MW00574286

Plant Invasions in Protected Areas

Invading Nature - Springer Series in Invasion Ecology

Volume 7

For further volumes:
http://www.springer.com/series/7228

Llewellyn C. Foxcroft • Petr Pyšek
David M. Richardson • Piero Genovesi
Editors

Plant Invasions in Protected Areas

Patterns, Problems and Challenges

 Springer

Editors
Llewellyn C. Foxcroft
South African National Parks
Centre for Invasion Biology
Conservation Services
Skukuza, South Africa

Petr Pyšek
Department of Invasion Ecology
Institute of Botany
Academy of Sciences of the Czech Republic
Průhonice, Czech Republic

David M. Richardson
Department of Botany and Zoology
Centre for Invasion Biology
Stellenbosch University
Stellenbosch, South Africa

Piero Genovesi
ISPRA – Institute for Environmental Protection
and Research, and Chair IUCN SSC Invasive
Species Specialist Group
Rome, Italy

ISBN 978-94-007-7749-1   ISBN 978-94-007-7750-7 (eBook)
DOI 10.1007/978-94-007-7750-7
Springer Dordrecht Heidelberg New York London

Library of Congress Control Number: 2013955739

© Springer Science+Business Media Dordrecht 2013
This work is subject to copyright. All rights are reserved by the Publisher, whether the whole or part of the material is concerned, specifically the rights of translation, reprinting, reuse of illustrations, recitation, broadcasting, reproduction on microfilms or in any other physical way, and transmission or information storage and retrieval, electronic adaptation, computer software, or by similar or dissimilar methodology now known or hereafter developed. Exempted from this legal reservation are brief excerpts in connection with reviews or scholarly analysis or material supplied specifically for the purpose of being entered and executed on a computer system, for exclusive use by the purchaser of the work. Duplication of this publication or parts thereof is permitted only under the provisions of the Copyright Law of the Publisher's location, in its current version, and permission for use must always be obtained from Springer. Permissions for use may be obtained through RightsLink at the Copyright Clearance Center. Violations are liable to prosecution under the respective Copyright Law.
The use of general descriptive names, registered names, trademarks, service marks, etc. in this publication does not imply, even in the absence of a specific statement, that such names are exempt from the relevant protective laws and regulations and therefore free for general use.
While the advice and information in this book are believed to be true and accurate at the date of publication, neither the authors nor the editors nor the publisher can accept any legal responsibility for any errors or omissions that may be made. The publisher makes no warranty, express or implied, with respect to the material contained herein.

Printed on acid-free paper

Springer is part of Springer Science+Business Media (www.springer.com)

*This book is dedicated to the
memory of our friend and colleague,
Professor Vojtěch Jarošík*

Foreword

When wise people first decided to proclaim portions of the Earth's surface as 'protected areas' for environmental conservation purposes in the late nineteenth century, the last thing they had on their minds were Invasive Alien Species. In fact at that time I doubt this human construct even existed in the minds of the most far-sighted of these wise people. Who could have foreseen that the greatest long-term threat to the integrity of the ecosystems 'protected' in these national parks, game reserves and state forests would turn out not to be the axeman, the hunter or the land developer, but rather the inexorable spread of alien species. But this is the situation that protected area managers throughout the world find themselves in today.

Fortunately, as the Invasive Alien Species (IAS) threat to protected areas has grown, the awareness of the scale of the problem has also grown. Fortunately, too, our understanding of the nature of the alien invasion phenomenon has also grown apace, as well as our understanding of the methods of combatting the phenomenon. The current volume sets out a great deal of the latest knowledge on this complex topic, and protected area managers will be well advised to read the relevant chapters with a view to applying the principles that are emerging in their own IAS prevention and management programmes.

In essence the problem is simple: natural ecosystems are being invaded by a host of alien species introduced outside their native ranges by human agency. These IAS vary in the severity of their impacts on the invaded ecosystems, but as a generalization, none of them should be tolerated in a protected area which should have the maintenance of its native biodiversity and natural ecosystem functioning as a top priority goal for the area's management. In reality, the scale of the problem is such that one of the managers' first tasks must be to prioritise on which IAS to focus their management efforts.

Unfortunately, that preeminent component of modern intelligent IAS management, 'prevention', is not optimally available to individual protected areas as these form parts of nations and subcontinents, and it is at the borders of such larger geographic and political entities that prevention strategies are generally best applied. However, it is still crucially important that all protected area IAS management strategies address this issue of preventing new alien species from entering the area. In this connection it is absurd that some protected areas still allow the cultivation of alien plant species in the developed areas within the protected area: this despite the ample historical evidence from protected areas all around the world that such introductions have frequently led to serious IAS problems.

Based on my own experience and my extensive reading of the experience of others in managing IAS in protected areas, a few simple points emerge: IAS management will only be successful if the nature of the IAS challenge to the protected area is well quantified, the appropriate IAS management strategy adopted, the optimum management measures employed and the actual field measures adequately monitored, documented and followed-up. It is crucially important at the outset to provide accurate estimates of the resources that will be required to implement the IAS management programme: under-resourced programmes are invariably doomed to failure and will only lead to 'decision makers' later shrugging their shoulders and declaring the problem insoluble.

Something that you are unlikely to read in the science-based chapters that comprise this volume is the overriding importance of 'commitment': an IAS management programme implemented half-heartedly by a management team that lacks a deep commitment to the programme will often fail miserably. The identical programme (as long as it is based on a sound understanding of the alien invasion to be managed and is adequately resourced) driven by a management team that is totally committed to making the programme work will generally succeed. 'Adaptive management' in which the lessons learned in initial management operations are rapidly fed back into an improved strategy and or improved tactical measures is essential.

Hopefully, reading the chapters of this book will lead to heightened levels of commitment by both decision-makers and managers to combatting the invasion of our priceless protected areas by alien species throughout the world. The editors and the chapter authors are to be commended for putting together this useful summary at a time when protected area managers throughout the world are in desperate need of such an up-to-date summary on this important management issue.

Ian A W Macdonald

Extraordinary Professor, Sustainability Institute, School of Public Leadership, Stellenbosch University, South Africa

Preface

> In the distribution of species over the Globe, the order of nature has been obscured through the interference of man. He has transported animals and plants to countries where they were previously unknown; extirpating the forest and cultivating the soil, until at length the face of the Globe itself is changed. To ascertain the amount of this interference, displaced species must be distinguished, and traced each to its original home.
>
> Charles Pickering, M.D. (*Chronological history of plants* 1879)

Interest in biological invasions has increased dramatically since introduced species were mentioned in faunas and floras in the late 1700s, briefly discussed in the works of Charles Darwin, Charles Pickering and others in the 1800s, and then brought to prominence in the mid-1900s through the work of, among others, Charles Elton in his 1958 book on *Invasions of Animals and Plants*.

Work directed at understanding the drivers and determinants of invasiveness of species and invasibility of habitats started in earnest, largely as a result of an international programme under the auspices of SCOPE (Scientific Committee on Problems of the Environment), in the late 1980s. Elucidation of the intricacies of the negative impacts of biological invasions has lagged behind. However, through initiatives such as the Millennium Ecosystem Assessment, there is now considerable awareness of the pervasiveness of invasive species and their role, often as part of a 'lethal cocktail' of factors, in driving ecosystem degradation. Considerable effort is now being devoted towards devising robust methods for forecasting and quantifying impacts, and developing effective prevention and management interventions. Dramatic evidence has emerged in recent decades that no ecosystems are free from invasive species and that even remote protected areas are being affected by many types of invasive species.

The growing recognition of the impacts of biological invasions on biodiversity has led conservation fora – such as the Convention on Biological Diversity and the IUCN World Commission on Protected Areas – to call on the global conservation community to strengthen prevention and response efforts for invasive species in protected areas. This became part of a more general process of recognising the urgent need of active conservation in response to the rapidly increasing pressures affecting protected areas. It has become obvious that more effective conservation action requires significant advances in invasion science. This is one of the reasons behind the increased number of scientific conferences, symposia and other fora related to invasive species organised in recent decades, such as the United Nations Conference on Alien Species in Trondheim, Norway, in 1996, and conferences such as BIOLIEF (World Conference on Biological Invasions and Ecosystem Functioning) and ICBI (International Conference on Biological Invasions).

The primary international forum for deliberations on plant invasions is the conference series on *Ecology and Management of Alien Plant Invasions* (EMAPI) which started in 1992. The concept of examining alien plant invasions in protected areas was initiated through a special session on the topic at the 10th EMAPI conference in Stellenbosch, South Africa, in 2009, and was followed up at the 11th EMAPI meeting in 2011 in Szombathely, Hungary. The seeds sown at these meetings grew into this book, which we hope presents a balanced synthesis of the current situation of invasive plants in protected areas and stimulates new work to deal with the massive challenges that lie ahead.

Skukuza, South Africa   Llewellyn C. Foxcroft
Průhonice, Czech Republic   Petr Pyšek
Stellenbosch, South Africa   David M. Richardson
Rome, Italy   Piero Genovesi
August 2013

Acknowledgements

We are grateful to the following reviewers for their time and providing helpful reviews which improved the chapters: Giuseppe Brundu, Laura Celesti-Grapow, Richard Cowling, Joe DiTomaso, Paul Downey, Essl Franz, Jack Ewel, Niek Gremmen, Vernon H Heywood, Patricia M Holmes, Philip E Hulme, Inderjit, Ingolf Kühn, Christoph Kueffer, Rhonda Loh, Lloyd L Loope, Margherita Gioria, John Mauremootoo, Laura Meyerson, Andrea Monaco, Jan Pergl, Peter G Ryan, Mathieu Rouget, Philip W Rundel, Peter Ryan, KfriSankaran, Dominique Strasberg, Federico Tomasetto, Brian van Wilgen, Nicola van Wilgen, John RU Wilson, Arne Witt.

We also thank Zuzana Sixtová for technical assistance with editing and other support.

We thank Prof. Dan Simberloff for supporting this book.

Llewellyn C. Foxcroft. I am most grateful to my wife, Sandra MacFadyen, for continuously and enthusiastically supporting me throughout working on this book. I also acknowledge the support of South African National Parks, the DST-NRF Centre of Excellence for Invasion Biology, Stellenbosch University, and National Research Foundation of South Africa.

David M. Richardson. I thank the DST-NRF Centre of Excellence for Invasion Biology and the National Research Foundation of South Africa (Grant 85417), and the Hans Sigrist Trust for financial support.

Petr Pyšek. I acknowledge financial support by Praemium Academiae award and long-term research development project no. RVO 67985939 from the Academy of Sciences of the Czech Republic, and from institutional resources of Ministry of Education, Youth and Sports of the Czech Republic.

Piero Genovesi. I thank Simon Stuart, chair of the IUCN Species Survival Commission for his constant support, Shyama Pagad, the members of the IUCN SSC Invasive Species Specialist Group and the subscribers of the Aliens list for the information, unpublished data and helpful suggestions provided.

Contents

**Part II   Regional Patterns: Mapping the Threats from Plant
            Invasions in Protected Areas**

Part IV   Conclusion

Contributors

Jake Alexander Institute of Integrative Biology – Plant Ecology, ETH Zurich, Zurich, Switzerland

Rachel Atkinson Charles Darwin Foundation, Santa Cruz, Galapagos Islands, Ecuador

Cláudia Baider The Mauritius Herbarium, Agricultural Services, (ex MSIRI-MCIA), Ministry of Agro-Industry and Food Security, Réduit, Mauritius

Stéphane Baret Parc national de La Réunion, La Plaine des Palmistes, La Réunion, France

Peter Bayliss National Environmental Research Program, Northern Australia Hub, Charles Darwin University, Darwin, NT, Australia

CSIRO, Cleveland, QLD, Australia

Maria José Bettencourt Direção Regional do Ambiente dos Açores, Rua Cônsul Dabney –Colónia Alemã, Horta, Portugal

Giuseppe Brundu Department of Science for Nature and Environmental Resources (DIPNET), University of Sassari, Sassari, Italy

Chris Buddenhagen Department of Biological Science, Florida State University, Tallahassee, FL, USA

Charles Darwin Foundation, Santa Cruz, Galapagos Islands, Ecuador

David F.R.P. Burslem Institute of Biological and Environmental Sciences, University of Aberdeen, Aberdeen, UK

Ramiro Bustamante Departamento de Ecología, Facultad de Ciencias, Universidad de Chile, Santiago, Chile

Instituto de Ecología y Biodiversidad (IEB), Santiago, Chile

Ted Center USDA, ARS, Invasive Plant Research, Fort Lauderdale, FL, USA

Hugo Costa CIBIO, Centro de Investigação em Biodiversidade e Recursos Genéticos, InBIO Laboratório Associado, Pólo dos Açores Departamento de Biologia, Universidade dos Açores, Ponta Delgada, Portugal

School of Geography, University of Nottingham, Nottingham, UK

Curt Daehler Department of Botany, University of Hawaii, Honolulu, HI, USA

Wayne Dawson Department of Biology, University of Konstanz, Constance, Germany

Michael M. Douglas National Environmental Research Program, Northern Australia Hub, Charles Darwin University, Darwin, NT, Australia

Paul O. Downey Parks and Wildlife Group, Office of Environment and Heritage, Hurstville, NSW, Australia

Institute for Applied Ecology, University of Canberra, Canberra, ACT, Australia

Ezekiel Edward Department of Forest and Landscape, Faculty of Life Sciences, University of Copenhagen, Copenhagen, Frederiksberg C, Denmark

Peter Edwards Institute of Integrative Biology – Plant Ecology, ETH Zurich, Zurich, Switzerland

Keith B. Ferdinands National Environmental Research Program, Northern Australia Hub, Charles Darwin University, Darwin, NT, Australia

Department of Land Resource Management, Weeds Management Branch, Palmerston, NT, Australia

Llewellyn C. Foxcroft Conservation Services, South African National Parks, Skukuza, South Africa

Centre for Invasion Biology, Department of Botany and Zoology, Stellenbosch University, Stellenbosch, South Africa

Nicol Fuentes Laboratorio de Invasiones Biológicas, Facultad de Ciencias Forestales, Universidad de Concepción, Concepción, Chile

Instituto de Ecología y Biodiversidad (IEB), Santiago, Chile

Mark R. Gardener Charles Darwin Foundation, Santa Cruz, Galapagos Islands, Ecuador

School of Plant Biology, University of Western Australia, Crawley, WA, Australia

Piero Genovesi ISPRA, Institute for Environmental Protection and Research, Rome, Italy

Chair IUCN SSC Invasive Species Specialist Group, Rome, Italy

Artur Gil Azorean Biodiversity Group, CITA-A, Departamento de Biologia, Universidade dos Açores, Ponta Delgada, Portugal

Sylvia Haider Institute of Biology/Geobotany and Botanical Garden, Martin Luther University Halle Wittenberg, Halle, Germany

Ruben Heleno Department of Life Sciences, Centre for Functional Ecology, University of Coimbra, Coimbra, Portugal

Charles Darwin Foundation, Santa Cruz, Galapagos Islands, Ecuador

Ankila J. Hiremath Ashoka Trust for Research in Ecology and the Environment 1, New Delhi, India

R. Flint Hughes Institute of Pacific Islands Forestry, USDA Forest Service, Hilo, HI, USA

Cang Hui Centre for Invasion Biology, Department of Botany and Zoology, Stellenbosch University, Stellenbosch, South Africa

Philip E. Hulme The Bio-Protection Research Centre, Lincoln University, Canterbury, New Zealand

Heinke Jäger Department of Ecology, Technische Universität Berlin, Berlin, Germany

Charles Darwin Foundation, Santa Cruz, Galapagos Islands, Ecuador

Alejandra Jiménez Laboratorio de Invasiones Biológicas, Facultad de Ciencias Forestales, Universidad de Concepción, Concepción, Chile

Instituto de Ecología y Biodiversidad (IEB), Santiago, Chile

Christoph Kueffer Institute of Integrative Biology – Plant Ecology, ETH Zurich, Zurich, Switzerland

Erwann Lagabrielle Université de La Réunion et Institut de Recherche pour le Développement - UMR 228 ESPACE-DEV, Sainte-Clotilde Cedex, La Réunion, France

Parc Technologique Universitaire, Sainte-Clotilde Cedex, La Réunion, France

Loralee Larios Department of Environmental Science, Policy and Management, University of California Berkeley, Berkeley, CA, USA

Lloyd L. Loope Formerly: U.S. Geological Survey, Pacific Island Ecosystems Research Center, Haleakala Field Station, Makawao (Maui), HI, USA

Current: Makawao, HI, USA

Wayne D. Lotter PAMS Foundation, Arusha, Tanzania

Sandra MacFadyen Conservation Services, South African National Parks, Skukuza, South Africa

Centre for Invasion Biology, Department of Botany and Zoology, Stellenbosch University, Stellenbosch, South Africa

Lori J. Makarick Grand Canyon National Park, Flagstaff, AZ, USA

Alicia Marticorena Departamento de Botánica, Facultad de Ciencias Naturales y Oceanográficas, Universidad de Concepción, Concepción, Chile

Keith McDougall Department of Environmental Management and Ecology, La Trobe University, Wodonga, VIC, Australia

Jeffrey A. McNeely Formerly: Gland Switzerland

Current: Petchburi, Thailand

Scott J. Meiners Department of Biological Sciences, Eastern Illinois University, Charleston, IL, USA

Jean-Yves Meyer Délégation à la Recherche, Government of French Polynesia, Papeete, Tahiti, French Polynesia

Laura A. Meyerson Department of Natural Resources Science, University of Rhode Island, Kingston, RI, USA

Institute of Botany, Department of Invasion Ecology, Academy of Sciences of the Czech Republic, Průhonice, Czech Republic

Ann Milbau Climate Impacts Research Centre – Department of Ecology and Environmental Science, Umeå University, Abisko, Sweden

Andrea Monaco ARP, Regional Parks Agency – Lazio Region, Rome, Italy

Catherine Parks Pacific Northwest Research Station, US Forest Service, La Grande, USA

Aníbal Pauchard Laboratorio de Invasiones Biológicas, Facultad de Ciencias Forestales, Universidad de Concepción, Concepción, Chile

Instituto de Ecología y Biodiversidad (IEB), Santiago, Chile

Jan Pergl Department of Invasion Ecology, Institute of Botany, Academy of Sciences of the Czech Republic, Průhonice, Czech Republic

Aaron M. Petty National Environmental Research Program, Northern Australia Hub, Charles Darwin University, Darwin, NT, Australia

Steward T.A. Pickett Cary Institute of Ecosystem Studies, Millbrook, NY, USA

Petr Pyšek Department of Invasion Ecology, Institute of Botany, Academy of Sciences of the Czech Republic, Průhonice, Czech Republic

Department of Ecology, Faculty of Science, Charles University in Prague, Prague, Czech Republic

Zafar A. Reshi Department of Botany, University of Kashmir, Srinagar, Jammu and Kashmir, India

Lisa J. Rew Land Resources and Environmental Sciences Department, Montana State University, Bozeman, MT, USA

John Richard Tanzania Forestry Research Institute, Lushoto, Tanzania

David M. Richardson Centre for Invasion Biology, Department of Botany and Zoology, Stellenbosch University, Stellenbosch, South Africa

Ramona A. Robison California State Parks, Sacramento, CA, USA

Mellesa Schroder NSW National Parks and Wildlife Service, Jindabyne, Australia

Tim Seipel Institute of Integrative Biology – Plant Ecology, ETH Zurich, Zurich, Switzerland

Samantha A. Setterfield National Environmental Research Program, Northern Australia Hub, Charles Darwin University, Darwin, NT, Australia

Justine D. Shaw Environmental Decision Group, School of Biological Sciences, The University of Queensland, St Lucia, Australia

Department of Sustainability, Environment, Water Population and Communities, Terrestrial Nearshore Ecosystems, Australian Antarctic Division, Hobart, Australia

Carlos M.N. Silva Sociedade Portuguesa para o Estudo das Aves SPEA, Lisboa, Portugal

Luís Silva CIBIO, Centro de Investigação em Biodiversidade e Recursos Genéticos, InBIO Laboratório Associado, Pólo dos Açores Departamento de Biologia, Universidade dos Açores, Ponta Delgada, Portugal

Daniel Simberloff Department of Ecology and Evolutionary Biology, University of Tennessee, Knoxville, TN, USA

Thomas J. Stohlgren US Geological Survey, Fort Collins Science Center, Fort Collins, CO, USA

Katharine N. Suding Department of Environmental Science, Policy and Management, University of California Berkeley, Berkeley, CA, USA

Bharath Sundaram Azim Premji University, PES Institute of Technology Campus, Bangalore, Karnataka, India

Joaquim Teodósio Sociedade Portuguesa para o Estudo das Aves SPEA, Lisboa, Portugal

Ann M. Thompson Department of Conservation, Wellington, New Zealand

Rosie Trevelyan Tropical Biology Association, Department of Zoology, Cambridge, UK

Mandy Trueman School of Plant Biology, University of Western Australia, Crawley, WA, Australia

Charles Darwin Foundation, Santa Cruz, Galapagos Islands, Ecuador

Mandy Tu Independent Consultant, Hillsboro, OR, USA

Alan Tye Charles Darwin Foundation, Santa Cruz, Galapagos Islands, Ecuador

Roy Van Driesche Department of Environmental Conservation, University of Massachusetts, Amherst, MA, USA

Carol J. West Department of Conservation, Wellington, New Zealand

Jan Wild Department of Invasion Ecology, Institute of Botany, Academy of Sciences of the Czech Republic, Průhonice, Czech Republic

Steve Winderlich Kakadu National Park, Jabiru, NT, Australia

Arne Witt CABI Africa, Nairobi, Gigiri, Kenya

Part I
Setting the Scene: Impacts, Processes and Opportunities

Chapter 1
Plant Invasions in Protected Areas: Outlining the Issues and Creating the Links

Llewellyn C. Foxcroft, David M. Richardson, Petr Pyšek, and Piero Genovesi

Abstract There are numerous excellent volumes on the topic of biological invasions, some of which deal with conservation-related issues to varying degrees. Almost 30 years since the last global assessment of alien plant invasions in protected areas during the SCOPE programme of the 1980s, the present book aims to provide a synthesis of the current state of knowledge of problems with invasive plants in protected areas. To set the scene we outline some of the major challenges facing the field of invasion biology. We discuss the extent and dimensions of problems that managers of protected areas deal with and what can be learnt from research and management interventions conducted in protected areas. A virtual tour through different regions of the world sheds light on the rapidly growing knowledge

L.C. Foxcroft (✉)
Conservation Services, South African National Parks, Private Bag X402, Skukuza 1350, South Africa

Centre for Invasion Biology, Department of Botany and Zoology, Stellenbosch University, Private Bag X1, Stellenbosch 7602, South Africa
e-mail: Llewellyn.foxcroft@sanparks.org

D.M. Richardson
Centre for Invasion Biology, Department of Botany and Zoology, Stellenbosch University, Private Bag X1, Stellenbosch 7602, South Africa
e-mail: rich@sun.ac.za

P. Pyšek
Department of Invasion Ecology, Institute of Botany, Academy of Sciences of the Czech Republic, Průhonice CZ 252 43, Czech Republic

Department of Ecology, Faculty of Science, Charles University in Prague, CZ 128 44 Viničná 7, Prague 2, Czech Republic
e-mail: pysek@ibot.cas.cz

P. Genovesi
ISPRA, Institute for Environmental Protection and Research, Via V. Brancati 48, I-00144 Rome, Italy

Chair IUCN SSC Invasive Species Specialist Group, Rome, Italy
e-mail: piero.genovesi@isprambiente.it

L.C. Foxcroft et al. (eds.), *Plant Invasions in Protected Areas: Patterns, Problems and Challenges*, Invading Nature - Springer Series in Invasion Ecology 7, DOI 10.1007/978-94-007-7750-7_1, © Springer Science+Business Media Dordrecht 2013

base in different socio-geographical settings, and applies such insights to the problems that managers face. We hope that this book captures the core concerns and creates the critical links that will be needed if the growing impacts of alien plant invasions on protected areas are to be managed effectively. We also aim to promote the role of protected areas as leaders and catalysts of global action on invasive species, and key study areas for basic and applied invasion science.

Keywords Conservation • Impact • Invasive alien plants • Management • Nature reserve

1.1 Protected Areas and Plant Invasions: History and Threats

The target of conserving 10 % of the world's ecological regions by 2010 was agreed to in 2004, at the seventh conference of the parties to the Convention on Biological Diversity (CBD 2004) and the CBD Strategic Plan for 2011–2020 raised this target to 17 % (Aichi Target 11). A recent summary estimates that there are about 157,000 terrestrial and marine areas that enjoy some form of legal status as protected areas (PAs) worldwide. These PAs cover more than 24 million km^2 (16 million km^2 terrestrial; IUCN and UNEP-WCMC 2012; http://www.wdpa.org/Statistics.aspx). The number of PAs grew tenfold between 1962, when there were approximately 10,000, and 2003 (the 5th World Parks Congress in Durban) when there were about 100,000 (Mulongoy and Chape 2004). Terrestrial PAs grew from about 3.5 % of the total land area in 1985 (Zimmerer et al. 2004), to 12.9 % in 2009 (Jenkins and Joppa 2009).

Protected areas are the foundation of national and international conservation initiatives, and are mandated with conserving biodiversity (Dudley and Parish 2006). They are designed to protect representative portions of natural landscapes, ensure the persistence of biodiversity and key ecosystem processes, provide ecosystem goods and services, and in many cases to contribute significant economic benefits (Barrett and Barrett 1997; Margules and Pressey 2000). The role that PAs can play in mitigating the impacts of global climate change is also increasing in importance (Conroy et al. 2011).

Empirical evidence of the overall contribution of PAs in conserving biodiversity is scarce. Nonetheless, and despite some conflicting case studies (Bruner et al. 2001; Mora et al. 2009; Butchart et al. 2012), there is little doubt that, globally, PAs buffer representative areas of biodiversity from many threatening processes (Gaston et al. 2008). Protected areas are, however, becoming increasingly isolated in a matrix of human-altered landscapes (Koh and Gardner 2010). Habitat fragmentation not only reduces the total amount of habitat and subdivides it into fragments, but also introduces new forms of land use (Bennett and Saunders 2010). These landscapes, modified to varying extents for different uses, differ in their conservation value, and in their compatibility with adjacent PAs. Moreover, PAs are faced with a number of threats, displacing the species and eroding the systems underpinning the reasons for their establishment. Within PAs the growing global impacts of habitat loss, fragmentation and over-exploitation are often eliminated or can be managed to some extent. Many anthropogenic threats to biodiversity are, however, not removed through formal protection. This is especially true for smaller PAs and those with larger edge/total area ratios. Biological invasions, one of the most pressing environmental concerns globally, are one such threat.

The concept of setting aside tracts of land for different forms of protection dates back thousands of years (Mulongoy and Chape 2004), with many being declared as sacred sites (Dudley et al. 2005). For instance in northern India (2,000 years ago) and Indonesia (1,500 years ago) areas were protected for religious beliefs and as homes of the Gods. Estimates suggest that there may be as many sacred sites as PAs, many of which fall outside formally listed PAs (Dudley et al. 2005). Modern philosophies behind conservation or protected areas were related to maintaining vast tracts of wilderness (of which John Muir was a major advocate; Devall 1982), a landscape ethic (Leopold 1949), the protection of fragments of habitats that were rapidly disappearing, or to support sustainable utilization or wildlife conservation (Meine 2010). Wildlife conservation has often focused on the preservation of single species at high risk of extinction and/or protecting dwindling herds of typically charismatic large mammals. For example, the preservation of rare or endangered species, which are also often charismatic, played a major role in leading to the promulgation in 1905 of Kaziranga National Park in India to protect the one-horned rhinoceros (*Rhinoceros unicornis*; Dudley and Parish 2006).

The first national park proclaimed globally (and the first formal use of the term 'national park'), primarily for protection of its scenic beauty (Dudley and Stolton 2012), was Yellowstone National Park in the United States, in 1872. The proclamation of Yellowstone National Park was followed shortly thereafter by national parks in a number of countries. By 2008, the US National Park system covers 338,000 km^2 of PAs, about 4 % of the country, including representative landscapes of all of the nation's biomes and ecosystems (Baron et al. 2008). Designation and management of PAs as an approach to preventing degradation of particular parcels of land continued with a focus on species populations, maintaining states in equilibrium (notions of the 'balance of nature') or agriculturally based concepts such as carrying capacity (Rogers 2003). In the 1990s, conservation practices and management approaches had started moving away from protection of single species and their habitats, towards the consideration of interactive networks of species and an ecosystem-based approach (Ostfeld et al. 1997). Species-centric approaches often developed into crisis-orientated approaches, whereas focusing on large-scale ecosystems and networks allows for the maintenance of the underlying requirements on which species depend (Fiedler et al. 1997; Ostfeld et al. 1997). There is also increasing acceptance by conservation agencies that systems are dynamic and heterogeneous, and that disturbance is both a driver and responder of system change (Pickett et al. 2003). Emerging concepts over the last decade include the growing understanding of the importance of ecosystem resilience for PAs (Wangchuk 2007; Baron et al. 2008; Hobbs et al. 2010) and that the interrelatedness of socio-ecological systems in the broader landscape are critical to long-term maintenance of PAs (Newton 2011). It is thus within this setting and new conservation paradigm that insights for invasion science may emerge.

In the USA, concern over alien species in the national parks was expressed by National Park Service scientists as early as in the 1930s (Houston and Schreiner 1995). Even earlier, however – shortly after the establishment of the Yosemite Valley state park in 1864, designated for public use and recreation – concerns about European weeds invading the park were raised (Randall 2011). In South Africa's Kruger National Park (established in 1898) the first official records of alien plants date to 1937, when six alien species were recorded during general botanical surveys (Foxcroft et al. 2003). In 1947 Bigalke, writing about the then National Parks Board of South Africa, published a strongly titled paper "The adulteration of the fauna and flora of our national parks". He stated that it should not be permissible to introduce animals and plants to a national park, and if the principle was not strictly adhered to the term 'national park' would have no meaning (Bigalke 1947). At a meeting of the American Association for the Advancement of Science (AAAS) in 1921, the council stated that it "...strongly opposes the introduction of non-native plants and animals into the national parks... and urges the National Park Service to prohibit all such introduction..." (Shelford 1926). Similar sentiments were expressed in Great Britain by the British Ecological Society in a report on nature conservation and nature reserves (British Ecological Society 1944).

Despite sentiments like these, some of the best-known examples of alien plant invasions come from PAs – and in some cases these are due to intentional introductions by park managers. For example, in Everglades NP, USA, *Melaleuca quinquenervia* (melaleuca) forms dense stands, replacing indigenous vegetation, altering habitats and fire regimes, and using large amounts of water (Schmitz et al. 1997). *Schinus terebinthifolius* (Brazilian pepper) has similar impacts, and has replaced *Cladium jamaicense* (saw grass) prairie and pineland with monospecific stands (Li and Norland 2001). *Mimosa pigra* (giant sensitive plant) is considered a major threat to Kakadu NP in Australia (Cowie and Werner 1993; Lonsdale 1993). Similarly, *Morella faya* (faya tree) in Hawaii Volcanoes NP has displaced the endemic *Metrosideros polymorpha* ('Ohi'a lehua) over large areas of protected land (Loope et al. 2014).

1.2 The SCOPE Programme on Biological Invasions in the 1980s

The last international research programme to focus specifically on invasive species in protected areas was a working group on invasions in nature reserves, initiated under the SCOPE (Scientific Committee on Problems of the Environment) programme on biological invasions in the 1980s (Wildlife Conservation and the Invasion of Nature Reserves by Introduced Species: a Global Perspective; Macdonald et al. 1989). The work on nature reserves culminated in a series of six papers published in the journal *Biological Conservation*, addressing invasions globally in nature reserves on islands (Brockie et al. 1988), on arid land (Loope et al. 1988), in tropical savannas and dry woodlands (Macdonald and Frame 1988), in Mediterranean-type climatic regions (Macdonald et al. 1988), and completed by a search for generalisations (Usher 1988). The central question posed by the working group on nature reserves was whether an undisturbed community could become invaded by alien species. The challenge, however, was to define such communities within which to work. It was felt that the best option would be in tracts of land that had been set aside to keep anthropogenic impacts on special features (e.g. wildlife and landscapes) to a minimum (Usher 1988). Using nature reserves as the sites most likely to accommodate these requirements, the working group aimed to (i) provide insights into differences between the extent to which natural and disturbed systems could become invaded; (ii) provide information on the consequences of invasions for indigenous species; and (iii) based on the outcomes, to provide management recommendations. The programme on nature reserves initially aimed to examine a larger list of biomes, but due to the lack of available information, work focussed on tropical and subtropical dry woodlands and savannas, Mediterranean-type shrublands and woodlands, arid lands, and oceanic islands. A total of 24 protected areas served as case studies.

Some findings from this SCOPE programme were that the nature and degree of invasions differ substantially between protected areas in different regions of the world. For example, it was suggested that nature reserves in arid regions of the tropics and sub-tropics have fewer invasive species (although notable exceptions were found); temperate regions in the northern hemisphere are relatively free of invasions, while reserves in the southern hemisphere were found to be severely impacted (Usher et al. 1988). All the nature reserves in the case studies included invasive vascular plants, comprising about 30 % of the flora on island reserves and about 5 % of all species in dry woodland and savanna (Usher 1988). Thus one of the most alarming generalisations of the programme at the time was the finding that all nature reserves contain invasive species and thus natural systems can indeed be invaded, some of them quite heavily. The authors also reported that invasions were found to impact both the structure and functioning of the ecosystems, and they recommended that priority should be given to species that threaten endemic species with extinction or those that have strong impacts at a landscape scale (Usher 1988). An important point was made that tourism poses dangers for invasions of reserves, as a positive correlation was detected between visitor numbers and numbers of introduced species (Usher 1988). This is obviously an increasingly concerning issue, as ecotourism is touted as a prime, low impact source of revenue in many parts of the world (see also Lonsdale 1999; Foxcroft et al. 2014).

Although the programme produced fundamental information on the invasibility of natural systems and the status of invasions across a number of regions globally, the six papers published in *Biological Conservation* have received less attention than deserved. Collectively the papers have been cited about 200 times, with half accruing to the synthesis paper (Usher 1988), Despite the growing intensity of research on biological invasions and the increasing focus on management issues (Richardson and

Pyšek 2008; Pyšek and Richardson 2010), there has been no follow-up synthesis on the topic of plant invasions in PAs in the last two decades. The question of whether natural systems can be invaded by alien plants has been answered, but many other issues have arisen.

1.3   Conservation and Policy Conventions

The World Conservation Strategy of 1980, developed jointly by the IUCN, UNEP and WWF (1980), had three main objectives: the maintenance of essential life support systems, the maintenance of natural diversity and the sustained utilization of species and ecosystems. Interestingly, although alien and invasive species (at the time 'exotic' species) were mentioned in the strategy document at various points, the problem was not listed as one of the 14 priority issues, on par with, for example, soil erosion and its role in the degradation of catchment areas and watersheds. The effects of invasive alien species (IAS) were listed as one of the threats to wild species, impacting on competition for space or food, predation, habitat destruction or degradation, and the transmission of diseases and parasites. The species of concern, however, did not include any alien plants, citing only trout, bass, goats and rabbits. Freshwater systems and islands were indicated as particularly vulnerable.

It was largely through the SCOPE programme in the late 1980s that a larger, more detailed body of knowledge began accumulating. This provided the foundations on which improved policies could be formulated, leading to the current situation where issues related to biological invasions are included at all levels, from local to international, and in almost all biodiversity or conservation conventions, specialist groups and non-governmental organisations.

We indicate key issues raised by some of these conventions as examples. Highlighting these initiatives provides an indication of the acceptance and growing importance of biological invasions as an agent of global environmental change. In particular, they show the increasing concern of the problems to biodiversity and conservation.

1.3.1   The Millennium Ecosystem Assessment

The Millennium Ecosystem Assessment (2005) provided a global account of the status and trends of the greatest threats to biodiversity and has gained high level attention. The work highlighted biological invasions as the second most important global driver of biodiversity loss, and – together with climate change – the most difficult to reverse. The study stressed the absence of an adequate regulation for several pathways of introductions and considered the adoption of measures to control major pathways as a fundamental goal to address the IAS threats to biodiversity (Goal 6).

1.3.2   Convention on Biological Diversity (CBD)

The adoption of the CBD by 101 countries in 1992 raised the political profile of IAS to an international level. The Convention (Article 8h) calls on contracting parties to "prevent the introduction of, control or eradicate those alien species which threaten ecosystems, habitats and species", with a number of key principles for addressing this threat being adopted (http://www.cbd.int/decision/cop/?id=7197). The CBD has also given much attention to the threat of invasive species to PAs. For example, the joint CBD and UNEP-WCMC report on PAs and biodiversity (Mulongoy and Chape 2004) stated that ". . . widespread threat is that of alien invasive species which may be released, deliberately or accidentally, within a protected area, or may move in from surrounding areas". At the 10th CBD-COP (in Nagoya, 2010), the threat of IAS to PAs (http://www.cbd.int/decision/cop/?id=12297) was again highlighted as an issue needing greater attention. Recognising the role of IAS as a key driver of biodiversity loss, the CBD invited the Parties to consider the role of IAS management as a cost-effective tool for the restoration and maintenance of protected areas and the ecosystem services they provide, and thus to include management of IAS in the action plans for implementation of the programme of work on PAs. At that occasion the CBD-COP adopted the Strategic Plan for Biodiversity 2011–2020, and 20 Aichi targets, including Target 9: "By 2020, invasive alien species and pathways are identified and prioritised, priority species are controlled or eradicated, and measures are in place to manage pathways to prevent their introduction and establishment".

1.3.3   International Union for Conservation of Nature (IUCN)

1.3.3.1   Protected Areas Programme

The 5th IUCN World Parks Congress in 2003 (Durban, South Africa) considered the need to manage IAS in PAs as an "emerging issue", stating that – "management of invasive alien species is a priority issue and must be mainstreamed into all aspects of protected area management". The Congress adopted a set of recommendations, including Recommendation I stating that pressures on PAs will increase as a result of global change, including invasions of alien species. The congress recognised and urged that "the wider audience of protected area managers, stakeholders and governments urgently need to be made aware of the serious implications for biodiversity, protected area conservation and livelihoods that result from lack of recognition of the IAS problem and failure to address it. Promoting awareness of solutions to the IAS problem and ensuring capacity to implement effective, ecosystem-based methods must be integrated into protected area management programmes. In addition to the consideration of benefits beyond boundaries, the impacts flowing into both marine and terrestrial PAs from external sources must be addressed" (https://cmsdata.iucn.org/downloads/emergingen.pdf).

1.3.3.2   World Commission Protected Areas Programme

The World Commission on Protected Areas (WCPA) is one of the five IUCN commissions, administered by IUCN's Global Programme on Protected Areas. It is a network of over 1,700 members, spanning 140 countries. The World Commission on Protected Areas aims to promote the establishment and effective management of a world-wide representative network of terrestrial and marine PAs as an integral contribution to IUCN's mission. To achieve this, WCPA supports planning of PAs and integrating them into all sectors, provides strategic advice to policy makers and strengthens capacity and investment in protected areas.

1.3.3.3   IUCN SSC Invasive Species Specialist Group

The Invasive Species Specialist Group (ISSG, http://www.issg.org/) is one of the five thematic specialist groups organised under the auspices of the Species Survival Commission (SSC) of the International Union for Conservation of Nature (IUCN). The ISSG, established in 1994, is a global network of scientific and policy experts on invasive species; it currently has about 200 core members from over 40 countries, and over 2,000 conservation practitioners and experts who contribute to its work. The three core activity areas of the ISSG are policy and technical advice, information exchange, and networking. The ISSG provides technical and scientific advice to, amongst others, the Convention on Biological Diversity, the Ramsar Convention, and the European Union. The ISSG promotes and facilitates the exchange of invasive species information, developing and managing the Global Invasive Species Database (GISD, http://www.issg.org/database/welcome), to provide information on the ecology of invasive species, their impacts and relevant management options. The GISD is cross-linked to the IUCN Red List of Threatened Species as well as the World Database on Protected Areas. The ISSG has worked with GISP to develop a scoping report on the threat of IAS to protected areas (De Poorter 2007). In 2012 a task force between IUCN SSC ISSG and the IUCN WCPA was established to produce guidelines for the management of IAS in PAs. In 2011 and 2012, IUCN and ISSG signed two Memoranda of Cooperation with the CBD Secretariat to provide support for the implementation of the Aichi targets in regard to the IAS issue.

1.3.4 Global Invasive Species Programme (GISP)

In 1996, concern that globalization was having negative consequences on the environment led the United Nations and the Government of Norway to convene the first international meeting IAS that was held in Trondheim, Norway (Sandlund et al. 1998). Participants concluded that IAS had become one of the most significant threats to biodiversity worldwide and recommended that a global strategy and mechanism to address the problem needed to be created immediately. In 1997, The Global Invasive Species Programme (GISP) was established. Working primarily at international and regional levels, GISP aimed to build partnerships, provide guidance, develop a supportive environment and build capacity for national approaches towards the prevention and management of invasive species by pursuing three key objectives: (i) facilitating information exchange; (ii) supporting policy and governance; and (iii) promoting awareness among key public and private sector decision makers (http://www.gisp.org/about/mission.asp). A GISP report on IAS in PAs (De Poorter 2007) identified the following key impediments or challenges to implementing invasive species management in PAs: (i) lack of capacity for mainstreaming of invasive alien species management into protected area management overall; (ii) lack of capacity for invasive alien species management at site level; (iii) lack of awareness of invasive alien species impacts on protected areas, of the options for fighting back, and of the urgency of prevention and early detection; (iv) lack of consolidated information on invasive alien species issue in protected areas at national, international, and global levels; (v) lack of information, at site level, on what alien species are present, what risks they pose and how to manage them; (vi) lack of funding and other resources; (vii) high level impediment, for example legal, institutional or strategic issues; and (viii) clashes of interests.

Unfortunately, due to a lack of financial resources the GISP Secretariat closed in March 2011.

1.3.5 Other International Conventions

The Ramsar Convention on Wetlands has, at different occasions, stressed the specific threat of invasive alien species to wetlands, and at the 10th COP (held in Korea, 2008) adopted The Ramsar Strategic Plan (2009–2015; http://www.ramsar.org/cda/en/ramsar-documents-resol/main/ramsar/1-31-107_4000_0). This document highlights IAS among the "challenges that still require urgent attention in order to achieve wetland wise use under the Convention". Ramsar has encouraged parties to develop national inventories of IAS impacting wetlands. Similarly, the Convention on the Conservation of Migratory Species of Wild Animals (in Bonn, 1979) recognised the threat posed by invasive species to migratory species in several provisions, and has included the struggle against IAS in the Strategic Plan for 2006–2014.

1.4 Why This Book?

Many books and syntheses have been written on biological invasions, covering all
dimensions of the discipline (see for example, Simberloff 2004; Cadotte et al. 2006;
Nentwig 2007; Davis 2009; Richardson 2011). Much work has also been done on
PAs since the SCOPE programme on nature reserves (Fig. 1.1), but the focus of the
work, and areas assessed, varies considerably in different parts of the world.
However, even with the progress in the field and the increasing number of publications, there has been no synthesis on the topic.

We set three main aims for this book:

(i)   To determine the status of knowledge on plant invasions in protected areas
      and synthesise these insights;
(ii) To integrate this with current models and theories of plant invasion ecology;
(iii) To determine key knowledge areas for informing the development of suc-
      cessful management strategies.

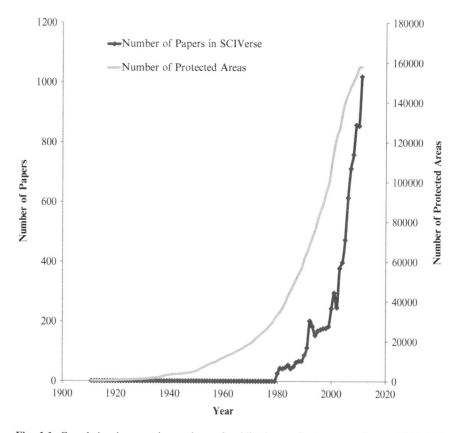

Fig. 1.1 Cumulative increase in numbers of publications referring to studies on biological
invasions in protected areas [from SCIVerse science direct; search: alien OR non-native OR
invasive OR biological invasion OR plant invasions AND protected area OR nature reserve OR
heritage site OR national park OR wilderness OR marine park], and in number of protected areas
(Data from IUCN and UNEP-WCMC 2012)

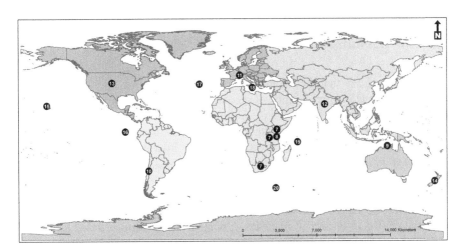

Fig. 1.2 Global map of regions discussed in the book. Numbers refer to the specific chapters

To achieve these objectives we aimed to cover a wide range of regions, as well as a variety of types of PAs and problems. We requested authors to specifically explore a number of issues pertinent to PAs. The book comprises three parts. The first section examines a number of general questions in invasion ecology; these are cross-cutting issues relevant to all regions. Here the authors discuss whether protected areas provide unique opportunities for gaining insights into these focal topics, and how work in PAs could provide further advances in the field. We also asked how PAs could be better used as model systems for future research. These topics include the role of PAs for developing an improved understanding of plant invasions and succession in natural systems, impacts of plant invasions, human dimensions of invasions, restoration, and large scale monitoring. The second part consists of case studies of plant invasions in PAs from 14 regions around the world (Fig. 1.2), including specific reference to about 135 protected areas. The case studies aim to capture experiences and to synthesise what has been done on invasive plants in PAs of different kinds and different sizes, in various environmental settings, and what has been learned from the research and management experiences in these areas. Case studies also explored the specific context of the systems and their unique attributes, and whether these aspects can provide natural laboratories for examining questions that cannot be studied in other regions. Specific attributes may include the modes and pathways of introduction and dispersal, impacts on biodiversity (whether species diversity, habitat structure or ecosystem function), the role of natural disturbance regimes (and whether these hold clues for understanding anthropogenic disturbance) and the usefulness of working in a range of sizes and types of PAs.

We also believe that these aims are crucial for providing knowledge that can contribute to meeting the Aichi target 9, and for ensuring full implementation of the provisions of Aichi target 11, which calls for effective management of the world's PAs. Moreover, effective management of IAPs is embodied in Aichi target 12, as being essential for reducing the rate of biodiversity loss.

1.5 Science in Protected Areas: Opportunities for the Future

The IUCN suggests that the main uses of protected areas are scientific research, wilderness protection, preservation of species and genetic diversity, maintenance of environmental services, protection of specific natural and cultural features, tourism and recreation, education, sustainable resource use and maintenance of cultural and traditional attributes (Mulongoy and Chape 2004). Invasive alien plants threaten, can impact on and respond to all these major attributes. As more reliance is being placed on PAs for ensuring the persistence of global biodiversity and related services, insight into how invasion processes progress and how systems function, or are likely to function, in an invaded state is essential for understanding the potential of PAs to fulfil their mandate. A number of other properties enhance the appeal of potential research sites: PAs cover a range of habitats and sizes, allowing for investigations at small plot scales, to large catchment type experiments. Many PAs are receiving increasing management attention, which can be factored into developing further understanding, and the outcomes can in return be implemented in management approaches directly, allowing for adaptation and further learning. With the increasing attention to PAs in general, a management body can protect or maintain research sites in the medium- to longer-term. This has already resulted in a number of PAs becoming focal points for research across a range of disciplines, which may allow for improved interaction and integration (for example, du Toit et al. 2003; Sinclair et al. 2008).

1.6 Management of Plant Invasions: The Future Roll of Protected Areas

Protected areas are crucial for protecting the global diversity and ecosystem services we all rely upon for our very existence. However, only evidence-based policy and management, developed through rigorous science, will allow us to respond appropriately to the growing environmental crisis. We believe that PAs can and should play a major role in combating invasions, not only by improving the efficacy of IAS management within their territories, but also raising awareness at all levels, improving the capacity of practitioners to deal with invaders, implementing site-based prevention efforts, enforcing early detection and rapid response frameworks, and catalysing action also beyond the park boundaries (Genovesi and Monaco 2014). Protected areas can thus be reservoirs of biodiversity, but also sentinels of invasions as well as of other emerging threats to biodiversity, champions of best practises, and catalysts of action also at a broader scale than that of the PAs.

1.7   Terminology

As with many fields in the conservation and ecological sciences, a plethora of terminology has arisen for describing issues relating to biological invasions. Much of the lexicon of invasions is heavily contested. For the purposes of this book we have adopted a generalised lexicon.

In defining invasions by alien plants we adopt the terminology associated with the introduced-naturalization-invasion continuum as elucidated by Richardson et al. (2011) and the proposed unified framework for biological invasions as set out by Blackburn et al. (2011). These frameworks provide the basis for the objective classification of the status of introduced species and for the related discussion of associated processes (see also Richardson and Pyšek 2012).

Many different terms and categories are used to define 'protected areas' in different parts of the world, reflecting the national objectives, societal needs and approaches to management. The IUCN definition of protected areas is: "A clearly defined geographical space, recognised, dedicated and managed, through legal or other effective means, to achieve the long-term conservation of nature with associated ecosystem services and cultural values" (Dudley 2008). The IUCN classifies PAs as one of six categories: Ia: strict nature reserve/wilderness protection area; Ib: wilderness area; II: national park; III: natural monument; IV: habitat/species management area; V: protected landscape/seascape; and VI: managed resource protected area. Similarly, but more simply, the term 'protected area' may be used to designate any area specifically designed or formally proclaimed for the protection of biodiversity, landscapes (natural or cultural) and processes therein. Different chapters and case studies refer to specific types of protected areas such as IUCN WDPA categories, nature reserves, heritage sites, Ramsar wetland sites, marine reserves or parks, wilderness areas and others.

Acknowledgments We thank Dan Simberloff for his support for this book. LCF thanks South African National Parks for supporting work on this book and for general support. LCF and DMR thank the Centre for Invasion Biology, the National Research Foundation (South Africa) and Stellenbosch University for support. PP was supported by long-term research development project no. RVO 67985939 (Academy of Sciences of the Czech Republic), institutional resources of Ministry of Education, Youth and Sports of the Czech Republic, and acknowledges the support by Praemium Academiae award from the Academy of Sciences of the Czech Republic. We thank Zuzana Sixtová for technical assistance with editing.

References

Baron JS, Allen CD, Fleishman E et al (2008) National parks. In: Adaptation options for climate-sensitive ecosystems and resources. The U.S. Climate Change Science Program, Washington, DC, p 35

Barrett NE, Barrett JP (1997) Reserve design and the new conservation theory. In: Pickett STA, Ostfeld RS, Shachak M et al (eds) The ecological basis of conservation. Heterogeneity, ecosystems and biodiversity. Chapman and Hall, New York, pp 236–251

Bennett AF, Saunders DA (2010) Habitat fragmentation and landscape change. In: Sodhi NS, Ehrlich PR (eds) Conservation biology for all. Oxford University Press, Oxford, pp 88–104

Bigalke R (1947) The adulteration of the fauna and flora of our national parks. S Afr J Sci 43:221–225

Blackburn TM, Pyšek P, Bacher S et al (2011) A proposed unified framework for biological invasions. Trends Ecol Evol 26:333–339

British Ecological Society (1944) Nature conservation and nature reserves. J Ecol 1:45–82

Brockie RE, Loope LL, Usher MB et al (1988) Biological invasions of island nature reserves. Biol Conserv 44:9–36

Bruner AG, Gullison RE, Rice RE et al (2001) Effectiveness of parks in protecting tropical biodiversity. Science 291:125–129

Butchart SHM, Scharlemann JPW, Evans MI et al (2012) Protecting important sites for biodiversity contributes to meeting global conservation targets. PLoS One 7:e32529

Cadotte MW, McMahon SM, Fukami T (2006) Conceptual ecology and invasion biology: reciprocal approaches to nature. Springer, Berlin

CBD (2004) Convention of biological diversity, CoP 7 decision VII/30. Strategic plan: future evaluation of progress. Goal 1 – promote the conservation of the biological diversity of ecosystems, habitats and biomes; Target 1.1. http://www.cbd.int/decision/cop/?id=7767. Accessed 17 Feb 2013

Conroy MJ, Runge MC, Nichols JD et al (2011) Conservation in the face of climate change: the roles of alternative models, monitoring, and adaptation in confronting and reducing uncertainty. Biol Conserv 144:1204–1213

Cowie ID, Werner PA (1993) Alien plant species invasive in Kakadu National Park, Tropical Northern Australia. Biol Conserv 63:127–135

Davis MA (2009) Invasion biology. Oxford University Press, Oxford

De Poorter M (2007) Invasive alien species and protected areas: a scoping report. Part 1. Scoping the scale and nature of invasive alien species threats to protected areas, impediments to invasive alien species management and means to address those impediments. Global Invasive Species Programme, Invasive Species Specialist Group. http://www.issg.org/gisp_publica tions_reports.htm

Devall B (1982) John Muir as deep ecologist. Environ Rev 6:63–86

du Toit JT, Rogers KH, Biggs HC (eds) (2003) The Kruger experience. Ecology and management of savanna heterogeneity. Island Press, Washington

Dudley N (ed) (2008) Guidelines for applying protected area management categories. IUCN, Gland. http://data.iucn.org/dbtw-wpd/edocs/PAPS-016.pdf

Dudley N, Parish J (2006) Closing the gap. Creating ecologically representative protected area systems: a guide to conducting the gap assessments of protected area systems for the convention on biological diversity. Secretariat of the Convention on Biological Diversity, Montreal

Dudley N, Stolton S (eds) (2012) Protected landscapes and wild biodiversity, vol 3. Values of protected landscapes and seascapes. Protected Landscapes Specialist Group of IUCN's World Commission on Protected Areas. International Union for Conservation of Nature, Gland

Dudley N, Higgins-Zogib L, Mansourian SM (2005) Beyond belief: linking faiths and protected areas to support biodiversity conservation. Arguments for Protection. WWF, Equilibrium and Alliance of Religions and Conservation (ARC). World Wide Fund for Nature, Gland

Fiedler PL, White PS, Leidy RA (1997) A paradigm shift in ecology and its implications for conservation. In: Pickett STA, Ostfeld RS, Shachak M et al (eds) The ecological basis of conservation. Heterogeneity, ecosystems, and biodiversity. Chapman and Hall, New York, pp 83–92

Foxcroft LC, Henderson L, Nichols GR et al (2003) A revised list of alien plants for the Kruger National Park. Koedoe 46:21–44

Foxcroft LC, Pyšek P, Richardson et al (2014) Chapter 2: Impacts of alien plant invasions in protected areas. In: Foxcroft LC, Pyšek P, Richardson DM, Genovesi P (eds) Plant invasions in protected areas: patterns, problems and challenges. Springer, Dordrecht, pp 19–41

Gaston KJ, Jackson SF, Cantú-Salazar L et al (2008) The ecological performance of protected areas. Ann Rev Ecol Evol Syst 39:93–113

Genovesi P, Monaco A (2014) Chapter 22: Guidelines for addressing invasive species in protected areas. In: Foxcroft LC, Pyšek P, Richardson DM, Genovesi P (eds) Plant invasions in protected areas: patterns, problems and challenges. Springer, Dordrecht, pp 487–506

Hobbs RJ, Cole DN, Yung L et al (2010) Guiding concepts for park and wilderness stewardship in an era of global environmental change. Front Ecol Environ 8:483–490

Houston DB, Schreiner EG (1995) Alien species in national parks: drawing lines in space and time. Conserv Biol 9:204–209

IUCN and UNEP-WCMC (2012) The World Database on Protected Areas (WDPA): February 2012 [On-line]. UNEP-WCMC, Cambridge. http://www.protectedplanet.net/. Accessed 21 Oct 2012

IUCN-UNEP-WWF (1980) World conservation strategy: living resource conservation for sustainable development. International Union for Conservation of Nature and Natural Resources, United Nations Environment Programme, World Wildlife Fund. doi:10.2305/IUCN.CH.1980.9.en

Jenkins CN, Joppa L (2009) Expansion of the global terrestrial protected area system. Biol Conserv 142:2166–2174

Koh LP, Gardner TA (2010) Conservation in human-modified landscapes. In: Sodhi NS, Ehrlich PR (eds) Conservation biology for all. Oxford University Press, Oxford, pp 236–258

Leopold A (1949) A sand county almanac and sketches here and there. Oxford University Press, New York

Li Y, Norland M (2001) The role of soil fertility in invasion of Brazilian pepper (*Schinus terebinthifolius*) in Everglades National Park, Florida. Soil Sci 166:400–405

Lonsdale WM (1993) Rates of spread of an invading species: *Mimosa pigra* in Northern Australia. J Ecol 81:513–521

Lonsdale WM (1999) Global patterns of plant invasions and the concept of invasibility. Ecology 80:1522–1536

Loope LL, Sanchez PG, Tarr PW et al (1988) Biological invasions of arid land nature reserves. Biol Conserv 44:95–118

Loope LL, Meyer J-Y, Hughes RF (2014) Chapter 15: Plant invasions in protected areas of tropical Pacific Islands, with special reference to Hawaii. In: Foxcroft LC, Pyšek P, Richardson DM, Genovesi P (eds) Plant invasions in protected areas: patterns, problems and challenges. Springer, Dordrecht, pp 313–348

Macdonald IAW, Frame GW (1988) The invasion of introduced species into nature reserves in tropical savannas and dry woodlands. Biol Conserv 44:67–93

Macdonald IAW, Graber DM, DeBenedetti S et al (1988) Introduced species in nature reserves in Mediterranean-type climatic regions of the world. Biol Conserv 44:37–66

Macdonald IAW, Loope LL, Usher MB et al (1989) Wildlife conservation and the invasion of nature reserves by introduced species: a global perspective. In: Drake J, Mooney HA, Di Castri F (eds) Biological invasions: a global perspective. Wiley, Chichester, pp 215–255

Margules CR, Pressey RL (2000) Systematic conservation planning. Nature 405:243–253

Meine C (2010) Conservation biology: past and present. In: Sodhi NS, Ehrlich PR (eds) Conservation biology for all. Oxford University Press, Oxford, pp 7–22

Millennium Ecosystem Assessment (2005) Ecosystems and human well-being: biodiversity synthesis. World Resources Institute, Washington, DC

Mora C, Myers RA, Coll M et al (2009) Management effectiveness of the world's marine fisheries. PLoS Biol 7:e1000131

Mulongoy KJ, Chape SP (eds) (2004) Protected areas and biodiversity: an overview of key issues. CBD Secretariat/UNEP-WCMC, Montreal/Cambridge

Nentwig W (ed) (2007) Biological invasions. Springer, Berlin

Newton AC (2011) Social-ecological resilience and biodiversity conservation in a 900-year-old protected area. Ecol Soc 16:13

Ostfeld RS, Pickett STA, Shachak M et al (1997) Defining the scientific issues. In: Pickett STA, Ostfeld RS, Shachak M (eds) The ecological basis of conservation. Heterogeneity, ecosystems, and biodiversity. Chapman and Hall, New York, pp 3–10

Pickett STA, Cadenasso ML, Benning TL (2003) Biotic and abiotic variability as key determinants of savanna heterogeneity at multiple spatiotemporal scales. In: du Toit JT, Rogers KH, Biggs HC (eds) The Kruger experience: ecology and management of savanna heterogeneity. Island Press, Washington, DC, pp 22–40

Pyšek P, Richardson DM (2010) Invasive species, environmental change and management, and health. Ann Rev Environ Res 35:25–55

Randall JM (2011) Protected areas. In: Simberloff D, Rejmánek M (eds) Encyclopaedia of biological invasions. University of California Press, Berkley/Los Angeles, pp 563–567

Richardson DM (ed) (2011) Fifty years of invasion ecology: the legacy of Charles Elton. Wiley-Blackwell, Oxford

Richardson DM, Pyšek P (2008) Fifty years of invasion ecology: the legacy of Charles Elton. Divers Distrib 14:161–168

Richardson DM, Pyšek P (2012) Naturalization of introduced plants: ecological drivers of bio-geographical patterns. New Phytol 196:383–396

Richardson DM, Pyšek P, Carlton JT (2011) A compendium of essential concepts and terminology in invasion ecology. In: Richardson DM (ed) Fifty years of invasion ecology: the legacy of Charles Elton. Wiley-Blackwell, Oxford, pp 409–420

Rogers KH (2003) Adopting a heterogeneity paradigm: implications for management of protected savannas. In: du Toit JT, Rogers KH, Biggs HC (eds) The Kruger experience: ecology and management of savanna heterogeneity. Island Press, Washington, DC, pp 41–58

Sandlund OT, Schei PJ, Åslaug V (eds) (1998) Invasive species and biodiversity management. Kluwer Academic Publishers, Dordrecht

Schmitz DC, Simberloff D, Hofstetter RH et al (1997) The ecological impact of nonindigenous plants. In: Simberloff D, Schmitz D, Brown T (eds) Strangers in paradise. Impact and management on nonindigenous species in Florida. Island Press, Washington, DC, pp 39–62

Shelford VE (1926) Naturalist's guide to the Americas. Williams and Wilkins, Baltimore

Simberloff D (2004) A rising tide of species and literature: review of some recent books on biological invasions. BioScience 54:247–254

Sinclair ARE, Packer C, Mduma SAR et al (eds) (2008) Serengeti III: human impacts on ecosystem dynamics. University of Chicago Press, Chicago

Usher MB (1988) Biological invasions of nature reserves: a search for generalizations. Biol Conserv 44:119–135

Usher MB, Kruger FJ, Macdonald IAW et al (1988) The ecology of biological invasions into nature reserves: an introduction. Biol Conserv 44:1–8

Wangchuk S (2007) Maintaining ecological resilience by linking protected areas through biological corridors in Bhutan. Trop Ecol 48:176–187

Zimmerer KS, Galt RE, Buck MV (2004) Globalization and multi-spatial trends in the coverage of protected-area conservation (1980–2000). Ambio 33:520–529

Chapter 2
The Bottom Line: Impacts of Alien Plant Invasions in Protected Areas

Llewellyn C. Foxcroft, Petr Pyšek, David M. Richardson, Jan Pergl, and Philip E. Hulme

Abstract Phrases like "invasive species pose significant threats to biodiversity..." are often used to justify studying and managing biological invasions. Most biologists agree that this is true and quantitative studies support this assertion. Protected areas are the foundation of conservation initiatives in many parts of the world, and are an essential component of an integrated approach to conserving biodiversity and the associated ecosystem services. The invasion of alien plants constitutes a

L.C. Foxcroft (✉)
Conservation Services, South African National Parks, Private Bag X402,
Skukuza 1350, South Africa

Centre for Invasion Biology, Department of Botany and Zoology,
Stellenbosch University, Private Bag X1, Stellenbosch 7602, South Africa
e-mail: Llewellyn.foxcroft@sanparks.org

P. Pyšek
Department of Invasion Ecology, Institute of Botany, Academy of Sciences
of the Czech Republic, Průhonice CZ 252 43, Czech Republic

Department of Ecology, Faculty of Science, Charles University in Prague,
CZ 128 44 Viničná 7, Prague 2, Czech Republic
e-mail: pysek@ibot.cas.cz

D.M. Richardson
Centre for Invasion Biology, Department of Botany and Zoology,
Stellenbosch University, Private Bag X1, Stellenbosch 7602, South Africa
e-mail: rich@sun.ac.za

J. Pergl
Department of Invasion Ecology, Institute of Botany, Academy of Sciences
of the Czech Republic, Průhonice CZ 252 43, Czech Republic
e-mail: pergl@ibot.cas.cz

P.E. Hulme
The Bio-Protection Research Centre, Lincoln University, PO Box 84,
Canterbury, New Zealand
e-mail: Philip.hulme@lincoln.ac.nz

L.C. Foxcroft et al. (eds.), *Plant Invasions in Protected Areas: Patterns, Problems and Challenges*, Invading Nature - Springer Series in Invasion Ecology 7, DOI 10.1007/978-94-007-7750-7_2, © Springer Science+Business Media Dordrecht 2013

substantial and growing threat to the ability of protected areas to provide this service. A large body of literature describes a range of impacts, but this has not been assessed within the context of protected areas. We do not aim to review the state of knowledge of impacts of invasive plants; rather, we collate examples of work that has been carried out in protected areas to identify important patterns, trends and generalities. We also discuss the outcomes of various studies that, while not necessarily undertaken in protected areas, are likely to become important for protected areas in the future. We discuss the range of impacts under five broad headings: (i) species and communities; (ii) ecosystem properties; (iii) biogeochemistry and ecosystem dynamics; (iv) ecosystem services; and (v) economic impacts.

Keywords Biogeochemistry • Conservation • Economic impact • Impact • Management • Nature reserve

2.1 Introduction: Why Are Impacts of Alien Plants in Protected Areas Especially Concerning?

Phrases like "invasive species pose significant threats to biodiversity…" are frequently used to justify the study and management of biological invasions. Most biologists agree that this is true and quantitative studies support this assertion (see e.g. Vilà et al. 2011; Pyšek et al. 2012; Simberloff et al. 2013 for recent reviews). Most ecologists and environmental managers agree that the diversity of life is in serious decline (Pimm et al. 2001; Pereira et al. 2010; Rudd et al. 2011), with some indicating that we are witnessing one of the greatest extinction events in our planet's history (e.g. Novacek and Cleland 2001). Protected areas (PAs) are part of an approach to conserve biodiversity and slow its loss (Hansen et al. 2010). Indeed, in a survey of 93 terrestrial PAs in 22 tropical countries, protected areas were shown to be effective in halting problems such as land clearing, logging, hunting, unplanned fires and overgrazing (Bruner et al. 2001); unfortunately impacts of invasive alien species were not included in the study.

The invasion of alien plants in PAs poses a serious concern for one of the most pressing conservation initiatives globally. The intensity of research on impacts of invasive plants varies among regions (Hulme et al. 2013), but there are some notable cases, for example in Hawaii Volcanoes National Park, where a substantial body of literature exists (see Loope et al. 2014). In this chapter we do not attempt a comprehensive review of what is known about the impacts of invasive plants in general, as many extensive reviews have been carried out on, for example, impacts of invaders on species, communities and ecosystems (Pyšek et al. 2012), soil nutrient cycling (Ehrenfeld 2003), mechanisms underlying impacts (Levine et al. 2003), ecosystem carbon and nitrogen cycling (Liao et al. 2007), hybridisation (Vilà et al. 2000), competition (Vilà et al. 2004), plant reproductive mutualisms (Traveset and Richardson 2006) and ecosystem services (Vilà et al. 2010). Rather, we examine what has been done within PAs, or what is specifically pertinent to them, due to their unique and essential conservation role.

Some studies have found that PAs contain fewer invasive species than their surrounds. A study of 184 PAs globally found about half the number of aliens inside the parks than outside (Lonsdale 1999). Similarly, across 302 nature reserves declared between 1838 and 1996 in the Czech Republic, significantly fewer alien species were found in the reserves (Pyšek et al. 2002). Further, the presence of intact natural vegetation appears to help slow the establishment of alien plants. A study examining the role of the boundary as a filter to alien plants in Kruger NP (South Africa) also showed that in areas where there was more than 90 % natural vegetation within a 5 km radius of the park, alien plants were significantly less likely to invade (Foxcroft et al. 2011; Jarošík et al. 2011).

Opposite trends are unfortunately frequently reported, for example, showing that alien plants can invade natural areas that have not experienced anthropogenic disturbances (e.g. Gros Morne NP in boreal Canada; Rose and Hermanutz 2004). As early as the 1980s, the SCOPE (Scientific Committee on Problems of the Environment) programme on biological invasions reported 1,874 alien invasive vascular plants from 24 case studies of nature reserves globally (Usher 1988; Macdonald et al. 1989). In southern Africa, only seven out of 307 PA managers that responded to a survey were of the opinion that no alien species were known to occur in their reserve (Macdonald 1986). In a 1980 report to Congress in the USA, 300 national park service areas reported 602 perceived threats to natural resources involving alien plants and animals (see Houston and Schreiner 1995). At around the same time, at least 115 invasive alien plant species that threaten natural areas, parks and other protected lands had been identified in Virginia, USA (Heffernan 1998). A decade later a study reported 20,305 alien plant species infestations, with 3,756 unique alien plants, totalling 7.3 million ha in 218 national parks in the USA (Allen et al. 2009). A Global Invasive Species Programme report (De Poorter 2007) identified 487 PAs where invasive alien species were recorded as a threat. More than 250 wildlife refuges and 145 National Parks in the USA were shown to have been invaded by invasive alien species (De Poorter 2007). The Nature Conservancy indicated that of 974 of their projects globally, about 60 % regard invasive alien plants to be the main threat (2009, unpublished data at http://conpro.tnc.org/reportThreatCount). In the US national parks, 61 % of 246 park managers indicated that alien plant invasions were moderate or major concerns (Randall 2011). An assessment of 110 PAs in South Africa's Ezemvelo KZN Wildlife conservation agency found that invasive alien plants represent the greatest threat to biodiversity in the province of KwaZulu-Natal (Goodman 2003). Based on results of an internet survey, it has recently been reported that managers of PAs in Europe perceive invasive species as the second greatest threat to their areas after habitat loss (Pyšek et al. 2014).

Without doubt, the threat, impact and management problems associated with alien plant invasions in PAs are increasingly being recognised as a major issue. Providing science-based evidence of the negative impacts of these invasions is becoming increasingly important in motivating for resources from frequently under-resourced conservation budgets. Protected areas face numerous challenges, including tourism-related issues, wildfire management, poaching and illegal harvesting of resources (Barber et al. 2004; Dudley et al. 2005; Alers et al. 2007), and climate change (Hannah et al. 2002; Huntley et al. 2011). Consequently, alien species control programmes must compete with these often emotive and charismatic management needs for resources. Managers often require evidence of potential problems within their area of concern and while localised empirical investigations can provide this information, collaboration across similar situations and systems can provide a much broader, synthetic understanding (Kueffer 2012).

Particular kinds of impacts are likely to be of more concern to different PAs than to other categories of land use, due to the specific objectives of PAs. Although the core function of many PAs is to conserve native 'biodiversity' in as natural a state as possible, the concept of 'biodiversity' is interpreted differently for different situations (Mayer 2006). Some PAs focus mainly on rare or single species protection, others on conservation of ecological processes, and yet others on landscapes, habitats or patch dynamics (Nott and Pimm 1997). Depending on the goal of the PA, where plant invasions threaten the specific entity of concern, different kinds of management approaches may be adopted. Outside PAs there is growing acceptance of the concept of 'novel ecosystems' which posits, among other things, that some ecosystems should be managed to ensure the continued delivery of particular services, irrespective of the composition of species in that system (native vs. alien) (Hobbs et al. 2006). This philosophy is unlikely to be widely adopted for PAs soon, except in very special cases, although tenets of the novel ecosystem philosophy will certainly be more widely discussed in general conservation forums in the future.

2.2 Impacts of Alien Plant Invasions: Species, Ecosystems, Processes and Economics

The search for general models for conceptualising and evaluating impacts of invasive alien species has been underway for many years. Early descriptions were mostly observational. For example Elton's widely acclaimed book, Ecology of invasions by animals and plants (Elton 1958), included many anecdotal observations. The SCOPE programme on nature reserves in the 1980s indicated that all case studies had examples of presumed effects of invasive species, but that it was difficult to clearly identify the cause of the observed impact (Usher 1988). Later, correlative approaches began being employed, comparing pre- and post-invasion sites, or sites with varying levels of abundance (for example, Parker and Reichard 1998). A generalised model for understanding ecological impacts (Parker et al. 1999) argued that the net impact of an invasive species should be conceptualised as the product of the geographic range of the invaders (area invaded), its abundance (density or biomass) and the per-capita or per-biomass effect. This model may provide PAs with a usable method for objectively assessing impacts of different species, especially where the distribution can be accurately

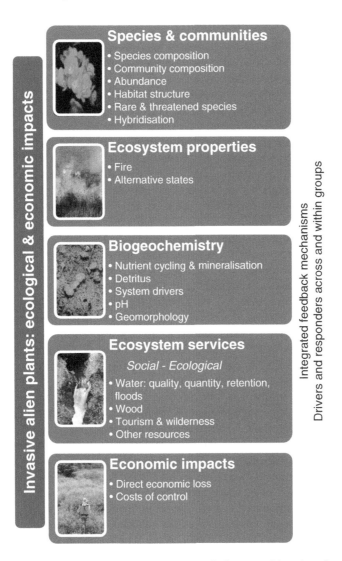

Fig. 2.1 Generalised outline of ecological and economic impacts of invasive alien plants in protected areas (Photos: Llewellyn C. Foxcroft, Navashni Govender (fire), Ezekiel Khoza (spraying of *Parthenium hysterophorus*))

mapped and abundance precisely estimated. Describing and quantifying the per-capita effect remains a challenge (Parker et al. 1999).

To avoid discussing the impacts of invasive plants as simply a list of examples, we used a general outline which clusters related issues (Fig. 2.1). The broad headings we use are (i) species and communities; (ii) ecosystem properties; (iii) biogeochemistry and ecosystem dynamics; (iv) ecosystem services; and (v) economic impacts.

2.2.1 Species and Communities

Global indicators for IAS under the Convention on Biological Diversity's 2010
Biodiversity Target show that invasive species are causing a decline in species
diversity of IUCN red-listed amphibians, birds and mammals (McGeoch et al.
2010). In an assessment of impacts on imperilled species in the USA, about 57 % of
1,055 listed plants were threatened by alien species (Wilcove and Chen 1998).
Although a number of such generalised lists indicate the threat from alien
plant invasions to native species (e.g. Mauchamp 1997; Pimentel et al. 2000), there
is still a shortage of quantitative data. Such information is particularly crucial for
advising policy makers, as the value of conservation is currently predominantly
measured by its ability to protect species richness (a common interpretation of
'biodiversity', Mayer 2006).

A review of about 150 studies provided a synthetic understanding of general
mechanisms underlying the impacts on plant and animal community structure,
nutrient cycling, hydrology and fire regimes (Levine et al. 2003). These authors
found that many studies examined the impacts of invasions on plant diversity and
composition, but fewer than 5 % test whether these effects arise through competition, allelopathy, alteration of ecosystem variables or other processes. Nonetheless,
competition was often hypothesised as a primary mechanism, and in nearly all
studies alien plants exhibited strong competitive effects over native species. In
contrast to studies of the impacts on plant community structure and higher trophic
levels, research examining impacts on nitrogen cycling, hydrology and fire regimes
is generally highly mechanistic, often driven by species-specific traits.

An early study that sought to quantify the impacts of alien plants in PAs was in
Theodore Roosevelt Island Nature Preserve in Washington, DC, USA. Here two
species of invasive vines were shown to inhibit the recruitment of native forest
species (Thomas 1980). *Lonicera japonica* (Japanese honeysuckle) inhibited the
reproduction of dominant forest trees such as *Liriodendron tulipifera* (tulip poplar),
Prunus serotina (wild black cherry) and *Ulmus americana* (American elm). *Hedera
helix* (English ivy) mainly inhibited the recruitment of herbaceous species. The
smaller plants were suppressed, and even established forest trees were eventually
killed through shading (Thomas 1980). Alien vine species may have advantages
due to altered phenologies (e.g. evergreen vs. deciduous) and can often invade low
light habitats (Gordon 1998). In this way, species with different life-forms can
cause patches of native plants to collapse and be completely replaced, thereby
altering, for example, community or species structure and light regimes (Gordon
1998). Floating species, such as *Eichhornia crassipes* (water hyacinth), native to
north-western Argentina, can invade the total surface area of a waterway,
completely preventing any light penetration (Ashton and Mitchell 1989).
Eichhornia crassipes is one of the world's worst aquatic invaders, and has been
reported to have invaded PAs in Asia, Australia, New Zealand, Africa, and the USA
(De Poorter 2007).

Biodiversity indicators, frequently using spiders or beetles, provide information on the presence of a set of other species in an area (McGeoch 1998) and are increasingly being used as proxies for quantifying impacts. In Hluhluwe-iMfolozi Game Reserve (South Africa), *Chromolaena odorata* (Siam weed) invasion altered native spider assemblages, with negative changes in abundance, diversity and estimated species richness. These changes were, however, reversed immediately following clearing (Mgobozi et al. 2008). In a similar study in Kruger NP, an assessment of the impact of *Opuntia stricta* (sour prickly pear) found that across a gradient of its density, species richness and species density for beetles and spiders did not change significantly (Robertson et al. 2011). Assemblages for spiders also did not differ across treatments, but beetle assemblages were significantly different. In South African National Parks as a whole, 663 alien plant species (813 alien species in total) have been recorded (Spear et al. 2011), but other than a few isolated projects, to date little work has been done on quantifying their impacts.

Impacts of invasive plants, primarily *C. odorata*, have also been reported on small and large mammals in Hluhluwe-iMfolozi GR (Dumalisile 2008). Small mammals showed both higher species richness and diversity in uninvaded sites compared to invaded sites, regardless of *C. odorata* density. Large mammals also decreased in richness and diversity as *C. odorata* invasion density increased. Invasive alien plants can also, perhaps unexpectedly, even threaten mega-herbivores. For example, Kaziranga NP in India is a vital habitat for the world's largest population of the great one-horned rhinoceros (*Rhinoceros unicornis*). The rhino is dependent on grasslands, which have been invaded by *Mimosa rubicaulis* (Himalayan mimosa), *M. diplotricha* (giant sensitive plant) and *Mikania micrantha* (mile-a-minute weed), hampering the growth of native palatable grasses (Lahkar et al. 2011). In Kenya, *Lantana camara* (lantana) invasions reportedly impact on the habitat of Sable antelope (*Hippotragus niger*; Steinfeld et al. 2006). Nile crocodile (*Crocodylus niloticus*) nesting habitat and sex ratios may be altered by invasions of *C. odorata* in KwaZulu-Natal, South Africa, due to shading and cooling of nesting sites by 5.0–6.0 °C. This can result in a female-biased sex ratio, with potentially adverse consequences for the population (Leslie and Spotila 2001). The wetlands of Kakadu NP in Australia, a world heritage and Ramsar site, are renowned for their high diversity and numbers of water birds, and are under threat from *Mimosa pigra* (sensitive plant) and *Urochloa mutica* (para grass; Setterfield et al. 2014).

By 2000, alien plants had invaded approximately 700,000 ha of US wildlife habitat per year (Babbitt 1998, as cited in Pimentel et al. 2000). In Great Smoky Mountains NP (USA) for example, 400 of the approximately 1,500 vascular plant species are alien, and 10 of these are currently displacing and threatening native plant species (Hiebert and Stubbendieck 1993). Overall, Hawaii is estimated to have lost about 8 % of its native plant species, with an additional 29 % still at risk (Loope 2004; Fig. 2.2).

Invasive plants can also contribute to an increased abundance of other invasive species, thus facilitating 'invasional meltdown' (sensu Simberloff and Von Holle 1999). In Hawaii Volcanoes NP, the widespread *Morella* (= *Myrica*) *faya* (faya tree) significantly increases the abundance of the alien insect *Sophonia rufofascia* (a leaf

Fig. 2.2 Sectional structure of *Psidium cattleianum* invasion in Hawaiian lowland rainforest (**a**) with closed canopy of *P. cattleianum* within the forest (**b**) uninvaded forest (**c**) aerial view of *P. cattleianum* invasion (Figures: Gregory Asner, Carnegie Airborne Observatory, Carnegie Institution for Science)

phloem-feeding insect of Asian origin). In areas where *M. faya* is present the abundance of *S. rufofascia* was up to 19 times more abundant than in areas where *M. faya* had been removed (Lenz and Taylor 2001). This is of substantial concern as the diversity of host plants fed upon by *S. rufofascia* is extremely broad, encompassing over 300 species from 87 different families. Among these, 67 species are endemic or native to Hawaii, and 14 are either endangered or candidates for listing (Lenz and Taylor 2001).

The loss of genetic purity of a species is an important concern, especially for those rare and or threatened species which may face extinction, and have been given sanctuary in PAs. The hybridisation between alien and native species can lead to genetic swamping and loss of native species' genetic diversity. These risks are increased when a rare species hybridises with an abundant species, producing fertile offspring that can back-cross (introgress; Rhymer and Simberloff 1996). Invasive species may swamp native species through hybridisation. For example, the native species *Hyacinthoides non-scripta* (bluebell), an iconic species in the British Isles, is being threatened by its conger *H. hispanica* and its hybrid with the native *Hyacinthoides × massartiana*. Both the introduced and hybrid species are naturalised, and are frequently found within 1 km of *H. non-scripta* (Kohn et al. 2009). Thus conservation programmes should strive to isolate rare species from cross-compatible congeners (Mooney and Cleland 2001).

Invasions can substantially influence plant reproductive mutualisms, while potentially disrupting mutualistic processes in invaded regions (Traveset and Richardson 2011). Good evidence exists for such impacts on pollination and reproductive success of native species (Traveset and Richardson 2011). For example, invasive plants that are highly attractive to pollinators can reduce overall visitation of native species (Morales and Traveset 2009; Gibson et al. 2012). These interactions can have consequences for whole communities due to the effects cascading through the network (Traveset and Richardson 2011). Such effects in PAs could be profound, by impacting directly on the biodiversity conservation objectives of the area.

Plant invasions have, to date, caused relatively few plant species to go extinct (Gurevitch and Padilla 2004; Sax and Gaines 2008). One reason for this is that, unlike the case with animal extinctions, plant extinctions can take decades or even centuries to play out (Gilbert and Levine 2013). However, plant invasions have led to the fragmentation of native plant communities worldwide, many of which currently survive as the 'living dead' (sensu Parker et al 1999, p. 12). This may be due to the persistence of native species in marginal habitats which, although still present, are reduced in abundance and distribution (Gilbert and Levine 2013). For example, serpentine soil landscapes in California include numerous rare and threatened plant species of high conservation concern. Invasions by European grasses impact on the area and quality of native species habitat, and may cause extinction hundreds of years after fully transforming the habitat (Gilbert and Levine 2013). Consequently, equating impact with numbers of absolute extinctions is misleading and inappropriate. Native species may still persist within an invaded area, but often be compromised or marginalised to such an extent that they no longer perform (to the same level, or at all) the functional roles they performed before they were affected by the invasive species (Wardle et al. 2011). Such changes are pervasive in ecosystems worldwide, and invasive plant species are increasingly prominent 'builders and shapers' of novel ecosystems in many regions (Richardson and Gaertner 2013). The examples discussed in this chapter show that invasive plant species very often drive ecosystems beyond thresholds at which ecological states are irreversibly altered. Such modified systems can sometimes be managed to deliver desired services, but such conditions are unacceptable in many PAs where the aim in to conserve species, community, structural and functional diversity.

2.2.2 Ecosystem Properties: Changes in Fire Regime

Ecosystems are the product of interactions between climatic conditions, resource availability and disturbance, of which the functional diversity of species is a major driver (Hooper et al. 2005). Biological invasions, often in concert with other global change drivers, have been shown in many cases to alter species diversity and community structure, thus having profound cascading effects on ecosystem functioning (Strayer 2012). The alternative states resulting from the relationship between invasive plants and fire, or the shading or smothering effect of vines and other species, are of some of the concerns facing PA managers, due in part to the irreversibility of such system changes.

Fire management has received substantial attention in PAs in many regions. This is because fire is a key driver of vegetation heterogeneity and patchiness in many systems, changing the structure, and relationships between trees and grasses (Bond et al. 2005), and acting as the primary diver of multiple ecosystem functions (Cole and Landres 1996; van Wilgen et al. 2003). Fire management has also attracted much attention because fire poses a hazard to infrastructure and human safety in PAs (e.g. Loehle 2004). Species, communities and even whole biomes have evolved with a particular tolerance to fire, including frequency, intensity, timing and vertical position (crown vs. ground). While the changing role of humans has received considerable attention in explaining and attempting to manage fire regimes generally (van Wilgen et al. 2003), increasing awareness is being given to changes in fire regimes, and consequently changes in ecosystem function, due to the widespread invasions of alien plants in many PAs.

Changes in fire regimes and ecosystem function due to invasions by alien plants have been documented from a range of habitats. One of the most frequently cited examples is related to the disruption of the grass-fire cycle (D'Antonio and Vitousek 1992). Invasive plants can increase vegetation flammability in areas where native species are poorly adapted and unable to cope in the presence of fire. In Hawaii Volcanoes NP at least one endangered plant and many of the dominant, poorly adapted, native species have been eliminated by fire (Hughes et al. 1991; D'Antonio and Vitousek 1992; Loope 2011). An increase in biomass of fine fuels can significantly increase the intensity or frequency of fires, or both. Introduced grasses that are fire adapted, or evolved in the presence of fire, are able to recover quickly after being burned, creating a positive feedback cycle that favours further invasion (D'Antonio and Vitousek 1992). In the Wildman Reserve in northern Australia, invasion of *Andropogon gayanus* (gamba grass) increased fuels loads by up to seven times, and increased fire intensity by up to eight times compared to areas with native grasses (Rossiter et al. 2003). Further, *A. gayanus* was shown to inhibit soil nitrification, thereby depleting total soil nitrogen from the already nitrogen-poor soils and promote fire mediated nitrogen loss (Rossiter-Rachor et al. 2009). Combined with the altered fire regime, it then forms self-perpetuating positive feedback loops (Rossiter-Rachor et al. 2009). In Kakadu NP, while *Urochloa mutica* produces dry season fuel loads similar to the native *Hymenachne acutigluma* (olive hymenchne), the fuel is drier and taller, increasing the fire intensity. Higher fire intensity and frequency may facilitate the displacement of *H. acutigluma*, which is fire sensitive, and damage other fire-sensitive woody vegetation (Setterfield et al. 2014).

In Mesa Verde NP (USA), successional pathways were altered following high intensity fires, with woodland-dominated systems being replaced by herbaceous species (Floyd-Hanna et al. 1993). Significant changes have been experienced in Dinosaur NP and Snake River Birds of Prey NP (USA), which have been invaded

Fig. 2.3 *Acacia paradoxa* thickets in Table Mountain National Park (Photo Rafael D. Zenni)

by *Bromus tectorum* (cheatgrass). The fire frequency has been changed from one in 60–100 year, to three in 3–5 year return cycles, converting native shrublands to alien dominated grasslands (Randall 2011). In the Florida Everglades, marshlands with sedges, grasses and herbs have been replaced by *Melaleuca quinquenervia* (Australian paper bark), creating large stands of swamp forests with little or no herbaceous understory. Moreover, *M. quinquenervia* promotes crown fires, whereas the native plants have evolved with higher frequency, low intensity surface fires (Randall 2011). Another invasive plant that has caused major changes to the fire regime in the Everglades is *Lygodium microphyllum* (Old World climbing fern), a vine-like fern that climbs on trees and shrubs, forming mats that cause canopy trees to collapse. Fires that would normally stop at the edge of native cypress sloughs, travel up the 'fire ladders' provided by dry fronds of *L. microphyllum* to kill tree canopies (Schmitz et al. 1997).

Should fire regimes be changed significantly, species may become globally, locally or functionally extinct in a PA. In Table Mountain NP in South Africa's fynbos region, fire plays a key role in the maintenance of ecosystems, and native plants are adapted to the fire regime (Forsyth and van Wilgen 2008). However, the most common invasive species in the park, a suite of Australian *Acacia* (Fig. 2.3) and *Hakea* species, are also fire adapted, and their ability to produce large numbers of seeds facilitates their prolific spread after fires (van Wilgen et al. 2012). These trees and shrubs increase biomass and add to fuel loads, leading to increased fire intensity and erosion (van Wilgen and Scott 2001). Due to uncontrolled fires combined with the effects of plant invasions, 13 endemic plant species are known to have gone extinct since European colonization, with many more facing imminent extinction (Trinder-Smith et al. 1996).

Where invasive plants provide positive feedback systems to enhance habitat invasibility by altering, for example, nutrient cycling and fire frequency and intensity (Rossiter et al. 2003), alternative ecosystem states may emerge (Richardson and Gaertner 2013). The ability of an ecosystem to recover from such states or severe degradation depends on the extent of change to functional and structural properties (Brooks et al. 2010). The degree to which invasion and degradation change the biotic and abiotic threshold determines the level of intervention required to return the system to a state allowing natural regenerative process to function (Brooks et al. 2010). Whether active or passive restoration is necessary may depend on the nature of these legacy effects (Larios and Suding 2014). In PAs specifically, preventing degradation to the point where alternative states emerge, should be a high priority, not only to prevent compromising the area integrity, but also to allow the resilience of the system to recover following control (Jäger and Kowarik 2010). This is also important due the substantial costs likely to be associated with resource demanding active restoration programmes (in addition to costs for removal of the invader only).

2.2.3   Biogeochemistry and Ecosystem Dynamics

The ecosystem-level energy budget and biogeochemical cycling involve complex interactions of many facets at multiple spatial scales. These very interactions provide the ecosystem services on which humans depend for their existence (Sekercioglu 2010). Invasive alien plants are implicated in driving substantial changes to biogeochemical cycling and ecosystem dynamics (Ehrenfeld 2011), by altering components of the soil carbon, nitrogen, water and other ecosystem cycles (Ehrenfeld 2003). Invasive plant species have been shown to increase biomass, net primary productivity and nitrogen availability in many areas. Nitrogen fixation rates are altered and litter with higher decomposition rates than that of co-occurring native species is produced (Ehrenfeld 2003). However, the trends are not always clear. In a review of 56 invasive plants, variations across sites, and even opposite trends, were found (Ehrenfeld 2003; see also Hulme et al. 2013). While this work can be generalised across some PAs in similar settings, it appears that less work has been done on investigating the effects in PAs. Examining site-specific cases not only contributes to a general understanding of invasion, but can provide detailed onsite information on the ecological integrity of a PA.

For example, in Hawaii Volcanoes NP, nitrogen-fixing species (*Morella faya* and *M. cereifera*) significantly increase soil nitrogen availability, by up to 400 % (Vitousek et al. 1987). These changes resulted in altered plant succession trajectories, promoted increases in populations of alien earthworms, which in turn increased nitrogen burial rates, thereby further changing soil nutrient cycles (Randall 2011). In contrast, in northern Australia *A. gayanus* inhibits soil nitrification, thereby depleting total soil nitrogen from the already nitrogen-poor soils (Rossiter-Rachor et al. 2009). In Picayune Strand State Forest in Florida, *Melaleuca quinquenervia* significantly altered both above- and belowground ecosystem components (Martin et al. 2009). However, the detectability of impacts on the changes in ecosystem dynamics remains difficult. By the time these changes have increased to the level where they can be quantified, severe impacts are likely to have already occurred (Vilà et al. 2011), especially on plant species and communities. This will require a detailed, long-term monitoring programme to detect changes and determine trends.

2.2.4 Ecosystem Services

Ecosystem services are the benefits or the range of ecosystem functions, on which human livelihoods and wellbeing depend (Millennium Ecosystem Assessment 2005; Sekercioglu 2010). The impacts of invasive alien plants on ecosystem services is gaining much interest (Charles and Dukes 2007), and increasing efforts are being made to understand which ecosystem processes are being disrupted or altered by biological invasions.

The Millennium Ecosystem Assessment (2005) classified a number of potential services in four broad classes, including provisioning, regulating, supporting and cultural services. In brief, provisioning services are those tangible products obtained from ecosystems, including food, freshwater, fibre, fuel and genetic resources. Regulating services relate to the governing functions of ecosystems in order to provide other kinds of resources, such as water regulation (timing and extent of flooding, runoff, and others), water purification and waste treatment. Cultural services are non-material benefits, including spiritual or religious values, cultural heritage, recreation/tourism, aesthetic values, and wilderness or values of a sense of place (Mulongoy and Chape 2004). Supporting services are required for the continued maintenance of globally encompassing functions, which include photosynthesis, primary production, nutrient cycling, water cycling and soil formation. While interactions are multi-faceted and complex, the ecosystem services are delivered by different taxa or trophic levels (e.g. from soil micro-organisms, to vegetation, mammals, or whole communities) and can be assessed in a range of functional groups (e.g. populations, ecosystems, species, ecosystems; Sekercioglu 2010).

Protected areas, besides their roles in conserving individual species and their habitats, can be important for maintaining ecosystem function. These functions underlie much of the ability of ecosystems to provide services. For example, where whole or large portions of water catchment areas can be protected from invasions or managed when invaded, the lower impacts on overall ecosystem cycles will allow for improved delivery of water-related services. The Sabie-Sand river is one of the healthiest rivers in the Kruger NP. However, by 2002 about 23 % of the upper catchment had been invaded to some degree, corresponding to a loss of about 9.4 % of the rivers natural flow (Le Maitre et al. 2002).

There are a growing number of cases where the importance of PAs in providing and maintaining ecosystems services have been realised. Baekdudaegan Mountains

Reserve, a unique forest protected area in Korea, explicitly aims to protect whole mountain ranges to maintain linkages, conserve biodiversity, sustain ecosystem services, and restore cultures and cultural values (IUCN 2009). Conservation planning frameworks are also beginning to be used to explore opportunities for aligning conservation goals for biodiversity with ecosystem services (Chan et al. 2006; Naidoo and Ricketts 2006).

2.2.5 Economic Impacts

Economic impacts are a crucial consideration in research on invasive alien species in general, and have been subject to intensive research in the last decade (e.g. Kasulo 2000; Pimentel et al. 2000; van Wilgen et al. 2008; Vilà et al. 2010). General studies on economic impacts by invasive alien species should provide compelling evidence for PA agencies as to the costs associated with inaction, and thus loss from resources as a result of invasion, and the costs associated with control, management or eradication. However, little work has been done on the economic costs of plant invasions in PAs specifically. This is probably because the situation in PAs is very different from other areas when it comes to economics. Because of the primary objective of PAs, which is most often to conserve biodiversity, standard economic models relating to production are often inappropriate.

For the above reason, economic assessments of plant invasions specifically related to PAs are scarce. However, some data are available from Europe. For example, the Czech regional offices for nature conservation of protected landscape areas spent about 1.8 million CZK (~US$100,000) per year on the management and eradication of the most important invasive plant species (Linc 2012; Fig. 2.4). Removal and management costs of *Prunus serotina* in conservation areas in Germany are estimated to be € 149 million (Reinhardt et al. 2003). The costs of controlling *Rhododendron ponticum* (rhododendron) invasion in the Snowdonia NP, Wales, was estimated to be £45 million (as at 2002; Gritten 1995; Pyšek et al. 2014).

In Kruger NP, between 1997 and 2011 the Working for Water programme spent about ZAR90 million (~US$10.7 million, as at September 2012) on control efforts. In 2008 the control of alien plants within Table Mountain NP cost approximately ZAR9 million (~US$1.08 million in 2008), with a focus of species of *Acacia*, *Hakea* and *Pinus* (Table Mountain National Park 2008). For the 2012–2013 financial year, the budgeted costs are approximately ZAR14 million (Table Mountain National Park US$1.7 million in September 2012; Foxcroft et al. 2014).

A potentially significant problem, and one in dire need of detailed assessments, is the relationship between tourism and PAs. Protected areas rely, to varying extents, on the revenue provided by (eco-)tourism for their long-term sustainability. Eco-tourism and PA visitation generates a significant proportion of the economic income of many countries (Eagles et al. 2002). Understandably, the development of infrastructure for tourism in PAs is being strongly promoted. However, two challenges arise; (i) invasive alien plants can impact on tourism experiences in various

Fig. 2.4 (**a**) *Impatiens glandulifera* (Himalayan balsam) is an annual plant native to Asia, currently invasive in many protected areas in Europe (see Pyšek et al. 2014). Introduced as a garden ornamental and still frequently planted, it spreads into semi-natural plant communities along water courses (Photo Jan Pergl), and (**b**) one of the top ten invasive plant species in European protected areas (see Pyšek et al. 2014), the hybrid taxon *Fallopia × bohemica* is a noxious invader in riparian habitats where it forms extensive continuous populations extending over large sections of river shores and outcompetes native flora (Photo Jan Pergl)

ways, and (ii) increased tourism increases the likelihood of new introductions into PAs.

2.3   Integration and Challenges

The disparity between the needs and focus of PA managers and scientists and the complexity of managing multiple drivers of invasions are key challenges for the management of invasive alien plants in PAs.

Scientific research on biological invasions has grown exponentially over the last decade, but the relevance of much of the research for solving the immediate problems of policy makers and managers has been questioned (e.g. Esler et al. 2010). In Spain, for example, an assessment of environmental managers revealed the concern that not enough attention was being paid to developing cost-efficient management approaches (Andreu et al. 2009).

With the limited resources available, managers of PAs need to prioritise all types of activities that are required, including the prioritisation of various control options across many invasive taxa and invaded areas (Pyšek et al. 2014). This includes taxa with many types of impacts on different attributes of biodiversity. Invasive alien plants can also be passengers, or secondary factors that take advantage of habitat change. For example, in the Haleakala and Hawaii Volcanoes NPs feral pigs are recognised as a keystone introduced species, as they are the single major factor contributing to the spread of many introduced plants. They not only create open habitats through digging, but also transport propagules in their hair and faeces (Stone and Loope 1987). In a similar case in Pasoh Forest Reserve, an undisturbed tropical forest on the Malaysian peninsular, *Clidemia hirta* (Koster's curse), which was considered unlikely to invade and alter forest regeneration, utilised patches where light was made available and soil disturbed by wild pigs, for establishment (Peters 2001). In Forty Mile Scrub NP, Australia, about 73 % of the dry rainforest and woodland savanna have been invaded with *Lantana camara*, with up to 5,000 individuals/ha being recorded (Fensham et al. 1994). It appears that root digging by pigs causes tree deaths, thereby allowing light penetration, which favours *L. camara*. The high level of invasion also causes substantially increased fuel loads and fires have killed canopy trees across a large area of the dry rainforest. In Gros Morne NP (Canada) moose (*Alces alces*), a non-native herbivore, appears to be the primary dispersal agents of alien plants, dispersing propagules and creating or prolonging disturbance by trampling (Rose and Hermanutz 2004).

The interaction of climate change and invasive alien plants is also becoming more concerning (see Dukes 2011 for a comprehensive review). One such concern is that the invasibility of habitats is likely to be increased. With the adaptability of introduced species to a wider range of climatic conditions, and the ability to rapidly exploit these changes, this may lead to an increase in distribution and abundance of invasive plants (Dukes 2011). There are a number of implications for PAs, with some suggesting that the impact may be greater in PAs than the broader landscape as the composition of species changes and vegetation types shift (e.g. Hannah et al. 2007; Gaston et al. 2008). While scientific models provide general recommendations, managers face the threats directly and are forced to develop and implement practical strategies, and thus need to be involved as collaborators in designing climate-change integrated conservation strategies (Hannah et al. 2002). One of the approaches recommend to enhance landscape connectivity between PAs against climate-change induced landscape and habitat shifts is implementing buffer zones around PAs, especially where conservation and compatible options are available (Hannah et al. 2002). However, many PAs already occur within a mosaic of highly transformed or disturbed landscapes, and heavily impacted areas can, for example, cause forest margins to retreat (Gascon et al. 2000), facilitating further invasions.

2.4   Conclusions

The range and severity of impacts of invasive plants in PAs is, in many areas, only starting to be realised. Examples of well documented impacts come from a few PAs, many of which are mentioned in this chapter. The extent of impacts are well documented for some of the best-studied PAs (e.g. Everglades NP and Hawaii Volcanoes NP, USA; Kakadu NP, Australia; Table Mountain NP, South Africa); these offer dire warnings that many types of invasive plants can cause many types of dramatic impacts. There is growing evidence that many other types of impacts are increasing in severity and extent in many other PAs; these include impacts that are driving the displacement of wildlife and changes to fire regimes, which undermine the justification for the existence of the PAs. More subtle effects of altered nutrient cycles and pollination and seed-dispersal networks are emerging more slowly, but will most likely result in significant and irreversible impacts.

Protected areas exclude different factors, to varying levels, extents and scales; this facilitates the examination of specific issues without the confounding effects that many factors associated with human dominated ecosystems typically bring to ecological studies. In the SCOPE programme (Usher et al. 1988) nature reserves were considered as useful outdoor laboratories where artificial impacts are minimised, and this is now even more important in a rapidly transforming world. While commendable efforts are being made to quantify the impacts in some PAs, improved knowledge transfer could certainly facilitate better uptake in management agendas worldwide. If PAs are to fulfil their role in the global conservation arena it is important that the mechanisms by which alien plant invasions degrade system attributes are understood, to enable appropriate response.

Acknowledgements LCF thanks South African National Parks for supporting work on this book and for general support. LCF and DMR thank the DST-NRF Centre of Excellence for Invasion Biology (C•I•B), the National Research Foundation (South Africa) and Stellenbosch University for support. This work benefitted from financial support from the C•I•B and the Working for Water Programme through their collaborative research project on "Research for Integrated Management of Invasive Alien Species". PP was supported by long-term research development project no. RVO 67985939 (Academy of Sciences of the Czech Republic), institutional resources of Ministry of Education, Youth and Sports of the Czech Republic, and acknowledges the support by Praemium Academiae award from the Academy of Sciences of the Czech Republic. We thank Zuzana Sixtová for technical assistance.

References

Alers M, Bovarnick A, Boyle T et al (2007) Reducing threats to protected areas lessons from the field. A joint UNDP and World Bank GEF lessons learned study. The Worldbank, Washington, DC   http://documents.worldbank.org/curated/en/2007/01/9532482/reducing-threats-protected-areas-lessons-field

Allen JA, Brown CS, Stohlgren TJ (2009) Non-native plant invasions of United States National Parks. Biol Invas 11:2195–2207

Andreu J, Vilà M, Hulme PE (2009) An assessment of stakeholder perceptions and management of alien plants in Spain. Environ Manag 43:1244–1255

Ashton P, Mitchell D (1989) Aquatic plants patters and modes of invasion, attributes of invading species and assessment of control programmes. In: Drake JA, Mooney H, di Castri F et al (eds) Biological invasions. A global perspective. Scope 37. Wiley, Chichester, pp 111–154

Barber CV, Miller KR, Boness M (eds) (2004) Securing protected areas in the face of global change: issues and strategies. IUCN, Gland/Cambridge

Bond WJ, Woodward FI, Midgley GF (2005) The global distribution of ecosystems in a world without fire. New Phytol 165:525–538

Brooks KJ, Setterfield SA, Douglas MM (2010) Exotic grass invasions: applying a conceptual framework to the dynamics of degradation and restoration in Australia's tropical savannas. Restor Ecol 18:188–197

Bruner AG, Gullison RE, Rice RE et al (2001) Effectiveness of parks in protecting tropical biodiversity. Science 291:125–129

Chan KMA, Shaw MR, Cameron DR et al (2006) Conservation planning for ecosystem services. PLoS Biol 4:2138–2152

Charles H, Dukes JS (2007) Impacts of invasive species on ecosystem services. In: Nentwig W (ed) Biological invasions, vol 193. Ecological studies. Analysis and synthesis. Springer, Berlin/Heidelberg, pp 217–237

Cole DN, Landres PB (1996) Threats to wilderness ecosystems: impacts and research needs. Ecol Appl 6:168–184

D'Antonio CM, Vitousek PM (1992) Biological invasions by exotic grasses, the grass/fire cycle, and global change. Annu Rev Ecol Syst 23:63–87

De Poorter M (2007) Invasive alien species and protected areas: a scoping report. Part 1. Scoping the scale and nature of invasive alien species threats to protected areas, impediments to invasive alien species management and means to address those impediments. Global Invasive Species Programme, Invasive Species Specialist Group. http://www.issg.org/gisp_publica tions_reports.htm

Dudley N, Mulongoy KJ, Cohen S et al (2005) Towards effective protected area systems. An action guide to implement the convention on biological diversity programme of work on protected areas. Secretariat of the Convention on Biological Diversity, Montreal

Dukes JS (2011) Responses of invasive species to a changing climate and atmosphere. In: Richardson DM (ed) Fifty years of invasion ecology: the legacy of Charles Elton. Wiley-Blackwell, Oxford, pp 345–357

Dumalisile L (2008) The effects of Chromolaena odorata on mammalian biodiversity in Hluhluwe-iMfolozi Park. University of Pretoria, Pretoria

Eagles PFJ, McCool SF, Haynes CDA (2002) Sustainable tourism in protected areas: guidelines for planning and management. IUCN, Gland/Cambridge

Ehrenfeld JG (2003) Effects of exotic plant invasions on soil nutrient cycling processes. Ecosystems 6:503–523

Ehrenfeld JG (2011) Transformers. In: Simberloff D, Rejmánek M (eds) Encyclopaedia of biological invasions. University of California Press, Berkley/Los Angeles, pp 667–670

Elton CS (1958) The ecology of invasions by animals and plants. University of Chicago Press, Chicago

Esler KJ, Prozesky H, Sharma GP et al (2010) How wide is the "knowing-doing" gap in invasion biology? Biol Invas 12:4065–4075

Fensham RJ, Fairfax RJ, Cannell RJ (1994) The invasion of Lantana camara L. in Forty Mile Scrub National Park, North Queensland. Aust J Ecol 19:297–305

Floyd-Hanna L, Romme D, Kendall D et al (1993) Succession and biological invasion at Mesa Verde National Park. Park Sci 9:16–18

Forsyth GG, van Wilgen BW (2008) The recent fire history of the Table Mountain National Park and implications for fire management. Koedoe 50:3–9

Foxcroft LC, Jarošík V, Pyšek P et al (2011) Protected-area boundaries as filters of plant invasions. Conserv Biol 25:400–405

Foxcroft LC, Witt A, Lotter WD (2014) Chapter 7: Invasive alien plants in African protected areas. In: Foxcroft LC, Pyšek P, Richardson DM, Genovesi P (eds) Plant invasions in protected areas: patterns, problems and challenges. Springer, Dordrecht, pp 117–143

Gascon C, Williamson GB, Da Fonseca GAB (2000) Receding forest edges and vanishing reserves. Science 288:1356–1358

Gaston KJ, Jackson SF, Cantú-Salazar et al (2008) The ecological performance of protected areas. Annu Rev Ecol Evol Syst 39:93–113

Gibson MR, Richardson DM, Pauw A (2012) Can floral traits predict an invasive plant's impact on native plant-pollinator communities? J Ecol 100:1216–1223

Gilbert B, Levine JM (2013) Plant invasions and extinction debts. Proc Natl Acad Sci USA 110:1744–1749

Goodman PS (2003) Assessing management effectiveness and setting priorities in protected areas in KwaZulu-Natal. BioScience 53:843–850

Gordon DR (1998) Effects of invasive, non-indigenous plant species on ecosystem processes: lessons from Florida. Ecol Appl 8:975–989

Gritten RH (1995) *Rhododendron ponticum* and some other invasive plants in the Snowdonia National Park. In: Pyšek P, Prach K, Rejmánek M et al (eds) Plant invasions: general aspects and special problems. SPB Academic Publishing, Amsterdam, pp 213–219

Gurevitch J, Padilla DK (2004) Are invasive species a major cause of extinctions? Trends Ecol Evol 19:470–474

Hannah L, Midgley GF, Millar D (2002) Climate change-integrated conservation strategies. Glob Ecol Biogeogr 11:485–495

Hannah L, Midgley GF, Andelman S et al (2007) Protected area needs in a changing climate. Front Ecol Environ 5:131–138

Hansen L, Hoffman J, Drews C et al (2010) Designing climate smart conservation: guidance and case studies. Conserv Biol 24:63–69

Heffernan KE (1998) Managing invasive alien plants in natural areas, parks, and small woodlands. Natural heritage technical report 98-25. Virginia Department of Conservation and Recreation, Division of Natural Heritage Program, Richmond, Virginia

Hiebert RD, Stubbendieck J (1993) Handbook for ranking exotic plants for management and control. U.S. Department of the Interior, National Park Service, Denver

Hobbs RJ, Arico S, Aronson J et al (2006) Novel ecosystems: theoretical and management aspects of the new ecological world order. Glob Ecol Biogeogr 15:1–7

Hooper DU, Chapin SF III, Ewel JJ et al (2005) Effects of biodiversity on ecosystem functioning: a consensus of current knowledge. Ecol Monogr 75:3–35

Houston DB, Schreiner EG (1995) Alien species in National Parks: drawing lines in space and time. Conserv Biol 9:204–209

Hughes F, Vitousek PM, Tunison T (1991) Alien grass invasion and fire in the seasonal submontane zone of Hawai'i. Ecology 72:743–746

Hulme PE, Pyšek P, Jarošík V et al (2013) Bias and error in current knowledge of plant invasions impacts. Trends Ecol Evol 28:212–218

Huntley B, Hole DG, Willis SG (2011) Assessing the effectiveness of a protected area network in the face of climatic change. In: Hodkinson TR, Jones MB, Waldren S et al (eds) Climate change, ecology and systematics. Cambridge University Press, Cambridge, pp 345–362

IUCN (2009) Proceedings of the international workshop on the future of the CBD programme of work on protected areas: Jeju Island, Republic of Korea, September 14–17, 2009. IUCN, Gland

Jäger H, Kowarik I (2010) Resilience of native plant community following manual control of invasive *Cinchona pubescens* in Galapagos. Restor Ecol 18:103–112

Jarošík V, Pyšek P, Foxcroft LC et al (2011) Predicting incursion of plant invaders into Kruger National Park, South Africa: the interplay of general drivers and species-specific factors. PLoS One 6:e28711

Kasulo V (2000) The impact of invasive species in African lakes. In: Perrings C, Williamson M, Dalmazzone S (eds) The economics of biological invasions. Edward Elgar, Cheltenham/Northampton, pp 183–207

Kohn D, Hulme PE, Hollingsworth P et al (2009) Are native bluebells (*Hyacinthoides non-scripta*) at risk from alien congenerics? Evidence from distributions and co-occurrence in Scotland. Biol Conserv 142:61–74

Kueffer C (2012) The importance of collaborative learning and research among conservationists from different oceanic islands. Rev Écol (Terre Vie) 11:125–135

Lahkar BP, Talukdar BK, Sarma P (2011) Invasive species in grassland habitat: an ecological threat to the greater one-horned rhino (*Rhinoceros unicornis*). Pachyderm 49:33–39

Larios L, Suding KN (2014) Chapter 28: Restoration within protected areas: when and how to intervene to manage plant invasions? In: Foxcroft LC, Pyšek P, Richardson DM, Genovesi P (eds) Plant invasions in protected areas: patterns, problems and challenges. Springer, Dordrecht, pp 621–639

Le Maitre DC, van Wilgen BW, Gelderblom CM et al (2002) Invasive alien trees and water resources in South Africa: case studies of the costs and benefits of management. For Ecol Manag 160:143–159

Lenz L, Taylor JA (2001) The influence of an invasive tree species (*Myrica faya*) on the abundance of an alien insect (*Sophonia rufofascia*) in Hawai'i Volcanoes National Park. Biol Conserv 102:301–307

Leslie AJ, Spotila JR (2001) Alien plant threatens Nile crocodile (*Crocodylus niloticus*) breeding in Lake St. Lucia, South Africa. Biol Conserv 98:347–355

Levine JM, Vilà M, D'Antonio CM et al (2003) Mechanisms underlying the impacts of exotic plant invasions. Proc R Soc Lond B 270:775–781

Liao C, Peng R, Luo Y et al (2007) Altered ecosystem carbon and nitrogen cycles by plant invasion: a meta-analysis. New Phytol 177:706–714

Linc O (2012) Efektivita likvidace invazních druhů v České republice na příkladu bolševníku velkolepého [The efficiency of eradication of invasive species in the Czech Republic exemplified by *Heracleum mantegazzianum*]. BSc thesis, University of Economics, Prague

Loehle C (2004) Applying landscape principles to fire hazard reduction. For Ecol Manag 198:261–267

Lonsdale WM (1999) Global patterns of plant invasions and the concept of invasibility. Ecology 80:1522–1536

Loope LL (2004) The challenge of effectively addressing the threat of invasive species to the national park system. Park Sci 22:14–20

Loope LL (2011) Hawaiian Islands: invasions. In: Simberloff D, Rejmánek M (eds) Encyclopaedia of biological invasions. University of California Press, Berkley/Los Angeles, pp 309–319

Loope LL, Hughes RF, Meyer J-Y et al (2014) Chapter 15: Plant invasions in protected areas of tropical Pacific Islands, with special reference to Hawaii. In: Foxcroft LC, Pyšek P, Richardson DM, Genovesi P (eds) Plant invasions in protected areas: patterns, problems and challenges. Springer, Dordrecht, pp 313–348

Macdonald IAW (1986) Invasive alien plants and their control in southern African nature reserves. In: Thomas LK (ed) Conference on science in the National Parks, vol 5. Management of exotic species in natural communities. Colorado State University, Fort Collins, pp 63–79

Macdonald IAW, Loope LL, Usher MB et al (1989) Wildlife conservation and the invasion of nature reserves by introduced species: a global perspective. In: Drake JA, Mooney H, di Castri F et al (eds) Biological invasions. A global perspective. Wiley, Chichester, pp 215–256

Martin MR, Tipping PW, Sickman JO (2009) Invasion by an exotic tree alters above and belowground ecosystem components. Biol Invas 11:1883–1894

Mauchamp A (1997) Threats from alien plant species in the Galápagos Islands. Conserv Biol 11:260–263

Mayer P (2006) Biodiversity: the appreciation of different thought styles and values helps to clarify the term. Restor Ecol 14:105–111

McGeoch MA (1998) The selection, testing and application of terrestrial insects as bioindicators. Biol Rev 73:181–201

McGeoch MA, Butchart SHM, Spear D et al (2010) Global indicators of biological invasion: species numbers, biodiversity impact and policy responses. Divers Distrib 16:95–108

Millennium Ecosystem Assessment (2005) Ecosystems and human well-being: biodiversity synthesis. World Resources Institute, Washington, DC

Mgobozi MP, Somers MJ, Dippenaar-Schoeman AS (2008) Spider responses to alien plant invasion: the effect of short- and long-term *Chromolaena odorata* invasion and management. J Appl Ecol 45:1189–1197

Mooney HA, Cleland EE (2001) The evolutionary impact of invasive species. Proc Natl Acad Sci USA 98:5446–5451

Morales CL, Traveset A (2009) A meta-analysis of impacts of alien vs. native plants on pollinator visitation and reproductive success of co-flowering native plants. Ecol Lett 12:716–728

Mulongoy KJ, Chape SP (eds) (2004) Protected areas and biodiversity: an overview of key issues. CBD Secretariat/UNEP-WCMC, Montreal/Cambridge

Naidoo R, Ricketts TH (2006) Mapping the economic costs and benefits of conservation. PLoS Biol 4:2153–2164

Nott PM, Pimm SL (1997) The evaluation of biodiversity as a target for conservation. In: Pickett STA, Ostfeld RS, Shachak M (eds) The ecological basis of conservation. Heterogeneity, ecosystems, and biodiversity. Chapman and Hall, New York, pp 125–135

Novacek MJ, Cleland EE (2001) The current biodiversity extinction event: scenarios for mitigation and recovery. Proc Natl Acad Sci USA 98:5466–5470

Parker IM, Reichard SH (1998) Critical issues in invasion biology for conservation science. In: Fiedler PL, Kareiva PM (eds) Conservation biology for the coming decade. Chapman and Hall, London, pp 283–305

Parker IM, Simberloff D, Lonsdale WM et al (1999) Impact: towards a framework for understanding the ecological effects of invaders. Biol Invas 1:3–19

Pereira H, Leadley P, Proenca V et al (2010) Scenarios for global biodiversity in the 21st century. Science 330:1496–1501

Peters HA (2001) *Clidemia hirta* invasion at the Pasoh Forest Reserve: an unexpected plant invasion in an undisturbed tropical forest. Biotropica 33:60–68

Pimentel D, Lach L, Zuniga R et al (2000) Environmental and economic costs of nonindigenous species in the United States. BioScience 50:53–65

Pimm SL, Ayres M, Balmford A et al (2001) Can we defy nature's end? Science 293:2207–2208

Pyšek P, Jarošík V, Kučera T (2002) Patterns of invasion in temperate nature reserves. Biol Conserv 104:13–24

Pyšek P, Jarošík V, Hulme PE et al (2012) A global assessment of invasive plant impacts on resident species, communities and ecosystems: the interaction of impact measures, invading species' traits and environment. Glob Change Biol 18:1725–1737

Pyšek P, Genovesi P, Pergl J et al (2014) Chapter 11: Invasion of protected areas in Europe: an old continent facing new problems. In: Foxcroft LC, Pyšek P, Richardson DM, Genovesi P (eds) Plant invasions in protected areas: patterns, problems and challenges. Springer, Dordrecht, pp 209–240

Randall JM (2011) Protected areas. In: Simberloff D, Rejmánek M (eds) Encyclopaedia of biological invasions. University of California Press, Berkley/Los Angeles, pp 563–567

Reinhardt F, Herle M, Bastiansen F et al (2003) Economic impact of the spread of alien species in Germany. Report No. UBA-FB. Biological and Computer Sciences Division; Department of Ecology and Evolution, Frankfurt am Main

Rhymer JM, Simberloff D (1996) Extinction by hybridization and introgression. Annu Rev Ecol Syst 27:83–109

Richardson DM, Gaertner M (2013) Plant invasions as builders and shapers of novel ecosystems. In: Hobbs RJ, Higgs EC, Hall CM (eds) Novel ecosystems: intervening in the new ecological world order. Wiley-Blackwell, Oxford, pp 102–114

Robertson MP, Harris KR, Coetzee JA et al (2011) Assessing local scale impacts of *Opuntia stricta* (Cactaceae) invasion on beetle and spider diversity in Kruger National Park, South Africa. Afr Zool 46:205–223

Rose M, Hermanutz L (2004) Are boreal ecosystems susceptible to alien plant invasion? Evidence from protected areas. Oecologia 139:467–477

Rossiter NA, Setterfield SA, Douglas MM et al (2003) Testing the grass-fire cycle: alien grass invasion in the tropical savannas of northern Australia. Divers Distrib 9:169–176

Rossiter-Rachor NA, Setterfield SA, Douglas MM et al (2009) Invasive *Andropogon gayanus* (gamba grass) is an ecosystem transformer of nitrogen relations in Australian savanna. Ecol Appl 19:1546–1560

Rudd MA, Beazley KF, Cooke SJ et al (2011) Generation of priority research questions to inform conservation policy and management at a national level. Conserv Biol 25:476–484

Sax DF, Gaines SD (2008) Species invasions and extinction: the future of native biodiversity on islands. Proc Natl Acad Sci USA 105:11490–11497

Schmitz D, Simberloff D, Hoffstetter R et al (1997) The ecological impact of non-indigenous plants. In: Simberloff D, Schmitz D, Brown T (eds) Strangers in paradise. Island Press, Washington, DC, pp 29–74

Sekercioglu CH (2010) Ecosystem services and functions. In: Sodhi NS, Erlich PR (eds) Conservation biology for all. Oxford University Press, Oxford, pp 45–67

Setterfield SA, Douglas MM, Petty AM et al (2014) Chapter 9: Invasive plants in the floodplains of Australia's Kakadu National Park. In: Foxcroft LC, Pyšek P, Richardson DM, Genovesi P (eds) Plant invasions in protected areas: patterns, problems and challenges. Springer, Dordrecht, pp 167–189

Simberloff D, Von Holle B (1999) Positive interactions of nonindigenous species: invasional meltdown? Biol Invas 1:21–32

Simberloff D, Martin J-L, Genovesi P et al (2013) Impacts of biological invasions: what's what and the way forward. Trends Ecol Evol 28:58–66

Spear D, McGeoch MA, Foxcroft LC et al (2011) Alien species in South Africa's National Parks. Koedoe 53:Art1032. doi:10.4102/koedoe. v53i1.1032

Steinfeld H, Gerber P, Wassenaar T et al (2006) Livestock's long shadow: environmental issues and options. Food and Agriculture Organization of the United Nations, Rome

Stone CP, Loope LL (1987) Reducing negative effects of introduced animals on native biotas in Hawaii: what is being done, what needs doing, and the role of National Parks. Environ Conserv 14:245–258

Strayer DL (2012) Eight questions about invasions and ecosystem functioning. Ecol Lett 15:1199–1210

Thomas LK (1980) The impact of three exotic plant species on a Potomac Island. National Park Service Scientific Monograph Series 13. U.S. Department of the Interior, Washington, DC, pp 1–179

Table Mountain National Park (2008) Park management plan. South African National Parks, Cape Town

Traveset A, Richardson DM (2006) Biological invasions as disruptors of plant reproductive mutualisms. Trends Ecol Evol 21:208–216

Traveset A, Richardson DM (2011) Mutualisms: key drivers of invasions... key casualties of invasions. In: Richardson DM (ed) Fifty years of invasion ecology: the legacy of Charles Elton. Wiley-Blackwell, Oxford, pp 143–160

Trinder-Smith TH, Cowling RM, Linder HP (1996) Profiling a besieged flora: endemic and threatened plants of the Cape Peninsula, South Africa. Biodiv Conserv 5:575–589

Usher MB (1988) Biological invasions of nature reserves: a search for generalizations. Biol Conserv 44:119–135

Usher MB, Kruger FJ, Macdonald IAW et al (1988) The ecology of biological invasions into nature reserves: an introduction. Biol Conserv 44:1–8

van Wilgen BW, Scott DF (2001) Managing fires on the Cape Peninsula: dealing with the inevitable. J Mediterr Ecol 2:197–208

van Wilgen B, Trollope WSW, Biggs HC et al (2003) Fire as a drive of ecosystem variability. In: du Toit JT, Rogers KH, Biggs HC (eds) The Kruger experience. Ecology and management of savanna heterogeneity. Island Press, Washington, DC, pp 149–170 van Wilgen BW, Reyers B, Le Maitre DC et al (2008) A biome-scale assessment of the impact of invasive alien plants on ecosystem services in South Africa. J Environ Manag 89:336–349

van Wilgen BW, Forsyth GG, Prins P (2012) The management of fire-adapted ecosystems in an urban setting: the case of Table Mountain National Park, South Africa. Ecol Soc 17:8. doi: org/10.5751/ES-04526-170108

Vilà C, Weber E, D'Antonio CM (2000) Conservation implications of invasion by plant hybridization. Biol Invas 2:207–217

Vilà M, Williamson M, Lonsdale M (2004) Competition experiments on alien weeds with crops: lessons for measuring plant invasion impact? Biol Invas 6:59–69

Vilà M, Basnou C, Pyšek P et al (2010) How well do we understand the impacts of alien species on ecosystem services? A pan-European, cross-taxa assessment. Front Ecol Environ 8:135–144

Vilà M, Espinar JL, Hejda M et al (2011) Ecological impacts of invasive alien plants: a meta-analysis of their effects on species, communities and ecosystems. Ecol Lett 14:702–708

Vitousek PM, Walker LR, Whiteacre LD et al (1987) Biological invasion by *Myrica faya* alters ecosystem development in Hawaii. Science 238:802–804

Wardle DA, Bardgett RD, Callaway RM et al (2011) Terrestrial ecosystem responses to species gains and losses. Science 332:1273–1277

Wilcove DS, Chen LY (1998) Management costs for endangered species. Conserv Biol 12:1405–1407

Chapter 3
Plant Invasions in Protected Landscapes: Exception or Expectation?

Scott J. Meiners and Steward T.A. Pickett

> *The disappearance of plants and animal species without*
> *visible cause despite efforts to protect them and the irruption*
> *of others as pests despite efforts to control them must in the*
> *absence of simpler explanations be regarded as symptoms of*
> *sickness in the land organism. Both are occurring too*
> *frequently to be dismissed as normal evolutionary events.*
>
> Aldo Leopold 1949 – A Sand County Almanac

Abstract While protection may alleviate some concerns within protected areas, should protection also be expected to mitigate the likelihood of invasion by alien species? Plant community dynamics may be thought of as generated by three broad classes of drivers: site availability and history, species availability and species performance. As all plant communities are dynamic and plant invasions may be involved in these dynamics, this framework may be useful in exploring the role of protection in invasion. In reviewing examples of individual drivers of plant community dynamics, the potential for mechanisms that lie outside local control to lead to invasion is found to be great. This suggests that alien plant invasions should be expected in most landscapes, even those with some level of protection. Linkages with management practices are also highlighted.

Keywords Invasion • Conceptual framework • Disturbance • Dispersal • Species pool • Herbivores • Nutrient deposition • Acid deposition • Climate change

S.J. Meiners (✉)
Department of Biological Sciences, Eastern Illinois University,
Charleston, IL 61920, USA
e-mail: sjmeiners@eiu.edu

S.T.A. Pickett
Cary Institute of Ecosystem Studies, Millbrook, NY 12545, USA
e-mail: picketts@caryinstitute.org

L.C. Foxcroft et al. (eds.), *Plant Invasions in Protected Areas: Patterns, Problems and Challenges*, Invading Nature - Springer Series in Invasion Ecology 7, DOI 10.1007/978-94-007-7750-7_3, © Springer Science+Business Media Dordrecht 2013

3.1 Introduction

Nature preserves typically revolve around the protection of a unique species, community or landscape that holds some biological or cultural value. This necessarily means that preserves are synonymous with threats; if the site were not threatened there would be no need for preservation or protection. As granting a site preserve status is largely a legal process that limits the types of activities that can occur within the area, it does remove some of the more immediate threats such as the anthropogenic conversion to non-habitat and resource harvesting. However, legal protection does nothing to ameliorate the context from which the site was originally threatened. There is no magic preservation bubble that appears to seal off the area from the outside influences that may continue to threaten the biological integrity of the system. In fact, threats to the system may already be present within the protected boundaries.

The threat of particular interest to this chapter and book is that of biological invasion. The focus of this chapter will be to highlight the ways in which alien species may colonise and come to dominate protected systems. To do this, we will first review the mechanisms of plant community dynamics using a hierarchical framework to aggregate these mechanisms into three primary drivers of community change. As biological invasions are also a type and mechanism of community dynamics, this framework should be useful in organizing mechanisms of invasion and in placing them in an appropriate ecological context. From there, we will visit examples of each major driver of community dynamics and explore how these may be related to biological invasions. At the end, we will return to the question posed in the title and discuss the utility of the hierarchical framework in guiding when we may expect biological invasions to occur in protected areas and plan for them accordingly.

3.2 A Hierarchical Framework for Community Change

Plant ecologists have identified a large number of individual drivers of community structure and dynamics over the past 100 years. These include such disparate topics as disturbance, competition, dispersal limitation, soil microbial interactions, herbivory, and resource availability among a long list of many, many others. An exhaustive list would be daunting, particularly if one was interested in using these potential drivers to establish a management plan for biological invasions. The laundry list approach is also not particularly useful in understanding the relationships among drivers. There are clear data in support of each of these controllers of community dynamics, but we would certainly not expect all to be equally important within a given community type or situation. What is necessary is a conceptual framework to help to organise these individual drivers into a logical order. One such framework is a hierarchical view that groups community drivers into three broad categories – site conditions and history, species

Table 3.1 The hierarchical conceptual framework for community dynamics. Community dynamics are generated by three broad drivers, which each contain many individual mechanisms. Variation within any of these broad drivers can result in changes in community composition or structure

Community dynamics		
Site conditions and history	Species availability	Species performance
Disturbance history	Vagility	Competition
Resource availability	Dispersal mode	Growth rates
Climatic conditions	Seed longevity	Reproductive rates
Soil structure	Behaviour of dispersers	Herbivore damage
	Location of fruiting plants	Interactions with soil biota
	Landscape connectivity	Chemical interactions
		Pathogens

availability and species performance (Pickett et al. 1987, 2011; Pickett and Cadenasso 2005). Each of these broad categories then contains individual drivers of community dynamics (Table 3.1). Variation in any of these processes, expressed as differentials, may then lead to changes in the composition and structure of a plant community.

The broad categories of drivers contain three discreet aspects of community dynamics. (i) Site conditions and history contain aspects of disturbance regime, resource availability, and site history. (ii) Species availability encompasses the ability of species disperse into a site or persist within that site by forming persistent seed banks. This driver also contains aspects of the landscape and the behaviour of dispersal agents. (iii) Species performance contains all aspects of the ecological sorting of species within a community. Variation in performance generates patterns of species dominance and microhabitat specificity within communities. This driver of community dynamics is by far the most active area of research and is populated by a large array of factors that may determine a species' success. The three broad drivers of dynamics appear to be conceptually complete as they encompass all of the known individual mechanisms of community change and should be sufficient to contain new developments. For example, this conceptual framework predates the more contemporary realization that feedbacks mediated by soil microbial communities can be very important in determining the structure and dynamics of plant communities (e.g. Bever 2003). This mechanism of species sorting clearly fits into the differential performance driver of community dynamics. Similarly the role of propagule pressure in biological invasions fits into species availability quite nicely.

Though there is no specific relationship among the broad categories of drivers suggested by the hierarchical framework, it can be manipulated to fit a range of systems. For biological invasions into protected areas, we may rearrange the hierarchical framework into a filter model of community dynamics that is more appropriate (Fig. 3.1). In this rearrangement, site conditions and history set the context for invasions, differential species availability across a site or landscape determines the pool of potential invaders, and differentials in species performance

Fig. 3.1 Rearrangement of
the hierarchical framework
into a filter model that
organises the broad drivers
of community dynamics to
illustrate key aspects of the
invasion process. The
dominant component
mechanisms within each of
the broad drivers will vary
based on the identity of the
system and the invader.
Similarly, not all of the filter
components will be
important in all invasions

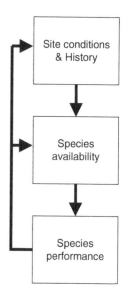

determine the success or failure of the invasion, the dominance achieved by the
invader, and its impacts on the recipient plant community. This directionality is not
absolute, however, as there may be interactions across community drivers. For
example, invasion by a nitrogen fixing legume may alter resource availability,
shifting the performance of other species in the community. Similarly, disturbances
in the protected landscape may alter the movement patterns of dispersal agents that
may increase or decrease invasion risk. It must be noted that this filter model only
deals with processes occurring within a given site or landscape and therefore
primarily reflects local processes of invasion. Other conceptual frameworks deal
with the broader geographic patterns of species movements, initial establishment
and spread in invaded areas (Catford et al. 2009; Blackburn et al. 2011; Foxcroft
et al. 2011). It should also be noted that while the conceptual framework presented
here was initially developed in the context of plant community dynamics, there is
no inherent reason that a similar structure would not be informative for animal
communities.

The following sections will provide a few examples of how these drivers may be
associated with alien species invasion in reserves and discuss their management
implications. These examples are illustrative only of some of the major invasion
opportunities for problematic species and are by no means an exhaustive list. There
are also clear linkages even among the brief set of examples discussed here. It is
impossible to fully separate such things as disturbance and herbivore activity from
resource availability. Likewise, global climate change will likely lead to cascading
impacts over longer time periods and may alter site conditions, disturbance
regimes, and populations of dispersers.

3.3 Site Availability and Conditions

In our rearrangement of the hierarchical framework, variation in site availability and conditions represents the first of the filters for invasion. The individual mechanisms under this broad driver of community dynamics set the context for the community and for any potential invasions. Species which disperse into the site as seeds or spores (species availability), but which cannot tolerate local conditions, will not become members of the community that differential performance can then act upon. Similarly, an alien species under one level of soil resources may be competitively inferior to the resident natives, but become invasive under another. In this section we will briefly discuss disturbance (contemporary and historical) and atmospheric deposition as factors that may mitigate invasion in protected areas, but which are not easily managed.

3.3.1 Disturbance History

Disturbance and the invasion of alien plant species have been mechanistically linked for decades and was one of the earliest noted causes of invasion (Fox et al. 1986; Hobbs and Drake 1989; Mack and D'Antonio 1998; Huston 2004; Moles et al. 2011). The linkage with disturbance may be related to the life history characteristics of alien plant species in the case of those species associated with agricultural, logging or other anthropogenic activities. In such cases, the presence of disturbed soils is necessary for germination, establishment and reproduction. For these alien species, mitigation can largely rely on the removal of anthropogenic disturbances (Meiners et al. 2002, 2007). However, disturbances can also provide opportunities in space and time for the initial establishment of an alien species into an otherwise closed community.

All plant communities naturally experience some level of disturbance, such as tree fall gaps in forests or soil disturbances from fossorial mammals in grasslands. These local disturbances may allow alien species to persist within the community and potentially displace similarly adapted native species. For species that require disturbance for establishment, the physical opportunity provided by a disturbance may be sufficient to gain entry into a community and then spread. For those alien species which can alter the disturbance regime, there can be large impacts on resident species as populations of the alien species increase. The conversion of tropical rainforests in Hawaii, South America and other areas to grazing lands planted with alien grasses brings fire into a system not adapted to fires (D'Antonio and Vitousek 1992; Mack and D'Antonio 1998; Rossiter et al. 2003). As fires occur, native plants susceptible to fires are killed, leading to the expansion of the alien species. These alien grass communities may persist even in the absence of fire as they are recalcitrant to tree colonization and establishment (Holl et al. 2000).

Disturbance is typically thought of as a factor that affects current interactions within the plant community – a tree falls and subordinate individuals compete to reach the canopy. However, there can also be important historical effects of disturbance, particularly when dealing with agricultural disturbances. Soils may bear the impacts of agriculture for decades to centuries (Koerner et al. 1997; Verheyen et al. 1999; Walker et al. 2010). The effects of disturbance history may then also be reflected in the plant communities that are supported by those soils (Foster 1993; Flinn and Marks 2007; Kuhman et al. 2011). Differences among sites in their disturbance history may therefore result in differentially invasible communities. For example, in the southern Appalachians of North America, agricultural areas that had been abandoned 100 years previously had higher abundances of several alien species than similar aged areas that had not been ploughed (Kuhman et al. 2011). Such long-term effects of disturbance on invasion are likely both species and system specific and provide real challenges to management.

Though disturbance can clearly lead to invasion, it is too simplistic to argue for the removal of disturbance from communities. Many native plant species are ruderals (sensu Grime 2001) and therefore depend on disturbances for regeneration. Furthermore, the preservation of some landscapes and species specifically require disturbances and the prevention of succession to a community type that may be more resistant to invasion. Such systems include: savannas, intermediate between grassland and forest; open shrublands, intermediate stages of succession in areas that support forest; and traditional agroecosystems, which often support species of interest as well as preserve cultural heritage. These systems often contain characteristic species that require intermediate successional communities to maintain their populations. Management decisions to incorporate disturbance will need to balance current threats to a community with the needs of native species in the system. For example, many butterfly species depend on early successional plants as larval food sources or for nectar. As agricultural disturbances are removed from systems, these species may decline along with their hosts. In Mediterranean landscapes where agriculture is being stopped, butterfly diversity is increased by removal of forest understory vegetation for fire management (Verdasca et al. 2012). Similarly, grazing in semi-natural European grasslands can be used to manage lepidopteran communities (Pöyry et al. 2004). While both of these management strategies increase early successional plant species, they would clearly also lead to opportunities for alien plant invasion. Ultimately, the decision to prevent or manage a particular invasion may lead to the decline or loss of other species from the system, just as management decisions to increase a native species may.

3.3.2 Nutrient/Acid Deposition

Anthropogenic effects are currently much broader than the physical footprint of human activity because of the regional alteration of rainfall chemistry by the combustion of fossil fuels. Nutrient deposition affects massive areas of nearly all continents and may dwarf pre-industrialization inputs into ecosystems. For nitrogen-limited terrestrial ecosystems, the addition of nitrogen through rainfall may alter competitive interactions and lead to changes in plant community composition (Stevens et al. 2006; Hautier et al. 2009). As resource availability is often linked with invasion (Burke and Grime 1996; Davis et al. 2000), anthropogenic nitrogen deposition may increase the potential area that meet an invader's N requirement or may alter existing communities to become more invasible. Nutrient deposition not only alters plant distribution, but may also affect interactions with soil microbial communities (Carreiro et al. 2000; Chung et al. 2007), mycorrhizal associations (Treseder 2004; Kjøller et al. 2012) and herbivory (Throop and Lerdau 2004). Together, these effects may cascade through the community, resulting in changes in plant community composition and opportunities for invasion.

Nutrient deposition occurs simultaneously with the acidification of soils and surface waters. While the fertilization effects of anthropogenic deposition receive most of the attention, acidification may also be problematic for terrestrial systems. Acidification has resulted in losses of cations, particularly Ca^+, from forest soils and may alter soil pH in areas where the buffering capacity has been exceeded. The implications of this nutrient loss are less well understood, but may also have large impacts on plant communities (DeHayes et al. 1999; Juice et al. 2006). There may also be cascading effects of calcium loss through altered soil fauna such as snails and the organisms that feed on them (Graveland et al. 1994).

Nutrient and acid deposition pose unique challenges for land managers. Accumulation of N in soils may ultimately shift a high quality protected area from being relatively resistant to invasion to susceptible once a threshold of nutrient availability has been reached. In habitat restoration, reduction in soil fertility can be achieved by removing plant biomass, removing topsoil or by chemical immobilization. This is particularly important in restoring plant communities on nutrient-poor soils where agricultural fertilizer application generates high fertility that persists after abandonment. In restoring heathland and other plant communities in The Netherlands, topsoil removal can lead to compositional shifts towards target communities (Verhagen et al. 2001). However, soil manipulations often occur before restoration is initiated. Reducing fertility in an intact plant community may be much more difficult, particularly as deposition is likely to continue. Acidification of soils may be ameliorated by addition of lime as is commonly done in agricultural ecosystems. Such amendments are likely only feasible in small, high priority areas.

3.4   Species Availability

Species availability encapsulates all of the ways in which a species can enter a plant community, whether from the seed bank, from seeds produced from a local individual, or from a long-distance dispersal event. For this reason, species availability combines species characteristics such as dormancy and longevity with inherently spatial issues such as where reproductive individuals are located in the landscape and the movement patterns of dispersal agents. This section will focus on the spatial aspects of species availability as this is where most of the management challenges for alien species invasions will lie.

3.4.1   The Matrix as a Source of Invasion

Protected areas lie within a matrix of non-protected areas that represent the most immediate source of alien invasive species to colonise a site. Seed dispersal of alien species is a spatially constrained process with most of the seeds that enter a site arriving from nearby areas. For this reason, the alien species pool of adjacent areas is the most important factor in determining the composition and number of colonizing individuals (Rose and Hermanutz 2004; Dawson et al. 2011). The original mode of introduction to an area may also be important in determining a species invasion potential as it should be strongly associated with species' characteristics. For simplicity, mode of introduction here is divided into accidental and deliberate.

The major source of anthropogenic disturbance in most landscapes is that associated with agricultural activities – crop production, livestock rearing and timber harvesting. These activities place very clear and strong selective filters on the alien species that can disperse into and spread within a site, even though the introductions are not deliberate. Species introduced accidentally through agricultural practices will predominately be classic 'weed' species (Baker 1965; Bazzaz and Mooney 1986). These species will depend on agricultural practices for their persistence in communities as they are typically shade intolerant and may require soil disturbances for establishment. Such agricultural weeds will likely decrease in abundance with succession and should not pose persistent problems in the system, though they may be maintained in the matrix and continually disperse into the protected area. Habitat management in this case may focus on enhancing successional transitions (Luken 1990; Walker et al. 2007). However, if the alien species are able to inhibit successional processes, they may require more direct intervention (Meiners et al. 2007).

Many alien plants are the result of deliberate introduction (Mack and Erneberg 2002) with ornamental species becoming an increasingly important component of the species pool the longer they are present in a region (Pyšek et al. 2003). Ornamental plantings represent a major source of alien species in many landscapes and the continued maintenance of these plants may be strongly linked with their ultimate invasion success (Mack 2000; Pyšek et al. 2009). Species deliberately introduced for ornamental purposes are often specifically selected for characteristics that may make them successful under local conditions. These traits often include adaptation to local climactic conditions, good growth rates (for horticultural production and success in plantings) and reproductive characteristics such as abundant flower production and attractive fruits (Dirr 1997; Li et al. 2004; Franco et al. 2006). These same traits may translate into species that are particularly well suited to becoming invasive. For example, alien woody species invasive in the metropolitan New York, USA area were more likely to have fleshy fruits (Aronson et al. 2007). Fruit displays, particularly large and showy ones, are commonly valued characters in selecting horticultural varieties and can lead to efficient spread of seeds across landscapes. The screening processes involved in selecting and breeding horticultural plants may enhance the potential for invasive alien plants that can persist and spread in protected areas (Li et al. 2004). The filtering processes involved in deliberate introductions may make horticultural species a much greater invasion risk.

From a management perspective, the reduction of alien plant species in the matrix that surrounds a protected area is largely unfeasible. The best strategy for dealing with the matrix is to use it to inform monitoring efforts. Knowing the composition of the potential invader pool and their abundance in the surrounding matrix should provide a reasonable estimate of the threats to the protected area. Particularly useful may be exploring natural habitats within the matrix to determine which alien species have the potential to establish local populations. For example, urban woodlots are often invaded by many of the species that are invasive in local forests (Godefroid and Koedam 2003; LaPaix et al. 2012). These urban populations indicate alien species that are not only suitable for the local environment, but are available within the local species pool. As such, they may be useful indicator habitats for predicting future plant invasions.

3.4.2 Dispersal in Fragmented Landscapes

To understand the role of dispersal in the invasion of protected areas, it is necessary to use a landscape perspective. Many protected areas, particularly large ones, are in landscapes with less anthropogenic disturbance. These areas are often available for conservation because they have little value for other activities (Pressey 1994; Scott et al. 2001). Areas preserved in this way may not represent the overall diversity of a region, but may face fewer invasion and other conservation issues. Smaller protected areas which occur in more modified landscapes with greater anthropogenic pressures may face a much greater risk of invasion.

While the species pool of the surrounding area determines the identity of potential invaders, landscape configuration is critical in determining which areas may become invaded. In areas of more or less contiguous habitat, the contrast between the protected and unprotected areas may be low. This may occur when a high quality patch is acquired for protection within a larger expanse of habitat that remains in private ownership. In the case of a forest reserve in a contiguous forest, the surrounding landscape may serve to buffer the protected area from invasion by non-forest species or at least minimise their spread. In many preserves, however, the matrix represents a structurally different habitat which generates a discernible edge with the protected area or the reserve itself may be a mosaic of habitat patches. Edges and habitat corridors may concentrate the activity of birds and other dispersers, allowing the initial colonization of alien species (Thompson and Willson 1978; Brothers and Spingarn 1992; Levey et al. 2005). Increased abundance at edges is not constrained to vertebrate-dispersed species as abiotic dispersal may also be highest at fragment edges (Cadenasso and Pickett 2001). Alien species that establish at edges may then expand from those initial populations into the interior of the protected areas as conditions allow.

The perimeter of the protected area is not the only conduit for alien plant invasion. Most reserves, particularly large ones, contain inholdings and other incursions into the interior of the reserve. These may be from roadways, existing private landholdings and previous land conversion. Incursions, and more importantly the species contained within them, represent opportunities for alien plant species to colonise the interior of the protected area. Continued land use activities may maintain alien species populations and the maintenance of alien horticultural species may bring new potential invaders into the protected area. Roads and pathways may also provide corridors that allow the spread of alien plants into the interior of protected areas (Benninger-Truax et al. 1992; Forman and Alexander 1998; Kuhman et al. 2011) though these species may remain restricted to habitats along the incursion (Pauchard and Alaback 2004; Flory and Clay 2006). In temperate parks in the Chilean Andes, alien plant species located in the surrounding matrix also occurred along roadsides within the protected areas (Pauchard and Alaback 2004). While the alien species present responded to elevational constraints, many appeared able to establish in protected areas composed of higher elevation forests. This study highlights the hierarchical controls on invasion imposed by site conditions (elevation) and species availability (landscape context).

From a practical perspective, dispersal is difficult to directly manage, but understanding it can provide guidance for monitoring and targeted intervention. Minimizing inholdings within protected landscapes may be helpful, but would be difficult in most situations. Similarly, as many protected areas serve the dual purposes of protecting habitats and allowing access, roads and trails are largely necessary. Monitoring alien plant species should focus on dispersal corridors, inholdings and habitat edges (Pauchard and Alaback 2004). Edge habitats, and potentially roadways may also be planted with native species to reduce seed dispersal and opportunities for establishment (Cadenasso and Pickett 2001).

3.5   Species Performance

This section will explore two mechanisms that may alter the differential performance of species, leading to the opportunity for invasion, or for the opportunity for a current alien species restricted to low abundance to increase and become problematic. For this section two processes have been chosen that will likely operate at spatial scales much larger than an individual nature reserve. Though the abundance of large herbivores is critical to many local plant communities, the controllers on herbivore abundance likely operate at landscape and regional scales. Similarly, global climate change operates at scales vastly larger than any nature preserve, but can subtly and profoundly change how species interact within local plant communities.

3.5.1 Alteration of Herbivore Abundances

Herbivores have an amazing ability to influence the structure and composition of plant communities through their feeding activity and can exert strong evolutionary pressures on plant species. Specifically, the selectivity of herbivores can differentially impact plant species and alter competitive hierarchies. While invertebrate and small mammalian herbivores are less affected by the human activities that may lead to the necessity for habitat preservation, large mammals are much more likely to be affected. Large herbivores require larger areas in which to forage and are controlled by larger predators, which are themselves often actively hunted to decrease real or perceived interactions with humans or their livestock. The movement, foraging, and other activity of large mammals may also be important in generating soil disturbances or in seed dispersal that also may impact plant communities (Myers et al. 2004; Rose and Hermanutz 2004; Bressette et al. 2012). Both increases and decreases in the abundance of large herbivores can change plant communities and lead to plant invasions.

Within eastern North America, the native whitetail deer (*Oidocoileus virginanus*) has gone through large changes in abundance over the last century. After suffering habitat loss and hunting to near extinction, the species has undergone a dramatic recovery following some early remediation. Because of the fragmentation of forest habitat, this species, which naturally favors edge habitats and diverse landscapes, has become overabundant in much of Eastern North America (Alverson et al. 1988; Côté et al. 2004). The lack of large predators such as wolves (*Canis lupus*) and coyotes (*Canis latrans*) and the lack of hunting in many suburban areas have also encouraged large populations. Overabundance of deer may lead to the loss of preferred forage species from forest understories, the reduction or elimination of tree regeneration (Côté et al. 2004), and to the synergistic invasion of alien species released by removal of potential competitors and disturbance of soil by trampling (Baiser et al. 2008; Eschtruth and Battles 2009). Preserves in areas with overabundant deer are equally subjected to deer pressure as unprotected areas in the same regions. Targeted hunting within reserves may temporarily reduce deer pressure but allows animals in surrounding areas to move in and increase local density once again. The fragmented context of most reserves makes adequate control of these herbivores problematic unless the site can be fenced, a large expense which excludes both deer and other mammals.

A related, but opposite issue is seen in the loss of wisents (*Bison bonasus*) from Europe's primeval forests. The wisent, a large herbivore of forested systems, once occurred across most of Europe and fed primarily on woody plants (Pucek et al. 2004; Kowalczyk et al. 2011). While this species has been reintroduced into Poland's Białowieża forest, the loss of wisent feeding patterns has probably had major impacts on deciduous forests across Europe. While overabundant herbivores may exert too much influence on plant communities, the lack of selective herbivores is equally problematic. Plant species that would normally be selectively fed upon may be released and increase abundance. This release may also include alien species or changes in forest structure that favor invasion.

Herbivore pressure, particularly feeding by large mammalian herbivores, is easily altered by anthropogenic activity. Any change in herbivore pressure can lead to shifts in competitive hierarchies and can provide opportunities for alien species invasion. While managing large herbivore pressure in small preserves may be difficult, some activity by these animals may be necessary to generate regeneration opportunities, reduce competitive exclusion, and generate heterogeneity (Knapp et al. 1999).

3.5.2   Global Climate Change and Invasion

Though we may include alien species invasions as a form of global change (e.g. Ricciardi 2007), invasions may themselves respond to climatic changes in complex ways (Bradley et al. 2010). The most obvious aspect of climate change is increasing temperature and its influence on growing season length and in limiting cold-intolerant species. However, to plant communities, changes in the seasonality and amount of rainfall, extreme high and low temperatures, increased severe weather events, and increased CO_2 may all be important in determining species abundances and success. Most of the effects of climate change are contained within the species performance component of the hierarchical model of community change, though it may impact all levels.

In response to the varied effects of global climate change, plant species are expected to migrate as their potential habitats shift (Iverson and Prasad 2002; Malcolm et al. 2002). Migration may be hampered by the rate at which species disperse across the landscape and the loss of native species from some communities may provide opportunities for alien plant invasion. The characteristics of alien plant species that make them good invaders may also allow them to preferentially expand in response to climate change (Dukes and Mooney 1999; Simberloff 2000). Pathogens and insect herbivores are often critical in regulating plant populations and relative abundances in plant communities. These species may also experience range changes, often through alleviation of cold limitation at higher latitudes (Bale et al. 2002; Jepsen et al. 2008; Dukes et al. 2009). Together, the combined and complex effects of global climate change make their influence on the interactions that regulate plant communities difficult to predict (Tylianakis et al. 2008).

Controlling global climate change is clearly beyond the ability of individual land managers, as is preventing the inevitable shifts in species abundance that occur with changing conditions. The challenge for land managers will be in minimizing the opportunities for alien plant invasions that result from climate change. Some native plant species will certainly decrease in abundance, while others will likely increase.

Opportunities for alien plant establishment will occur during such species transitions (Davis et al. 2000). When species are lost from a system and not replaced by resident natives, assisted migration may be required to ensure that an appropriate suite of native species colonise the site to maintain ecosystem integrity. It is also likely that climactic changes may increase the suitability of the environment for alien species that are not currently problematic, allowing them to expand their populations (Simberloff 2000). With the exception of insect or pathogen outbreaks and catastrophic disturbances, the changes brought about by climate change should be gradual and allow for the development of an adaptive management plan.

3.6 Conclusions

The examples listed here illustrate that the potential for alien plant invasion within protected areas is great. While the examples above involve various aspects of plant community dynamics, they are similar in that they involve factors that are either external to the protected system or are beyond the control of local land managers. It is also important to note, that while these examples are discussed within one of three broad drivers of community dynamics, there are interactions among these drivers, as indicated by the interconnecting arrows in Fig. 3.1. For example, the manipulation of herbivore density most directly alters competitive hierarchies and so belongs within species performance. Herbivores also may act as seed dispersers, generate local disturbances, and alter rates of nutrient cycling. Shifts in the composition of the plant community from herbivory may have similar effects on rates of nutrient cycling, and the abundances of other consumers. The connections among drivers and implications of changes to any one part can lead to complex dynamics.

Using the conceptual framework of community dynamics to understand plant invasions provides a useful way to organise the many mechanisms by which alien plants may come to invade and dominate protected areas. In fact, one can easily replace the individual drivers of plant community dynamics with the many mechanisms that have been proposed as reasons for the invasion and success of alien species (Table 3.2). All of these invasion hypotheses neatly fit within the three broad categories that capture plant community dynamics. Within site conditions and history are contained in hypotheses dealing with resource availability and disturbance. Species availability contains ideas such as propagule pressure, reproductive output and dispersal ability. Finally, species performance contains most of the interaction-based mechanisms of invasion such as competitive ability, plasticity and natural enemies. In total, the components of the hierarchical model of invasion are remarkably similar to the components of the conceptual model for community dynamics. The applicability of the same conceptual framework to both community dynamics and plant invasions suggests that both processes are essentially the same. Alien species invasions are merely special cases of community dynamics, much in the same way that succession is a special case of the broader community dynamics (Pickett and Cadenasso 2005).

Table 3.2 Individual mechanisms of alien species invasion mapped onto the hierarchical framework of community dynamics

Invasion dynamics		
Site conditions and history	Species availability	Species performance
Disturbance history/regime	Vagility	Competitive ability (EICA)
Resource availability	Propagule pressure	Growth rates
Nutrient deposition	Species persistence	Reproductive rates
Cultivation	Behaviour of dispersers	Escape from natural enemies
	Location of fruiting plants	Novel traits
	Landscape connectivity	Phenotypic plasticity
	Dispersal mode	Allelopathy
		Phenology
		Soil microbial feedbacks

EICA evolution of increased competitive ability

We need to view our protected areas as dynamic plant communities that are constantly shifting in response to contemporary and historical influences. Among contemporary influences, alien plant invasions are a potentially critical driver. While protection removes some threats, particularly anthropogenic disturbances, it does little to remove regional influences on the plant community. The natural dynamics of communities in protected areas will generate opportunities for invasion regardless of the type or intensity of management. Therefore, we should expect and plan for plant invasions within protected areas. This should entail identification of local threats, monitoring susceptible areas for invasion, and mitigation of invasions when warranted.

This chapter started with a quote from an early North American champion of conservation and land management, Aldo Leopold. He placed plant invasions into a broader context of what he termed 'land sickness'. While this view is a bit too Clementsian for most ecologists today, it does suggest that plant invasion as a process is integrated with many other challenges to plant communities. Whether alien plant invasions are the cause of Leopold's land sickness or merely an accompanying symptom, their remedy clearly needs to involve the whole system.

References

Alverson WS, Waller DM, Solheim SL (1988) Forests too deer: edge effects in northern Wisconsin. Conserv Biol 2:348–358

Aronson MFJ, Handel SN, Clemants SE (2007) Fruit type, life form and origin determine the success of woody plant invaders in an urban landscape. Biol Invasion 9:465–475

Baiser B, Lockwood J, La Puma D et al (2008) A perfect storm: two ecosystem engineers interact to degrade deciduous forests of New Jersey. Biol Invasion 10:785–795

Baker HG (1965) Characteristics and modes of origin of weeds. In: The genetics of colonizing species. Academic, New York, pp 147–172

Bale JS, Masters GJ, Hodkinson ID et al (2002) Herbivory in global climate change research: direct effects of rising temperature on insect herbivores. Glob Chang Biol 8:1–16

Bazzaz FA, Mooney HA (1986) Life history characteristics of colonizing plants: some demographic, genetic, and physiological features. In: Ecology of biological invasions of North America and Hawaii. Springer, New York, pp 96–110

Benninger-Truax M, Vankat JL, Schaefer RL (1992) Trail corridors as habitat and conduits for movement of plant species in Rocky Mountain National Park, Colorado, USA. Landsc Ecol 6:269–278

Bever JD (2003) Soil community feedback and the coexistence of competitors: conceptual frameworks and empirical tests. New Phytol 157:465–473

Blackburn TM, Pyšek P, Bacher S et al (2011) A proposed unified framework for biological invasions. Trends Ecol Evol 26:333–339

Bradley BA, Blumenthal DM, Wilcove DS et al (2010) Predicting plant invasions in an era of global change. Trends Ecol Evol 25:310–318

Bressette JW, Beck H, Beauchamp VB (2012) Beyond the browse line: complex cascade effects mediated by white-tailed deer. Oikos 121:1749–1760

Brothers TS, Spingarn A (1992) Forest fragmentation and alien plant invasion of central Indiana old-growth forests. Conserv Biol 6:91–100

Burke MJW, Grime JP (1996) An experimental study of plant community invasibility. Ecology 77:776–790

Cadenasso ML, Pickett STA (2001) Effect of edge structure on the flux of species into forest interiors. Conserv Biol 15:91–97

Carreiro MM, Sinsabaugh RL, Repert DA et al (2000) Microbial enzyme shifts explain litter decay responses to simulated nitrogen deposition. Ecology 81:2359–2365

Catford JA, Jansson R, Nilsson C (2009) Reducing redundancy in invasion ecology by integrating hypotheses into a single theoretical framework. Divers Distrib 15:22–40

Chung H, Zak DR, Reich PB et al (2007) Plant species richness, elevated CO_2, and atmospheric nitrogen deposition alter soil microbial community composition and function. Glob Chang Biol 13:980–989

Côté SD, Rooney TP, Tremblay J-P et al (2004) Ecological impacts of deer overabundance. Annu Rev Ecol Evol Syst 35:113–147

D'Antonio CM, Vitousek PM (1992) Biological invasions by exotic grasses, the grass/fire cycle, and global change. Annu Rev Ecol Syst 23:63–87

Davis MA, Grime P, Thompson K (2000) Fluctuating resources in plant communities: a general theory of invasibility. J Ecol 88:528–534

Dawson W, Burslem DFRP, Hulme PE (2011) The comparative importance of species traits and introduction characteristics in tropical plant invasions. Divers Distrib 17:1111–1121

DeHayes DH, Schaberg PG, Hawley GJ et al (1999) Acid rain impacts on calcium nutrition and forest health. BioScience 49:789–800

Dirr MA (1997) Dirr's hardy trees and shrubs: an illustrated encyclopedia. Timber Press, Portland

Dukes JS, Mooney HA (1999) Does global change increase the success of biological invaders? Trends Ecol Evol 14:135–139

Dukes JS, Pontius J, Orwig D et al (2009) Responses of insect pests, pathogens, and invasive plant species to climate change in the forests of northeastern North America. Can J For Res 39:231–248

Eschtruth AK, Battles JJ (2009) Acceleration of exotic plant invasion in a forested ecosystem by a generalist herbivore. Conserv Biol 23:388–399

Flinn KM, Marks PL (2007) Agricultural legacies in forest environments: tree communities, soil properties, and light availability. Ecol Appl 17:452–463

Flory S, Clay K (2006) Invasive shrub distribution varies with distance to roads and stand age in eastern deciduous forests in Indiana, USA. Plant Ecol 184:131–141

Forman RTT, Alexander LE (1998) Roads and their major ecological effects. Annu Rev Ecol Syst 29:207–231

Foster DR (1993) Land-use history (1730–1990) and vegetation dynamics in central New England, USA. J Ecol 80:753–771

Fox MD, Fox BJ, Groves RH (1986) The susceptibility of natural communities to invasion. In: Ecology of biological invasions. Cambridge University Press, Cambridge, pp 57–66

Foxcroft LC, Pickett STA, Cadenasso ML (2011) Expanding the conceptual frameworks of plant invasion ecology. Perspect Plant Ecol Evol Syst 13:89–100

Franco JA, Martinez-Sanchez JJ, Fernandez JA et al (2006) Selection and nursery production of ornamental plants for landscaping and xerogardening in semi-arid environments. J Hort Sci Biotechnol 81:3–17

Godefroid S, Koedam N (2003) Distribution pattern of the flora in a peri-urban forest: an effect of the city-forest ecotone. Landsc Urban Plann 65:169–185

Graveland J, van der Wal R, van Balen JH et al (1994) Poor reproduction in forest passerines from decline of snail abundance on acidified soils. Nature 368:446–448

Grime JP (2001) Plant strategies, vegetation processes, and ecosystem properties. Wiley, Chichester

Hautier Y, Niklaus PA, Hector A (2009) Competition for light causes plant biodiversity loss after eutrophication. Science 324:636–638

Hobbs RJ, Drake JA (1989) The nature and effects of disturbance relative to invasions. In: Biological invasions: a global perspective. Wiley, Chichester, pp 389–405

Holl KD, Loik ME, Lin EHV et al (2000) Tropical montane forest restoration in Costa Rica: overcoming barriers to dispersal and establishment. Restor Ecol 8:339–349

Huston MA (2004) Management strategies for plant invasions: manipulating productivity, disturbance, and competition. Divers Distrib 10:167–178

Iverson LR, Prasad AM (2002) Potential redistribution of tree species habitat under five climate change scenarios in the eastern US. For Ecol Manag 155:205–222

Jepsen JU, Hagen SB, Ims RA et al (2008) Climate change and outbreaks of the geometrids *Operophtera brumata* and *Epirrita autumnata* in subarctic birch forest: evidence of a recent outbreak range expansion. J Anim Ecol 77:257–264

Juice SM, Fahey TJ, Siccama TG et al (2006) Response of sugar maple to calcium addition to northern hardwood forest. Ecology 87:1267–1280

Kjøller R, Nilssom LO, Hansen K et al (2012) Dramatic changes in ectomycorrhizal community composition, root tip abundance and mycelial production along a stand-scale nitrogen deposition gradient. New Phytol 194:278–286

Knapp AK, Blair JM, Briggs JM et al (1999) The keystone role of bison in North American tallgrass prairie. BioScience 49:39–50

Koerner W, Dupouey JL, Dambrine E et al (1997) Influence of past land use on the vegetation and soils of present day forest in the Vosges Mountains, France. J Ecol 85:351–358

Kowalczyk R, Taberlet P, Coissac E et al (2011) Influence of management practices on large herbivore diet – case of European bison in Białowieża primeval forest (Poland). For Ecol Manag 261:821–828

Kuhman TR, Pearson SM, Turner MG (2011) Agricultural land-use history increases non-native plant invasion in a southern Appalachian forest a century after abandonment. Can J For Res 41:920–929

LaPaix R, Harper K, Freedman B (2012) Patterns of exotic plants in relation to anthropogenic edges within urban forest remnants. Appl Veg Sci 15:525–535

Leopold A (1949) A sand county almanac: and sketches here and there. Oxford University Press, New York

Levey DJ, Bolker BM, Tewksbury JJ et al (2005) Effects of landscape corridors on seed dispersal by birds. Science 309:146–148

Li Y, Cheng Z, Smith WA et al (2004) Invasive ornamental plants: problems, challenges, and molecular tools to neutralize their invasiveness. Crit Rev Plant Sci 23:381–389

Luken JO (1990) Directing ecological succession. Chapman and Hall, London

Mack RN (2000) Cultivation fosters plant naturalization by reducing environmental stochasticity. Biol Invasive 2:111–122

Mack MC, D'Antonio CM (1998) Impacts of biological invasions on disturbance regimes. Trends Ecol Evol 13:195–198

Mack RN, Erneberg M (2002) The United States naturalized flora: largely the product of deliberate introductions. Ann Mo Bot Gard 89:176–189

Malcolm JR, Markham A, Neilson RP et al (2002) Estimated migration rates under scenarios of global climate change. J Biogeogr 29:835–849

Meiners SJ, Pickett STA, Cadenasso ML (2002) Exotic plant invasions over 40 years of old field succession: community patterns and associations. Ecography 25:215–223

Meiners SJ, Cadenasso ML, Pickett STA (2007) Succession on the Piedmont of New Jersey and its implication for ecological restoration. In: Cramer VA, Hobbs RJ (eds) Old fields: dynamics and restoration of abandoned farmland. Island Press, Washington, DC, pp 145–161

Moles AT, Flores-Moreno H, Bonser SP et al (2011) Invasions: the trail behind, the path ahead, and a test of a disturbing idea. J Ecol 100:116–127

Myers JA, Vellend M, Gardescu S et al (2004) Seed dispersal by white-tailed deer: implications for long-distance dispersal, invasion, and migration of plants in eastern North America. Oecologia 139:35–44

Pauchard A, Alaback PB (2004) Influence of elevation, land use, and landscape context on patterns of alien plant invasions along roadsides in protected areas of South-Central Chile. Conserv Biol 18:238–248

Pickett STA, Cadenasso ML (2005) Vegetation dynamics. In: van der Maarel E (ed) Vegetation ecology. Blackwell Publishing, Malden, pp 172–198

Pickett STA, Collins SL, Armesto JJ (1987) A hierarchical consideration of causes and mechanisms of succession. Plant Ecol 69:109–114

Pickett STA, Meiners SJ, Cadenasso ML (2011) Domain and propositions of succession theory. In: Scheiner SM, Willig MR (eds) The theory of ecology. University of Chicago Press, Chicago, pp 185–216

Pöyry J, Lindgren S, Salminen J et al (2004) Restoration of butterfly and moth communities in semi-natural grasslands by cattle grazing. Ecol Appl 14:1656–1670

Pressey RL (1994) Ad hoc reservations: forward or backward steps in developing representative reserve systems? Conserv Biol 8:662–668

Pucek Z, Belousova IP, Krasinska M et al (eds) (2004) European bison. Status survey and conservation action plan. IUCN, Gland

Pyšek P, Sádlo J, Mandák B et al (2003) Czech alien flora and the historical pattern of its formation: what came first to central Europe? Oecologia 135:122–130

Pyšek P, Křivánek M, Jarošík V (2009) Planting intensity, residence time, and species traits determine invasion success of alien woody species. Ecology 90:2734–2744

Ricciardi A (2007) Are modern biological invasions an unprecedented form of global change? Conserv Biol 21:329–336

Rose M, Hermanutz L (2004) Are boreal ecosystems susceptible to alien plant invasion? Evidence from protected areas. Oecologia 139:467–477

Rossiter NA, Setterfield SA, Douglas MM et al (2003) Testing the grass-fire cycle: alien grass invasion in the tropical savannas of northern Australia. Divers Distrib 9:169–176

Scott JM, Davis FW, McGhie RG et al (2001) Nature reserves: do they capture the full range of America's biological diversity? Ecol Appl 11:999–1007

Simberloff D (2000) Global climate change and introduced species in United States forests. Sci Total Environ 262:253–261

Stevens CJ, Dise NB, Gowing DJG et al (2006) Loss of forb diversity in relation to nitrogen deposition in the UK: regional trends and potential controls. Glob Chang Biol 12:1823–1833

Thompson JN, Willson MF (1978) Disturbance and the dispersal of fleshy fruits. Science 200:1161–1163

Throop HL, Lerdau MT (2004) Effects of nitrogen deposition on insect herbivory: implications for community and ecosystem processes. Ecosystems 7:109–133

Treseder KK (2004) A meta-analysis of mycorrhizal responses to nitrogen, phosphorus, and atmospheric CO_2 in field studies. New Phytol 164:347–355

Tylianakis JM, Didham RK, Bascompte J et al (2008) Global change and species interactions in terrestrial ecosystems. Ecol Lett 11:1351–1363

Verdasca MJ, Leitãoa AS, Santana J et al (2012) Forest fuel management as a conservation tool for early successional species under agricultural abandonment: the case of Mediterranean butterflies. Biol Conserv 146:14–23

Verhagen R, Klooker J, Bakker JP et al (2001) Restoration success of low-production plant communities on former agricultural soils after top-soil removal. Appl Veg Sci 4:75–82

Verheyen K, Bossuyt B, Hermy M et al (1999) The land use history (1278–1990) of a mixed hardwood forest in western Belgium and its relationship with chemical soil characteristics. J Biogeogr 26:1115–1128

Walker LR, Walker J, Hobbs RJ (2007) Linking restoration and ecological succession. Springer, New York

Walker LR, Wardle DA, Bardgett RD et al (2010) The use of chronosequences in studies of ecological succession and soil development. J Ecol 98:725–736

Chapter 4
Global Efforts to Address the Wicked Problem of Invasive Alien Species

Jeffrey A. McNeely

Abstract As the globalization of trade continues to expand, species inevitably have more opportunities to spread. Some are intentionally traded and become established in new ecosystems, thereby potentially changing the functioning of the now-modified ecosystem. Invasive alien species are gaining greater public attention, especially when their spread becomes dramatic and threatens ecosystems that people value. But by the time an invasive alien species becomes a problem, it may be too late to effectively respond. Perhaps worse, some people may welcome the alien species, making it even more difficult to implement eradication or control measures. This can make invasive species a 'wicked problem', challenging the building of the consensus necessary to reach a solution. But building public awareness about the impacts of invasive species can help to provide the necessary support, drawing on broad government support through international agreements such as the Convention on Biological Diversity. Practical measures to address the problems posed to native biodiversity (at the level of genes, species, and ecosystems) by invasive species of plants are the subject of active research and practical experience, posing some hope that even a problem as 'wicked' as invasive alien species can be successfully addressed.

Keywords Everglades • Global change • Management responses • International conventions • Wicked problems

J.A. McNeely (✉)
Formerly: Gland Switzerland

Current: Petchburi, Thailand
e-mail: jam@iucn.org

L.C. Foxcroft et al. (eds.), *Plant Invasions in Protected Areas: Patterns, Problems and Challenges*, Invading Nature - Springer Series in Invasion Ecology 7, DOI 10.1007/978-94-007-7750-7_4, © Springer Science+Business Media Dordrecht 2013

4.1   Introduction

In July 2012, the popular press in the United States grabbed onto the news that a
Burmese python 5.20 m long and weighing 74.5 kg had been killed in Florida's
Everglades National Park (ENP), the largest Burmese python yet reported from the
park. Even more worrying to some was that it was a female and had 87 eggs in its
oviduct, ready to be laid. But this should have come as no surprise, as Burmese
pythons (*Python molurus bivittatus*) have been known in ENP for over 30 years
(Meshaka et al. 2000). While the capture of such a large individual was exciting for
readers, it is a disaster for the Everglades ecosystem, which is included on the
UNESCO World Heritage List for the outstanding universal value of its native
biodiversity. Since the species first arrived (probably released by a pet owner), the
populations of potential prey species have been declining rapidly. Marsh rabbits
(*Sylvilagus palustris*), cottontails (*Sylvilagus floridiana*), grey foxes (*Urocyon
cinereoargenteus*), red foxes (*Vulpes vulpes*, itself an invasive alien species) and
opossums (*Didelphis marsulialis*, also invasive alien) are now seldom seen (Dorcas
et al. 2012). The pythons are also known to feed on American alligators (*Alligator
mississippiensis*), crocodiles (*Crocodylus acutus*), endangered Florida panthers
(*Puma concolor coryi*), and the endangered wood stork (*Mycteria americana*).

   While this is a dramatic story, it is only a symptom of a much larger problem.
Over 137 species of animals are recognised as invasive in Florida (Krysko
et al. 2011) and ENP supports five other invasive alien reptiles besides the Burmese
python, as well as nine invasive mammals and eight invasive birds. Florida also has
76 species of invasive alien plants (IAPs), with another 76 plant species established
and potentially invasive. The Everglades alone has 69 species of IAPs, including
many that are widespread and invasive in other parts of the world, for example,
Eichhornia crassipes (water hyacinth), two species of *Casuarina*, *Lantana camara*
(lantana), *Melaleuca quinquenervia* (melaleuca) and *Mimosa pigra* (giant sensitive
plant; FLEPPC 2011). Established to conserve native species, the Everglades has
instead become a magnet for invasive species of alien plants and animals.

   Other states, such as California (California Invasive Plant Council 2006) and
Hawaii (Loope et al. 2013), have fared far worse than Florida, and few of the
world's protected areas (PAs) are immune to the problem. The continuing spread of
invasive alien species (IAS; specifically those alien species that damage native
ecosystems and cause ecological or economic damage), is the result of globalization gone wild. As the case of the Everglades dramatically illustrates, IAS do not
respect the boundaries of PAs. Plants such as *Pueraria montana* (kudzu) invade so
quickly that PA managers seem helpless to prevent their invasion into PAs such as
the Great Smoky Mountains National Park and Shenandoah National Park in the US
(National Park Service 2012). Even World Heritage sites noted for their wildlife
populations are suffering from IAPs that reduce the habitats available for their main
tourist attractions. Examples include *Argemone mexicana* (Mexican poppy) in
Serengeti National Park and Ngorongoro Conservation Area in Tanzania (Henderson
2002), and *Mikania micrantha* (mile-a-minute weed), *Chromolaena odorata*

(Siam weed), and *L. camara* in Chitwan National Park in Nepal, noted for its greater one-horned rhinoceros (*Rhinoceros unicornis*) populations and is also included on the World Heritage List (Raj et al. 2012).

It is sometimes said that change is the only constant, but in today's world the rates of many changes seem to be unprecedented and constantly accelerating. The global problem of the increasing spread of IAS is too often lost among concerns about climate change, growing population, and accelerating economic stress, yet it is linked to all of these.

Invasive alien species provide an example of a 'wicked problem' (see Conklin 2006 on 'wicked problems'), which is a problem that cannot be easily defined and all interested parties often cannot even agree on the problem to be solved. In the case of IAS, these include horticulturalists, gardeners, farmers, pet dealers, importers, resource managers, politicians, biologists, and so forth. Objective solutions may therefore be impossible, especially when the problem is not recognised until the alien species is well established (as in the case of the Burmese python and *M. quinquenervia* in the Everglades).

Resource managers today are forced to address the problem of IAS as best they can, with the incomplete information at hand, which makes the problem 'wicked' for them. Further, PA managers must be prepared to adapt to further invasions that may appear in the course of seeking to manage ecosystems sustainably (an end point that can never be definitively reached because ecosystems are constantly evolving; Hill et al. 1999). The 'wickedness' of the problem makes it even more challenging because sufficient resources required to address it are seldom forthcoming.

Humanity and nature are intimately linked, yet people are living within the 'environmentalist's paradox' (Raudsepp-Hearne et al. 2010): human wellbeing continues to improve (for at least some people, and perhaps even a majority), while many indicators of the state of the environment, such as the increasing numbers and impacts of IAS, continue to worsen. Further, PAs continue to expand but biodiversity continues to decline while invasions by alien species become epidemic in scale (Cox and Underwood 2011). Reasons that have been proposed for this paradox include everything from lag times in biodiversity's response to human actions, and to technologies that promote a decoupling of the fundamental relationship between people and nature. The irony is that economic growth has actually made people more dependent on healthy ecosystems, even as their actions often lead to species invasions that undermine the very ecosystems that they value highly, of which the Everglades is but one example among thousands (Guo et al. 2010).

4.2   Invasive Species as a Global Change Driver Affecting Ecosystems

It is now well recognised that Planet Earth is being affected by a combination of unprecedented ecological, socioeconomic, and institutional changes (UNDP 2003; Millennium Ecosystem Assessment 2005). While this brief review focuses on the ecological changes brought about by IAS, these are intimately related to socioeconomic changes (such as economic growth, global trade, energy demand, human rights, consumption patterns, demographics, settlement patterns, and land use) and institutional changes (laws, regulations, technological innovation, governance, and so forth). The various kinds of global changes are not independent variables; instead, they interact in complex ways that are not always well understood and in any case are dynamic. Such variable factors are often both drivers and consequences of change, with feedback loops that may accelerate or decelerate the spread of IAS. This brief assessment will therefore incorporate at least some of the human dimensions, while leaving it to others to explore these in more detail (see for example McNeely 2001). It will also recognise that the problems of IAS are related to other global changes, requiring a systemic approach that at least recognises these links. Climate change, for example, may lead to the spread of alien species, as well as affecting the distribution of native species (Hellmann et al. 2008).

While many ecosystems are losing native species, many are adding alien species of which only some are invasive. For example, only about 200 of California's 1,800 alien plants are considered invasive by the California Invasive Plant Council (CIPC 2006). Dealing with the responses of ecosystems to such species gains is a major focus of many ecosystem managers, not least in PAs (Mooney et al. 2005; Tu 2009).

A major concern is that the ecosystems of the world are becoming more homogenised, with cosmopolitan species dominating and many native species, especially those endemic to relatively small areas, declining towards extinction (IUCN 2012). While relatively few species introductions are detrimental to ecosystems or human wellbeing, some alien species, when imported into new habitats, are highly destructive to native species, thereby fundamentally changing ecosystems. Three examples illustrate the concerns:

(i) An assessment of 90 alien and 75 native plants in eastern Australia concluded that the alien species were better able to assimilate nitrogen and phosphorus, thereby generating higher primary productivity than the natives and successfully out-competing them. As a result, they profoundly influenced the properties of the ecosystems in which they were found (Leischman et al. 2007).
(ii) The invasion of at least 11 non-native species of earthworms into the forests in the north-eastern part of North America has led to fundamental changes in plant community structure (Nuzzo et al. 2009). Eradicating these earthworms is impossible, so this ecosystem change is essentially permanent.
(iii) As many as 1,500 species of birds (about 20 % of the world's bird species at that time) were driven to extinction on Pacific islands following their settlement by the Polynesian pioneers who first arrived on these islands, accompanied by rats (*Ratus* spp.) and pigs (*Sus scrofa*; Steadman 1995). The ecosystem function on many Pacific islands has been permanently altered as a result, with the spread of non-native plants further disrupting the native ecosystems. Tahiti, for example, is dominated by an invasive tree species, *Miconia calvescens* (miconia), which was introduced in 1937 and now covers over 60 % of the island, threatening some 25 % of the native species and causing major ecosystem disruption (Meyer et al. 2010).

The Parties to the Convention on Biological Diversity (CBD 2010) have agreed that the alien species that threaten ecosystems, habitats, or native species should be controlled or eradicated, or, better yet, prevented from being introduced (CBD article 8(h)). Such species are considered 'invasive alien species' and their impact is increasing as a function of globalization (Meyerson and Mooney 2007); the substantial expansion of trade over the past several decades provides many more opportunities for species to be spread, either intentionally or inadvertently. The damage caused by IAS is widely considered one of the five major drivers of ecosystem disruption (habitat change, climate change, pollution, over-exploitation; Millennium Ecosystem Assessment 2005).

Some alien species have been shown to have some specific conservation values, for example through providing habitat or food resources to threatened or rare species (Schlaepfer et al. 2011). They may also serve as functional substitutes for extinct taxa and help carry out important ecological functions that otherwise might have led to extinctions of other species. One example is the replacement of an extinct subspecies of the Aldabra tortoise with a related subspecies of *Aldabrachelys gigantea* from another island, which led to the recovery of *Diospyros egrettarum* (ebony tree) on the Ile aux Aigrettes in the Seychelles. Seeds that had passed through the digestive systems of tortoises germinated much faster than those that had not been so blessed (Griffiths et al. 2009).

Others have argued that the problem of IAS is not a problem at all, but rather part of the natural process of ecosystem change, and actually increases species diversity (Davis et al. 2011). This position has been vigorously opposed by many ecologists who are concerned about IAS that damage ecosystems (Simberloff et al. 2011). Invasive alien species may also have less genetic diversity than native species, because they typically spread from a very restricted number of founders. Others argue that the study of invasive species may provide important insights into the effects of new species on ecological functioning, thereby providing ecosystem managers with useful tools for responding to the ecosystem changes that are expected to result from the various human impacts that are discussed in the scientific literature (Lockwood et al. 2011).

It may well be that the problem of IAS will never be 'solved' because biota will continue to be moved around the world by international trade. However, greater attention to the problem by ecosystem managers may keep its symptoms within ecologically acceptable bounds (seeking to reach agreement on such bounds helps to make this a wicked problem).

4.3   Management Responses to Invasive Species

Ecosystem management responses to invasive species can take place at multiple scales, from local to global. Most of the global responses are at the policy level (see the following section), but the actual management responses will typically take place at national or lower levels where the action is appropriate to the scale at which the impact of the IAS is perceived.

While management interventions aimed at reducing the impacts of IAS can be as diverse as the species themselves, some generalities based on experience have been identified. Examples include:

(i) Establish explicit goals and objectives for the management intervention.
(ii) Focus on maintaining ecosystem functioning and biodiversity (diversity at the level of genes, species, and ecosystems, as defined in the Convention on Biological Diversity), rather than on species richness (which can lead to promoting alien species).
(iii) Adopt comprehensive approaches that enable the participation of a broad range of interest groups (in political terms, build the strongest possible constituency for the management action of eradicating invasive species or at least reducing their impact).
(iv) Recognise the value of taking a diversity of approaches to various dimensions of species invasions.
(v) Identify the causes of the fundamental problems and design management to address these (such as the link to global trade), rather than just treating their symptoms (though symptoms such as invasive snakes in the Everglades also need to be treated).
(vi) Build understanding of ecosystem management that incorporates the complexity and dynamism of ecosystems, how these are affected by IAS, and human interactions with them.

While these are reasonable guiding concepts, they are hardly comprehensive. Additional management concepts will surely continue to emerge as greater experience is gained (see Wittenburg and Cock (2001), and Genovesi and Monaco (2013) for general guidelines, Simberloff (2013) for eradication, Van Driesche and Center (2013) for biological control, and Meyerson and Pyšek (2013) for prevention).

Humans, like many other animals, are ecosystem managers, seeking to provide forms of stewardship that will enable adaptations to rapidly-changing conditions (also called resilience). General strategies include reducing the magnitude of known stresses, thereby reducing exposure and sensitivity to these changes, developing policies that will shape change rather than responding to effects that have already taken place, and avoiding or escaping from socio-ecological traps that are unsustainable (Chapin et al. 2009). But the growing challenges posed by the role of IAS in ecosystem management, such as those outlined above, have also generated some new management approaches, or modifications of traditional approaches, some of which are discussed below.

4.4   Controlling the Spread of Invasive Alien Species

With the recognition of the close link between global trade and the spread of IAS, coupled to increasing concern about the negative impacts on ecosystems, and the success of at least some eradication efforts (Simberloff 2013), the issue has begun to receive growing international attention. Institutional and legal frameworks for dealing with IAS are now available (Shine et al. 2000), along with tools for assessment and control of biological invasion risks (Koike et al. 2006) and identification of best prevention and management practices (Wittenburg and Cock 2001).

Application on the ground has followed, including the eradication of 27 species of invasive plants and animals from Galapagos Islands (Simberloff et al. 2011). Many other successes have been reported from islands belonging to New Zealand, Australia, Seychelles, Nauru, Mauritius, USA, British West Indies, French Southern Territories, and Northern Ireland (Veitch and Clout 2002), but considerable challenges remain on larger land masses and waters.

These promising efforts are being supplemented by stronger border controls among countries that are especially concerned about invasive species of plant pathogens or about IAS in general. Much can also be learned from efforts to address the problem of invasive species of disease organisms, a problem that poses severe threats to human health and therefore earns considerable public attention (and investment). But with the continuing growth of international travel and trade, dealing with potentially invasive species will require eternal vigilance. Preventing entry is far less expensive than trying to eradicate a species that has become established (arguing for quick response when a potentially invasive species is first noticed, as has been learned from invasive disease vectors), but ecosystem management may be the only available response if an invasive species has become a part of the native ecosystem.

4.5 Policies to Support Control of Invasive Alien Species

The management responses discussed above are practical applications of more general policies that have been put into place to address environmental problems. These policies have been established at global, national, and local levels, by governments, the private sector, farmers, and others. The following examines some of the most relevant global policy responses to the challenges of species invasions.

Ecosystems provide both global and local goods that often span geopolitical boundaries. In dealing with global changes through ecosystem management, governments have realised that global responses, or at least cooperation, will be necessary. Many ecosystems cross international borders: water often flows between countries and managing international rivers (which may support invasive species of fish and other aquatic species) requires cooperation among the riparian states; processes like climate change affect all countries and all ecosystems; and invasive species are intimately linked to global trade that is a major political and economic concern (Lausche 2008).

International environmental law gained momentum at the International Conference on Environment and Development, held in Stockholm, Sweden, in 1972. Only a few of the international agreements emerging from that meeting specifically addressed ecosystem management and thus, at least indirectly, invasive species. The Convention on Wetlands of International Importance (also called the Ramsar Convention, after the Iranian city where the convention was agreed in 1973) is notable in this regard, with resolutions calling for Parties to address the problem of invasive species in Ramsar Sites (Ramsar 2002). Others, such as the World Heritage Convention (agreed in 1975) and the Convention on International Trade in Endangered Species of Flora and Fauna (CITES, agreed in 1973), deal with some PAs and species but with insufficient attention given to invasive species (though, for example, Everglades is now on the List of World Heritage in Danger, at least partly due to the problem of invasive species).

A second generation of conventions were agreed at the so-called 'Earth Summit', held in Rio de Janeiro, Brazil, in 1992. These were 'framework conventions', requiring further agreements between governments on the details of implementation, such as the Kyoto Protocol under the UN Framework Convention on Climate Change and the Biosafety Protocol under the Convention on Biological Diversity (CBD 2010). The CBD specifically deals with invasive species in its Article 8(h), which calls on Parties to "prevent the introduction of, control or eradicate those alien species which threaten ecosystems, habitats, or species". Its 2007 Conference of Parties, held in Kuala Lumpur, issued a detailed decision on alien invasive species (Decision VII/13), helping to put this issue on the conservation agenda (though as one issue among many). At the same meeting, the Conference of Parties agreed a Programme of Work on Protected Areas (Decision VII/28) that includes many elements that could be applied to invasive alien species in PAs (though they do not explicitly do so) (SCBD 2010).

In 2010, the tenth Conference of Parties of the CBD approved a Strategic Plan for the coming decade (2011–2020), which specifically addressed invasive species in its Target 9: "By 2020, invasive alien species and their pathways are identified and prioritised, priority species are controlled or eradicated, and measures are in place to manage pathways to prevent their introduction and establishment". This is certainly a useful step, especially because prior to the CBD meeting, other multilateral environmental conventions agreed to adopt the CBD Strategic Plan as the common basis for their own work in the coming years, taking a significant advance towards coherence in obligations, data collection and reporting.

Another important recent policy outcome that could help address the problem in IAS is the 2010 UN General Assembly resolution that supported the establishment of an Intergovernmental Platform on Biodiversity and Ecosystem Services (IPBES). It is expected to further strengthen the credibility, legitimacy and saliency of the information exchange processes between the scientific community and policy makers in areas relating to biodiversity and ecosystem services, which certainly includes IAS (though not automatically; vigilance is still required to ensure that invasive species are not left lurking in the background).

Dealing with governments can be frustrating, because many of the decisions are reached by a consensus that is strongly affected by political factors. But the successes of the international structure that supports ecosystem management indicate that considerable benefits can arise from seeking consensus on the priority actions that need to be taken to address the problems posed by IAS.

4.6   Conclusions

The ecological changes brought about by IAS are best understood as a package of variable changes, with the relative impacts differing from site to site and depending on numerous external factors, ranging from geographical location to human demography. An agent of global change, such as the spread of species, can harm some components of an ecosystem but can benefit others, further complicating the challenge of making broad generalisations. That said, this review has reached the following conclusions:

(i) All efforts to manage IAS need to be based on the best available information, including solid science, political reality, and local knowledge.

(ii) Few IAS have yet generated the appropriate institutional responses, at least partly because they tend to have complex drivers that are the responsibility of many different actors. Also, their symptoms are not always immediately apparent to the public and policy-makers, and many of the major issues are influenced more by social, economic, and political factors than scientific ones. This calls for renewed efforts to strengthen the science component of responses to IAS and seek more effective means of reaching policy-makers through collaboration with other sectors of the economy.

(iii) Many management responses to invasive species will involve trade-offs, as the different interest groups negotiate for favourable outcomes to suit their own interests. This argues for including the key interest groups in negotiating the trade-off's, and addressing some of their major concerns, particularly the economic and social dimensions of invasive species.

(iv) With budgets becoming increasingly tight in many countries, creative ways to enforce regulations regarding invasive species will be required. For example, citizen scientists can help to provide early warning for IAS (Acevedo-Gutierrez et al. 2011).

Our planet is experiencing multiple wicked problems that are undermining the ecosystems upon which humanity depends. Increased investments in managing IAS should be seen as part of a package of investments that will be required to ensure a healthy and prosperous future for all of humanity and the rest of the living world.

References

Acevedo-Gutierrez A, Acevedo L, Boren L (2011) Effects of the presence of official-looking volunteers on harassment of New Zealand fur seals. Conserv Biol 25:623–627

California Invasive Plant Council (2006) California invasive plant inventory. http://www.cal.ipc. org. Accessed 28 Oct 2012

CBD (Convention on Biological Diversity) (2010) The Aichi biodiversity targets. Secretariat of the Convention on Biological Diversity, Montreal

Chapin FS III, Carpenter SR, Kofinas GP et al (2009) Ecosystem stewardship: sustainability strategies for a rapidly changing planet. Trends Ecol Evol 25:241–249

Conklin J (ed) (2006) Dialogue mapping: building shared understanding of wicked problems. Wiley, West Sussex

Cox R, Underwood C (2011) The importance of conserving biodiversity outside of protected areas in Mediterranean ecosystems. PLoS One 6(1):e14508

Davis MA, Chew MJ, Hobbs RJ et al (2011) Don't judge species on their origins. Nature 474:153–154

Dorcas M, Willson JD, Reed RN et al (2012) Severe mammal declines coincide with proliferation of invasive Burmese pythons in Everglades National Park. Proc Natl Acad Sci U S A 109:2418–2422

FLEPPC (Florida Exotic Pest Plant Council) (2011) Florida exotic pest plant database 2011. http://www.fleppc.org/list/11list.html. Accessed 18 May 2013

Genovesi P, Monaco A (2013) Chapter 22: Guidelines for addressing invasive species in protected areas. In: Foxcroft LC, Pyšek P, Richardson DM, Genovesi P (eds) Plant invasions in protected areas: patterns, problems and challenges. Springer, Dordrecht, pp 487–506

Griffiths CJ, Jones CG, Hansen DM et al (2009) The use of extant non-indigenous tortoises as a restoration tool to replace extant ecosystem engineers. Restor Ecol 18:1–7

Guo Z, Zhang L, Li Y (2010) Increased dependence of humans on ecosystem services and biodiversity. PLoS One 5(10):e13113

Hellmann JJ, Beyers JE, Bierwagen BG et al (2008) Five potential consequences of climate change for invasive species. Conserv Biol 22:534–543

Henderson L (2002) Problem plants in Ngorongoro Conservation Area. Final report to the Ngorongoro Conservation Area Authority. Arusha, Tanzania

Hill SB, Vincent C, Chouinard G (1999) Evolving ecosystems approaches to fruit insect pest management. Agric Ecosyst Environ 73:107–110

IUCN (International Union for Conservation of Nature) (2012) IUCN red list. http://www.iucnredlist.org. Accessed 28 Oct 2012

Koike F, Clout MN, Kawamichi M et al (2006) Assessment and control of biological invasion risks. Shoukadoh Book Sellers/International Union for Conservation of Nature, Kyoto/Gland

Krysko KL, Enge KM, Moler PE (2011) Verified non-indigenous amphibians and reptiles in Florida from 1863 to 2010. Outlining the invasion process and identifying invasion pathways and stages. Zootaxa 3028:1–64

Lausche B (2008) Weaving a web of environmental law. Erich Schmidt Verlag, Berlin

Leischman MR, Haselhurst T, Areas A et al (2007) Leaf trait relationships of native and invasive plants: community and global-scale comparisons. New Phytol 176:635–643

Lockwood J, Hoopes M, Marchetti M (2011) Non-natives: plusses of invasion ecology. Nature 475:36

Loope L, Flint Hughes R, Meyer J-Y (2013) Chapter 15: Plant invasions in protected areas of tropical Pacific Islands, with special reference to Hawaii. In: Foxcroft LC, Pyšek P, Richardson DM, Genovesi P (eds) Plant invasions in protected areas: patterns, problems and challenges. Springer, Dordrecht, pp 313–348

McNeely JA (ed) (2001) The great reshuffling: human dimensions of invasive alien species. International Union for Conservation of Nature, Gland

Meshaka WE, Loftus WF, Steiner T (2000) The herpetofauna of Everglades National Park. Florida Sci 63:84–103

Meyer JY, Fourdrigniez M, Taputuarai R (2010) The recovery of the native and endemic flora after the introduction of a fungal pathogen to control the invasive tree *Miconia calvescens* in Tahiti, French Polynesia. Biol Control Nat 3:1–21

Meyerson LA, Mooney HA (2007) Invasive alien species in an era of globalization. Front Ecol Environ 5:199–208

Meyerson LA, Pyšek P (2013) Chapter 21: Manipulating alien plant species propagule pressure as a prevention strategy for protected areas. In: Foxcroft LC, Pyšek P, Richardson DM, Genovesi P (eds) Plant invasions in protected areas: patterns, problems and challenges. Springer, Dordrecht, pp 473–486

Millennium Ecosystem Assessment (2005) Ecosystems and human well-being: biodiversity synthesis. World Resources Institute, Washington, DC

Mooney HM, Mack RN, Mcneely JA et al (eds) (2005) Invasive alien species: a new synthesis. Island Press, Washington, DC

National Park Service (ed) (2012) Great Smoky Mountains National Park. United States National Park Service, Washington, DC

Nuzzo V, Maerz JC, Blossey B (2009) Earthworm invasion as the driving force behind plant invasion and community change in Northeastern North American forests. Conserv Biol 23:966–974

Raj RK, Scarborough H, Subedi N et al (2012) Invasive plants: do they devastate or diversify rural livelihoods. J Nat Conserv 20:170–176

Ramsar (2002) Decision VIII.18. Invasive species and wetlands. 8th meeting of the conference of parties to the convention on wetlands. Ramsar Secretariat, Gland

Raudsepp-Hearne C, Peterson GD, Tengö M et al (2010) Untangling the environmentalist's paradox: why is human well-being increasing as ecosystem services degrade? BioScience 60:576–589

SCBD (Secretariat of the Convention on Biological Diversity) (2010) Handbook of the convention on biological diversity. United Nations Environment Program, Montreal

Schlaepfer MA, Sax D, Olden JD (2011) The potential conservation value of non-native species. Conserv Biol 25:428–437

Shine C, Williams N, Gundling L (2000) A guide to designing legal and institutional frameworks on alien invasive species. International Union for Conservation of Nature, Gland

Simberloff D (2013) Chapter 25: Eradication – pipe dream or real option? In: Foxcroft LC, Pyšek P, Richardson DM, Genovesi P (eds) Plant invasions in protected areas: patterns, problems and challenges. Springer, Dordrecht, pp 549–559

Simberloff D et al (2011) Non-natives: 141 scientists object. Nature 475:36

Steadman DW (1995) Prehistoric extinctions of Pacific island birds: biodiversity meets zooarcheology. Science 267:1123–1131

Tu M (2009) Assessing and managing invasive species within protected areas. The nature conservancy, convention on biological diversity, and IUCN World Commission on Protected Areas. Washington, DC

UNDP (United Nations Development Program) (2003) Human development report 2003. United Nations, New York

Van Driesche R, Center T (2013) Chapter 26: Biological control of invasive plants in protected areas. In: Foxcroft LC, Pyšek P, Richardson DM, Genovesi P (eds) Plant invasions in protected areas: patterns, problems and challenges. Springer, Dordrecht, pp 561–597

Veitch CR, Clout MN (eds) (2002) Turning the tide: the eradication of invasive species. Proceedings of the international conference on eradication of island invasives. International Union for Conservation of Nature, Gland, Switzerland

Wittenburg R, Cock MJW (2001) Invasive alien species: a toolkit of best prevention and management practices. CAB International, Wallingford

Chapter 5
A Cross-Scale Approach for Abundance Estimation of Invasive Alien Plants in a Large Protected Area

Cang Hui, Llewellyn C. Foxcroft, David M. Richardson, and Sandra MacFadyen

Abstract Efficient management of invasive alien plants requires robust and cost-efficient methods for measuring the abundance and spatial structure of invasive alien plants with sufficient accuracy. Here, we present such a monitoring method using ad hoc presence-absence records that are routinely collected for various management and research needs in Kruger National Park, South Africa. The total and local abundance of all invasive alien plants were estimated using the area-of-occupancy model that depicts a power-law scaling pattern of species occupancy across scales and a detection-rate-based Poisson model that allows us to estimate abundance from the occupancy, respectively. Results from these two models were consistent in predicting a total of about one million invasive alien plant records for the park. The accuracy of log-transformed abundance estimate improved significantly with the increase of sampling effort. However, estimating abundance was shown to be much more difficult than detecting the spatial structure of the invasive alien plants. Since management of invasive species in protected areas is often hampered by limited resources for detailed surveys and monitoring, relatively simple and inexpensive monitoring strategies are important. Such data should also be appropriate for multiple purposes. We therefore recommend the use of the scaling pattern of species distribution as a method for rapid and robust

C. Hui (✉) • D.M. Richardson
Centre for Invasion Biology, Department of Botany and Zoology, Stellenbosch University, Private Bag X1, Stellenbosch 7602, South Africa
e-mail: chui@sun.ac.za; rich@sun.ac.za

L.C. Foxcroft • S. MacFadyen
Conservation Services, South African National Parks, Private Bag X402, Skukuza 1350, South Africa

Centre for Invasion Biology, Department of Botany and Zoology, Stellenbosch University, Private Bag X1, Stellenbosch 7602, South Africa
e-mail: Llewellyn.foxcroft@sanparks.org; sandramf@live.co.za

L.C. Foxcroft et al. (eds.), *Plant Invasions in Protected Areas: Patterns, Problems and Challenges*, Invading Nature - Springer Series in Invasion Ecology 7, DOI 10.1007/978-94-007-7750-7_5, © Springer Science+Business Media Dordrecht 2013

monitoring of invasive alien plants in protected areas. Not only do these approaches provide valuable tools for managers and biologists in protected areas, but this kind of data, which can be collected as part of routine activities for a protected area, provides excellent opportunities for researchers to explore the status of aliens as well as their assemblage patterns and functions.

Keywords Invasive plants • Kruger National Park • *Opuntia stricta* • *Lantana camara* • Optimum monitoring strategy

5.1   Introduction

Population size (abundance) is a measure of the extent and impact of biological invasions (Parker et al. 1999). Therefore, to assess the impact of invasive alien species and propose efficient and timely control strategies, we need to have a cost-efficient and robust approach for estimating abundance. Without this it is difficult to assess the status and change in invasions, and to propose refinements to existing management strategies. To maximise the application of this approach, we cannot only utilise datasets collected and recorded for specific purposes (e.g. relative abundances recorded along transects), but those in widely-accessed format collected in the field by conservation and alien plant control managers. To quickly assess the invasion extent, ground managers often accumulate localities of a focal invader found in a haphazard fashion or reported by others. To this end, designing methods that can utilise these geo-referenced presence-only or presence-absence records and robustly estimate the abundance at a satisfactory level could provide substantial benefits and opportunities for control and risk assessment, and thus deserves closer attention.

Many types of models can be used to estimate abundance from presence-absence records (e.g. Wright 1991; MacKenzie et al. 2002; He and Gaston 2003; Royle et al. 2005; Hui et al. 2009, 2011a). These models can be clustered into three groups. First, models based on the abundance-occupancy relationship require a robust representation of the focal species' occupancy based samplings of spatially equal effort (e.g. He and Gaston 2003). These types of models often assume that the environment is homogenous (Hui et al. 2009) and thus can only be used at local scales. Second, models incorporating imperfect detection rate often require repeated samples (MacKenzie et al. 2002; Hui et al. 2011a); this often not feasible when working over large areas. Moreover, using this type of model, and also interpreting the model predictions, usually requires knowledge of fairly advanced mathematical methods. This has hindered their wide application in the field. Moreover, both types of models require systematic sampling and are only appropriate for studies of particular species at local sites. They fall short when a large-scale quick assessment is needed.

We here recommend the use of a third kind of model, namely the area-of-occupancy (AOO), which does not involve abstruse mathematics. They are also suitable for rapid, large-scale assessments using records. A power-law pattern of

AOOs across scales has been widely observed for species, across taxa (Kunin 1998). The power law AOO could reflect the fractal structures of species distributions driven by the fractal structures of environmental variables and multiple ecological processes. Although the exact reason behind a power-law AOO is still unknown and widely debated (e.g. Halley et al. 2004 and references therein), this specific form of AOO model has been suggested as an elegant way for extrapolating species occupancy across scales (Wilson et al. 2004) and estimating abundance with reasonable accuracy over large, heterogeneous environments (Hartley and Kunin 2003; Hui et al. 2009). Due to the lack of suitable large-scale monitoring approaches, incorporating the AOO model into monitoring programmes for alien and invasive plants is an attractive option. Here, we aim to (i) explore the presence records collected by rangers during their routine patrols in protected areas and (ii) demonstrate how the AOO method can help us to estimate the abundances and distributions of top plant invaders in the protected areas.

5.2   Kruger National Park as a Model System

We demonstrate the AOO approach for the abundance estimation of invasive alien plants (IAPs) in a large protected area, which is actively managed for wildlife protection and biodiversity conservation – South Africa's Kruger National Park (KNP). Kruger National Park (c. 20,000 km^2, the size of Israel) is situated in the north-eastern part of South Africa, in a semi-arid savanna system (Fig. 5.1a). A full account of the KNP's biophysical landscape and its rich 100-year management and research history has been given in Du Toit, Rogers and Biggs (2003). Unfortunately KNP also contains a large number of alien species, with 348 invasive alien plant (IAP) species identified to date (Foxcroft et al. 2003; Spear et al. 2011).

In 2004 an electronic handheld data capturing system – CyberTracker – was initiated. The CyberTracker system is a customised, icon driven programme run on a personal digital assistant (PDA) device, and is used during routine patrols by rangers, to capture distribution records of a number of features such as animal sightings, ephemeral water distribution, carcasses, alien species and many others (full details on the CyberTracker programme are given in Kruger and MacFadyen 2011). Examples of other uses of CyberTracker data are rich in literature (Dietemann et al. 2006; Foxcroft et al. 2009; Hui et al. 2011a). Approaches with which the abundance of IAPs may be estimated using the CyberTracker records are thus not only of management value for KNP, but also provide an opportunity to assess, refine and provide guidelines for designing similar monitoring programmes for other protected areas.

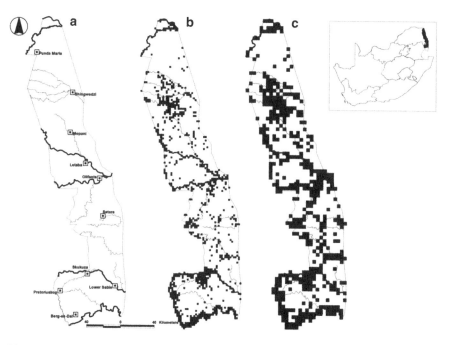

Fig. 5.1 (**a**) The spatial geography of Kruger National Park and presence-absence maps of invasive alien plants at a resolution of 2 × 2 km (**b**) and 4 × 4 km (**c**) of the Kruger National Park

5.3   Power Law Area-Of-Occupancy

To estimate the total number of all IAPs in the KNP, we first divided the landscape into lattices with the grain size of each cell a km^2 (Fig. 5.1b, c). Let $u_a(i)$ denote the number of records in the cell i, in which a number of $u_a^+(i)$ records are presence and the rest absence. The occupancy at the scale of a can thus be defined as (Eq. 5.1),

$$P_a = \sum_i a \cdot \varphi_a(i) \qquad (5.1)$$

where $\varphi_a(i) = 1$ if $u_a^+(i) \geq 1$; otherwise, $\varphi_a(i) = 0$. The power-law AOO describes how the occupancy P_a changes across scales, specifically as a function of grain a (Kunin 1998; Hartley and Kunin 2003; Gaston and Fuller 2009; Hui et al. 2009). Therefore, there are a number of $N_a = P_a/a$ cells occupied by the IAPs at the scale of a; that is, the number of presences at the scale a is (Eq. 5.2),

$$N_a = \sum_i \varphi_a(i) \qquad (5.2)$$

When the grain is small enough to only hold at most one IAP individual (i.e. the grain equals the individual size, $a = \delta$), the number of occupied cells N_δ is thus equal to the abundance of the IAPs (Hartley and Kunin 2003). The individual size is not the canopy size of a plant but equals the reciprocal of local density; that is, the minimum land size that can support the growth of a plant. For plant species that are unfeasible for identifying individuals, we can use the reciprocal of the density of plant patches to define δ; of course, N_δ will not represent the abundance but rather the number of plant patches of a focal species. A power-law form of the AOO has been confirmed for the distribution of plants (Kunin 1998; Hartley and Kunin 2003; Hui et al. 2011b) and butterflies (Wilson et al. 2004),

$$P_a = c \cdot a^d \qquad (5.3)$$

where c and d are constants (Eq. 5.3). Parameter c represents the occupancy when the grain equals 1 km^2, whilst d denotes the exponent of the power law and is proportional to how fast the occupancy changes with the spatial scales when the grain is around 1 km^2. We chose this power-law AOO for the abundance estimation because of its simplicity and empirical support of plant distributions. The power-law AOO also indicates a self-similar and fractal nature of species distribution (Hui and McGeoch 2008).

The power-law AOO is a powerful predictive tool, as once the occupancies at two (or more) different scales have been determined, the parameters c and d can be estimated and therefore the total abundance N_δ can be calculated. For a robust calculation we divided the landscape into lattices with 64 different grains, with the width of the cells ranging from 125 m to 8 km. The power law AOO was fitted using linear regression on log-log transformed axes for the occupancies at these 64 scales. The number of occupied cells was also calculated for seven classes to understand the species-level assembly of IAPs: *Opuntia stricta* (sour prickly pear), *Lantana camara* (lantana), *Opuntia* spp. (all other *Opuntia* records combined, except those specified as *O. stricta*), *Chromolaena odorata* (Chromolaena/Siam weed), *Pistia stratiotes* (water lettuce), *Parthenium hysterophorus* (parthenium) and others (the combination of species with less than 200 records). Lennon et al.'s (2007) χ^2 test of the difference in maximum log-likelihoods was used to verify this power-law AOO.

To measure the performance of using the power-law AOO for total abundance estimation under different sampling efforts, we calculated the occupancies P_a and P_{4a} (by combining the four adjacent a-size cells to form a $4a$-size cell; specifically $a = 4 \times 4$ km and thus $4a = 8 \times 8$ km), from which the parameters c and d can be calculated and the total abundance can be estimated. The status (presence or absence) of boundary cells was determined solely on the records within the park. Since the occupancy of a $4a$-size cell could be derived from the combined information of less than four a-size cells (e.g. cells along the boundaries), the abundance estimates at coarse scales were likely to be overestimated. We propose a general relationship between the abundance estimation and sampling effort (Eq. 5.4),

$$\ln N(s) = \ln N_\delta \cdot \left(1 - \exp\left(-a \cdot s^\beta\right)\right) \tag{5.4}$$

where $N(s)$ is the abundance estimation under the sampling effort s; α is a measure of the converging speed from the abundance estimation $N(s)$ to its limit N_δ; β is a scaling parameter for sampling effort. Let $A = 1\text{-Abs[pred-obs]/pred}$ denote the accuracy of the prediction (Hui et al. 2006), the above *abundance-effort relationship* would allow us to estimate the minimum sampling effort (s^*) for estimating the abundance at a satisfactory level (say, $A = 0.95$).

5.4 Poisson Occupancy-Abundance Model

For the calculation of the local abundance, i.e. the abundance of IAP in specific cell, we first estimated the proportion of presences in each cell, $D_a(i) = u_a^+(i)/u_a(i)$. Second, if we assume that CyberTracker records occur randomly and independently of one another within the cell, we can estimate the local abundance of IAPs according to the Poisson occupancy-abundance model (Eq. 5.5; Wright 1991),

$$n_a(i) = -U \times \ln(1 - D_a(i)) \tag{5.5}$$

where $1/U$ is the minimum size of the area that one record can represent; that is, the size of the detection area when a ranger stands still. In practice, U is estimated by the size of grain divided by the maximum number of records in cells. The local abundance was calculated at a grain of 4×4 km for demonstration. More sophisticated approaches that deal with pseudo-absence dilemma and zero-inflation problem are also available (e.g. Bayesian estimation model; Hui et al. 2011a) but we prefer this simple Poisson model here that also provides a reasonable estimate of local abundance. The AOO method was not used for calculating the local abundance because additivity does not apply for the power law (e.g. $x^{1.5}+y^{1.5} < (x+y)^{1.5}$; Cohen et al. 2005) and thus the summation of local abundances often underestimate the total abundance.

To measure the performance of the above approach under different sampling efforts for estimating the spatial structure of IAPs, the normal approach would be to calculate the discrepancy between the frequency distributions of the predicted local abundance and observed local abundance, using chi-square statistics. However, these statistics encounter problems when the frequency for some categories is lower than five (Quinn and Keough 2002), which is often the case during the simulation. Instead, we use the sum of squared errors for the distance (or deviation) between the original spatial structure and the spatial structure from sampling (Eq. 5.6),

$$d = \sum_i \left(\log_2 D(i) - \log_2 D_a'(i) \right)^2 \tag{5.6}$$

where $D_a'(i)$ is the estimate of the proportion of presences in the cell i under a specific sampling effort. The sum and mean of squared errors have often been used in comparing the similarity of two images in the field of image processing (e.g. Wang et al. 2004). The reason for the log-transformation here is because it can largely normalise the observed frequency distribution of the proportion of presences.

The significant distance can be determined by a randomization test (also called a permutation test; Sokal and Rohlf 1995). For this test we reshuffled the observed proportions of presences for the cells with at least one presence record 5,000 times. After each run, we calculated the above distance between the reshuffled spatial structure and the original observed spatial structure. We then built a probability distribution of these 5,000 distances in order to proceed with a test of significance. We found that the observed probability distribution of these distance was not different from a normal distribution (Shapiro-Wilk test; SW-W $= 0.9995$, p $= 0.217$), from which we identified a one-tail critical value with p $= 0.05$, $d_{0.05} = 5{,}027.5$. With certain amount of sampling effort, we can predict the spatial structure of local abundances of IAPs. The deviation of this sampling-effort-dependent structure from the real structure can be captured by d; if we have $d < d_{0.05}$, the predicted spatial structure is not significantly different from the observed structure (p < 0.05); otherwise ($d > d_{0.05}$), the spatial structure prediction is significantly different from that which was observed. The minimum sampling effort was defined as the sampling effort so that the spatial structure from simulations is not significantly different from the observed spatial structure.

5.5   Invasion Status of Kruger National Park

We used data from 2004 to 2007, comprising 2.4 million records with 27,777 presences and the rest absences (absences were taken from records of other features that were collected; see Hui et al. 2011a for issues with pseudo-absences). The extremely low occurrence (1.15 %) of IAPs does not necessarily indicate insufficient sampling intensity but could suggest that most species are currently: (i) at an early phase of invasion, with high potential for rapid expansion, (ii) are being maintained at the current state of low abundance through on-going management, or (iii), for some species at least, especially those introduced from regions with a completely different climate, have already occupied the full extent of their potential range in the area.

There were 167 IAP species represented by the presence records, with most records for *O. stricta* (72.1 %) and *L. camara* (8.4 %), and 119 species with fewer

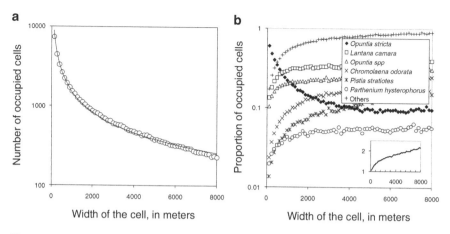

Fig. 5.2 (a) The scaling pattern of the number of occupied cells of all the invasive alien plants in Kruger National Park. The *solid line* indicates the regression from all 64 grains; the *solid circles* indicate the number of occupied cells at 64 spatial scales (from 125 × 125 m to 8 × 8 km). The *fitted lines* have been extrapolated over two orders of magnitude to the scale of 1 × 1 m. (b) The proportion of cells occupied by each species in the cells with records of IAPs. The inset indicates the sum of the proportion of all species approaches one when scaling down

than ten records. The maximum number of records per cell for the grain of 4 × 4 km² is $U = 72,203$, indicating a detection area of 221.6 m² with a diameter of 16.8 m. The mean canopy size of all IAPs weighted by the number of records for each species is 0.748 (0.743–0.754) m² (a mean radius of 0.49 m). For simplicity, we added a buffer zone beyond the canopy (an extension of 0.076 m) to make the individual size to be exactly 1 m² (i.e. the total abundance estimates equal the number of occupied unit-size cells).

The power-law form of the area-of-occupancy (AOO) provided an accurate description of the scaling pattern of IAP's distribution (Fig. 5.2a). The maximal log-likelihood ($= -6,160.38$) of Lennon et al.'s (2007) scale-dependent model showed no significant difference ($\chi^2 = 3.72$; p > 0.05) from the power law AOO (log-likelihood$_{max} = -6,162.24$). When considering all 64 scales, we had $N_a = e^{13.2}a^{-0.425}$ ($R^2 = 0.995$; solid line in Fig. 5.2a), indicating a total number of 552,820 IAPs with a box-counting fractal dimension $D = 2 \times 0.425 = 0.85$. We further calculated the AOO for 16 smaller scales (from 125 × 125 m to 2 × 2 km; $N_a = e^{12.7}a^{-0.385}$, $R^2 = 0.999$, $D = 0.77$; a total number of 313,461 IAPs) and for 49 larger scales (from 2 × 2 km to 8 × 8 km; $N_a = e^{14.2}a^{-0.485}$, $R^2 = 0.97$, $D = 0.97$; a total number of 1,450,685 IAPs).

Species-level partitioning of the occupied cells showed a decrease of overlapping among species when scaling down, with the sum of occupied cells for each species approaching the number of occupied cells when combining all IAP species together (Fig. 5.2b), suggesting a reliable estimation of the total number of IAPs. Although the log-transformed abundances are compatible, disparities do appear among these estimates – a typical problem of any scaling method (Lennon et al. 2007). Therefore, the abundance estimate must be verified by the sum of local abundance. When the width of the grain cell is greater than 4 km, the proportion of occupied cells for each species is fairly stable (i.e. scale-insensitive; Fig. 5.2b). However, it starts to group into three different levels at finer scales: *C. odorata*, *P. stratiotes* and *P. hysterophorus* converges to only occupying 1 % of the total occupied cells of IAPs; *L. camara*, *Opuntia* spp. and other species combined converges to occupy 10 % of the occupied cells; *O. stricta* converges to occupying a majority of all occupied cells (Fig. 5.2b). Evidently, most IAP species have a fairly wide range but are locally rare, whilst *O. stricta* has a moderate range but are locally extremely abundant.

We calculated the local abundance at the scale (grain) of 4 × 4 km. Because the Poisson model can only be applied for cells with presence records (Fig. 5.1), the estimation of local abundance were literally for those areas only. Abundance of IAPs in absence cells was thus considered extremely low and was neglected in the further analyses. The total abundance of IAPs calculated by summing up the local abundances of all cells is 1,033,969, which is half way between the AOO estimates from all scales (552,820) and from large scales (1,450,685).

5.6   Assemblage Patterns of Aliens

Preston (1948) first identified the log-normal form of the "species" and "individual" curves, i.e. the frequency distribution of species and individuals falling in any given octave class. Based on his finding, we expect a log-normal form of the frequency distribution of local abundance in the cells (i.e. the number of IAPs in each cell). Even though the $\log_2$-transformed frequency distributions of the proportion of presences and local abundance have similar shape (Fig. 5.3), neither followed a strict normal distribution (SW-W = 0.963, p < 0.01; SW-W = 0.957, p < 0.01; Fig. 5.3).

If we consider each cell to be identical in hosting individuals (i.e. a homogenous landscape), a neutral model prediction that considers the birth and death events within each cell, should be expected (Volkov et al. 2005). This is equal to switching the concept of a species' abundance in a community by the number of IAPs in a cell; that is, the number of individuals of a neutral species behaves similarly to the number of individuals of a homogenous landscape site. Therefore, we can test whether the frequency of cells with n number of IAPs (F_n) follows Volkov et al.'s (2005) neutral model, $F_n = \theta x^n/(n + c)$, where x and c are regression parameters, and θ a normalization constant to ensure the number of cells sensible). Parameter x indirectly represents the increase rate of IAPs, and c controls the strength of the density dependence. Using maximum log-likelihood method for parameterization, we also found a significant distinction between the neutral model prediction from the local abundance estimates (log-likelihood = -1599.1, $x = 0.99992$, $c = 33$, $\theta = 0.1865$, $\chi^2 = 75.8$, $df = 10$, p < 0.01).

Frequency

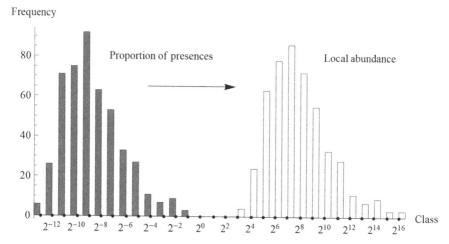

Fig. 5.3 Frequency distributions of the proportion of presences (*grey bars*) and the local abundance (i.e. the number of IAP individuals per cell; *white bars*) of the Kruger National Park at the 4 × 4 km scale

Evidently, besides the overall log-normal shape of the local-abundance frequency distribution (Fig. 5.3), it differs from both log-normal shape and neutral model predictions. This indirectly suggests (i) that the KNP landscape is heterogeneous (thereby supporting other work, for example Pickett et al. 2003) and (ii) that IAPs in KNP have not reached their demographic equilibrium, indicating that the invasions are at an early stage or that, in some areas at least, IAPs are being maintained at their current state through management efforts. Of course, other interpretation of this result may also exist. For instance, the maximum likelihood estimate of parameter c is much greater than those estimated for tropical forests (Volkov et al. 2005), suggesting there could be a strong density dependence of the IAPs (positively or negatively). This could lead to strong spatial autocorrelation of the number of IAPs between cells and violate the assumptions of neutral models, suggesting the possible hotspots of high invasibility of certain areas in the KNP.

5.7 Optimal Sampling and Monitoring Effort

To determine the efficient sampling scheme, we first let s denote the total number of records reported (a sum of presences and absences), i.e. the sampling effort. For each unit of sampling effort, only one cell can be visited and the chance of reporting a presence record is equal to the proportion of presences of the cell. Five sampling schemes are examined, including random, systematic, addictive, elusive and random-walk. In a random sampling scheme, the cell visited each time is randomly chosen. In a systematic sampling scheme, all cells will be visited an equal number of times. In an addictive sample scheme, the ranger tends to visit the cells having more presence records, i.e. the probability of visiting the cell i depends on its recording history (Eq. 5.7):

$$v(i)/\sum_j v(j) \tag{5.7}$$

where $v(i)$ and $v(j)$ are one plus the current number of presence records of cell i and j, respectively. Plus one is to ensure cells currently with no presence records can still have a chance to be visited. In an elusive sampling scheme, the ranger will try to avoid visiting the cells with presence records, with the probability of visiting cell i being (Eq. 5.8),

$$(1/v(i))/\sum_j (1/v(j)) \tag{5.8}$$

In a random-walk sampling scheme, the ranger will randomly choose a cell adjacent to the cell visited at the last time, i.e. randomly choosing among the four neighbouring cells of the current visiting cell. If the cell is at the border of KNP, the choice of adjacent cells will only be among those within the KNP.

We then simulated the sampling process according to the above schemes on a 4×4 km resolution for the whole KNP. Twenty-six simulations with different scenarios of sampling efforts ($s = 512, 724, 1024, \ldots, 3 \times 10^6$) were chosen for each random, addictive and random-walk schemes; 22 simulations with different sampling efforts ($1333, 2666, \ldots, 3 \times 10^6$) for systematic scheme; 17 simulations with different sampling efforts ($512, 724, 1024, \ldots, 131072$) for elusive scheme. The maximum number of sampling effort was constrained by the computational capacity. Each simulation was then repeated five times to reduce the effect of stochasticity, and, thus, a total number of 585 simulations were run, with the maximum sampling effort in a single simulation reaching three million. The results from these simulations allowed us to further compare the performance from each sampling scheme and calculate the minimum sampling effort in the estimation of total abundance and the spatial structure of IAPs in the KNP.

The Poisson occupancy-abundance model explained a significant amount of variance in the relationship between abundance estimation and sampling effort (F-ratio, p < 0.01; Fig. 5.4a). In terms of accuracy, we compared the limit of the logarithmic abundance estimations ($\ln N_\delta$) with the AOO abundance estimations (High: $\ln(1450685) = 14.188$;   Middle: $\ln(552820) = 13.22$;   Low: $\ln(313461) = 12.66$). At least 13.7 % of current sampling effort (i.e. 3×10^5 records) is needed for reaching the 0.95-level accuracy when using the random sampling scheme. For detecting the spatial structure of IAPs in KNP (the spatial distribution of local abundance), the minimum sampling effort with satisfactory similarity ($d < d_{0.05}$) was attained at a mere level of 6.7×10^4 records for random sampling (2.78 % of current sampling effort; Fig. 5.4b). Current sampling efforts (121 records per km^2) could be reduced to 15.3 records per km^2 for an acceptable abundance estimate, and to 3.4 records per km^2 for distribution detection. Detecting the overall

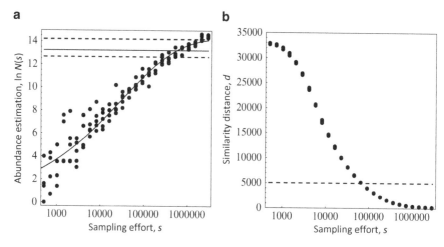

Fig. 5.4 (a) Abundance estimation and, (b) similarity distance as a function of sampling effort (s) for random sampling scheme. The *solid* and *dashed straight lines* in (a) correspond to the area-of-occupancy estimation of abundance (see Fig. 5.2a); the *dashed line* in (b) indicates the 0.05 critical level $d_{0.05}$ of the similarity distance. Similar plots for other sampling schemes are not shown due to resembling patterns

Table 5.1 Effects of sampling schemes and effort on detecting total abundance and spatial structure of invasive alien plants in Kruger National Park

Sampling scheme	$\ln N_\delta$	α	β	A_H	A_M	A_L	s^*	s^{**}
Random	14.24 ± 0.29	0.029	0.34	>0.99	0.93	0.89	15.77	3.46
Systematic	15.58 ± 0.45	0.059	0.26	0.91	0.85	0.81	20.94	3.44
Addictive	7.92 ± 0.10	0.002	0.81	0.21	0.33	0.40	–	–
Elusive	10.80 ± 0.36	0.007	0.55	0.69	0.78	0.83	–	3.45
Random-walk	15.16 ± 0.79	0.066	0.27	0.94	0.87	0.84	12.13	2.24

$\ln N_\delta$ indicates the converging limit of ln-transformed total abundance estimation; α and β are two model parameters; A_H, A_M and A_L the accuracy for abundance estimation with respect to the high, middle and low estimates; s^* indicates the minimum sampling effort (records per km^2) for $A_M = 0.95$; s^{**} is the minimum sampling effort (records per km^2) for the detected spatial structure not significantly different from observations ($d_s < d_{0.05}$); '–' indicates schemes failed to reach the 0.95 accuracy

spatial pattern of the IAP distribution takes much less effort than having an accurate assessment of abundance (Joseph et al. 2006). The accuracy A of the random sampling scheme showed consistently top ranking for abundance estimation, followed by random-walk, systematic, elusive and addictive (Table 5.1).

5.8   Cross-Scale Monitoring and Management

Efficient monitoring programmes for protected areas require robust methods for estimating target species abundance and distribution, and detecting changes thereof over time. Area-of-occupancy (AOO), using the scaling pattern of occupancy demonstrated here, fulfils the requirements of such a programme. Traditional mensuration methods of abundance estimating such as systematic or cluster sampling are only useful at local scales (e.g. $0.1–10$ km^2 for complete counts) due to the method of data collection and costs. There is increasing interest in using binary (presence/absence) data for large scale surveys (Brotons et al. 2004; Joseph et al. 2006). In this regard, two categories of abundance estimation models have been developed. First, the intraspecific occupancy-abundance relationship is grounded in the ubiquitous positive correlation between species abundance and range size (Gaston and Blackburn 2000; He and Gaston 2003). Second, the scaling pattern of occupancy describes how adjacent occupied cells merge with increasing grain (Hartley and Kunin 2003; Hui et al. 2006; Lennon et al. 2007; Gaston and Fuller 2009). A multi-criteria test suggests the supremacy of the scaling pattern of occupancy models over the occupancy-abundance relationship models in estimating abundance and yielding macroecological patterns (Hui et al. 2009). Indeed, Kunin's (1998) power-law AOO requires only one-tenth of the current sampling effort for a robust estimate of IAP abundance in the KNP (Table 5.1). Furthermore, scaling pattern of occupancy models provide a framework for further analysing biodiversity patterns across scales (e.g. Hurlbert and Jetz 2007; Foxcroft et al. 2009). We thus suggest that both these methods are useful for estimating abundance and distribution in large-scale monitoring programmes in protected areas.

Even though the AOO method can capture the essence of species distributions across a range of scales, we found a slight change (decline) in the slope of the AOO, calibrated from large to small scales (Fig. 5.2). The box-counting fractal dimension also declines from 0.97 to 0.77, indicating a more scattered, less structured distribution at finer scales (Lennon et al. 2007). Two reasons for this are plausible. First, AOO only considers the scaling pattern of presence records, not absence records. The status of absence for cells at small scales is inferred by either without records or with a low number of absence records which result in an overestimation of absence cells in the small scaled grid cells (i.e. a low omission error; Pearce and Ferrier 2000; Anderson et al. 2003). This points to an underestimation of abundance when using small-scale AOO models due to the concave shape of the scaling pattern of occupancy. However, the high sampling intensity in the KNP can largely reduce the influence of such nonlinearity in the AOO that the power-law exponent and thus the fractal dimension declines when the spatial resolution increases (i.e. scaling down to finer scales). Second, this declining of fractal dimensions when scaling down reflects that the fractal structure of species distribution breaks down at small scales (Hartley et al. 2004; Hui and McGeoch 2007; Lennon et al. 2007). This, however, suggests an overestimation of abundance when using large-scale AOO. These two effects can compensate each other and allow for a reasonably robust estimate from the AOO method. In addition, the fractal dimension at large scales could reflect the multidirectional range expansion through continuous habitat (Hui 2011). Furthermore, a fractal dimension close to one could indicate a linearised distribution of IAPs which can help to identify linear-shaped habitat (e.g. spread via rivers and roads) that determines the pathways of range expansion.

As with all methods for inference and extrapolation, the AOO method has its own limitations. First, uncertainty is often high for species with a low number of records. To have a fairly good estimate of abundance or coverage, we need at least 15–20 records per km^2. In addition, cells with a low number of records will also suffer from a high risk of a false categorization of non-detected species as absence. Other methods should thus also be consulted if this is suspected (e.g. MacKenzie et al. 2002; Hui et al. 2011a). Second, in general, species with a moderate range but which are locally abundant will have a high power-law exponent of the AOO across scales (i.e. steep scaling pattern of AOOs). However, the structures of species distribution across scales (e.g. at regional and local scales) are diverse (Gaston 1994). How these diverse structures of species distributions are related to the forms of AOOs across scales certainly needs more investigation. Finally, extrapolation across orders of magnitude carries inherent risks as each ecological process only works at a specific range of spatial scales. The power-law patterns can break up at very fine scales (e.g. Hui and McGeoch 2007). In this regard, extrapolation should only be used for scales close to the calibration range of the AOOs. However, this is often not possible for the purpose of rapid assessment of the invasion status. Managers have to base their decision on the trade-off between invasion risks and the uncertainty of inference. Extrapolation to the detection range (16 × 16 m) could be reliable (25 % of the AOO scale ranges from 125 m to 8 km in cell width) but the abundance estimates at 1 m^2 scale are crude and can only be used as an indicator for rapid risk assessment. Once the rapid assessment is done, managers can then choose more sophisticated approaches to investigate, and if needed, more detailed species-specific population-level structures and derive local-scale management plans.

In conclusion, to better manage biodiversity in protected areas, implementing knowledge gained in science into conservation management action is essential for conservation agencies. The use of the AOO model provides managers with a rigorous assessment of the options available for monitoring invasive alien plants over a large area.

Acknowledgments We are grateful to G. Blanchet, G. Cruz-Piñón, F. He, K. J. Gaston, I. Kühn, W. E. Kunin, S. Hartley and anonymous reviewers for comments and logistic help, and SANParks and the Centre for Invasion Biology, Stellenbosch University, for financial support. C.H., L.C.F and D.M.R. acknowledge financial support from the NRF. We thank the Kruger National Park rangers for collecting the CyberTracker data and Zuzana Sixtová for technical assistance with editing.

References

Anderson RP, Lew D, Peterson AT (2003) Evaluating predictive models of species' distributions: criteria for selecting optimal models. Ecol Model 162:211–232

Brotons L, Thuiller W, Araújo MB et al (2004) Presence-absence versus presence-only modelling methods for predicting bird habitat suitability. Ecography 27:437–448

Cohen JE, Jonsson T, Muller CB, Godfray HCJ, Savage VM (2005) Body sizes of hosts and parasitoids in individual feeding relationships. Proc Natl Acad Sci U S A 102:684–689

Dietemann V, Lubbe A, Crewe RM (2006) Human factors facilitating the spread of a parasitic honey bee in South Africa. J Econ Entomol 99:7–13

Du Toit JT, Rogers KH, Biggs HC (eds) (2003) The Kruger experience. Ecology and management of savanna heterogeneity. Island Press, Washington, DC

Foxcroft LC, Henderson L, Nichols GR et al (2003) A revised list of alien plants for the Kruger National Park. Koedoe 4:21–44

Foxcroft LC, Richardson DM, Rouget M et al (2009) Patterns of alien plant distribution at multiple spatial scales in a large national park: implications for ecology, management and monitoring. Divers Distrib 15:367–378

Gaston KJ (1994) Rarity. Chapman and Hall, London

Gaston KJ, Blackburn TM (2000) Pattern and process in macroecology. Blackwell Science, Oxford

Gaston KJ, Fuller RA (2009) The sizes of species' geographic ranges. J Appl Ecol 46:1–9

Halley JM, Hartley S, Kallimanis AS et al (2004) Uses and abuses of fractal methodology in ecology. Ecol Lett 7:254–271

Hartley S, Kunin WE (2003) Scale dependency of rarity, extinction risk, and conservation priority. Conserv Biol 17:1559–1570

Hartley S, Kunin WE, Lennon JJ et al (2004) Coherence and discontinuity in the scaling of species' distribution patterns. Proc R Soc B-Biol Sci 271:81–88

He F, Gaston KJ (2003) Occupancy, spatial variance, and the abundance of species. Am Nat 162:366–375

Hui C (2011) Forecasting population trend from the scaling pattern of occupancy. Ecol Model 222:442–446

Hui C, McGeoch MA (2007) Modeling species distributions by breaking the assumption of self-similarity. Oikos 116:2097–2107

Hui C, McGeoch MA (2008) Does the self-similarity species distribution model lead to unrealistic predictions? Ecology 89:2946–2952

Hui C, McGeoch MA, Warren M (2006) A spatially explicit approach to estimating species occupancy and spatial correlation. J Anim Ecol 75:140–147

Hui C, McGeoch MA, Reyers B et al (2009) Extrapolating population size from the occupancy-abundance relationship and the scaling pattern of occupancy. Ecol Appl 19:2038–2048

Hui C, Foxcroft LC, Richardson DM et al (2011a) Defining optimal sampling effort for large-scale monitoring of invasive alien plants: a Bayesian method for estimating abundance and distribution. J Appl Ecol 48:768–776

Hui C, Richardson DM, Robertson MP et al (2011b) Macroecology meets invasion ecology: linking the native distributions of Australian acacias to invasiveness. Divers Distrib 17:872–883

Hurlbert AH, Jetz W (2007) Species richness, hotspots, and the scale dependence of range maps in ecology and conservation. Proc Natl Acad Sci U S A 104:13384–13389

Joseph LN, Field SA, Wilcox C et al (2006) Presence-absence versus abundance data for monitoring threatened species. Conserv Biol 20:1679–1687

Kruger JM, MacFadyen S (2011) Science support within the South African National Parks adaptive management framework. Koedoe 53(2):Art. #1010, 7 pages. doi:10.4102/koedoe.v53i2.1010

Kunin WE (1998) Extrapolating species abundance across spatial scales. Science 281:1513–1515

Lennon JJ, Kunin WE, Hartley S et al (2007) Species distribution patterns, diversity scaling and testing for fractals in southern African birds. In: Storch D, Marquet PA, Brown JH (eds) Scaling biodiversity. Cambridge University Press, Cambridge, pp 51–76

MacKenzie DI, Nichols JD, Lachman GB et al (2002) Estimating site occupancy rates when detection probabilities are less than one. Ecology 83:2248–2255

Parker IM, Simberloff D, Lonsdale WM et al (1999) Impact: toward a framework for understanding the ecological effects of invaders. Biol Invasive 1:3–19

Pearce J, Ferrier S (2000) Evaluating the predictive performance of habitat models developed using logistic regression. Ecol Model 133:225–245

Pickett STA, Cadenasso ML, Benning TL (2003) Biotic and abiotic variability as key determinants of savanna heterogeneity at multiple spatiotemporal scales. In: Du Toit JT, Rogers KH, Biggs HC (eds) The Kruger experience: ecology and management of savanna heterogeneity. Island Press, Seattle, pp 22–40

Preston FW (1948) The commonness, and rarity, of species. Ecology 29:254–283

Quinn GP, Keough MJ (2002) Experimental design and data analysis for biologists. Cambridge University Press, Cambridge

Royle JA, Nichols JD, Kery M (2005) Modelling occurrence and abundance of species when detection is imperfect. Oikos 110:353–359

Sokal RR, Rohlf FJ (1995) Biometry: the principles and practice of statistics in biological research. W. H. Freeman and Company, New York

Spear D, McGeoch MA, Foxcroft LC et al (2011) Alien species in South Africa's national parks. Koedoe 53:Art. #1032, 4 pages. doi:10.4102/koedoe.v53i1.1032

Volkov I, Banavar JR, He F et al (2005) Density dependence explains tree species abundance and diversity in tropical forests. Nature 438:658–661

Wang Z, Bovik AC, Sheikh HR et al (2004) Image quality assessment: from error visibility to structural similarity. IEEE Trans Image Process 13:600–612

Wilson RJ, Thomas CD, Fox R et al (2004) Spatial patterns in species distributions reveal biodiversity change. Nature 432:393–396

Wright DH (1991) Correlations between incidence and abundance are expected by chance. J Biogeogr 1:463–466

Chapter 6
Plant Invasions into Mountain Protected Areas: Assessment, Prevention and Control at Multiple Spatial Scales

Christoph Kueffer, Keith McDougall, Jake Alexander, Curt Daehler, Peter Edwards, Sylvia Haider, Ann Milbau, Catherine Parks, Aníbal Pauchard, Zafar A. Reshi, Lisa J. Rew, Mellesa Schroder, and Tim Seipel

Abstract Mountains are of great significance for people and biodiversity. Although often considered to be at low risk from alien plants, recent studies suggest that mountain ecosystems are not inherently more resistant to invasion than other types of ecosystems. Future invasion risks are likely to increase greatly, in particular due to climate warming and increased human land use (e.g. intensification of human activities, human population growth, and expansion of tourism). However, these risks can be reduced by minimising anthropogenic disturbance in and around protected areas, and by preventing the introduction of potentially invasive alien plants into these areas, particularly at high elevations. Sharing information and experiences gained in different mountainous areas is important for devising effective management strategies. We review current knowledge about plant invasions into mountains, assembling evidence from all continents and across different climate zones, and describe experiences at local to global scales in preventing and managing plant invasions into mountain protected areas. Our findings and recommendations are also relevant for managing native species that expand to higher elevations.

Keywords Alpine • Altitude • Arctic • Climate change • Cold climate • Elevation gradient • Global • Invasibility • Mountain • Non-native • Ornamental plant trade • Precautionary principle • Tourism

C. Kueffer (✉) • J. Alexander • P. Edwards • T. Seipel
Institute of Integrative Biology – Plant Ecology, ETH Zurich, CH-8092 Zurich, Switzerland
e-mail: kueffer@env.ethz.ch; jake.alexander@env.ethz.ch; peter.edwards@env.ethz.ch;
t.seipel@env.ethz.ch

K. McDougall
Department of Environmental Management and Ecology, La Trobe University,
PO Box 821, Wodonga, VIC, Australia 3689
e-mail: k.mcdougall@latrobe.edu.au

L.C. Foxcroft et al. (eds.), *Plant Invasions in Protected Areas: Patterns, Problems and Challenges*, Invading Nature - Springer Series in Invasion Ecology 7, DOI 10.1007/978-94-007-7750-7_6, © Springer Science+Business Media Dordrecht 2013

6.1 Introduction

Mountains are of great significance for people and biodiversity (Messerli and Ives 1997). Not only do they support very diverse ecological communities, including many endemic species (Körner 2003), but they also have great value for historic, aesthetic and economic reasons. In order to protect these values, many mountain systems worldwide have been designated as protected areas (PAs). Indeed, according to the IUCN-WCPA Mountain Protected Areas Network, one third of the world's PAs are in mountainous regions (http://conservationconnectivity.org/mountains-wcpa/down loads/IUCNMountainsFacts2004.htm).

Despite these attempts to protect mountain ecosystems, they are exposed to increasing pressures from climate change, atmospheric nitrogen deposition and changing land use (Price 2006; Spehn et al. 2006). The growth in tourism, for example, entails greater disturbance, much of it associated with constructing new

C. Daehler
Department of Botany, University of Hawaii, 3190 Maile Way, Honolulu, HI 96822, USA
e-mail: daehler@hawaii.edu

S. Haider
Institute of Biology/Geobotany and Botanical Garden,
Martin Luther University Halle Wittenberg, D-06108 Halle (Saale), Germany
e-mail: sylvia.haider@botanik.uni-halle.de

A. Milbau
Climate Impacts Research Centre – Department of Ecology and Environmental Science,
Umeå University, SE-981 07 Abisko, Sweden
e-mail: ann.milbau@emg.umu.se

C. Parks
Pacific Northwest Research Station, US Forest Service, 1401 Gekeler Lane,
La Grande OR 97850, USA
e-mail: cparks01@fs.fed.us

A. Pauchard
Laboratorio de Invasiones Biológicas, Facultad de Ciencias Forestales,
Universidad de Concepción, Casilla 160-C, Concepción, Chile

Instituto de Ecología y Biodiversidad (IEB), Santiago, Chile
e-mail: pauchard@udec.cl

Z.A. Reshi
Department of Botany, University of Kashmir, Srinagar 190006, Jammu and Kashmir, India
e-mail: zreshi@kashmiruniversity.ac.in

L.J. Rew
Land Resources and Environmental Sciences Department, Montana State University,
Bozeman, MT 59717, USA
e-mail: lrew@montana.edu

M. Schroder
NSW National Parks and Wildlife Service, PO Box 2298, Jindabyne, Australia 2627
e-mail: Mel.Schroder@environment.nsw.gov.au

roads, trails and accommodations. And climate change is also affecting plant communities – in part directly, through effects upon the phenology, competitive balance, productivity and distribution of plant species, and in part indirectly, through changes to hydrology, disturbance regimes (including fires or landslides), the incidence of pests and diseases, and the abundance of herbivores (e.g. Theurillat and Guisan 2001; Klanderud 2004; Price 2006; Spehn et al. 2006; Inouye 2008).

Although high mountains are generally considered to be at low risk from alien plants (Millennium Ecosystem Assessment 2005), there is growing evidence that this risk – even if lower than in some other ecosystems – can be significant (Pauchard et al. 2009; McDougall et al. 2011b). Indeed, the currently low number of alien species to be found in mountains may be related more to low human activity at higher elevations – resulting in low alien propagule pressure and habitat disturbance – than to any inherent resistance of mountain ecosystems to invasion (Pauchard et al. 2009). Mountains might also be relatively uninvaded because in the past most alien plants were introduced to the lowlands, and few of the introduced species have succeeded in spreading far along steep climate gradients (Alexander et al. 2011). However, it may be merely a matter of time before some lowland species do spread to higher elevations. And the risk of invasion may also increase if more species are directly introduced to high elevations, and if anthropogenic disturbance and a changing climate make conditions more favourable for lowland species (Pauchard et al. 2009; Petitpierre et al. 2010; Kueffer 2010a, 2011; McDougall et al. 2011b).

Our aim in this chapter is to draw together information on plant invasions and management options in mountainous regions throughout the world. We begin by reviewing the extent of plant invasions in mountains on different continents and in different climate zones. We then consider two questions particularly relevant for managing invasive species: (i) how do alien plants enter mountain areas (introduction pathways)? (ii) are cold, high elevation habitats resistant to invasion (invasibility)? Finally, we review the management responses that have so far been taken and discuss possible management strategies for mountain protected areas, focusing especially on measures to prevent potentially invasive alien plants from being introduced. This information is important because mountain ecosystems remain among the few terrestrial ecosystems not severely affected by plant invasions, giving managers of PAs a unique opportunity to take preventive action against an emerging threat.

6.2   How Widespread and Problematic Are Alien Plants in Mountains?

Alien mountain floras have been well documented in several regions, including North America, Europe, Australia, New Zealand, South Africa, and some oceanic islands (e.g. Baret et al. 2006; Kalwij et al. 2008; Kueffer 2010a; McDougall et al. 2011a). However, for most other parts of the world – including large parts of Asia, Africa and South America – data on alien plant distributions are limited or entirely lacking. Preliminary data from terrestrial ecosystems in the Himalayan region indicate that invasive alien plants are common in subtropical lower elevations, but that few species occur at higher elevations (Khuroo et al. 2007; Weber et al. 2008; Kosaka et al. 2010). However, this may change rapidly, given the pace of economic development in the region, especially in China and India (Ding et al. 2008).

Alien species richness typically decreases strongly with elevation, with the number and abundance of alien species being usually very low at the highest sites (Pauchard and Alaback 2004; Kalwij et al. 2008; Kosaka et al. 2010; Seipel et al. 2012). Therefore, mountain PAs at higher elevation in most regions are not yet faced with problematic invasions by alien plants (Table 6.1), with infestations usually being concentrated along roads and on disturbed sites. Nevertheless, more than 100 invasive alien species, many of them deliberately introduced as pasture plants, timber trees or ornamentals (McDougall et al. 2011a), are now regarded as requiring management in different mountain regions around the world (McDougall et al. 2011a). These problematic species include *Acacia* spp., *Anthoxanthum odoratum* (sweet vernal grass), *Bromus* spp., *Carduus* spp., *Centaurea* spp., *Cirsium* spp., *Cytisus scoparius* (Scotch broom), *Hieracium* spp., *Lepidium draba* (whitetop), *Leucanthemum vulgare* (oxeye daisy), *Linaria* spp., *Pinus* spp., *Poa* spp., *Potentilla recta* (sulfur cinquefoil), *Salix* spp., *Taraxacum officinale* (dandelion), *Ulex europaeus* (common gorse), and *Verbascum thapsus* (woolly mullein; Fig. 6.1).

Some of these invasions have been investigated in detail, including *Pinus* spp. invading montane and subalpine meadows and woodlands in Hawaii and Chile (Daehler 2005; Peña et al. 2008), *T. officinale* colonizing cushion plant microhabitats in the alpine zone of the Andes (Cavieres et al. 2005), and invasions of grasslands and abandoned pastures by *Hieracium* spp. in New Zealand (Treskonova 1991), *V. thapsus* in Hawaii (Ansari and Daehler 2010), *L. vulgare* in the Australian Alps (Benson 2012) and Kashmir (Khuroo et al. 2010), and species of *Bromus*, *Centaurea*, *Linaria*, and *Potentilla* in the US Pacific Northwest and Intermountain West (Parks et al. 2005).

6.3 Introduction Pathways of Alien Plant Species to Mountain Areas

The spread of invasive species is influenced by many factors operating at differing spatial scales; these include global transportation, regional land use and environmental conditions, and patterns of local disturbance (Seipel et al. 2012). Global studies of alien plants in mountainous regions show that the majority, at least in the New World, are short-lived, of Eurasian origin, and linked historically with the European tradition of pastoralism and agriculture established by colonial settlers

Table 6.1 Alien plant diversity and management in selected mountain protected areas

Protected area	Latitude	Area (ha)	Elevation range (m)	Native vascular taxa (no.)	Alien vascular taxa (no.)	Species being managed (no.)	References
Australia							
Ben Lomond	−41°30	16,520	600–1,572	222	9	0	Ben Lomond National Park (1998)
Kosciuszko	−36°30	690,000	250–2,228	991	321	22	Doherty M, Wright G, Duncan A and McDougall K, unpublished data
India							
Dachigam	34°10	14,100	1,660–4,300	Uncertain	216	0	Reshi ZA, unpublished data
New Zealand							
Arthurs Pass	−43°00	114,350	240–2,400	668	154	12	Burrows (1986), Department of Conservation (2007)
Sweden							
Abisko	68°20	7,725	340–1,000	171	4	0	Aronsson (2002)
USA							
Yellowstone	44°30	898,300	1,600–3,462	1,350	218	47	http://www.nps.gov/yell/naturescience/plants.htm (accessed 4 March 2012); http://www.nps.gov/yell/naturescience/upload/exoticveg_9_13_07.pdf (accessed 4 March 2012)
Great Smoky	35°40	211,000	270–2,025	1,573	341	35	Whipple (2001); http://www.nps.gov/grsm/naturescience/non-natives.htm (accessed 4 March 2012)
Glacier	48°30	410,000	1,000–3,190	1,131	126	20	Whipple (2001); http://www.nps.gov/glac/naturescience/ccrlc-citizen-science_weeds.htm (accessed 4 March 2012)
Denali	63°00	2,458,500	170–6,195	794	22	0	Roland (2004)

The number of alien plants per protected area ranges widely. In most protected areas currently only a small proportion of alien species is being managed

Fig. 6.1 Examples of plant invasions in mountain protected areas. (**a**) *Verbascum thapsus* invading subalpine grasslands on the island of Hawaii (Hawaii, USA); (**b**) *Pinus contorta* and *P. sylvestris* invasion in an *Araucaria araucana* forest, Malalcahuello National Reserve in the Andes of south-central Chile; (**c**) the white flowers of *Leucanthemum vulgare* cover a subalpine grassland in Kosciuszko National Park (Australia) (photo credits: (**a**) Curt Daehler, (**b**) Aníbal Pauchard, (**c**) Mellesa Schroder)

(McDougall et al. 2011a; Seipel et al. 2012). These studies also show that most alien species were introduced to the lowlands, where human land use was most intense (McDougall et al. 2011a), and that some subsequently spread to higher elevations, either naturally or by human agency along roads and other transport corridors. This spread from lowland source populations was accompanied by a process of ecological filtering, resulting from the varying climatic tolerances of different alien species (Alexander et al. 2011). As a consequence, most alien plants currently found at high elevations are species with a wide ecological amplitude, being able to grow under both warm, lowland conditions and under the colder conditions at high elevations (Alexander et al. 2011; Haider et al. 2011). Thus, contemporary patterns of plant invasion in mountains largely reflect historical pathways of introduction, with ecological filtering and declining propagule pressure accounting for the typical decline in alien species richness towards the highest elevations (Alexander et al. 2011; Pyšek et al. 2011; Seipel et al. 2012).

But will alien species continue to expand their elevational ranges? It has been noted that some species reach similar maximum elevations on different continents, suggesting that they have reached their climatic limits (Alexander et al. 2009, and MIREN, unpublished data). Meanwhile, other studies show maximum elevation increasing with the time since introduction, suggesting that some species are still spreading upwards (Becker et al. 2005; Haider et al. 2010). Such spread could indicate either that a species has not yet reached its ecological limit, or that local adaptation is occurring, enabling the species to extend its ecological range (Allan and Pannell 2009; Haider et al. 2012). In addition to phenotypic plasticity, evolutionary changes may be important in enabling species to spread, or perhaps in increasing the performance of populations once they have established a foothold at high elevations. In support of the latter possibility, genetically-based differences have been demonstrated between high and low elevation populations of several introduced species in traits such as phenology and plant size (e.g. Leger et al. 2009; Monty and Mahy 2009; Alexander 2010; Haider et al. 2012).

Among native floras, several cases are known of plants extending their elevational range in response to climate change (e.g. Lenoir et al. 2008), and there is increasing quantitative evidence that alien species are also responding to climatic change. In general, we would expect the risk of invasions at higher elevations to increase as the climate becomes warmer (Walther 1999; Petitpierre et al. 2010; Bromberg et al. 2011). This outcome seems probable in temperate Europe, where plausible climate change scenarios suggest that all major invaders currently present in the lowlands could expand into high elevation mountain ecosystems (Petitpierre et al. 2010); and an increased vulnerability to invasions has also been predicted for some other cold temperature regions, including Antarctica (Chown et al. 2012). In hotter or more arid regions, however, changes in precipitation or evapotranspiration may prove more important than temperature changes, and could lead to reduced elevational ranges for some alien species (Jakobs et al. 2010; Petitpierre et al. 2010; Juvik et al. 2011).

Introduction pathways of alien plants into mountains may also change in the future, with more cold-adapted mountain specialists being deliberately introduced to high elevations – for instance, through intensification of agriculture, or for gardens or forestry or the restoration of ski runs. If this were to occur on a large scale, it could substantially increase invasion risks at high elevations and radically change the composition of alien floras (Kueffer 2010b; McDougall et al. 2011a). Land use change may also alter human-assisted dispersal at regional and local scales and thereby affect alien propagule pressure into mountain ecosystems.

6.4 How Vulnerable Are Arctic/Alpine Ecosystems to Plant Invasions?

An important question is how vulnerable are cold-temperature ecosystems to plant invasions (Pauchard et al. 2009). The invasibility of alpine ecosystems has scarcely been investigated, and we will therefore also refer to studies conducted in Arctic ecosystems, while recognising that there are important differences between the two kinds of ecosystems (Körner 2003). For example, alpine ecosystems have larger daily temperature fluctuations during the growing season than arctic ecosystems, as well as more intense solar radiation during the day, while alpine soils are usually more drained than those of arctic tundra.

A harsh climate is assumed to limit plant invasions in both mountain and arctic ecosystems, although several alien plant species have been found to establish and spread under these conditions (Rose and Hermanutz 2004; Morgan and Carnegie 2009; Pauchard et al. 2009; Chown et al. 2012; Ware et al. 2012). These plants clearly have the physiological capacity to grow at sites with a short growing season, low temperatures and strong winds, suggesting that, given time and a sufficient supply of propagules (e.g. Quiroz et al. 2011), new species may colonise.

In mountain landscapes, alien species are often restricted to human-modified habitats such as roadsides, ruderal sites, settlements, pastures, and disturbed forests and plantations, indicating – as in other ecosystems – that anthropogenic disturbance is the most important factor permitting invasion (Pauchard and Alaback 2004; Rose and Hermanutz 2004; Arévalo et al. 2005; Daehler 2005; McDougall et al. 2005; Parks et al. 2005; Kalwij et al. 2008; Haider et al. 2010; Seipel et al. 2012). However, various types of natural disturbance – landslides, avalanches, insect outbreaks and fires – are also common in mountain ecosystems and could also facilitate the entry of alien species. And at a smaller scale, invasibility may be increased by animals that graze, trample or burrow (e.g. by lemmings, voles, rabbits and wombats), and by cryogenic processes that create cracks and bare patches in the soil. Whether the invasibility of mountain ecosystems will increase or decrease in the future might therefore depend on how regimes of both natural and anthropogenic disturbance change. While the former are difficult to predict, there is little doubt that anthropogenic pressures due to activities such as hiking and skiing will increase in many regions (Johnston and Pickering 2001; Morgan and Carnegie 2009).

The invasibility of undisturbed alpine and arctic plant communities varies widely (Milbau et al. 2013). Mountain and arctic systems characterised by high nutrient levels, soil moisture, and pH appear to be the most vulnerable (Welling and Laine 2002; Forbis 2003; Graae et al. 2011), although some species – such as *Hieracium lepidulum* (tussock hawkweed) in New Zealand (Radford et al. 2006) – have been shown to be particularly invasive under nutrient-poor conditions. Moreover, species composition of the native vegetation is also important, since some species may strongly inhibit colonization. For example, leaf litter of the dense, low-growing shrub *Empetrum hermaphroditum* (mountain crowberry) has a high phenolic content that greatly restricts the establishment of seedlings (Pellissier et al. 2010). *Empetrum hermaphroditum* dominated dwarf-shrub heath is the commonest vegetation type in northern mountain tundra regions, and it is also the least colonised and invaded (Aerts 2010; Pellissier et al. 2010). Indeed, important changes in the species composition of such vegetation can only be expected when *E. hermaphroditum* is very heavily damaged, which could occur if winters were to become warmer (Aerts 2010).

Depending on the overall harshness of the environment, the role of biotic interactions for invasions can range from positive (facilitation) to negative (competition) (Callaway et al. 2002). A good example of facilitation during invasion is the improved establishment and performance of the alien herb *T. officinale* in cushions of *Azorella monantha* in the high Andes of central Chile (Cavieres et al. 2005). However, plant competition frequently reduces the invasibility of such ecosystems, despite slow plant growth under cold conditions (e.g. Klanderud and Totland 2007; Eskelinen 2010). Indeed, several studies in arctic and alpine communities have shown recruitment and colonization to be significantly increased when competitors are removed (Graae et al. 2011). All that can be safely said at present is that the balance between facilitation through improved microclimatic conditions and competition varies widely according to species, life stage, site conditions and climate (including season; Dona and Galen 2006).

6.5   Regional Differences in Current Management Efforts

Strategies for dealing with invasive plants in mountain PAs vary widely, from no management, to a focus on a few species of economic importance, to elaborate integrated pest management programmes (McDougall et al. 2011b). To characterise this diversity, we describe four contrasting situations that reflect the varying ecological and socioeconomic contexts in different regions. These are: (i) No problem, no active management, (ii) No resources, no action; (iii) Increasing problem, recent action; (iv) Recognised problem, integrated management plans and preventive action in surrounding lowlands.

6.5.1   No Problem Yet, and No Active Management: European Alps

In the European Alps, between 450 and 500 alien vascular plant species have been recorded, amounting to approximately 10 % of the total flora of the Alps or 15–20 % of all alien plant species in Europe (Kueffer 2010a). Most of these, however, are confined to lower elevations (Becker et al. 2005); indeed, only 23 of the most important invasive species in the European lowlands occur in the montane zone, and only nine reach to the subalpine zone (Kueffer 2010a). Furthermore, most of the species that do reach high altitudes are largely restricted to disturbed sites, and none is known to harm biodiversity or habitat quality within a PA. However, given predictions that many lowland species could expand into the subalpine or alpine zone under plausible scenarios of climate change (Petitpierre et al. 2010), precautionary measures are needed. Fortunately, awareness is increasing throughout the Alpine region, and a preliminary list of likely invaders of high altitude zones was recently produced (Kueffer 2010a).

6.5.2 Lack of Resources, Lack of Action: Perspectives from Developing Countries

Options for managing invasive species in PAs are limited in developing countries by two important factors: limited funding and expertise, and the need to balance the needs of biodiversity with those of local communities (Nuñez and Pauchard 2010). Very often, local people depend upon resources that are produced within PAs, and it is important to ensure that their livelihoods are not harmed by policies to control invasive species (e.g. Kull et al. 2007). For example, people living in and around the Golden Gate Highlands and Table Mountain National Parks in South Africa use alien woody species of the genera *Acacia, Eucalyptus, Hakea* and *Pinus* for food, fuel and building materials (Shackleton et al. 2007). However, these species are highly invasive, and the park authorities would like to see them removed. A study by Le Maitre et al. (1996) demonstrated that clearing these plants would increase the water supply for arid lowland agriculture, and that this increase would outweigh the economic value of alien woody species to the mountain communities. It was subsequently possible to implement this finding, offsetting the negative impact of removing these plants by employing people from the mountain communities to do this work (Working for Water – van Wilgen et al. 2001).

Protected areas at lower elevations in the Himalayan State of Jammu and Kashmir, India, harbour a large number of alien plant species. However, lack of awareness and resources means that no strategy has been implemented to control these species. On the other hand, alien aquatic plants are removed from some mountain lakes in Kashmir because it benefits the tourist industry, as well as providing employment to local people (McDougall et al. 2011b).

In Chile, the spread of pine species from plantations is harming mountain PAs (e.g. *Pinus contorta*, lodgepole pine, Peña et al. 2008) (see Case Study 1), but conflicts of interest with the forestry industry must be solved before any management can be implemented. Recently, forest certification standards such as FSC (Forest Stewardship Council) have introduced a requirement for forestry companies to control tree invasions on their lands. However, no such regulations apply to publicly owned natural areas, though this may change with the introduction of a new law recognising services provided by biodiversity and PAs (Pauchard et al. 2011).

6.5.2.1 Case Study 1: Introduced Conifers in Chilean Protected Areas: A Solution That Turned into a Problem

During the early 1900s, the forests of south-central Chile were severely degraded by logging and by fires used for clearing grazing land. By the 1950s, a number of conifer species (e.g. *P. contorta; Pinus ponderosa*, ponderosa pine; *Pinus sylvestris,* Scots pine; *Pinus monticola*, western white pine; *Pseudotsuga menziesii*, Douglas fir; *Larix occidentalis*, western larch) had been introduced for timber because they grew faster and suffered lower seedling mortality than the native *Nothofagus* species. Even PAs were subject to government afforestation schemes, both to combat soil erosion and as experimental field trials for new conifer cultivars. Unfortunately, no protocol was established to monitor these plantations and field trials, and no plans prepared for restoring native vegetation once the vegetation cover had recovered. After 30 or 40 years of planting, many conifer species are now invading into the surrounding vegetation; in some places this includes native forests such as those dominated by *Araucaria araucana* (monkey puzzle tree), highly valued for their emblematic biodiversity and scenic value (Peña et al. 2008; Fig. 6.1).

Currently, alien conifers are only controlled sporadically, and the Chilean Forest Service (CONAF) and other authorities have no clear mandate to deal with this problem (Pauchard and Villarroel 2002). However, a new alliance between the Laboratory of Biological Invasions (LIB; www.lib.udec.cl) and the Forest Service aims to improve the control of invasive plants, especially conifers, in PAs. Nonetheless, the goal of replacing conifer plantations with native ecosystems and preventing further invasion in PAs remains a major challenge for both research and management.

6.5.3   Sudden Increased Awareness Triggered by Recent Invasions: Australian Alps

Until recently, the alpine and subalpine zones of Australia were thought to be at low risk of invasion because of their cold climate and limited accessibility (Costin 1954). The only frequent alien species were a few, relatively uncompetitive ruderal species such as *Hypochaeris radicata* (hairy cat's ear), and *Rumex acetosella* (common sheep sorrel). Since 1999, however, several species known to be problematic in other mountain regions (e.g. *Hieracium* spp., *L. vulgare*) have been spreading, and the management of invasive plants has been given higher priority (see Case studies 2 and 3). However, the chances of controlling these plants, let alone of eradicating them, have diminished due to spiralling costs and the spread of species into inaccessible areas. Furthermore, recent introductions of cold adapted horticultural plants to ski resorts pose new risks to the PAs (McDougall et al. 2005, 2011b).

6.5.3.1   Case Study 2: Early Detection and Eradication in a Mountainous Landscape: The Australian Experience

In 1999, *Hieracium aurantiacum* (orange hawkweed) was discovered in a ski village adjoining the Alpine National Park in Australia. The species was recognised as posing a significant threat to biodiversity in mountain areas of New Zealand and North America (Morgan 2000), being highly competitive and requiring no major disturbance to establish in natural vegetation. Resort and park managers therefore took prompt action to eradicate the initial population and search for new

Fig. 6.2 Early detection and eradication of emerging alien plants in mountain protected areas can be very resource-demanding as shown by the experience in Kosciuszko National Park in the Australian Alps. (**a**) volunteers searching for *Hieracium aurantiacum*, (**b**) the extent of volunteer searches (*brown lines*) in relation to the known occurrences of *H. aurantiacum* (*orange dots*) (Photo and map: Jo Caldwell, NSW National Parks and Wildlife Service)

populations, some of which were detected a few years later in Kosciuszko National Park, more than 100 km to the north (Fig. 6.2).

In the following years, populations in both areas were sprayed with herbicide and the plants marked to determine whether they had been killed. It quickly became obvious that the species was spreading faster than it was being eradicated, and a more systematic programme of search and control was introduced, together with regular monitoring of treated areas (Fig. 6.2). Associated with these measures, research was undertaken related to the detection (Moore et al. 2011), dispersal (Williams et al. 2008) and ecology of the species. The first results suggest that the programme has been effective in reducing the area actually occupied by *H. aurantiacum*, and because the soil seed reserve is small, managers remain hopeful that the species can be eradicated from the PAs. However, monitoring has shown that plants are reproductive in their first year of growth, and any plants remaining undetected in that year can become the source of new populations, which may explain why the total extent continues to increase (Fig. 6.2). Furthermore, plants are difficult to locate in dense native vegetation, and the species is increasingly found in rugged and remote parts of Kosciuszko National Park. With the need to revisit treated populations and survey a larger area each year, monitoring costs are escalating, and the work would be impossible without a large number of volunteers.

The programme to eradicate *H. aurantiacum* is also put at risk by the need to allocate resources to controlling other species. Managers in Alpine National Park are attempting to eradicate three other invasive species detected within the last decade – *Hieracium praealtum* (king devil hawkweed), *Hieracium pilosella* (mouse-ear hawkweed) and *Myosotis laxa* (bay forget-me-not) (Charlie Pascoe, Parks Victoria, pers. comm.) – while managers in Kosciuszko National Park are facing a seemingly greater threat from *L. vulgare* (Fig. 6.1). This latter species was initially confined to road verges, and a decade ago it could perhaps have been eradicated. Now that it has invaded so much natural vegetation, however, a combination of containment and asset protection is considered a more realistic management goal. The recent proliferation of highly disruptive invasive species probably reflects a shift in land use in the Alps from summer grazing to tourism and hydroelectricity production, which commenced during the 1960s and brought increasing roads, infrastructure and amenity plantings to the Alps. After a brief establishment phase, many new introductions now have a sufficient foothold to spread rapidly, which reduces the chances of eradicating them. Nevertheless, the experience gained with *H. aurantiacum* may prepare managers better for tackling future invasions.

6.5.3.2   Case Study 3: Containment of Scotch Broom (*Cytisus scoparius*) in Barrington Tops National Park, New South Wales, Australia

Cytisus scoparius at varying densities infests over 10,000 ha of the Barrington Tops sub-alpine plateau (1,400–1,580 m a.s.l.). The plateau is geographically isolated from other high altitude areas, which may account for its many endemic plant and animal species (DECC 2010). The infestation originated in the mid-nineteenth century from a garden planting in the north of the plateau, and by the 1950s it had become widespread. Gazetting of the national park in 1969 altered land use practices by halting grazing and summer burning, which resulted in a rapid increase in the abundance and extent of *C. scoparius* (Waterhouse 1988).

To manage the species effectively, it was critical to understand the causes and extent of the infestation, which in turn required better knowledge of the species' ecology. And because *C. scoparius* seeds have a long viability, it was clear that a long-term commitment to managing the species would be needed (Sheppard and Hosking 2000). In 1989, the extent of the infestation was mapped from the air, and a containment strategy was devised to prevent the species from spreading into unaffected catchments and to limit its spread to adjoining uninvaded areas of the plateau. The containment lines made use of existing roads and natural barriers such as heavily canopied creek lines. Biennial control with herbicides and physical removal is now undertaken along tracks, roads and camping areas within the main area of infestation. Whilst the parent plant is easy to kill, follow-up measures to prevent seedling recruitment are essential. The activities coordinated by the programme include community education, measures to control vertebrate pests, and a biological control programme involving the release of four potential control agents (DECC 2010). Despite these activities, however, it was soon recognised that Scotch broom was not only increasing in density within the main infestation area, but also invading wetlands and grassland frost hollows where it had previously been absent. To protect the ecological value of the national park it was important to identify which endemic plant and animal species might be most at risk from the invasion (Hosking and Schroder, unpublished data). Following this work, control programmes have been initiated within the most vulnerable plant communities.

6.5.4 Integrated Management Plans Are in Place: Western United States

A large proportion of the land in the mountainous areas of the Western United States is controlled and managed by the Federal Government. While the specific missions of different government entities vary, all are bound by a Presidential Order that directs them to prevent and respond to new invaders (Executive Order 13112 of February 3, 1999). These agencies also work with private landowners in 'weed management associations (WMA)' to manage alien species, particularly those on a legally defined 'noxious' weed list (http://plants.usda.gov/java/noxiousDriver, accessed on August 23 2012). Integrated management approaches, which involve inventory/survey, prioritization, control, monitoring, prevention, outreach and co-operation with stakeholders, are encouraged. The most appropriate control techniques are determined on a case-by-case basis, and may include biological control, herbicide application, burning, prescribed grazing, or pulling by hand. To reduce their spread within national parks and forests, many designated noxious species are sprayed with herbicide along roads and near human infrastructure, but completely extirpating a species is usually difficult or impossible.

Preventive measures include a certified weed-free forage programme to prevent people who use stock animals for recreational riding or hunting from bringing hay infested with invasive plants into mountain areas (McDougall et al. 2011b). The USDA Forest Service requires the use of mobile vehicle wash units to clean all vehicles arriving and leaving wildfire staging areas, and other agencies are beginning to follow this example. Many alpine areas in designated wilderness areas or within national parks remain isolated from propagule sources, and have therefore escaped invasion. The greatest concerns for the future are new introductions of species adapted to wildfire and capable of growing under cold, dry conditions.

6.5.5   Preventive Action in the Surrounding Lowlands: Oceanic Islands

Most high-elevation ecosystems on oceanic islands lie within PAs. However, many of these parks extend to lower elevations where invasive species already pose a major problem. This is one reason why managers of PAs on some oceanic islands haven chosen to take preventive action against potential invaders. For instance, in the 1990s managers of Haleakala National Park in the Hawaiian islands identified which invasive species were most likely to threaten higher areas of the park (e.g. *Verbascum thapsus*), and they now provide funding to the Maui Invasive Species Committee (an interagency management body) to eradicate or control them at lower elevations outside the park (McDougall et al. 2011b). In La Réunion, modelling studies have demonstrated the need for preventive measures against future invasion risks, especially in mountain ecosystems, but concrete measures have yet to be introduced (Strasberg et al. 2005; Baret et al. 2006).

6.6   Prevention of Plant Invasion into Mountain Protected Areas Across Multiple Spatial Scales

As discussed above, most alien species were introduced to lowland regions, which explains why propagule pressure tends to be low at higher elevations and why potentially problematic alien mountain specialists are rarely found (Alexander et al. 2011; McDougall et al. 2011a). However, invasion risks are expected to increase greatly in the future, and strategies should be developed now to prevent potentially problematic species reaching high elevation sites, especially in PAs. In the following section we argue that success in preventing new introductions will depend on measures taken at scales ranging from the global to the local (Fig. 6.3). At a global scale, plants that pose a particular risk to mountain ecosystems should be identified, the transport of such species to mountain areas regulated, and mountain managers informed about the risks. At a regional scale, the spread of alien plants from lowland source pools to high elevations must be contained. And at local scales at high elevations, steps should be taken to minimise the risk of propagules being dispersed from disturbed to undisturbed sites.

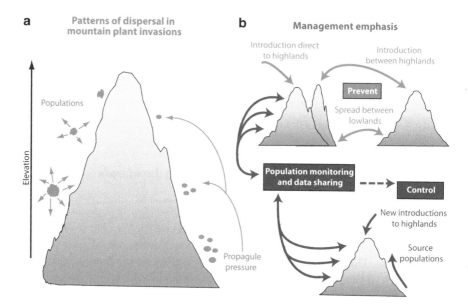

Fig. 6.3 A diagram representing a multi-scale management approach for the prevention of plant invasions into mountain protected areas. The historical introduction of alien plants into mountain areas occurred predominately at low elevations. (**a**) The spread of these alien plant species follows an elevation and environment gradient from the lowlands to the highlands. As elevation and environmental stress increase, both population size and propagule pressure decrease. For some species, populations at low elevation are source populations (indicated with *arrows*) and as elevation increases, populations lose source strength or become demographic sinks. (**b**) Using this model, management of invasive alien species in mountain ecosystems should focus on prevention, monitoring, data sharing, and control. Prevention aims to reduce new introductions, particularly of alien mountain specialists directly to the highlands and between highlands. Monitoring aims to determine where populations of particular species are being most invasive both within and between regions by sampling populations along the elevation gradient. Afterwards, data on species invasiveness can be shared within and between regions, along with information on the effectiveness of different control practices

6.6.1 Global Networking: The Approach of the Mountain Invasion Research Network (MIREN)

A particular challenge in managing invasive species is to link processes operating at local and larger scales. The invasive behaviour and resulting negative impacts of some alien plants become evident at a local scale, but preventive action depends upon sharing information at a global scale. Mountains throughout the world have many similarities, a generally harsh climate at high elevations, isolation as topographic islands, and ecologically distinctive floras and faunas, and it can be expected that they present similar management challenges (McDougall et al. 2011a, b). Therefore, global networking of invasive species managers and experts is particularly important for PAs in mountains.

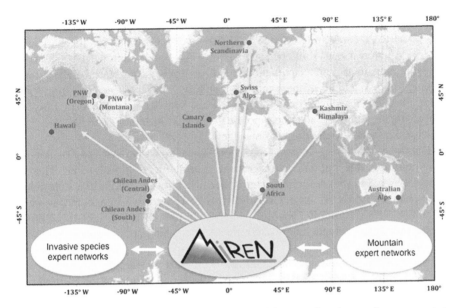

Fig. 6.4 The Mountain Invasion Research Network (MIREN) networks science and management at local to global scales. MIREN aims to increase knowledge about plant invasions in mountains and simultaneously connect experts of different topics – invasive species and mountain ecosystems – and from different areas. MIREN encompasses 11 core sites that are situated in different climate zones on all continents (except Antarctica) and some oceanic islands. MIREN works closely with existing regional and global networks of invasive species and mountain ecosystem experts

The Mountain Invasion Research Network (MIREN, www.miren.ethz.ch) is a network established to increase knowledge about plant invasions in mountains and simultaneously to create information links that build awareness and capacity amongst managers in different regions (Dietz et al. 2006). MIREN, which started in 2005 with six mountain regions, now encompasses 11 regions across the globe and links different expert groups in several ways (Fig. 6.4). On the one hand, it provides a bridge between experts on mountain ecosystems and invasion biologists, and on the other hand it links the local and international scales. The mountain research and management community has a long history of international collaboration (Messerli 2012), and MIREN complements existing initiatives. At an international level, for instance, MIREN is associated with the Global Mountain Biodiversity Assessment (GMBA, Diversitas, http://gmba.unibas.ch/index/index.htm) and the Mountain Research Initiative (MRI, http://mri.scnatweb.ch/). MIREN also regularly publishes in mountain-related policy documents, newsletters, and scientific journals. At regional scales, MIREN interacts with agencies such as the Alpine Network of Protected Areas (ALPARC, http://www.alparc.org/) in the European Alps and the Consortium for Integrated Climate Research in Western Mountains (CIRMOUNT, http://www.fs.fed.us/psw/cirmount/) in the western North American mountains. And MIREN members are equally well integrated into the international community of invasion biologists.

The link between the international level and local mountain PAs is organised via the 11 MIREN core sites, situated in different climate zones on all continents (except Antarctica) and some oceanic islands (Fig. 6.4). Some MIREN members are professionally associated with mountain PAs (e.g. Kosciuszko and Yellowstone National Parks), and all of them interact regularly with representatives from such areas. Through these local contacts, expertise and knowledge from PAs in different biogeographic, climatic and socio-political settings feeds into MIREN research, which is fine-tuned to the needs of managers.

Databases providing information about alien and invasive species are an important tool for building management capacity at a global level. This is especially true in mountain ecosystems, because climatic similarities determine that species invasive in one region are likely to be invasive in another, as well. Therefore, this information enables managers to predict which species entering the surrounding lowlands may pose a threat, and which species already present require urgent action. Threats may also come from human-mediated dispersal between distant mountains, such as the accidental movement of alien seeds between holiday resorts or the deliberate introduction of plants for horticulture (McDougall et al. 2005, 2011b), forestry (Daehler 2005; Peña et al. 2008) or ski-run restoration (Burt 2012). MIREN has developed a global database of alien species in mountain regions currently containing more than 1,300 species and 2,500 records (http://www.miren.ethz.ch/database/index.html). The MIREN database will be crucial in developing weed risk assessment systems (WRA) specifically for mountain ecosystems.

6.6.2 Monitoring and Control of Alien Plant Invasion in the Landscape

Limiting the spread of existing invasive populations along elevational gradients is an important management goal (Fig. 6.3). At a regional scale, regular surveys can be used to monitor distributional changes of established species and to detect the establishment of new species. Because alien plants often establish first at ruderal and highly disturbed sites, an effective approach is to record populations along roadsides (Pauchard and Alaback 2004). However, monitoring should be stratified by habitat type and include samples away from roadsides (Rew et al. 2006, 2007), because some species do spread into more natural habitats (e.g. McDougall et al. 2005; Pollnac et al. 2012). In principle, this approach can be used to monitor changes in the elevational limits of alien plants (Seipel 2011), though the data must be interpreted with care because the highest populations are often temporary, representing demographic sinks that are recolonised from time to time by propagules from source populations in the lowlands (Seipel 2011). Thus, the best way to quantify species spread may be to investigate how the probability of occurrence changes over time along gradients of elevation and disturbance. If sufficient resources are available, it is helpful to identify source and sink populations by monitoring the demography of populations and linking such information to species distribution models (Maxwell et al. 2009; Giljohann et al. 2011).

The introduction pathway of alien plants to high elevations, usually through spread from low elevation sources, has important implications for designing containment strategies at regional scales. High elevation populations of many species probably depend on occasional re-colonization from lowland populations (e.g. after a harsh winter). In such cases, an effort to also control the larger lowland infestations could prove more effective than attempting to locate and destroy isolated plants at the elevational limit.

6.6.3   Preventing Dispersal of Alien Plants to Mountain Protected Areas at Local Scales

Efforts to detect and monitor alien plants at local scales should focus on those places known to be most prone to invasion, including disturbed and open environments such as river flood plains and trails (Rose and Hermanutz 2004), and areas of high human use such as trails and rock climbing sites (McMillan and Larson 2002).

While propagules may be dispersed in many ways, humans and their vehicles are often especially important for transporting them over large distances to new areas. For example, in a study conducted in a PA in the Australian Alps, 27 alien species were germinated from soil collected in a car park (Mallen-Cooper 1990) that did not occur in the surrounding subalpine vegetation, of which 20 had not previously been recorded at such a high elevation. In another study, vehicles were found to deposit between 2 and 78 seeds per kilometre, with type of vehicle, type of road surface and weather conditions accounting for much of this variation (Rew et al., unpublished data). Similar factors have also been found to affect the distance that seeds are transported before falling from the vehicle; for example, under dry conditions, only 1–14 % of the seeds were lost over a distance of 256 km (Taylor et al. 2012).

Seeds of alien species can also be transported in substantial numbers in people's clothing (Mount and Pickering 2009), and regulations requiring hikers to clean their shoes and clothes before entering a PA can be important, especially in remote areas. Such regulations must, of course, also be observed by the staff who work in these areas, and it may also be necessary to prevent seed from being introduced in machinery and materials used for constructing infrastructure. For example, in Yellowstone National Park gravel used on roads (e.g. to improve traction on icy surfaces) has to be certified as weed-free, while in the Kosciuszko National Park (Australian Alps) contractors are required to remove all soil from earth-moving machinery before entering natural areas. Unfortunately, gravel of lowland origin is still used for constructing alpine walking tracks in Kosciuszko, which may explain why alien species such as *Echium vulgare* (viper's bugloss), *Melilotus alba* (white sweet clover) or *Juncus effusus* (common rush) have been detected even in remote areas.

A particular challenge for the managers of PAs is to prevent the deliberate introduction of cold-climate adapted alien species to neighbouring areas such as hotel gardens. This requires engaging with the relevant stakeholders, for example from tourism, forestry, and horticulture, to build awareness of the potential problems of invasive species. In Kosciuszko National Park, for instance, a cooperative programme has been established between the National Parks and Wildlife Service and ski resort lodges that aims to remove alien ornamentals from ski resorts and replace them with local native species.

6.7 Conclusions

Thanks to generally low levels of anthropogenic disturbance and propagule pressure, and a harsh climate, most mountain regions and habitats have not been heavily invaded by alien plants. However, mountain ecosystems are not invulnerable to plant invasions, and the risk is likely to increase with climate warming and more intensive use of these areas. When invasions do occur, managers of mountain PAs have little time to react, especially since invasions in rugged terrain soon become unmanageable. For effective management, information is needed about the ecology of the species and how it behaves elsewhere in its range, which in turn requires an effective dialogue between research and management at both local and global scales.

In most mountain PAs, people are the cause of the problem of plant invasions, and they should also be part of the solution. For example, the risk of new invasions can be reduced by encouraging people to use native rather than introduced species for amenity planting; and in cases where the invasive species have some economic value, it may be possible to mitigate the social impact of control measures by employing local people to perform this work.

In this synthesis, we have focused on the invasion risks of alien plant species. However, anthropogenic changes are also affecting the distribution and abundance of native species (e.g. Gottfried et al. 2012), with some lowland species spreading to higher altitudes (e.g. Lenoir et al. 2008). In many arctic and mountain ecosystems, similar processes of changing land use, climate warming and nitrogen deposition are causing the spread of woody species (Tape et al. 2006). In Dachigam National Park (Indian Kashmir), for instance, a native shrub *Strobilanthes urticifolia* (blue nettle) has almost completely invaded the PA (Zafar A. Reshi, personal communication). Whether native or alien, it will be important to understand which species threaten endangered organisms and ecosystem functioning. With this information, it will be possible to focus management efforts upon preventing those species from spreading, or managing their impacts so as to minimise unwanted vegetation change.

Acknowledgements This chapter is based on research and ideas of the Mountain Invasion Research Network Consortium (MIREN, www.miren.ethz.ch/people). We would in particular like to acknowledge the important role of Hansjörg Dietz as initiator of MIREN. Inputs from two anonymous reviewers helped improve the manuscript. AP was funded by ICM P02-005 and CONICYT PFB-23.

References

Aerts R (2010) Nitrogen-dependent recovery of subarctic tundra vegetation after simulation of extreme winter warming damage to *Empetrum hermaphroditum*. Glob Change Biol 16:1071–1081

Alexander J (2010) Genetic differences in the elevational limits of native and introduced *Lactuca serriola* populations. J Biogeogr 37:1951–1961

Alexander JM, Naylor B, Poll M et al (2009) Plant invasions along mountain roads: the altitudinal amplitude of alien Asteraceae forbs in their native and introduced ranges. Ecography 32:334–344

Alexander JM, Kueffer C, Daehler CC et al (2011) Assembly of non-native floras along elevational gradients explained by directional ecological filtering. Proc Natl Acad Sci U S A 108:656–661

Allan E, Pannell JR (2009) Rapid divergence in physiological and life-history traits between northern and southern populations of the British introduced neo-species, *Senecio squalidus*. Oikos 118:1053–1061

Ansari S, Daehler CC (2010) Life history variation in a temperate plant invader, *Verbascum thapsus* along a tropical elevational gradient in Hawaii. Biol Invasion 12:4033–4047

Arévalo JR, Delgado JD, Otto R et al (2005) Distribution of alien vs. native plant species in roadside communities along an altitudinal gradient in Tenerife and Gran Canaria (Canary Islands). Perspect Plant Ecol Evol Syst 7:185–202

Aronsson M (2002) Torneträskområdets kärlväxter checklist, 7th edn. Private Publishing, Bålsta

Baret S, Rouget M, Richardson DM et al (2006) Current distribution and potential extent of the most invasive alien plant species on La Réunion (Indian Ocean, Mascarene islands). Austral Ecol 31:747–758

Becker T, Dietz H, Billeter R et al (2005) Altitudinal distribution of alien plant species in the Swiss Alps. Perspect Plant Ecol Evol Syst 7:173–183

Ben Lomond National Park (1998) Ben Lomond National Park management plan. Parks and Wildlife Service Department of Primary Industries. Water and Environment, Hobart

Benson JS (2012) Ox-eye daisy: an expanding weed on the tablelands. Nat N S W 56:24–25

Bromberg JE, Kumar S, Brown CS et al (2011) Distributional changes and range predictions of downy brome (*Bromus tectorum*) in Rocky Mountain National Park. Invasion Plant Sci Manage 4:173–182

Burrows CJ (1986) Botany of Arthur's Pass National Park South Island, New Zealand. I. History of botanical studies and checklist of the vascular flora. N Z J Bot 24:9–68

Burt JW (2012) Developing restoration planting mixes for active ski slopes: a multi-site reference community approach. Environ Manage 49:636–648

Callaway RM, Brooker RW, Choler P et al (2002) Positive interactions among alpine plants increase with stress. Nature 417:844–848

Cavieres LA, Quiroz CL, Molina-Montenegro MA et al (2005) Nurse effect of the native cushion plant *Azorella monantha* on the invasive non-native *Taraxacum officinale* in the high-Andes of Central Chile. Perspect Plant Ecol Evol Syst 7:217–226

Chown SL, Huiskes AHL, Gremmen NJM et al (2012) Continent-wide risk assessment for the establishment of nonindigenous species in Antarctica. Proc Natl Acad Sci U S A 109:4938–4943

Costin AB (1954) A study of the ecosystems of the Monaro Region of New South Wales. Government Printer, Sydney

Daehler CC (2005) Upper-montane plant invasions in the Hawaiian Islands: patterns and opportunities. Perspect Plant Ecol Evol Syst 7:203–216

DECC (2010) Barrington Tops National Park, Mount Royal National Park and Barrington Tops State Conservation Area plan of management. Department of Environment and Climate Change, New South Wales. http://www.environment.nsw.gov.au/resources/planmanagement/final/20100833BarringtonMtRoyalFinal.pdf, Sydney

Department of Conservation (2007) Arthur's Pass National Park management plan. Department of Conservation, Hokitika

Dietz H, Kueffer C, Parks CG (2006) MIREN: a new research network concerned with plant invasion into mountain areas. Mt Res Dev 26:80–81

Ding JQ, Mack RN, Lu P et al (2008) China's booming economy is sparking and accelerating biological invasions. Bioscience 58:317–324

Dona AJ, Galen C (2006) Sources of spatial and temporal heterogeneity in the colonization of an alpine krummholz environment by the weedy subalpine plant *Chamerion angustifolium* (fireweed). Can J Bot 84:933–939

Eskelinen A (2010) Resident functional composition mediates the impacts of nutrient enrichment and neighbour removal on plant immigration rates. J Ecol 98:540–550

Forbis TA (2003) Seedling demography in an alpine ecosystem. Am J Bot 90:1197–1206

Giljohann KM, Hauser CE, Williams SG et al (2011) Optimizing invasive species control across space: willow invasion management in the Australian Alps. J Appl Ecol 48:1286–1294

Gottfried M, Pauli H, Futschik A et al (2012) Continent-wide response of mountain vegetation to climate change. Nat Clim Chang 2:111–115

Graae BJ, Ejrnæs R, Lang SI et al (2011) Strong microsite control of seedling recruitment in tundra. Oecologia 166:565–576

Haider S, Alexander J, Dietz H et al (2010) The role of bioclimatic origin, residence time and habitat context in shaping non-native plant distributions along an altitudinal gradient. Biol Invasion 12:4003–4018

Haider S, Alexander J, Kueffer C (2011) Elevational distribution limits of non-native species: combining observational and experimental evidence. Plant Ecol Divers 4:363–371

Haider S, Kueffer C, Edwards PJ et al (2012) Genetic differentiation in growth of multiple non-native plant species along a steep environmental gradient. Oecologia 170:89–99

Inouye DW (2008) Effects of climate change on phenology, frost damage, and floral abundance of montane wildflowers. Ecology 89:353–362

Jakobs G, Kueffer C, Daehler CC (2010) Introduced weed richness across altitudinal gradients in Hawai'i: humps, humans and water-energy dynamics. Biol Invasion 12:4019–4031

Johnston FM, Pickering CM (2001) Alien plants in the Australian Alps. Mt Res Dev 21:284–291

Juvik JO, Rodomsky BT, Price JP et al (2011) The upper limits of vegetation on Mauna Loa, Hawai'i: a fiftieth-anniversary reassessment. Ecology 92:518–525

Kalwij JM, Robertson MP, van Rensburg B (2008) Human activity facilitates altitudinal expansion of exotic plants along a road in montane grassland, South Africa. Appl Veg Sci 11:491–498

Khuroo AA, Rashid I, Reshi Z et al (2007) The alien flora of Kashmir Himalaya. Biol Invasion 9:269–292

Khuroo AA, Malik AH, Reshi ZA et al (2010) From ornamental to detrimental: plant invasion of *Leucanthemum vulgare* Lam. (Ox-eye daisy) in Kashmir Valley, India. Curr Sci 98:600–602

Klanderud K (2004) Climate change effects on species interactions in an alpine plant community. J Ecol 93:127–137

Klanderud K, Totland O (2007) The relative role of dispersal and local interactions for alpine plant community diversity under simulated climate warming. Oikos 116:1279–1288

Körner C (2003) Alpine plant life: functional plant ecology of high mountain ecosystems, 2nd edn. Springer, Berlin

Kosaka Y, Saikia B, Mingki T et al (2010) Roadside distribution patterns of invasive alien plants along an altitudinal gradient in Arunachal Himalaya, India. Mt Res Dev 30:252–258

Kueffer C (2010a) Alien plants in the Alps: status and future invasion risks. In: Price MF (ed) Europe's ecological backbone: recognising the true value of our mountains. European Environment Agency (EEA), Copenhagen, pp 153–154, EEA Report No 6/2010

Kueffer C (2010b) Transdisciplinary research is needed to predict plant invasions in an era of global change. Trends Ecol Evol 25:619–620

Kueffer C (2011) Neophyten in Gebirgen – Wissensstand und Handlungsbedarf. Gesunde Pflanzen 63:63–68

Kull CA, Tassin J, Rangan H (2007) Multifunctional, scrubby, and invasive forests? Wattles in the highlands of Madagascar. Mt Res Dev 27:224–231

Le Maitre DC, van Wilgen BW, Chapman RA et al (1996) Invasive plants and water resources in the Western Cape Province, South Africa: modelling the consequences of a lack of management. J Appl Ecol 33:161–172

Leger EA, Espeland EK, Merrill KR et al (2009) Genetic variation and local adaptation at a cheatgrass (Bromus tectorum) invasion edge in Western Nevada. Mol Ecol 18:4366–4379

Lenoir J, Gégout JC, Marquet PA et al (2008) A significant upward shift in plant species optimum elevation during the 20th century. Science 320:1768–1771

Mallen-Cooper PJ (1990) Introduced plants in the high altitude environment of Kosciusko National Park. PhD thesis. The Australian National University, Canberra

Maxwell BD, Lehnhoff E, Rew LJ (2009) The rationale for monitoring invasive plant populations as a crucial step for management. Invasion Plant Sci Manage 2:1–9

McDougall KL, Morgan JW, Walsh NG et al (2005) Plant invasions in treeless vegetation of the Australian Alps. Perspect Plant Ecol Evol Syst 7:159–171

McDougall K, Alexander J, Haider S et al (2011a) Alien flora of mountains: global comparisons for the development of local preventive measures against plant invasions. Divers Distrib 17:103–111

McDougall KL, Khuroo AA, Loope LL et al (2011b) Plant invasions in mountains: global lessons for better management. Mt Res Dev 31:380–387

McMillan MA, Larson DW (2002) Effects of rock climbing on the vegetation of the Niagara Escarpment in Southern Ontario, Canada. Conserv Biol 16:389–398

Messerli B (2012) Global change and the world's mountains. Where are we coming from, and where are we going to? Mt Res Dev 32:S55–S63

Messerli B, Ives JD (eds) (1997) Mountains of the world: a global priority. Parthenon, New York

Milbau A, Shevtsova A, Osler N et al (2013) Plant community type and small-scale disturbances, but not altitude, influence the invasibility in subarctic ecosystems. New Phytol 197:1002–1011

Millennium Ecosystem Assessment (2005) Ecosystems and human well-being: biodiversity synthesis. World Resources Institute, Washington, DC

Monty A, Mahy G (2009) Clinal differentiation during invasion: Senecio inaequidens (Asteraceae) along altitudinal gradients in Europe. Oecologia 159:305–315

Moore JL, Hauser CE, Bear JL et al (2011) Estimating detection–effort curves for plants using search experiments. Ecol Appl 21:601–607

Morgan JW (2000) Orange hawkweed Hieracium aurantiacum L.: a new naturalised species in alpine Australia. Vic Nat 117:50–51

Morgan JW, Carnegie V (2009) Backcountry huts as introduction points for invasion by non-native species into subalpine vegetation. Arct Antarct Alp Res 41:238–245

Mount A, Pickering CM (2009) Testing the capacity of clothing to act as a vector for non-native seed in protected areas. J Environ Manag 91:168–179

Nuñez MA, Pauchard A (2010) Biological invasions in developing and developed countries: does one model fit all? Biol Invasion 12:707–714

Parks CG, Radosevich SR, Endress BA et al (2005) Natural and land-use history of the Northwest mountain ecoregions (USA) in relation to patterns of plant invasions. Perspect Plant Ecol Evol Syst 7:137–158

Pauchard A, Alaback PB (2004) Influence of elevation, land use, and landscape context on patterns of alien plant invasions along roadsides in protected areas of South-Central Chile. Conserv Biol 18:238–248

Pauchard A, Villarroel P (2002) Protected areas in Chile: history, current status, and challenges. Nat Areas J 22:318–330

Pauchard A, Kueffer C, Dietz H et al (2009) Ain't no mountain high enough: plant invasions reaching new elevations. Front Ecol Environ 7:479–486

Pauchard A, García R, Langdon B et al (2011) The invasion of non-native plants in Chile and their impacts on biodiversity: history, current status, and challenges for management. In: Figueroa E (ed) Biodiversity conservation in the Americas: lessons and policy recommendations. Editorial FEN-Universidad de Chile, Santiago, pp 133–165

Pellissier L, Anne Bråthen K, Pottier J et al (2010) Species distribution models reveal apparent competitive and facilitative effects of a dominant species on the distribution of tundra plants. Ecography 33:1004–1014

Peña E, Hidalgo M, Langdon B et al (2008) Patterns of spread of *Pinus contorta* Dougl. ex Loud. Invasion in a natural reserve in southern South America. For Ecol Manage 256:1049–1054

Petitpierre B, Kueffer C, Seipel T et al (2010) Will the risk of plant invasions into the European Alps increase with climate change? In: Kollmann J, van Mölken T, Ravn HP (eds) Biological invasions in a changing world – from science to management. Neobiota book of abstracts. Department of Agriculture & Ecology, University of Copenhagen, Copenhagen, p 7

Pollnac F, Seipel T, Repath C et al (2012) Plant invasion at landscape and local scales along roadways in the mountainous region of the greater yellowstone ecosystem. Biol Invasion 14:1753–1763

Price ME (ed) (2006) Global change in mountain regions. Sapiens Publishing, Duncow

Pyšek P, Jarošík V, Pergl J et al (2011) Colonization of high altitudes by alien plants over the last two centuries. Proc Natl Acad Sci U S A 108:439–440

Quiroz CL, Cavieres LA, Pauchard A (2011) Assessing the importance of disturbance, site conditions, and the biotic barrier for dandelion invasion in an alpine habitat. Biol Invasion 13:2889–2899

Radford IJ, Dickinson KJM, Lord JM (2006) Nutrient stress and performance of invasive *Hieracium lepidulum* and co-occurring species in New Zealand. Basic Appl Ecol 7:320–333

Rew LJ, Maxwell BD, Dougher FL et al (2006) Searching for a needle in a haystack: evaluating survey methods for non-indigenous plant species. Biol Invasion 8:523–539

Rew LJ, Lehnhoff EA, Maxwell BD (2007) Non-indigenous species management using a population prioritization framework. Can J Plant Sci 87:1029–1036

Roland CA (2004) The vascular plant floristics of Denali National Park and Preserve: a summary, including the results of inventory fieldwork 1998–2001. Denali National Park and Preserve, Denali Park

Rose M, Hermanutz L (2004) Are boreal ecosystems susceptible to alien plant invasion? Evidence from protected areas. Oecologia 139:467–477

Seipel T (2011) Distributions and demographics of non-native plants in mountainous regions. PhD thesis, ETH Diss. Nr. 20031. ETH Zurich, Zurich

Seipel T, Kueffer C, Rew LJ et al (2012) Processes at multiple spatial scales determine non-native plant species richness and similarity in mountain regions around the world. Glob Ecol Biogeogr 21:236–246

Shackleton CM, McGarry D, Fourie S et al (2007) Assessing the effects of invasive alien species on rural livelihoods: case examples and a framework from South Africa. Hum Ecol 35:113–127

Sheppard AW, Hosking JR (2000) Broom management. Proceedings of a workshop held at Ellerston and Moonan on 16–17 November 1998. Plant Prot Q 15:133–186

Spehn EM, Libermann M, Körner C (eds) (2006) Land use change and mountain biodiversity. CRC Press, Andover

Strasberg D, Rouget M, Richardson DM et al (2005) An assessment of habitat diversity and transformation on La Réunion Island (Mascarene Islands, Indian Ocean) as a basis for identifying broad-scale conservation priorities. Biodiver Conserv 14:3015–3032

Tape K, Sturm M, Racine C (2006) The evidence for shrub expansion in Northern Alaska and the Pan-Arctic. Glob Change Biol 12:686–702

Taylor K, Brummer TJ, Taper ML et al (2012) Human-mediated long-distance dispersal: an empirical evaluation of seed dispersal by vehicles. Divers Distrib 18:942–951

Theurillat JP, Guisan A (2001) Potential impact of climate change on vegetation in the European Alps: a review. Clim Change 50:77–109

Treskonova M (1991) Changes in the structure of tall tussock grasslands and infestation by species of Hieracium in the MacKenzie country, New Zealand. N Z J Ecol 15:65–78 van Wilgen BW, Richardson DM, Le Maitre DC et al (2001) The economic consequences of alien plant invasions: examples of impacts and approaches to sustainable management in South Africa. Environ Dev Sustain 3:145–168

Walther G-R (1999) Distribution and limits of evergreen broad-leaved (laurophyllous) species in Switzerland. Bot Helv 109:153–167

Ware C, Bergstrom D, Müller E et al (2012) Humans introduce viable seeds to the Arctic on footwear. Biol Invasion 14:567–577

Waterhouse BM (1988) Broom (*Cytisus scoparius*) at Barrington Tops, New South Wales. Aust Geogr Stud 26:239–248

Weber E, Sun SG, Li B (2008) Invasive alien plants in China: diversity and ecological insights. Biol Invasion 10:1411–1429

Welling P, Laine K (2002) Regeneration by seeds in alpine meadow and heath vegetation in sub-arctic Finland. J Veg Sci 13:217–226

Whipple JJ (2001) Annotated checklist of exotic vascular plants in Yellowstone National Park. West N Am Nat 61:336–346

Williams NSG, Hahs AK, Morgan JW (2008) A dispersal-constrained habitat suitability model for predicting invasion of alpine vegetation. Ecol Appl 18:347–359

Part II
Regional Patterns: Mapping the Threats from Plant Invasions in Protected Areas

Chapter 7
Icons in Peril: Invasive Alien Plants in African Protected Areas

Llewellyn C. Foxcroft, Arne Witt, and Wayne D. Lotter

Abstract Protected areas in Africa are global conservation icons, attracting millions of tourists a year. However, these areas are being threatened by a growing human population making increasing demands on the natural capitol being conserved. Moreover, global environmental change, of which biological invasions are a key concern, pose significant threats to the function of ecosystems and their constituents. Other than in a few regions, primarily in South Africa, little is known about alien plant invasions in protected areas across the continent. In order to present a first approximation of the threat of plant invasions to protected areas across Africa, we present the information we could find by drawing on published literature, grey literature and personal observations. We also present six case studies from prominent protected areas across Kenya, Tanzania and South Africa. These case studies aim to illustrate what is known in different regions and the key concerns and management approaches, thereby providing examples that may facilitate shared learning. Where information is available it suggests that some species are likely to be widespread, impacting severely on indigenous species diversity. If protected areas are to be successful in carrying out their mandate of biodiversity conservation, and increasingly, revenue creation, long-term management of invasive plants is essential. However, in developing countries, which

L.C. Foxcroft (✉)
Conservation Services, South African National Parks,
Private Bag X402, Skukuza 1350, South Africa

Centre for Invasion Biology, Department of Botany and Zoology,
Stellenbosch University, Private Bag X1, Stellenbosch 7602, South Africa
e-mail: Llewellyn.foxcroft@sanparks.org

A. Witt
CABI Africa, ICRAF Complex, United Nations Avenue, Nairobi, Gigiri, Kenya
e-mail: a.witt@cabi.org

W.D. Lotter
PAMS Foundation, Arusha, Tanzania
e-mail: wayne@pamsfoundation.org

L.C. Foxcroft et al. (eds.), *Plant Invasions in Protected Areas: Patterns, Problems and Challenges*, Invading Nature - Springer Series in Invasion Ecology 7, DOI 10.1007/978-94-007-7750-7_7, © Springer Science+Business Media Dordrecht 2013

characterise much of Africa, resources are severely lacking. Where funds are available for conservation these are often channelled to other aspects of protected area management, such as anti-poaching. Protected areas in Africa include a number of unique attributes that can provide natural laboratories for research on basic ecological principles of invasions, while the research can, in turn, contribute directly to the needs of the protected area agencies.

Keywords Alien plant distribution • Biodiversity • Biological invasions • Conservation • National Parks

7.1 Introduction

Protected areas (PAs) in Africa include some of the world's best known, iconic national parks. Indeed the mention of national parks in Africa conjures romanticised images of imposing lions (*Panthera leo*), large herds of elephants (*Loxodonta africana*), and the annual migration of over 1.5 million wildebeest (*Connochaetes taurinus*) in the Serengeti – Maasai Mara ecosystem. It is unlikely that many tourists pay much attention to the increasing pressures being placed on the processes underpinning the functioning of these PAs and the species that depend on them. Fewer are likely to be aware of widespread invasions that may be within or encroaching on the PAs' boundaries. Indeed, little appears to be known on plant invasions in Africa's PAs in general.

Protected areas, defined by the IUCN as "A clearly defined geographical space, recognised, dedicated and managed, through legal or other effective means, to achieve the long-term conservation of nature with associated ecosystem services and cultural values" (Dudley 2008), are regarded as one of the most important approaches for conserving biodiversity globally (Chape et al. 2005; Gaston et al. 2008). In Africa, PAs cover about 12.2 % of the continental landmass, equating to just less than 17 million km^2 (Fig. 7.1; IUCN 2012). The percentage of PAs relative to the entire terrestrial region of Northern Africa was about 4.0 % in 2010 (3.7 % in 2000; 3.3 % in 1990) and approximately 11.8 % in sub-Saharan Africa that same year (11.3 % in 2000; 11.1 % in 1990). However, many PAs are likely to be contested as being 'paper parks' (i.e. areas that have been proclaimed but have little or no management, and are therefore ineffective in fulfilling their mandate; Erwin 1991).

Most PAs, however large, are islands in a sea of differing land uses. Some forms of neighbouring land use types are compatible to an extent, but the management objectives of many are completely contradictory to the concept of biodiversity conservation (Newmark 2008). These different forms of land use bring with them different levels of associated pressures and risks. Protected areas in developing countries are further challenged with rapid population growth, high levels of poverty and for some, political instability (Naughton-Treves et al. 2005). Thus today, more so than in the past, PAs are also expected to contribute directly to

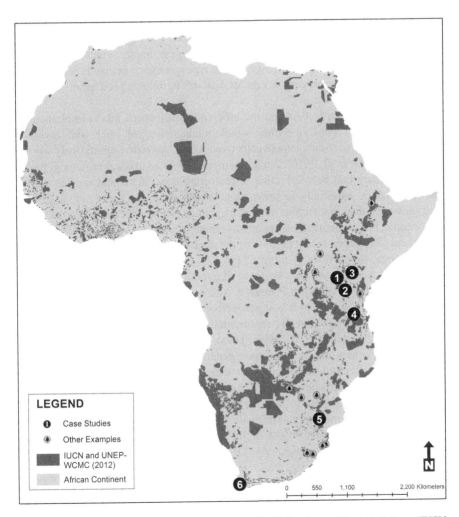

Fig. 7.1 Protected areas in Africa, according to the World Database of Protected Areas (IUCN and UNEP-WCMC 2012). In Africa, protected areas cover about 12.2 % of the continental landmass, equating to just less than 17 million km². Some of these protected areas are likely however to be contested as 'paper parks' (i.e. areas that have been proclaimed but have little or no management, and are therefore ineffective in fulfilling their mandate; Erwin 1991). The numbered points refer to the six case studies discussed in the chapter, while the yellow circles indicate other protected areas that are mentioned. (*1*) Serengeti – Maasai Mara ecosystem, (*2*) Ngorongoro Conservation Area, (*3*) Nairobi National Park, (*4*) Mikumi National Park, (*5*) Kruger National Park, (*6*) Table Mountain National Park (Fig. Sandra MacFadyen)

poverty reduction (Naughton-Treves et al. 2005). In many areas population expansion is taking place adjacent to PAs, leading to increased levels of land degradation and as a result increasing the demands made on these areas for access to natural resources. Access has also opened to remote areas, driven by global demands for oil extraction, mining and logging (Naughton-Treves et al. 2005), while locally, communities require access to resources such as fuel wood, medicinal plants, grazing, timber and expansion of agricultural areas. Having just travelled extensively in Zambia and Tanzania – one of the biggest threats without a doubt is charcoal production. These different land use types, resource demands and impacts all present PAs with varying levels of risk of introducing different kinds of potentially invasive species.

The attention given to PAs and the efficiency with which this is implemented provides important context within which management of alien and invasive plants are considered. Management effectiveness of PAs varies significantly worldwide (Leverington et al. 2010). An assessment of PAs in tropical regions globally showed that they had been effective in protecting ecosystems and species within their boundaries, especially in preventing habitat clearing (Bruner et al. 2001). It was however clear that various other management problems require improvement, for example illegal hunting (invasions by alien species were not included in the assessment). However, recent reviews of the management of PAs in general showed that in Africa, proportionally more of the PAs have little effective management in place and are in need of assistance (Leverington et al. 2010). Through the CBD Programme of Work on Protected Areas and IUCN World Commission on Protected Areas, much effort is being placed on developing "comprehensive, effectively managed and ecologically-representative national and regional systems of protected areas" (Dudley et al. 2005). These programmes acknowledge that while the extent of PAs has increased, few have general management structures and are, in many cases, ineffectual in carrying out their mandate. As one of the goals for developing effective PAs networks, mechanisms to identify and mitigate the impacts of key threats to PAs were to have been in place by 2008 (Dudley et al. 2005). This included taking measures to mitigate the risks posed by invasive alien species.

There has been no assessment of the state of alien plant invasions in most countries in Africa, let alone across PAs. If PAs are to fulfil their mandate successfully in the long-term, basic information on the alien species that are present, and those likely to become highly invasive, is a prerequisite. We aim to provide a synthesis of what is currently known, by providing a review of the available literature, adding recent observations and providing case studies to illustrate various problems and potential solutions.

7.2   State of Knowledge

Although we carried out an extensive review, little information is available for most areas. While we acknowledge that we are most familiar with South Africa and parts of East Africa, we searched for material across the continent, and are confident that the information is a reasonably detailed assessment of what is generally known. We suggest that the lack of information is an indication of the level of resources and attention that has been given to this problem, and thus not the accessibility of information. Much work has been done on collecting and synthesising data on species patterns and invasion processes the tropical forests in the East Usambara

Mountains (Eastern Arc Mountains) in Tanzania. However, as an in-depth review is given by Hulme et al. 2014, we have not included a discussion on the region here. Due to the availability of information, we use six case studies as examples to illustrate how different approaches to managing, surveying, monitoring and research have been undertaken, and how these may provide insight into other situations. These case studies include different regions and biomes of Africa: grasslands, dense to open savanna type ecosystems (Serengeti National Park – Maasai Mara National Reserve, Mikumi National Park, Ngorongoro Conservation Area and Kruger National Park) and two urban parks (Table Mountain National Park and Nairobi National Park). These PAs also include different histories of invasion and management. For example, Table Mountain NP, a Mediterranean type biome and a global centre of endemism, is surrounded by the city of Cape Town and has had a history of plant introductions dating back to the 1650s with European colonisation, and has detailed records from the early 1800s. Nairobi NP, which includes mega-herbivores, covers an area of 117 km^2 and is 7 km from Nairobi city centre, has virtually no records of plant invasions. Kruger NP, a 20,000 km^2 savanna PA, and proclaimed more than 100 years ago, has records of alien plants introductions from the 1930s. The Serengeti, Maasai Mara and Ngorongoro PAs in Tanzania and Kenya are global icons, but they, and Mikumi National Park, have had a shorter history of alien plant introductions and attention paid thereto.

The SCOPE (Scientific Committee on Problems of the Environment) programme of the 1980s assessed the invasion of alien species in nature reserves, using these as examples of systems having been protected from anthropogenic impacts (Usher et al. 1988). The aim was to test whether natural systems could be invaded, as opposed to disturbed systems, which was considered necessary for colonisation by alien plant species. The programme reviewed a number of biomes globally, including five PAs across Africa: savanna regions included Serengeti and Ngorongoro in Tanzania, and Kruger National Park and Hluhluwe-iMfolozi Game Reserve in South Africa (Macdonald and Frame 1988). For arid regions, the Skeleton Coast in Namibia (Loope et al. 1988), and for Mediterranean type systems, the Cape of Good Hope Nature Reserve (now incorporated into Table Mountain National Park) in South Africa (Macdonald et al. 1988), were included. Within these parks 80 alien plants (73 invasive) were reported for the Cape of Good Hope, 12 species in Serengeti-Ngorongoro (12 invasive), 156 species in Kruger (113 invasive), 74 species in Hluhluwe-iMfolozi (71 invasive) and seven species Skeleton Coast (7 invasive). It would be important to note however that the terminology used to indicate invasiveness has changed considerably since these studies were conducted, and species may be categorised differently under more recent approaches (see Richardson et al. (2011) for a discussion on terminology). Nevertheless, these lists still provide insight into the species already present at the time. It also led to an important and necessary finding, and which came to play a role in shaping future questions around the susceptibility of ecosystems to invasion: the realisation that all nature reserves (and thus 'natural' systems) appear to contain invasive species. The programme revealed that the nature and degree of invasions differed substantially between PAs in different regions of the world (Usher 1988). For example, it was suggested that PAs in arid regions of the tropics and sub-tropics have fewer invasive species (although notable exceptions were found); temperate regions in the northern hemisphere are relatively free of invasions, while the southern hemisphere has been severely impacted by invasions (Usher 1988). An important point raised by the programme was that tourism poses a substantial threat for invasions of PAs, as a positive correlation between visitor numbers and numbers of introduced species was found (Usher 1988), most likely due to associated infrastructure (lodges, roads). This is obviously an increasingly important issue as eco-tourism is a major source of revenue, and frequently, the primary motivation for continued existence and expansion of the global PA network (Barber et al. 2004). Moreover, this is likely to become especially important in Africa, as the quality of resources available for the tourism industry is high, and for most countries the development of their full tourism potential is in its infancy (Christie and Crompton 2001).

A recent review of plant invasions on tropical and sub-tropical savannas, while not explicitly focusing on PAs, found that much of the available information was also based on work and collections in PAs (see references in Foxcroft et al. 2010). In this work only a few species had been reported in published literature, some of which are ruderal or roadside weeds and not invasive. In Uganda these include six species in Lake Mburo NP, 15 species in Murchison Falls NP and 26 in Queen Elizabeth NP. In Mkomazi NP, Tanzania, eight species had been listed. However, as is discussed later, recent observations have substantially changed these figures.

While information is available and increasing in South Africa and parts of East Africa, little is known about the status of alien plant invasions in West Africa, especially from PA's. Most work in the region has focused on invasive alien plants (IAPs) impacting on agricultural systems and water resources, and to a large extent on the biological control thereof. In a workshop on prevention and management of alien plant invasions in the West African region (CABI 2004), 19 taxa were listed, of which eight were considered priority species. However, no information was given on their status in PAs. More recently, a brief survey of eight protected areas in Burkina Faso and Ghana (IUCN/PACO 2013) listed 26 species, including many that are highly invasive in similar biomes elsewhere. These include *Cardiospermum* sp. (balloon vine), *Chromolaena odorata* (chromolaena/triffid weed), *Lantana camara* (lantana), *Leucaena leucocephala* (leucaena), *Mimosa pigra* (giant sensitive plant) and four species of *Senna*. The aquatic species *Eichhornia crassipes* (water hyacinth) and *Pistia stratiotes* (water lettuce), which are probably of the most problematic in Africa, are also present (IUCN/PACO 2013).

7.2.1   East Africa

The *Nairobi Prevention Protocol Concerning Protected Areas and Wild Fauna and Flora in the Eastern African Region*, Article 7 (21 June 1985), requires the contracting parties to take all appropriate measures to prohibit the intentional or accidental introduction of alien species which may cause significant or harmful changes to the East African region. Although already having been in place for

25 years, the protocol is still largely unknown and as such not adhered to. Moreover, a lack of capacity and resources, especially with regard to the identification and monitoring of alien plants, has largely prevented implementation thereof.

There are no documented records of the total number of alien plant species present in PAs in East Africa. Lists that have been compiled have been done for other specific purposes and often mention alien plants incidentally. A few plant identification guides (e.g. Ivens 1967; Terry 1984; Terry and Michieka 1987) have been published on the common agricultural weeds in the region, of which many are alien species. Checklists of regional floras, such as the Flora of Tropical East Africa, and others for more specific areas, such as Pemba Island (Williams 1949; Koenders 1992) have been compiled, in which alien species are included. The Kenyan Horticultural Society produced a book of the most common ornamental plants (Hobson 1995), many of which are alien, potentially invasive or have already become invasive. A series of books lists the most useful trees and shrubs, of which many are, again, naturalised or invasive in Kenya (Maundu and Tengnas 2005), Uganda (Katende et al. 1995) and Tanzania (Mbuya et al. 1994). Other illustrated field guides include some alien species, many of which are invasive (e.g. Blundell et al. 2003; Birnie and Noad 2011; Dharani 2011). A comprehensive list of more than 500 alien plant species, which were introduced in the early 1900s to the Amani Botanical Gardens in Tanzania, was compiled in the 1930s (Greenway 1934) and recently revisited by Dawson et al. (2008). Thus, while there are many botanical publications, no work has been directed at compiling a comprehensive list of alien and invasive plants until recently, where a database providing 'fact sheets' on vertebrate pests and 100 of the most invasive IAPs known to occur in East Africa has been compiled (http://keys.lucidcentral.org/keys/v3/eafrinet/plants.htm).

Introductions of alien plants for ornamentation (via nurseries and other means) are a well-known and effective pathway of introduction in general (Reichard and White 2001), and even for PAs specifically (e.g. in Kruger NP, Foxcroft et al. 2008). As such, due to the use of ornamental species being one of the few pathways that can potentially be managed, special attention should be given to preventing new introductions of potentially invasive ornamental plants. An additional pathway of concern is the intentional introduction of agro-forestry species, for example *Prosopis juliflora* (mesquite), *Leucaena* spp., Australian *Acacia* spp., *Calliandra calothyrsus* (calliandra), by Non-Governmental Organisations (NGOs) and other agencies. Additional efforts need to be made to manage this process and pathway more effectively. Many of the species invading PAs are also widely utilised as hedge plants by rural communities, such as *L. camara*, various *Opuntia* species, *Tithonia diversifolia* (Mexican sunflower), *Caesalpinia decapetala* (Mauritius thorn), *Thevetia peruviana* (yellow oleander) and *Brugmansia suaveolens* (angel trumpet), and collaboration with communities to explore the use of indigenous alternatives or non-invasive alien species should be sought.

7.2.1.1   Case Study 1: Serengeti – Maasai Mara Ecosystem: Serengeti National Park (Tanzania) and Maasai Mara National Reserve (Kenya)

The Serengeti – Maasai Mara ecosystem is an area of approximately 25,000 km^2 spanning the border between Tanzania and Kenya. It is renowned for the annual wildebeest and zebra migration between the southern plains of the Serengeti NP (SNP) and the northern grasslands of Maasai Mara National Reserve (MMNR). The Kenyan section of the total ecosystem is about 6,000 km^2 in size, of which 1,510 km^2 consists of the MMNR. The adjacent SNP covers an area of 14,763 km^2. The area surrounding Serengeti – Maasai Mara consists of a mixture of private nature reserves/conservancies or unprotected communally owned land.

Invasive alien plants were generally considered of low importance in the SNP, with one report indicating that there was no sign of introduced weeds colonizing natural disturbances or invading the undisturbed grassland community (Belsky 1987). In the late 1980s 12 species of alien plants were recorded (Macdonald and Frame 1988). Four of the listed species were considered to have substantial ecological impacts. *Tagetes minuta* (khaki weed) appeared to thrive in disturbed areas along roadsides and under high grazing pressure from livestock. Additionally, heavy disturbance from the indigenous mole rat (*Tachyoryctes daemon*) provided habitat in which the plants proliferated, displacing indigenous grasses (Macdonald and Frame 1988). Of the additional species considered to have a high impact, *Rorippa nasturtium-aquaticum* (watercress) was intentionally introduced by European settlers for food. *Medicago laciniata* (cutleaf medick), an annual herb of Mediterranean origin, appears to have been introduced accidentally, with seeds on army coats sold to Maasai pastoralists after World War II (Macdonald and Frame 1988). *Euphorbia tirucalli* (Indian spurge) was also listed, but has recently been shown to be indigenous to east and southern Africa (Foxcroft et al. 2010). In 2003, ten additional alien species were reported (Foxcroft 2003a), including *Opuntia stricta* (sour prickly pear; Fig 7.2a), *O. monacantha* (drooping prickly pear) and *P. stratiotes*. Although much research has been done in the Serengeti NP for a number of decades (Sinclair and Norton-Griffiths 1979; Sinclair and Arcese 1995; Sinclair et al. 2008) no research programme has been focused specifically on IAPs.

Very little is known about the alien plant species present in the MMNR, although some studies have been undertaken at a landscape level in an attempt to identify the main drivers of vegetation change since the early 1900s (e.g. Glover and Trump 1970; Dublin 1986). Ironically an invasive alien species, rinderpest (an acute, usually fatal disease of ruminant animals), together with fire and elephants, have been identified as being some of the main drivers responsible for changes in the grass and woodland cover. Invasive plants may however become the next major driver of change, as a number of invasive plant species have recently been identified (Witt and Sospeter, unpubl.). The main pathways for their introduction, both accidental and intentional, have no doubt been the extensive road network, rivers and streams originating outside the MMNR, and lodge gardens.

Fig. 7.2 (**a**) *Opuntia stricta* in Kruger National Park, South Africa. *Opuntia stricta* has also been recorded in Serengeti National Park, Tsavo East National Park and a number of conservancies in Kenya. (**b**) *Parthenium hysterophorus* being controlled along the Crocodile River, Kruger National Park. *Parthenium hysterophorus* has been recorded widely across north-eastern South Africa, Mozambique and East Africa, including Maasai Mara National Reserve, Ngorongoro Conservation Area, Nairobi National Park and Awash National Park, Ethiopia (Photograph **a** Llewellyn C Foxcroft, **b** Ezekiel Khoza)

The first intentional introduction of alien plant species into the MMNR was probably associated with the development of the first tourist lodge, Keekorok lodge, in 1965. Since then more than 70 introduced ornamental species have been introduced to lodge gardens in the MMNR (Witt and Sospeter, unpubl.). Many of the current 24 permanent camps in and around the MMNR host a number of well-known invasive plant species such as *L. camara*, *Tithonia diversifolia* (Mexican sunflower), *T. rotundifolia* (red sunflower), *O. monacantha*, and *Anredera cordifolia* (Madeira vine). With the exception of *Parthenium hysterophorus* (parthenium; Fig 7.2b), a number of other alien plant species, most of them relatively benign, have also been inadvertently introduced and are widely established along the extensive road network. The extensive network of formal roads and informal jeep tracks facilitates the movement of invasive plants and is likely to be a major pathway for the dispersal of *P. hysterophorus*.

There have been no official attempts to manage any invasive plant species in the MMNR until very recently. This is probably due to the combination of a lack of awareness and that alien plant invasions are likely to have been relatively recent. The discovery of *P. hysterophorus* in the MMNR in November 2010 led to the first official attempt to manage an invasive plant in the Reserve. With funding from the Australian High Commission in Kenya, and working together with the Kenya Wildlife Service, community members were employed to manually remove *P. hysterophorus* from the MMNR. All visible plants were manually removed, but unfortunately there was insufficient funding for follow-up activities. However, management of *P. hysterophorus* (initially manually, but more recently using herbicides) has continued in the Mara Triangle, which is managed by the Mara Conservancy. Whether the control has been successful in the long-term may be too soon to determine. A compounding problem is that *P. hysterophorus* is not being managed on the adjacent property.

7.2.1.2   Case Study 2: Ngorongoro Conservation Area, Tanzania

As with Serengeti NP, *Rorippa nasturtium-aquaticum* was probably intentionally introduced by European settlers into the Ngorongoro crater for food. It was considered likely to have displaced the indigenous aquatic plant, *Crassula granvikii* (Macdonald and Frame 1988).

More recently an additional 43 alien plant species were listed for the Ngorongoro crater (Lyons and Miller 1999; Henderson 2002). A large number of alien ornamental species that have potential to become invasive were also observed at the lodges surrounding the Ngorongoro crater (Henderson 2002). The most important species, and for which management recommendations have been suggested include *Acacia mearnsii* (black wattle) and *C. decapetala* (Henderson 2002; Foxcroft 2003b; Lotter 2004), *Melia azedarach* (syringa), *Azolla filiculoides* (red water fern), *L. camara*, *Jacaranda mimosifolia* (jacaranda) and *P. hysterophorus* (Clark et al. 2011). Manual control has been conducted for some species including *A. mearnsii*, *C. decapetala* and *P. hysterophorus*. Two species previously considered alien and requiring management were *Bidens schimperi* (yellow-flowered blackjack) and *Gutenbergia cordifolia*. However, although these have been confirmed as indigenous species (Henderson 2002) they are still being managed to improve forage on the crater floor.

7.2.1.3 Case Study 3: Nairobi National Park (Kenya)

The Nairobi National Park (NNP), only 7 km from the Nairobi city centre, is considered to be unique in that it is the only urban National Park in the world that has lion, rhino, leopard, and buffalo in such close proximity to a city. Formally established in 1946 as Kenya's first national park, it is approximately 117 km^2 in extent. It is bounded by an electric fence on three sides but is open to the Kitengela Conservation Area, located to the south of the park.

There are virtually no published accounts of the flora of NNP, and as such no formal records of IAPs. The only references to some alien plant species were made by Heriz-Smith (1962) who lists a number of ruderal/roadside weeds, and *Dovyalis caffra* (kei apple), *Ageratum conyzoides* (invading ageratum), *Opuntia* spp., *Eucalyptus* spp., and *Schinus molle* (pepper tree). Recent informal surveys revealed that *L. camara* is the most abundant IAP in the park with infestations mainly confined to the dry forest, and the Mbagathi and other river valleys. The cactus species, *O. ficus-indica* (sweet prickly pear), *O. monacantha*, *Austrocylindropuntia subulata* (long-spine cactus) and *Cereus* sp. (queen of the night) are present on the open grass plain, together with *D. caffra*, which also occurs in the forest. *Parthenium hysterophorus* has recently invaded the park, while *C. decapetala*, *Bryophyllum delagoense* (chandelier plant), *B. fedtschenkoi* (kalanchoe stonecrop), *Agave sisalana* (sisal) and *A. americana* (American agave) are localised within the PA. There are a large number of introduced roadside or ruderal plants that appear not to have impacted on the indigenous flora. Despite the presence of a large number of other invasive and potentially IAPs that have been grown as ornamentals for more than 100 years in and around the city of Nairobi, since its establishment in 1899, the park is still relatively uninvaded, compared to others such as Kruger National Park.

There have been very few attempts to control invasive plants in NNP. In 1998 funds were made available by the David Sheldrick Wildlife Trust to control some of the invasive cactus species in NNP. As herbicides could not be used, plants were removed manually and dumped elsewhere. Unfortunately funds were insufficient to eradicate the plants. More recently KWS, with funding from USAID, has started with the manual control of *L. camara*.

7.2.1.4 Case Study 4: Mikumi National Park (Tanzania)

Mikumi National Park (MNP) was established in 1964 and although it is 3,230 km^2 in extent, it is unfenced and adjoins the Selous Game Reserve World Heritage Site, which is more than 43,000 km^2 in extent. A survey in 2009 listed ten species of potentially invasive alien plant species within MNP, including *L. camara* and *P. stratiotes* (Clark and Lotter 2009). Fortunately, manual control measures were already in place by Tanzanian National Parks to minimise their impact and spread. Apart from the national public road traversing the park and rivers flowing into MNP from outside its borders, lodges and camps were also observed to be important seed source sites. Of additional concern was the observation of 11 invasive species occurring outside, but in the immediate vicinity of MNP. These species included *C. decapetala*, *M. azedarach* and *Psidium guajava* (guava). *Mimosa pigra*, also found outside of MNP, has already invaded the Selous Game Reserve, where, unfortunately, no control measures are in place (Clark and Lotter 2009).

7.2.2 Southern Africa

In Zimbabwe, 1449 introduced and naturalised species have been listed (Maroyi 2006), of which 391 have been analysed for their mode or purpose of introduction and their invasion status. This study indicates 153 (39.1 %) species as casual aliens, 154 (39.4 %) as naturalised and 84 (21.5 %) as invasive species (Maroyi 2012). Of the most invasive species countrywide, *J. mimosifolia* and *M. azedarach*, have also been long recorded as invaders in Matopos, Hwange and Kyle National Parks (Southern Rhodesia Commission Forestry 1956). In the 1980s concerns were raised in Nyanga National Park where the park estates are either adjacent to commercial forestry plantations, or had their own plantings in the early 1920s, with 20 % of the park reportedly affected by invasive alien tree species (Nyoka 2003). The alien species were planted to provide fuel wood (mainly *Pinus* spp. and *Acacia* spp.), construction timber, and to 'beautify' (Nyoka 2003) the parks with ornamentals such as *J. mimosifolia* and *M. azedarach*. Additionally, Chimanimani National Park in the eastern highlands of Zimbabwe has also been heavily invaded (Nyoka 2003).

South Africa has a well-documented history of alien plant introductions, with about 750 tree species and around 8 000 shrubby, succulent and herbaceous species having been introduced (van Wilgen et al. 2001). Of these, 198 have been declared by legislation (Conservation of Agricultural Resources Act; Act 43 of 1983) as alien weeds or invasive plants (Henderson 2001), and about 240 species recommended for listing under the National Environmental Management: Biodiversity Act (Act 10 of 2004). In South Africa's KwaZulu-Natal province, IAPs were rated as the most serious threat to biodiversity in a management effectiveness assessment of all the PAs, including both Natural World Heritage Sites, uKhahlamba/Drakensberg Park and iSimangaliso/Greater St Lucia Wetland National Park (Goodman 2003). In 1983, 20 alien plant species were reported in Hluhluwe-iMfolozi (Macdonald 1983), increasing to 74 species by 1988 (Macdonald and Frame 1988). The most important species were regarded to be *C. odorata* and *M. azedarach*, the former of which has become a major problem, costing millions of ZAR in control annually (Lotter and Clark 2008).

A total of 663 alien plants (all alien plants, not only invasive species) have been listed for South Africa's 19 national parks. Approximately 20 % are considered invasive (Spear et al. 2011). Kruger National Park (KNP) has the highest number of alien plants listed with 350 alien plants, followed by Table Mountain National Park (TMNP) with 239 (Spear et al. 2011). We discuss Table Mountain and Kruger National Park further as specific case studies.

7.2.2.1 Case Study 5: Kruger National Park (South Africa)

Detailed accounts of the introduction and management of alien plants in Kruger National Park have been given in a number of publications (Foxcroft 2001; Foxcroft and Richardson 2003; Freitag-Ronaldson and Foxcroft 2003; Foxcroft and Freitag-Ronaldson 2007; Foxcroft and Downey 2008). Here we provide a synopsis of the key issues: introduction of alien plants as ornamental plants, early control efforts and current management initiatives, research, and lessons learned.

The southern region of KNP was proclaimed in 1898. The Park extends 360 km from north to south, covers 20,000 km^2, and is bisected by seven major river systems, which originate in the highlands to the west of the park and drain a combined area of about 88,600 km^2. The first six alien plants were recorded in 1937 (Obermeijer 1937; Foxcroft et al. 2003), increasing to 350 (Spear et al. 2011). There was however little support for alien species management until the late 1990s (Foxcroft and Freitag-Ronaldson 2007). The tourist camps and staff villages have been landscaped to varying degrees, with 258 alien plant species being recorded in 36 camps or staff villages, many of which were intentionally introduced (Foxcroft et al. 2008), even though they are now known to be invasive elsewhere. In 1957 KNP prohibited planting ornamental plants in tourist camps and staff villages, and although the park policy on the control of ornamental species was updated periodically thereafter, it was largely disregarded. In 2004 the policy was again revised and the species list expanded. Species were prioritised on the basis of their potential to invade KNP as indicated by their invasiveness in similar habitats elsewhere, and on the prevailing national legislation. These species were removed manually and follow-up control still continues. Once the priority species were deemed to be under control, less invasive species were targeted. All alien species were removed from vacated houses (as staff retired or transferred elsewhere in the park), regardless of their potential for invasion (Foxcroft et al. 2008).

Management of alien plants began in 1956 on *M. azedarach* (Foxcroft and Freitag-Ronaldson 2007) on a relatively small scale. From 1982 the park dedicated one team of ten general workers to the control of IAPs. Biological control of *P. stratiotes* was initiated with the introduction of the snout weevil *Neohydronomus affinis* in 1985 (Cilliers et al. 1996). The insects were successfully introduced and resulted in a complete reduction of the populations and abundance of *P. stratiotes* in the park. *Opuntia stricta*, one of the parks worst invasive plant problems, was initially introduced as an ornamental garden plant in Skukuza in the early 1950s (Lotter and Hoffmann 1998). In 1980, *O. stricta* was estimated to have invaded an area of only 100 km², but by the mid to late 1990s it was estimated that *O. stricta* covered an area of 300–400 km². In 1996 the revised management plan delineated an area of 670 km², with the plants covering about 530 km² (although the patches are widespread, with a few high density patches interspersed with isolated plants; Foxcroft et al. 2004). Control by means of herbicides was initiated in the mid-1980s and in 1987 followed with the introduction of the biocontrol agent *Cactoblastis cactorum* 'stricta' biotype (Hoffmann et al. 1998a, b). In 1997 the cochineal *Dactylopius opuntiae* was released, resulting in rapid control of large dense patches of *O. stricta*. Following this success, a hot-house was erected for rearing the cochineal on a continuous basis, and cladodes of *O. stricta* covered in cochineal were distributed across the region, thereby facilitating widespread dispersal of the insect. *Opuntia stricta* is now considered under control, with little resurgence of previously densely invaded areas (see Figure 3 in Paterson et al. 2011). *Lantana camara*, one of KNPs most widespread invasive species occurs along most of the parks major rivers. It was first recorded in KNP in 1940, and was planted in tourist camps in the early 1950s (Vardien et al. 2012). Together with *M. azedarach*, it became the focus of early control efforts in the late 1950s. Upper catchment areas are heavily invaded, forming a substantial supply of propagules, which are dispersed along the rivers. *Lantana camara* is likely to remain in the system and should follow-up clearing be ceased, will most likely rapidly reinvade. Although the threat of invasions from ornamental plants has virtually been removed, pressure from outside the park remains high where neighbouring populations are found on the parks periphery (Spear et al. 2013). *Parthenium hysterophorus* probably represents one of the most important emerging invasive species. Invading from the south-east of the park, the plants have shown a steady increase in their distribution even though control efforts are underway (Fig 7.2b).

In 1997 the KNP Working for Water project was initiated. The catalyst for the national programme was the ability to combine (i) the need to manage IAPs that impact on already stressed water resources and biodiversity, and (ii) the high levels of poverty in many parts of the country. The project was initiated in 1997 with the sponsorship of ZAR3 million (~ US$650,000) for a period of three years from the Royal Netherlands Government, and ZAR6 million (~ US$1.3 million) from the Poverty Relief Fund of the South African Government. This allowed the employment of up to 1,000 people, focusing solely on the control of IAPs. The Working for Water – poverty relief programme has continued to the present, with a total of around ZAR90 million (~ US$10.7 million; as at September 2012 values) having been spent on control efforts in the park up to the end of the 2010/2011 financial year.

Although the park has a long tradition of research in general, with various programmes dating back to the 1950s (for example on fire, large mammals, carnivores, Du Toit et al. (2003), there was no research programme dedicated to the study of invasive alien species until about 1996. Research that was conducted was driven by external scientists, focusing largely on post-release evaluation of biological control agents. These studies included biological control of *P. stratiotes* (Cilliers et al. 1996), and *O. stricta* (Hoffmann et al. 1998a, b; Lotter and Hoffmann

1998; Reinhardt et al. 1999; Paterson et al. 2011). This has expanded in the last decade to include, for example, work on risk assessment (Foxcroft et al. 2007), ornamental plants (Foxcroft et al. 2008), processes and patterns of invasion (e.g. Foxcroft et al. 2004, 2009, 2011) and invasion potential and allelopathy of priority species such as *P. hysterophorus* (van der Laan et al. 2008).

The intentional introduction of alien plants as ornamental species has been one of the most important pathways of introduction, with many species that are now considered as some of the parks most problematic species (e.g. *O. stricta, L. camara* and *P. stratiotes*) now incurring substantial costs to control. While this pathway is in effect closed, the upper catchment of the KNP produces an unending supply of propagules, which will require on-going commitment to minimise. While KNP has itself invested in on-going control for a number of decades, due to the scale of the problem, it is unlikely that without the collaboration with the Working for Water programme, the current low density of invasive species would be possible. Preventative measures have generally been poor, and more strategic placement of control teams requires improved insights into likely sources of invasion from adjacent areas. Recent work has aimed to quantify the role that the boundary plays in filtering invasions from adjacent areas (Foxcroft et al. 2011; Jarošík et al. 2011) in order to develop buffer zones.

7.2.2.2 Case Study 6: Table Mountain National Park (South Africa)

The Cape of Good Hope Nature Reserve, which now forms the southern section of Table Mountain National Park (TMNP), was proclaimed in 1939 (Macdonald et al. 1988). Through the consolidation of a number of smaller nature reserves in 1998, TMNP now covers 471 km^2. The park is situated within the Cape Town metropolitan area and was declared a Natural World Heritage Site in 2003. TMNP also falls within the Cape Floral Kingdom, providing sanctuary for 2285 indigenous plant species, of which 90 are endemic (TMNP 2008). The fynbos biome (Mediterranean-climate type shrublands) in which TMNP is situation, has a rich history of botanical study (Cowling et al. 1997; Gelderblom et al. 2003), however invasions by alien plants have become an increasingly important component thereof (van Wilgen 2012). By the mid-1990s IAPs were already considered as one of the key threats to TMNPs ecosystem integrity (Richardson et al. 1996).

Alien plants have long been introduced into the Western Cape region of South Africa, for example, *Pinus pinaster* (cluster pine) in 1680, *Hakea sericea* (silky hakea) in 1830 (Macdonald and Richardson 1986), and *Acacia longifolia* (long-leaved wattle), which was reportedly introduced by Kew Gardens' collector James Bowie in 1827 (Stirton 1978). Introduced in the mid-1880s, *Acacia cyclops* (red eye) was considered the most important species 100 years later (Macdonald et al. 1988). By the late 1800s and early 1900s, botanists Peter MacOwan (in 1888) and Rudolf Marloth (in 1908) both raised concerns that IAPs would replace the natural vegetation (Stirton 1978). In 1945 plant invasions had reached alarming proportions, leading to the statement by a prominent naturalist that "one of the greatest, if not the greatest, threats to which the Cape vegetation is exposed, is suppression through the spread of vigorous exotic plant species" (Wicht 1945).

In an assessment of threats to the biodiversity of the region in 1996, dense patches of invasive plants were shown to be impacting on a third of the known localities of threatened taxa, with only about 10 % of the sites occurring only within areas already invaded (Richardson et al. 1996).

Another key problem coupled to the impact of IAPs in the fynbos region is that of fire and the alteration of fire frequency, seasonal incidence, size and intensity (Macdonald and Richardson 1986). The fynbos system is predisposed to fire as an ecosystem driver (Forsyth and Wilgen 2008), but usually to longer return intervals and lower intensities (van Wilgen 2009). Invasive alien plants are often more competitive under fire regimes that differ from those that the indigenous biota evolved with (Forsyth and Wilgen 2008). The most common invasive species in TMNP are also fire adapted, and their ability to produce large numbers of seeds facilitates their prolific spread after fires (van Wilgen et al. 2012). These trees (e.g. wattles and conifers; van Wilgen and Richardson 2012) and shrubs significantly increase biomass and add to fuel loads, leading to increased fire intensity and erosion (van Wilgen and Scott 2001).

Control operations started in 1941 but up until the 1970s were unsuccessful, largely due to the lack of a coordinated management plan and poor understanding of the biology of the target plants (Macdonald et al. 1988). In 1984/1985 the programme cost ZAR154,000 ($\sim$ US$100,000 at that time) and consumed about 40 % of the Cape of Good Hope Nature Reserve annual budget. These efforts were unfortunately not sufficient to contain the problem (Macdonald et al. 1988). In 2008 the control of IAPs within TMNP cost approximately ZAR9 million ($\sim$ US$1.08 million in 2008), with priority species including *Acacia*, *Hakea* and *Pinus* spp. (TMNP 2008). For the 2012–2013 financial year, the budgeted costs are approximately ZAR14 million ($\sim$ US$1.7 million in September 2012).

A challenge that has arisen over the last few years with regards to management of IAPs in TMNP, and from which valuable insights may be gained, is that of reconciling environmental imperatives and personal value systems (van Wilgen 2012). Although in keeping with TMNPs status as a world heritage site and its primary function of conserving the biodiversity rich region, IAP clearing programmes have become highly controversial. Arguments against removing IAPs include, for example, perceptions that alien trees contribute positively to water and soil resources (van Wilgen 2012). More difficult to manage however are the arguments related to aesthetics, recreational and ethical values. An assessment of the challenges (van Wilgen 2012) provides important lessons for the future, as these kinds of situations can conceivably increase in multi-use PA landscapes. Appreciation of the biodiversity values of the PA, and the adoption of policy, based on international conventions and best practice, is a key element to ensure a support base. A sound scientific understanding of the ecology and impacts of plant invasions is required to provide evidence based management and policy input. The lack thereof was shown to be a contributing factor to the unsuccessful management attempts between the 1940s and 1970s. The partnerships between academic researchers-managers-policy makers-funders facilitated an integrated strategic approach to implementation of control programmes and opportunities for co-learning. The management of IAPs in South Africa has however been facilitated by the renowned Working for Water programme, which has successfully been able to provide employment and economic outcomes, while integrating essential ecological needs (van Wilgen 2012). This kind of clearing programme may not be feasible in all situations, but alternative mechanisms need to be investigated to provide this level of support.

7.3 Impacts: Empirical Evidence and Anecdotal Observations

A wide range of impacts from plant invasions have been reported globally, with some originating from work in PAs (Foxcroft et al. 2014). Of the total body of literature on impacts of plant invasions, little of this work has been specifically directed at PAs in Africa, except perhaps for some work done in South Africa. We do not provide a comprehensive synthesis of the impacts of IAPs, but rather highlight some examples from PAs in Africa. A more in-depth review is provided elsewhere in this volume (Foxcroft et al. 2014).

Some examples include impacts on beetles and spiders as biodiversity indicators in Kruger National Park (Robertson et al. 2011), and Hluhluwe-iMfolozi Game Reserve (Mgobozi et al. 2008). In both studies using beetle and spiders as indicators, either the assemblage patterns, abundance, diversity and estimated species richness were changed. In Hluhluwe-iMfolozi small mammal species richness and diversity were also shown to have decreased (Dumalisile 2008). Due to the role that large- to mega-herbivores play as ecosystem drivers in most African PAs, an improved understanding of the effects of them being displaced on other species or ecosystems is also required. Dense patches of C. odorata were shown to impact on forage availability and access, causing spatial reorganization of the black rhino (Diceros bicornis) population (Howison 2009). This led to the hypothesis that C. odorata may have been partly responsible for a decline in the black rhino population in Hluhluwe-iMfolozi in the mid-1990s (Howison 2009). Shading of crocodile (Crocodylus niloticus) nesting sites by the invasive plant C. odorata in Lake Saint Lucia (iSimangaliso Wetland Park) has resulted in a female-biased sex ratio (Leslie and Spotila 2001). In the fynbos biome, of which TMNP plays a crucial role in protecting the species rich region, invasions could reduce species richness by between 45 and 67 % (Richardson et al. 1989). By 1996, due to uncontrolled fires, combined with the effects of plant invasions, many species were considered to be facing imminent extinction (Trinder-Smith et al. 1996). This is further exacerbated by the impact of fire on soil erosion (van Wilgen and Scott 2001). Soil loss following fires typically amounts to 0.1 tonnes/ha in fynbos habitat, but when fuelled by high biomass from IAPs, increases to 6 tonnes/ha in patches invaded by *Pinus* (Scott et al. 1998). Other studies have provided documented evidence of modified nutrient regimes due to either increased nitrogen fixation or increased decomposing biomass (Musil and Midgley 1990; Yelenik et al. 2004) and impacts on seed banks of native fynbos species associated with *Acacia saligna* (Port Jackson willow) invasions (Holmes and Cowling 1997).

Water loss due to increased use by invading trees is also of concern as few PAs contain entire water catchments within their boundaries, and thus widespread invasion and habitat alteration may have severe consequences for downstream users. For example, a study in 2002 showed the upper reaches of the Sabie-Sand river catchment, which flows through the Kruger NP, to be about 23 % invaded, with a corresponding 9.4 % loss in natural river flow (Le Maitre et al. 2002).

There have been very few scientific studies on the impacts of IAPs in East Africa, although various studies have been carried out in the Eastern Arc Mountains of East Africa (Hulme et al. 2014). Considered by some to be an introduced invasive plant, the impacts by the toxic *Solanum campylacanthum* (bitter apple), were investigated in Hells Gate (HGNP) and Nakuru NP, Kenya. In Nakuru NP, *S. campylacanthum* invaded all grassland vegetation types, with densities of up to 3,334 plants ha (Ng'weno et al. 2010), decreasing forage quantity significantly with increased invasion (Ng'weno et al. 2010).

Parthenium hysterophorus (Terfa 2009) and *P. juliflora* (Demissie 2009) were shown to displace native plant species in Awash National Park (Ethiopia). The impact of *M. pigra*, which has invaded Lochinvar National Park (Zambia) was assessed using fixed transects along the shoreline. Bird species diversity was reduced by almost 50 %, with only 314 individual birds recorded in invaded sites, compared to 19,265 in uninvaded open floodplains (Shanungu 2009).

7.4 Management: What Is Being Done and What Can Be Done?

There is a wealth of information widely available on the techniques of managing alien plants, including for example, A Toolkit of Best Management Practices (Wittenberg and Cock 2001), Turning the Tide: the Eradication of Invasive Species (Veitch and Clout 2002), Assessing and Managing Invasive Species within Protected Areas (Tu 2009) and Tools and Techniques for Use in Natural Areas (Tu et al. 2001). Also, a number of chapters in this volume provide further information on various aspects of alien plant management (e.g. Meyerson and Pyšek 2014; Simberloff 2014; Tu and Robison 2014). Therefore we do not discuss the tools of management (except for a short discussion on biological control; see Van Driesche and Center 2014 for a detailed discussion), but rather examples being used in different settings in PAs in Africa currently.

Management of alien plant invasions in Africa's PAs is generally inadequate. The extent to which management interventions can be implemented is limited due to both the scarcity of resources for PAs, and competition for allocation of resources within overall park management programmes. For example, resources may need to be allocated to anti-poaching, fire management, animal capture and relocation, monitoring of threatened species, infrastructure maintenance, community relations and invasive species management. Additionally, the lack of capacity, and awareness of the problem, and PA risk assessments to prioritise threats undermines the importance of invasive species management. Where control operations are in place, objective management effectiveness evaluations would assist in continuous improvement. Innovative approaches are required if progress is to be made in the long-term. It is also highly unlikely that one management approach will suit all situations. However, examples do exist that may be tailored to provide unique solutions. These may include, for example, the South African Working for Water model or local community driven approaches.

South Africa, as with most other African states, is characterised by high levels of unemployment (Buch and Dixon 2008). In redressing the high levels of poverty, the Working for Water programme uses control of invasive plants as a means to achieve social, economic and biodiversity outcomes (Van Wilgen et al. 2010). Nationally, this programme provides employment for up to 20,000 people per annum, while simultaneously treating large areas invaded by invasive plants. Conservation agencies are supported by the national programme to enable them to maintain their objectives, which in the absence of this funding would be unlikely in most PAs. Due cognizance does however need to be given to potential challenges of such public-works programmes (Buch and Dixon 2008), such as the importance of being economically sustainable for the individual employees who are usually employed on a short-term contractual basis. Also, even within a large programme such as Working for Water, on-going reassessment of priorities and approaches is required. For example, a review of the programme 15 years after its initiation suggests that the distribution of some invasive plants has still increased (although would be significantly worse in the absence of the programme), requiring reprioritisation of species and areas to be cleared (van Wilgen et al. 2012).

An integrated community approach advocated by the PAMS Foundation (http://www.pamsfoundation.org/) has been employed in a few village communities in southern Tanzania. It focuses on education and a strong awareness campaign to create an appreciation of the negative impacts of IAPs. Additionally, it aims to promote alternative indigenous plants that would be beneficial to use. As the communities and local authorities develop an understanding of the problem and request assistance, empowerment programmes are initiated. While assistance, including the provision of seedlings of desirable species and the necessary equipment is provided to commence the programme, most manual labour is voluntary due to the benefits to the community itself. These projects suggest that this bottom-up approach is more cost effective in areas where specialised felling and herbicide treatments are not essential, or in other words where the use of well-trained contractors is not required. It also has a higher chance of being sustainable, due to the ownership of the programme by land users/owners from the start of the programme. The passing of by-laws prohibiting the use of certain species forms part of the programme from the early stages as well.

In another example, Kenyan Wildlife Service is making progress in developing an invasive species management strategy for PAs. Resources have also for the first time been allocated to management. In addition, the revised draft of the Wildlife Bill of 2011 makes special reference to invasive species, to the extent that any person who introduces an invasive species into wildlife conservation areas, or fails to comply with the measures prescribed by the Cabinet Secretary regarding invasive alien species, commits an offence and is liable to prosecution. A number of alien species are listed under the Bill, including 12 invasive plant species. It is hoped that the list will be expanded to include all invasive and potentially invasive species present in East Africa.

Unfortunately the biological control of terrestrial invasive species has not been practiced widely in Africa, outside of South Africa. This despite the fact that many of the species on which biocontrol agents have been released in South Africa, or even Australia, are invasive in other parts of the continent. Opposition is often based on a poor understanding of the theory and practice of biocontrol. However in the long-term, biological control (Van Driesche and Center 2014) has the potential to become one of the best management options available to PA managers in Africa, because most countries do not have the resources to implement costly mechanical and chemical control programmes. The main benefits of biocontrol are (i) that the agents establish self-perpetuating populations and can often establish in areas that are not accessible for chemical or mechanical control, (ii) control of the target species is permanent, (iii) there are no negative impacts on the environment, (iv) the cost of biocontrol is low relative to other approaches and usually requires a once-off investment, and (v) benefits can be reaped by many stakeholders independent of their financial status, and irrespective if they contributed to the initial research (Greathead 1995). Biological control of invasive plants at a global level has been completely successful in about 25 % of cases, which means that the target weeds have been totally suppressed by the agents themselves with no need for further chemical or mechanical control interventions (Cruttwell McFadyen 1998). In the majority of other cases, chemical and/or mechanical inputs have been considerably reduced. An analysis of some biocontrol research programmes in South Africa found that the benefit:cost ratios ranged from 50:1 for tropical woody shrubs (de Lange and van Wilgen 2010; van Wilgen and de Lange 2011), to 34:1 for *L. camara* and 4331:1 for *Acacia pycnantha* (van Wilgen et al. 2004). The value of the South African biocontrol programme has been estimated at ZAR840 million for fire-adapted trees, ZAR104 billion for invasive Australian trees, ZAR37 billion for succulents and ZAR2.5 billion for subtropical woody shrubs (1 US\$ = ZAR7.7 in January 2010; de Lange and van Wilgen 2010). It was also estimated that by 1998, biocontrol agents present in South Africa had already reduced the financial costs of mechanical and chemical control by more than 20 % (Versfeld et al. 1998).

While no single management model will fit all situations, similar kinds, or a combination of approaches, may provide useful for specific cases. The South

African Working for Water model, while successful in a country focusing on creation of employment opportunities linked to water conservation, is unlikely to be successful in other areas where national needs, as well as available budgets, are different. In certain circumstances implementation of local scale projects may prove more successful for a specific PA, especially if the actions provide beneficial spinoffs for the related community. However, if control initiatives are to be successful in the long-term, stakeholder support from the level of local communities to national government is essential.

7.5 Conclusions

An assessment of regional contributions to invasion ecology science found a low representation of developing countries with Asia and Africa (with the exception of South Africa) severely understudied (Pyšek et al. 2008). In a literature review of ecology and biodiversity conservation, only 15.8 % of all published papers related to alien species had authors from developing countries, and only 6.5 % had authors solely from developing countries (Nuñez and Pauchard 2010).

The clear lack of information on plant invasions in PAs across much of the continent is a critical shortfall in the long-term security of increasingly important biodiversity rich areas. The examples of the more data rich areas do however provide examples of approaches that may be adopted and modified elsewhere, reducing some of the need for costly and time-consuming relearning. Trends indicate that as further surveys are conducted in other PAs and sampling intensity is increased, the number of species and distribution is likely to be substantially more than currently estimated. If Africa's unique ecosystems and biodiversity are to be safeguarded in the long-term, substantial work needs to be done to collect this information, which is essential to the design and implementation of control programmes. However, as with most countries (especially developing countries in Africa), the scarcity of resources is likely to inhibit any large scale, long-term management, and innovative approaches will be required. Although little research has been carried out in Africa's PAs, the work that has been done has revealed that the selected biodiversity indicators and other ecosystem properties are being impacted upon. However, Africa's PAs offer many opportunities for science that can both develop a basic understanding of invasion biology and contribute directly to real world management problems.

Acknowledgements LCF thanks South African National Parks, the Centre for Invasion Biology, Stellenbosch University, and the National Research Foundation (South Africa) for support. We thank Sandra MacFadyen for Fig. 7.1 and Zuzana Sixtová for technical assistance with editing. AW acknowledges the Australian High Commission Kenya, UNEP-GEF and CABI for financial support.

References

Barber CV, Miller KR, Boness M (eds) (2004) Securing protected areas in the face of global change: issues and strategies. IUCN, Gland/Cambridge

Belsky AJ (1987) Revegetation of natural and human-caused disturbances in the Serengeti National Park, Tanzania. Vegetation 70:51–60

Birnie A, Noad T (2011) Trees of Kenya – an illustrated field guide, 3rd edn. Koeltz Scientific Books, Nairobi

Blundell AG, Scatena FN, Wentsel R et al (2003) Ecorisk assessment using indicators of sustainability – invasive species in the Caribbean National Forest of Puerto Rico. J For 101:14–19

Bruner AG, Gullison RE, Rice RE et al (2001) Effectiveness of parks in protecting tropical biodiversity. Science 291:125–129

Buch A, Dixon AB (2008) South Africa's Working for Water programme: searching for win–win outcomes for people and the environment. Sustain Dev 17:129–141

CABI (2004) Prevention and management of alien invasive species: forging cooperation throughout West Africa. In: Proceedings of a workshop held in Accra, Ghana, 9–11 March, 2004. CABI, Nairobi, Kenya & CABI, Accra, Ghana

Chape S, Harrison J, Spalding M et al (2005) Measuring the extent and effectiveness of protected areas as an indicator for meeting global biodiversity targets. Phil Trans Biol Sci 360:443–455

Christie IT, Crompton DE (2001) Tourism in Africa. Africa Region working paper series no 12. The World Bank Group, Africa Region

Cilliers CJ, Zeller DA, Strydom G (1996) Short- and long-term control of water lettuce (*Pistia stratiotes*) on seasonal water bodies and on a river system in the Kruger National Park, South Africa. Hydrobiologia 340:173–179

Clark K, Lotter WD (2009) Preliminary invasive alien plant survey of Mikumi National Park. PAMS Foundation, Arusha

Clark K, Lotter WD, Runyoro VA (2011) Ngorongoro conservation area invasive alien plant strategic management plan. Ministry of Natural Resources and Tourism, Tanzania

Cowling RM, Richardson DM, Mustart PJ (1997) Fynbos. In: Cowling RM, Richardson DM, Pierce SM (eds) Vegetation of Southern Africa. Cambridge University Press, Cambridge, pp 99–130

Cruttwell McFadyen RE (1998) Biological control of weeds. Annu Rev Entomol 43:369–393

Dawson W, Mndolwa AS, Burslem D et al (2008) Assessing the risks of plant invasions arising from collections in tropical botanical gardens. Biodivers Conserv 17:1979–1995

de Lange WJ, van Wilgen BW (2010) An economic assessment of the contribution of biological control to the management of invasive alien plants and to the protection of ecosystem services in South Africa. Biol Invasion 12:4113–4124

Demissie H (2009) Invasion of *Prosopis juliflora* (Sw.) DC. into Awash National Park and its impact on plant species diversity and soil characteristics. Addis Ababa University, Ethiopia

Dharani N (2011) Field guide to common trees and shrubs of East Africa. Struik Nature, Cape Town

Du Toit JT, Rogers KH, Biggs HC (eds) (2003) The Kruger experience. Ecology and management of savanna heterogeneity. Island Press, Washington, DC

Dublin HT (1986) Decline of the Mara woodlands: the role of fire and elephants. University of British Columbia, Vancouver

Dudley N (ed) (2008) Guidelines for applying protected area management categories. IUCN, Gland. http://data.iucn.org/dbtw-wpd/edocs/PAPS-016.pdf

Dudley N, Mulongoy KJ, Cohen S et al (2005) Towards effective protected area systems. An action guide to implement the convention on biological diversity programme of work on protected areas. Secretariat of the Convention on Biological Diversity, Montreal

Dumalisile L (2008) The effects of *Chromolaena odorata* on mammalian biodiversity in Hluhluwe-iMfolozi Park, South Africa. University of Pretoria, Pretoria

Erwin TL (1991) An evolutionary basis for conservation strategies. Science 253:750–752

Forsyth GG, Wilgen BW (2008) The recent fire history of the Table Mountain National Park and implications for fire management. Koedoe 50:3–9

Foxcroft LC (2001) A case study of human dimensions in invasion and control of alien plants in the personnel villages of Kruger National Park. In: McNeely JA (ed) The great reshuffling. Human dimensions of invasive alien species. IUCN, Gland, pp 127–134

Foxcroft LC (2003a) Observations and recommendations for the management of invasive alien plant species in the Serengeti National Park. South African National Parks, Skukuza

Foxcroft LC (2003b) Recommendations for the management of invasive alien plants and problematic indigenous weed species in the Ngorongoro Conservation Area Authority. South African National Parks, Skukuza

Foxcroft LC, Downey PO (2008) Protecting biodiversity by managing alien plants in national parks: perspectives from South Africa and Australia. In: Tokarska-Guzik B, Brock JH, Brundu G et al (eds) Plant invasions: human perception, ecological impacts and management. Backhuys Publishers, Leiden, pp 387–403

Foxcroft LC, Freitag-Ronaldson S (2007) Seven decades of institutional learning: managing alien plant invasions in the Kruger National Park, South Africa. Oryx 41:160–167

Foxcroft LC, Richardson DM (2003) Managing alien plant invasions in the Kruger National Park, South Africa. In: Child L, Brock J, Brundu G et al (eds) Plant invasions: ecological threats and management solutions. Backhuys Publishers, Leiden, pp 385–403

Foxcroft LC, Henderson L, Nichols GR et al (2003) A revised list of alien plants for the Kruger National Park. Koedoe 46:21–44

Foxcroft LC, Rouget M, Richardson DM et al (2004) Reconstructing 50 years of Opuntia stricta invasion in the Kruger National Park, South Africa: environmental determinants and propagule pressure. Divers Distrib 10:427–437

Foxcroft LC, Rouget M, Richardson DM (2007) Risk assessment of riparian plant invasions into protected areas. Conserv Biol 21:412–421

Foxcroft LC, Richardson DM, Wilson JRU (2008) Ornamental plants as invasive aliens: problems and solutions in Kruger National Park, South Africa. Environ Manage 41:32–51

Foxcroft LC, Richardson DM, Rouget M et al (2009) Patterns of alien plant distribution at multiple spatial scales in a large national park: implications for ecology, management and monitoring. Divers Distrib 15:367–378

Foxcroft LC, Richardson DM, Rejmánek M et al (2010) Alien plant invasions in tropical and sub-tropical savannas: patterns, processes and prospects. Biol Invasion 12:3913–3933

Foxcroft LC, Jarošík V, Pyšek P et al (2011) Protected-area boundaries as filters of plant invasions. Conserv Biol 25:400–405

Foxcroft LC, Pyšek P, Richardson DM et al (2014) Chapter 2: Impacts of alien plant invasions in protected areas. In: Foxcroft LC, Pyšek P, Richardson DM, Genovesi P (eds) Plant invasions in protected areas: patterns, problems and challenges. Springer, Dordrecht, pp 19–41

Freitag-Ronaldson S, Foxcroft LC (2003) Anthropogenic influences at the ecosystem level. In: du Toit JT, Rogers KH, Biggs HC (eds) The Kruger experience. Ecology and management of savanna heterogeneity. Island Press, Washington, DC, pp 392–421

Gaston KJ, Jackson SF, Cantú-Salazar L et al (2008) The ecological performance of protected areas. Annu Rev Ecol Evol Syst 39:93–113

Gelderblom CM, van Wilgen BW, Nel JL et al (2003) Turning strategy into action: implementing a conservation action plan in the Cape Floristic Region. Biol Conserv 112:291–297

Glover PE, Trump EC (1970) An ecological survey of the Narok District of Kenya Masailand. The vegetation. Kenya National Parks, Nairobi

Goodman PS (2003) Assessing management effectiveness and setting priorities in protected areas in KwaZulu-Natal. BioScience 53:843–850

Greathead DJ (1995) Benefits and risks of classical biological control. In: Hokkanen HMT, Lynch JM (eds) Biological control: benefits and risks. Cambridge University Press, Cambridge, pp 53–63

Greenway PJ (1934) Report of a botanical survey of the indigenous and exotic plants in cultivation at the East African Agricultural Research Station, Amani, Tanganyika Territory. East African Agricultural Research Station, Amani, Tanganyika

Henderson L (2001) Alien weeds and invasive plants. Plant Protection Research Institute Handbook No. 12. Agricultural Research Council, Pretoria

Henderson L (2002) Problem plants in Ngorongoro Conservation Area. Agricultural Research Council – Plant Protection Research Institute, Pretoria

Heriz-Smith S (1962) The wild flowers of the Nairobi Royal National Park. D.A. Hawkins, Nairobi

Hobson B (1995) Gardening in East Africa. Horticultural Society of Kenya, Nairobi

Hoffmann JH, Moran CC, Zeller DA (1998a) Evaluation of *Cactoblastis cactorum* (Lepidoptera: Phycitidae) as a biological control agent of *Opuntia stricta* (Cactaceae) in the Kruger National Park, South Africa. Biol Control 12:20–24

Hoffmann JH, Moran CC, Zeller DA (1998b) Long-term population studies and the development of an integrated management programme for control of *Opuntia stricta* in Kruger National Park, South Africa. J Appl Ecol 35:156–160

Holmes PM, Cowling RM (1997) The effects of invasion by *Acacia saligna* on the guild structure and regeneration capabilities of South African fynbos shrublands. J Appl Ecol 34:317–332

Howison RA (2009) Food preferences and feeding interactions among browsers, and the effect of an exotic invasive weed *Chromolaena odorata* on the endangered Black Rhino in an African savanna. University of Kwazulu-Natal, Westville

Hulme PE, Burslem DFRP, Dawson W et al (2014) Chapter 8: Aliens in the Arc: are invasive trees a threat to the montane forests of East Africa? In: Foxcroft LC, Pyšek P, Richardson DM, Genovesi P (eds) Plant invasions in protected areas: patterns, problems and challenges. Springer, Dordrecht, pp 145–165

IUCN and UNEP-WCMC (2012) The World Database on Protected Areas (WDPA): February 2012 [On-line]. UNEP-WCMC, Cambridge. http://www.protectedplanet.net/. Accessed 21 Oct 2012

IUCN/PACO (2013) Invasive plants affecting protected areas of West Africa. Management for reduction of risk for biodiversity. IUCN/PACO, Ouagadougou

Ivens G (1967) East African weeds and their control. Oxford University Press, Nairobi

Jarošík V, Pyšek P, Foxcroft LC et al (2011) Predicting incursion of plant invaders into Kruger National Park, South Africa: the interplay of general drivers and species-specific factors. PLoS One 6:e28711

Katende AB, Birnie A, Tengnas B (1995) Useful trees and shrubs for Uganda. Identification, propagation and management for agricultural and pastoral communities. Regional Soil Conservation Unit (RELMA), Swedish International Development Cooperation Agency (SIDA)

Koenders L (1992) Flora of Pemba Island. A checklist of plant species, 2nd edn. Wildlife Conservation Society of Tanzania, Dar es Salaam

Le Maitre DC, van Wilgen BW, Gelderblom CM et al (2002) Invasive alien trees and water resources in South Africa: case studies of the costs and benefits of management. For Ecol Manage 160:143–159

Leslie AJ, Spotila JR (2001) Alien plant threatens Nile crocodile (*Crocodylus niloticus*) breeding in Lake St. Lucia, South Africa. Biol Conserv 98:347–355

Leverington F, Costa KL, Courrau J et al (2010) Management effectiveness evaluation in protected areas – a global study, 2nd edn. University of Queensland, Brisbane

Loope LL, Sanchez PG, Tarr PW et al (1988) Biological invasions of arid land nature reserves. Biol Conserv 44:95–118

Lotter WD (2004) Progress report and updated recommendations for the management of invasive alien plants in the Ngorongoro Conservation Area. Ngorongoro Conservation Area Authority, Ngorongoro, Tanzania

Lotter WD, Clark K (2008) Invasive alien species strategic management plan for Ezemvelo KwaZulu-Natal Wildlife, South Africa. Ezemvelo KwaZulu-Natal Wildlife, Pietermaritzburg

Lotter WD, Hoffmann JH (1998) An integrated management plan for the control for *Opuntia stricta* (Cactaceae) in the Kruger National Park, South Africa. Koedoe 41:63–68

Lyons EE, Miller SE (eds) (1999) Invasive species in eastern Africa. Proceedings of a workshop held at ICIPE, July 5–6, 1999, ICIPE Science Press, Nairobi

Macdonald IAW (1983) Alien trees, shrubs and creepers invading indigenous vegetation in the Hluhluwe-Umfolozi Game Reserve Complex in Natal. Bothalia 14:949–959

Macdonald IAW, Frame GW (1988) The invasion of introduced species into nature reserves in tropical savannas and dry woodlands. Biol Conserv 44:67–93

Macdonald IAW, Richardson DM (1986) Alien species in terrestrial ecosystems of the fynbos biome. In: Macdonald IAW, Kruger FJ, Ferrar A (eds) The ecology and management of biological invasions in southern Africa. Oxford University Press, Cape Town, pp 93–108

Macdonald IAW, Graber DM, Debenedetti S et al (1988) Introduced species in nature reserves in Mediterranean-type climatic regions of the world. Biol Conserv 44:37–66

Maroyi A (2006) Preliminary checklist of introduced and naturalised plants in Zimbabwe. Kirkia 18:177–247

Maroyi A (2012) The casual, naturalised and invasive alien flora of Zimbabwe based on herbarium and literature records. Koedoe 54:Art. #1054, 6 p. doi:10.4102/koedoe.v54i1.1054

Maundu P, Tengnas B (2005) Useful trees and shrubs of Kenya. World Agroforestry Centre, Nairobi

Mbuya LP, Msanga HP, Ruffo CK et al (1994) Useful trees and shrubs for Tanzania. Identification, propagation and management for agricultural and pastoral communities. Regional Soil Conservation Unit (RSCU). Swedish International Development Authority (SIDA), Embassy of Sweden, Kenya

Meyerson LA, Pyšek P (2014) Chapter 25: Manipulating alien species propagule pressure as a prevention strategy in protected areas. In: Foxcroft LC, Pyšek P, Richardson DM, Genovesi P (eds) Plant invasions in protected areas: patterns, problems and challenges. Springer, Dordrecht, pp 549–559

Mgobozi MP, Somers MJ, Dippenaar-Schoeman AS (2008) Spider responses to alien plant invasion: the effect of short- and long-term *Chromolaena odorata* invasion and management. J Appl Ecol 45:1189–1197

Musil CF, Midgley GF (1990) The relative impact of invasive Australian acacias, fire and season on the soil chemical status of a sand plain lowland fynbos community. S Afr J Bot 56:419–427

Naughton-Treves L, Holland MB, Brandon K (2005) The role of protected areas in conserving biodiversity and sustaining local livelihoods. Annu Rev Environ Res 30:219–252

Newmark WD (2008) Isolation of African protected areas. Front Ecol Environ 6:321–328

Ng'weno CC, Mwasi SM, Kairu JK (2010) Distribution, density and impact of invasive plants in Lake Nakuru National Park, Kenya. Afr J Ecol 48:905–913

Nuñez MA, Pauchard A (2010) Biological invasions in developing and developed countries: does one model fit all? Biol Invasion 12:707–714

Nyoka BI (2003) Biosecurity in forestry: a case study on the status of invasive forest trees species in Southern Africa. FAO Forest Biosecurity Working Paper FBS/1E. Forestry Department, Rome

Obermeijer AA (1937) A preliminary list of the plants found in the Kruger National Park. Ann Transvaal Mus 17:185–227

Paterson ID, Hoffmann JH, Klein H et al (2011) Biological control of Cactaceae in South Africa. Afr Entomol 19:230–246

Pyšek P, Richardson DM, Pergl J et al (2008) Geographical and taxonomic biases in invasion ecology. Trends Ecol Evol 23:237–244

Reichard SH, White P (2001) Horticulture as a pathway of invasive plant introductions in the United States. BioScience 51:103–113

Reinhardt CF, Rossouw L, Thatcher L et al (1999) Seed germination of *Opuntia stricta*: implications for management strategies in the Kruger National Park. S Afr J Bot 65:295–298

Richardson DM, Macdonald IAW, Forsyth GG (1989) Reductions in plant species richness under stands of alien trees and shrubs in the fynbos biome. S Afr For J 149:1–8

Richardson DM, van Wilgen BW, Higgins SI et al (1996) Current and future threats to plant biodiversity on the Cape Peninsula, South Africa. Biodivers Conserv 5:607–647

Richardson DM, Pyšek P, Carlton JT (2011) A compendium of essential concepts and terminology in invasion ecology. In: Richardson DM (ed) Fifty years of invasion ecology – the legacy of Charles Elton. Wiley-Blackwell, Oxford, pp 409–420

Robertson MP, Harris KR, Coetzee JA et al (2011) Assessing local scale impacts of *Opuntia stricta* (Cactaceae) invasion on beetle and spider diversity in Kruger National Park, South Africa. Afr Zool 46:205–223

Scott DF, Versfeld DB, Lesch W (1998) Erosion and sediment yield in relation to afforestation and fire in the mountains of the Western Cape Province, South Africa. S Afr Geogr J 80:52–59

Shanungu GK (2009) Management of the invasive *Mimosa pigra* L. in Lochinvar National Park, Zambia. Biodiversity 10:56–60

Simberloff D (2014) Chapter 25: Eradication – pipe dream or real option? In: Foxcroft LC, Pyšek P, Richardson DM, Genovesi P (eds) Plant invasions in protected areas: patterns, problems and challenges. Springer, Dordrecht, pp 549–559

Sinclair ARE, Arcese P (eds) (1995) Serengeti II: dynamics, management, and conservation of an ecosystem. University of Chicago Press, Chicago

Sinclair ARE, Norton-Griffiths M (eds) (1979) Serengeti: dynamics of an ecosystem. University of Chicago Press, Chicago

Sinclair ARE, Packer C, Mduma SAR et al (eds) (2008) Serengeti III: human impacts on ecosystem dynamics. University of Chicago Press, Chicago

Southern Rhodesia Commission Forestry (1956) Exotic forest trees in the British Commonwealth, Southern Rhodesia. Government Printers, Rhodesia

Spear D, McGeoch MA, Foxcroft LC et al (2011) Alien species in South Africa's National Parks (SANParks). Koedoe 53:Art. #1032, 4 p. doi:10.4102/koedoe.v53i1.1032

Spear D, Foxcroft LC, Bezuidenhout H et al (2013) Human population density explains alien species richness in protected areas. Biol Conserv 159:137–147

Stirton CH (1978) Plant invaders: beautiful, but dangerous. Department of Nature and Environmental Conservation, Cape Town

Terfa AE (2009) Impact of *Parthenium hysterophorus* L. (Asteraceae) on herbaceous plant biodiversity in the Awash National Park (ANP). Addis Ababa University, Ethiopia

Terry PJ (1984) A guide to weed control in East African crops. Kenya Literature Bureau, Nairobi

Terry PJ, Michieka RW (1987) Common weeds of East Africa/Magugu ya Afrika Mashariki. Food and Agricultural Organization of the United Nations, Rome

TMNP (2008) Table Mountain National Park – park management plan. South African National Parks, Cape Town

Trinder-Smith TH, Cowling RM, Linder HP (1996) Profiling a besieged flora: endemic and threatened plants of the Cape Peninsula, South Africa. Biodivers Conserv 5:575–589

Tu M (2009) Assessing and managing invasive species within protected areas. In: Ervin J (ed) Protected area quick guide series. The Nature Conservancy, Arlington, 40 pp

Tu M, Robison RA (2014) Chapter 24: Overcoming barriers to the prevention and management of alien plant invasions in protected areas: a practical approach. In: Foxcroft LC, Pyšek P, Richardson DM, Genovesi P (eds) Plant invasions in protected areas: patterns, problems and challenges. Springer, Dordrecht, pp 529–547

Tu M, Hurd C, Randall JM (2001) Weed control methods handbook: tools and techniques for use in natural areas. The Nature Conservancy, Wildland Invasive Species Team, Arlington

Usher MB (1988) Biological invasions of nature reserves – a search for generalizations. Biol Conserv 44:119–135

Usher MB, Kruger FJ, Macdonald IAW et al (1988) The ecology of biological invasions into nature reserves – an introduction. Biol Conserv 44:1–8

van der Laan M, Reinhardt CF, Belz RG et al (2008) Interference potential of the perennial grasses *Eragrostis curvula*, *Panicum maximum* and *Digitaria eriantha* with *Parthenium hysterophorus*. Tropic Grassl 42:88–95

Van Driesche R, Center T (2014) Chapter 26: Biological control of plant invasions in protected areas. In: Foxcroft LC, Pyšek P, Richardson DM, Genovesi P (eds) Plant invasions in protected areas: patterns, problems and challenges. Springer, Dordrecht, pp 561–597

van Wilgen BW (2009) The evolution of fire and invasive alien plant management practices in fynbos. S Afr J Sci 105:342

van Wilgen BW (2012) Evidence, perceptions, and trade-offs associated with invasive alien plant control in the Table Mountain National Park, South Africa. Ecol Soc 17:23

van Wilgen BW, de Lange WJ (2011) The costs and benefits of biological control of invasive alien plants in South Africa. Afr Entomol 19:504–514

van Wilgen BW, Richardson DM (2012) Three centuries of managing introduced conifers in South Africa: benefits, impacts, changing perceptions and conflict resolution. J Environ Manage 106:56–68

van Wilgen BW, Scott DF (2001) Managing fires on the Cape Peninsula: dealing with the inevitable. J Medit Ecol 2:197–208

van Wilgen BW, Richardson DM, Le Maitre DC et al (2001) The economic consequences of alien plant invasions: examples of impacts and approaches to sustainable management in South Africa. Environ Dev Sustain 3:145–168

van Wilgen BW, de Wit MP, Anderson HJ et al (2004) Costs and benfits of biological control of invasive alien plants: case studies from South Africa. S Afr J Sci 100:113–122

van Wilgen BW, Khan A, Marais C (2010) Changing perspectives on managing biological invasions: insights from South Africa and the working for water programme. In: Richardson DM (ed) Fifty years of invasion ecology – the legacy of Charles Elton. Wiley-Blackwell, Oxford, pp 377–393

van Wilgen BW, Forsyth GG, Prins P (2012) The management of fire-adapted ecosystems in an urban setting: the case of Table Mountain National Park. S Afr Ecol Soc 17:8

Vardien W, Le Roux JJ, Richardson DM et al (2012) The introduction history, spread, and current distribution of Lantana camara in South Africa. S Afr J Bot 81:81–94

Veitch CR, Clout MN (eds) (2002) Turning the tide: the eradication of invasive species. IUCN SSC Invasive Species Specialist Group, IUCN, Gland/Cambridge

Versfeld DB, Le Maitre DC, Chapman RA (1998) Alien invading plants and water resources in South Africa: a preliminary assessment. CSIR, Stellenbosch

Wicht CL (1945) Report of the Committee on the preservation of the vegetation of the South Western Cape. Royal Society of South Africa, Cape Town

Williams RO (1949) Useful and ornamental plants of Zanzibar and Pemba. Zanzibar Protectorate, Zanzibar

Wittenberg R, Cock MJW (eds) (2001) Invasive alien species: a toolkit of best prevention and management practices. CABI, Wallingford

Yelenik SG, Stock WD, Richardson DM (2004) Ecosystem level impacts of invasive Acacia saligna in the South African fynbos. Restor Ecol 12:44–51

Chapter 8
Aliens in the Arc: Are Invasive Trees a Threat to the Montane Forests of East Africa?

Philip E. Hulme, David F.R.P. Burslem, Wayne Dawson, Ezekiel Edward, John Richard, and Rosie Trevelyan

Abstract Although plant invasions are often regarded as a significant threat to global biodiversity, current understanding of the vulnerability of tropical forests to invasion or the factors that lead to alien species becoming invasive in the tropics remains limited. Here, we synthesise available information on plant invasions in protected areas for the most ecologically important montane forests of East Africa. We undertake a hierarchical analysis to explore patterns across the entire mountain chain with those within an individual mountain block down to a single nature

P.E. Hulme (✉)
The Bio-Protection Research Centre, Lincoln University, Canterbury,
PO Box 84, New Zealand
e-mail: Philip.hulme@lincoln.ac.nz

D.F.R.P. Burslem
Institute of Biological and Environmental Sciences, University of Aberdeen,
Cruickshank Building, Aberdeen AB24 3UU, UK
e-mail: d.burslem@abdn.ac.uk

W. Dawson
Department of Biology, University of Konstanz, Universitaetstrasse 10,
D 76464 Konstanz, Germany
e-mail: wayne.dawson@uni-konstanz.de

E. Edward
Department of Forest and Landscape, Faculty of Life Sciences, University of Copenhagen,
Rolighedsvej 23, DK-1958 Copenhagen, Frederiksberg C, Denmark
e-mail: ezedwa@yahoo.com

J. Richard
Tanzania Forestry Research Institute, PO Box 95, Lushoto, Tanzania
e-mail: jorijomb@yahoo.com

R. Trevelyan
Tropical Biology Association, Department of Zoology,
Downing Street, Cambridge CB2 3EJ, UK
e-mail: rjt34@cam.ac.uk

L.C. Foxcroft et al. (eds.), *Plant Invasions in Protected Areas: Patterns, Problems and Challenges*, Invading Nature - Springer Series in Invasion Ecology 7, DOI 10.1007/978-94-007-7750-7_8, © Springer Science+Business Media Dordrecht 2013

reserve. A common feature of the occurrence of alien trees in the Eastern Arc Mountains is the overwhelming importance of propagule pressure in the representation of species found colonising forests. The patterns observed emphasise the need for scientifically sound advice regarding not only the potential impact of an alien species on native biodiversity but also an assessment of which mitigation strategies might be most appropriate and highlights the research, control and social challenges of managing invasive agroforestry trees in the tropics.

Keywords Biological invasions • Botanic garden • Fragmentation • Propagule pressure • Tropical forest

8.1 Introduction

Although plant invasions are often regarded as a significant threat to global biodiversity, current understanding of the vulnerability of tropical forests to invasion or the factors that lead to alien species becoming invasive in the tropics remains limited (Lugo 2004; Dawson et al. 2008). Compared to other biomes, tropical forest ecosystems appear less vulnerable to invasions by alien plants, at least judged by the number of naturalised taxa in floras (Rejmánek 1996). The presence of fast-growing multi-layered vegetation has been proposed as the main mechanism by which undisturbed tropical forests are resistant to invasions (Rejmánek et al. 2005). Fine (2002) reviewed instances of alien plant invasion into tropical forests, and concluded that the low number of invasion events observed globally may be attributed to low levels of human disturbance, and a lack of life history traits required for success in forests among introduced alien species, such as shade tolerance. However, in contrast to studies of plant invasions on tropical islands, knowledge of these problems in protected areas of continental tropical forests is poor. Here, we synthesise available information on plant invasions in protected areas for the most ecologically important montane forests of East Africa and undertake a hierarchical analysis to explore patterns across the entire mountain chain with those within an individual mountain block down to a single nature reserve. We use this system to illustrate that the vulnerability of tropical forests to invasion is intimately linked to their: (i) marked man-made disturbance; (ii) fragmentation; and (iii) exposure to high propagule pressure of alien species. These human pressures are not unique to continental tropical forests but, faced with high population densities and low incomes, pose a significant challenge in the montane forest of East Africa.

8.2 Plant Invasions and Conservation in the Eastern Arc Mountains

The term 'Eastern Arc Mountains' describes a chain of 13 forest-capped ancient crystalline mountain blocks that stretch for some 900 km from the Makambako Gap, southwest of the Udzungwa Mountains in southern Tanzania to the Taita Hills

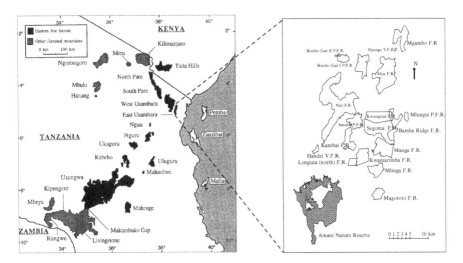

Fig. 8.1 Map of East Africa showing the locations of the 13 blocks that comprise the Eastern Arc Mountains with an inset of the East Usambara mountains depicting the locations and size of the individual forest reserves (Adapted from Doody et al. 2001c)

in south-coastal Kenya (Lovett and Wasser 1993; Fig. 8.1). The Eastern Arc Mountains rise from the coastal plain to 2,635 m in altitude at the highest point (Kimhandu Peak in the Ulugurus). Above about 2,200 m the vegetation is elfin woodland/forest, montane grassland and bog while forest extends down from the frost line at c. 2,400 m to the base of the mountains at around 300 m (Burgess et al. 2007). Four forest formations can be distinguished with upper montane (1,800–2,635 m), montane (1,500–1,800 m) and sub-montane (800–1,500 m) forest while at lower elevations, the forest grades into lowland coastal forest typical of the eastern seaboard of Africa (Lovett and Wasser 1993). Geologically the mountains are formed mainly from Pre-Cambrian basement rocks uplifted about 100 million years ago (Griffiths 1993). Their proximity to the Indian Ocean ensures high rainfall (3,000 mm/year on the eastern slopes of the Ulugurus, falling to 600 mm/year in the western rain shadow). Climatic conditions are believed to have been more-or-less stable for at least the past 30 million years (Axelrod and Raven 1978). The high rainfall and long-term climatic stability, together with the isolation of the individual mountain blocks, have resulted in forests that are both ancient, biologically diverse and exhibit high rates of local endemism. The Eastern Arc Mountains are home to at least 1,500 endemic plant species, 10 endemic mammals, 19 endemic birds, 29 endemic reptiles and 38 endemic amphibians (Lovett and Wasser 1993; Myers et al. 2000; Burgess et al. 2007). Because of the relatively small area of the region the densities of endemic species are among the highest in the world and the region is considered as one of 11 'hyperhot' priorities for conservation investment (Brooks et al. 2002).

The original forest cover (2,000 years ago) on the Eastern Arc Mountains has been estimated to have been around 23,000 km^2, of which approximately 15,000 km^2 remained by 1900, at most only 5,340 km^2 were present by the

Table 8.1 Summary data on forest area, distance to coast, altitude range, forest loss between 1970 and 2000 (%) and the number of endemic vertebrates and trees in each of the 13 major blocks of the East Arc Mountains

Mountain block	Forest area (ha)	Distance to coast (km)	Altitude range (m)	Endemic vertebrates	Endemic Trees	Forest loss 1970–2000
Taita Hills	300	165	1,500–2,140	8	8	N/A
North Pare	2,500	220	1,300–2,113	5	0	12 %
South Pare	13,540	150	820–2,463	8	1	15 %
West Usambara	26,500	100	1,200–2,200	22	27	31 %
East Usambara	25,800	50	130–1,506	35	40	21 %
Nguu	24,900	150	1,000–1,550	9	6	15 %
Nguru	34,000	150	400–2,000	20	25	6 %
Ukaguru	17,400	220	1,500–2,250	10	4	3 %
Rubeho	47,400	300	520–2,050	12	0	9 %
Uluguru	27,000	180	300–2,400	45	26	13 %
Malundwe Hill	450	270	1,200–1,275	0	4	N/A
Mahenge	1,940	300	460–1,040	2	5	5 %
Udzungwa	102,400	300	300–2,580	41	37	5 %

Data from Newmark (2002), Burgess et al. (2007) and Hall et al. (2009)

mid-1990s and by 2005 this had fallen further to only 3,241 km^2 (Newmark 1998; Hall et al. 2009; Table 8.1). Thus, although the degree of loss varies considerably across the mountain blocks, on average around only 15 % of the original forest remains and has become highly fragmented, with a mean patch size of approximately at 10 km^2 (Newmark 1998). Most of the remaining natural habitat on the mountains is found within approximately 150 Government Forest Reserves, with 107 of these managed nationally for water catchment where forest exploitation is not allowed based on the need to preserve the water supply. These catchment forest reserves are found at higher altitudes and thus much of the lower forests have not had protection, such that forest loss has been greatest below 1,000 m altitude. As a result the upper montane zone (>1,800 m) has lost 52 % of its paleoecological forest area, 6 % since 1955, whereas the figures for the submontane habitat (800–1,200 m) are 93 and 57 %, respectively (Hall et al. 2009).

The Tanzanian government has classified 8 Nature Reserves and 83 Forest Reserves (covering 656,815 ha) in the Eastern Arc Mountains in accordance with the IUCN Protected Areas Management Categories system (URT 2010). Three Nature Reserves, Uzungwa Scarp and Kilombero in the Udzungwa, as well as both Amani and Nilo in the East Usambara, have been allocated to IUCN Category Ib (Wilderness Area). Category Ib protected areas are usually large unmodified or slightly modified areas, retaining their natural character and influence, without permanent or significant human habitation, which are protected and managed so as to preserve their natural condition. The remaining Forest Reserves and the four other Nature Reserves, Mkingu in the Nguru, Magamba in the West Usambara,

Uluguru in the Uluguru and Chome in the South Pare have been allocated to IUCN Category IV (Habitat/Species Management Area). Category IV protected areas aim to protect particular species or habitats and management reflects this priority that will need regular, active interventions to address the requirements of particular species or to maintain habitats. The forests of Udzungwa Mountains National Park fall under the IUCN category II protected area (national park). Category II protected areas are large natural or near natural areas set aside to protect large-scale ecological processes, along with the complement of species and ecosystems characteristic of the area, which also provide a foundation for environmentally and culturally compatible spiritual, scientific, educational, recreational and visitor opportunities.

Outside these reserves most forest has been cleared, except in small village burial sites, a few village Forest Reserves, and inaccessible areas. In most Eastern Arc Mountains the local populations respect the reserve boundaries (where they are clear), but about 40 % of total household consumption in some forest adjacent communities is accounted for by forest and woodland products such as firewood, construction material, medicinal herbs, wild fruits and other food materials, thus some forests are heavily degraded. Fire is also a problem and can damage these forests during the dry season (Newmark 2002).

The global biodiversity value of the region is widely recognised (Lovett 1998; Myers et al. 2000; Brooks et al. 2002; Burgess et al. 2007). The Eastern Arc Mountains are one of the 200 Global Ecoregions identified by WWF (Olson and Dinerstein 1998), part of an Endemic Bird Area of BirdLife International (Stattersfield et al. 1998), and a major component of the East Afromontane biodiversity hotspot of Conservation International (Mittermeier et al. 2004). It is also one of the regions of the world facing the most urgent conservation threat and likely to suffer the highest number of plant and vertebrate extinctions for a given loss of habitat (Brooks et al. 2002; Ricketts et al. 2005). As expected, the big forest blocks (Usambara, Uluguru and Udzungwa) are more species-rich than the smaller blocks and have higher numbers of endemic vertebrates and trees (Table 8.1). The Eastern Arc Mountains also contain a number of valuable timber species, such as *Ocotea usambarensis*, *Beilschmiedia kweo* and *Podocarpus* spp., which have been logged on these mountains for more than a century. While all logging is illegal, it has proved difficult to eliminate. Other threats to these forests include the collection of firewood, charcoal production, hunting, and gathering plants for medicine. Intentional burning has been responsible for converting much of the forests in the region to grassland and scrub-grassland. In addition, artisanal mining for gold, rubies and garnets poses a threat to some areas (CEPF 2003).

Not surprisingly, in the face of these major challenges the potential threats posed by alien species to the Eastern Arc Mountains have received relatively limited attention. A prioritised list of the overall threats affecting the Eastern Arc Mountains has been developed through an extensive stakeholder participation process that classified threats in terms of their extent, severity and urgency (URT 2010). Invasive species ranked lowest among ten main pressures believed to threaten biodiversity in the Eastern Arc Mountains (URT 2010), with fire and agricultural expansion being seen as the most important threats, but this view may reflect the limited knowledge regarding the threat of biological invasions.

Systematic assessments of plant invasions have not been undertaken across the Eastern Arc Mountains and current information is usually anecdotal, qualitative and incomplete. Nevertheless, some general patterns do emerge from what information is currently available. On the drier slopes of the Taita Hills, *Cinnamomum camphora*, *Caesalpinia decapetala* and *Acacia mearnsii* are problematic and colonise gaps and edges of the montane forest (Spanhove and Lehouck 2008). The latter two species are also a problem in the drier South Pare mountains, which are vulnerable to colonisation by *Eucalyptus* species if the forest is disturbed or burnt. *Eucalyptus* species are also a problem in the Mahenge as well as the West Usambara mountains where *Leucaena leucocephala* and *Caesalpinia decapetala* have also established in lowland forest. Further south, alien *Rubus* spp. are invading areas of heavily disturbed forest on the Uluguru and western Udzungwa mountains, while the forest edges of the latter have been colonised by *Tectona grandis*. While these patterns appear idiosyncratic there are several species found across a number of different mountain blocks including *Lantana camara* (found in most mountain blocks), *Maesopsis eminii* (Taita Hills, East and West Usambara, Ukaguru, Uluguru mountains), *Cedrela odorata* (Mahenge, East Usambara, Uluguru mountains) and alien *Rubus* spp. (South Pare, Ukaguru, Uluguru, Udzungwa mountains). The need to meet Government targets for forest expansion and land allocation for rapid enhancement of water catchment value in the Eastern Arc Mountains has meant that the establishment of plantations of exotic tree species has been strongly encouraged. Unfortunately, the fast growing species selected for plantations e.g. *Eucalyptus* spp., *Grevillea robusta*, *A. mearnsii*, *Casuarina* spp. have a tendency to naturalise in disturbed forest (Cronk and Fuller 1995). Furthermore, alien species such as *Eucalyptus* spp., *Cedrela odorata*, and *T. grandis* have been used to demarcate forest reserve boundaries and subsequently colonise forest edges. However, while it appears most Eastern Arc Mountain blocks do face some issues with alien plants naturalising in forest edges and gaps following disturbance most evidence to date points to one block, the East Usambara Mountains, being the scene of invasion by multiple alien plant species (Sheil 1994; Dawson et al. 2008). It is therefore to the East Usambaras that we turn our focus to understand better the threats posed by invasive alien plants to the Eastern Arc Mountains. The following sections examine the circumstances underpinning plant invasions in the area, the character of the main invasive species involved, the conflicts such species generate and the scope for management.

8.3   Plant Naturalisation in the East Usambara Mountains, Tanzania

The East Usambara Mountains occupy an area of approximately 86 km^2 in the Tanga Region, Tanzania (Fig. 8.1) and in terms of their species endemism and biodiversity are considered to be one of the most important forest blocks in Africa

(Rodgers and Homewood 1982; Lovett 1989). As in the other blocks of the Eastern Arc Mountains, much of the original forest cover has been lost (Table 8.1). The East Usambaras have been inhabited for more than 2,000 years and the area supports a relatively high human population (101,767 people) distributed across 61 villages. Although forests cover almost half of the East Usambara mountains, much of this is has been exploited for timber or firewood or under-planted with crops (e.g. cardamom) such that only around 25 % of the original dense lowland and sub-montane forest remains (Johansson and Sandy 1996). Dense forest occurs almost exclusively (~95 %) within a network of 17 Forest Reserves and the Amani Nature Reserve. With the exception of a few Village Forest Reserves and those parcels of forest that remain in private hands, outside of the Forest Reserves most of the forest has been cleared for small scale village farms, forestry plantations (mostly alien species: *Eucalyptus* spp., *M. eminii*, *Terminalia ivorensis*) and commercial estates growing cash crops (Johansson and Sandy 1996). The most common estate crops are alien species such as sisal (*Agave sisalana*), tea (*Camellia sinensis*), and cocoa (*Theobroma cacao*). Estate crops are also grown by local villagers as cash crops but a much wider variety of, mostly alien, species are grown in home gardens and include: cardamom (*Elettaria cardamomum*), black pepper (*Piper nigrum*), sugarcane (*Saccharum officinarum*), cloves (*Syzygium aromaticum*), cinnamon (*Cinnamomum verum*), coconuts (*Cocos nucifera*), and coffee (*Coffea* spp.). The food crops cultivated are mainly maize (*Zea mays* spp.), cassava (*Manihot esculenta*), rice (*Oryza sativa*), and yams (*Ipomoea batatas*). Banana (*Musa* spp.) is the most common food and cash crop in the area (Reyes et al. 2005).

Given the dramatic land-use change experienced within this region, the fragmented nature of much of the forest, its continued encroachment by the human population, and the dependence on alien plants in both commercial and local agriculture, we might expect to find a wide variety of naturalised alien plants in the East Usambara mountains. A number of studies have expressed concern regarding the threat that invasive alien plants may represent to biodiversity in the indigenous forests of the East Usambara mountains (Hamilton and Bensted-Smith 1989; Sheil 1994; Dawson et al. 2008). However, a systematic assessment across the East Usambara mountains has yet to be undertaken.

The most comprehensive assessment of the flora of the East Usambara mountains, indicates approximately 270 naturalised plants occur within the region, the vast majority being herbaceous and of neotropical origin (Fig. 8.2a, b, Iversen 1991). As might be expected, given the disturbed nature of the landscape, most naturalised plants are found on waste ground or along roadsides (Fig. 8.2c) and these species include many cosmopolitan or pantropical herbaceous weeds (e.g. *Bidens pilosa*, *Tithonia diversifolia*) or neotropical species that have become pantropical weeds (e.g. *Mimosa pudica*, *Stachytarpheta jamaicensis*). By comparison, relatively few naturalised species are found in forests and indeed alien plants (often shrubs and small trees) are more frequently found in forest edges, clearings or in secondary growth (e.g. *Lantana camara*, *Clidemia hirta*). The relative low occurrence of naturalised alien species in forests is all the more striking given that, even with deforestation, forests of some description still encompass one third of the land cover in the East Usambara mountains.

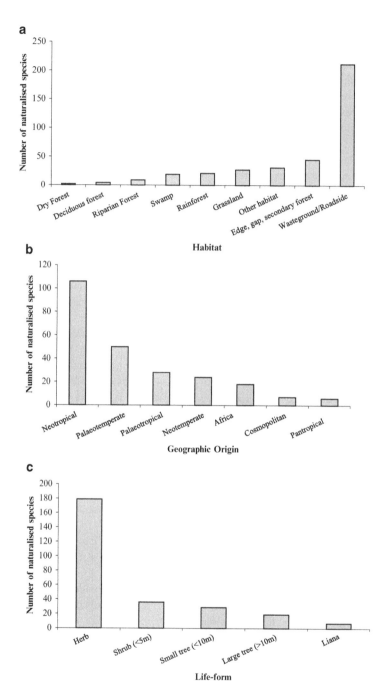

Fig. 8.2 Description of the alien plant species naturalised in the East Usambara mountains in relation to (**a**) habitat, (**b**) geographic origin and (**c**) life-form (Data from Iversen 1991)

A more detailed assessment of naturalised alien plants in forests can be gleaned from systematic vegetation surveys within Amani Nature Reserve and the other Forest Reserves (Table 8.2). Between 1996 and 2002, each protected area was divided into a 450 m × 900 m grid within which one, 50 m × 20 m plot was sampled in each grid square (Huang et al. 2003). Within each sample plot, every tree with a diameter at breast height (dbh) <10 cm was identified. The regeneration layer was recorded within nested 3 m × 3 m and 6 m × 6 m subplots at the centre of each vegetation plot and all plants with a dbh <10 cm were recorded. Although the sampling approach had an intensity of only 0.25 % it provides a means of comparing the prevalence of naturalised alien plants within forests. Unfortunately, there was no systematic attempt to identify the native or alien status of species and conflicting categorisations occur among different surveys as well as between the surveys and checklists of the East Usambara flora e.g. *Mimusops kummel*, *Millettia dura* (Ruffo et al. 1989; Iversen 1991). Three broad groups of taxa can be discerned among the most frequent aliens recorded in the reserves (Table 8.2): fruit crops arising from home gardens (*Artocarpus heterophyllus*, *Citrus aurantium*, *Mangifera indica*, *Psidium guajava*), multi-purpose trees planted around villages (*Albizia chinensis*, *Castilla elastica*, *Leucaena leucocephala*, *Manihot glaziovii*) and escapes from commercial timber plantations (*Maesopsis eminii*).

The mean level of invasion across the reserves was 2.10 % but marked differences existed among reserves with four reserves recording no alien trees whereas over 8.78 % of species in Segoma were alien (Table 8.2). Species richness was positively associated with both the area and elevational range of the reserves, although natives showed the strongest relationship with elevation (Pearson's r = 0.76, df 16, P < 0.01) while aliens were more strongly correlated with area (r = 0.60, df 16, P < 0.01). However, neither of these covariates explained significant variation in the level of invasion. Given the putative sources of many of the aliens observed in the reserves, and the idiosyncratic species composition, it is likely that proximity to villages, agricultural fields and plantations may prove to be better explanatory variables of the level of invasion. For example, the relatively large number of alien species recorded in Segoma is in part due to plants originating from abandoned plantations immediately to the south-east, which primarily consist of citrus and oil palm (*Elais guinensis*) with a number of rubber (*C. elastica*), cocoa and jack fruit (*A. heterophyllus*) trees. A similar explanation exists for Amani Nature Reserve which not only abuts many horticultural plantations but also includes several within the actual reserve boundaries. Nevertheless, a number of crop/ornamental trees appear to regenerate in these reserves such as *Citrus* spp., *M. glaziovii*, *Senna siamea* in Segoma, and *A. pinnata*, *Citrus* spp., *C. elastica*, *C. odorata*, *M. azedarach*, *M. eminii* and *S. campanulata* in Amani. Understanding the potential role of such propagule pressure, and its interaction with species traits, on the likelihood of alien taxa colonising forests requires a detailed history of the location and size of species introductions into the areas surrounding the reserves. For most reserves this information is not available, but for the Amani Nature Reserve a unique opportunity exists to explore these issues through the detailed examination of escapes from the Amani Botanical Garden.

P.E. Hulme et al.

Table 8.2 Details of the 18 forest reserves in the East Usambara mountains with data for reserve area, altitude range, the number of native and alien tree/shrub species and a list of alien species recorded in systematic surveys of the reserves undertaken by Frontier Tanzania

Forest Reserve	Area (ha)	Altitude range (m)	Native tree	Alien tree	Alien tree/shrub species
Amani Nature Reserve	8,380	190–1,130	246	15	*Arenga pinnata, Artocarpus heterophyllus, Castilla elastica, Casuarina equisetifolia, Cedrela odorata, Cinchona succirubra, Citrus* spp., *Eucalyptus saligna, Mangifera indica, Maesopsis eminii, Melia azedarach, Pentadesma butyracea, Spathodea campanulata, Spondia lutea, Toona ciliata*
Bamba Ridge	1,131.5	150–1,033	167	5	*A. heterophyllus, M. indica, Manihot glaziovii, Senna spectabilis, Psidium guajava*
Bombo East I	448.0	220–620	59	0	
Bombo East II	404.0	440–840	47	1	*Ceiba pentandra*
Kambai	1,046.3	200–870	162	0	
Kwamarimba	887.4	95–445	165	1	*Elaeis guineensis*
Kwamgumi	1,708.4	150–915	192	2	*C. pentandra, Citrus aurantium*
Longuza	1,579.9	95–345	106	0	
Magoroto	591.0	650–770	109	3	*Clidemia hirta, M. eminii, Syzygium malaccensis,*
Manga	1,616.0	120–360	115	3	*C. pentandra, M. glaziovii, Leucaena leucocephala*
Mpanga	24.0	650–920	73	3	*C. aurantium, Citrus sinensis, Lantana camara*
Mgambo	1,346.0	320–820	101	1	*A. heterophyllus*
Mlinga	890.0	220–1,069	134	3	*Albizia chinensis, A. heterophyllus, M. eminii*
Mlungui	1,046.3	200–450	56	0	
Mtai	3,107.0	180–1,016	223	2	*A. heterophyllus, M. eminii*
Nilo	6,025.0	400–1,506	207	3	*A. chinensis, A. heterophyllus, C. pentandra, M. eminii, M. glaziovii*
Segoma	1,877.3	80–920	148	13	*Samanea saman, A. heterophyllus, C. elastica, C. aurantium, Citrus limon, E. guineensis, L. leucocephala, M. eminii, P. guajava, Senna siamea, Theobroma cacao*
Semdoe	980.0	95–520	81	1	*C. pentandra*

Cunneyworth (1996a, b, c), Cunneyworth and Stubblefield (1996a, b, c), Doggart et al. (1999a, b, c, 2001), Beharell et al. (2001), Doody et al. (2001a, b, c), Hall et al. (2002), Oliver et al. (2002), and Staddon et al. (2002a, b)

8.4   Amani Botanical Garden and the Escape of Alien Plants into Amani Nature Reserve

While the effects of disturbance (e.g. Colon and Lugo 2006), fragmentation (e.g. Muthuramkumar et al. 2006) and species richness (e.g. Zimmerman et al. 2008) on invasion can often be assessed *post hoc*, the effects of propagule pressure are often difficult to discern since knowledge of the original numbers of individuals introduced is usually not known or difficult to estimate. Few studies, particularly in the tropics, have been able to account for propagule pressure when assessing the drivers of plant invasion. To address this challenge, we describe a unique natural experiment to assess the spread of introduced trees, shrubs and lianas from botanic garden plantations into a surrounding mosaic of differentially disturbed humid forest.

The Amani Botanical Garden (ABG) is situated in the lowland and submontane rainforests of the East Usambara Mountains and effectively embedded within the Amani Nature Reserve. The ABG was formally established under the German administration in 1902, though after the First World War, the British managed the gardens for agricultural research until the early 1950s, when the research station closed and the herbarium was moved to Nairobi (Iversen 1991). Over 600 species (mostly woody) were planted at ABG over a 30-year period from 1902 to 1930 in a series of trial plantations (Dawson et al. 2008). The majority of species were introduced for potential commercial gain, with economic development of the area being the central goal (Iversen 1991). The botanical gardens are spread over some 300 ha, and originally consisted of 20 plantation blocks, divided into 141 compartments; these compartments vary in shape and size, from 0.1 to 7 ha, and originally contained almost 2,000 species plots of varying size (Greenway 1934). However, the collections were subject to near abandon between 1948 and 1993. Accordingly many accessions have been lost over the years and today, approximately only one third of species and plots remain.

Detailed surveys of the original plantations were undertaken by Greenway (1934) who recorded plot location, species survival and, in many cases, the planting date as well as the numbers of individuals planted. Subsequent surveys of extant plots and compartments were undertaken in 1963, 1997 and 2005 (Dawson et al. 2008). This most recent survey found that out of the 214 alien plant species surviving from the original plantings in the early twentieth century over half showed no evidence of regeneration, 35 had set seed and produced seedlings with no recruitment of adult trees, 38 had naturalised within the vicinity of the original plot, while 16 had naturalised widely and were found in many of the botanical garden compartments. Several of these species were also found established within neighbouring secondary forest. Species were significantly more likely to occur in forests if they had naturalised widely rather than locally ($\chi^2 = 15.53$, df 1, $P < 0.0001$) and the latter species were also mostly restricted to forest edges. Of the 11 widely naturalised tree/palm species capable of naturalising in forest, seven were also found in the other forest reserves across the East Usambaras (Table 8.3).

Table 8.3 Alien trees and palms that were introduced to Amani Botanical Garden and have subsequently naturalised widely and established in forest. Information is provided on the species, family and life-form

Species	Family	Life-form
Elaeis guineensis	Arecaceae	Palm
Arenga pinnata	Arecaceae	Palm
Cordia alliodora	Boraginaceae	Tree
Cedrela odorata	Meliaceae	Tree
Toona ciliata	Meliaceae	Tree
Castilla elastica	Moraceae	Tree
Psidium cattleianum	Myrtaceae	Tree
Psidium guajava	Myrtaceae	Tree
Syzygium jambos	Myrtaceae	Tree
Piper aduncum	Piperaceae	Tree
Maesopsis eminii	Rhamnaceae	Tree

Adapted from Dawson et al. (2008)

Thus overall around 7 % of species introduced in ABG satisfy the criteria of being invasive (sensu Richardson et al. 2000) through being able to disperse widely and establish in semi-natural vegetation. This base rate is significantly higher than the 1–2 % observed in many floras (Hulme 2012) and may be indicative of a species pool biased towards species able to establish in tropical forest environments. The proportion of species that had been recorded elsewhere in the world as naturalised/invasive was related to their status in ABG, with 94 % of widely and 79 % of locally naturalising species being recorded as naturalised or invasive in another global region, compared to 57 % of species that were only found regenerating in plots and 49 % of species that showed no evidence of successful regeneration (Dawson et al. 2008).

What determines these differences among species in their likelihood of naturalisation and invasion? Further analysis on the patterns of naturalisation of species in ABG has revealed that species with native ranges centred in the tropics and with larger seeds were more likely to regenerate, whereas local naturalization success was explained by longer residence time, faster growth rate, fewer seeds per fruit, smaller seed mass and shade tolerance (Dawson et al. 2009a). Naturalised species that spread more widely from original plantings tended to have more seeds per fruit, often were dispersed by canopy-feeding animals and had their native ranges centred on the tropics. Species dispersed by canopy feeding animals and with greater seed mass were more likely to be established in closed forest. There was no indication that species succeeding or failing to establish in either disturbed or intact forest differed in the leaf traits that might be associated with shade tolerance (Dawson et al. 2011). There was also no relationship between the degree of herbivory observed on leaves in the field and the likelihood that a species would naturalise widely (Dawson et al. 2009b). However, species establishing in disturbed forest were planted in twice as many plantations while establishment in intact forest was more likely for species planted closer to forest edges. Thus it appears leaf, life-history and dispersal traits may be less important in the colonization of tropical forest than introduction characteristics. Given sufficient propagule pressure or proximity to forest, alien species are much more likely to become established.

In general, the importance of propagule pressure underscores the perils faced by small founder populations early on in the establishment process in which stochasticity in demography, the environment and/or genetics as well as Allee effects will all reduce the probability of establishment of small founder populations (Hulme 2011a). However, a further insight into propagule pressure effects can be gleaned from these studies in Amani Nature Reserve. The Australian Weed Risk Assessment (WRA) protocol was found to successful predict the extent of naturalisation of 214 alien plant species introduced into ABG, rejecting 83 % of widely naturalised species but accepting 74 % of species failing to regenerate (Dawson et al. 2009c). Overall, the WRA score was a good indication of whether a species would spread into open/disturbed habitats but performed poorly at predicting which species might colonise forest. Further examination of these data shows that the WRA score was also significantly correlated with the number of plots planted (Pearson r = 0.15, df 192, P <0.05). Thus the curators preferentially introduced species with higher invasion potential, pre-selecting those that were known to grow well in similar tropical climates and planting them more frequently, compounding the likelihood that the species would be more likely to naturalise. This appears to be a common finding for botanic gardens where species with known invasion histories tend to be found more frequently in collections than more benign species (Hulme 2011b).

Above the threshold introduction effort necessary for population persistence, the relationship between propagule pressure and establishment success should be a function of both how well matched the species is to the prevailing environment and the ecological resistance of the recipient ecosystem. Indeed, propagule pressure may reach such levels that it overcomes ecological resistance (Von Holle and Simberloff 2005). However, disturbance and other habitat attributes are often confounded with propagule pressure, since degraded areas tend to be close to the site of invasion. To explore this interaction, a more detailed study in ABG addressed the importance of propagule pressure and anthropogenic disturbance (evidence of pole cutting, tree stumps etc.) by examining the spread of an intro-duced tree, *Cordia alliodora*, from a single plantation of 210 trees into the sur-rounding mosaic of humid forest (Edward et al. 2009). By assessing vulnerability to invasion along transects radiating from the plantation, the effects of distance (a measure of potential propagule pressure), and forest disturbance were discerned. For all life stages, distance from source population was the strongest correlate of density. A marked influence of disturbance was only found for seedlings. The evidence suggests that propagule pressure was a more important determinant of *C. alliodora* density than disturbance. If this is true for other alien tree species in tropical forests, controlling for introduction effort is essential when assessing the relative importance of different drivers of plant naturalisation.

The foregoing highlights that ABG is the primary source for the introduction of more than one dozen species into closed canopy forests in Amani Nature Reserve. While this is a small proportion of the total number of extant species in ABG, a larger number are colonists of forest edges, riparian vegetation and disturbed forest. It would seem logical that any management plan to address plant invasions into

Amani Nature Reserve would need to address the future of ABG. This is probably true of all plantations proximate to the few remaining fragments of forest in the East Usambara mountains. The scientific value of the living collections in ABG is limited since the majority of collections are of unknown provenance. Thus their value for crop breeding and research is greatly reduced while the extensive collections of coffee and tea have largely eroded with only residual samples surviving. The living collections are of little value to the local villages, apart from some timber harvesting (e.g. *Caesalpinia echinacea*) and the collection of resin (e.g. *Canarium* and *Araucaria*). The main value of the living collections appears to be aesthetics and the opportunity to attract foreign income from tourists who might visit ABG. However, such a rationale is not consistent with aspirations to conserve biodiversity in the Amani Nature Reserve. Thus, in the case of ABG the usual context of botanic gardens supporting conservation seems clearly outweighed by the risks it poses with regarding to escape of alien plants (Hulme 2011c). Nevertheless, the focus on the rehabilitation of ABG remains an important goal in the management plan for Amani Nature Reserve (Sandy et al. 1997; Doody et al. 2001c). A potential solution may be to use tools such as the Australian WRA to identify high risk taxa and prioritise selective removal of these plantations and their 'feral' trees. However, attempts to date to take a pro-active line of attack against invasive alien plants have highlighted conflicts in such an approach and the difficulty in assessing long-term impacts of plant invasions in the East Usambara mountains.

8.5 Is *Maesopsis eminii* a Threat to the East Usambara Mountain Forests?

Although a variety of species have spread into the forests of the Eastern Arc Mountains and could potentially pose a threat to the integrity of these tropical forests, one species in particular, *M. eminii*, has received the lion's share of attention. Although introduced to the Eastern Arc Mountains, *M. eminii* is among the most widely distributed of African lowland forest tree species occurring from West Africa in Togo and Nigeria, to Congo and southern Sudan while in East Africa it occurs naturally from West and Southern Uganda to the north-western regions of Tanzania and western Kenya. While primarily a lowland forest species found in forest gaps and secondary regrowth, *M. eminii* is usually an uncommon tree with the exception of the forest-savannah boundary in Uganda where it may become dominant. Outside of Africa, *M. eminii* has been widely disseminated across the tropics in timber plantations (Puerto Rico and Fiji), as a shade tree for coffee plantations (India, Indonesia) and elsewhere in Asia as an agroforestry tree (Buchholz et al. 2010).

Attributes that probably facilitate *M. eminii* naturalisation and spread include its fast growth and survival on poor soils, ability to reproduce after 4–6 years, and the attractiveness of its fruit to a wide range of vertebrate seed dispersal agents

(Bingelli and Hamilton 1993). The species has naturalised in most regions where it is grown but has only been reported as an invasive of concern in Tanzania (Eastern Arc Mountains and Pemba island) and Puerto Rico (Hall 2010). In the East Usambara mountains, *M. eminii* was initially planted in the early twentieth century at Amani and Longuza from which spread was noted within a decade (Moreau 1935). Having been found to grow well at both sites it was widely planted after logging in Kwamkoro forest and used as a nursery tree for plantations of the endemic timber tree *Cephalosphaera usambarensis* (Bingelli and Hamilton 1993). As primarily a pioneer species, *M. eminii* colonises forest gaps and secondary forests and is believed to be extremely competitive in this context in comparison with other gap species in the East Usambara mountain forests (Binggeli 1989; Cordeiro et al. 2004). The species is now widespread in the East Usambara mountain forests and has been found to comprise over 30 % of large trees (>20 m) in secondary forest and 6 % in more pristine forest (Hall et al. 2011). As a result much of the focus on its potential impact has addressed the question as to whether by colonising forest gaps it is capable of altering forest structure.

If *M. eminii* is able to retain sites following colonisation then the potential exists for it to alter forest succession by limiting opportunities for late-successional native species. The evidence indicates that while *M. eminii* is able to establish under its own canopy through seedling regeneration and epicormic shoots (Binggeli 1989) this is not to the exclusion of late-successional native species (Viisteensaari et al. 2000). Although *M. eminii* was widespread in disturbed forests in the 1980s (Binggeli 1989) more recent evidence suggests that following a subsequent reduction in timber extraction and disturbance, the species has become less abundant though is still common (Cordeiro et al. 2004; Hall et al. 2011). Thus predictions that *M. eminii* would establish a novel type of secondary forest (Binggeli 1989) to the exclusion of many native species have been proved incorrect. However, on the other hand, evidence that native species can regenerate under a *M. eminii* canopy is not in itself sufficient to support the claim made by Viisteensaari et al. (2000) that the alien pioneer poses no risk to the Eastern Arc Mountain forests. Although native primary forest species can establish under the *M. eminii* canopy, whether the community regenerating under these circumstances is similar to that found under native pioneers remains unknown. Reductions in soil quality as a result of high leaf litter decomposition and changes in soil pH have been noted for *M. eminii* (Hall 2010) and may alter which native late successional species successfully colonise these sites. Furthermore, while most attention has been on the regeneration of late successional species, *M. eminii* may have significant impact on the recruitment of native pioneers. Neither of these aspects has been explored but the existing plantations in the Eastern Arc Mountains provide a substantial seed source that may strongly bias the propagule pool colonising forest gaps towards *M. eminii*. Such propagule pressure will be enhanced by the regular and copious seed production combined with effective long-distance seed dispersal by silver-cheeked hornbills (*Ceratogymna brevis*), that can disseminate seeds up to 4 km from source plants (Cordeiro et al. 2004).

Given this scenario, is management to reduce the propagule pressure of *M. eminii* feasible or even desired? Since the declaration of Forest/Nature Reserves was primarily for ensuring their protection in perpetuity, future management plans may consider the felling of *M. eminii*, which might be a solution with benefits to both conservation and local communities. However, currently any harvesting of trees is prohibited in these protected areas. Felling itself causes problems and can facilitate further colonisation of felled area by *M. eminii* and other alien species (Binggeli 1989) and thus should only be considered if followed by enrichment planting with native taxa. Alternatives to felling, such as ring-barking, have been trialled but take more than 4 years to kill trees (Hall 2010), would be labour intensive and provide little net economic benefit to local villagers. Furthermore, national forest policies restricting the use of native hardwood trees has led to farmers planting *M. eminii* as a source of timber and firewood such that they may facilitate its regeneration within their agroforestry systems (Hall et al. 2011). Although *M. eminii* may now be too widespread for eradication or even local control to be feasible, the foregoing emphasises the need for scientifically sound advice regarding not only the potential impact of an alien species on native biodiversity but also an assessment of which mitigation strategies might be most appropriate. Thus, while the direct economic benefits of *M. emini* have not been gauged the species highlights the research, control and social challenges of managing an invasive agroforestry tree in the tropics.

8.6 Aliens in Protected Areas: Lessons Learned from the Eastern Arc Mountains

A common, albeit unsurprising, feature of the occurrence of alien trees in the Eastern Arc Mountains is the overwhelming importance of propagule pressure in the representation of species found colonising forests. Although many species appear to be restricted to disturbed forests, evidence from *C. alliodora* highlights that high propagule pressure may overcome the ecological resistance of relatively undisturbed forest. Given an annual population growth rate of about 3.5 %, equivalent to the population doubling every 20 years, *C. alliodora* poses a potentially significant threat to the East Usambara as well as other humid forests where it is promoted for agroforestry (Edward et al. 2009). Yet similar dire predictions have been made for other alien trees such as *M. eminii*, and this species has revealed that a far greater understanding of the demography and community dynamics is required to assess possible long-term impacts. Such information is urgently required and there is therefore a critical need for building local capacity in the assessment and management of biological invasions in East Africa. However, while resources are low, the Eastern Arc Mountain forests have more staff per hectare and more donor support than any other natural forests in Tanzania (Rodgers 1998; CEPF 2003). Better targeting of capacity building of local rather than overseas staff and towards management rather than cataloguing biodiversity would seem to be a priority. A further, overwhelming challenge may be the conflict between managing invasive species and local livelihoods since, as the studies of Amani Botanical Garden have shown, many agroforestry trees are also likely to colonise secondary and even relatively pristine forest. This conflict is likely to be played out across most tropical regions where low incomes and high population growth rates coincide with high biodiversity. Resolving these conflicts requires not only recognition that alien plants may pose environmental problems in the region but also a level of capacity among local managers and their institutions that will allow them to make their own decisions on managing and monitoring alien plants. This could include retraining of field staff, new research and extension packages as well as revisions to existing management plans regarding alien plants and their management. Capacity building activities related to invasive plant management in the Eastern Arc Mountains should prioritise:

(i) Establishment of research programmes to understand more about invasive plants, their distribution and abundance, temporal trends and assessment of their impact in order to manage them from an informed perspective. (ii) Development, trialling and dissemination of sustainable methods and technical advice for the removal of invasive alien plants that will limit indirect impacts on native biodiversity, re-invasion and water quality issues (soil erosion, herbicides etc.). (iii) Consideration of the potential for using biological control agents to limit spread of invasive alien agroforestry species where such agents do not negatively affect the economics of production (e.g. seed feeders/pathogens where seeds are not a valued crop). (iv) Identification, screening and promotion of alternative species in agroforestry that are preferably native or, if alien, have a low risk of invasion and are easy to contain within agricultural systems. (v) Production and dissemination of information to raise public awareness and capability regarding the identification, impact assessment and management of invasive plant species. (vi) Data management, archiving and sharing to ensure efforts across the Eastern Arc Mountains can be coordinated and lessons learnt in one region transferred to other areas.

A first step would be to consider conducting customised training programmes (stand alone short courses, workshops, and elements in existing courses, etc.) for different stakeholders e.g. policy-makers, scientists, extension workers and affected communities to raise capacity. These training programmes would aim to increase capability in topics such as invasive plant awareness, risk analysis, alien plant identification, invasive species management, data management, accessing and using global invasive species information sources, communication and teaching of invasive species issues, and promotion of both compliance as well as enforcement of invasive species guidelines. Furthermore, there remains a requirement for the Tanzanian government to facilitate the implementation of effective invasive species management programmes including setting of invasive plant species classification standards, public awareness, advocacy and organisational skills. Thus there is also a need to strengthen the enabling policy environment for invasive alien species management. It has long been recognised that government and community partnerships, supported by international, national and local NGOs will be essential to resolve large scale environmental problems in the Eastern Arc Mountains (Rodgers 1998). While these partnerships now exist in the region they have not been mobilised to address the conflicts arising from the use of alien species. Hopefully, our appraisal of the current situation regarding alien plant species in the Eastern Arc Mountains will provide the stimulus and impetus for greater momentum in initiatives to address plant invasions in the region.

References

Axelrod DI, Raven PH (1978) Late Cretaceous and Tertiary vegetation history of Africa. In: Werger MJA (ed) Biogeography and ecology of Southern Africa. Dr. W. Junk Publications, The Hague, pp 77–130

Beharrell NK, Fanning E, Howell KM (eds) (2001) Nilo Forest Reserve: a biodiversity survey. East Usambara Conservation Area Management Programme, technical paper no. 53. Frontier Tanzania, Dar es Salaam

Binggeli P (1989) The ecology of *Maesopsis* invasion and dynamics of the evergreen forest of the East Usambaras, and their implications for forest conservation and forest practices. In: Hamilton AC, Bensted-Smith R (eds) Forest conservation in the East Usambara Mountains, Tanzania. IUCN, Gland, pp 269–300

Binggeli P, Hamilton AC (1993) Biological invasions by *Maesopsis eminii* in the East Usambara forests, Tanzania. Opera Bot 121:229–235

Brooks TM, Mittermeier RA, Mittermeier CG et al (2002) Habitat loss and extinction in the hotspots of biodiversity. Conserv Biol 16:909–923

Buchholz T, Tennigkeit T, Weinreich A (2010) Single tree management models: *Maesopsis eminii*. In: Bongers F, Tennigkeit T (eds) Degraded forests in Eastern Africa: management and restoration. Earthscan, London, pp 247–266

Burgess ND, Butynski TM, Cordeiro NJ et al (2007) The biological importance of the Eastern Arc Mountains of Tanzania and Kenya. Biol Conserv 134:209–231

CEPF (2003) Ecosystem profile: Eastern Arc Mountains and coastal forests of Tanzania and Kenya biodiversity hotspot. Critical Ecosystem Partnership Fund, Washington, DC

Colon SM, Lugo AE (2006) Recovery of a subtropical dry forest after abandonment of different land uses. Biotropica 38:354–364

Cordeiro NJ, Patrick DAG, Munisi B et al (2004) Role of dispersal in the invasion of an exotic tree in an East African submontane forest. J Trop Ecol 20:449–457

Cronk Q, Fuller J (1995) Plant invaders: the threat to natural ecosystems. Chapman and Hall, New York

Cunneyworth P (ed) (1996a) Kwamarimba Forest Reserve: a biodiversity survey. East Usambara Conservation Area Management Programme, technical paper no. 33. Frontier Tanzania, Dar es Salaam

Cunneyworth P (ed) (1996b) Longuza (north) Forest Reserve: a biodiversity survey. East Usambara Conservation Area Management Programme, technical paper no 34. Frontier Tanzania, Dar es Salaam

Cunneyworth P (ed) (1996c) Kambai Forest Reserve: a biodiversity survey. East Usambara Conservation Area Management Programme, technical paper no. 35. Frontier Tanzania, Dar es Salaam

Cunneyworth P, Stubblefield L (eds) (1996a) Magaroto Forest: a biodiversity survey. East Usambara Conservation Area Management Programme, technical paper no. 30. Frontier Tanzania, Dar es Salaam

Cunneyworth P, Stubblefield L (eds) (1996b) Bamba Ridge Forest Reserve: a biodiversity survey. East Usambara Conservation Area Management Programme, technical paper no 31. Frontier Tanzania, Dar es Salaam

Cunneyworth P, Stubblefield L (eds) (1996c) Mlungui Proposed Forest Reserve: a biodiversity survey. East Usambara Conservation Area Management Programme, technical paper no 32. Frontier Tanzania, Dar es Salaam

Dawson W, Mndolwa AS, Burslem DFRP et al (2008) Assessing the risks of plant invasions arising from collections in tropical botanical gardens. Biodivers Conserv 17:1979–1995

Dawson W, Burslem DFRP, Hulme PE (2009a) Factors explaining alien plant invasion success in a tropical ecosystem differ at each stage of invasion. J Ecol 97:657–665

Dawson W, Burslem DFRP, Hulme PE (2009b) Herbivory is related to taxonomic isolation, but not to invasiveness of tropical alien plants. Divers Distrib 15:141–147

Dawson W, Burslem DFRP, Hulme PE (2009c) The suitability of weed risk assessment as a conservation tool to identify invasive plant threats in East African rainforests. Biol Conserv 142:1018–1024

Dawson W, Burslem DFRP, Hulme PE (2011) The comparative importance of species traits and introduction characteristics in tropical plant invasions. Divers Distrib 17:1111–1121

Doggart N, Dilger M, Cunneyworth P, Fanning E (eds) (1999a) Kwamgumi Forest Reserve: a biodiversity survey. East Usambara Conservation Area Management Programme, technical paper no. 40. Frontier Tanzania, Dar es Salaam

Doggart N, Dilger M, Kilenga R, Fanning E (eds) (1999b) Mtai Forest Reserve: a biodiversity survey. East Usambara Conservation Area Management Programme, technical paper no. 39. Frontier Tanzania, Dar es Salaam

Doggart N, Joseph L, Bayliss J, Fanning E (eds) (1999c) Manga Forest Reserve: a biodiversity survey. East Usambara Conservation Area Management Programme, technical paper no. 41. Frontier Tanzania, Dar es Salaam

Doggart NH, Doody KZ, Howell KM, Fanning E (eds) (2001) Semdoe Forest Reserve: a biodiversity survey. East Usambara Conservation Area Management Programme, technical paper no. 42. Frontier Tanzania, Dar es Salaam

Doody KZ, Beharrell NK, Howell KM, Fanning E (eds) (2001a) Mpanga Village Forest Reserve: a biodiversity survey. East Usambara Conservation Area Management Programme, technical paper no 51. Frontier Tanzania, Dar es Salaam

Doody KZ, Fanning E, Howell KM (eds) (2001b) Segoma Forest Reserve: a biodiversity survey. East Usambara Conservation Area Management Programme, technical paper no. 50. Frontier Tanzania, Dar es Salaam

Doody KZ, Howell KM, Fanning E (eds) (2001c) Amani Nature Reserve: a biodiversity survey. East Usambara Conservation Area Management Programme, technical paper no 52. Frontier Tanzania, Dar es Salaam

Edward E, Munishi PKT, Hulme PE (2009) Relative roles of disturbance and propagule pressure on the invasion of humid tropical forest by *Cordia alliodora* (Boraginaceae) in Tanzania. Biotropica 41:171–178

Fine PVA (2002) The invasibility of tropical forests by exotic plants. J Trop Ecol 18:687–705

Greenway PJ (1934) Report of a botanical survey of the indigenous and exotic plants in cultivation at the East African Agricultural Research Station, Amani, Tanganyika Territory. Unpublished typescript, Royal Botanic Gardens, Kew

Griffiths CJ (1993) The geological evolution of Eastern Africa. In: Lovett JC, Wasser SK (eds) Biogeography and ecology of the rain forest of Eastern Africa. Cambridge University Press, Cambridge, pp 9–21

Hall JB (2010) Future options for *Maesopsis*: agroforestry asset or conservation catastrophe? In: Bongers F, Tennigkeit T (eds) Degraded forests in Eastern Africa: management and restoration. Earthscan, London, pp 221–246

Hall SM, Fanning E, Howell KM, Pohjonen V (eds) (2002) Mlinga Forest Reserve: a biodiversity survey. East Usambara Conservation Area Management Programme, technical paper no. 56. Frontier Tanzania, Dar es Salaam

Hall J, Burgess N, Lovett JC et al (2009) Conservation implications of deforestation across an elevational gradient in the Eastern Arc Mountains, Tanzania. Biol Conserv 142:2510–2521

Hall JM, Gillespie TW, Mwangoka M (2011) Comparison of agroforests and protected forests in the East Usambara Mountains, Tanzania. Environ Manag 48:237–247

Hamilton AC, Bensted-Smith R (1989) Forest conservation in the East Usambara Mountains, Tanzania. IUCN, Gland

Huang WD, Pohjonen V, Johansson S et al (2003) Species diversity, forest structure and species composition in Tanzanian tropical forests. For Ecol Manag 173:11–24

Hulme PE (2011a) Biosecurity: the changing face of invasion biology. In: Richardson DM (ed) Fifty years of invasion ecology – the legacy of Charles Elton. Blackwells, Oxford, pp 301–314

Hulme PE (2011b) Addressing the threat to biodiversity from botanic gardens. Trends Ecol Evol 26:168–174

Hulme PE (2011c) Botanic garden benefits do not repudiate risks: a reply to Sharrock et al. Trends Ecol Evol 26:434–435

Hulme PE (2012) Weed risk assessment: a way forward or a waste of time? J Appl Ecol 49:10–19

Iversen ST (1991) The Usambara mountains, NE Tanzania: phytogeography of the vascular plant flora. Symbolae Botanicae Upsaliensis 29:1–234

Johansson SG, Sandy R (1996) Protected areas and public lands - land use in the East Usambara. East Usambara Catchment Forest Project technical paper no 28. Forestry and Beekeeping Division, Dar es Salaam

Lovett JC (1989) The botanical importance of the East Usambara forests in relation to other forests in Tanzania. In: Hamilton AC, Bensted-Smith R (eds) Forest conservation in the East Usambara Mountains, Tanzania. IUCN, Gland, pp 207–212

Lovett JC (1998) Importance of the Eastern Arc Mountains for vascular plants. J East Afr Nat Hist 87:59–74

Lovett JC, Wasser SK (1993) Biogeography and ecology of the rain forests of Eastern Africa. Cambridge University Press, Cambridge

Lugo AE (2004) The outcome of alien tree invasions in Puerto Rico. Front Ecol Environ 2:265–273

Mittermeier RA, Robles Gil P, Hoffmann M et al (2004) Hotspots revisited. CEMEX, Mexico

Moreau RE (1935) A synecological study of Usambara, Tanganyika territory, with particular reference to birds. J Ecol 23:1–43

Muthuramkumar S, Ayyappan N, Parthasarathy N et al (2006) Plant community structure in tropical rain forest fragments of the Western Ghats, India. Biotropica 38:143–160

Myers N, Mittermeier RA, Mittermeier CG et al (2000) Biodiversity hotspots for conservation priorities. Nature 403:853–858

Newmark WD (1998) Forest area, fragmentation and loss in the Eastern Arc Mountains: implications for the conservation of biological diversity. J East Afr Nat Hist 87:29–36

Newmark WD (2002) Conservation biodiversity in Eastern African forests: a study of the Eastern Arc Mountains, Ecological studies 155. Springer, Berlin

Oliver SA, Bracebridge CE, Fanning E, Howell KM (eds) (2002) Mgambo Forest Reserve: a biodiversity survey. East Usambara Conservation Area Management Programme, technical paper no. 59. Frontier Tanzania, Dar es Salaam

Olson DM, Dinerstein E (1998) The Global 200: a representation approach to conserving the Earth's most biologically valuable ecoregions. Conserv Biol 12:502–515

Rejmánek M (1996) Species richness and resistance to invasions. In: Orians RD, Dirzo R, Cushman JH (eds) Diversity and processes in tropical forest ecosystems. Springer, New York, pp 153–172

Rejmánek M, Richardson DM, Higgins SI et al (2005) Ecology of invasive plants: state of the art. In: Mooney HA, McNeely JA, Neville L et al (eds) Invasive alien species: a new synthesis. Island Press, Washington, DC, pp 104–162

Reyes T, Quiroz R, Msikula S (2005) Socio-economic comparison between traditional and improved cultivation methods in agroforestry systems, East Usambara Mountains, Tanzania. Environ Manag 36:682–690

Richardson DM, Pyšek P, Rejmánek M et al (2000) Naturalization and invasion of alien plants: concepts and definitions. Divers Distrib 6:93–107

Ricketts TH, Dinerstein E, Boucher T et al (2005) Pinpointing and preventing imminent extinctions. Proc Natl Acad Sci U S A 102:18497–18501

Rodgers WA (1998) An introduction to the Eastern Arc Mountains. J East Afr Nat Hist 87:7–18

Rodgers WA, Homewood KM (1982) The conservation of the East Usambara Mountains, Tanzania: a review of biological values and land use pressures. Biol J Linn Soc 24:285–304

Ruffo CK, Mmari C, Kibuwa SP et al (1989) A preliminary list of plant species recorded from the East Usambara forests. In: Hamilton AC, Bensted-Smith R (eds) Forest conservation in the East Usambara Mountains, Tanzania. IUCN, Gland, pp 156–184

Sandy RF, Boniface G, Rajabu I (1997) A survey and inventory of the Amani Botanical Garden. East Usambara Catchment Forest Project, technical paper no. 38. Forestry and Beekeeping Division, Dar es Salaam

Sheil D (1994) Naturalised and invasive plant species in the evergreen forests of the East Usambara Mountains, Tanzania. Afr J Ecol 32:66–71

Spanhove T, Lehouck V (2008) Don't miss the invasions! A note on forest health monitoring in the Taita Hills, Kenya. J East Afr Nat Hist 97:255–256

Staddon S, Fanning E, Howell KH (eds) (2002a) Bombo East I forest reserve: a biodiversity survey. East Usambara Conservation Area Management Programme, technical paper no. 57. Frontier Tanzania, Dar es Salaam

Staddon S, Fanning E, Howell KM (eds) (2002b) Bombo East II Forest Reserve: a biodiversity survey. East Usambara Conservation Area Management Programme, technical paper no. 58. Frontier Tanzania, Dar es Salaam

Stattersfield AJ, Crosby MJ, Long AJ et al (1998) Endemic bird areas of the world: priorities for biodiversity conservation. BirdLife International, Cambridge

URT (2010) Nomination of properties for inclusion on the World Heritage List serial nomination: Eastern Arc Mountains forests of Tanzania. UNESCO Convention Concerning the Protection of the World Cultural and Natural Heritage. United Republic of Tanzania, Ministry of Natural Resources and Tourism, Dar es Salaam, Tanzania

Viisteensaari J, Johansson S, Kaarakka V et al (2000) Is the alien tree species Maesopsis eminii Engl. (Rhamnaceae) a threat to tropical forest conservation in the East Usambaras, Tanzania? Environ Conserv 27:76–81

Von Holle B, Simberloff D (2005) Ecological resistance to biological invasion overwhelmed by propagule pressure. Ecology 86:3213–3218

Zimmerman N, Hughes RF, Cordell S et al (2008) Patterns of primary succession of native and introduced plants in lowland wet forests in eastern Hawai'i. Biotropica 40:277–284

Chapter 9
Invasive Plants in the Floodplains
of Australia's Kakadu National Park

Samantha A. Setterfield, Michael M. Douglas, Aaron M. Petty,
Peter Bayliss, Keith B. Ferdinands, and Steve Winderlich

Abstract Kakadu National Park is Australia's premier protected area and one of the
few World Heritage areas listed for both its natural and cultural heritage values.
Kakadu National Park encompasses vast areas of seasonally inundated wetlands that
support an outstanding abundance of biodiversity, particularly birds and fish. The
wetlands provide critical resources for the Indigenous landowners and are also a major
tourist attraction. The international importance of Kakadu National Parks' wetlands is
also reflected by their listing under the Ramsar Wetlands Convention. Unfortunately,
these wetlands are under substantial threat from a range of high impact invasive alien
plants. The response of managers to different invasive alien plants has varied substantially. For example, the response by Kakadu National Park managers to the threat
from the alien shrub *Mimosa pigra* has widely been used as a case study of best

S.A. Setterfield (✉) • M.M. Douglas • A.M. Petty
National Environmental Research Program, Northern Australia Hub,
Charles Darwin University, Darwin, NT 0909, Australia
e-mail: samantha.setterfield@cdu.edu.au; michael.douglas@cdu.edu.au;
aaron.petty@cdu.edu.au

P. Bayliss
National Environmental Research Program, Northern Australia Hub,
Charles Darwin University, Darwin, NT 0909, Australia

CSIRO, GPO Box 2583, Brisbane, QLD 4001, Australia
e-mail: peter.bayliss@csiro.au

K.B. Ferdinands
National Environmental Research Program, Northern Australia Hub,
Charles Darwin University, Darwin, NT 0909, Australia

Department of Land Resource Management, Weeds Management Branch,
Palmerston, NT 0831, Australia
e-mail: keith.ferdinands@nt.gov.au

S. Winderlich
Kakadu National Park, PO Box 71, Jabiru, NT 0886, Australia
e-mail: steve.winderlich@environment.gov.au

L.C. Foxcroft et al. (eds.), *Plant Invasions in Protected Areas: Patterns, Problems*   167
and Challenges, Invading Nature - Springer Series in Invasion Ecology 7,
DOI 10.1007/978-94-007-7750-7_9, © Springer Science+Business Media Dordrecht 2013

practice. The response was rapid, appropriately resourced, consistent over time and well-monitored. In contrast, the response to two aquatic invasive alien grass species, *Hymenachne amplexicaulis* and *Urochloa mutica*, has been relatively poor. Subsequently, whereas *M. pigra* remains under control, with a limited number of small infestations, the alien grasses have spread extensively in recent years and now pose a substantial threat. This chapter explores the history, invasion and management response to invasive alien grass management in Kakadu National Park. We suggest actions that should commence immediately to avoid wasting the past efforts made to save Kakadu National Park's wetland ecosystems from *M. pigra*, and prevent their conversion into invasive alien grass dominated systems.

Keywords Australia • Best practice • *Hymenachne amplexicaulis* • Kakadu National Park • *Mimosa pigra* • *Urochloa mutica*

9.1 Introduction

A protected area (PA) is defined "as a geographical space, recognised, dedicated and managed, through legal or other effective means, to achieve the long-term conservation of nature with associated ecosystem services and cultural values" (Dudley 2008). In Australia, this includes mainland and offshore PAs managed by the Federal Government, and PAs within each of the six States and two Territories of Australia. The nationwide network of Australia's parks and reserves is called the National Reserve System, which was established in 1992. It brings together the range of parks, reserves, private PAs and Indigenous PAs, into a single system to conserve Australia's unique biodiversity (Sattler and Taylor 2008). The system now covers nearly 13 % of the Australian land mass. Across such a large network of PAs, managers have to respond to numerous threats, including climate change, invasive alien plants (IAPs) and animals, fire and mining. In this chapter we focus on the threat and management of IAPs on the extensive floodplains within Australia's most iconic PA, Kakadu National Park (KNP).

9.2 Kakadu National Park: Australia's Premier Protected Area

Kakadu National Park (Fig. 9.1), located approximately 200 km east of Darwin in Australia's Northern Territory, is the country's largest national park (Wellings 2007). It is one of the world's premier PAs and also one of the few World Heritage areas listed for both its natural and cultural values (Wellings 2007). Kakadu National Park was declared in three stages between 1979 and 1991 (Press and Lawrence 1995). Prior to becoming a national park, parts of KNP were under grazing lease for cattle (*Bos primigenius*). Large herds of feral buffalo (*Bubalus*

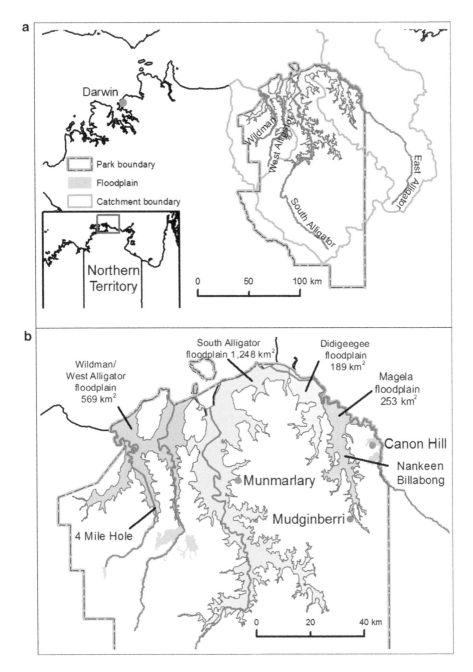

Fig. 9.1 Location of (**a**) Kakadu National Park, 200 km east of Darwin in Australia's Northern Territory and (**b**) the location of the major seasonally inundated rivers and their floodplain: West Alligator, Wildman, South Alligator and East Alligator

Fig. 9.2 Magela floodplain, inundated during the wet season, showing typical patchy mosaic of mixed grasses, sedges and water lily communities (Photo Michael Douglas)

bubalis) were also exploited for meat, hides and horns (Levitus 1995). As a result, some of the management problems in KNP, such as ecosystem disturbance and invasions of some alien plants, are a legacy of the previous land use and management practices (Cowie and Werner 1993).

Kakadu National Park is often described as a living cultural landscape because it has been continuously inhabited for at least 50,000 years (Roberts et al. 1993; Roberts and Jones 1994). Approximately half of KNP is Aboriginal land under the Aboriginal Land Rights (Northern Territory) Act 1976, and most of the remaining area of land is under claim by Aboriginal people. Management of KNP occurs under a co-management arrangement between the Aboriginal traditional owners (Bininj/Munguy) and the Australian Government (Director of National Parks), through a joint Board of Management (Wellings 2007).

Kakadu National Park contains a range of landscapes, from the coastal estuaries in the north, through extensive seasonally flooded wetlands, to vast areas of *Eucalyptus*-dominated tropical savanna forests and woodlands, through to the Arnhem Land sandstone plateau in the east (Russell-Smith 1995). Many of the ecological features of KNP result from its location within the monsoonal tropics, with its distinct high-rainfall wet season followed by extended drought in the dry season (Taylor and Tulloch 1985). Kakadu National Park receives 1,300–1,500 mm of rain, with most occurring between November and April. This concentrated period of rain in an area of low topographic relief has led to the formation of the extensive wetlands in the region (Finlayson et al. 1990). Each wet season, water spills over the river levees, slows and spreads out to fill their vast, shallow floodplains to a depth of several meters (Douglas et al. 2005; Fig. 9.2). The

Fig. 9.3 Cracking clay
soils of the floodplain
during the extended dry
season (Photo Aaron Petty)

Fig. 9.3 Cracking clay soils of the floodplain during the extended dry season (Photo Aaron Petty)

floodwaters gradually recede or evaporate over the dry season so that most of the floodplains have no surface water from August to December (Douglas et al. 2005; Fig. 9.3). The KNPs wetlands occur within the catchments of the Wildman, West Alligator, South Alligator and East Alligator Rivers (Fig. 9.1).

9.3   Ecology, Function and Significance of Kakadu National Park's Floodplains

As the floodplains become inundated, there is rapid aquatic plant growth. The extensive floodplains support a diverse mosaic of grass and/or sedge dominated communities reflecting geomorphic features, including drainage depressions, permanent and semi-permanent swamps and billabongs, and a wide range of more briefly inundated systems (Finlayson et al. 1990; Cowie et al. 2000). Many plant species are widely distributed in other tropical regions of the world (Pettit et al. 2011), with common species including *Oryza* spp. (wild rice), *Eleocharis* spp. (spike-rush), *Hymenachne acutigluma* (native hymenachne), *Pseudoraphis spinescens* (water couch) and water lilies (*Nymphaea* spp. and *Nymphoides* spp.; Finlayson 2005). As the floodwater recedes many plants senesce, often surviving the dry season as dormant seeds, tubers or corms (Finlayson 2005). The Northern Territory floodplain flora is generally divided into three broad groupings; saline/semi-saline, dry freshwater and wet freshwater (Wilson et al. 1991; Cowie et al. 2000). The invasive alien grasses discussed in this chapter preferentially invade and transform the freshwater communities. These communities are typically dominated by native grass, for example, *Leersia hexandra* (southern cut grass), *Oryza* spp., *P. spinescens*, *H. acutigluma* and *Eleocharis dulcis* (water chestnut – sedges; Cowie et al. 2000).

The seasonal wetting and drying of the KNP floodplains supports a high diversity of species (Finlayson et al. 2006). Freshwater fish abundance and diversity in KNP are both high by Australian standards (Allen et al. 2002), with the greatest fish diversity within KNP supported by the river channel and floodplain environments (Cowie et al. 2000). The floodplains are internationally known for the number and diversity of water birds that they support, providing food resources for many of these species, and are also important breeding sites for five migratory species, magpie geese (*Anseranas semipalmata*), plumed whistling-duck (*Dendrocygna eytoni*), wandering whistling-duck (*Dendrocygna arcuata*), radjah shellduck (*Tadorna radjah*) and comb-crested jacana (*Irediparra gallinacean*; Bayliss and Yeomans 1990; Chatto 2006; Finlayson et al. 2006). Over 172,000 water birds were counted within the upstream South Alligator River floodplains during October 2001 (Chatto 2006). Up to 27 % of the Northern Territory's breeding water bird population is within KNP, with the South Alligator floodplains regarded as the third most important nesting habitat area in the Northern Territory (Bayliss and Yeomans 1990). The transformation of these wetlands thus poses substantial ramifications for a range of species.

As indicated, KNP is jointly managed by its Indigenous traditional owners. The integrity of the wetlands is also important because they provide traditional foods ('bush tucker'), such as magpie goose, fish, turtles (*Chelodina rugosa*), *Nelumbo nucifera* (lotus lily), and other products such as wood and fibre for weapons, utensils, weaving, and traditional medicines (Lucas and Russell-Smith 1993). In addition, there are culturally significant sites on the floodplains and access to these sites and their maintenance is important. Kakadu National Park attracts national and international tourists, making it an important economic asset to the Northern Territory and Australia (Tremblay 2007).

9.4 The Threat from Invasive Alien Plants

The seasonally inundated floodplain ecosystems are being substantially impacted by three major IAPs: *M. pigra*, a woody shrub, and two grasses, *U. mutica* (para grass) and *Hymenachne amplexicaulis* (olive hymenachne, Figs. 9.4a–c, and 9.5). *Mimosa pigra* (giant sensitive plant) and *H. amplexicaulis* are included in the list of Australia's Weeds of National Significance (Thorp and Lynch 2000). *Urochloa mutica* and *H. amplexicaulis* are listed as part of a group of five species recognised as a Key Threatening Process under Australia's Environmental Protection and Biodiversity Conservation Act (EPBC; Anonymous 2009; Department of Sustainability, Water, Population and Communities 2009). We describe the differences in the introduction, spread and management response to these species, and the likely consequences over the next few decades if current management approaches do not change substantially.

Fig. 9.4 (a) Infestations of
Mimosa pigra outside
Kakadu National Park
convert grassland to
shrubland (**b**) *Urochloa
mutica* forms dense
monocultures displacing
native grasses and sedges
(**c**) *Hymenachne
amplexicaulis* also forms
dense monocultures. It is
listed as one of Australia's
weeds of national
significance. This photo is
from the Mary River
floodplain, neighbouring
Kakadu National Park, and is
an analogue of the invasion
that may occur in KNP if this
species is not controlled
(Photo (**a**) Michael Douglas
(**b**) Michael Douglas
(**c**) Aaron Petty)

174 S.A. Setterfield et al.

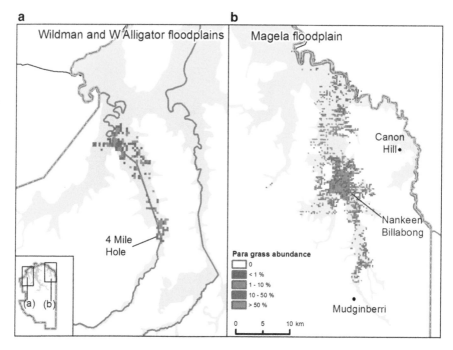

Fig. 9.5 Distribution of *Urochloa mutica* on the (**a**) Wildman and West Alligator floodplains, and (**b**) Magela floodplain in Kakadu National Park, recorded from helicopter survey (250 × 250 m pixels)

9.4.1 Case Study 1: Control of Mimosa pigra: *An Example of Best Practice Invasive Alien Plant Management*

Mimosa pigra is an invasive woody shrub, native to tropical America, which was probably first introduced to the Botanical Gardens in Darwin in the mid-nineteenth century (Miller and Lonsdale 1987). It is a hard-seeded, thorny, leguminous shrub which grows up to 6 m in northern Australia; substantially taller than in its native range where it grows to 1–2 m. It is considered to have demonstrated a 'textbook' lag-phase of IAP invasion. There was reportedly relatively little spread around Darwin for 60–80 years after its introduction, followed by a large infestation being discovered in 1952 in the upper reaches of the Adelaide River, 100 km south of Darwin (Miller et al. 1981). Despite *M. pigra's* limited distribution, its potential impact was recognised and it was declared a noxious weed in the Northern Territory in 1966 (Miller et al. 1981). However, irrespective of its status as "a plant that should be eradicated" (Miller et al. 1981) and the limited number of populations, the resources available for management were insufficient and sporadic. As a result, *M. pigra* soon spread downstream of the upper reaches of the Adelaide River resulting in "plants scattered over approximately 4,000 ha" in 1980 (Miller et al. 1981). Between 1980 and 1989 *M. pigra* spread from 4,000 ha to 80,000 ha in the Northern Territory (NT Government 1997). Although the spread of *M. pigra* has been reduced by the use of an integrated management approach, including herbicide, physical and biological control (Paynter and Flanagan 2004), it is now estimated that approximately 140,000 ha of the Northern Territory have been invaded (Burrows and Lukitsch 2012). This includes significant areas of floodplains bordering Kakadu; approximately 10,000 ha in the Mary River on the western boundary and 4,000 ha in the East Alligator/Murganella-Cooper East Alligator (Walden et al. 2004).

When comparing the successful management of *M. pigra* in KNP to other IAPs, it is important to recognise that when the first *M. pigra* plants were discovered in Kakadu in 1981, it was already a declared noxious weed and land managers had a legislative requirement to manage it (Miller et al. 1981). It was unequivocally considered an IAP in Australia with detrimental impacts for all users of the floodplains. Additionally, evidence had been documented to demonstrate that control was extremely difficult to manage once infestations were well established (Miller and Lonsdale 1987). As such, the threat to KNPs wetlands was clearly recognised and park management immediately implemented a strong management response. Initial *M. pigra* infestations were located by staff or visitors, but a systematic survey and eradication programme was soon implemented (Cook et al. 1996). In 1984 two people were employed, and within 2 years this increased to a team of four people dedicated to *M. pigra* control (Cook et al. 1996). In a continuing example of best practice management, the four-person team continues to operate to this day, locating, mapping and eradicating new populations (Hunter et al. 2010). The team currently monitors approximately 260 plots, visiting each three times a year.

Kakadu National Park's approach to *M. pigra* control formed part of an effective regional strategy because the KNP infestations were the satellite outliers, or the 'nascent foci' (sensu Moody and Mack 1988) of large main infestations in Oenpelli floodplain to the east and Mary River floodplain to west (Cook et al. 1996). Initially, control efforts on the floodplains neighbouring KNP focused on controlling the large established populations with limited success, as the outliers expanded rapidly. For example, in the Oenpelli floodplain, *M. pigra* expanded from several plants to nearly 6,000 ha between 1984 and 1991, doubling in area every 1.4 years (Cook et al. 1996). Consequently, by the time control was attempted at Oenpelli it required millions of dollars to implement. Between 1991 and 1996, nearly $7 million of Australian Government funds (along with $2.1 million of NT Government in-kind support) were committed to the control of the Oenpelli infestation, with over 60 tonnes of herbicide applied over 5 years (Walden et al. 2004). On-going funds for follow-up programmes on the Oenpelli floodplain have also been provided, and are particularly important given the longevity (up to 20 years) of the seed (Lonsdale et al. 1988).

The contrast between the approach in Oenpelli and KNP is significant. Storrs et al. (1999) calculated the cost of control in KNP to approximately $2 ha^{-1} year^{-1}, whereas the spray programme to control the large *M. pigra* infestation at Oenpelli cost \$220 ha^{-1} year^{-1} for 5 years, with follow-up control requiring funds of at least the same cost as the KNP programme. The KNP control approach became a model approach throughout the region (Storrs et al. 1999).

9.4.2   Case Study 2: Management of Urochloa mutica: A Legacy of Past Land Use

The management response to *M. pigra* in KNP and the broader region of northern Australia provides a stark contrast to that for *U. mutica*. It also provides a valuable lesson for the management of multiple IAPs in other high value conservation areas.

Urochloa mutica is a semi-aquatic stoloniferous grass that was introduced to Queensland, Australia, from Brazil or North Africa in the 1880s (Douglas and O'Connor 2004). The first record of *U. mutica* in the Northern Territory was from the Darwin Botanical Gardens in the late 1800s (Wesley-Smith 1973). *Urochloa mutica* was promoted as a desirable replacement to native grasses on the floodplains because of a number of attributes, such as rapid spread, high yield, tolerance of waterlogging and drought, and recovery from heavy grazing (Anning and Hyde 1987). It is promoted as being highly palatable and capable of supporting a larger carrying capacity than native pastures (Cameron and Lemcke 1996), although this has been contested (Calder 1981; Rea and Storrs 1999). More recently, *U. mutica* was also used to suppress the regrowth of *M. pigra* following chemical treatment (Miller and Lonsdale 1992; Grace et al. 2004) and to stabilise earthworks, such as small dams constructed to provide water for cattle and fire fighting (Beggs 2012).

The negative impacts of *U. mutica* were first recognised at least 70 years ago in agricultural systems. Winders (1937) noted that "para grass has come to be looked upon with disfavour in certain areas (particularly on the Upper Tweed River in New South Wales), because of its habit of invading irrigation channels and small streams and impeding the flow of water. The grass may also assume pest proportions". By the 1960s, it was considered to be one of the worst IAPs in irrigation channels in the tropical Kimberley region of north-west Australia, and trials were undertaken to test the effectiveness of chemical control methods (van Rijn 1963). Before KNP was declared a national park, *U. mutica* was planted on a small scale at Canon Hill in the 1930s (Christian and Aldrick 1977) and in the Wildman and East Alligator catchments in the 1960s and 1970s (Salau 1995).

In Australia there was little attention or resources directed to areas set aside for protection of conservation values at this time, let alone preventing the use of agricultural plants (Cook and Dias 2006). However, in the late 1980s concerns were raised at land management meetings (Cowie and Werner 1987). In 1991, a report to the Federal Government identified IAPs that posed major threats to natural habitats, stated that "some serious and very serious weeds are deliberately planted" (Humphries et al. 1991). The report listed *U. mutica* as one of the 18 most serious environmental weeds in Australia. Unfortunately, the report did not result in a strong management response by the NT Government or PA managers in northern Australia. *Urochloa mutica* continued to be promoted by government agriculture departments and other non-government land management organisations, with no restrictions on locations for planting or precautions to prevent spread (Friedel et al. 2010). During this period, *U. mutica* was planted in the wetlands on KNPs eastern and western boundaries for pasture production and as part of the *M. pigra* control strategy (Cameron and Lemcke 2003), ignoring warnings that it may prove an even more intractable problem than *M. pigra* (Cook and Setterfield 1995).

9.4.2.1   The Spread of *Urochloa mutica* Within Kakadu National Park

In the 1970s *U. mutica* was planted at Four Mile Hole Billabong in KNPs Wildman catchment (Fig. 9.1), and in the early 1990s the distribution in KNP was still relatively limited. In 1995 the invasion was described as "satellite infestations scattered for approximately 2 km down the floodplain" (Salau 1995; Walden and Bayliss 2003). By 2012, a systematic aerial survey of 95,700 ha of the Wildman and West Alligator floodplains was conducted, assessing 250 × 250 m (6.25 ha) quadrants. The results showed that 3,440 ha contained low levels of *U. mutica*, with 440 ha being more than 50 % invaded.

   Urochloa mutica was first reported in KNPs Magela Creek catchment in 1946 on the Cannon Hill floodplain (Christian and Aldrick 1977, Fig. 9.1). In 1968/69, *U. mutica* was one of a range of species trialled by the NT Government within paddocks at Mudginberri in the upper Magela floodplain which was a pastoral property at the time (Miller 1979). This planting was limited in extent (<100 ha) and not considered very successful due to the significant effort required to plant runners and limited establishment success (Geoff Cross, pers. comm.). However, *U. mutica* spread from this site and by the mid-1990s, invasion was described as spreading from Mudginberri to the north of Nankeen billabong (Salau 1995), some 20 km away. The middle reaches of the Magela floodplain were highly suitable for *U. mutica*. Knerr (1996) used aerial photographs to determine the change in distribution of *U. mutica* and estimated that it expanded from 132 to 422 ha between 1991 and 1996. The increase in the area of *U. mutica* caused a corresponding decrease in area of wild rice (*Oryza meridionalis*) (Knerr 1996). In 2008/2009, we completed systematic aerial survey of 35,700 ha of the Magela Floodplain, assigning each 250 × 250 m (6.25 ha) into a 5-point cover class as described previously. This survey showed that 8,412 ha contained some invasion by *U. mutica*, with 1,637 ha in the densest cover class (>50 %). Taking cover class into account, over 2,200 ha of *U. mutica* now occurs on the floodplain.

Fig. 9.6 High intensity fires that are fuelled by *Urochloa mutica* cause death of *Melaleuca* spp. overstory (Photo Damien McMaster)

9.4.2.2 Negative Impacts of *Urochloa mutica* Invasion

In response to the concerns of the *U. mutica* invasion, a detailed study of the impacts of *U. mutica* on aquatic ecosystems in KNP (Douglas et al. 2001) showed significant negative effects on native vegetation. Compared to native *Oryza* spp. and *H. acutigluma* communities, *U. mutica* communities had significantly lower plant biodiversity, particularly during the dry season. In a comparison of the relative impacts of *M. pigra* and *U. mutica* invasion (Bayliss et al. 2012), re-analysis of experimental data from the Oenpelli floodplain (Cook 1992) showed that in sites with 100 % cover of *M. pigra*, 14 % of native floodplain plant species remained, whereas with 100 % cover of *U. mutica* resulted in complete loss of native plant species (Bayliss et al. 2006, 2012). *Urochloa mutica* does not support the fledging growth rates of magpie goose goslings obtained from a diet of native grasses and this can be fatal (Whitehead and Dawson 2000; Whitehead et al. 2000). Magpie geese may also be detrimentally affected by *U. mutica* invasion as they preferentially nest in *Eleocharis* and wild rice (Bayliss and Yeomans 1990; Corbett and Hertog 1996).

Urochloa mutica also has different fuel characteristics compared to the native grasses it replaces (Douglas and O'Connor 2004). For example, *U. mutica* produces approximately twice the dry season fuel load of *Oryza* spp. While *U. mutica* has a similar fuel load to the native perennial *H. acutigluma* it produces taller, drier fuel, increasing fire intensity. This may facilitate the displacement of *H. acutigluma*, which is fire sensitive, and damages other fire-sensitive woody vegetation including *Melaleuca* (paperbark; Figs. 9.6 and 9.7) and rainforest. On the Magela floodplain,

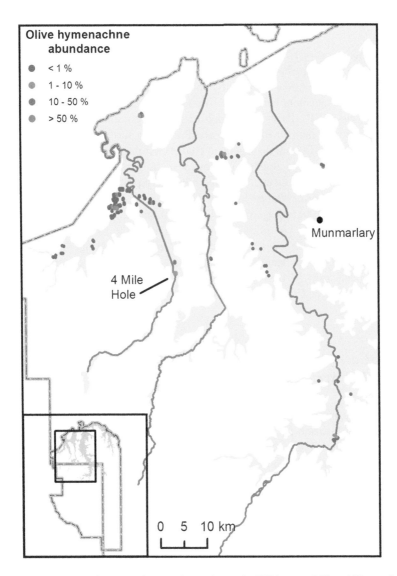

Fig. 9.7 Distribution of *Hymenachne amplexicaulis* on the Wildman and West Alligator flood-plains in Kakadu National Park, recorded from helicopter survey (250 × 250 m pixels)

U. mutica is responsible for a reduction in the area of monsoon vine forest adjacent to the floodplain. More intense fires also pose a threat to turtles that aestivate in the floodplain soil during the dry season, presumably as they are unable to dig through the dense fuel to the soil, or are killed by fire even while buried.

9.4.2.3   *Urochloa mutica* Management

Whereas managers of KNP responded immediately to the invasion of *M. pigra*, the response to *U. mutica* was far more delayed and was allocated significantly fewer resources. A 'grassy weeds team' of two people was established in 2003 to support the mimosa management team and KNP rangers to undertake work on *U. mutica* and other grassy IAPs (Parks Australia 2012). At that point there were already large invasions across a number of catchments. Preventing further spread of *U. mutica* would have been a difficult management goal even if the team was dedicated to just this species and on one floodplain. However, the grassy weeds team was also responsible for the management of three other very problematic terrestrial grassy IAPs, *Pennisetum pedicellatum* (annual mission grass), *P. polystachion* (perennial mission grass) and *Andropogon gayanus* (gamba grass). This means that the staff allocated to *U. mutica* control was effectively one-eighth of that allocated to *M. pigra*. In reality the other species, particularly *P. pedicellatum* and *P. polystachion*, took the majority of the team's time. The result of these resource limitations was that early detection of new populations and rapid management response was significantly less for *U. mutica* than it was for *M. pigra*.

9.4.3   Case Study 3: Hymenachne amplexicaulis: *The Accidental Tourist*

Unfortunately, managers of KNP and other PAs in northern Australia have also had to respond to invasion by another high impact, semi-aquatic grass. *Hymenachne amplexicaulis* is a relatively recent introduction into northern Australia. It was introduced as part of an on-going programme to find and use new species for primary production, despite the early and consistent warnings about the potential spread of introduced pasture grasses to non-pastoral areas (Clarkson 1991, 1995; Cook and Setterfield 1995; Csurhes et al. 1999). *Hymenachne amplexicaulis* is a perennial grass that commonly grows to between 1 and 2.5 m. The (incorrect) assessment that *U. mutica* was limited to habitats with water depths less than 50–60 cm was used as a rationale for introducing *H. amplexicaulis* and *Echinochloa polystachya* (aleman grass), another semi-aquatic grass, as they were suited to growing in seasonally inundated areas up to 2 m deep (Pittaway and Chapman 1996). The first experimental planting occurred in Queensland from plant material introduced in 1983 (Wearne et al. 2010). In 1988, *Hymenachne amplexicaulis* cv. 'Olive' was approved for release by the Queensland Herbage Plants Liaison Committee (Wildin 1989). Initially, *H. amplexicaulis* was primarily promoted for use in 'ponded pastures', that is, artificially created wetlands for grazing purposes. This new species allowed grazers to build and graze deeper ponds than they were able to with *U. mutica* or native grasses. In the Northern Territory, *H. amplexicaulis* was also being promoted as a more productive pasture species for floodplain areas

(Cameron 1999) and was planted for this purpose on the Adelaide, Daly, Finniss and Mary River floodplains, and at Arafura Swamp in northern central Arnhem Land. It was also used to suppress seedling growth of *M. pigra* (Paynter 2004).

Concerns were raised soon after the release of *H. amplexicaulis* because it showed evidence of spread from planted areas (Clarkson 1991). Like *U. mutica*, *H. amplexicaulis* was clearly another 'conflict' species that was promoted for pastoral use across northern Australia despite concerns for conservation areas or other users. However, there were some important differences to the introduction and spread of *U. mutica*. By the time of *H. amplexicaulis's* release, there were already scientists, conservation managers and others raising concerns about pasture grasses, particularly these aquatic pasture species. Calls were made for the need for restrictions on the use of *H. amplexicaulis*, such as localised planting in sensitive areas, monitoring spread and management approaches to limit spread (Clarkson 1991; Humphries et al. 1991; Csurhes et al. 1999). During the 1980–1990s, IAP monitoring and reporting became increasingly systematic as GPS units and other technological devices were developed. As a result, documented evidence of populations of *H. amplexicaulis* in non-pastoral areas were widely reported and further concerns raised nationally. Importantly, early sites of invasion by *H. amplexicaulis* had an impact on another powerful agricultural sector, the sugar industry, by blocking water flow in irrigation channels. In 1999, barely a decade after its official release, *H. amplexicaulis* was listed as one of Australia's 20 Weeds of National Significance (WoNS, Thorp and Lynch 2000). It took another 4 years before it was declared a noxious weed in Queensland (July 2003) and 6 years (November 2005) before it was declared as a weed in the Northern Territory.

By the time the first infestations were located in KNP in 2001, *H. amplexicaulis* was officially considered one of Australia's worst IAPs. Unfortunately, before it was declared a noxious weed in the NT, plantings had occurred in many catchments near KNP. These included the Adelaide River and Mary River floodplains, and Arafura Swamp in northern central Arnhem Land. *Hymenachne amplexicaulis* was first reported in KNP in August 2001 at two separate locations on the South Alligator floodplain (Wearne et al. 2010). Despite control efforts by KNP weed management teams, there has been a steady increase in the reported occurrence of new populations. Populations of *H. amplexicaulis* have been recorded over 23 km^2 in the West Alligator and Wildman catchments, and scattered populations continue to be reported from these catchments and the South and East Alligator Rivers.

9.5 Understanding the Factors Influencing Management Responses

The case studies above demonstrate three very different responses and outcomes for three of the aquatic IAPs threatening KNP. Given this varied management responses, it is worth reflecting on the reasons why KNP is a model of best practice

182   S.A. Setterfield et al.

Table 9.1 Factors that potentially affecting the difference in management response for *Urochloa mutica, Hymenachne amplexicaulis* and *Mimosa pigra*

Factors	Urochloa mutica	Hymenachne amplexicaulis	Mimosa pigra
Year that weed was first recorded in region, and the number of years either before or after the formation of KNP	1930s; >40 years before KNP formation	2001; 22 years after KNP formation	1981; 2 years after KNP formation
Declared as a Weed of National Significance (year)?	No	Yes (1999)	Yes (1999)
Declared as a weed in the Northern Territory (year)?	No	Yes (2005)	Yes (1981)
Recognised threat or benefit to cattle industry?	Benefit	Benefit	Threat
Recognised as environmental threat when it first occurred	No	Yes	Yes
Rating from Northern Territory weed risk assessment	Very high	Very high	Very high
Recognised nationally as key threatening process (year)	Yes (2011)	Yes (2011)	No
Obvious transformation of vegetation structure	No	No	Yes (grass to shrubland)
Current distribution in			
(i) East Alligator	High	Low-medium	Low
(ii) South Alligator	Low	Low	Low
(iii) West Alligator	Medium	Medium	Low
Rate of spread (years to double in area)	Medium (5) Bayliss et al. (2012)	Medium (unknown)	High (1.4) Cook et al. (1996)
Potential for unassisted spread from adjacent catchments	Low	High	Medium
Seed bank longevity	Low	Low	High
Staff dedicated to control (no. of full-time staff per species)	Low (<1)	Low (<1)	High (4)

IAP management for *M. pigra*, yet appears to be losing the fight against aquatic grassy IAPs (Table 9.1).

An important factor in the management response appears to have been the IAP status of the plant when it was first detected in KNP. *Mimosa pigra* and *H. amplexicaulis* had both already been formally recognised as significant IAPs at a national level. Consequently, park management had a high level of awareness of their negative impacts and a responsibility to manage these species within the requirements of the national strategy. In contrast, *U. mutica* is still not a declared weed in any State or Territory, although it was one of the five invasive grass species identified as a Key Threatening Process under the Federal EPBC Act (Anonymous 2009).

Part of the reason for the lack of declaration of *U. mutica* by Government weed management agencies is strong opposition from the cattle grazing industry and the government of primary industry departments and ministers themselves (Ferdinands et al. 2005; Friedel et al. 2010). These parties view *U. mutica* as a beneficial species and have consistently disputed claims of its ability to spread and the negative environmental impacts (Rea and Storrs 1999; Whitehead and Wilson 2000). For example, Cameron and Lemke (2003) states that within the government's agricultural advice to growers is that *U. mutica* has been "present on floodplains for considerable time without taking over". No management approaches such as restrictions on planting or formal management plans have been introduced despite formal risk assessment processes suggesting they should (Clarkson et al. 2010). The IAP risk assessment in the Northern Territory resulted in it being ranked a high risk species based on its invasiveness, impacts and potential distribution (Friedel et al. 2010), and Queensland's pest risk assessment stated that *U. mutica* has significant "negative impacts in Queensland, despite its use as pasture" and that there remain important areas at risk in Cape York, where substantial areas of wetlands are still free of *U. mutica* and perhaps some other wetlands in North Queensland (Hannan-Jones and Csurhes 2012). The Northern Territory Department of Primary Industries still promote *U. mutica* as a beneficial grass without acknowledging the IAP issues for other properties within a catchment. For example, the NT Government (2012) advise that best practice floodplain management includes the use of introduced grasses for re-vegetation and IAP control on the sub-coastal floodplains, and the use of introduced grasses and banks to retain water to improve productivity on floodplains. The four species listed to have been successfully planted are *U. mutica*, *H. amplexicaulis*, *E. polystachya* and *Setaria sphacelata* (Kazungula setaria). Only *H. amplexicaulis* is given the warning that it is a WoNS and a declared noxious weed and therefore cannot be planted. A clear contrast occurred with *M. pigra*, where the potential invasion and loss of floodplain pasture was seen as having major economic impacts and there was no opposition to its listing as an IAP (Miller et al. 1981).

Like *U. mutica*, *H. amplexicaulis* was also promoted as a pasture grass and this potential beneficial use contributed to the delay in its listing as an IAP in the NT. Its relatively recent introduction, rapid spread, vigorous growth and capacity to establish in deep water appears to have made it easier to argue the case for its listing as a serious environmental IAP. However formal risk assessments have shown *H. amplexicaulis* and *U. mutica* pose a very similar environmental risk (as demonstrated by the outcomes of weed risk assessments in the NT) and research documenting the environmental impacts of *U. mutica* was in fact used to justify the case for listing *H. amplexicaulis* as a WoNS (Thorp and Lynch 2000).

Another potential factor that made the invasion by *M. pigra* prompt immediate attention was that it causes dramatic and very visible structural change on the floodplains, changing the grass/sedgelands to shrublands (Cook et al. 1996). By contrast, *U. mutica* and *H. amplexicaulis* invasion are only noticeable to an expert observer. For example, one of the largest areas in KNP invaded by *U. mutica* is visible from a popular tourist destination, yet there is almost no public awareness that the vista is mostly occupied by a weed. This means that there is little public pressure on the park to manage it.

All these factors may have contributed to an initial false sense of security and masked the potential impact and threat posed by the spread of *U. mutica* in KNP, resulting in a delayed management response. However, these perceptions should have changed during the 1990s, when the risk of spread and knowledge of impacts on biodiversity and cultural assets were reported in many forums. At that time, *U. mutica* distribution was still limited in the Wildman/West Alligator and South Alligator catchments, even though it had spread significantly in the East Alligator. Whereas KNP managers were applauded for their best practice management of *M. pigra*, they failed to adequately address the threat posed by *U. mutica*. The 'grassy weeds' team was eventually established, but the level of resourcing and operation of this group has proven inadequate to successfully manage either *U. mutica* or *H. amplexicaulis*.

The pattern of spread, based on recent surveys, suggests no new invasions of *U. mutica* entering KNP since it was originally introduced in the 1990s, thus eradication of satellite populations would be feasible. Seeds remain viable in the soil for a much shorter period than *M. pigra* and would therefore require far less follow-up control and monitoring. This is in marked contrast to *H. amplexicaulis*, where the pattern of spread and new invasions suggest it is being spread readily by animal vectors (e.g. birds and pigs) from adjacent catchments, resulting in many new populations across the park over the past decade. Given that, in most instances, these invasion vectors are not being controlled outside KNP, this represents an on-going source of invasion and a major challenge to future management.

9.6   Conclusion: The Future of Managing Multiple Invasive Alien Plant Threats in Multiple Catchments

The management of these IAPs in KNP is at a critical phase. *Mimosa pigra* remains under control but invasive alien grasses are continuing to expand in both area invaded and in the number of satellite populations. As Bayliss et al. (2012) point out, the benefits of the sustained allocation of resources to *M. pigra* will be lost if the floodplains remain free of *M. pigra* but become invaded by alien grasses.

Urgent decisions need to be made about protecting the floodplains' key biological and cultural assets. This includes modeling future spread patterns, costs of control and evaluating management strategies and potential returns on money invested (economic, cultural and environmental). We suggest some actions that the park managers should commence immediately to minimise the risks posed by IAPs. In essence these relate to applying the same or similar approach to that which were successfully applied to managing the risks posed to KNP by *M. pigra*. These are to (i) adopt a strategic approach to the management of *U. mutica* and *H. amplexicaulis* by taking into account the potential impact, reinvasion and seed longevity, (ii) set clear and quantitative targets for the control of these IAPs taking a multi-catchment, park-wide approach, (iii) give the highest priority to eradicating the source populations of *U. mutica* invasion on the South Alligator floodplain and containing spread from the West Alligator floodplain onto the South Alligator, (iv) review the level of resources currently allocated to the management of floodplain IAPs, bearing in mind the future financial savings accrued from eradicating small, satellite populations before they become large infestations where eradication is no longer viable and on-going control is required, and (v) acknowledge the very serious threat that these grassy IAPs pose to KNP, relative to the other threats (see Bayliss et al. 2012) and afford them a higher level of priority than there has been to date.

Acknowledgements We thank Anne O'Dea, Buck Salau, Bert Lukitsch for information, data and advice.

References

Allen GR, Midgley SH, Allen M (2002) Field guide to the freshwater fishes of Australia. Western Australian Museum, Perth

Anning P, Hyde R (1987) Ponded para grass in North Queensland. Qld Agric J 113:171–180

Anonymous (2009) Listing advice for key threatening process. Invasion of northern Australia by gamba grass and other introduced grasses. Australian Department of Sustainability, Environment, Water, Population and Communities, Canberra

Bayliss P, Yeomans KM (1990) Seasonal distribution and abundance of magpie geese, *Anseranas semipalmata* Latham, in the Northern Territory, and their relationship to habitat, 1983–86. Aust Wildl Res 17:15–38

Bayliss P, van Dam R, Boyden J et al (2006) Ecological risk assessment of Magela floodplain to differentiate mining and non-mining impacts. In: Evans KG, Rovis-Hermann J, Webb A et al (eds) Eriss research summary 2004–2005, Supervising scientist report 189. Office of Supervising Scientist, Darwin, pp 172–185

Bayliss P, van Dam R, Bartolo R (2012) Quantitative ecological risk assessment of the Magela Creek floodplain in Kakadu National Park, Australia: comparing point source risks from the Ranger Uranium Mine to diffuse landscape-scale risks. Human Ecol Risk Assess 18:115–151

Beggs KE (2012) Effects of exotic pasture grasses on biodiversity in the Mary River Catchment, Northern Territory. Charles Darwin University, Darwin

Burrows N, Lukitsch B (2012) Biological control agents are observed on Mimosa pigra six and 12 years after their release in the Northern Territory, Australia. In: Eldershaw V (ed) Developing solutions to evolving weed problems. Proceedings of the 18th Australasian Weeds Conference. Weed Society of Victoria, Melbourne, pp 347–348

Calder GJ (1981) *Hymenachne acutigluma* in the Northern Territory, Technical bulletin no 46. Department of Primary Production, Darwin

Cameron A (1999) Management of grazing on NT floodplains. Agnote 104. Department of Primary Industries, Darwin

Cameron AG, Lemcke B (1996) Management of improved grasses on NT floodplains. Agnote No. E17. Darwin

Cameron AG, Lemcke B (2003) Floodplain grazing management. Agnote, vol E54. Department of Primary Industry, Fisheries and Mines, Darwin

Chatto R (2006) The distribution and status of waterbirds around the coast and coastal wetlands of the Northern Territory. Parks and Wildlife Commission of the Northern Territory, Palmerston

Christian CS, Aldrick JM (1977) Alligator Rivers study: a review report of the Alligator Rivers region environmental fact-finding study. AGPS, Canberra

Clarkson JR (1991) The spread of pondage species beyond the pasture system – the risk and associated ecological consequences. In: Anonymous (ed) Proceedings of the probing ponded pastures workshop, Rockhampton, 16–18 July 1991. University of Central Queensland, Rockhampton, pp 1–6

Clarkson J (1995) Ponded pastures: a threat to wetland biodiversity. In: Finlayson CM (ed) Wetland research in the wet-dry tropics of Australia, Supervising Scientist Report 101. Supervising Scientist, Canberra, pp 206–211

Clarkson JR, Grice AC, Friedel MH et al (2010) The role of legislation and policy in dealing with contentious plants. In: Zydenbos SM (ed) New frontiers in New Zealand: together we can beat the weeds. Proceedings of the 17th Australasian Weeds, Christchurch, 2010. New Zealand Plant Protection Society, pp 474–477

Cook GD (1992) Control of *Mimosa pigra* at Oenpelli: research and monitoring program. Preliminary report to the Steering Committee of the program to control mimosa on Aboriginal Lands in the Northern Territory. CSIRO Division of Wildlife & Ecology, Canberra

Cook GD, Dias L (2006) It was no accident: deliberate plant introductions by Australian government agencies during the 20th century. Aust J Bot 54:601–625

Cook GD, Setterfield SA (1995) Ecosystem dynamics and the management of environmental weeds in wetlands. In: Finlayson CM (ed) Wetland research in the wet-dry tropics of Australia. Supervising Scientist, Jabiru, pp 200–205

Cook GD, Setterfield SA, Maddison JP (1996) Shrub invasion of a tropical wetland: implications for weed management. Ecol Appl 6:531–537

Corbett L, Hertog AL (1996) An experimental study of the impact of feral swamp buffalo *Bubalus bubalis* on the breeding habitat and nesting success of magpie geese *Anseranas semipalmata* in Kakadu National Park. Biol Conserv 76:227–287

Cowie ID, Werner PA (1987) Weeds in Kakadu National Park: a survey of alien plants. Unpublished final report to the Australian National Parks and Wildlife Service. CSIRO, Darwin

Cowie ID, Werner PA (1993) Alien plant species invasive in Kakadu National Park, tropical Northern Australia. Biol Conserv 63:127–136

Cowie ID, Short PS, Osterkamp Madsen M (2000) Floodplain flora: a flora of the coastal floodplains of the Northern Territory, Australia. Australian Biological Resources Study, Canberra

Csurhes SM, Mackey AP, Fitzsimmons L (1999) Hymenachne (*Hymenachne amplexicaulis*) in Queensland, Pest status review series. Department of Natural Resources and Mines, Brisbane

Department of Sustainability, Environment, Water, Population and Communities (2009) Invasion of northern Australia by gamba grass and other introduced grasses. Department of Sustainability, Environment, Water, Population and Communities, Canberra. http://www.environ ment.gov.au/biodiversity/threatened/ktp/northern-australia-introduced-grasses.html. Accessed 22 July 2012

Douglas MM, O'Connor RA (2004) Effects of para grass (*Urochloa mutica* (Forssk.) Q. Nguyen) invasion on terrestrial invertebrates of a tropical floodplain. In: Sindel BM, Johnson SB (eds) Weed management: balancing people, planet, profit. Proceedings of the 14th Australian Weeds Conference, Wagga Wagga, New South Wales, Australia, 6–9 September 2004. Weed Society of New South Wales, Sydney, pp 153–156

Douglas MM, Bunn SE, Pidgeon RJW et al (2001) Weed management and the biodiversity and ecological processes of tropical wetlands. National Wetlands Research and Development Program, Canberra

Douglas MM, Bunn SE, Davies PM (2005) River and wetland food webs in Australia's wet-dry tropics: general principles and implications for management. Mar Freshw Res 56:329–342

Dudley N (ed) (2008) Guidelines for applying protected area management categories. IUCN, Gland

Ferdinands K, Beggs K, Whitehead P (2005) Biodiversity and invasive grass species: multiple-use or monoculture? Wildl Res 32:447–457

Finlayson CM (2005) Plant ecology of Australia's tropical floodplain wetlands: a review. Ann Bot 96:541–555

Finlayson CM, Bailey BJ, Cowie ID (1990) Characteristics of a seasonally flooded freshwater system in monsoonal Australia. In: Whigham DF, Goode RE, Kvet J (eds) Wetland ecology and management—case studies. Kluwer Academic Publishers, Dordrecht, pp 141–162

Finlayson CM, Lowry J, Grazia Bellio M et al (2006) Biodiversity of the wetlands of the Kakadu region, Northern Australia. Aquat Sci 68:374–399

Friedel MH, Grice AC, Clarkson JR et al (2010) How well are we currently dealing with contentious plants? In: Zydenbos SM (ed) New frontiers in New Zealand: together we can beat the weeds Proceedings of the 17th Australasian Weeds Conference. New Zealand Plant Protection Society, Christchurch, pp 470–473

Grace BS, Gardener MR, Cameron AG (2004) Pest or pasture? Introduced pasture grasses in the Northern Territory. In: Sindel BM, Johnson SB (eds) Weed management: balancing people, planet, profit. Proceedings of the 14th Australian Weeds Conference, Wagga Wagga, New South Wales, Australia, 6–9 September 2004. Weed Society of New South Wales, Sydney, pp 157–160

Hannan-Jones M, Csurhes SM (2012) Invasive species risk assessment: para grass (Urochloa mutica). Queensland Department of Agriculture, Fisheries and Forestry, Brisbane

Humphries SE, Groves RH, Mitchell DS (1991) Plant invasions and Australian ecosystems: a status review and management directions. In: Longmore R (ed) Plant invasions. The incidence of environmental weeds in Australia, Kowari 2:1–134 Australian National Parks and Wildlife Service, Canberra, pp 1–127

Hunter F, Ibbett M, Salau B (2010) Weed management in Kakadu National Park. In: Winderlich S (ed) Kakadu National Park Landscape Symposia Series 2007–2009. Symposium 2: weeds management. Jabiru Field Station, 2007. Supervising Scientist, pp 22–28

Knerr NJA (1996) Grassland community dynamics of a freshwater tropical floodplain: invasion of Brachiaria mutica (para grass) on the Magela floodplain, Kakadu National Park. University of New England, Armadale

Levitus R (1995) Social history since colonisation. In: Press AJ, Lea DM, Webb A et al (eds) Kakadu: natural and cultural heritage and management. Australian Nature Conservation Agency, Darwin, pp 74–93

Lonsdale WM, Harley KLS, Gillett JD (1988) Seed bank dynamics of Mimosa pigra, an invasive tropical shrub. J Appl Ecol 25:963–976

Lucas DE, Russell-Smith J (1993) Traditional resources of the South Alligator floodplain: utilisation and management. Final consultancy report to the Australian Nature Conservation Agency. Australian Nature Conservation Agency, Jabiru

Miller IL (1979) Para grass. Turnoff 2:2

Miller IL, Lonsdale WM (1987) Early records of Mimosa pigra in the Northern Territory. Plant Protect Q 2:140–142

Miller IL, Lonsdale WM (1992) The use of fire and competitive pastures to control Mimosa pigra. In: Harley KLS (ed) A guide to the management of Mimosa pigra. Commonwealth Scientific and Industrial Research Organisation, Canberra, pp 104–106

Miller IL, Nemestothy L, Pickering SE (1981) Mimosa pigra in the Northern Territory, Northern Territory Department of Primary Production technical bulletin no. 51. Department of Primary Production Division of Agriculture and Stock, Darwin

Moody ME, Mack RN (1988) Controlling the spread of plant invasions: the importance of nascent foci. J Appl Ecol 25:1009–1021

NT Government (1997) Report of inquiry into matters relating to the occurrence, spread, impact and future management of Mimosa pigra in the Northern Territory, April 1997. Sessional Committee on the Environment, Darwin

NT Government (2012) Cattle and land management best practices in the Top End region, 2011. Northern Territory Government. Department of Resources, Darwin

Parks Australia (2012) Draft. Kakadu National Park weed management plan 2012–17. Jabiru

Paynter Q (2004) Evaluating *Mimosa pigra* biocontrol in Australia. In: Julien M, Flanagan G, Heard T et al (eds) Research and management of *Mimosa pigra*. CSIRO, Canberra, pp 141–148

Paynter Q, Flanagan GJ (2004) Integrating herbicide and mechanical control treatments with fire and biological control to manage an invasive wetland shrub, *Mimosa pigra*. J Appl Ecol 41:615–629

Pettit N, Townsend S, Dixon I et al (2011) Plant communities of aquatic and riverine habitats. In: Pusey BJ (ed) Aquatic biodiversity in Northern Australia: patterns, threats and future. Charles Darwin University Press, Darwin, pp 37–50

Pittaway PA, Chapman DG (1996) The downstream benefits of ponded pastures. In: Hunter HM, Eyles AG, Rayment GE (eds) Proceedings of the national conference on downstream effects of land use, Rockhampton, Queensland, 26–28 April 1995. Department of Natural Resources, Queensland, pp 297–299

Press AJ, Lawrence D (1995) Kakadu National Park: reconciling competing interests. In: Press AJ, Lea DM, Webb A et al (eds) Kakadu: natural and cultural heritage and management. Australian Nature Conservation Agency, Darwin, pp 1–14

Rea N, Storrs MJ (1999) Weed invasions in wetlands of Australia's Top End: reasons and solutions. Wetl Ecol Manage 7:47–62

Roberts RG, Jones R (1994) Luminescence dating of sediments: new light on the human colonisation of Australia. Aust Aborig Stud 2:2–17

Roberts R, Jones R, Smith MA (1993) Optical dating at Deaf Adder George, Northern Territory, indicates human occupation between 53,000 and 60,000 years ago. Austr Archaeol 31:58–59

Russell-Smith J (1995) Flora. In: Press AJ, Lea DM, Webb A et al (eds) Kakadu: natural and cultural heritage and management. Australian Nature Conservation Agency, Darwin, pp 127–166

Salau R (1995) Para grass in Kakadu National Park. Report to Natural Resources Section, Parks Australia. Parks Australia, Jabiru

Sattler PS, Taylor MFJ (2008) Building nature's safety net 2008. Progress on the directions for the National Reserve System. WWF-Australia, Sydney

Storrs M, Ashley M, Brown M (1999) Aboriginal community involvement in the management of mimosa (*Mimosa pigra*) on the wetlands of the Northern Territory's Top End. In: Bishop AC, Boersma M, Barnes CD (eds) Weed management into the 21st Century: Do we know where we're going? Proceedings of the 12th Australian weeds conference, Hobart, 1999. Weed Society of Tasmania, Hobart, pp 562–565

Taylor JA, Tulloch D (1985) Rainfall in the wet-dry tropics: extreme events at Darwin and similarities between years during the period 1870–1983 inclusive. Aust J Ecol 10:281–295

Thorp JR, Lynch R (2000) The determination of weeds of national significance. National Weeds Strategy Executive Committee, Launceston

Tremblay P (2007) Economic contribution of Kakadu National Park to tourism in the Northern Territory. Sustainable Tourism CRC, Darwin

van Rijn PJ (1963) Chemical weed control in irrigation channels at the Kimberley Research Station, Western Australia. Austr J Exp Agric Anim Husb 3:170–172

Walden D, Bayliss P (2003) An ecological risk assessment of the major weeds on the Magela Creek floodplain, Kakadu National Park, Internal report 439. Supervising Scientist, Darwin

Walden D, van Dam R, Finlayson M et al (2004) A risk assessment of the tropical wetland weed *Mimosa pigra* in northern Australia. Supervising Scientist Report 177. Darwin

Wearne LJ, Clarkson JR, Vitelli JS (2010) The biology of Australian weeds. 56. *Hymenachne amplexicaulis* (Rudge) Nees. Plant Protect Q 25:146–161

Wellings P (2007) Joint management: aboriginal involvement in tourism in the Kakadu world heritage area. In: Bushell R, Eagles P (eds) Tourism in protected areas: benefits beyond boundaries. CABI, Wallingford, pp 89–100

Wesley-Smith RN (1973) Para grass in the Northern Territory – parentage and propagation. Tropic Grassl 7:249–250

Whitehead PJ, Dawson T (2000) Let them eat grass! Nat Austr 26:46–55

Whitehead P, Wilson C (2000) Exotic grasses in Northern Australia: species that should be sent home. In: Proceedings of the Northern Grassy Landscapes Conference, Katherine, August 29–31, 2000. Tropical Savannas CRC, Darwin, pp 83–87

Whitehead PJ, Dawson T, McLean A et al (2000) Digestive function in Australian magpie geese. Austr J Zool. Tropical Savannas CRC, Darwin, 48:265–279

Wildin JH (1989) Register of Australian herbage plant cultivars. A. Grasses. 24. Hymenachne, (a) *Hymenachne amplexicaulis* (Rudge) Nees (hymenachne) cv. Olive. Reg. No. 1-24a-1. Austr J Exp Agric 29:293

Wilson BA, Whitehead PJ, Brocklehurst PS (1991) Classification, distribution and environmental relationships of coastal floodplain vegetation, Northern Territory, Australia, March-May 1990. Technical memorandum. Conservation Commission of the NT, Darwin

Winders CW (1937) Sown pastures and their management. Qld Agric J 48:258–280

Chapter 10
Alien Plants Homogenise Protected Areas: Evidence from the Landscape and Regional Scales in South Central Chile

Aníbal Pauchard, Nicol Fuentes, Alejandra Jiménez, Ramiro Bustamante, and Alicia Marticorena

Abstract Protected areas are generally considered 'remnant islands' of relatively natural ecosystems and are thus less susceptible to plant invasions than the anthropogenic matrix. However, there is increasing evidence that some invasive species are capable of invading more isolated natural landscapes, higher elevations and relatively undisturbed ecosystems. With an increasing influx and establishment of alien species into protected areas, we should expect a decrease in biotic differentiation and a homogenization of plant communities. Protected areas of south-central Chile have been shown to contain a significant number of alien species and can serve as a good model to test for homogenization processes due to the broad latitudinal, elevational and disturbance gradients. In this chapter, we use a comprehensive floristic survey (n = 165), collected across ten protected areas of central and south central Chile, to test whether alien plant species have contributed to the homogenization of plant communities. We test this homogenization using changes in Jaccard similarity index with the addition of alien species at local and regional scales. By analysing this case study, we expect to shed light into broader questions about the effectiveness of protected areas in filtering out alien plant invasions and to

A. Pauchard (✉) • N. Fuentes • A. Jiménez
Laboratorio de Invasiones Biológicas, Facultad de Ciencias Forestales,
Universidad de Concepción, Casilla 160-C, Concepción, Chile

Instituto de Ecología y Biodiversidad (IEB), Santiago, Chile
e-mail: pauchard@udec.cl; nfuentes@udec.cl; aljimene@udec.cl

R. Bustamante
Departamento de Ecología, Facultad de Ciencias, Universidad de Chile,
Santiago, Chile

Instituto de Ecología y Biodiversidad (IEB), Santiago, Chile
e-mail: rbustama@uchile.cl

A. Marticorena
Departamento de Botánica, Facultad de Ciencias Naturales y Oceanográficas
Universidad de Concepción, Concepción, Chile
e-mail: amartic@udec.cl

L.C. Foxcroft et al. (eds.), *Plant Invasions in Protected Areas: Patterns, Problems and Challenges*, Invading Nature - Springer Series in Invasion Ecology 7, DOI 10.1007/978-94-007-7750-7_10, © Springer Science+Business Media Dordrecht 2013

provide recommendations for specific management actions to reduce the threat of plant invasions in protected areas. We found that distance to the road (roadside vs. interior), context (protected area vs. anthropogenic matrix) were the only factors significantly associated with alien species richness. Higher alien species richness was found in transects located in matrices as compared to those in protected areas. Roadside transects showed higher alien species richness both in protected areas and matrices. On the other hand, no significant association to any environmental variable was detected for native species. Delta similarity in the Jaccard index was positive, indicating homogenization, for most of the 165 transects at all scales of comparisons. Overall floristic similarity is higher when alien and native species are included in the composition matrix compared to the only-native species matrix. Our study highlights the importance of complete floristic surveys and inventories in protected areas and their surrounding matrices, and the value of establishing effective monitoring networks, which can facilitate the large management challenge of reducing the threat posed by alien species to protected areas biodiversity.

Keywords Exotic plant invasions • Floristic similarity • Mountains • Nature reserves • Roads

10.1 Introduction

Protected areas (PAs) are generally considered less susceptible to invasion by alien plants than their surrounding areas or matrices. Low rates of invasibility are a product of the history of protection from anthropogenic land-use types, low disturbance levels, isolation and in many cases, an association with higher elevation environments, which increases the climatic barrier for alien species (Pauchard et al. 2009; Foxcroft et al. 2011). However, recently, there is mounting evidence that some invasive species are capable of crossing these barriers reaching more isolated natural landscapes, higher elevations and relatively undisturbed ecosystems (Pauchard et al. 2009; Seipel et al. 2012). The effects of plant invasions into these natural areas are still not completely understood, but it may jeopardise the main reason why protected areas are created, which is to conserve biodiversity.

Protected areas are 'remnant islands' of relatively natural ecosystems in a matrix of anthropogenic land-uses (Cameron 2006; Foxcroft et al. 2011). The difference from oceanic islands is that the rate of biotic exchange with the surrounding matrix is much higher. Therefore, most alien plant species that are abundant in the matrix can percolate or invade PAs across their boundaries, especially if there are vectors such as humans, livestock and wildlife that freely cross those boundaries (Allen et al. 2009; Dovrat et al. 2012). In addition, the movement of alien species is greatly enhanced through roads and other human and natural corridors. Roads have been shown to promote species establishment in natural areas (Pauchard and Alaback 2004; Seipel et al. 2012) both by creating and maintaining roadside disturbance, but also because they serve as conduits for

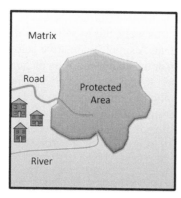 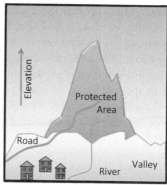

Fig. 10.1 Interaction between landscape context and elevation defines plant invasions in protected areas. The matrix, which is usually heavily affected by human activities, is a source of alien species that disperse into the protected area. Roads, rivers and other anthropogenic or natural corridors can increase the percolation of alien species into protected areas, while elevation tend to filter out species not adapted to harsher environmental conditions

more rapid movement of propagules aided by vehicles and road machinery. Because most PAs occur in rugged, higher elevation terrain, elevation and access reduce the potential for plant invasions (Fig. 10.1). A history of anthropogenic land-use may also favour plant invasions (McKinney 2002), creating a legacy of invasions even if disturbances are currently reduced. In summary, PAs although not completely invulnerable, they are still disturbed less by human activities, and thus more shielded from invasions than surrounding areas.

The relative resistance that PAs show against alien plant species does not ensure halting biotic homogenization processes caused by the addition of new species. Most alien plants are homogenizing disturbed anthropogenic environments at a high rate, which can lead to increased percolation of alien species into more pristine environments. Homogenization is not only occurring at broad intercontinental scales but also at local or regional scales, where biotic singularity is being lost due to the rapid movement of widely distributed species (Sax and Gaines 2003). In mountains for example, native species tend to show a very clear stratification along elevational gradients due to a long history of natural selection (Korner 2000). Alien species on the other hand, show a generalist strategy, occupying broad elevational bands, thereby reducing floristic distinctiveness (Alexander et al. 2011). In fact, most alien species in mountains are also invaders in lowlands. The reduction in the number of alien species with elevation is just a process of filtering out species that are not physiologically tolerant of harsher mountain conditions (Alexander et al. 2011). The evidence also suggests that given enough time, floristic similarity between non-protected lowland areas and higher elevation protected areas will increase, especially if propagule pressure, changes in disturbance regimes and climate change lead to increased stress on native species (McDougall et al. 2011).

With increasing introduction and establishment of alien species in protected areas, we should expect a decrease in biotic differentiation and a homogenization of plant communities, especially in the more frequently disturbed landscape elements such as roadsides (Arevalo et al. 2010). Unfortunately, this hypothesis is difficult to test without having long-term monitoring data. A simple alternative is to test how alien species contribute to the differentiation or the homogenization of current plant communities. If alien species behave similarly to native species, there should be no difference in the similarity of plant communities with or without alien plants. However, if similarity increases with the addition of alien plants, we can conclude that these species may be homogenizing the flora. Evidently, local extinction cannot be accounted for with this method and therefore the results may be considered rather conservative. Although several methods have been developed to test for homogenization, a simple comparison of similarity indexes across multiple communities to estimate the change in similarity has been widely accepted (Cassey et al. 2007). To conduct such analysis, a comprehensive dataset of alien and native plant presence and abundance in a hierarchical multi-scale design may increase the chances for detection of subtle homogenization or for the differentiation of trends.

Although Chile has a long history of species introduction, it is however relatively low compared to other climatically similar regions (Jimenez et al. 2008). Continental Chile has about 743 alien plants species representing about 15 % of all plants (native and non-native) (Fuentes et al. 2013). Most species were introduced during the Spanish colonization, but with the arrival of new immigrants and the internationalization of the country there have been a greater number of introductions in the nineteenth and twentieth century (Fuentes et al. 2008). In the past decade, invasive plant research has shown that the impacts of these species on biodiversity in Chile are not trivial and their importance has been neglected simply because of the lack of scientific evidence (Quiroz et al. 2009). There is no evidence or theory that suggests that non-native plants will decrease in coming years. The rate of trade and commerce in Chile continues to grow because of free trade and economic consolidation of the country, which will most likely lead to even greater alien species invasions (Fuentes et al. 2008; Nunez and Pauchard 2010).

Protected areas of south-central Chile have been shown to contain a significant number of alien species. These PAs are located across a wide climatic and latitudinal gradient (from 35°25′ S to 38°37′ S) (Table 10.1) and ranging from the warm and dry Mediterranean, to cool and wet temperate rainforest climates (Arroyo et al. 1995a, b). This area is considered to be a hotspot for global biodiversity due to the remarkably high levels of endemism and biogeographic isolation (Myers et al. 2000). The national protected areas system (SNASPE) in the Mediterranean region of Chile covers less than 5 % of the landscape (Pauchard and Villarroel 2002). Paradoxically, most of the alien plants present in Chile are concentrated in this region (Arroyo et al. 2000; Fuentes et al. 2008), representing a high risk to biodiversity conservation efforts (Arroyo et al. 2000; Fuentes et al. 2010). Land-use changes have fragmented and isolated natural vegetation in central Chile (Aguayo et al. 2009), a trend that has affected landscapes surrounding PAs. For south-central Chile, Pauchard and Alaback (2004) described a clear pattern of alien species reduction towards the higher zones of PAs and a very limited percolation of alien species into forested environments, reflecting the contrast between the anthropogenic matrix and protected area.

Table 10.1 Study sites (n = 11) in ten protected areas of central and south-central Chile. The number of transects, location, elevation, area of the protected area and the total number of native and alien species in all transects for that site is presented

National Park (NP)							
Natural Reserve (NR)	Transects	Latitude	Longitude	Area (ha)	Altitude (m a.s.l.)	Native species richness	Alien species richness
Radal 7 Tazas NR	16	35°25' S	71°00' W	5.026	766–1,093	102	68
Nahuelbuta Np	16	37°44' S	72°55' W	6.832	1,013–1,253	160	43
Los Ruiles NR	9[a]	35°37' S	72°21' W	45	165–332	93	52
Los Queules NR	12[a]	35°58' S	72°42' W	147	223–514	83	55
Laguna Del Laja NP	16	37°23' S	71°24' W	11.880	870–1,207	118	41
Ralco NR	16	37°55' S	71°25' W	12.492	952–1,109	67	26
Villarrica NP (A)	16	39°21' S	71°27' W	63.000	567–1,122	84	19
Villarrica NP (B)	16	39°21' S	71°27' W	63.000	599–1,051	64	38
Rio Clarillo NP	16	33°43' S	70°28' W	13.085	770–928	83	30
Rio Los Cipreces NR	16	34°18' S	70°30' W	36.882	926–1,169	52	28
Conguillio NP	16	38°37' S	71°46' W	60.080	764–1,276	69	17

[a]Areas with lower number of transects

Protected areas and their adjacent matrices in central and south-central Chile can serve as a good model to test for homogenization processes due to the broad latitudinal, elevational and disturbance gradients. We can also use this system to test how effective protected areas are in filtering the influx of alien plants. In this chapter, we use a comprehensive floristic survey, collected across ten protected areas of central and south central Chile, to test whether alien plant species have contributed to the homogenization of plant communities. We further test whether this homogenization is associated with landscape context, road proximity, land-use and elevation. By analysing this case study, we expect to shed light into broader questions about the effectiveness of protected areas in filtering out alien plant invasions and to provide recommendations for specific management actions to reduce the threat of plant invasions in PAs.

10.2 Methods

10.2.1 Data Collection

We selected a total of 11 sites in 10 PAs (Table 10.1) (Jiménez et al. 2013). Each site included a PA and its surrounding landscape. In each site, eight floristic sampling points were located along the access road, separated 1 km (following the road) from each other (Fig. 10.2). Four points were located inside the PA and four in the adjacent matrix (Fig. 10.1). At each point, we set a pair of transects (50 × 2 m) parallel to the road, one was located right along the roadside and the other 50 m away from the road (interior habitat). In two sites, not all transects were established due to the lack of roads in the interior of the protected area and inaccessibility (Table 10.1). A total of 165 transects were sampled during the 2005 and 2006 growing season.

In each transect, five continuous sub-transects (2 × 10 m) were used to record all native and alien vascular plant species. Species presence and their abundance was estimated using seven cover classes (0–1, 1–5, 5–15, 15–25, 25–50, 50–75, 75–100 %) and was recorded for each sub-transect. The data was later aggregated by transect. For each transect, the following landscape and environmental variables were recorded: latitude, elevation, land-use, distance to the road (0, roadside or 50 m, interior), and landscape context (PA or matrix).

10.2.2 Data Analysis

A univariate GLM model with fixed factors and covariates was developed to test the association between environmental and landscape variables with alien species richness. We tested for differences in mean native and alien species richness between parks and matrices and between roadsides and interior habitats using Mann-Whitney test.

We adapted the method used by Cassey et al. (2007) to test for homogenization at multiple scales. Instead of using a temporal dynamic of native and alien plant species in plant communities, we consider a hypothetical stage where only native species were present in the communities (i.e. initial model) and a stage where both native and alien species were present (i.e. current model). Two presence/absence matrices were built containing all transects and using: (i) only native species and (ii) native and alien species. Similarity matrices were calculated using Jaccard similarity index that ranges between 0 and 100, where 0 represents no species in common and 100, all species shared between the two communities (Cassey et al. 2007). Initial mean similarity (S0) for each plot was calculated using the native species abundance matrix. Current mean similarity (S1) for each plot was calculates using the native and alien species abundance matrix. These parameters were calculated at three scales of comparisons (Fig. 10.3): (i) total (St), considering

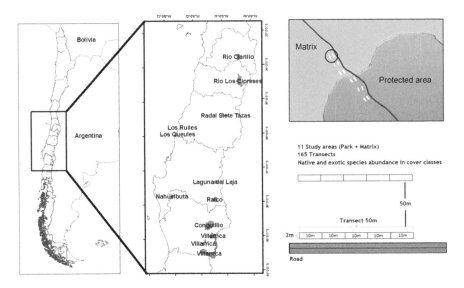

Fig. 10.2 Sampling scheme for transects in protected areas and their matrices in South-Central Chile

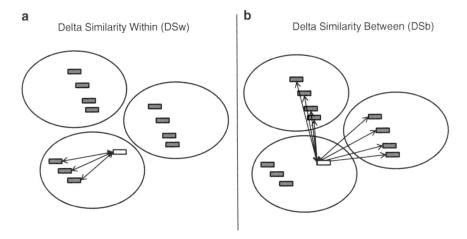

Fig. 10.3 Schematic diagram of the calculation of the delta similarity using Jaccard similarity index. An initial mean similarity value is calculating using only native species and then a final similarity index is calculating with both native and alien species, the difference between both values is the Delta Similarity (DS). The DSW scale indicates mean Jaccard similarity value of that transect with all transects in the study area of the specific transect (**a**). The DSb scale indicates mean Jaccard similarity value of that transect with all the transects in other study sites and not with transects in that site (**b**). Total similarity is the mean Jaccard similarity value of that transect with all transects independent of their location

the mean similarity between each transect and all other transects; (ii) between (Sb), considering the mean similarity between each transect and all other transects not located in that site; and (iii) within (Sw), considering the mean similarity between each transect and all transects in that site. A delta similarity (DS) was calculated for each transect subtracting S0 from S1 (DS $= $ S1$-$S0), and each scale of comparison (DSt, DSb, DSw). Positive DS values indicate homogenization and negative values indicated differentiation for that particular transect due to the addition of alien species in the plant communities. DSb and DSw values were plotted to detect differences in homogenization at the different scales of comparisons. The association between DSt and environmental variables was tested using a GLM and mean comparisons (Mann-Whitney test).

To test for differences in overall community composition caused by alien species, we used Analysis of Similarities (ANOSIM) using species composition at the two scenarios (initial, current). Multi-Dimensional Scaling (MDS) was used to create a graphic ordination of the data using a combined composition matrix with all transects (n $=$ 165) at the two scenarios (n-total $=$ 330). All analyses were run in Primer 6.0 and SPSS 12.0.

10.3   Results

10.3.1   Alien and Native Species Richness

Distance to the road (roadside vs. interior), context (PA vs. matrix) were the only environmental factors significantly associated with alien species richness. No significant interaction between these factors was detected (GLM, n $=$ 165, F $=$ 34.322, $R^2 = 0.390$, $P < 0.001$). Higher alien species richness was found in transects located in matrices as compared to those in PAs (Mann-Whitney test, $P < 0.01$, Fig. 10.4). In addition, roadside transects showed higher alien species richness both in PAs and matrices (Mann-Whitney test, $P < 0.01$, Fig. 10.4). On the other hand, no significant association to any environmental variable was detected for native species (GLM).

10.3.2   Homogenization

Delta similarity (DS) using the Jaccard index was positive for most of the 165 transects at all scales of comparisons (DSt (125), DSb (125), DSw (102)). Overall, the mean delta similarity was positive for all comparisons (DSt $= 1.65 \pm 0.14$, DSb $= 1.59 \pm 0.12$, DSw $= 2.17 \pm 0.40$). A positive relationship was found between DSb and DSw ($R^2 = 0.446$, $P < 0.001$, Fig. 10.5). At both scales, 100 transects showed homogenization (DSt $= 2.55 \pm 0.15$, DSb $= 2.34 \pm 0.13$,

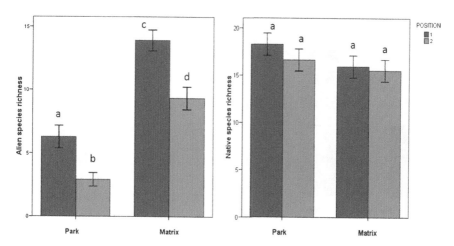

Fig. 10.4 Mean alien plant species richness (±SE) and mean native plant species richness (±SE) in transects located in protected areas (Park) and their matrices, in both roadsides (*blue*) and interior (*green*) habitats ($n = 165$). Letters indicate significant differences ($P < 0.01$, Mann-Whitney test)

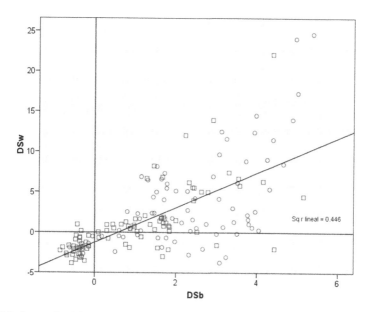

Fig. 10.5 Scatterplot delta similarity at the between (DSb) and the within (DSw) scales based on Jaccard Similarity Index for all transects ($n = 165$). Negative values indicate differentiation, while positive values indicate homogenization. *Squares* are transects in protected areas and *circles* transects in matrices

200   A. Pauchard et al.

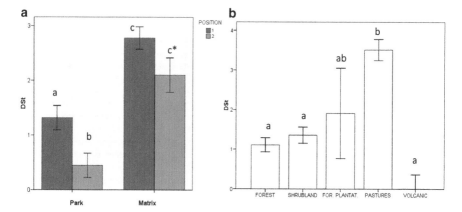

Fig. 10.6 Mean delta in similarity total (DSt) (±SE) based on changes in Jaccard Similarity Index in transects located in protected areas (Park) and their matrices, in both roadsides (*blue*) and interior (*green*) habitats (**a**) and in transects classified by vegetation type (**b**) ($n = 165$). Letters indicate significant differences ($P < 0.01$, Mann-Whitney test; *: $P < 0.05$)

DSw $= 4.70 \pm 0.52$), while 38 showed differentiation but of lower mean magnitude (DSt $= -0.53 \pm 0.03$, DSb $= 0.40 \pm 0.03$, DSw $= -2.02 \pm 0.14$).

DSt was associated to position (roadside vs. interior), context (protected area vs. matrix) and latitude (GLM, n $= 165$, F $= 26.026$, $R^2 = 0.327$, $P < 0.001$). DSt was higher in matrices than in PAs, with relatively higher homogenization occurring at roadsides compared to the interior habitat. In PAs, this was highly significant for roadsides ($P < 0.01$), while in the matrices the effect was less pronounced, but still significant ($P < 0.05$) (Mann-Whitney test, Fig. 10.6). Pastures were the most sensitive to homogenization (Mann-Whitney test, Fig. 10.6). DSt decreases with elevation for most sites indicating less homogenization in higher areas (9 of 11 sites, $P < 0.01$). A positive relationship between alien species and DSt was found for all transects ($R^2 = 0.652$, $P < 0.001$, Fig. 10.7).

Overall, the floristic similarity is higher when alien and native species are included in the composition matrix, compared to the native-only species matrix (Jaccard Similarity Matrix, ANOSIM, $P < 0.01$, Fig. 10.8).

10.4 Discussion

Our results indicate three major trends in plant invasions in PAs of Chile. First, plant invasions are strongly influenced by landscape context, road proximity and land-use. Second, plant invasions have, for the most part, homogenised plant communities at local and regional scales. Third, the fact that the number of alien species is positively correlated with homogenization of the communities indicate that most alien plants are generalists that tend to occupy broader ranges than the native plant species.

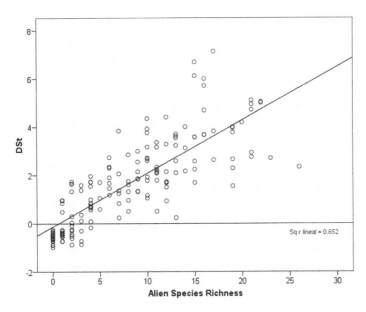

Fig. 10.7 Linear Regression between alien species richness and delta of similarity total (DSt) based on Jaccard Similarity Index ($n = 165$, $R^2 = 0.652$, $P < 0.001$). Negative values for transects with 0 alien species are a result of a differentiation to the other plots were alien species have been added

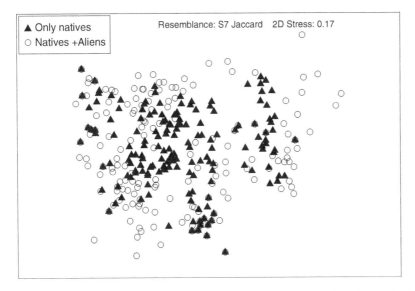

Fig. 10.8 Multi-Dimensional Scaling (*MDS*) using Jaccard Similarity Index. A combination matrix of transects including both native and alien species and transects where alien species were discarded ($n = 165$). There is a significant difference between the similarity among the two groups (ANOSIM, $P < 0.01$, Global R = 0.128)

Protected areas are less invaded than their surrounding matrices in central and south-central Chile. It appears that because of a combination of factors (e.g. land-use, history and isolation), plant communities in PAs show a lower rate of invasion. This pattern applies to the interior environments as well as roadside communities, showing that landscape context is a major driver of plant invasion. Our results confirm an increasing body of literature that emphasises the role of roads and other corridors as conduits for invasions (Pauchard and Alaback 2004; Arteaga et al. 2009; Seipel et al. 2012). Therefore, transportation design and management of roads and other corridors should be the first targets of a management strategy in PAs (Foxcroft et al. 2007, 2011). However, this is good news for managers as roads and other corridors are a small fraction of the PAs.

The fact that some land-use types concentrate alien species, especially anthro-pogenically generated pastures, where these land-uses are less common they play a role in the overall low rates of invasion into PAs. However, it is interesting to see that some natural vegetation types are more prone to invasion than others: volcanic areas do not seem to be suited for species invasions due to limited soil development and poor nutrient availability, while shrublands contain a higher number of alien species because of the open canopy conditions, which are favourable to ruderal species. On the other hand, native species do not respond to these anthropogenic gradients in a predictable manner and show a much more site specific response. Alien species that invade PAs in central and south-central Chile seem to be related to agricultural and forestry activities. Interestingly, in our dataset we only found very limited occurrence of the most invasive tree and shrub species, which seem to have not yet invaded more isolated natural ecosystems (e.g. *Acacia* spp., *Pinus* spp., *Cytisus* spp.). Therefore, we could expect an increase in the invasion of highly invasive lowland species and other escapees from less traditional uses (e.g. ornamentals, biofuels), especially around heavily developed areas in central Chile (Pauchard et al. 2011).

We found that from the ten PAs analysed in this study, *Rosa rubiginosa* (Fig. 10.9a) is present in all of them, *Hypochaeris radicata* (b) and *Rumex acetosella* (c) in nine, and *Prunella vulgaris* (d), *Hypericum perforatum* (e) and *Plantago lanceolata* (f) in eight. Most of these invasive plants are herbs dispersed by livestock from adjacent matrix into PAs. These invasive plants were introduced since 1851 (oldest record) to 1916 (newest record), and currently are widely extended in Chile. According to Fuentes et al. (2010), these plants have a high invasive potential based on the weed risk assessment protocol applied for alien plant present in Chile. Amongst the biological characteristics explaining the remarkable success of this pool of invasive plants there are short life cycle (with exception of *Rosa rubiginosa*, being shrub plant), large productions of seeds and tolerance of a wide range of climatic conditions. Certainly, at the country scale (i.e. 13 administrative regions), these invasive plants occur in more than 60 % of the territory, with *Rumex acetosella* and *Plantago lanceolata* occurring in all of them. This pool of invasive plants is contributing to the homogenization effect of PAs, as well as other ecosystem in Chile. Surprisingly, in this study we found very limited occurrence of the most invasive trees and shrubs such as *Acacia dealbata*, *Cytisus*

Fig. 10.9 Examples of the most widely distributed alien plants in protected areas of central Chile. (a) *Rosa rubiginosa*, (b) *Hypochaeris radicata*, (c) *Rumex acetosella*, (d) *Prunella vulgaris*, (e) *Hypericum perforatum*, (f) *Plantago lanceolata*

striatus, *Teline monspessulana*, *Rubus ulmifolius*, and *Pinus* spp. Protected Areas with its biotic and abiotic conditions may be acting as a filter of these pool of invasive plants. However, as human activities increases in and around PAs these filters may be not as effective in reducing the arrival of highly invasive species, which can quickly dominate landscapes with higher ecological impacts than the alien species currently widely distributed.

Overall, there is a trend towards homogenization of plant communities due to invasion of alien species in and around PAs. The addition of new species into the system does not follow the same distinctiveness as native plant assemblages. Our results confirm what has been found for mountains around the world, which is that alien plants tend to be more generalists than resident native plants (Alexander et al. 2011). In fact, the most abundant and widely distributed plants have life history traits associated with a generalist pioneer strategy adapted to high frequency or intensity disturbance.

Larger changes in similarity were observed at the within-site scale, indicating a stronger signal of homogenization at local scales. Alien species show less patchiness at the landscape scale compared to native species, and therefore they tend to homogenise plant communities within sites. Although lower in magnitude, the occurrence of regional homogenization (between sites) indicates that alien species respond differently to native species with larger regional filters (e.g. climate, dispersal). How the homogenization patterns change their magnitude at even larger scales (e.g. intercontinental scales) remains an interesting question to address in order to gain insight into the overall effect of alien species on biodiversity (sensu Sax and Gaines 2003). McDougall et al. (2011, b) showed that alien species across mountain regions are more similar to alien floras of their own region rather than to other mountain floras. However, this observation does not inform the rate at which homogenization is occurring at higher elevation areas or in mountain PAs. Clearly, lag phases may be obscuring the fact that invasive species are already present in these isolated environments and that we cannot detect them due to their low abundance (Rew et al. 2006).

The higher the number of alien plant species in a community, the higher the rate of homogenization. Therefore, not surprisingly, homogenization is associated with variables that are most closely correlated with alien species richness. Protected areas tend to have lower levels of homogenization and therefore we can assume that they are achieving their conservation goal of maintaining local and regional diversity. However, the trends towards increasing numbers of alien species with increasing development of PAs, road construction and visitation, may change these scenarios over a relatively short period of time (Pauchard et al. 2009). Even very remote sites can decrease their uniqueness/distinctiveness by the addition of just a few generalist species. It is difficult to quantify the ecosystem effects of such a phenomenon, as relatively few studies have addressed the impacts of species additions to natural systems. For example, do these generalist alien species replace more specialist native species, causing local extinctions, or do they just add new species in the community (Sax and Gaines 2003)? Long-term monitoring programmes in protected areas may help to address these questions, with which the kinds of data presented in this study are impossible to address (Chong et al. 2001; Graham et al. 2007).

10.5   Conclusion: Risk Analysis and Management

Our study highlights the importance of complete floristic surveys and inventories in PAs and their surrounding matrices, and the value of establishing effective monitoring networks in these critical areas. Data collected using a multi-scale approach, as shown here, can be used to assess the risk of species introductions before they reach a PA. These results are of significant relevance because they can be applied to identify alien plants proven to be invasive inside and outside PAs (Fig. 10.10). A risk assessment approach can be developed using these findings, based on assessments of PA and the anthropogenic matrix, levels of homogenisation across multiple spatial scales, identifying corridors of invasion into PAs and then defining management actions (Table 10.2).

A multi-scale approach is useful for spatially explicit modeling and other techniques that can better inform decision makers on the risk of alien plant invasions, and

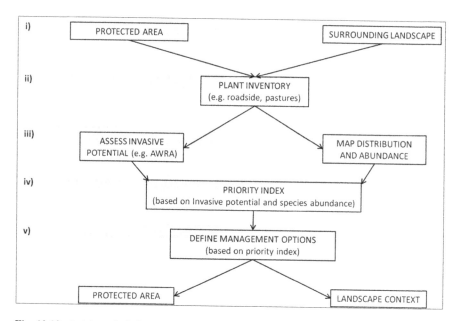

Fig. 10.10 A risk analysis framework for protected areas

Table 10.2 Stages for risk assessment and management of alien plant invasions in protected areas

Stages	Activities/management actions
Define the area of primary interest	Include both the PA and surrounding landscape (i.e. matrix), because many alien plants occur outside the PA, increasing propagule pressure and acting as a potential source of invasive alien plant to the PA
Identify major corridors of alien plant introduction	Identify those corridors by which alien plants are introduced from one location to another (source-sink dynamics). In PAs increase in tourists will boost the role of corridors for the dispersal of alien plants. Therefore, efforts to increase the monitoring along corridors should be included into a comprehensive scheme for appropriate management responses
Assess species invasiveness	Assess the chances that a given alien plant may become invasive can be determined using approaches such as the Australian Weed Risk Assessment (AWRA; Pheloung et al. 1999; see Fuentes et al. 2010 for its application in Chile). The AWRA is based on 49 questions about biogeography, biology/ecology, and undesirable traits of the species under scrutiny. Based on the answers a numeric scores is calculated, whereby it is possible to identify species that pose a negligible invasion risk (i.e. accepted, sensu AWRA), species with a clear invasion potential (rejected, sensu AWRA), and species that requires further evaluation before a decision can be made (i.e. further evaluation, sensu AWRA). Additionally, based on the score

(continued)

Table 10.2 (continued)

Stages	Activities/management actions
	of each alien plant, it is possible to identify the type of impact (e.g. agricultural impacts, ecological impacts). Subsequently, mapping alien plant distribution and its abundance will help to assess the extent and magnitude of the problem, given better basis for management and monitoring programmes. The resolution of the data should be appropriate to the scale of the study area
Calculate priority indices	Calculate priority indices based on the invasiveness of alien plants and its corresponding abundance in the area of interest (i.e. PA and matrix). Land managers rarely have sufficient resources to eradicate or control all invasive plants, especially in developing countries. Therefore, realistic priorities must be set. To generate an objective method to establish priorities for alien species management or control, we propose the scheme of Daehler et al. (2004), which combine alien species abundance with risk assessment score, based on Australian method: Priority $= S/A$; where S is the alien plant score, and A is the abundance of the respective alien plant. Alien plants with the highest priority will be those having a high score and low abundance. These species are likely to pose a significant threat to natural ecosystems, and they must be controlled. Species with high score and high abundance will demand a large amount of resources to control; therefore will have lower priority (Daehler et al. 2004)
Explore management options	Define management options based on the priority index. First, efforts should be directed in targeting alien plants with a high invasive potential but low abundance in the area of interest. Examine each species more closely to identify others factors that might help in making prioritization decisions, such as the availability of cost-effective control methods and the availability of native species to replace alien species (Daehler et al. 2004)

also to establish priorities for management actions for PAs. By presenting data in a more compelling way with, for example, local, regional and national scenarios of alien plant invasions in PAs, government agencies are becoming increasingly engaged in the process. Recently, in order to develop a comprehensive national plan the Chilean Forest Service (CONAF) is collaborating with the Laboratory of Biological Invasions (Universidad de Concepción) to address the threat of alien plant invasions. However, challenges remain due to the unprecedented task of assessing and managing the threat of plant invasions across a diverse range of PAs (Kueffer et al. 2013). Generalities drawn from ecological studies will clearly facilitate the unprecedented management challenge of controlling invasive plants across PAs at national and global scales.

Acknowledgements We thank Paul Alaback for insightful comments on the manuscript and Jocelyn Esquivel for maps and GIS work. This research was conducted in the Laboratory of Biological Invasions (LIB-www.lib.udec.cl) and was funded by Fondecyt 1040528, Conicyt Basal Funding PFB-23 and CONICYT ICM P05-002. NF holds a postdoctoral grant FONDECYT-3120125.

References

Aguayo M, Pauchard A, Azocar G et al (2009) Land use change in the south central Chile at the end of the 20(th) century. Understanding the spatio-temporal dynamics of the landscape. Rev Chil Hist Nat 82:361–374

Alexander JM, Kueffer C, Daehler CC et al (2011) Assembly of nonnative floras along elevational gradients explained by directional ecological filtering. Proc Natl Acad Sci U S A 108:656–661

Allen JA, Brown CS, Stohlgren TJ (2009) Non-native plant invasions of United States National Parks. Biol Invasions 11:2195–2207

Arevalo J, Otto R, Escudero C et al (2010) Do anthropogenic corridors homogenize plant communities at a local scale? A case studied in Tenerife (Canary Islands). Plant Ecol 209:23–35

Arroyo M, Cavieres L, Marticorena C et al (1995a) Convergence in the Mediterranean floras in central Chile and California: insights from comparative biogeography. In: Arroyo M, Zedler P, Fox M (eds) Ecology and biogeography of Mediterranean ecosystems in Chile, California, and Australia. Springer, New York, pp 43–88

Arroyo M, Riveros M, Peñaloza A et al (1995b) Phytogeographic relationships and regional richness patterns of the cool temperate rainforest flora of southern South America. In: Lawford R, Alaback P, Fuentes E (eds) High-latitude rainforests and associated ecosystems of the West Coasts of the Americas. Climate, hydrology, ecology and conservation. Springer, New York, pp 164–172

Arroyo MTK, Marticorena CM, Matthei O et al (2000) Plant invasions in Chile: present patterns and future predictions. In: Mooney HA, Hobbs R (eds) Invasive species in a changing world. Island Press, Covelo, pp 385–421

Arteaga MA, Delgado JD, Otto R et al (2009) How do alien plants distribute along roads on oceanic islands? A case study in Tenerife, Canary Islands. Biol Invasions 11:1071–1086

Cameron RP (2006) Protected area – working forest interface: ecological concerns for protected areas management in Canada. Nat Areas J 26:403–407

Cassey P, Lockwood JL, Blackburn TM et al (2007) Spatial scale and evolutionary history determine the degree of taxonomic homogenization across island bird assemblages. Divers Distrib 13:458–466

Chong G, Reich R, Kalkhan M et al (2001) New approaches for sampling and modeling native and exotic plant species richness. West North Am Nat 61:328–335

Daehler CC, Denslow JS, Ansari S, Kuo H-C (2004) A risk assessment system for screening out invasive pest plants from Hawai'i and other Pacific Islands. Conserv Biol 18:360–368

Dovrat G, Perevolotsky A, Ne'eman G (2012) Wild boars as seed dispersal agents of exotic plants from agricultural lands to conservation areas. J Arid Environ 78:49–54

Foxcroft LC, Rouget M, Richardson DM (2007) Risk assessment of riparian plant invasions into protected areas. Conserv Biol 21:412–421

Foxcroft LC, Jarošík V, Pyšek P et al (2011) Protected-area boundaries as filters of plant invasions. Conserv Biol 25:400–405

Fuentes N, Ugarte E, Kuhn I et al (2008) Alien plants in Chile: inferring invasion periods from herbarium records. Biol Invasions 10:649–657

Fuentes N, Ugarte E, Kuhn I et al (2010) Alien plants in southern South America. A framework for evaluation and management of mutual risk of invasion between Chile and Argentina. Biol Invasions 12:3227–3236

Fuentes N, Pauchard A, Sánchez P, Esquivel J, Marticorena A (2013) A new comprehensive database of alien plant species in Chile based on herbarium records. Biol Invasions 15:847–858. doi:10.1007/s10530-012-0334-6

Graham J, Newman G, Jarnevich C et al (2007) A global organism detection and monitoring system for non-native species. Ecol Inform 2:177–183

Jimenez A, Pauchard A, Cavieres LA et al (2008) Do climatically similar regions contain similar alien floras? A comparison between the mediterranean areas of central Chile and California. J Biogeogr 35:614–624

Jiménez A, Pauchard A, Marticorena A et al (2013) Patrones de distribución de plantas introducidas en áreas silvestres protegidas y sus áreas adyacentes del centro-sur de Chile. Gayana Botánica 70:110–120

Korner C (2000) Why are there global gradients in species richness? Mountains might hold the answer. Trends Ecol Evol 15:513–514

Kueffer C, McDougall K, Alexander J (2013) Chapter 6: Plant invasions into mountain protected areas: assessment, prevention and control at multiple spatial scales. In: Foxcroft LC, Pyšek P, Richardson DM, Genovesi P (eds) Plant invasions in protected areas: patterns, problems and challenges. Springer, Dordrecht, pp 89–113

McDougall KL, Alexander JM, Haider S et al (2011) Alien flora of mountains: global comparisons for the development of local preventive measures against plant invasions. Divers Distrib 17:103–111

McKinney ML (2002) Influence of settlement time, human population, park shape and age, visitation and roads on the number of alien plant species in protected areas in the USA. Divers Distrib 8:311–318

Myers N, Mittermeier R, Mittermeier C et al (2000) Biodiversity hotspots for conservation priorities. Nature 403:853–858

Nunez M, Pauchard A (2010) Biological invasions in developing and developed countries: does one model fit all? Biol Invasions 12:707–714

Pauchard A, Alaback PB (2004) Influence of elevation, land use, and landscape context on patterns of alien plant invasions along roadsides in protected areas of south-central Chile. Conserv Biol 18:238–248

Pauchard A, Villarroel P (2002) Protected areas in Chile: history, current status, and challenges. Nat Areas J 22:318–330

Pauchard A, Kueffer C, Dietz H et al (2009) Ain't no mountain high enough: plant invasions reaching new elevations. Front Ecol Environ 7:479–486

Pauchard A, García R, Langdon B et al (2011) The invasion of non-native plants in Chile and their impacts on biodiversity: history, current status, and challenges for management. In: Figueroa E (ed) Biodiversity conservation in the Americas: lessons and policy recommendations. Editorial FEN-Universidad de Chile, Santiago, pp 133–165

Pheloung PC, Williams PA, Halloy SR (1999) A weed risk assessment model for use as a biosecurity tool evaluating plant introductions. J Environ Manag 57:239–251

Quiroz C, Pauchard A, Cavieres L et al (2009) Quantitative analysis of the research in biological invasions in Chile: trends and challenges. Rev Chil Hist Nat 82:497–505

Rew LJ, Maxwell BD, Dougher FL et al (2006) Searching for a needle in a haystack: evaluating survey methods for non-indigenous plant species. Biol Invasions 8:523–539

Sax DF, Gaines SD (2003) Species diversity: from global decreases to local increases. Trends Ecol Evol 18:561–566

Seipel T, Kueffer C, Rew LJ et al (2012) Processes at multiple scales affect richness and similarity of non-native plant species in mountains around the world. Glob Ecol Biogeogr 21:236–246

Chapter 11
Plant Invasions of Protected Areas in Europe: An Old Continent Facing New Problems

Petr Pyšek, Piero Genovesi, Jan Pergl, Andrea Monaco, and Jan Wild

Abstract Europe has a particularly long history of land protection measures, and is the region of the world with the largest number of protected areas, which has grown rapidly over the last decades. This was to a large extent due to the Natura 2000 programme of the European Union which focused on extending the existing network of legally protected areas with other habitats of conservation value. As a result, Europe has over 120,000 nationally designated protected sites (the most in the world) and 21 % of the continent area (1,228,576 km^2) currently enjoys some form of legal protection. Despite these impressive statistics, the effectiveness of the existing network in protecting biodiversity is constrained by habitat fragmentation and other factors. Despite the generally high awareness of the importance of biodiversity protection in Europe, invasive alien species are not perceived as the most

P. Pyšek (✉)
Department of Invasion Ecology, Institute of Botany, Academy of Sciences
of the Czech Republic, Průhonice CZ 252 43, Czech Republic

Department of Ecology, Faculty of Science, Charles University in Prague,
CZ 128 44 Viničná 7, Prague 2, Czech Republic
e-mail: pysek@ibot.cas.cz

P. Genovesi
ISPRA, Institute for Environmental Protection and Research,
Via V. Brancati 48, I-00144 Rome, Italy

Chair IUCN SSC Invasive Species Specialist Group, Rome, Italy
e-mail: piero.genovesi@isprambiente.it

J. Pergl • J. Wild
Department of Invasion Ecology, Institute of Botany, Academy of Sciences
of the Czech Republic, Průhonice CZ 252 43, Czech Republic
e-mail: pergl@ibot.cas.cz; wild@ibot.cas.cz

A. Monaco
ARP, Regional Parks Agency – Lazio Region, Via del Pescaccio 96,
I-00166 Rome, Italy
e-mail: amonaco@regione.lazio.it

L.C. Foxcroft et al. (eds.), *Plant Invasions in Protected Areas: Patterns, Problems
and Challenges*, Invading Nature - Springer Series in Invasion Ecology 7,
DOI 10.1007/978-94-007-7750-7_11, © Springer Science+Business Media Dordrecht 2013

pressing problem by the public. This is in contrast with the fact that many of them have serious impacts on biodiversity and ecosystem functioning in protected areas. Among these, *Ailanthus altissima, Fallopia* taxa, *Heracleum mantegazzianum, Impatiens glandulifera* and *Robinia pseudoacacia* are considered as top invaders by managers of protected areas. Surprisingly, continent-wide rigorous data on the distribution and abundance of invasive alien species are lacking and there is an urgent need for collating checklists of alien species using standardised criteria to record their status. With the exception of very few regions such information is missing, or incomplete, based on varying criteria and scattered in grey literature and unpublished reports. To put the management on a more scientific basis the collection and curation of better data is an urgent priority; this could be done by using existing instruments of the EU as a convenient platform. As found by means of a web survey reported here, managers of protected areas in Europe are well aware of the seriousness of the problem and threats imposed by invasive plant species but are constrained in their efforts by the lack of resources, both staff and financial, and that of rigorous scientific information translated into practical guidelines.

Keywords European Union • Natura 2000 • Neophytes • Propagule pressure • Species distribution

11.1   Introduction

Europe, and in particular the European Union (EU) which comprises 27 out of the total of 52 European countries, is one the regions of the world with the highest number of protected areas (PAs), and the number has grown rapidly in recent decades (Lockwood 2006; Gaston et al. 2008). Europe has more than 120,000 nationally designated sites[1], of which 105,000 are located in the 39 member as well as collaborating countries associated with the European Environment Agency (EEA). European PAs represent 69 % of the records in the World Database on Protected Areas managed by UNEP-WCMC (European Environment Agency 2012). Protected Areas in the EU cover 15.3 % of the total surface (661,692 km^2), or even 25 % (1,081,195 km^2) if sites implemented as part of the Natura 2000 scheme (Natura 2000 Networking Programme 2007; Gaston et al. 2008) are considered. In the 39 EEA member and collaborating countries the proportion of protected land is 13.7 % (801,500 km^2), or 21 % (1,228,576 km^2) if Natura 2000 sites are included. These figures are well above the world average where the total land area under any legal protection was recently reported as 12.9 %, with only 5.8 % under strict protection for biodiversity (Jenkins and Joppa 2009). Since 1995, the Natura 2000 network has grown to 26,400 sites with a total surface area of about 986,000 km^2, now accounting for nearly

---

[1] A given area can be designated under several designations, often with different boundaries. By 'site' we mean each individual record of a given area under a specific designation type.

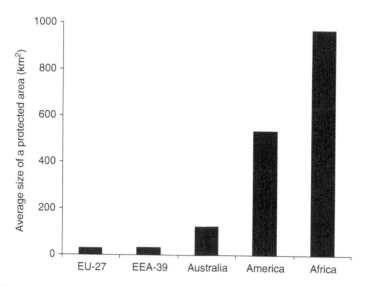

Fig. 11.1 Average size of terrestrial nationally designated protected areas in different regions of the world. EU-27 includes member states of European Union, EEA-39 includes 32 European Environment Agency member countries and seven collaborating countries (http://www. eea.europa.eu/data-and-maps/figures/political-map-of-eea-member-and-collaborating-countries. Taken from UNEP-WCMC 2011)

768,000 km^2 of land, and 218,000 km^2 of sea (European Environment Agency 2012).

In Europe, the term 'protected area' covers a wide variety of designations. Protected areas in this continent are characterised by quite different management regimes, from highly protected sites with limited access to visitors, to parks with a high numbers of visitors, and large areas with rather intense human presence, including dwellings and important economic activities within the borders of the PAs. Such intense human presence in some European PAs is reflected by the large extension of agro-ecosystems, accounting for over 28 % of PAs (European Environment Agency 2006).

The strong influence of humans on nature in Europe began as early as the Neolithic (ca. 3000–1100 BC), and over the centuries has radically altered the natural ecosystems of this region, through for example the harvesting of natural resources, the establishment of settlements, and the cultivation of land. As a consequence, Europe is characterised by a particularly high human density (the average for EU member states is 112 inhabitants per km^2), much higher than that recorded in most other regions of the world. Such density is associated with extensive urbanization, high levels of transport infrastructures and a high degree of fragmentation of the land. As a result of all these characteristics, European PAs are, on average, very small in size compared to other regions of the world (Fig. 11.1; see also Gaston et al. 2008 for more detailed data on selected countries). Most PAs in Europe (90 %) are smaller than 1,000 ha and 65 % range between 1 and

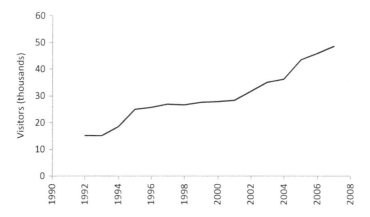

Fig. 11.2 Increase in the number of visitors to Finnish national parks between 1992 and 2007, expressed as the average number of annual visits per park (Based on data from Puhakka 2008)

100 ha; the largest PA is the Yugyd Va National Park in Russia which covers 1,891,700 ha. The high and still growing level of fragmentation of natural areas also brings about concerns about whether the existing PA systems can maintain their biodiversity values under the likely impacts of climate change (Gaston et al. 2008).

On the other hand, Europe is a continent characterised by relative political stability, low levels of poverty and slow human population growth (see e.g. Naughton-Treves et al. 2005; Foxcroft et al. 2014b). Furthermore, Europe is also characterised by a high level of attention paid by the general public to nature, as illustrated by the numbers of visitors to Natura 2000 sites (1.2–2.2 billion visitor days per annum; Gantioler et al. 2010), and by an increasing interest in PAs. For example in Finland the visitation rate to national parks more than tripled between 1992 and 2007 (Fig. 11.2). Tourism plays a key role in regional development, and many PAs have become attractive tourist destinations (Puhakka 2008). The positive trend recorded in nature-based tourism – one of the fastest growing economic sectors globally – contrasts with the declining numbers of visitors in other regions of the world such as United States or Japan (Balmford et al. 2009). On the other hand, increase in tourism also has negative connotations especially in southern and Mediterranean Europe, where extensive coastalization, landscaping of hotels and increasing urbanization, with the demand for more non-native ornamental plants, increases the threat of problem with invasive plants.

11.2 History and Legislation

The history of PAs in Europe is particularly long. It starts with monarchs, who used areas they owned for their personal benefit, for example to harvest game or wood, and prevented the rest of the society from accessing and using these areas. The first example of this kind of land protection in Europe can be dated to 1066, when William the Conqueror created the first hunting forests in Britain, declaring the first game-keeping forest in 1087. Similar legislation, aimed to protect game and forests as a symbol of royal power, was introduced repeatedly at least until the sixteenth century (Welzholz and Johann 2007). Another example of early PAs used by monarchs as hunting preserves is Coto Doñana (Spain) where Alfonso X set up a Hunting Palace in 1262. Since the seventeenth and eighteenth centuries, with the emergence of landscape gardening the interest in natural areas started to shift from the resources they contained to their natural beauty, creating the foundation of modern nature conservation. This aesthetic view was taken up by the European Romantic movement and became one important ideal of Romanticism, which placed great importance on the beauty of such untamed places (European Environment Agency 2012).

In the nineteenth century, the protection of land started to be driven also by private associations that purchased parcels of land for the intrinsic value of nature present in those sites. It was as early as in the 1820s that the first formal PAs were declared in Germany, followed by the creation of PAs in what was then the Austro-Hungarian Empire (present day Austria, Czech Republic, Hungary and Slovakia; European Environment Agency 2012).

In the twentieth century the ownership of natural areas in many cases shifted to the state, and after the Second World War European society started to value the maintenance of biodiversity in PAs. Following the establishment of national parks in North America, many European countries created similar institutions in their colonies. The first country to establish national parks that were owned by the state was Sweden in 1909, and Switzerland followed soon after, in 1914. However, most European national parks were set up after the Second World War, and only in the past 30 years has a broader vision of PAs emerged in Europe, whereby such areas are valued for multiple reasons such as their beauty, their role as repositories of biodiversity, and as potential sources of economic wealth. In this period planners and managers of PAs started to give attention to a proper management of the sites, to involving local communities, and to the need to establish networks of PAs. Protected areas in Europe are currently seen not only as reservoirs for habitats and species, but also as nodes of environmental resilience (European Environment Agency 2010). Furthermore, the economic benefits that PAs can provide are now valued much more than in the past, and Europeans now expect these sites to attract tourists, supply natural resources, and in general to provide the key ecosystem services that are crucial for their livelihood (CREDOC 2008; European Environment Agency 2010).

European legislation on PAs is extensive, complex and continuously evolving. At EU level, several directives have been particularly important for the creation of PAs. Both the EU Birds Directive and the Habitats Directive envisage the creation of PAs as a means of achieving their objectives (see e.g. Gaston et al. 2008 for evaluation of their effectiveness and state of the art). The Special Protected Areas (SPAs) classified under the Birds Directive, and the Special Areas of Conservation (SACs) designated under the Habitats Directive form the Natura 2000 network.

It must be stressed that the establishment of this network (but also its close relative, the Emerald Network[2]) was a turning point in the history of European PAs which contributed to the considerable expansion of the existing system. Through this tool Europe has created the most extensive PA system in the world, which currently (as of the end of 2012) comprises more than 26,000 sites.

In 2001, as part of its commitments to the CBD, the European Commission adopted the biodiversity strategy "Our life insurance, our natural capital: an EU biodiversity strategy to 2020", that, among other targets, commits to improve the conservation status of species and habitats by 2020 and to maintain, enhance and restore ecosystems and their services by the same date. No specific target on the coverage of PAs was included in the Europe Biodiversity Strategy, while CBD target 12 calls to conserve by 2020 "at least 17 % of terrestrial and inland water, and 10 % of coastal and marine areas (. . .) through effectively and equitably managed, ecologically representative and well-connected systems of protected areas . . ." This European decision reflects a shift from designation of new PAs toward a full implementation and enforcement of protection of species and habitats included in the Habitat Directive (in fact, Action 1 of the EU Strategy calls to a full establishment and good management of Natura 2000 sites); in this regard it must be stressed that the need to pass from legal protection to the effective management of PAs is indeed considered a crucial advancement at the global scale (Jenkins and Joppa 2009). Moreover, it must be recalled that Target 5 of the EU Biodiversity Strategy strives to identify pathways of invasions for improving prevention, and to prioritise invasive alien species (IAS) for control. As far as the legal framework on IAS is concerned, the EU has committed itself by a decision of the European Union Council of June 25th 2009, to develop a dedicated legislative instrument on the issue, also mentioned in the above mentioned EU Biodiversity Strategy. However, at this stage the scope and coverage of the instrument are not yet clear.

11.3 Big Picture: Continental Patterns

Globally, a call for developing lists of alien species in PAs was made already in the late 1990s. Usher (1988) summarised available information from 24 nature reserves all over the world, but this data set included only two PAs from Europe (Isle of Rhum, UK and Salvage Islands, Portugal). The most complete data exist for the

[2] The Emerald Network, now under development as part of the Bern Convention, is conceptually similar to the Natura 2000 network, but it incorporates more countries. As the European Union is also a signatory to the Bern Convention, the Natura 2000 network can be considered as the contribution of the EU to the Emerald Network. The Emerald Network works as an extension to non-EU countries of Natura 2000. At present, non-EU countries engaged in the constitution of the Emerald Network are Albania, Armenia, Azerbaijan, Belarus, Bosnia and Herzegovina, Croatia, Georgia, Iceland, Moldova, Montenegro, Norway, the Russian Federation, Serbia, Switzerland, Turkey, Ukraine, and the former Yugoslav Republic of Macedonia (European Environment Agency 2012).

Table 11.1 Summary of available data on representation of alien species in European protected areas

Protected area	Country	Area (ha)	Total species number	Number of alien species	Inclusion criteria	% of alien	reference
Thayatal-Podyjí National Park	Austria/Czech Republic	22,700	1,287	116	neophyte	9.0	Grulich (1997)
Mt Medvednica Nature Park	Croatia	22,826	1,352	27	invasive	2.0	Vuković et al. (2010), native: Dobrovic et al. (2006)
Maksimir Park	Croatia	407	443	16	invasive	3.6	B. J. Hutinec, unpublished data
Gajna Protected Area	Croatia	1,500	235	13	invasive	5.5	Kumbarić (1999)
Labské pískovce Landscape Protected Area	Czech Republic	25,000	1,300–1,400	107	naturalised aliens	8.2–8.9	AOPK (2009)
various (n = 302)	Czech Republic	36,500	2,152 (total)	312 (total)	alien	6.1 (mean)	Pyšek et al. (2004a)
La Brede: reserve naturelle geologique de Saucats	France	75	366	15 (11)	aliens (invasive)	4.1 (3.0)	C. Gréaume, unpublished data
Etang des Landes	France	100	414	3	alien	0.7	K. Guerbaa, unpublished data
Réserve Naturelle Nationale du Bois du Parc	France	45	242	0	alien	0.0	RN Bois du Parc administration, unpublished
Réserve naturelle nationale de la Sangsurière et de l'Adriennerie	France	396	270	2	invasive	0.7	C. Binet, unpublished data
Eifel National Park	Germany	11,000	920	140 (19)	alien (invasive)	17.9 (3.3)	A. Pardey, unpublished data
Asinara National Park	Italy	74,653	n.a.	78	common aliens	n.a.	Camarda et al. (2002)
Arcipelago di La Maddalena National Park	Italy	5,134	n.a.	113	common aliens	n.a.	Camarda et al. (2002)
Selva del Lamone Natural Reserve	Italy	2,002	870	14	alien	1.6	L. Carotenuto, unpublished
Salvage Islands National Park	Portugal	400	107	15	alien	14.0	Brockie et al. (1988)
Sefton Coast, North Merseyside	UK	2,100	1,327	530	alien	40.0	Smith (2012)
Isle of Rhum National Reserve	UK	10,650	520	60	alien	11.5	Brockie et al. (1988)
Mid Yare Reserve	UK	780	572	54	alien	9.4	T. Strudwick, unpublished data

Czech Republic, the only country where a thorough analysis of plant invasions in nature reserves has been published (see Case study 1). However, even this study did not distinguish between invasive and non-invasive aliens and analysed the patterns of species richness and their determinants for all alien plants (Pyšek et al. 2002). The only other summarizing studies are the one on 10 PAs in Slovenia, which focuses on 32 selected alien species (Veenvliet and Humar 2011), and the recent account of the mapping invasive species in PAs on Mediterranean islands (Brundu 2014).

Besides the few published studies listed in Table 11.1, and unpublished reports, some continent-wide data are available from a recent web survey aimed at managers in European PAs that yielded 138 responses from 21 European countries (A. Monaco and P. Genovesi, unpublished). These data provide insights on the quality of information currently available for Europe. Of the total responses received, 95 (79 %) indicated that they have some list of alien plant species available, but the vast majority of lists included only a few invasive plant species of concern. Also, there is much variation in how invasive species are defined, with many data sets not being based on standard scientific criteria (Fabiszewski and Brej 2008; Lamprecht 2008; Schiffleithner and Essl 2010). Data resulting from the survey (Table 11.1) clearly indicates that in some PAs the focus is only on invaders that have some impact on native species and ecosystems, or are otherwise considered important. The proportions of alien/invasive species are therefore not comparable among individual regions. The data nevertheless suggest that the overall variation is within the global range of 5–30 % representation of alien species reported by Usher (1988). In general, the percentage of all aliens in European PAs is within the range of 6–18 %, while that of invasive species, where given and bearing in mind differences in definitions adopted by individual PAs, varies between 2.0 and 5.5 % (Table 11.1). A high number and proportion of aliens in the Sefton Coast PA from where 40 % alien plants are reported is caused by including also casual alien species (P. Smith, unpublished).

From the above it follows that the information on plant invasions in European PAs is surprisingly scarce and mostly scattered in unpublished reports and the grey literature (e.g. AOPK 2009; Bacchetta et al. 2009). Compared to other regions of the world, Europe does not have a comprehensive list of alien plants at least for some kinds of PAs such as national parks in USA (McKinney 2002) and South Africa (Spear et al. 2011), or a subset of PAs delimited by habitat in New Zealand (Timmins and Williams 1991). That invasions in European PAs are seriously understudied is rather surprising since in general terms, this continent is among those where plant invasions are most intensively studied (Pyšek et al. 2008). Furthermore, the majority of available reports are of little use for robust comparison or analysis of factors that determine the levels of invasion due to their incompleteness and selectivity about which species to include. This is also reflected in how often studies on impact are conducted in PAs compared to non-protected areas. In this respect, Europe also lags behind other regions of the world, with little focus on studying impacts directly in PAs (Hulme et al. 2013; Foxcroft et al. 2014a).

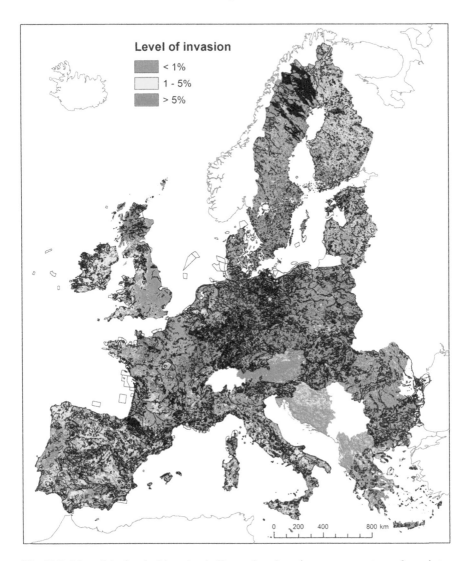

Fig. 11.3 Map of the level of invasion in Europe based on the mean percentage of neophytes (plant species introduced to Europe after ca 1500 A.D.) in vegetation plots corresponding to individual CORINE land-cover classes. Within the mapping limits, areas with non-available land-cover data or insufficient vegetation-plot data are blank. Taken from Chytrý et al. (2009); reproduced courtesy of Blackwell Scientific Publications. The position of Natura 2000 sites is shown in black (not available for some countries)

An indirect insight into the threat from plant invasions in PAs in Europe is provided by comparing the geographical distribution of Natura 2000 sites with the level of invasion in European regions derived from the map of plant invasions at the continent (Fig. 11.3). The overall picture reveals that areas of conservation interest, represented by the Natura 2000 network[3], are located in areas less threatened by invasions; this is most obvious in UK, the Mediterranean region and southern Europe. The map, however, also indicates that in many regions, namely in central and Eastern Europe, Natura sites are often located in landscapes that are heavily invaded. Overall, areas containing Natura 2000 sites are about half as invaded (containing on average 1.8 % alien species in vegetation plots; see Chytrý et al. 2009 for details on calculations) as areas without Natura 2000 sites (3.5 %).

11.3.1   Case Study 1. Regional Patterns Illustrated by Protected Areas in the Czech Republic

Surprisingly, the only comprehensive study dealing with a detailed pattern of plant invasions into European PAs seems to be the one from the Czech Republic (Pyšek et al. 2002, 2003a). These data can be used to illustrate regional patterns of invasions into natural temperate plant communities.

Based on over 300 PAs of various status and size (representing 17 % of the number of PAs in the country and 44 % in terms of protected land area), the study showed that the level of PAs invasion by neophytes (modern invaders introduced after the end of the Medieval Period; Pyšek et al. 2004b, 2012a, b) was determined by an interplay of environmental and human-related factors, the most important being climate (decreasing level of invasion with increasing altitude due to colder conditions), propagule pressure (increasing in areas with a high human population density and indirectly pointing to previously reported effects of visitation; Usher 1988; Macdonald et al. 1989; Lonsdale 1999; McKinney 2002; Mortensen et al. 2009), and habitat heterogeneity (indicated by the positive relationship with the number of native species; cf. Timmins and Williams 1991). On average only 6.1 % of plant species recorded in a PA were alien. However, there was a great variation in the level of invasion among particular sites and in some PAs the proportion of alien species was as high as 25 %. Of the two standardly distinguished categories of alien plants in Europe, based on the time of immigration, neophytes were less represented, only 2 % of the total number of species (Pyšek et al. 2002). As neophytes are the group from which the majority of important invaders are recruited in Europe (Lambdon et al. 2008), the data indicate that the overall threat from alien species in Czech nature reserves was relatively minor.

However, the overall level of invasion based on the presence of all aliens is only part of the story. Although this measure was shown to be correlated with the presence of

---

[3] The degree of overlap between nationally designated PAs and Natura 2000 sites is variable; in some countries, as Malta, Estonia, Latvia, there is no overlap because there was no developed pre-existing national system of PAs (Gaston et al 2008). In other countries (e.g. Cyprus, Bulgaria, Denmark, Ireland) the overlap is more than 80 %. Other figures include, e.g.: Italy and France >40 %; Poland and Spain >60 %, Germany ~10 %, Czech Republic >20 % (source European Topic Centre on Biological Diversity – ETC/BD, 2009).

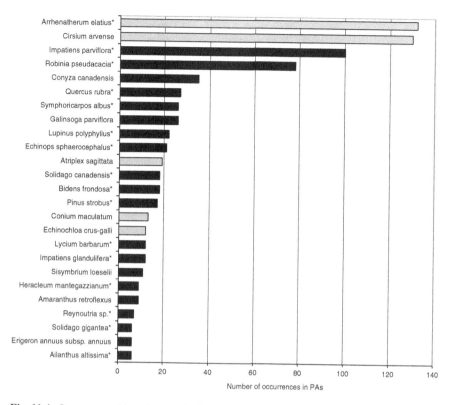

Fig. 11.4 Occurrence of invasive species in protected areas in the Czech Republic, showing the number of protected areas in which the species was recorded. Based on data in Pyšek et al. 2002, with the delimitation of an invasive species following Pyšek et al. 2012b (as invasive are considered only those that currently spread in the country). Archaeophytes (introduced to the country before 1500 A.D.) are shown as *grey bars*, neophytes (introduced after that date) in *black*. Species that are not restricted to disturbed sites and are capable of invading natural vegetation are marked with *asterisks*

major invasive plants at the continental scale of Europe (Chytrý et al. 2012), focusing on invasive species (in the sense of Richardson et al. 2000; Blackburn et al. 2011) provides deeper insights into the real threats posed by invasions. Of the total number of 50 neophytes and 11 archaeophytes currently considered as invasive in an updated national checklist (Pyšek et al. 2012a, b), 31 and 8, respectively, were recorded in the investigated sample of PAs. This corresponds to rather high percentage of the total, 62 and 72 %, respectively, a much higher figure than for all neophytes (Pyšek et al. 2002). While many invasive aliens occur only occasionally, 25 were recorded in at least five PAs and some are rather widespread – the top four species were present in at least 25 % of the PAs studied (Fig. 11.4). This group includes some neophytes invading also in semi-natural habitats, such as *Robinia pseudoacacia* (black locust), *Quercus rubra* (northern red oak), *Lupinus polyphyllus* (large-leaved lupin), *Solidago canadensis* (Canada goldenrod), *Pinus strobus* (eastern white pine), *Heracleum mantegazzianum* (giant hogweed), or *Fallopia* (knotweed) taxa. The impact of these

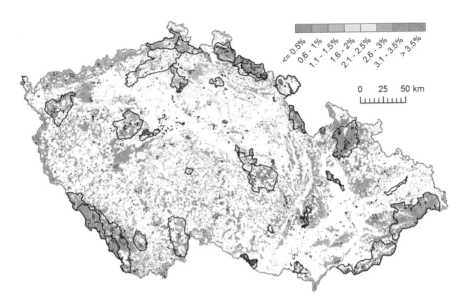

Fig. 11.5 Map of the level of invasion in seminatural habitats in the Czech Republic based on a quantitative assessment of the proportion of neophytes among the total number of species in vegetation plots located in 35 terrestrial habitat types at different altitudes (see Chytrý et al. 2009 for details on methods). The network of protected areas, as of 1994, is displayed as *black areas*; large areas are outlined. Based on the map published in Chytrý et al. (2009), reproduced courtesy of the Czech Botanical Society

species on invaded communities has been documented in the national and continental literature (DAISIE 2009; Hejda et al. 2009; Pyšek et al. 2012b, c) and the group includes a number of woody species, invasions of which are known to be especially devastating (Richardson and Rejmánek 2011). Similar results pointing to relatively high potential for future invasions emerged from a finer-scale study of 48 urban reserves in the city of Prague; these reserves harbour a comparable figure of about two-thirds of invasive neophytes recorded in the country, and many of the most invasive species are shrubs and trees (Jarošík et al. 2011). This indicates that overall levels of invasions derived from numbers of all aliens may not provide a reliable picture about the magnitude of threat from invasions, and management plans need to be designed for individual reserves based on the occurrences of major invasive species.

That the degree of threat of invasions in PAs varies at the country scale, depending on geographic conditions, climate, altitude and intensity of human influence, can be illustrated by a national map of plant invasions, based on a quantitative assessment of the proportion of neophytes among the total number of species in vegetation plots located in particular habitat types. Generally, mountain reserves are little affected but some PAs are located in heavily invaded regions such as lowland sandy areas and river corridors (Fig. 11.5).

At the global scale, nature reserves are invaded about half as much as sites outside reserves (Lonsdale 1999). This difference seems to be even more pronounced at the regional scale of the Czech Republic: the network of PAs analysed in the Czech data set captured about 80 % of the country's native flora but less than 20 % of neophytes (Pyšek et al. 2002). The mechanism underlying this phenomenon is that it is more difficult for alien species to colonise PAs than corresponding sections of non-protected landscape because natural vegetation acts as a buffer against plant invasions (see also Foxcroft et al. 2011). Among PAs in the Czech Republic those established long ago harbour significantly fewer neophytes than those established more recently, and the neophytes from a rapidly increasing pool of species in the surrounding landscapes were not captured over time in the PA network any faster than were native species from the pool available at the time of establishment of the first reserves (Pyšek et al. 2003b). This suggests that natural vegetation in nature reserves creates an effective barrier against the establishment of alien species (see also Meyerson and Pyšek 2014).

11.4   The Most Invasive Plant Species in European Protected Areas

At the species level, what information is available on the most serious plant invaders in European PAs and how does this continent stand compared to others? A summary is provided by De Poorter (2007) in her scoping report produced for the World Bank as a contribution to the Global Invasive Species Programme (GISP). This study emphasises that there is a shortage of consolidated information at global, international and/or regional level, on IAS impacts, threats and management in PAs. It also found that a wealth of information is available at site and national levels, but that it is very dispersed and not standardised, which makes it difficult to get a balanced global picture (De Poorter 2007).

De Poorter (2007) list 58 significant invasive plant and animal species for Europe, the criterion for inclusion being that they have impact and represent threat to PAs. This number, although the comparison is biased by different sizes of regions, places Europe in the middle of the range given for continents; the number of plants and animals these authors list as invasive in PAs in USA and Canada is 109 (84 of which are plant species), in Australia and New Zealand 87 (57), Africa 58 (47), Asia 43 (30), Oceania 19 (13), and South and Central America and Mexico 18 (10). Among the 58 European invaders there are 37 plants. The list includes 25 trees and shrubs, eight perennials, and four annuals. Although the results of the survey were influenced by the limited information accessible (De Poorter 2007), the species perceived as problematic in European PAs nevertheless overlap to a large degree with those known to be invasive in Europe in general, i.e. also outside PAs (DAISIE 2009).

A more detailed picture of how the major invasive plants are distributed in European PAs can be inferred from the above mentioned web survey (A. Monaco and P. Genovesi, unpublished) in which managers reported species they consider most harmful to their areas (Table 11.2). Among the 378 taxa listed at least once, the top invasive species are *Fallopia japonica* (Japanese knotweed, which most

Table 11.2 Plant species reported as most harmful in European protected areas by managers

Taxon	LH	Origin	Number of PAs	Number of European regions
Fallopia japonica et sp.	p	Asia	48	36
Impatiens glandulifera	a	Asia	29	34
Robinia pseudoacacia	t	N America	26	42
Ailanthus altissima	t	Asia	16	36
*Heracleum mantegazzianum**	p	Asia	11	25
Ambrosia artemisiifolia	a	N America	10	33
*Solidago canadensis**	p	N America	9	36
Solidago gigantea	p	N America	8	32
Amorpha fruticosa	s	N America	7	17
*Elodea canadensis**	p	N America	6	38
Acer negundo	t	N America	6	33
Acer pseudoplatanus	t	Europe	6	19
Prunus serotina	s	N America	5	24
*Baccharis halimifolia**	s	N America	4	6
*Buddleia davidii**	s	Asia	4	23
Caulerpa racemosa	al	Africa	4	15
*Echinocystis lobata**	a	N America	4	15
*Heracleum sosnowskyi**	p	Asia	4	7
*Impatiens parviflora**	a	Asia	4	31
*Opuntia ficus-indica**	p	C America	4	13
*Phytolacca americana**	p	N America	4	29
Carpobrotus edulis	p	Africa	4	22
*Asclepias syriaca**	p	N America	3	18
*Datura stramonium**	a	N America	3	42
Rhododendron ponticum	s	Europe, Asia	3	10
*Senecio inaequidens**	a	Africa	3	26
Xanthium italicum	a	N America	3	20

Based on web survey (A. Monaco and P. Genovesi, unpublished; see text for details)

Number of PAs (n = 118) in which the species ranked among the most harmful invasive plants is shown and compared with the overall distribution in Europe, expressed as the number of regions in which it occurs, based on DAISIE database (DAISIE 2009)

Species missing from the European list presented in a global survey of De Poorter (2007, see text) are marked with asterisk

LH life history, *a* annual, *p* perennial, *s* shrub, *t* tree, *al* alga

likely includes other European taxa of this genus such as *F. sachalinensis*, giant knotweed, and the hybrid *F. ×bohemica*; Pyšek 2009), reported to have impact in 41 % of the total number of surveyed PAs, *Impatiens glandulifera* (Himalayan balsam; 25 %), *R. pseudoacacia* (22 %), *Ailanthus altissima* (tree of heaven; 14 %), *H. mantegazzianum*[4] (9 %), and *Ambrosia artemisiifolia* (common ragweed; 9 %). Interestingly, a number of species perceived as the top invaders at the site level in European PAs are not listed for this continent in the above mentioned global survey

---

[4] This may include also related species *H. sosnowskyi* and *H. persicum* (Jahodová et al. 2007).

for Europe (De Poorter 2007; Table 11.2) which supports the concerns about the quality of information available. Missing from the global list are some species whose absence can be attributed to taxonomic issues (e.g. *S. canadensis*, *Opuntia ficus-indica*, prickly pear).

11.5 Magnitude of the Problem: Impact and Management

The screening conducted by De Poorter (2007) revealed that IAS are reported as having impact in 144 PAs surveyed, located in 29 European countries. Those numbers are, in absolute terms, higher than in other regions of the world, for the number of PA sites approximately twice as many as in Africa, Asia, Americas and Australia with New Zealand. Globally, the study showed a significant number of PAs where IAS have been recorded as an issue (De Poorter 2007). However, a rigorous data set recently assembled on global impacts of invasive plants on species, communities and ecosystems (Vilà et al. 2011; Pyšek et al. 2012c) indicates that in Europe, studies on ecological impacts are conducted in PAs disproportionally less frequently than on other continents. Europe contributes only 5 % to the total number of impacts tested in PAs but 31 % to those resulting from studies conducted outside PAs (P. Hulme et al. 2013).

An insight into how impacts of invasive plant species are perceived by the administration of PAs in Europe is provided by the web survey (A. Monaco and P. Genovesi, unpublished). In general, merging both plants and animals, competition with native species and changes imposed to the habitats and ecosystem functioning are considered as the most serious impacts of invasive species in European PAs (Fig. 11.6). Interestingly, a comparison with rigorous data available from the recent analysis of impacts of invasive plants reveals that the ranking of impacts perceived by managers corresponds well to the ranking of impacts resulting from published scientific studies, as reported in Pyšek et al. (2012c). The impacts that can be largely attributed to competition, i.e. those on richness, diversity and abundance of resident species, are most likely to be significant, and those affecting habitats, i.e. mainly on soil properties, come second.

Closely linked with impacts are management options that PA managers in Europe consider to be most effective (Fig. 11.7). They perceive eradication and control to be the best approaches for dealing with invasive species. The fact that European PA managers consider these two measures more important than prevention, education or public involvement, probably reflects the approach often adopted in PAs that tend to focus more on responding to invasions than working on prevention, although prevention is increasingly viewed as the best management option (Pyšek and Richardson 2010; Meyerson and Pyšek 2014). Interestingly, if available management options are compared with what is actually being implemented (Fig. 11.7), several issues emerge. The most frequently implemented action against alien species in PAs is monitoring. Both active management options, eradication and control, are in reality highly under-represented compared with how

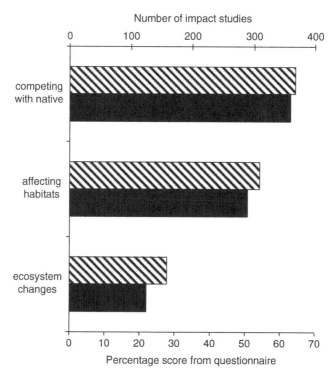

Fig. 11.6 Comparison of the most serious impacts of plant invasions as perceived by managers of protected areas in Europe (based on a web survey of A. Monaco and P. Genovesi, unpublished) with data from studies rigorously testing impact in European PAs. The former measure (hatched columns) is a percentage score calculated from the received responses. Managers were asked to rank the five most serious impacts in their PA on a semi-quantitative scale, and these were scored from 5 to 1; the full score (100 %) would therefore correspond to an impact perceived by all managers in all PAs as the most serious. The latter measure (*black columns*) are percentages of significant impacts among all tested, as addressed by studies conducted within protected areas, based on 574 individual cases (see Pyšek et al. 2012c for details on primary data, and Hulme et al. 2013, for the frequency of studies conducted in PAs). Tested impacts were grouped so as to correspond to the categories delimited by the web survey

frequently they are suggested by managers as being the best strategy. The same is true for prevention which is assumed to be most effective and cheapest measure. Well represented by real activities is part of prevention devoted to education and public involvement where communication towards public is highly used. A clear message from such comparisons is to act as early as possible when the infestations are relatively small to be effectively manageable, instead of relying on long-term monitoring (Mack and Lonsdale 2002; Simberloff 2003; Pluess et al. 2012).

Unfortunately, the ability of many European PAs to withstand biological invasions is limited by inadequate management, not only concerning biological invasions but also in general terms. As a result of this inadequacy, it has been estimated that less than 20 % of the species and habitats listed by the Habitats Directive have a

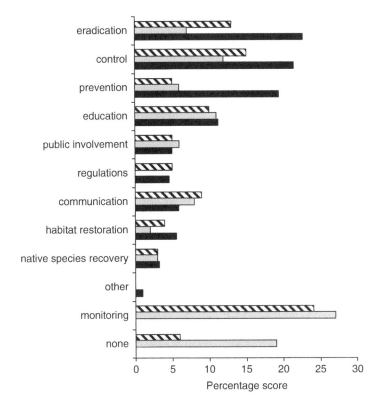

Fig. 11.7 Comparison of the management options considered by managers of protected areas as most effective (*black columns*, measured as the cumulative score from the questionnaire for both plants and animals) with the frequency of how actual measures are implemented (hatched for plants, *grey* for animals). Based on a web survey (A. Monaco and P. Genovesi, unpublished); see Fig. 11.6 for details on calculation of the percentage score

favourable conservation status in Europe (European Environment Agency 2012), similar to the situation globally (Bertzky et al. 2012). According to the adaptive management framework approach (Walters 1986; Foxcroft 2004), knowledge and expertise improve by practice, i.e. actually doing things. Therefore management actions in European PAs are often being undertaken even though the full extent of the problem is not known, especially if only a few invasive species need to be eradicated or contained rapidly. So, in a number of PAs in Europe actions were taken to control or eradicate invasive plants, for example, *Amorpha fruticosa* (desert false indigo) in Croatia (Council of Europe 2011), or management actions against invasive plants in national parks in Poland (M. Opęchowska, unpublished). These efforts were often part of LIFE projects aimed at ecological restoration within Natura 2000 sites (Scalera and Zaghi 2004; Table 11.3; see also Case study 2).

A major challenge to the management of European PAs relates to on-going global change, especially climate change. There is strong evidence in the literature that climate change is likely to result in changes in species' distributions. However,

Table 11.3 Selected LIFE projects conducted in Europe aimed to control or eradicate invasive and alien plants in Natura 2000 sites

LIFE project no.	Name	Period	Country	Major target IAS	Habitat
LIFE95 ENV/F/000782	Control of the *Caulerpa taxifolia* extension in the Mediterranean Sea	1996–1999	France	*Caulerpa taxifolia*	Posidonia beds
LIFE96 NAT/E/3180	Restoration and integrated management of the Island of Buda	1996–2000	Spain	*Phoenix* spp., *Washingtonia* spp.	Atlantic insular ecosystems
LIFE97 NAT/UK/004244	Restoration of Atlantic Oakwoods	1997–2001	UK	*Rhododendron ponticum*, non-native conifers	Atlantic forest
LIFE97 NAT/UK/004242	Securing Natura 2000 objectives in the New Forest	1997–2001	UK	*Rhododendron ponticum*, non-native conifers	Heathland
LIFE97 NAT/P/004082	Management and conservation of the Laurisilva Forest of Madeira	1998–2000	Portugal	*Hedychium gardnerianum*	Laurel forest
LIFE97 NAT/IT/4134	Restoration of alluvial woods in the Ticino Park	1997–2000	Italy	*Prunus serotina, Robinia pseudoacacia, Ailanthus altissima, Quercus rubra*	Wet woodlands
LIFE98 NAT/A/5418	Pannonian sand dunes	1998–2002	Austria	*Ailanthus altissima, Robinia pseudoacacia*	Relict dunes
LIFE99 NAT/E/6392	Restoration of the islets and cliffs of Famara (Lanzarote Island)	1999–2002	Spain	*Nicotiana glauca*	Atlantic insular ecosystems
LIFE00 NAT/UK/007074	Woodland habitat restoration: Core sites for a forest habitat network	2001–2005	UK	*Rhododendron ponticum*	Atlantic forest
LIFE00 NAT/E/007339	Dunas Albufera: Model of restoration of dunes habitats in 'L'Albufera de Valencia'	2001–2004	Spain	*Carpobrotus edulis*	Coastal dunes
LIFE00 NAT/E/7355	Conservation of areas with threatened flora on the Island of Minorca	2001–2004	Spain	*Carpobrotus edulis*	Mediterranean insular ecosystems
LIFE03 NAT/IT/000139	RETICNET 5 SCI for the conservation of wetlands and main habitats	2003–2006	Italy	Unspecified	Alpine wetland

Code	Project description	Years	Country	Species	Habitat type
LIFE03 NAT/FIN/000039	Lintulahdet: Management of wetlands along the gulf of Finland migratory flyway	2003–2007	Finland	*Phragmites australis*	Wetland
LIFE04 NAT/ES/000044	Recovery of the littoral sand dunes with *Juniper* spp. in Valencia	2004–2007	Spain	*Carpobrotus edulis, Agave americana*	Mediterranean dunes
LIFE04 NAT/CY/000013	Conservation management in Natura 2000 sites of Cyprus	2004–2008	Cyprus	*Robinia pseudoacacia, Eucalyptus regnans*	Mattoral
LIFE05 NAT/D/000051	Large herbivores for maintenance and conservation of coastal heaths	2005–2009	Germany	*Prunus serotina*	Heathland
LIFE05 NAT/NL/000124	Dutch coastal dunes: Restoration of dune habitats along the Dutch coast	2005–2011	The Netherland	Unspecified	Coastal dunes
LIFE05 NAT/IT/000037	DUNETOSCA: Conservation of ecosystems in northern Tuscany	2005–2009	Italy	*Yucca gloriosa, Amorpha fruticosa*	Mediterranean coastal ecosystems
LIFE05 NAT/IRL/000182	Restoring priority woodland habitats in Ireland	2006–2009	Ireland	*Picea abies, Picea sitchensis, Larix decidua, Pinus radiata, Fagus sylvatica, Acer pseudoplatanus, Aesculus hippocastanum, Laurus nobilis, Fallopia japonica, Rhododendron ponticum*	Woodland
LIFE05 TCY/CRO/000111	IBM, Central Posavina: Wading toward integrated basin management	2006–2008	Croatia	*Amorpha fruticosa, Xanthium spp.*	Floodplain ecosystem
LIFE08 NAT/IT/000353	Montecristo 2010	2010–2014	Italy	*Ailanthus altissima, Carpobrotus spp., Pinus halepensis, Acacia pycnantha*	Mediterranean insular ecosystems

the current networks of PAs may no longer be effective in conserving biodiversity under rapidly changing climatic conditions because they were designed on the basis of a paradigm of long-term stability of species' geographical distributions (Huntley et al. 2011). It is widely acknowledged that PAs will soon have to face changes due to global change. This also has important implications for the design of new PAs in that core areas need to be secured to accommodate predicted changes (Hannah 2001; Hannah et al. 2007). However, information on how these processes could influence the distribution/abundance of plant invasions in PAs is very scarce (see Kleinbauer et al. 2010; Case study 3).

11.5.1 Case Study 2. Management of Rhododendron ponticum in Protected Areas in the UK

An alien plant species that is causing major conservation problems in European PAs is *Rhododendron ponticum* (rhododendron). This densely branched evergreen shrub produces several million seeds per bush; the seeds are dispersed over long distances (up to 100 m) by wind and water under favourable open conditions, but over shorter distances in closed canopy forest, and remain viable in soil for several years. The plants are also capable of limited branch rooting in contact with soil, usually only at forest edges, and sprout vigorously after cutting (Stout 2007; Hulme 2009). It is unpalatable to vertebrates and few insects feed on the plant. The species represents an invader that is native to part of Europe but is an invasive alien outside its native range (see Lambdon et al. 2008). Formerly widely distributed throughout Europe during the Tertiary, the extant native range is disjunct with *R. ponticum* subsp. *baeticum* occurring in Iberian Peninsula, and subsp. *ponticum* occurring around the Black Sea. The species was introduced to parts of Europe where it is now invasive as an ornamental plant (Milne and Abbott 2000; Erfmeier and Bruelheide 2010) and it is still available from nurseries. It is naturalised in the British Isles and western continental Europe, and the extent of invasion is increasing. The shrubs often completely dominate the invaded area, accumulate thick litter layers allowing a few plants to survive under the canopy (Hulme 2009), and integrate into existing pollination networks (Vilà et al. 2009). The invasion success of *R. ponticum* has been attributed to the greater environmental suitability of the new regions, a wider range of favourable habitat types (Shaw 1984; Erfmeier and Bruelheide 2010), but also to genetic shift in invasive populations towards an increased investment in growth and a faster germination rate, and genotype × environment interactions play a major role during the invasion process. Both hybridization and ecological release from constraints experienced in the native range are plausible explanations for its success (Erfmeier and Bruelheide 2005).

After the initial introduction to UK in 1763, it was introduced to many private estates, parks and woodlands for game cover, mainly from Spanish populations. It was recorded as naturalised by the late nineteenth century and spread widely in the

Fig. 11.8 Invasion of *Rhododendron ponticum* in Killarney National Park, Ireland (Photo: P. Pyšek)

twentieth century (Scalera and Zaghi 2004). The species became highly invasive in semi-natural forests and woodlands, but also heaths, bogs and sand dunes on a wide range of damp acid substrates over British Isles (Fig. 11.8; Cross 1975; Foley 1990; Gritten 1995), negatively impacting numerous habitats identified for protection under the EU Habitats Directive).

In Snowdonia National Park, Wales, it was first planted as an ornamental shrub in large estates, and extensively used as rootstock for grafting ornamental varieties (Gritten 1995). Control efforts started to appear in 1980s (e.g. Shaw 1984; Gritten 1992, 1995), but by the time it has been generally accepted that control was needed, the extent of the effort required was huge, over landscapes (Scalera and Zaghi 2004). For example, the costs of control at a park-wide scale were estimated at £45 million at 1992 prices (Gritten 1995). Since 1997 the EU has co-financed five LIFE Nature projects (http://ec.europa.eu/environment/life; Table 11.3) to tackle the problem of invasion by rhododendron in England, Scotland and Wales, with focus on eradicating rhododendron populations from the core Natura 2000 areas, providing a *cordon sanitaire* around the treated sites and to work with private landowners and communities to seek support for a coordinated programme. The techniques employed involve the widespread removal of plants using mechanical methods, burning the cut plants, removing root mats to expose fresh soil, and controlling re-growth with herbicides. One of the LIFE projects was aimed at the large-scale removal of rhododendron from woodlands at Loch Sunart, western Scotland. The project mobilised private landowners by covering 95 % of the costs, and higher payment rates were offered for work within Natura 2000. In a follow-up project in Loch Sunart it appeared that until all sources of seed are tackled the threat of reinvasion persisted and not all landowners supported the continuation of eradication programme. Another project resulted in clearing of 110 ha of rhododendron-invaded heaths and woods in the New Forest National Park, followed by monitoring to ensure that any new foci were quickly targeted for management. As an additional constraint, the eradication projects faced obstacles due to rhododendron's popularity as a garden plant. Overall, despite the long-term effort and the huge investment of resources, rhododendron remains a problem in and near PAs and Natura 2000 sites. Besides a call for a general change in attitude, the replacement of cultivated stock by planting of dwarf sterile rhododendron hybrids has been suggested as a solution (Scalera and Zaghi 2004).

Rhododendron ponticum is an example of a highly invasive species whose distribution across Europe, including PAs, is rather localised (Lambdon et al. 2008; DAISIE 2009). Nonetheless, it is one of the most invasive species in European PAs, incurs great economic costs, and attempts to undertake coordinated control efforts have been ineffective. For example, in Scotland it still considered one of the most noxious plant invaders, with £1.6 million spent on its management in 2011–2012 in several districts (Forestry Commission Scotland 2012).

11.5.2 Case Study 3. Climate Warming Will Drive the Invasive Tree Robinia pseudoacacia into Nature Reserves

Robinia pseudoacacia is among the most widespread invasive plant species in Europe (Lambdon et al. 2008), in central Europe it is among those with the broadest habitat range (Chytrý et al. 2005), and it is invasive in most countries (Essl and Rabitsch 2002; Pyšek et al. 2009b, 2012a, b; Medvecká et al. 2012). This deciduous tree, native to central and eastern North America, is up to 30 m tall and grows as an early successional species in open and disturbed habitats. It has a good regeneration capacity, resprouting well from roots and stumps. It was introduced to Europe in 1601 as an ornamental species, and was later used for timber production and erosion control (Başnou 2009; Pyšek et al. 2012a). As a nitrogen fixing species, it can achieve early dominance on open sites where nitrogen is limiting to other species, and ecological impacts of this species on biodiversity and ecosystem functioning are well documented (Kowarik 2003; Rice et al. 2004).

In a study on the current and future distribution of *R. pseudoacacia* in Austria, using niche-based predictive modelling, Kleinbauer et al. (2010) investigated whether the predicted dynamics might represent invasion threat to the existing network of Natura 2000 sites. By doing this, the study addressed potential problems resulting from the static nature of the PAs network that disregards the dynamics of species ranges; this issue will be increasingly relevant under on-going environmental change, including global warming, that will potentially shift the distribution of suitable habitats for many species that are nowadays protected in areas designated based on their distribution in the past (Parmesan and Yohe 2003; Parmesan 2006). Climate change can thus not only drive potentially endangered species out of the boundaries of existing reserves but also facilitate colonization of reserves by invasive plant species (Kleinbauer et al. 2010).

The study showed that current distribution of the species was strictly controlled by temperature constraints and predicted an increase in the area invaded by *R. pseudoacacia* under warmer climate. This is a general phenomenon seen for many invasive plant species in central Europe that originate from areas climatically warmer than is this target region (Pyšek et al. 2003b). The predictions differed among the 13 forest and grassland habitats that were included in the study as potentially invasible by *R. pseudoacacia*. Therefore, the risk of invasion into legally protected areas and habitats that are vulnerable to colonization by *R. pseudoacacia* is likely to increase with climate warming, with the threat being most pronounced for endangered habitats of a high conservation value. Moreover, the study predicts not only an increase in area invaded but also that in the abundance of *R. pseudoacacia* populations, exacerbating their impact on invaded ecosystems. These results point strongly to the necessity of proactive management of PAs whereby consideration should be given to different facets of global change in a more explicit manner. They also suggest that reducing propagule pressure by avoiding the establishment of plantations close to endangered reserves and habitats is the most straightforward way to prevent further invasion under a warmer climate (Kleinbauer et al. 2010).

11.6 Challenges, Solutions and Strategy: Towards the Brighter Future of European Protected Areas?

Europe has lagged behind other regions of the world in the struggle against IAS (Genovesi and Shine 2004), largely because of the limited awareness of the European society on this issue. Despite the fact that 96 % of Europeans consider the protection of the environment to be important and 84–93 % perceive the loss of biodiversity as a serious problem, only 2–3 % of European citizens acknowledge IAS as a significant threat (Gallup Organisation 2007; Hulme et al. 2009). The limited level of understanding and concern regarding IAS is indeed a major obstacle to more effective policies on biological invasions (Brunel et al. 2013). It is therefore urgent to inform the public better on this issue. Protected areas can play a pivotal role in this regard because these institutions have a direct link with their visitors, and enjoy a high credibility in public opinion. Better information and education of the public could be ensured also by directly involving them in activities such as monitoring or

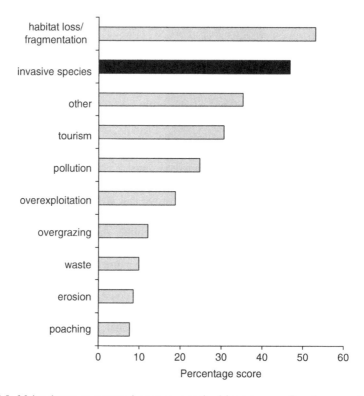

Fig. 11.9 Major threats to protected areas as perceived by managers. Based on a web survey (A. Monaco and P. Genovesi, unpublished), see Fig. 11.6 for details on calculation of the percentage score. Other threats listed with low frequency are shown together and include: human conflicts, climate change, lack of resources, ecological instability and lack of political support

management, as in the case of the "balsam blitz" at the Pembrokeshire Coast National Park (Wales), where volunteers are engaged in controlling *I. glandulifera* (NewsWales 2011). Another example is the on-going eradication of *Lysichiton americanus* (American skunk cabbage) in the Taunus Nature Park (Germany). This project involves over 100 volunteers and is planned take at least 10 years to complete it (B. Albertemst, personal communication). Such initiatives provide opportunities to launch far-reaching awareness campaigns.

Fortunately, the generally limited awareness of biological invasions by the public does not extend to the managers of PAs, who have a high concern about the threats posed by IAS (Scalera and Zaghi 2004). Based on the results of the web survey (A. Monaco and P. Genovesi, unpublished), IAs are now considered by managers and administrators of PAs the second most serious threat, after habitat loss and fragmentation – and more important than tourism (Fig. 11.9). This specific attention of European PAs on biological invasions probably reflects the direct experience of managers with the impacts caused by IAS whose numbers are constantly and rapidly growing in all European environments and regions (Hulme et al. 2009). The perception by European PA managers of these main threats

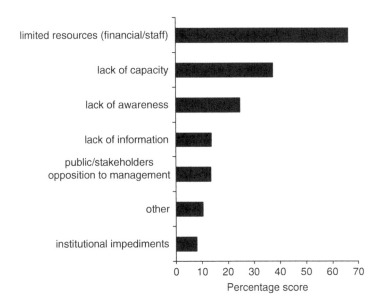

Fig. 11.10   Key impediments to dealing with the spread of invasive species in European protected areas as perceived by managers. Based on a web survey (A. Monaco and P. Genovesi, unpublished); see Fig. 11.6 for details on calculation of the percentage score

also corresponds reasonably well with general global threats as identified by e.g. Millennium Ecosystem Assessment (2005) – that underlined the need to improve the management of PAs, to mitigate the impacts of development, over harvesting, unsustainable tourism, invasive species, and climate change – or specifically for nature reserves and island ecosystems (Robertson et al. 2011).

The magnitude of impacts of invasive species in European PAs is expected to grow rapidly in the near future, with severe effects on biodiversity and ecosystem services. Besides increasing numbers of introductions to the continent, PAs are under increasing pressures from the continuous growth of tourism (Fig. 11.2) which is a major source of propagules of alien species (Usher 1988; Lonsdale 1999; Pretto et al. 2012). Furthermore, on-going climate change may aggravate the impact from invasions (e.g. Kleinbauer et al. 2010, see Case study 3 above).

However, the ability to develop more effective and science-based responses to this threat in European PAs is constrained by several factors. These include limited support from the rest of the society (including decision makers), the inadequate legal framework reflecting the European context, the lack of early warning rapid response frameworks (Brunel et al. 2013), the lack of specific financial mechanisms, including those for contingency actions, and – last but not least – the lack of data on invasive species in PAs (Fig. 11.10). Regarding the last mentioned aspect, inventories of invasive species in PAs, using standard scientific criteria, are urgently needed to support European PAs in their efforts to prevent and control invasions (e.g. Pyšek et al. 2009a). This can only be achieved through coordinated international cooperation for which the system of projects funding within EU provides a suitable framework.

To achieve this aim, and in order to improve the knowledge base of actual distribution of IAS, managers of protected areas should make more use of the 'citizen science' or 'citizen as a sensor' approach, where the public is involved in monitoring of natural resources for improving management and/or research, often allowing scientists to accomplish studies that would otherwise be unfeasible. In general the 'citizen science' approach can also be aimed at promoting public engagement, information and education. Properly trained volunteers could be effectively involved in inventories and monitoring programmes of IAS distribution and could play a fundamental role in the surveillance of new IAS arrivals, to support an early warning and rapid response system (Genovesi et al. 2010; Gallo and Wait 2011). The initiative of the EEA "Eye on Earth" is an interesting example of the involvement of the public for recording data on IAS (http://www.eyeonearth.org/en-us/Pages/Home.aspx, section nature), for which aim a number of open platforms can provide valuable tools, such as the applications developed for several electronic devices such as mobile phones, tablets, etc. (e.g. "Aliens Among Us app"; http://www.royalbcmuseum.bc.ca/TravellingExhibitions/default.aspx; "iAs_sess", http://ias-ess.org).

More generally, in order to prevent further impacts by IAS, European PAs should give priority to the prevention of new introductions, by identifying the priority pathways of IAS introduction, and addressing them through a balanced and regulatory approach. European PAs, compared to other regions of the world, are characterised by widespread presence of human activities within their borders or in their immediate surroundings[5]. It is important to extend such a precautionary approach outside the borders of PAs, and to discuss with competent authorities – not only at the local or national scale, but also at the European level – ways of preventing introductions of alien plants by forestry, horticulture, or via botanical and zoological gardens. Protected areas should be involved in fostering more responsible behaviour by private individuals and industries, for example by promoting the adoption of agreed standards, best-practice guidelines, or codes of conduct that are being developed by European institutions (Heywood and Brunel 2009; Heywood 2011, 2012; Scalera et al. 2012). Furthermore, it is crucial to improve the ability of European PAs to promptly detect new invasions, and to enforce effective response measures. European institutions should consider the adoption of specific financial mechanisms to allow for prompt response to and development of contingency plans for new invasions, based on a rapid evaluation process.

---

[5] This is confirmed by the analysis on PAs coverage per IUCN category (EEA 2012) that highlights that the most represented IUCN category by surface (about 50 %) in European PAs is the V, that is "... protected area where the interaction of people and nature over time has produced an area of distinct character with significant ecological, cultural and scenic value ..." (IUCN, http\\www.iucn.org).

11.7 Conclusions

Several issues that emerge from our synthesis can be summarised as follows (see also Gaston et al. 2008):

- In Europe there is a well-developed system of PAs that was expanded by the Natura 2000 scheme, and a high level of interest of the public in nature protection. However, PAs suffer from habitat fragmentation and diverse human pressures, and are facing pressure due to climate change. Despite the extensive legal framework for conservation planning, quantitative goals and indicators of the effectiveness of biodiversity protection within PAs are still needed.
- Despite generally high levels of awareness of Europeans regarding the importance of biodiversity protection, IAS are not perceived as a key conservation issue by the public, and there is a need for more education of visitors to PAs about the threats resulting from invasions. The problem of IAS is likely to accelerate in the near future as on-going environmental change may potentially facilitate colonization of PAs by invasive species that are currently kept out due to climatic constraints. The dynamics of species ranges need to be incorporated in proactive management in a more explicit manner.
- The management of many PAs is still inadequate in general. Among other things, collecting standardised continent-wide data on the distribution and abundance of IAS is an urgent priority. The lack of such data is surprising given that Europe is one of the continents where biological invasions are most intensively studied. There is an urgent need to coordinate such systematic data collation and to integrate this element in the reporting instruments of EU, such as the reporting requested by the Habitat Directive, the Bird Directive, the Marine Strategy, and the Water Directive. It is advisable to employ citizen science in schemes aimed at improving the availability of IAS data. Also, European institutions need to support the implementation of a dedicated regional information system, as was requested by the European Union Council of June 25th, 2009.
- PA managers in Europe are aware of the seriousness of the problem and threats imposed by IAS but are constrained in their efforts to deal with them by the lack of staff and budgetary resources, the inadequate legal context, and the lack of rigorous scientific information translated into practical guidelines. European institutions should develop specific financial mechanisms to react promptly to new incursions, and PAs should establish contingency plans for invasions. It is crucial for Pan-European and national institutions to address the legal constraints to achieve more effective management of IAS.

Acknowledgements We thank three anonymous reviewers for comments on the manuscript, and Dave Richardson for improving our English. PP and JP were supported by grant no. P504/11/1028 (Czech Science Foundation), long-term research development project no. RVO 67985939 (Academy of Sciences of the Czech Republic). PP was also supported by institutional resources of Ministry of Education, Youth and Sports of the Czech Republic, and acknowledges the support by

Praemium Academiae award from the Academy of Sciences of the Czech Republic. We thank V. Andrić, A. Badré, P. Bauer, C. Binet, J. Bride, V. Buskovic, L. Carotenuto, B. Fritsch, C. Greaume, K. Guerbaa, B.J. Hutinec, G. Kupczac, A. Labouille, A. Pardey, A. Pringarbe, P. Smith and T. Strudwick for providing us the data for PAs, and Zuzana Sixtová for technical assistance. The survey on invasive species in European PAs, that provided key information for the present chapter, was made possible by the support of the Council of Europe, and by the help of many experts, protected areas staff, and organizations, including in particular Europarc, IUCN World Commission on Protected Areas, IUCN Regional Office for Europe, IUCN Med Office, the Group of Experts of the Bern Convention on Protected Areas and Ecological Networks.

References

AOPK (2009) Rozbory Chráněné krajinné oblasti Labské pískovce k 31. 10. 2009. AOPK ČR, Správa CHKO Labské pískovce, Czech Republic

Bacchetta G, Mayoral Garcia Berlanga O, Podda L (2009) Catálogo de la flora exótica de La Isla De Cerdeña (Italia). Flora Montiberica 41:35–61

Balmford A, Beresford J, Green J et al (2009) A global perspective on trends in nature-based tourism. PLoS Biol 7:e1000144

Başnou C (2009) *Robinia pseudoacacia* L., black locust (Fabaceae, Magnoliophyta). In: DAISIE (ed) Handbook of alien species in Europe. Springer, Dordrecht, p 357

Bertzky B, Corrigan C, Kemsey J et al (2012) Protected planet report 2012: tracking progress towards global targets for protected areas. IUCN/UNEP-WCMC, Gland/Cambridge

Blackburn TM, Pyšek P, Bacher S et al (2011) A proposed unified framework for biological invasions. Trends Ecol Evol 26:333–339

Brockie RE, Loope LL, Usher MB et al (1988) Biological invasions of island nature reserves. Biol Conserv 44:9–36

Brundu G (2014) Chapter 18: Invasive alien plants in protected areas in Mediterranean islands: knowledge gaps and main threats. In: Foxcroft LC, Pyšek P, Richardson DM, Genovesi P (eds) Plant invasions in protected areas: patterns, problems and challenges. Springer, Dordrecht, pp 395–422

Brunel S, Fernández-Galiano E, Genovesi P et al (2013) Invasive alien species: a growing but neglected threat? In: Late lessons from early warning: science, precaution, innovation. Lessons for preventing harm. EEA Report 1/2013, Copenhagen

Camarda I, Brundu G, Carta M et al (2002) Invasive alien plants in the national parks of Sardinia. In: Camarda I et al (eds) Global challenges of parks and protected area management: proceedings of the 9th ISSRM: October 10–13, 2002, La Maddalena, Sardinia

Chytrý M, Pyšek P, Tichý L et al (2005) Invasions by alien plants in the Czech Republic: a quantitative assessment across habitats. Preslia 77:339–354

Chytrý M, Wild J, Pyšek P et al (2009) Maps of the level of invasion of the Czech Republic by alien plants. Preslia 81:187–207

Chytrý M, Wild J, Pyšek P et al (2012) Projecting trends in plant invasions in Europe under different scenarios of future land-use change. Glob Ecol Biogeogr 21:75–87

Council of Europe (2011) Final report of group of experts on invasive alien species. Directorate of Culture and Cultural and Natural Heritage, Convention on the Conservation of European Wildlife and Natural Habitats. T-PVS/(2011)6. http://www.coe.int/t/dg4/cultureheritage/nature/bern/IAS/default_en.asp

CREDOC (2008) La valeur économique et sociale des espaces protégés. Cahier de recherche n. 247, Novembre 2008, Paris

Cross JR (1975) Biological flora of the British Isles. *Rhododendron ponticum* L. J Ecol 63:345–364

DAISIE (2009) Handbook of alien species in Europe. Springer, Berlin

De Poorter M (2007) Invasive alien species and protected areas: a scoping report. Part 1. Scoping the scale and nature of invasive alien species threats to protected areas, impediments to invasive alien species management and means to address those impediments. Global Invasive Species Programme, Invasive Species Specialist Group. http://www.issg.org/gisp_publica tions_reports.htm

Dobrovic I, Nikolic T, Jelaska SD et al (2006) An evaluation of floristic diversity in Medvednica Nature Park (northwestern Croatia). Plant Biosyst 140:234–244

Erfmeier A, Bruelheide H (2005) Invasive and native *Rhododendron ponticum* populations: is there evidence for genotypic differences in germination and growth? Ecography 28:417–428

Erfmeier A, Bruelheide H (2010) Invasibility or invasiveness? Effects of habitat, genotype, and their interaction on invasive *Rhododendron ponticum* populations. Biol Invasions 12:657–676

Essl F, Rabitsch W (eds) (2002) Neobiota in Österreich. Umweltbundesamt GmbH, Wien

European Environment Agency (2006) CORINE land cover 2006. http://www.eea.europa.eu/data-and-maps/data/corine-land-cover-2006-raster

European Environment Agency (2010) Protected areas. "10 messages for 2010". EEA, Copenhagen

European Environment Agency (2012) Protected areas in Europe: an overview. EEA Report 5/2012, Copenhagen

Fabiszewski J, Brej T (2008) Ecological significance of some kenophytes in lower Silesian national parks. Acta Soc Bot Polon 77:167–174

Foley C (1990) *Rhododendron ponticum* at Killarney National Park. In: Proc. The biology and control of invasive plants. Univ Wales Cardiff 20–21. 8. 1990. Industrial Ecology Group, BES, pp 62–63

Forestry Commission Scotland (2012) NFE aims to be rhododendron free (in 15 years)! News release no. 14783. http://www.forestry.gov.uk/newsrele.nsf/WebNewsReleases/E9AA5F5F7817A56E802578CC0046B2C2

Foxcroft LC (2004) An adaptive management framework for linking science and management of invasive alien plants. Weed Technol 18:1275–1277

Foxcroft LC, Jarošík V, Pyšek P et al (2011) Protected-area boundaries as filters of plant invasions. Conserv Biol 25:400–405

Foxcroft LC, Genovesi P, Pyšek P et al (2014a) Chapter 2: Impacts of alien plant invasions in protected areas. In: Foxcroft LC, Pyšek P, Richardson DM, Genovesi P (eds) Plant invasions in protected areas: patterns, problems and challenges. Springer, Dordrecht, pp 19–41

Foxcroft LC, Witt A, Lotter WD (2014b) Chapter 7: Invasive alien plants in African protected areas. In: Foxcroft LC, Pyšek P, Richardson DM, Genovesi P (eds) Plant invasions in protected areas: patterns, problems and challenges. Springer, Dordrecht, pp 117–143

Gallo T, Wait D (2011) Creating a successful citizen science model to detect and report invasive species. BioScience 61:459–465

Gallup Organisation (2007) Attitudes of Europeans towards the issue of biodiversity. European Commission, Brussels

Gantioler S, Rayment M, Bassi S et al (2010) Costs and socio-economic benefits associated with the Natura 2000 Network. Final report to the European Commission, DG Environment on Contract ENV.B.2/SER/2008/0038, Institute for European Environmental Policy/GHK/Ecologic, Brussels

Gaston KJ, Jackson SF, Nagy A et al (2008) Protected areas in Europe. Ann N Y Acad Sci 1134:97–119

Genovesi P, Shine C (2004) European strategy on invasive alien species. Nat Environ 137:1–67

Genovesi P, Scalera R, Brunel S et al (2010) Towards an early warning and information system for invasive alien species (IAS) threatening biodiversity in Europe. EEA technical report n. 5/2010

Gritten RH (1992) The control of *Rhododendron ponticum* in the Snowdonia National Park. Aspects Appl Biol 29:279–286

238 P. Pyšek et al.

Gritten RH (1995) *Rhododendron ponticum* and some other invasive plants in the Snowdonia National Park. In: Pyšek P, Prach K, Rejmánek M et al (eds) Plant invasions: general aspects and special problems. SPB Academic Publishing, Amsterdam, pp 213–219

Grulich V (1997) Atlas rozšíření cévnatých rostlin národního parku Podyjí. Verbreitungsatlas der Gefäßpflanzen des Nationalparks Thayatal. Masarykova Univerzita, Brno

Hannah L (2001) The role of a global protected areas system in conserving biodiversity in the face of climate change. In: Visconti G, Benniston M, Iannorelli ED et al (eds) Global change and protected areas. Kluwer, New York, pp 413–422

Hannah L, Midgley G, Andelman S et al (2007) Protected area needs in a changing climate. Front Ecol Environ 5:131–138

Hejda M, Pyšek P, Jarošík V (2009) Impact of invasive plants on the species richness, diversity and composition of invaded communities. J Ecol 97:393–403

Heywood V (2012) European code of conduct for botanic gardens on invasive alien species. Council of Europe Document T-PVS/Inf (2012)1, Strassbourg

Heywood V, Brunel S (2009) Code of conduct on horticulture and invasive alien plants. Nat Environ 155:1–35

Heywood VH (2011) The role of botanic gardens as resource and introduction centres in the face of global change. Biodivers Conserv 20:221–239

Hulme PE (2009) *Rhododendron ponticum* L., rhododendron (Ericaceae, Magnoliophyta). In: DAISIE (ed) Handbook of alien species in Europe. Springer, Berlin, p 356

Hulme PE, Pyšek P, Pergl J et al (2013) Greater focus needed on plant invasion impacts in protected areas. Conserv Lett doi:10.1111/conl.12061 (in press)

Hulme PE, Pyšek P, Nentwig W et al (2009) Will threat of biological invasions unite the European Union? Science 324:40–41

Huntley B, Hole DG, Willis SG (2011) Assessing the effectiveness of a protected area network in the face of climatic change. In: Hodkinson TR, Jones MB, Waldren S et al (eds) Climate change, ecology and systematics. Cambridge University Press, Cambridge, pp 345–362

Jahodová Š, Trybush S, Pyšek P et al (2007) Invasive species of *Heracleum* in Europe: an insight into genetic relationships and invasion history. Divers Distrib 13:99–114

Jarošík V, Pyšek P, Kadlec T (2011) Alien plants in urban nature reserves: from red-list species to future invaders. NeoBiota 10:27–46

Jenkins CN, Joppa L (2009) Expansion of the global terrestrial protected area system. Biol Conserv 142:2166–2174

Kleinbauer I, Dullinger S, Peterseil J et al (2010) Climate change might drive the invasive tree *Robinia pseudacacia* into nature reserves and endangered habitats. Biol Conserv 143:382–390

Kowarik I (2003) Biologische Invasionen: Neophyten und Neozoen in Mitteleuropa. E. Ulmer Verlag, Stuttgart

Kumbarić A (1999) Flora of Gajna protected area. Faculty of Science, University of Zagreb, Zagreb

Lambdon PW, Pyšek P, Basnou C et al (2008) Alien flora of Europe: species diversity, temporal trends, geographical patterns and research needs. Preslia 80:101–149

Lamprecht A (2008) Die Verbreitung invasiver und potenziell invasiver Neophyten im Nationalpark oberösterreichische Kalkalpen sowie Notwendigkeit und Möglichkeiten ihrer Bekämpfung. Thesis, Fachhochschule Weihenstephan, Germany

Lockwood M (2006) Global protected area framework. In: Lockwood M, Worboys GL, Kothari A (eds) Managing protected areas: a global guide. Earthscan, London, pp 73–100

Lonsdale WM (1999) Global patterns of plant invasions and the concept of invasibility. Ecology 80:1522–1536

Macdonald IAW, Loope LL, Usher MB et al (1989) Wildlife conservation and the invasion of nature reserves by introduced species: a global perspective. In: Drake JA, Mooney HA, di Castri F et al (eds) Biological invasions: a global perspective. Wiley, Chichester, pp 215–255

Mack RN, Lonsdale WM (2002) Eradicating invasive plants: hard-won lessons from islands. In: Veitch CR, Clout M (eds) Turning the tide: the eradication of invasive species. IUCN SSC Invasive Species Specialist Group, Gland/Cambridge, pp 164–172

McKinney ML (2002) Influence of settlement time, human population, park shape and age, visitation and roads on the number of alien plant species in protected areas in the USA. Divers Distrib 8:311–318

Medvecká J, Kliment J, Májeková J et al (2012) Inventory of the alien flora of Slovakia. Preslia 84:257–309

Meyerson LA, Pyšek P (2014) Chapter 21: Manipulating alien species propagule pressure as a prevention strategy in protected areas. In: Foxcroft LC, Pyšek P, Richardson DM, Genovesi P (eds) Plant invasions in protected areas: patterns, problems and challenges. Springer, Dordrecht, pp 473–486

Millennium Ecosystem Assessment (2005) Ecosystems and human well-being. World Resources Institute, Washington, DC

Milne RI, Abbott RJ (2000) Origin and evolution of invasive naturalized material of *Rhododendron ponticum* L. in the British Isles. Mol Ecol 9:541–556

Mortensen DA, Rauschert ESJ, Nord AN et al (2009) Forest roads facilitate the spread of invasive plants. Inv Plant Sci Manag 2:191–199

Natura 2000 Networking Programme (2007) Natura 2000. http://www.natura.org

Naughton-Treves L, Holland MB, Brandon K (2005) The role of protected areas in conserving biodiversity and sustaining local livelihoods. Annu Rev Environ Res 30:219–252

NewsWales (2011) Volunteer for balsam blitz in Pembrokeshire National Park. http://www.newswales.co.uk/index.cfm?section=Environment&F=1&id=21327

Parmesan C (2006) Ecological and evolutionary responses to recent climate change. Annu Rev Ecol Evol Syst 37:637–669

Parmesan C, Yohe G (2003) A globally coherent fingerprint of climate change impacts across natural systems. Nature 421:37–42

Pluess T, Jarošík V, Pyšek P et al (2012) Which factors affect the success or failure of eradication campaigns against alien species? PLoS One 7:e48157

Pretto F, Celesti-Grapow L, Carli E et al (2012) Determinants of non-native plant species richness and composition across small Mediterranean islands. Biol Invasions 14:2559–2572

Puhakka R (2008) Increasing role of tourism in Finnish national parks. Fennia 186:47–58

Pyšek P (2009) *Fallopia japonica* (Houtt.) Ronse Decr., Japanese knotweed (Polygonaceae, Magnoliophyta). In: DAISIE (ed) Handbook of alien species in Europe. Springer, Berlin, p 348

Pyšek P, Richardson DM (2010) Invasive species, environmental change and management, and health. Annu Rev Environ Res 35:25–55

Pyšek P, Jarošík V, Kučera T (2002) Patterns of invasion in temperate nature reserves. Biol Conserv 104:13–24

Pyšek P, Jarošík V, Kučera T (2003a) Inclusion of native and alien species in temperate nature reserves: an historical study from Central Europe. Conserv Biol 17:1414–1424

Pyšek P, Sádlo J, Mandák B et al (2003b) Czech alien flora and a historical pattern of its formation: what came first to Central Europe? Oecologia 135:122–130

Pyšek P, Kučera T, Jarošík V (2004a) Druhová diverzita a rostlinné invaze v českých rezervacích: co nám mohou říci počty druhů? Příroda 21:63–89

Pyšek P, Richardson DM, Rejmánek M et al (2004b) Alien plants in checklists and floras: towards better communication between taxonomists and ecologists. Taxon 53:131–143

Pyšek P, Richardson DM, Pergl J et al (2008) Geographical and taxonomic biases in invasion ecology. Trends Ecol Evol 23:237–244

Pyšek P, Hulme PE, Nentwig W (2009a) Glossary of the main technical terms used in the handbook. In: DAISIE (ed) Handbook of alien species in Europe. Springer, Berlin, pp 375–379

Pyšek P, Křivánek M, Jarošík V (2009b) Planting intensity, residence time, and species traits determine invasion success of alien woody species. Ecology 90:2734–2744

Pyšek P, Chytrý M, Pergl J et al (2012a) Plant invasions in the Czech Republic: current state, introduction dynamics, invasive species and invaded habitats. Preslia 84:576–630

Pyšek P, Danihelka J, Sádlo J et al (2012b) Catalogue of alien plants of the Czech Republic (2nd edition): checklist update, taxonomic diversity and invasion patterns. Preslia 84:155–255

Pyšek P, Jarošík V, Hulme PE et al (2012c) A global assessment of invasive plant impacts on resident species, communities and ecosystems: the interaction of impact measures, invading species' traits and environment. Glob Change Biol 18:1725–1737

Rice KS, Westerman B, Federici R (2004) Impacts of the exotic, nitrogen-fixing black locust (*Robinia pseudoacacia*) on nitrogen-cycling in a pine-oak ecosystem. Plant Ecol 174:94–107

Richardson DM, Rejmánek M (2011) Trees and shrubs as alien invasive species: a global review. Divers Distrib 17:788–809

Richardson DM, Pyšek P, Rejmánek M et al (2000) Naturalization and invasion of alien plants: concepts and definitions. Divers Distrib 6:93–107

Robertson P, Bainbridge I, de Soye Y (2011) Priorities for conserving biodiversity on European islands. Strassbourg

Scalera R, Zaghi D (2004) LIFE focus/Alien species and nature conservation in the EU: the role of the LIFE program. European Commission, Office for Official Publications of the European Communities, Luxembourg. http://195.207.127.31/comm/environment/life/infoproducts/alienspecies_en.pdf

Scalera R, Genovesi P, De Man D et al (2012) European code of conduct on zoological gardens and acquaria and invasive alien species. Council of Europe Document T-PVS/Inf (2011) 26 rev., Strassbourg

Schiffleithner V, Essl F (2010) Untersuchung ausgewählter Neophyten im NP Thayatal im Jahr 2010: Verbreitung und Evaluierung von Managementmaßnahmen. Institut für angewandte Biologie und Umweltbildung, Wien

Shaw MW (1984) *Rhododendron ponticum*: ecological reasons for the success of an alien species in Britain and features that may assist in its control. Aspects Appl Biol 5:231–242

Simberloff D (2003) How much information on population biology is needed to manage introduced species? Conserv Biol 17:83–92

Smith PH (2012) Inventory of vascular plants for the Sefton Coast. Unpublished report to the Sefton Coast Partnership

Spear D, McGeoch MA, Foxcroft LC et al (2011) Alien species in South Africa's National Parks (SANParks). Koedoe 53:1032–1034

Stout JC (2007) Reproductive biology of the invasive exotic shrub, *Rhododendron ponticum* L. (Ericaceae). Bot J Linn Soc 155:373–381

Timmins SM, Williams PA (1991) Weed numbers in New Zealand's forest and scrub reserves. N Z J Ecol 15:153–162

UNEP-WCMC (2011) The world database of protected areas. Cambridge. http://www.wdpa.org/Statistics.aspx

Usher MB (1988) Biological invasions of nature reserves: a search for generalisation. Biol Conserv 44:119–135

Veenvliet JK, Humar M (2011) Tujerodne vrste na zavarovanih območjih [Alien species in protected areas]. Report on capacity building activity in the framework of the WWF project Dinaric Arc Ecoregion, Ministry of Environment, Slovenia

Vilà M, Bartomeus I, Dietzsch AC et al (2009) Invasive plant integration into native plant-pollinator networks across Europe. Proc R Soc B 276:3887–3893

Vilà M, Espinar JL, Hejda M et al (2011) Ecological impacts of invasive alien plants: a meta-analysis of their effects on species, communities and ecosystems. Ecol Lett 14:702–708

Vuković N, Bernardic A, Nikolic T et al (2010) Analysis and distributional patterns of the invasive flora in a protected mountain area: a case study of Medvednica nature park (Croatia). Acta Soc Bot Polon 79:285–294

Walters C (1986) Adaptive management of renewable resources. MacMillan, New York

Welzholz JC, Johann E (2007) History of protected forest areas in Europe. In: Frank G, Parviainen J, Vandekerhove K et al (eds) COST Action E27. Protected forest areas in Europe – analysis and harmonisation (PROFOR): results, conclusions and recommendations, pp 17–40

Chapter 12
Invasive Plant Species in Indian Protected Areas: Conserving Biodiversity in Cultural Landscapes

Ankila J. Hiremath and Bharath Sundaram

Abstract Invasive plant species in Indian protected areas have received relatively little attention until recently. This may partly be due to a historical emphasis on wildlife protection, rather than on a broader science-based approach to conservation of biodiversity and ecosystem functioning. A literature review of invasive plant species in India showed that nearly 60 % of all studies have been done since 2000, and only about 20 % of all studies are from protected areas. Studies from protected areas have largely focused on a small subset of invasive alien plants, and almost half these studies are on a single species, *Lantana camara*, probably reflecting the species' ubiquitous distribution. The spread of alien plants in India has been both ecologically and human mediated. Efforts to manage plant invasions have, in the past, been diluted by the ambivalence of managers attempting to find beneficial uses for these species. Despite growing knowledge about the harmful impacts of certain invasive plants on native species and ecosystems, their deliberate spread has continued, even till quite recently. And, despite the successful implementation of management initiatives in some protected areas, these efforts have not expanded to other areas. The lack of a national coordinated effort for invasive species monitoring, research, and management largely underlies this.

Keywords Invasive alien species • *Lantana camara* • Management • *Prosopis juliflora*

A.J. Hiremath (✉)
Ashoka Trust for Research in Ecology and the Environment 1, K-Block Commercial Complex (2nd floor), Jangpura Extension, New Delhi 110014, India
e-mail: hiremath@atree.org

B. Sundaram
Azim Premji University, PES Institute of Technology Campus, Electronics City, Hosur Road, Bangalore 560100, Karnataka, India
e-mail: bharath.sundaram@azimpremjifoundation.org

L.C. Foxcroft et al. (eds.), *Plant Invasions in Protected Areas: Patterns, Problems and Challenges*, Invading Nature - Springer Series in Invasion Ecology 7, DOI 10.1007/978-94-007-7750-7_12, © Springer Science+Business Media Dordrecht 2013

12.1   Introduction

Two decades ago, Usher (1991) ventured tentatively that it was unlikely tropical nature reserves (as with reserves elsewhere) were free from alien invasive species. With the tremendous growth in global trade and travel, and with increasing landscape fragmentation, this can now be categorically stated (Mack and Lonsdale 2001; Denslow and deWalt 2008; Weber and Li 2008). This is likely to be especially true in a country like India, with its network of relatively small protected areas (PAs) set in a matrix of altered, human-dominated landscapes.

Worldwide there is a growing catalogue of the potential impacts of invasive species on native species, wildlife habitats, disturbance regimes, and ecosystem services (e.g. Pyšek et al. 2011; Foxcroft et al. 2014; Simberloff et al. 2013). Yet in Indian PAs invasive plant species have received relatively little attention until recently, whether from researchers, managers, or the general public. This neglect may, at least in part, lie in the history of forest management and conservation, and in the genesis of PAs in India.

In this chapter we review the available scientific literature on invasive alien plant species (IAPs) in Indian PAs. We then trace the history of introductions of the better-known invasive species that have been reported from Indian PAs. Using two examples of widespread invasive plant species in India, we assess their impacts. Finally, we look at patterns regarding the drivers of invasion that are starting to emerge from these studies. These findings provide valuable insights for future management of invasive species in these PAs, with their long and continuing history of human habitation and use.

12.2   India's Protected Areas

India's 668 PAs account for about 4.9 % of the country's geographic area (Krishnan et al. 2012). The categories of PAs under the Indian Wildlife Protection Act (WLPA) include national parks, wildlife sanctuaries, conservation reserves, and community conserved areas, varying in the degree of human use permitted within them. Indian parks are generally small relative to PAs in some other parts of the world, being, on average, on the order of a few 100 km^2. Many PAs have had a long history of forest management and use by communities that lived in these forests prior to their notification. Historical management and use by these communities included shifting cultivation, burning, hunting, grazing, and fuel-wood and non-timber-forest-product (NTFP) collection. Other PAs had historically been managed for the harvest of timber. Thus, these PAs are cultural landscapes as much as they are natural landscapes.

Even today a large proportion of PAs have resident forest-dependent communities. Other than in national parks, forest-dwelling communities have rights to NTFP and fuel-wood collection and grazing in protected areas, though shifting cultivation, hunting, and burning have been curtailed (Krishnan et al. 2012). An amendment to the WLPA in 2002 (Government of India 2002) banned NTFP collection for commercial use, permitting only subsistence collection. The exception to this general pattern is the case of tiger reserves, which are a subset of PAs especially earmarked for tiger (*Panthera tigris*) protection. In tiger reserves the most recent amendment to the WLPA (Government of India 2006) mandates the setting aside of a core inviolate zone, the critical tiger habitat, which is to be free from all human habitation and use.

Initially, the modern era of forest management in India (1864 onwards) was dominated by production forestry. The first PAs were established only at the turn of the twentieth century. They owed their origins to diminishing populations of valuable game animals as a result of overhunting and on-going habitat transformation. As animal numbers declined, the numbers of hunters-turned-conservationists, and their influence, increased (Burton 1953; Rangarajan 2001). Prominent amongst them was Colonel James Corbett, of the eponymous national park, and the first such park in India (Rangarajan 2001). Other protected areas owed their origins to erstwhile princely rulers who were prescient in setting aside portions of their hunting preserves – *shikargahs* – for the protection of valuable endangered species (Rangarajan 2001; Krishnan et al. 2012). One example is that of Gir in the western Indian state of Gujarat, and the last remaining home of the Asiatic lion (*Panthera leo*), which was protected by the rulers of Junagadh. Another example is Bandipur in the southern Indian state of Karnataka, where the ruler of Mysore had set aside tiger protection blocks in which hunting of wildlife was strictly prohibited (Rangarajan 2001).

Having begun in this manner, conservation in India had a single-minded focus, the protection of charismatic mammals, as is evident from the earliest reserves (e.g. Kaziranga established for the one-horned rhinoceros, *Rhinoceros unicornis*, and Kanha for the tiger; Krishnan et al. 2012). Many protected areas continue to be synonymous with individual species, for example, Gir (the Asiatic lion) and Corbett (tiger). Management in this context was largely focused on inventorying and maintaining stocks, and preventing illegal activities such as hunting and poaching (Burton 1953; Stracey 1960). This preoccupation with numbers only increased in the 1970s; a nationwide tiger census in 1972 showed that its numbers had declined markedly, leading to the initiation of Project Tiger, under which a series of tiger reserves were established (Rangarajan 2001). Even when the emphasis of conservation broadened to include other species in the landscape, management remained largely unchanged. The argument was that conservation of big mammals, so-called umbrella species, automatically guaranteed conservation of all other plants and animals, though this is not always the case (e.g. Das et al. 2006).

It is only in the last three or four decades that the focus of conservation in India has broadened to include not only species, but unique habitats and ecosystems. A biogeographic assessment of the country in 1992, under the National Wildlife Action Plan of 1983, led to the identification of gaps in the PA network, and the setting up of new reserves (Krishnan et al. 2012). Other initiatives during this time have included the establishment of biosphere reserves and world heritage sites under the aegis of UNESCO, as part of a global programme to conserve unique landscapes. In the last couple of decades the growing global focus on biodiversity hotspots (biodiversity rich areas with high degrees of endemism and threats; Myers et al. 2000) has led to recognition of the conservation value of large regions like the Eastern Himalayas and the Western Ghats (Critical Ecosystems Partnership Fund 2013). Even so, the management of PAs has remained largely unchanged, with its emphasis on protection from illegal activities, and on inventorying and maintaining wildlife numbers, with the role of science continuing to be largely insignificant (Madhusudan et al. 2006).

Given this history of forest management and conservation, it is not surprising that the insidious spread of invasive plant species in Indian PAs went largely unnoticed until quite recently. Although IAPs have occasionally been included as part of habitat management plans in PAs (e.g. *Lantana camara* in the Melghat Tiger Reserve; Sawarkar 1984), these have tended to be isolated instances. It is only in the past decade or so that both managers and researchers have become increasingly interested in the issue of IAPs in India's PAs.

12.3 Invasive Plants in Indian Protected Areas: An Overview

From the list of 100 of the world's worst invasive species (see Lowe et al. 2000, for criteria they use), 11 plant species occur in India and several of these occur in PAs. These 11 species comprise *Acacia mearnsii* (black wattle), *Arundo donax* (giant cane), *Chromolaena odorata* (Siam weed), *Clidemia hirta* (Koster's curse), *Imperata cylindrica* (cogon grass), *Lantana camara* (lantana), *Leucaena latisiliqua* (= *Leucaena leucocephala*, false koa), *Mikania micrantha* (mile-a-minute weed), *Opuntia stricta* (prickly pear), *Ulex europaeus* (gorse), and *Sphagneticola trilobata* (= *Thelechitonia trilobata*, *Wedelia trilobata*, Singapore daisy). These are a subset of the 225 alien plant species in India that Khuroo et al. (2012) have classified as invasive, using a modification of the classification proposed by Pyšek et al. (2004). They also recognise an additional 134 species as naturalised, but with potential to become invasive in the near future.

Our review of IAPs in India is based on a variety of sources. The Thomson-Reuters ISI Web of Science database was searched using the following search string: (exotic OR invasive OR invad* OR alien OR non*native OR weed) AND (India). The same search string was used to search the Agricola database. In addition to these two sources, we also searched the available archives for the journal *Tropical Ecology*, which is not indexed in either of the databases searched, and relied on our knowledge of the Indian literature (especially articles in other journals not indexed in either of the two databases, e.g. the Journal of the Bombay Natural History Society, Conservation and Society, and Indian Forester). We supplemented our findings with other relevant articles and reports cited in records retrieved from these searches. Overall, this constitutes a reasonably complete review of the published literature (excluding book chapters) on terrestrial and aquatic invasive

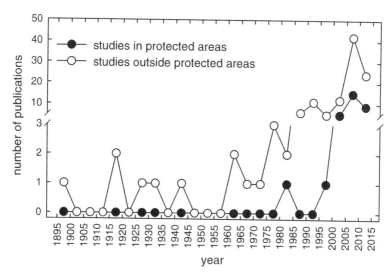

Fig. 12.1 Temporal trends in the published information on invasive alien plants in India, showing the relative numbers of studies conducted in protected areas compared to studies outside protected areas (Publications for the period 2010–2015 are still accruing)

plant species in India, but by no means is it a complete review of the grey literature. Most reports of IAP occurrences and accounts of management undertaken by the Forest Department are almost certain to remain in departmental reports and forest management plans that are difficult to access; very few such accounts make their way into the public domain (in this case, the Indian Forester, which is the journal of the Indian Forest Department). Abstracts of the initial list of articles retrieved were screened to eliminate studies that exclusively pertained to weeds of agricultural systems, though articles pertaining to IAPs in shifting cultivation systems were retained. Only articles pertaining to alien plant species that are recognised to be invasive (i.e. those that are widespread and dominant, Colautti and MacIsaac 2004) were retained. Those that dealt with other introduced or naturalised species, for example alien species in plantation forests or agro-ecosystems, were excluded. Finally, studies that were not typically ecological in their emphasis (e.g. those looking at chemical or molecular characteristics of particular species) were also excluded.

The most noteworthy finding to emerge from this review was how few studies exist on IAPs in India. We did not restrict the search to any particular time period, so the earliest report dates back to the end of the nineteenth century (Anon 1895), long before there was widespread interest in the biology of species invasions (generally attributed to Elton 1958, but see Chew 2011). For a period spanning more than a century, the search yielded less than 150 studies, with more than 60 % of these studies after the year 2000.

A second noteworthy finding from this review is how few studies are from PAs. The research on invasive plant species in PAs accounts for barely a fifth of all IAP research in India (Fig. 12.1). Considering that Indian PAs constitute only a small fraction of the country's area, making them valuable repositories of the country's biological diversity, this result is striking.

Third, the work on invasive plant species in Indian PAs has focused only on a handful of species, with *L. camara* being the subject of almost half of these studies (Table 12.1). *Lantana camara* is also the most studied invasive plant species in India, and is the focus of over a third of all studies generally. This may be a reflection of its status globally. Cronk and Fuller (1995) classify it as one of the most ubiquitous invasive plant species worldwide, ranging from tropical to subtropical and warm temperate regions of the world. Based on the current state of knowledge in India, an account of IAPs in Indian PAs is largely, though not exclusively, an account of *L. camara*.

12.4   A History of Introductions

Studies of IAPs in Indian PAs repeatedly mention only a small subset of species. One reason for this could be that most of the species in this subset have been in India for at least 100 years, presumably long enough to have become invasive (Wilson et al. 2007; Table 12.2). Second, at least two of these species (*L. camara* and *Prosopis juliflora*, mesquite) are very widespread. *Lantana camara*, particularly, occurs in a wide variety of different ecosystems (Table 12.1). *Prosopis juliflora*, on the other hand, forms extensive and conspicuous stands, even though it is restricted to the arid and semi-arid regions of the country (Saxena 1998). Third, most of the species in this subset tend to represent a new life form in the systems they have invaded. Whether it is the shrubby or clambering *L. camara* invading relatively open deciduous forests or woodland savannas, or *P. juliflora* (a tree), *Cytisus scoparius* (Scotch broom, a shrub) or *Mimosa diplotricha* (giant sensitive plant, a clambering vine) invading grasslands, the impacts of these species are more readily visible than they would be if the alien invader merely resulted in more of the same, for example, an alien grass invading a grassland.

In India, as with other regions of the world, invasive species have arrived in a variety of ways. Most alien plant species that are known to be invasive in PAs in India were first introduced into the country as garden ornamentals. Other reasons why alien species were introduced intentionally was to meet fuel-wood requirements, to prevent desert spread, and for commercial cultivation (Table 12.2). The most unusual, and perhaps apocryphal, example of an intentional introduction is that of *M. micrantha*, a climber known for its rapid growth in humid tropical environments. *Mikania micrantha* is thought to have been introduced by the allied forces during World War II to camouflage airfields built along the Indo-Burmese border as a defence against the advancing Japanese forces (Randerson 2003).

An example of an accidentally introduced species that has become invasive is *Parthenium hysterophorus* (congress grass). Reports suggest that it arrived in India as a contaminant of imported wheat in the mid-1950s, though there is evidence that it may already have been in India as early as 1810 (Paul 2010). Attempting to

Table 12.1 Protected areas in India for which published information on invasive alien plants is available. Note the disproportionate number of studies on *Lantana camara* compared with other invasive plant species, and the range of ecosystem types in which *L. camara* occurs

Protected area	Ecosystem type	Invasive species reported	Source
Kalakad Mundanturai Tiger Reserve	Tropical evergreen forest	*Lantana camara*, *Chromolaena odorata*	Chandrasekaran and Swamy (2002, 2010)
Protected forest, Anamalais	Tropical evergreen forest	*Coffea arabica*, *Coffea canephora*	Joshi et al. (2009)
Greater Nicobar Biosphere Reserve	Tropical evergreen forest	*Mikania micrantha*, *Chromolaena odorata*, *Lantana camara*, *Ageratina* spp., *Merremia peltata*	Babu and Leighton (2004)
Southern Western Ghats (no specific protected area mentioned)	Tropical evergreen forest	*Mikania micrantha*	Sankaran and Srinivasan (2001)
North-eastern India (no specific protected area mentioned)	Tropical evergreen forest	*Mikania micrantha*	Gogoi (2001)
Achanakmar-Amarkantak Biosphere Reserve	Tropical moist-deciduous forest	*Lantana camara*	Sahu and Singh (2008), Shukla et al. (2009)
Mudumalai National Park	Tropical moist-deciduous, dry deciduous forest, scrub forest	*Lantana camara*, *Chromolaena odorata*, *Opuntia stricta* var. *dillenii* (=*O. dillenii*)	Mahajan and Azeez (2001), Ramaswami and Sukumar (2011)
Biligiri Rangaswamy Temple Tiger Reserve	Tropical moist-deciduous, dry deciduous forest	*Lantana camara*, *Chromolaena odorata*	Murali and Setty (2001), Sundaram and Hiremath (2012)
Bandipur National Park	Tropical dry deciduous forest	*Lantana camara*, *Chromolaena odorata*	Puyravaud et al. (1995), Prasad (2009, 2010, 2012)
Chinnar Wildlife Sanctuary	Tropical dry deciduous forest, scrub forest	*Lantana camara*, *Ageratum houstonianum*	Chandrashekara (2001)
Melghat Tiger Reserve	Tropical dry deciduous forest	*Lantana camara*	Sawarkar (1984)
Tadoba-Andhari Tiger Reserve	Tropical dry deciduous forest	*Lantana camara*, *Hyptis suaveolens*, *Parthenium hysterophorus*	Giradkar and Yeragi (2008)
Kumbalgarh Wildlife Sanctuary	Tropical dry deciduous forest	*Prosopis juliflora*	Waite et al. (2009)
Ranthambore National Park	Tropical dry deciduous forest	*Prosopis juliflora*	Dayal (2007)
Corbett Tiger Reserve	Subtropical moist-deciduous, dry deciduous forest	*Lantana camara*	Babu et al. 2009; Love et al. (2009)

(continued)

Table 12.1 (continued)

Protected area	Ecosystem type	Invasive species reported	Source
Rajaji National Park	Subtropical moist deciduous, dry deciduous forest	*Lantana camara*	Rishi (2009), Kimothi and Dasari (2010), Kimothi et al. (2010)
Valley of Flowers National Park	Alpine meadow	*Polygonum polystachyum*	Saberwal et al. (2000), Kala and Shrivastava (2004)
Mukurti National Park	Montane forest and grassland (grassland-shola mosaic)	*Cytisus scoparius, Chromolaena odorata, Ulex europaeus, Acacia mearnsii*	Zarri et al. (2006), Srinivasan et al. (2007), Srinivasan (2011)
Kaziranga National Park	Floodplain grassland	*Mimosa invisa* (= *Mimosa diplotricha*), *Mikania micrantha*	Vattakkavan et al. (2005), Lahkar et al. (2011)
Orang National Park	Floodplain grassland	*Mimosa diplotricha, Mikania micrantha, Chromolaena odorata*	Lahkar et al. (2011)
Pabitora Wildlife Sanctuary	Floodplain grassland	*Mikania micrantha, Ipomoea carnea*	Lahkar et al. (2011)
Manas National Park	Floodplain grassland	*Mikania micrantha, Chromolaena odorata*	Lahkar et al. (2011)
Jaldapara Wildlife Sanctuary	Floodplain grassland	*Mikania micrantha*	Lahkar et al. (2011)
Garumara Wildlife Sanctuary	Floodplain grassland	*Mikania micrantha, Chromolaena odorata*	Lahkar et al. (2011)
Gulf of Mannar Marine Biosphere Reserve	Marine	*Kappaphycus alvarezii*	Bagla (2008), Chandrasekaran et al. (2008), Namboothri and Shankar (2010)

reconcile these disparate reports, Kohli et al. (2006) suggest that it may have arrived in the nineteenth century, but only became widespread in the mid-twentieth century.

The introduction of the seaweed *Kappaphycus alvarezii* for the commercial production of carrageenan deserves special mention. It was first introduced in 1993 to the Central Salt and Marine Chemicals Research Institute in western India. From there it was introduced into the Palk Bay at the southern tip of India in 2001, even though it was known to be invasive in other analogous environments (in Hawaii and the Caribbean; Namboothri and Shankar 2010). It has since spread to the Gulf of Mannar Marine Biosphere Reserve, where it is now rapidly growing over coral colonies, forming dense mats and smothering the corals below (Chandrasekaran et al. 2008).

Table 12.2 The subset of invasive alien plants in India that have been reported from protected areas, with the motives for their introduction and their source regions

Invasive species	Year of introduction	Source region	Reason for introduction	Source
Acacia mearnsii	1840s	Australia	Intentional (fuelwood)	Nair (2010)
Ageratina spp.	1800s	Mexico	Intentional (ornamental)	Muniappan et al. (2009)
Ageratum conyzoides	prior to 1882	South America	Possibly intentional (ornamental)	Kohli et al. (2006)
Ageratum houstonianum	a	Mexico	Possibly intentional (ornamental)	Khuroo et al. (2012)
Chromolaena odorata	1800s	Central, South America	Intentional (ornamental)	Bingelli et al. (1998)
Coffea arabica, C. canephora	1500s	Yemen	Intentional (cultivation)	Coffee Board of India (www.indiacoffee.org)
Cytisus scoparius	Prior to 1930	United Kingdom/ Europe	Intentional (ornamental)	Zarri et al. (2006), Srinivasan et al. (2007)
Hyptis suaveolens	a	Central, South America	a	Raizada (2006)
Ipomoea carnea	Late 1800s	South America	Intentional (ornamental)	Chaudhuri et al. (1994)
Kappaphycus alvarezii	1993	Philippines	Intentional (commercial)	Namboothri and Shankar (2010)
Lantana camara	1809, introduced several times during nineteenth century	South America (via Europe)	Intentional (ornamental)	Anon (1895), Iyengar (1933), Thakur et al. (1992), Day et al. (2003), Kannan et al. (2013)
Merremia peltata	a	Indo-Pacific region	a	Paynter et al. (2006)
Mikania micrantha	1940s	Central, South America	Intentional (camouflage)	Randerson (2003)
Mimosa invisa (syn. Mimosa diplotricha)	1960s	South America via Southeast Asia	Intentional (soil improvement)	Vattakkavan et al. (2005)
Opuntia stricta var. dillenii (= O. dillenii)	a	Mexico	a	Khuroo et al. (2012)
Parthenium hysterophorus	1950s (or prior to 1810; see text)	Latin America	Accidental	Kohli et al. (2006)

(continued)

Table 12.2 (continued)

Invasive species	Year of introduction	Source region	Reason for introduction	Source
Polygonum polystachyum	–	Indigenous weed	–	Kala and Shrivastava (2004)
Prosopis juliflora	1857, 1878	Central & South America (Mexico, Jamaica, Peru, Argentina, Uruguay)	Intentional (to halt desertification, for fuelwood)	Pasiecznik et al. (2001)
Ulex europaeus	Prior to 1910	Europe	Intentional (ornamental)	Bingelli et al. (1998)

[a]Denotes lack of information

12.5 Introduction, Invasiveness and Impacts: The Example of Two Widespread Invasive Species

In this section, the invasion history, invasiveness and impacts of IAPs in India's PAs is illustrated by using examples of two widespread invasive plant species, *L. camara* and *P. juliflora*.

12.5.1 Case Study 1: Lantana camara

12.5.1.1 Introduction and Spread

The European 'plant hunters' of the seventeenth and eighteenth centuries brought back a number of botanically and horticulturally interesting plants from their voyages and introduced them to botanical gardens across Europe, from where they travelled to other parts of the world. One such species was *L. camara*, whose earliest recorded introduction to India was in 1809 as an ornamental plant brought to the Calcutta Botanical Gardens (Thakur et al. 1992; but see also Kannan et al. 2013). There are also other later accounts of *L. camara* arriving in India, for example, in Coorg around 1865 (Anon 1895), and in peninsular India via Sri Lanka (Iyengar 1933). By the time *L. camara* was introduced into the old world tropics it had already been in cultivation as a garden ornamental in Europe since the mid- to late-seventeenth century (Day et al. 2003; Kannan et al. 2013). Plants that were introduced from Europe were thus likely to have been a complex of hybrids, which then hybridised further in their introduced environments (Day et al. 2003). This may be what underlies *L. camara's* wide ecological amplitude both in India and elsewhere (e.g. Vardien et al. 2012). Today it occurs in a variety of habitats across India, from tropical forests in the south all the way up to the subtropical and warm temperate lower reaches of the Himalayas in the north (e.g. Table 12.1;

a   b

Fig. 12.2 *Lantana camara* in the Biligiri Rangaswamy Temple Tiger Reserve, India. *Lantana camara* exhibits substantial morphological variation, (**a**) forming dense thickets 3–4 m tall, or (**b**) clambering up into tree crowns (Photo: (**a**) AJ Hiremath, (**b**) B Sundaram)

Kannan et al. 2013). It also manifests tremendous morphological variability (see Fig. 12.2). The extent to which these differences are genotypic or phenotypic is unknown, though the tools to investigate these differences now exist (Ray et al. 2013).

The earliest reports of *L. camara* spread date back to the late nineteenth century (Anon 1895; Kannan et al. 2013). Between 1917 and 1931 it was recorded to have spread at the rate of 600–1,280 ha per year across four forest ranges in North Salem, southern India, going from 3 % to 42 % of all forests in the district during this period (Iyengar 1933). The report does not, however, mention the abundance of *L. camara*, or how exactly this spread was measured. Another account indicates that *L. camara* spread at the rate of more than 2 km/year between 1911 and 1930, from a location where it was introduced in the Himalayan foothills (Hakimuddin 1930).

A recent account of *L. camara* from the Biligiri Rangaswamy Temple Tiger Reserve (BRT) constitutes perhaps the first systematic, long-term monitoring record of an invasive species' spread in a PA in India (Sundaram and Hiremath 2012). Over an 11-year period, the frequency of *L. camara* occurrence doubled across the 540 km^2 of this reserve. In 1997 *L. camara* was encountered in about 40 % of all plots surveyed, while by 2008 it was encountered in over 80 % of these plots (Fig. 12.3). The increase in spatial extent was accompanied by a commensurate, and disproportionate, increase in density. *Lantana camara* increased from one in every 20 stems in 1997, to one in every three stems by 2008. This increase in *L. camara* density was accompanied by a reduction in stems of native species, because there was no overall increase in the total numbers of stems recorded.

252   A.J. Hiremath and B. Sundaram

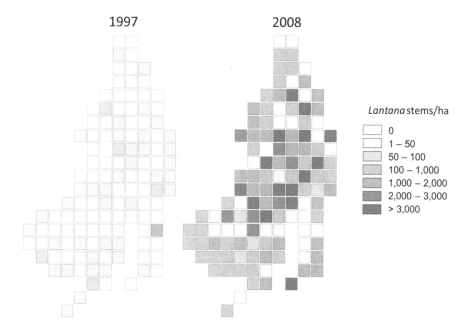

Fig. 12.3 *Lantana camara's* spread in the Biligiri Rangaswamy Temple Tiger Reserve, India, between 1997 and 2008. The squares represent cells of a 2 × 2 km grid across the 540 km² park. *Lantana camara* density is depicted based on vegetation surveys in a 400 m² plot at the centre of each cell (Sundaram 2011, with permission)

12.5.1.2 Invasiveness and Impacts

There have been several recent reviews of hypotheses pertaining to species traits that make them invasive, or community characteristics that make them invasible (Catford et al. 2009; Gurevitch et al. 2011; Jeschke et al. 2012). Several of these alternative mechanisms appear to play a role in the successful invasion of *L. camara*. One explanation is that of enemy release, which argues that species tend to grow uncontrolled in introduced environments in the absence of herbivores or pathogens that would keep them in check in their native environments (Keane and Crawley 2002). *Lantana camara* is not palatable to herbivores and thus does not appear to be preferentially browsed in Indian forests. Another explanation is that invasive species are able to take more rapid advantage of available resources and at the same time use nutrients more efficiently in low resource environments, when compared with native species (Funk and Vitousek 2007), a combination of characteristics typically thought to trade off against one another (Berendse and Aerts 1987). *Lantana camara* has been shown to be efficient at nutrient uptake and use (Bhatt et al. 1994), which would potentially give it a competitive advantage over other species, especially on nutrient poor soils.

It has also been suggested that invasive species typically produce large numbers of fruits that are widely dispersed, as do pioneer species, thus enabling them to exert propagule pressure (Lockwood et al. 2005). *Lantana camara* flowers and fruits year round, and in southern Indian deciduous forests it has been estimated to produce on the order of 10^4 fruits per individual during a single fruiting season (M. Kaushik, unpublished data). A combination of prolific fruiting and dispersal aided by avian frugivores can result in *L. camara* dominating the soil seed bank. In BRT, Sundaram (2011) found over 600 seeds/m^2 in the top 10 cm of soil, which is more than twice the number of seeds of all other native woody species (i.e. trees, shrubs, lianas) combined. These characteristics could enable *L. camara* to pre-emptively take advantage of opportunities to germinate and establish. Indeed, it has been shown to effectively colonise edges of fragmented forests (Sharma and Raghubanshi 2010) and colonise rapidly after disturbances (Duggin and Gentle 1998).

Apart from the ecological reasons, there are also human-mediated reasons underlying the successful establishment and spread of invasive species in Indian PAs. In the case of *L. camara*, despite extensive documentation of its spread and harmful impacts on agriculture and forestry, early work was focused on its potential uses (Hakimuddin 1930; Iyengar 1933), diluting attempts at control (e.g. Tireman 1916). While this was in keeping with the production-oriented approach to forest management of the time, surprisingly, this ambivalence continues. Soni et al. (2006), for example, list potential economically beneficial uses of *L. camara*, despite the accumulating ecological literature on its harmful impacts.

Invasive alien plants can have impacts at multiple scales (Parker et al. 1999). They may not be a significant cause of species extinction, other than in very specific environments like oceanic islands (Davis et al. 2011). However, there are other well-documented types of impacts on, for example, community structure and composition (e.g. Hejda et al. 2009), plant-animal interactions (e.g. Ghazoul 2004), disturbance regimes (e.g. D'Antonio and Vitousek 1992), and ecosystem processes (e.g. Vitousek and Walker 1989; Le Maitre et al. 2001). In Indian PAs these potential effects of IAPs take on added significance, especially considering the small area of remaining natural ecosystems that these PAs represent, relative to the country as a whole.

Studies indicate that *L. camara* invasion is correlated with changes in nitrogen cycling (Sharma and Raghubanshi 2009). This has been attributed to changes in litter quality and turnover under *L. camara* compared to background levels. There are also indications that *L. camara* is correlated with changes in community structure and composition (Sharma and Raghubanshi 2010; Sundaram and Hiremath 2012; Prasad 2012). The mechanism by which this happens may be the suppression of native regeneration. Although seedlings of native trees are found beneath *L. camara*, very few appear to recruit into the sapling stage (R. Ganesan, unpublished data). *Lantana camara* presence also appears to be associated with adult tree mortality (Prasad 2009; Sundaram and Hiremath 2012). A plausible explanation for this, based on observation, is that *L. camara* alters fuel characteristics (see also Berry et al. 2011), leading to fires that are more intense and severe than they would be in its absence (Tireman 1916; Hiremath and Sundaram 2005). In the long-term this could drastically alter the physiognomy of *L. camara*-invaded forests, with dire consequences for the wildlife they are meant to conserve.

In addition to changes in plant community structure and composition, there is growing evidence to suggest that *L. camara* may also have cascading trophic impacts. Increased abundance of unpalatable *L. camara* has been correlated with reduced abundances of native species. This means increased susceptibility of native vegetation to browsing, forage scarcity for herbivores, and, in turn, implications for predators like the tiger (Prasad 2010). *Lantana camara's* prolific fruiting attracts large numbers of frugivores, especially birds, potentially disrupting native plant-frugivore interactions (M. Kaushik, unpublished), and altering bird community composition, especially the abundance of certain feeding guilds (Aravind et al. 2010). Finally, the suppression of natural regeneration by *L. camara* can also have detrimental demographic consequences for important non-timber-forest-product species and for forest-dependent communities that harvest these fruits (e.g. *Phyllanthus emblica* and *P. indoficheri*, collectively known as the Indian gooseberry; Ticktin et al. 2012).

12.5.2 Case Study 2: Prosopis juliflora

12.5.2.1 Introduction and Spread

Prosopis juliflora was first brought to India in the latter half of the nineteenth century. There are at least two different accounts of its introduction, in 1857 and 1878, to halt the spread of the Thar desert in Northwest India, and for use as a fuel-wood species in peninsular India. There are indications that the sources of these two introductions differed, with seeds of the former coming from Mexico, and the latter from Jamaica. There are also records of subsequent introductions of seeds from Peru, Argentina and Uruguay (Pasiecznik et al. 2001). Following the initial success of *P. juliflora*, it was planted on a large scale in the western Indian states of Gujarat (in 1894), Rajasthan (in 1913) and Maharashtra (in 1934) (Tiwari 1999). In 1940 *P. juliflora* was even declared a "Royal Plant", and given special protection in the erstwhile princely kingdom of Jodhpur in Rajasthan (Pasiecznik et al. 2001).

Prosopis juliflora was also introduced to several PAs as a way to alleviate pressure for fuel wood from local forest-dependent communities (Robbins 2001), from where it has spread rapidly. Two prominent examples of this are Keoladeo Ghana (Anoop 2010), and Ranthambore (Dayal 2007), both in the desert state of Rajasthan in Northwest India. Yet, while it is widely recognised as a problem in PAs in Rajasthan – Keoladeo Ghana, Ranthambore, and also Kumbalgarh (Robbins 2001) – in other parts of the country *P. juliflora* is still cited as a successful and desirable example of afforestation. In the neighbouring state of Gujarat, for instance, Saxena (1998) talks of how the entire region of Kutch was successfully converted to *P. juliflora* woodland in just a 30-year period. In the Banni grasslands, which form part of Kutch, *P. juliflora*'s rate of spread between 1980 and 1992 was estimated (using remote sensing) to be as much as 25.5 km^2 per year (Jadhav et al. 1993; Tewari et al. 2000).

12.5.2.2 Invasiveness and Impacts

Unlike *L. camara*, much less is known about what contributes to *P. juliflora's* success as an invasive species. In a study comparing it with its only native congener, *P. cineraria* (khejri), Sharma and Dakshini (1996) suggest that its seed characteristics enable it to establish and grow faster than the native species. *Prosopis juliflora* has also been shown to be tolerant of drought and salinity (Pasiecznik et al. 2001). These characteristics, in combination with its low palatability, probably give it an advantage over native species.

The impacts of *P. juliflora* in PAs have not been as well documented as those of *L. camara*. The invasion of *P. juliflora* has, however, been shown to be replacing natural habitat in India's premier bird reserve, the Keoladeo Ghana (Anoop 2010), and to have allelopathic impacts on native vegetation (Kaur et al. 2012). Others have documented its impacts on native vegetation and forest-dependent communities in and around Kumbalgarh wildlife sanctuary, where *P. juliflora* invasion, accompanied by other un-palatable shrubby species (including *L. camara*), has led to the exclusion of important fodder grasses (Robbins 2001). *Prosopis juliflora* has also been documented to be encroaching on unique habitats for grassland birds such as the rare Houbara bustard (*Chlamydotis undulata*; Tiwari 1999).

In a bio-economic analysis of Ranthambore National Park, Dayal (2007) examined the impacts of *P. juliflora* on different stakeholders, namely, wildlife managers and local villagers (a composite of fuelwood collectors, cattle grazers, and goat owners). Findings suggest that the spread of *P. juliflora* is potentially reducing forage availability for wild herbivores as well as for cattle, though not for goats, which browse on the fruits and help to disperse seeds. The detrimental impacts of the tree on wild herbivores could, in turn, have bottom-up consequences for their predators, the tiger. Different scenarios for managing *P. juliflora* may potentially lead to very different outcomes for each of the stakeholders. In the context of this biological and socio-economic complexity that characterises many of India's PAs, a key question may be whether certain types of *P. juliflora* utilization (e.g. fuel wood collection, but not browsing by goats) could also further conservation goals by benefiting both managers and villagers (Dayal 2007).

With the exception of *L. camara* and *P. juliflora,* other invasive species in Indian PAs have barely been studied. Evidence is gradually accruing to suggest that *M. diplotricha* is encroaching floodplain habitat to which rhinoceros are restricted (Lahkar et al. 2011). Also, that the spread of *C. scoparius* in the montane grasslands of the Nilgiris (southern India), is altering community composition in these unique ecosystems (Srinivasan et al. 2007). However, these examples are probably just the tip of the iceberg (see Table 12.3). For most PAs in India even basic information regarding invasive species presence or absence is lacking, while information about their impacts is virtually non-existent.

Table 12.3 The subset of invasive alien plant species in Indian protected areas for which there is documented information on impacts

Invasive species	Impact at the population or community level	Impact at the ecosystem level	Source
Coffea canephora	Correlated with altered plant community composition	[a]	Joshi et al. (2009)
Chromolaena odorata	Invading floodplain grasslands, reducing habitat for the rhinoceros	[a]	Talukdar et al. (2008), Lahkar et al. (2011)
Cytisus scoparius	Correlated with altered plant community composition	[a]	Srinivasan et al. (2007)
Ipomoea carnea	Invading floodplain grasslands, reducing habitat for the rhinoceros	[a]	Lahkar et al. (2011)
Kappaphycus alvarezii	Forms dense mats over corals, eventually killing them	[a]	Chandrasekaran et al. (2008)
Lantana camara	Correlated with altered plant community composition, altered bird community composition; potential impacts on higher trophic levels (due to impacts on herbivores)	Increased soil nitrogen cycling, change in fire regime	Tireman (1916), Prasad (2009, 2010, 2012), Sharma and Raghubanshi (2009, 2010), Aravind et al. (2010), Sundaram and Hiremath (2012)
Mikania micrantha	Smothers other vegetation, eventually killing it; invasion of floodplain grasslands, reducing habitat for the rhinoceros	[a]	Gogoi (2001), Sankaran and Srinivasan (2001), Talukdar et al. (2008), Lahkar et al. (2011)
Mimosa diplotricha	Invading floodplain grasslands, reducing habitat for the rhinoceros, toxic to herbivores	[a]	Talukdar et al. (2008), Lahkar et al. (2011)
Prosopis juliflora	Replacing grasslands, reducing habitat for grassland birds; correlated with altered plant community composition; allelopathic	Increased soil organic nitrogen, organic carbon, exchangeable phosphorus, accumulation of phenolics	Tiwari (1999), Robbins (2001), Kaur et al. (2012)

[a]Denotes lack of information

12.6 Drivers of Invasion

12.6.1 Fire

Findings are starting to emerge from Indian PAs that seem to counter prevailing theoretical (Davis et al. 2000) and empirical (e.g. Myers 1983; Hobbs 1989; Larson et al. 2001) evidence for disturbed systems being more vulnerable to invasion than undisturbed systems. In BRT, for example, anecdotal evidence suggests that initial *L. camara* colonization and establishment had been preceded by widespread fires following bamboo flowering and dieback. More recent observations suggest that *L. camara* is able to recover from fire faster than the surrounding native vegetation, leading to a self-perpetuating fire-*Lantana* cycle (proposed by Hiremath and Sundaram 2005), analogous to the invasive grass-fire cycle reported from the Americas (D'Antonio and Vitousek 1992).

In a study aimed at testing this *L. camara*-fire cycle hypothesis, Sundaram (2011) found, contrary to expectation, that areas that had burned more frequently over an 11 year period (1997–2008) showed less abundant *L. camara* than areas that burned less frequently over the same period. It is possible that this may just reflect the time since fires, with areas that burned more frequently still recovering. However, a related ethnographic study suggests otherwise. Sundaram et al. (2012) found that the local inhabitants of the BRT, the *Soliga*, date the beginning of *L. camara's* spread in these forests to about 40 years ago. This roughly coincides with the notification of BRT as a PA and the cessation of the local community's traditional forest management practices (including cool early season ground fires, termed *'taragu benki'*). The *Soliga* maintain that the annual occurrence of *taragu benki* helped to suppress *L. camara*.

Needless to say, *L. camara* is now so abundant that the occurrence of cool ground fires or *taragu benki* would be impossible; fires would today rapidly become crown fires, causing widespread mortality of native vegetation, as witnessed annually during the dry season. Thus, with the spread of *L. camara* in BRT, these forests appear to have changed from a state where fire possibly halted the spread of *L. camara*, to a state where fire promotes the spread of *L. camara* (or at least causes damage to native vegetation, which could benefit *L. camara*), all in the space of about 40 years.

12.6.2 Cessation of Disturbance: A Paradoxical Driver of Invasions

Lantana camara in BRT is not an isolated example of a possible link between cessation of a particular disturbance regime and the spread of an IAP. Studies on *C. scoparius* in the montane grasslands of the Mukurti National Park in the Nilgiris

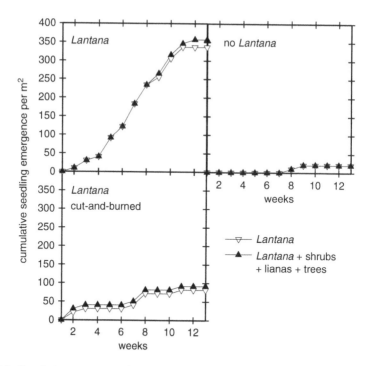

Fig. 12.4 Cumulative emergence of seeds of *Lantana camara* and other woody species from the soil seedbank over a 13-week period. Samples are from the dry season, when there are fewer viable seeds in the soil than at other times. Note the relatively large number of *L. camara* seedlings emerging in invaded sites compared with uninvaded sites. Also note the reduction in numbers of *L. camara* seedlings emerging in invaded sites that were burned

point to a similar situation. Srinivasan et al. (2012) suggest that suppression of fires may be the proximate cause of the spread of *C. scoparius* populations in Mukurti over the past few decades. This region has been home to Toda pastoralists, a community that traditionally practised fire management. The earliest records of the Todas in Mukurti go back to around 1117 A.D., suggesting that until burning ceased there had been an almost 900-year history of anthropogenic fires (Noble 1967). Similarly, in the Valley of Flowers National Park in the Himalayas, the spread of *Polygonum polystachyum* (Himalayan knotweed, a 'native invader') has been related to the cessation of grazing by nomadic pastoralists when the area became a national park (Naithani et al. 1992; Saberwal et al. 2000; but see Kala and Shrivastava 2004).

For *L. camara* in BRT, experiments suggest that fires kill seeds in the soil seed bank (Sundaram and Hiremath, unpublished; Fig. 12.4). This may be one mechanism by which *taragu benki* suppressed *L. camara* when it was not yet as widespread as it is today. Another mechanism may have been that fire helped to maintain the largely grassy understory of these deciduous woodland-savanna forests and that fire suppression enabled *L. camara* to out-compete grasses.

The cessation of a historical management regime (whether fire or grazing) could be considered equivalent to the removal of top-down control, thus exposing weaker competitors to competitive exclusion by stronger competitors (Paine 1966; Miller et al. 2001). This may be especially important where native vegetation has historically evolved with a prolonged anthropogenic disturbance regime. Anthropogenic fires may thus have played a role in preventing *C. scoparius* from becoming dominant in the montane grasslands of the Nilgiris, or in preventing *L. camara* from saturating the soil seedbank in BRT. Likewise, livestock grazing may have played a role in keeping *P. polystachyum* in check in the Valley of Flowers. Unfortunately, we only have observational evidence for these patterns. Additionally, two or three examples are insufficient to extract widespread trends. But these observations suggest that understanding the dynamics of IAPs in Indian PAs is unlikely to be complete without factoring in the ubiquitous human influence, both historical and on-going.

12.7 Conclusions and Implications for Management

Of the more than 200 IAPs and almost 700 PAs in India, information has been published on only a few invasive species and from only about 20 PAs. Despite the growing worldwide awareness of alien species invasions, India still lacks specific legislation to screen and regulate the introductions of potentially invasive species into the country. There is also no action plan for IAP management, no coordinated national research programme, and not even a repository for information on the distribution, extent and impact of even the better known invasive species (Khuroo et al. 2011). This has resulted in, for example, the seaweed *Kappaphycus alvarezii* being introduced into the vicinity of a marine biosphere reserve as recently as 2001, in spite of knowledge about its invasiveness in other similar environments.

Lantana camara and *P. juliflora* are excellent examples of the shortcomings that need to be addressed in managing invasive species in India's PAs, as well as of the opportunities that exist to do so. To develop creative solutions for long-term management and monitoring, PAs in India need to develop an adaptive management plan that promotes collaboration between researchers and managers. The lack of shared information on the distribution and impact of IAPs underlies the ambivalence with which forest managers have treated invasive species, and continue to do so. Thus, for instance, *P. juliflora* continues to be planted extensively (e.g. Saxena 1998), even as intensive efforts are in place to remove it in other areas (Anoop 2010).

In a recent review of *L. camara*, Bhagwat et al. (2012) described its management as a lost battle, suggesting that attempts to eradicate it have failed and ways of adaptively managing it need to be developed. Given how ubiquitous *L. camara* is today, it would be hard to disagree; any attempt to completely eradicate it is bound to fail. In the context of human-dominated landscapes, utilizing *L. camara* to enhance livelihoods and offset some of its costs may be one of the few viable options available (e.g. Uma Shaanker et al. 2009). However, in the context of high conservation value landscapes, there is a strong case to be made for *L. camara* control. Indeed, there are examples to demonstrate that this can be a realistic goal if researchers and managers collaborate to integrate the growing ecological understanding of invasive species with attempts to manage them, as in the case of the Corbett Tiger Reserve (Babu et al. 2009; Love et al. 2009). A similar successful collaborative programme between forest managers, researchers, and non-governmental conservation organizations is the monitoring and removal of *M. diplotricha* (= *M. invisa*) in Kaziranga National Park (Vattakkavan et al. 2005).

Another example of a successful attempt to control an invasive species in a PA comes from Keoladeo Ghana (Anoop 2010). Here park management has taken advantage of an existing poverty alleviation programme, the Mahatma Gandhi National Rural Employment Guarantee Scheme, to employ local villagers for *P. juliflora* removal (cf. the Working for Water programme in South Africa). The villagers use the *P. juliflora* for fuel wood and are employed to continue monitoring. However, despite the success of this initiative, it has not yet been expanded to other PAs across the country.

Invasive species management in PAs in India needs to move beyond just invasive plant removal. It needs to include an ecosystem approach that also considers drivers of invasion. Understanding the interaction between IAPs and the long-term anthropogenic disturbance regimes that these landscapes have evolved with may be as important to their management as understanding the biology and impacts of IAPs (Hobbs et al. 2011). Moles et al. (2012) have suggested that it is not disturbance, per se, but rather a change in disturbance regime, including the cessation of past disturbance, which may better explain ecosystem invasibility. Sharp changes in disturbance or management regimes have historically accompanied PA creation, with strict protection replacing past management (e.g. grazing, burning, etc.). Such practice is rooted in what Hobbs et al. (2006) argue is a "one-dimensional dichotomy between natural and human-dominated". They go on to suggest that we need to move away from these simplistic depictions to a more realistic understanding of how human beings interact with nature. Though they were referring to contemporary landscapes that are increasingly human-modified, it would be equally relevant in the context of PAs in India. Neither scientists nor managers can neglect the historical and on-going role of people in shaping Indian PAs. Engaging with this management history, rather than its abrupt cessation, may be a critical element in the management of IAPs in these landscapes.

Acknowledgments We thank the editors for the invitation to write this chapter and for input that helped to improve it. The work in BRT was supported by the Department of Science and Technology, India, and by the International Foundation for Science, Sweden. R. Geeta provided early assistance in the search for literature on invasive species in India. This manuscript benefited from comments provided by G. Joseph, A. Prasad, R. Ostertag, and two anonymous reviewers.

References

Anon (1895) Is *Lantana* a friend or an enemy? Ind For 21:454–460

Anoop KR (2010) Progress of *Prosopis juliflora* eradication work in Keoladeo National Park. Unpublished Report, Rajasthan Forest Department

Aravind NA, Rao D, Ganeshaiah KN et al (2010) Impact of the invasive plant, *Lantana camara*, on bird assemblages at Malé Mahadeshwara Reserve Forest, South India. Trop Ecol 51:325–338

Babu S, Leighton DP (2004) The Shompen of Greater Nicobar Island (India) – between "development" and disappearance. Policy Matter 13:198–211

Babu S, Love A, Babu CR (2009) Ecological restoration of *Lantana*-invaded landscapes in Corbett Tiger Reserve, India. Ecol Restor 27:468–478

Bagla P (2008) Seaweed invader elicits angst in India. Science 320:1271–1271

Berendse F, Aerts R (1987) Nitrogen-use efficiency: a biologically meaningful definition? Funct Ecol 1:293–296

Berry ZC, Wevill K, Curran TJ (2011) The invasive weed *Lantana camara* increases fire risk in dry rainforest by altering fuel beds. Weed Res 51:525–533

Bhagwat SA, Breman E, Thekaekara T et al (2012) A battle lost? Report on two centuries of invasion and management of *Lantana camara* L. in Australia, India and South Africa. PLoS One 7:e32407

Bhatt YD, Rawat YS, Singh SP (1994) Changes in ecosystem functioning after replacement of forest by *Lantana* shrubland in Kumaun Himalaya. J Veg Sci 5:67–70

Bingelli P, Hall JB, Healey JR (1998) An overview of invasive woody plants in the tropics. Publication no. 13. School of Agricultural and Forest Sciences, University of Wales, Bangor

Burton RW (1953) The preservation of wildlife in India. A compilation with a summarized index of contents. Bangalore Press, Bangalore

Catford JA, Jansson R, Nilsson C (2009) Reducing redundancy in invasion ecology by integrating hypotheses into a single theoretical framework. Divers Distrib 15:22–40

Chandrasekaran S, Swamy PS (2002) Biomass, litterfall and aboveground net primary productivity of herbaceous communities in varied ecosystems at Kodayar in the Western Ghats of Tamil Nadu. Agric Ecosyst Environ 88:61–71

Chandrasekaran S, Swamy PS (2010) Growth patterns of *Chromolaena odorata* in varied ecosystems at Kodayar in the Western Ghats, India. Acta Oecol 36:383–392

Chandrasekaran S, Nagendran NA, Pandiaraja D et al (2008) Bioinvasion of *Kappaphycus alvarezii* on corals in the Gulf of Mannar, India. Curr Sci 94:1167–1172

Chandrashekara UM (2001) *Lantana camara* in Chinnar Wildlife Sactuary, Kerala, India. In: Sankaran KV, Murphy ST, Evans HC (eds) Alien weeds in moist tropical zones: banes and benefits. KFRI/CABI Bioscience, Kerala/Ascot, pp 56–63

Chaudhuri H, Ramaprabhu T, Ramachandran V (1994) *Ipomea carnea* Jacq. A new aquatic weed problem in India. J Aquat Plant Manag 32:37–38

Chew MK (2011) Invasion biology: historical precedents. In: Simberloff D, Rejmanek M (eds) Encyclopedia of biological invasions. University of California Press, Berkeley, pp 369–375

Colautti RI, MacIsaac HJ (2004) A neutral terminology to define 'invasive' species. Divers Distrib 10:135–141

Critical Ecosystems Partnership Fund (2013) Global map. http://www.cepf.net/where_we_work/Pages/map.aspx. Accessed 14 Feb. 2013

Cronk QCB, Fuller JL (1995) Plant invaders. The threat to natural ecosystems. Chapman and Hall, London

D'Antonio CA, Vitousek PM (1992) Biological invasions by exotic grasses, the grass-fire cycle, and global change. Annu Rev Ecol Syst 23:63–87

Das A, Krishnaswamy J, Bawa KS et al (2006) Prioritisation of conservation areas in the Western Ghats, India. Biol Conserv 133:16–31

Davis MA, Grime JP, Thompson K (2000) Fluctuating resources in plant communities: a general theory of invasibility. J Ecol 88:528–536

Davis MA, Chew MK, Hobbs RJ et al (2011) Don't judge species on their origins. Nature 474:153–154

Day MD, Wiley CJ, Playford J et al (2003) *Lantana*. Current management status and future prospects. ACIAR Monograph 102, Canberra

Dayal V (2007) Social diversity and ecological complexity: how an invasive tree could affect diverse agents in the land of the tiger. Environ Dev Econ 12:1–19

Denslow J, deWalt S (2008) Exotic plant invasions in tropical forests: Patterns and hypotheses. In: Carson WP, Schnitzer S (eds) Tropical forest community ecology. Blackwell Scientific, Oxford, pp 409–426

Duggin JA, Gentle CB (1998) Experimental evidence on the importance of disturbance intensity for invasion of *Lantana camara* L. in dry rainforest – open forest ecotones in north-eastern NSW, Australia. For Ecol Manage 109:279–292

Elton CS (1958, 2000) The ecology of invasions by animals and plants. University of Chicago Press, Chicago

Foxcroft LC, Pyšek P, Richardson DM et al (2014) Chapter 2: Impacts of alien plant invasions in protected areas. In: Foxcroft LC, Pyšek P, Richardson DM, Genovesi P (eds) Plant invasions in protected areas: patterns, problems and challenges. Springer, Dordrecht, pp 19–41

Funk J, Vitousek PM (2007) Resource-use efficiency and plant invasion in low-resource systems. Nature 446:1079–1081

Ghazoul J (2004) Alien abduction: disruption of native plant-pollinator interactions by invasive species. Biotropica 36:156–164

Giradkar PG, Yeragi SG (2008) Flora of Tadoba National Park. Ind For 134:263–269

Gogoi AK (2001) Status of *Mikania micrantha* infestation in Northeastern India: management options and future research thrust. In: Sankaran KV, Murphy ST, Evans HC (eds) Alien weeds in moist tropical zones: banes and benefits. KFRI/CABI Bioscience, Kerala/Ascot, pp 77–79

Government of India (2002) The wildlife protection (Amendment) Act 2002. http://www.indiaenvironmentportal.org.in/node/258103. Accessed 12 Mar. 2013

Government of India (2006) The wildlife protection (Amendment) Act 2006. http://www.indiaenvironmentportal.org.in/reports-documents/wild-life-protection-amendment-act-2006. Accessed 12 Mar 2013

Gurevitch J, Fox GA, Wardle GM et al (2011) Emergent insights from the synthesis of conceptual frameworks for biological invasion. Ecol Lett 14:407–418

Hakimuddin M (1930) *Lantana* in northern India as a pest and its probable utility in solving the cowdung problem. Ind For 56:405–410

Hejda M, Pyšek P, Jarošík V (2009) Impact of invasive plants on the species richness, diversity and composition of invaded communities. J Ecol 97:393–403

Hiremath AJ, Sundaram B (2005) The fire-*Lantana* cycle hypothesis in Indian forests. Conserv Soc 3:26–42

Hobbs RJ (1989) The nature and effects of disturbance relative to invasions. In: Drake JA, Mooney HA, di Castri RH et al (eds) Biological invasions: a global perspective. SCOPE 37. Wiley, New York, pp 389–405

Hobbs RJ, Arico S, Aronson J et al (2006) Novel ecosystems: theoretical and management aspects of the new ecological world order. Glob Ecol Biogeogr 15:1–7

Hobbs RJ, Hallett LM, Ehrlich PR et al (2011) Intervention ecology: applying ecological science in the twenty-first century. BioScience 61:442–450

Iyengar AVV (1933) The problem of the *Lantana*. Curr Sci 21:266–269

Jadhav RN, Kimothi MM, Kandya AK (1993) Grassland mapping/monitoring of Banni, Kachchh (Gujarat) using remotely-sensed data. Int J Remote Sens 14:3093–3103

Jeschke JM, Aparicir LG, Haider S et al (2012) Support for major hypotheses in invasion biology is uneven and declining. Neobiota 14:1–20

Joshi AA, Mudappa D, Shankar Raman TR (2009) Brewing trouble: coffee invasion in relation to edges and forest structure in tropical rainforest fragments of the Western Ghats, India. Biol Invasions 11:2387–2400

Kala CP, Shrivastava RJ (2004) Successional changes in Himalayan alpine vegetation: two decades after removal of livestock grazing. Weed Technol 18:1210–1212

Kannan R, Shackleton CM, Uma Shaanker R (2013) Reconstructing the history of introduction and spread of the invasive species, *Lantana*, at three spatial scales in India. Biol Invasions 15:1287–1302

Kaur R, Gonzáles WL, Llambi LD et al (2012) Community impacts of *Prosopis juliflora* invasion: biogeographic and congeneric comparisons. PLoS One 7:e44966

Keane RM, Crawley MJ (2002) Exotic plant invasions and the enemy release hypothesis. Trends Ecol Evol 17:164–170

Khuroo AA, Reshi ZA, Rashid I et al (2011) Towards an integrated research framework and policy agenda on biological invasions in the developing world: a case-study of India. Environ Res 111:999–1006

Khuroo AA, Reshi ZA, Malik AH et al (2012) Alien flora of India: taxonomic composition, invasion status and biogeographic affiliations. Biol Invasions 14:99–113

Kimothi MM, Dasari A (2010) Methodology to map the spread of an invasive plant (*Lantana camara* L.) in forest ecosystems using Indian remote sensing satellite data. Int J Remote Sens 31:3273–3289

Kimothi MM, Anitha D, Vasistha HB et al (2010) Remote sensing to map the invasive weed, *Lantana camara* in forests. Trop Ecol 51:67–74

Kohli RK, Batish DR, Singh HP et al (2006) Status, invasiveness and environmental threats of three tropical American invasive weeds (*Parthenium hysterophorus* L., *Ageratum conyzoides* L., *Lantana camara* L.) in India. Biol Invasions 8:1501–1510

Krishnan P, Ramakrishnan R, Saigal S et al (2012) Conservation across landscapes: India's approaches to biodiversity governance. United Nations Development Programme, New Delhi

Lahkar BP, Talukdar BK, Sarma P (2011) Invasive species in grassland habitat: an ecological threat to the greater one-horned rhino (*Rhinoceros unicornis*). Pachyderm 49:33–39

Larson DL, Anderson PJ, Newton W (2001) Alien plant invasion in mixed-grass prairie: effects of vegetation type and anthropogenic disturbance. Ecol Appl 11:128–141

Le Maitre DC, van Wilgen BW, Gelderblom CM et al (2001) Invasive alien trees and water resources in South Africa: case studies of the costs and benefits of management. For Ecol Manage 160:143–159

Lockwood JL, Cassey P, Blackburn T (2005) The role of propagule pressure in explaining species invasions. Trends Ecol Evol 20:223–228

Love A, Babu S, Babu CR (2009) Management of *Lantana*, an invasive alien weed in forest ecosystems of India. Curr Sci 97:1421–1429

Lowe S, Browne M, Boudielas S (2000) 100 of the world's worst invasive alien species. A selection from the Global Invasive Species Database. The Invasive Species Specialist Group (ISSG) of the World Conservation Union (IUCN), Gland

Mack RN, Lonsdale WM (2001) Humans as global plant dispersers: getting more than we bargained for. BioScience 51:95–102

Madhusudan ND, Shanker K, Kumar A et al (2006) Science in the wilderness: the predicament of scientific research in India's wildlife reserves. Curr Sci 91:1015–1019

Mahajan M, Azeez PA (2001) Distribution of selected exotic weeds in Nilgiri Biosphere Reserve. In: Sankaran KV, Murphy ST, Evans HC (eds) Alien weeds in moist tropical zones: banes and benefits. KFRI, India & CABI Bioscience, UK, pp 46–55

Miller B, Dugelby B, Foreman D et al (2001) The importance of large carnivores to healthy ecosystems. Endanger Species Update 18:202–210

Moles A, Flores-Moreno H, Bonser SP et al (2012) Invasion: the trail behind, the path ahead, and a test of a disturbing idea. J Ecol 100:116–127

Muniappan R, Raman A, Reddy GVP (2009) *Ageratina adenophora* (Sprengel) King and Robinson (Asteraceae). In: Muniappan R, Reddy GVP, Raman A (eds) Biological control of tropical weeds using arthropods. Cambridge University Press, Cambridge, pp 63–73

Murali KS, Setty RS (2001) Effect of weeds *Lantana camara* and *Chromolaena odorata* growth on the species diversity, regeneration and stem density of tree and shrub layer in BRT sanctuary. Curr Sci 80:675–678

Myers RL (1983) Site susceptibility to invasion by the exotic tree *Melaleuca quinquenervia* in southern Florida. J Appl Ecol 20:645–658

Myers N, Mittermeier RA, Mittermeier CG et al (2000) Biodiversity hotspots for conservation priorities. Nature 403:853–858

Nair KPP (2010) The agronomy and economy of important tree crops of the developing world. Elsevier, Amsterdam

Naithani HB, Negi JDS, Thapliyal RC et al (1992) Valley of flowers: needs for conservation or preservation. Ind For 118:371–378

Namboothri N, Shankar K (2010) Corals or coke: between the devil and the deep blue sea. Curr Conserv 4:18–20

Noble WA (1967) The shifting balance of grasslands, shola forests, and planted trees on the upper Nilgiris, southern India. Ind For 93:691–693

Paine RT (1966) Food web complexity and species diversity. Am Nat 100:65–75

Parker IM, Simberloff D, Lonsdale WM et al (1999) Impact: toward a framework for understanding the ecological effects of invaders. Biol Invasions 1:3–19

Pasiecznik NM, Felker P, Harris PJC et al (2001) The *Prosopis juliflora–Prosopis pallida* complex: a monograph. HDRA, Coventry

Paul TK (2010) The earliest record of *Parthenium hysterophorus* L. (Asteraceae) in India. Curr Sci 98:1272

Paynter Q, Harman H, Waipara N (2006) Prospects for biological control of *Merremia peltata*. Landcare Research Contract Report: LC0506/177, New Zealand

Prasad AE (2009) Tree community change in a tropical dry forest: the role of roads and exotic plant invasion. Environ Conserv 36:201–207

Prasad AE (2010) Effects of an exotic plant invasion on native understory plants in a tropical dry forest. Conserv Biol 24:747–757

Prasad AE (2012) Landscape-scale relationships between the exotic invasive shrub *Lantana camara* and native plants in a tropical deciduous forest in southern India. J Trop Ecol 28:55–64

Puyravaud JP, Shridhar D, Gaulier A et al (1995) Impact of fire on a dry deciduous forest in the Bandipur National Park, southern India – preliminary assessment and implications for management. Curr Sci 68:745–751

Pyšek P, Richardson DM, Rejmanek M et al (2004) Alien plants in checklists and floras: towards better communication between taxonomists and ecologists. Taxonomy 53:131–143

Pyšek P, Jarošík V, Hulme PE et al (2011) A global assessment of invasive plant impacts on resident species, communities, and ecosystems: the interaction of impact measures, invading species' traits, and environment. Glob Chang Biol 18:1725–1737

Raizada P (2006) Ecological and vegetative characteristics of a potent invaders, *Hyptis suaveolens* Poit. from India. Lyonia 11:115–120

Ramaswami G, Sukumar R (2011) Woody plant seedling distribution under invasive *Lantana camara* thickets in a dry-forest plot in Mudumalai, southern India. J Trop Ecol 27:365–373

Randerson J (2003) Fungus in your tea sir? New Sci 2401:10

Rangarajan M (2001) India's wildlife history. An introduction. Permanent Black, Delhi

Ray A, Sumangala RC, Ravikanth G et al (2013) Isolation and characterization of polymorphic microsatellite loci from the invasive plant *Lantana camara* L. Conserv Genet Res 4:171–173

Rishi V (2009) Wildlife habitat enrichment for mitigating human-elephant conflict by biological displacement of *Lantana*. Ind For 135:439–448

Robbins P (2001) Tracking invasive land covers in India, or why our landscapes have never been modern. Annu Assoc Am Geogr 91:637–659

Saberwal V, Rangarajan M, Kothari A (2000) People, parks and wildlife: towards coexistence. Orient Longman, New Delhi

Sahu PK, Singh JS (2008) Structural attributes of *Lantana*-invaded forest plots in Achanakmar-Amarkantak Biosphere Reserve, Central India. Curr Sci 94:494–500

Sankaran KV, Srinivasan MA (2001) Status of *Mikania* infestation in the Western Ghats. In: Sankaran KV, Murphy ST, Evans HC (eds) Alien weeds in moist tropical zones: banes and benefits. KFRI/CABI Bioscience, Kerala/Ascot, pp 67–76

Sawarkar VB (1984) *Lantana camara* in wildlife habitats with special reference to the Melghat Tiger Reserve. Cheetal 26:24–38

Saxena SK (1998) Ecology of *Prosopis juliflora* in the arid regions of India. In: Tewari JC, Pasiecznik NM, Harsh LN et al (eds) *Prosopis* species in the arid and semi-arid zone of India. Proceedings of a conference held at the Central Arid Zone Research Institute, Jodhpur, Rajasthan, India. November 21–23, 1993. The Prosopis Society of India and The Henry Doubleday Research Association, pp 17–20. http://www.fao.org/docrep/006/AD321E/ad321e04.htm#bm04.1. Accessed 11 Mar. 2012

Sharma R, Dakshini KMM (1996) Ecological implications of seed characteristics of the native *Prosopis cineraria* and the alien *P juliflora*. Vegetatio 124:101–105

Sharma GP, Raghubanshi AS (2009) *Lantana* invasion alters soil nitrogen pools and processes in the tropical dry deciduous forest of India. Appl Soil Ecol 42:134–140

Sharma GP, Raghubanshi AS (2010) How *Lantana* invades dry deciduous forest: a case study from Vindhyan highlands, India. Trop Ecol 51:305–316

Shukla AN, Singh KP, Singh JS (2009) Invasive alien species of Achanakmar-Amarkantak Biosphere Reserve, Central India. Proc Natl Acad Sci Ind Sect B-Biol Sci 79:384–392

Simberloff D, Martin J-L, Genovesi P et al (2013) Impacts of biological invasions: what's what and the way forward. Trends Ecol Evol 28:58–66

Soni PL, Naithani S, Gupta PK et al (2006) Utilization of economic potential of *Lantana camara*. Ind For 132:1625–1630

Srinivasan MP (2011) The ecology of disturbance and global change in the montane grasslands of the Nilgiris, South India. University of Kentucky Doctoral Dissertations, Lexington, Paper 213

Srinivasan MP, Shenoy K, Gleason SK (2007) Population structure of Scotch broom (*Cytisus scoparius*) and its invasion impacts on the resident plant community in the grasslands of Nilgiris, India. Curr Sci 93:1108–1113

Srinivasan MP, Kalita R, Gurung IK et al (2012) Seedling germination success and survival of the invasive shrub Scotch broom (*Cytisus scoparius*) in response to fire and experimental clipping in the montane grasslands of the Nilgiris, South India. Acta Oecol 38:41–48

Stracey PD (1960) Wildlife management. Leaflet no. 3. Indian Board for Wildlife, Ministry of Food and Agriculture, New Delhi

Sundaram B (2011) Patterns and processes of *Lantana camara* persistence in South Indian tropical dry forests. Doctoral Dissertation, Ashoka Trust for Research in Ecology and the Environment and Manipal University

Sundaram B, Hiremath AJ (2012) *Lantana camara* invasion in a heterogeneous landscape: patterns of spread and correlation with changes in native vegetation. Biol Invasion 14:1127–1141

Sundaram B, Krishnan S, Hiremath AJ et al (2012) Ecology and impacts of the invasive species, *Lantana camara*, in a social-ecological system in South India: perspectives from local knowledge. Hum Ecol 40:931–942

Talukdar BK, Emslie R, Bist SS et al (2008) *Rhinoceros unicornis* (greater one-horned rhino, great Indian rhinoceros, Indian rhinoceros). http://www.iucnredlist.org/details/19496/0. Accessed 1 Jan 2013

Tewari JC, Harris PJC, Harsh LN et al (2000) Managing *Prosopis juliflora* (vilayti babul) – a technical manual. CAZRI, Jodhpur

Thakur ML, Ahmad M, Thakur RK (1992) Lantana weed (*Lantana camara* var. *aculeata* Linn.) and its possible management through natural insect pests in India. Ind For 118:466–486

Ticktin T, Ganesan R, Paramesha M et al (2012) Disentangling the effects of multiple anthropogenic drivers on the decline of two tropical dry forest trees. J Appl Ecol 49:774–784

Tireman H (1916) *Lantana* in the forests of Coorg. Ind For 42:384–392

Tiwari JWK (1999) Exotic weed *Prosopis juliflora* in Gujarat and Rajasthan, India: boon or bane? Tigerpaper 26:21–25

Uma Shaanker R, Joseph G, Aravind NA et al (2009) Invasive plants in tropical human-dominated landscapes: need for an inclusive management strategy. In: Perrings C, Mooney H, Williamson M (eds) Bioinvasions and globalization. Ecology, economics, management, and policy. Oxford University Press, Oxford, pp 202–219

Usher MB (1991) Biological invasion into tropical nature reserves. In: Ramakrishnan PS (ed) Ecology of biological invasions in the tropics. International Scientific Publications, New Delhi

Vardien W, Richardson DM, Foxcroft LC et al (2012) Invasion dynamics of *Lantana camara* L. (sensu lato) in South Africa. S Afr J Bot 81:81–94

Vattakkavan J, Vasu NK, Varma S et al (2005) Silent stranglers. Eradication of *Mimosa* in Kaziranga National Park, Assam. Wildlife Trust of India, New Delhi

Vitousek PM, Walker LR (1989) Biological invasion by *Myrica faya* in Hawai'i: plant demography, nitrogen fixation, and ecosystem effects. Ecol Monogr 59:247–265

Waite TA, Corey SJ, Campbell LG et al (2009) Satellite sleuthing: does remotely sensed land-cover change signal ecological degradation in a protected area? Divers Distrib 15:299–309

Weber E, Li B (2008) Plant invasions in China: what is to be expected in the wake of economic development. BioScience 58:437–444

Wilson JRU, Richardson DM, Rouget M et al (2007) Residence time and potential range: crucial considerations in modelling plant invasions. Divers Distrib 13:11–22

Zarri AA, Rahmani AA, Behan MJ (2006) Habitat modification by Scotch broom *Cytisus scoparius* invasion of the grasslands of Upper Nilgiris in India. J Bombay Nat Hist Soc 103:356–365

Chapter 13
Invasive Plants in the United States National Parks

Thomas J. Stohlgren, Lloyd L. Loope, and Lori J. Makarick

Abstract Natural area parks managed by the United States National Park Service were established to protect native species and historical (but living) landscapes and scenery, and to provide public enjoyment of the same, as long as the natural area remained "unimpaired for future generations." A growing human population, and a 40-fold increase global trade and transportation may provide the most significant challenge to Park Service management: the invasion of alien plants, animals, and diseases into so called 'protected areas'. General ecological theory suggests that an increase in biological diversity should increase the overall stability of an ecosystem. We provide examples of plant invasions in U.S. National Parks to show that, despite increases in plant diversity, alien plant species are capable of greatly affecting the native species that the parks were established to protect. Furthermore, alien plant species are affecting natural patterns of grazing, disturbance, and nutrient cycling, resulting in decreased habitat quality, perhaps providing less resistance and resilience to natural and anthropogenic stresses such as climate change, land use change, recreation, and future invasions. The primary mission of the National Park Service, protecting native species and ecosystems for present and future generations, may be increasingly difficult due to the continuing invasions

T.J. Stohlgren (✉)
US Geological Survey, Fort Collins Science Center,
2150 Centre Ave. Bldg. C, Fort Collins, CO 80526, USA
e-mail: stohlgrent@usgs.gov

L.L. Loope (retired)
Formerly: USGS Pacific Island Ecosystems Research Center, Haleakala Field Station,
Makawao (Maui), HI 96768, USA

Current: 751 Pelenaka Place, Makawao, HI 96768, USA
e-mail: lll@aloha.net

L.J. Makarick
Grand Canyon National Park, 1824 S. Thompson Street Suite 200, Flagstaff,
AZ 86001, USA
e-mail: Lori_Makarick@nps.gov

L.C. Foxcroft et al. (eds.), *Plant Invasions in Protected Areas: Patterns, Problems and Challenges*, Invading Nature - Springer Series in Invasion Ecology 7, DOI 10.1007/978-94-007-7750-7_13, © Springer Science+Business Media Dordrecht 2013

of alien organisms. Key elements of an effective invasive plant management programme are identified.

Keywords Alien plant species • Early detection • Globalisation • Invasive species control • Non-native species • Preservation • Prevention

13.1 Introduction

The United States National Park Service manages 400 land areas totalling 33,993,594 ha (84,000,000 acres) of land and 1,822,155 ha (4,502,644 acres) of oceans, lakes, reservoirs (http://www.nps.gov/aboutus/index.htm; accessed July, 30 2012). Over 87 % of the protected areas are in the western states, but because a vast majority of native species and ecosystems are represented in the National Park system, these protected areas serve as premier global example of natural resource conservation (Burns et al. 2003).

Because of their beauty, uniqueness, and American historical value, the National Park Service was created to protect natural resources and cultural features (National Park Service Organic Act 1916; 16 U.S.C. 1 2 3, and 4). The legislation is quite clear. The National Park Service is directed: "to conserve the scenery and the natural and historic objects and the wild life therein and to provide for the enjoyment of the same in such manner and by such means as will leave them unimpaired for the enjoyment of future generations." However, a brief history of invasive species management in national parks, adapted from Houston and Schreiner (1995) and Drees (2004), and expanded here, shows how the National Park Service may be falling further behind each year.

Eighty years ago, Wright et al. (1933) suggested that alien animals posed a significant threat to the national parks. The authors criticised the purposeful introduction of game birds and fishes, accidental releases of alien species, and human-caused range extensions of alien species as wholly inappropriate for nature preserves. In the 1960s, two influential reports asserted that the introduction of alien species was inappropriate for areas set aside to preserve natural resources (Leopold et al. 1963; Robbins et al. 1963). In 1968, the national park policy stated that "nonnative species may not be introduced into natural areas. Where they have become established or threaten invasion of a natural area, an appropriate management plan should be developed to control them, where feasible. . ." and that "nonnative species of plants and animals will be eliminated where it is possible to do so by approved methods which will preserve wilderness qualities" (National Park Service 1968).

By the 1990s, scientists more strongly urged policy makers on the potential for alien species to displace native species and change 'protected' landscapes (Usher et al. 1988; Vitousek 1990; Mack et al. 2000; Pimentel et al. 2000; Myers and Bazely 2003). Meanwhile, field studies in national parks from around the world reported that nature reserves were hotbeds of invasions (Loope et al. 1988; Macdonald and Frame 1988; Macdonald et al. 1989; Lonsdale and Lane 1994). In addition, Stohlgren et al. (1998, 1999) reported that hotspots of native plant diversity coincided with hotspots of alien species invasions. Awareness of the problem grew.

Drees (2004) reported that, of the 34 million ha managed by the National Park Service, 1.1 million ha were infested by alien plants and animals. Many alien animals, including feral pigs (*Sus scrofa*) and goats (*Capra hircus*), hemlock woolly adelgid (*Adelges tsugae*), New Zealand mudsnail (*Potamopyrgus antipodarum*) and African oryx (*Oryx beisa*), were rampant in the Parks System. The National Park Service estimated that staffing and funding for a viable invasive species programme might cost US$80 million per year, but the budgetary response was considerably less. In 2003, Exotic Plant Management Teams received US$2.8 million for weed control in national parks (Drees 2004). While scientists at the time urged that invasive species science and management remained fragmentary and underfunded relative to the scope of the problem (Vitousek 1990; Mack et al. 2000; Pimentel et al. 2000; Myers and Bazely 2003; Loope 2004), some Park Service officials remained optimistic. "If we stay the course, management of invasive species in parks is within our grasp" (Drees 2004). That same year, Loope (2004) was far more pessimistic, stating that, "Major breakthroughs in science, policy, and management will likely be needed to address the complex and important issue of biological invasions if substantial impairment of the parks is to be averted" (pg. 7). Loope may have been more correct.

In 2006, the National Park Service clarified its policy regarding invasive species (NPS 2006; Section 4.4.4.2 Removal of Exotic Species Already Present): "All exotic plant and animal species that are not maintained to meet an identified park purpose will be managed – up to and including eradication – if (1) control is prudent and feasible, and (2) the exotic species:

- interferes with natural processes and the perpetuation of natural features, native species or natural habitats, or
- disrupts the genetic integrity of native species, or
- disrupts the accurate presentation of a cultural landscape, or damages cultural resources, or
- significantly hampers the management of park or adjacent lands, or
- poses a public health hazard as advised by the U.S. Public Health Service (which includes the Centres for Disease Control and the NPS public health programme), or
- creates a hazard to public safety."

Valiant, but modest, efforts continue to be made by the National Park Service in containing invasive species. Exotic Plant Management Teams continue to control invasive weeds in many natural areas parks. Volunteers/citizen scientists work alongside park staff to assist with invasive species control, donating tens of thousands of hours each year. There are many other examples of locally-effective invasive species management in the national parks, but recent studies suggest it may be time to re-evaluate the efforts made relative to the increasing scope and severity of the problem (Pimentel et al. 2005; Stohlgren et al. 2006, 2011, 2013).

13.2 Recent Studies: Examples of a Growing Problem

13.2.1 Alien Plant Invasions Have Accelerated

An alarming study by Allen et al. (2009) reported 20,305 alien plant species infestations, with 3,756 unique alien plants totalling 7.3 million ha in 218 national parks in the USA. Many of the 254 natural area parks had inadequate data, or mapped data for only a small fraction of their invasive species. Still, many national parks reported hundreds of alien plant species (Fig. 13.1). Chesapeake and Ohio Canal National Historical Park reported 483 alien plant species. Alien plant species richness was significantly associated with the richness of rare plant species and native plant richness.

A recent assessment of alien plant species from the National Park Service Biodiversity Database (IRMA version. https://irma.nps.gov/Species.mvc/Welcome; (accessed 15 March 2012) showed that 127 national park units contained 50 or more alien plant species (Table 13.1). These numbers correspond to other recent studies of widespread alien plant species in the United States (Stohlgren et al. 2011).

Upon closer examination of the patterns of alien plant invasions, the investigators found strong positive associations between alien species richness and both the number of visitors and the length of the trail systems in the parks. This confirmed earlier studies in other countries (Lonsdale and Lane 1994; Lonsdale 1999) and elsewhere in the United States (Tyser and Worley 1992; Harrison et al. 2002; Gelbard and Belnap 2003) that roads and trails may be important conduits of invasive species. Likewise, visitation rates on island reserves were strongly correlated to invasion (Brockie et al. 1988).

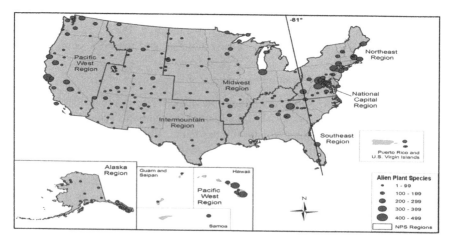

Fig. 13.1 Map of the National Park Service regional boundaries, alien plant species richness, and study park distribution (Adapted from Allen et al. 2009)

Table 13.1 Alien plant species that occur in 50 or more U.S. National Park units

Rumex crispus	181	Portulaca oleracea	76
Taraxacum officinale	180	Potentilla recta	76
Verbascum thapsus	172	Amaranthus retroflexus	75
Trifolium repens	168	Commelina communis	75
Melilotus officinalis	162	Rumex obtusifolius	75
Medicago lupulina	157	Malva neglecta	74
Capsella bursa-pastoris	153	Dianthus armeria	73
Plantago lanceolata	153	Verbascum blattaria	73
Cirsium vulgare	145	Hedera helix	71
Lactuca serriola	143	Nepeta cataria	71
Poa pratensis	142	Trifolium campestre	71
Stellaria media	142	Linaria vulgaris	70
Bromus tectorum	136	Thlaspi arvense	70
Dactylis glomerata	136	Polygonum convolvulus	69
Convolvulus arvensis	135	Agrostis stolonifera	68
Sonchus asper	133	Elymus repens	68
Rumex acetosella	131	Albizia julibrissin	67
Poa annua	130	Berberis thunbergii	67
Phleum pratense	129	Tribulus terrestris	66
Trifolium pratense	127	Anagallis arvensis	65
Chenopodium album	120	Hemerocallis fulva	65
Erodium cicutarium	115	Holcus lanatus	65
Poa compressa	111	Lespedeza cuneata	65
Tragopogon dubius	111	Lolium pratense	65
Echinochloa crus-galli	108	Achillea millefolium	63
Cynodon dactylon	106	Ulmus pumila	63
Arctium minus	104	Carduus nutans	62
Bromus japonicus	104	Alliaria petiolata	60
Plantago major	104	Elaeagnus angustifolia	60
Daucus carota	103	Cardamine hirsuta	59
Veronica arvensis	102	Cerastium glomeratum	59
Lolium perenne	101	Digitaria ischaemum	58
Asparagus officinalis	99	Lamium purpureum	58
Cirsium arvense	99	Lepidium campestre	58
Lonicera japonica	98	Microstegium vimineum	58
Polygonum aviculare	98	Conium maculatum	57
Leucanthemum vulgare	96	Digitaria sanguinalis	57
Medicago sativa	96	Duchesnea indica	57
Sisymbrium altissimum	96	Lolium arundinaceum	57
Ailanthus altissima	95	Solanum dulcamara	57
Melilotus alba	95	Triticum aestivum	57
Morus alba	94	Allium vineale	56
Cichorium intybus	91	Trifolium dubium	56
Setaria viridis	89	Cerastium fontanum ssp. vulgare	55
Lamium amplexicaule	88	Hesperis matronalis	55
Barbarea vulgaris	87	Arabidopsis thaliana	54
Rosa multiflora	87	Bromus commutatus	54

(continued)

Table 13.1 (continued)

Sorghum halepense	87	Trifolium arvense	54
Agrostis gigantea	84	Chenopodium ambrosioides	53
Sonchus oleraceus	84	Lotus corniculatus	53
Bromus inermis	83	Paspalum dilatatum	53
Polygonum persicaria	83	Arenaria serpyllifolia	52
Descurainia sophia	82	Bromus catharticus	52
Marrubium vulgare	82	Kochia scoparia	52
Polypogon monspeliensis	81	Lathyrus latifolius	52
Saponaria officinalis	80	Ligustrum vulgare	52
Glechoma hederacea	79	Malus pumila	52
Vinca minor	79	Syringa vulgaris	52
Anthoxanthum odoratum	78	Leonurus cardiaca	51
Eleusine indica	78	Populus alba	51
Hypericum perforatum	78	Pyrus communis	51
Salsola tragus	78	Veronica serpyllifolia	51
Trifolium hybridum	77	Acer platanoides	50
Eragrostis cilianensis	76		

Data from the U.S. National Park Service (March 2011), presented as genus, species, and number of park units

A recent study on global patterns of invasion showed that a 40-fold increase in trade (imports) in the past 50 years may be responsible for another alarming pattern: about half of the most widely distributed plant species in the United States are from other countries (Stohlgren et al. 2011). It is now common in national parks to see plant species from Europe and Asia (e.g. annual grasses, asters, tamarisk, thistles, etc.) dominating landscapes. Thus, external forces (globalisation and trade) are combining with internal forces (transportation networks and park visitation) to greatly increase the dispersal of alien propagules. Urbanisation adjacent to the parks and ground-disturbing projects within them, combined with the landscaping and nursery trade (Reichard and White 2001), may increasingly add to the number and genetic diversity of alien plant invaders in U.S. National Parks.

13.2.2 Site-Specific Plant Invasions Increasing Over Time

In many national parks and protected areas in the United States, detailed surveys over time are lacking and complete inventories are rare. However, an illustrated example from the Grand Canyon National Park, Arizona, is available. The park maintains a herbarium of plant collections, with the oldest specimen dating back to 1913. The list of alien plants recorded to date (2011) contains 197 species, although the nativity of one species of *Galium* is still being examined. Although survey efforts have increased over time in the park, they have been reasonably consistent in the past few decades in their spatial extent, intensity, and training of personnel involved in collections. A time series graph of alien plant species in the park (Fig. 13.2) depicts increases during the 1990s from literature reviews, herbaria,

GRAND CANYON NATIONAL PARK, ARIZONA

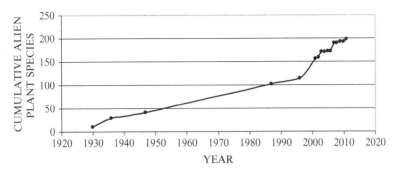

Fig. 13.2 Cumulative alien plant species reported in Grand Canyon National Park over time

surveys, and a master's thesis on the topic (Makarick 1999). Most surprising to park staff, was the rapid increase attributed to new species arriving in the park in recent years. Five alien plant species were thought to arrive in 2011, based on field surveys in previously well-surveyed areas. One species, *Scorzonera lacinata* (cutleaf vipergrass), had only one previous specimen collected in Arizona, but based on its rapid spread in other states, it is likely to keep expanding its range.

Verifying the exact date of species arrival is problematic. However, repeat surveys decades apart provide corroborating evidence that alien species may be continually arriving in various areas around the country. For example, Hoagland (2007) repeated a complete floristic survey of the Ozark Plateau in Oklahoma. The survey was first conducted by an equally competent botanist, Charles Wallis in 1959. Wallis recorded 134 alien plant species in 1959 (Wallis 1959). Hoagland documented 188 alien plant species over the same area by 2007, an increase of about one plant species per year over the course of the two surveys. Reports of new alien species in national parks are consistent with increases in trade and transportation (Stohlgren et al. 2013).

Documenting the actual spread of alien plant species is also difficult. However, modeling the distributions of *Bromus tectorum* (cheatgrass) in Rocky Mountain National Park, Colorado, may provide insights. Bromberg et al. (2011) used vegetation plot data from 1996, 2003, and 2007 to model habitat suitability of *B. tectorum* in the park. The habitat models were generally able to predict new occurrences of *B. tectorum* based on previous data. Subsequent field surveys in 2008 confirmed the spread of *B. tectorum* into some areas modeled to be highly suitable (J. Bromberg, personal communication, 2012).

13.2.3 Invasions Changing Landscapes

Many national parks and forests are also affected by a steady influx of competing tree species from other countries. Top invaders include: *Acer platanoides* (Norway

Fig. 13.3   *Miconia calvescens* in Tahiti, and the threat to Hawaiian National Parks

maple), *Ailanthus altissima* (tree-of-heaven), *Schinus terebinthifolius* (Brazilian peppertree), *Morella faya* (firetree), *Robinia pseudoacacia* (black locust), *Melia azedarach* (Chinaberry tree), *Albizia julibrissin* (mimosa), *Paulownia tomentosa* (royal paulownia), *Populus alba* (white poplar), and *Triadica sebifera* (tallow tree). Like most introduced woody species in the United States (Reichard and White 2001), many of these trees are commercially available, and many of their seeds are dispersed by wind and birds far beyond their intended garden use.

One of the most dangerous plant invaders in Hawaii is *Miconia calvescens* (miconia), a small, large-leaved tree native to Central and South America. To assess the potential impact in Hawaii, we may look at what the species does on climatically similar Tahiti. *Miconia calvescens* was intentionally introduced to Tahiti (French Polynesia, South Pacific Ocean) in 1937 as an ornamental plant (Meyer and Florence 1996). As horrifying as it sounds, dense monocultures of *M. calvescens* now cover over two-thirds of that island (Meyer and Florence 1996). Because it forms impenetrable monocultures that allow little light to reach the forest floor, the tree has displaced Tahiti's native forests, placing 40–50 native plant species close to extinction. Almost half the island's 107 endemic plant species may disappear in a human lifetime. This serves as an urgent warning because *M. calvescens* is now on the doorstep of Haleakala National Park, on the island of Maui in Hawaii (Fig. 13.3).

The news of *M. calvescens'* impact in Tahiti reached Maui, where *M. calvescens* was discovered 8 km from Haleakala NP in 1988; its initial aggressive behaviour in Maui rainforests indicated a clear threat to the park and adjacent watershed lands. A major interagency campaign was launched in the 1990s to contain this invasion until a biological control programme could be developed (Medeiros et al. 1997). Biological control testing is progressing, but no biocontrol agent has yet been released. The containment effort has been a success to date, but at great cost (over US$1 million per year), an obvious challenge to sustain.

Working in concert with plant invasions, is the invasion of forest pests that weaken native plant species, and drastically changing landscapes. More than 450 alien forest insects are established in the United States (Aukema et al. 2011), and the national parks are not immune to their ecological effects. Native tree composition in most national parks in the eastern United States has been radically altered by Dutch elm disease (*Ceratocystis ulmi*), Asian gypsy moth (*Lymantia dispar*), chestnut blight (*Cryphonecttria parasitica*), balsam woolly adelgid (*Adelges piceae*), emerald ash borer (*Agrilus planipennis*) and many other species of alien forest pests (Lovett et al. 2006). Likewise, the forest composition of national parks in the western United States has been altered by white pine blister (*Cronartium ribicola* rust), sudden oak death (*Phytophthora ramorum*) and many other forest pests (Wittenberg and Cock 2001).

13.2.4 Changing Disturbance Regimes

Many native species in national parks have been affected by past management actions (e.g. predator control, water diversions, fire suppression, and grazing activities). Restoring more natural disturbance regimes, for example, using prescribed fire, may have unintended repercussions due to alien plant species present in the new species pool.

Many riparian systems in national parks have been affected by adding campgrounds, roads, and visitor facilities. Human disturbances have often facilitated invasions of riparian systems, and restoring these ecosystems may be difficult and expensive. A recent study by Reynolds and Cooper (2011) in Canyon de Chelly National Monument, Arizona, showed that removing *Tamarix* spp. (tamarisk) and *Elaeagnus angustifolia* (Russian olive) resulted in replacement by upland species rather than desired riparian species (e.g. *Populus* spp., cottonwoods). Controlling floods may have long term deleterious effects on cottonwood regeneration, with or without alien species. However, research supports the notion that restoring floods (natural water regimes) may greatly benefit the native *Populus deltoides* (cottonwood tree), while decimating the alien *T. ramosissima* (Sher and Marshall 2003).

Plant invasions also affect disturbance regimes, as shown by *Pennisetum ciliare* (buffelgrass) invasion in Saguaro National Park, Arizona (Brooks et al. 2004). Sonoran Desert vegetation is not historically adapted to broad-scale wildfire, due to patchy fuels and sparse vegetation (http://www.nps.gov/sagu/naturescience/

276

T.J. Stohlgren et al.

invasive-plants.htm; accessed July 2012). As a result, some endemic plant species may be particularly vulnerable to intense fire, including the *Carnegiea gigantea* (saguaro cactus) and foothill tree *Cercidium microphyllum* (palo verde). Thus, just in the past decade, the recent invasion of alien *P. ciliare* in desert regions in the south-western United States poses a significant management challenge to managers of protected areas. *Pennisetum ciliare*, an introduced forage crop native to Asia and Africa, is a fast growing, perennial grass that creates a vast flammable fuel base. One 138-ha fire in Saguaro National Park 1994, for example, resulted in 24 % mortality of *C. gigantea* and 73 % mortality of *C. microphyllum* 6 years after the fire (Esque et al. 2004). Other protected areas at risk from *P. ciliare* invasion include Organ Pipe Cactus National Monument and Ironwood Forest National Monument in Arizona; and parks and wildlife refuges in Texas such as Big Bend National Park, Santa Ana National Wildlife Refuge, Lower Rio Grande Valley NWR, and Laguna Atascosa NWR (Esque et al. 2007). Wildfires threaten habitat for other rare species including desert tortoises (*Gopherus agassizii*), ocelots (*Leopardus pardalis*), and jaguarundis (*Herpailurus yaguarondi cacomitli*).

Another example comes from the foothills of Sequoia and Kings Canyon National Parks, California. For more than a century, the native annual and perennial grasslands have been replaced by alien grass species (Burcham 1957), leaving a grassland ecosystem in Sequoia National Park dominated primarily by alien grasses (e.g. *Avena fatua*, common wild oat; *Bromus mollis = Bromus hordeaceus*, soft brome; and *Bromus diandrus*, ripgut brome; Parsons and Stohlgren 1989). Tall, highly flammable fuel loads from the alien annual grasses coincide with a hot, dry Mediterranean climate to create a self-replacing fire cycle in the blue oak woodland (*Quercus douglasii*) and surrounding grasslands and chaparral. Burning appears to increase the number and biomass of both alien and native forb species, but few native grass species become established after fire (Parsons and Stohlgren 1989). With up to 90 % of live plant biomass taken up by alien plant species, this remains the most altered vegetation assemblage in the Park.

Fire-prone invaders provide a particular challenge to the National Park Service. For example, in Hawaii, *Schizachyrium condensatum* (Colombian bluestem), an alien perennial grass invaded native shrublands and woodlands (D'Antonio and Vitousek 1992). The newly added fuel promoted more frequent and larger fires. The fires killed many native trees and shrubs, but promoted *S. condensatum* and another flammable alien perennial grass, *Melinis minutiflora* (molasses grass; Hughes et al. 1991).

13.2.5 Invasive Plants Facilitating Other Invasions

Invasive plants arriving to a protected area may facilitate others, by improving the suitability of the habitat for the subsequent invaders. The classic example is that of *Myrica faya*, a nitrogen-fixing invasive tree in Hawaii, improving impoverished soil for the invasion of the alien tree, *Psidium cattleianum* (strawberry guava; Vitousek

1986; Vitousek and Walker 1989). Altering soil nutrient dynamics, common to many nitrogen-fixing weeds such as *E. angustifolia* and *E. umbellata* (Autumn olive) in western parks, *Pueraria sp.* (kudzu) in southern parks, *Cytisus scoparius* (Scotch broom) in California parks, and *Melilotus alba* (white sweetclover) in northern parks from Kentucky to Alaska, and *M. officinalis* (yellow sweetclover) in many national parks across the Unites States, serve as alien species facilitators.

An alien plant species need not be a nitrogen fixer to facilitate invasions. Invasive *Tamarix* sp. (salt cedar) provides shade facilitating *E. angustifolia* seedling establishment, outcompeting native *Populus* spp. seedlings (Reynolds and Cooper 2010).

Trampling by domestic stock animals, humans, and invasive animals such as wild hog, may further facilitate plant invasions. For example, Hernandez and Sandquist (2011) showed that biological soil crusts (i.e. bacteria, algae, fungi, and lichens found on soil surfaces in arid and semiarid ecosystems) are easily disturbed and heavily impacted by trampling and grazing. These activities significantly increased (>3 times) alien plant emergence, while insignificantly affecting native species emergence after disturbance (Hernandez and Sandquist 2011).

Alien species interactions also facilitate plant invasions in protected areas. Wild hogs and escaped domestic grazers can disturb the ground to facilitate the invasion of alien annual grasses. Horses and pack stock used for recreation can distribute alien plant seeds after defecating. And nitrogen-fixing weeds can improve soil conditions for subsequent invaders.

13.3   Discussion

Given the prospects of ever-increasing invasions of alien species, the National Park Service may be wise to adapt a risk assessment approach to manage invasive species (Stohlgren and Schnase 2006). Briefly discussed below are key elements of an effective invasive plant management programme including prevention, early detection and rapid response, a triage approach to containment, and restoration.

13.3.1   Prevention

Prevention begins by identifying probable invaders from regional watch lists (Drucker et al. 2008), and evaluating probable pathways of invasion including wind, animal vectors, vehicles, and even visitors. While preventing wind and animal vectors is difficult, cooperative invasive species management with surrounding land owners and agencies may help. Visitors may be important vectors of invasion. A recent study of alien plant seeds in Antarctica serves as a reminder that even our areas that are most protected areas are not immune from invasion: about 40,000 human visitors per year carried, on average, 9.5 seeds per person in their clothing (Chown et al. 2012). Multiply that effect times up to four million visitors per year to some U.S. National Parks, with far more seeds carries by cars, trucks, boats, pets and livestock. Educating visitors on their responsibilities in reducing plant invasions remains an important challenge.

If your door mat is muddy, you can't keep a clean house for long. External forces will make it increasingly difficult to protect national parks (and all protected areas) from the invasion of plant species. Global trade has increased 40-fold in the past 60 years (Stohlgren et al. 2013). Trade and transportation are directly linked to species invasions (Stohlgren et al. 2011). Horticulture and landscaping throughout America will continue to add sources of alien plant materials that will find their way to the national parks and other protected areas (Reichard and White 2001). What goes on in Vegas doesn't stay in Vegas. Wind, water, birds, and other animals will continue to serve as primary dispersers of many plant species. The National Park Service may have to reach far beyond their borders to protect the valuable resources within them. Similar studies in central Europe have reached similar conclusions (Pyšek et al. 2002, 2003). Economists in Australia demonstrated that increasing inspections for invaders at the ports of entry may save money in the long run (McAusland and Costello 2004). It may behove the National Park Service to actively participate in a global strategy to reduce the costs of future invasions.

13.3.2   Early Detection and Rapid Response

Early warning systems also may require a 'risk assessment' approach to biological invasions (Stohlgren and Schnase 2006; Jarnevich and Stohlgren 2009). Modeling the potential distributions of invasive species with Geographic Information Systems is an important component of an early warning system (Jarnevich et al. 2010). Such models are often data intensive, meaning that a large number of field observations may be required inside and adjacent to the national park for the best results. Developing and training teams of park visitors as 'citizen scientists' may increase awareness and help contain invasive species problems (Newman et al. 2010). Especially for extremely invasive plant species, control efforts must be swift (Rejmánek and Pitcairn 2002). A strategy similar to the one used to contain wildfires may be useful. Such a strategy might include interagency centres and specifically trained crews for containing alien plant species early in the invasion process. However, some new invasions will likely remain undetected. A long-term commitment to invasive species containment will likely be necessary.

13.3.3   A Triage Approach to Containment

Quantifying patterns of plant invasions in and adjacent to national parks is an essential first step (Drucker et al. 2008). For example, in Rocky Mountain National Park, Colorado, Stohlgren et al. (1997) found hot spots of native plant species were also hotspots for alien plant species. In Yosemite National Park, Underwood et al. (2004) found a similar story, alien plant species occurred most frequently at low- to mid-elevations, in flat areas with other herbaceous species. This information can help target coordinated control efforts inside and outside parks.

Many parks now have access to Exotic Plant Management Teams to control major weed infestations. However, even back in 2004, the National Park Service admitted that "the funding needs for a viable invasive species programme was estimated at $80 million per year," a far cry from the amount allocated to invasive species issues. In 2003, the Exotic Plant Management Teams received $2.8 million (Drees 2004). We were unable to find precise budget numbers for invasive species expenditures by the National Park Service. However, there is strong evidence that the invasive species issues have grown, probably well beyond the $80 million per year estimate in 2004.

In a triage approach, it also may be important to consider alien plants, animals, and diseases. Is it more important and feasible to contain this alien plant, or this alien animal or this alien disease first? Of course, such decisions should be made after scientific risk assessments of many potential target species (Stohlgren and Schnase 2006). Targeting potentially dominant and ecosystem-altering alien plant species may be a first step. Protecting parks from likely forest pests may also be wise. Establishing early detection programmes for the nun moth (*L. monacha*) from northwest forests, or Asian Long-horned beetle (*Anoplophora glabripennis*) from mid-western and eastern forests, or other harmful insects, nematodes and fungi may save money, species, and landscapes, in the long term.

Control efforts must be cognizant of native species and non-target effects whether or not they are in native species-rich areas (Stohlgren et al. 1999). It may not be enough to track 'the number of acres treated'.

13.3.4 Restoration

It may become increasingly important to track the size if the area treated, the effects of treatment on target and non-target species, and the size of the area restored and hardened against future re-invasion by the previous target species, or invasion by newly arrived alien species. Large-scale restoration of the pristine, native-species-only landscapes may not be feasible. However, demonstration projects may be a good first step.

13.4 Conclusions

It may be time for the National Park Service to convene a panel of experts to develop a synthesis report similar to the Leopold and Robbins reports of 1963, focused specifically of issues related to invasive species in national parks.

The committee might view the challenge of protecting parks from 'biological pollution' in the same way the National Park Service has worked with state and federal agencies to protect air quality standards in the parks. Obvious 'point sources' for botanical pollutants may include ports of entry, commercial nurseries, and botanical gardens. Non-point sources of botanical pollutants may include land surface disturbances, urbanisation and landscaping efforts, vehicles, wildlife, and wind from areas adjacent to parks. New weed screening tools are available for immediate use (Koop et al. 2012). Genetically Modified Organisms (GMOs), especially herbicide-resistant GMOs, may pose additional future risks to national park units. The committee might also be charged with the task of evaluating and prioritizing multiple threats to park resources and native wildlife species, including invasive species, land use change, contaminants, recreationists as vectors, and climate change. It may be important to revise current science and management priorities based on a 'triage approach' to internal and external threats. The committee might begin by asking, "What are the proximate causes of native population declines, degradation of habitat quality, and species extirpations?" In many cases, the answer may lead to invasive alien species. National park managers may find it increasingly challenging to maintain native species and landscapes, because alien plant species continue to invade, reproduce, and spread when budgets are insufficient.

References

Allen JA, Brown CS, Stohlgren TJ (2009) Non-native plant invasions of the United States National Parks. Biol Invasions 11:2195–2207

Aukema JE, Leung B, Kovacs K et al (2011) Economic impacts of non-native forest insects in the continental United States. PLoS One 6:e24587

Brockie RE, Loope LL, Usher MB et al (1988) Biological invasions of island nature reserves. Biol Conserv 44:9–36

Bromberg JE, Kumar S, Brown CS et al (2011) Distributional changes and range predictions of downy brome in Rocky Mountain National Park. Invasive Plant Sci Manag 4:173–182

Brooks ML, D'Antonio CM, Richardson DM et al (2004) Effects of invasive alien plants on fire regimes. BioScience 54:677–688

Burcham LT (1957) California range land. California Department of Forestry, Sacramento

Burns CE, Johnston KM, Schmitz OJ (2003) Global climate change and mammalian species diversity in U.S. National Parks. Proc Natl Acad Sci 100:11474–11477

Chown SL, Huiskes AHL, Gremmen NJM et al (2012) Continent-wide risk assessment for the establishment of nonindigenous species in Antarctica. Proc Natl Acad Sci 109:4938–4943

D'Antonio CM, Vitousek PM (1992) Biological invasions by exotic grasses, the grass/fire cycle, and global change. Annu Rev Ecol Syst 23:63–87

Drees L (2004) A retrospective on NPS invasive species policy and management. Park Sci 22:21–26

Drucker HR, Brown CS, Stohlgren TJ (2008) Developing regional invasive species watch lists: Colorado as a case study. Invasive Plant Sci Manag 1:390–398

Esque TC, Schwalbe CR, Haines DF et al (2004) Saguaros under siege: invasive species and fire. Desert Plants 20:49–55

Esque TC, Schwalbe CR, Lissow JA et al (2007) Buffelgrass fuel loads in Saguaro National Park, Arizona, increase fire danger and threaten native species. Park Sci 24:33–56

Gelbard JL, Belnap J (2003) Roads as conduits for exotic plant invasions in a semiarid landscape. Conserv Biol 17:420–432

Harrison S, Hohn C, Ratay S (2002) Distribution of exotic plants along roads in a peninsular nature reserve. Biol Invasions 4:425–430

Hernandez RR, Sandquist DR (2011) Disturbance of biological soil crust increases emergence of exotic vascular plants in California sage scrub. Plant Ecol 212:1709–1721

Hoagland BW (2007) A checklist of the vascular flora of the Ozark Plateau in Oklahoma. Okla Native Plant Rec 7:21–53

Houston DB, Schreiner EG (1995) Alien species in national parks: drawing lines in space and time. Conserv Biol 9:204–209

Hughes F, Vitousek PM, Tunison T (1991) Alien grass invasion and fire in the seasonal submontane zone of Hawai'i. Ecology 72:743–746

Jarnevich CS, Stohlgren TJ (2009) Near term climate projections for invasive species distributions. Biol Invasions 11:1373–1379

Jarnevich CS, Holcombe T, Barnett D et al (2010) Forecasting weed distributions using climate data: a GIS early warning tool. Invasive Plant Sci Manag 3:365–375

Koop AL, Fowler L, Newton LP et al (2012) Development and validation of a weed screening tool for the United States. Biol Invasions 14:273–294

Leopold AS, Cain SA, Cottam CM et al (1963) Wildlife management in the national parks: advisory board on wildlife management appointed by Secretary of the Interior. Transact N Am Wildl Nat Res Conf 28:29–44

Lonsdale WM (1999) Global patterns of plant invasions and the concept of invasibility. Ecology 80:1522–1536

Lonsdale WM, Lane AM (1994) Tourist vehicles as vectors of weed seeds in Kakadu National Park, Northern Australia. Biol Conserv 69:277–283

Loope LL (2004) The challenge of effectively addressing the threat of invasive species to the National Park System. Park Sci 22:14–20

Loope LL, Sanchez PG, Loope WL et al (1988) Biological invasions of arid land nature reserves. Biol Conserv 44:95–118

Lovett GM, Canham CD, Arthur MA et al (2006) Forest ecosystem responses to exotic pests and pathogens in Eastern North America. BioScience 56:395–405

Macdonald IAW, Frame GW (1988) The invasion of introduced species into nature reserves in tropical savannas and dry woodlands. Biol Conserv 44:67–93

Macdonald IAW, Loope LL, Usher MB et al (1989) Wildlife conservation and the invasion of nature reserves by introduced species: a global perspective. In: Drake JA, Mooney HA, di Castri F et al (eds) Biological invasions: a global perspective. Wiley, Chichester, pp 215–255

Mack RN, Simberloff D, Lonsdale WM et al (2000) Biotic invasions: causes, epidemiology, global consequences, and control. Ecol Appl 10:689–710

Makarick LJ (1999) Exotic plant species management plan for Grand Canyon National Park. U.S. Department of the Interior, National Park Service, Grand Canyon

McAusland C, Costello C (2004) Avoiding invasives: trade-related policies for controlling unintentional exotic species introductions. J Environ Econ Manag 48:954–977

Medeiros AC, Loope LL, Conant P et al (1997) Status, ecology, and management of the invasive tree Miconia calvescens DC (Melastomataceae) in the Hawaiian Islands. In: Evenhuis NL, Miller SE (eds) Records of the Hawaii biological survey for 1996. Bishop Museum occasional papers no 48, pp 23–35

Meyer J-Y, Florence J (1996) Tahiti's native flora endangered by the invasion of Miconia calvescens DC. (Melastomataceae). J Biogeogr 23:775–781

Myers JH, Bazely DR (2003) Ecology and control of introduced plants. Cambridge Press, Cambridge, p 313

National Park Service (1968) Administrative policies for natural areas of the national park system. U.S. Government Printing Office, Washington, DC, pp 16–21

National Park Service (2006) Management policies 2006: the guide to managing the national park system. U.S. Department of the Interior, Washington, DC

Newman G, Laituri M, Graham J et al (2010) Teaching citizen science skills online: implications for invasive species training programs. Appl Environ Educ Commun 9:276–286

Parsons DJ, Stohlgren TJ (1989) Effects of varying fire regimes on annual grasslands in the Southern Sierra Nevada of California. Madroño 36:154–168

Pimentel D, Lach L, Zuniga R et al (2000) Environmental and economic costs of nonindigenous species in the United States. BioScience 50:53–65

Pimentel D, Zuniga R, Morrison D (2005) Update on the environmental and economic costs associated with alien-invasive species in the United States. Ecol Econ 52:273–288

Pyšek P, Jarošík V, Kučera T (2002) Patterns of invasions in temperate nature reserves. Biol Conserv 104:13–24

Pyšek P, Jarošík V, Kučera T (2003) Inclusion of native and alien species in temperate nature reserves: an historical study from Central Europe. Conserv Biol 17:1414–1424

Reichard SH, White P (2001) Horticulture as a pathway of invasive plant introductions in the United States. BioScience 51:103–113

Rejmánek M, Pitcairn MJ (2002) When is eradication of exotic pest plants a realistic goal? In: Veitch CR, Clout MN (eds) Turning the tide: the eradication of invasive species. IUCN SSC Invasive Species Specialist Group, Gland, pp 249–253

Reynolds LV, Cooper DJ (2010) Environmental tolerance of an invasive riparian tree and its potential for continued spread in the Southwestern US. J Veg Sci 21:733–743

Reynolds LV, Cooper DJ (2011) Ecosystem response to removal of exotic riparian shrubs and a transition to upland vegetation. Plant Ecol 212:1243–1261

Robbins WJ, Ackerman EA, Bates M et al (1963) A report by the Advisory Committee to the National Park Service on research. National Academy of Sciences, National Research Council, Washington, DC

Sher AA, Marshall DL (2003) Seedling competition between native *Populus deltoides* (Salicaceae) and exotic *Tamarix ramosissima* (Tamaricaceae) across water regimes and substrate types. Am J Bot 90:413–422

Stohlgren TJ, Schnase JL (2006) Risk analysis for biological hazards: what we need to know about invasive species. Risk Anal 26:163–173

Stohlgren TJ, Chong GW, Kalkhan MA et al (1997) Rapid assessment of plant diversity patterns: a methodology for landscapes. Environ Monit Assess 48:25–43

Stohlgren TJ, Bull KA, Otuski Y et al (1998) Riparian zones as havens for exotic plant species in the central grasslands. Plant Ecol 138:113–125

Stohlgren TJ, Barnett D, Chong GW et al (1999) Exotic plants species invade hot spots of native plant diversity. Ecol Monogr 69:25–46

Stohlgren TJ, Barnett D, Flather C et al (2006) Species richness and patterns of invasion in plants, birds, and fishes in the United States. Biol Invasions 8:427–447

Stohlgren TJ, Pyšek P, Kartesz J et al (2011) Widespread plant species: natives vs. aliens in our 532 changing world. Biol Invasions 13:1931–1944

Stohlgren TJ, Pyšek P, Kartesz J et al. (2013) Globalization effects on common plant species. In: Levin SA (ed) Encyclopaedia of biodiversity, vol 3, 2nd edn. Academic, Waltham, pp 700–706. doi:10.1016/B978-0-12-384719-5.00239-2

Tyser RW, Worley CA (1992) Alien flora in grasslands adjacent to roads and trail corridors in Glacier National Parks, Montana (USA). Conserv Biol 6:253–262

Underwood EC, Klinger R, Moore PE (2004) Predicting patterns of non-native plant invasions in Yosemite National Park, California, USA. Divers Distrib 10:447–459

Usher MB, Kruger FJ, Macdonald IAW et al (1988) The ecology of biological invasions into nature reserves: an introduction. Biol Conserv 44:1–8

Vitousek PM (1986) Biological invasions and ecosystem properties: can species make a difference? In: Mooney HA, Drake JA (eds) Ecology of biological invasions of North America and Hawaii. Springer, New York, pp 163–176

Vitousek PM (1990) Biological invasions and ecosystem processes: towards integration of population biology and ecosystem studies. Oikos 57:7–13

Vitousek PM, Walker LR (1989) Biological invasion by *Myrica faya* in Hawai'i: plant demography, nitrogen fixation, ecosystem effects. Ecol Monogr 59:247–265

Wallis CS (1959) Vascular plants of the Oklahoma Ozarks. Oklahoma State University, Oklahoma

Wittenberg R, Cock MJW (eds) (2001) Invasive alien species: a toolkit of best prevention and management practices. CAB International, Wallingford, p 228

Wright GM, Dixon JS, Thompson BH (1933) Fauna of the national parks of the United States: a preliminary survey of faunal relations in National Parks, Fauna series 1. U.S. Government Printing Office, Washington, DC

Chapter 14
Small, Dynamic and Recently Settled: Responding to the Impacts of Plant Invasions in the New Zealand (Aotearoa) Archipelago

Carol J. West and Ann M. Thompson

Abstract New Zealand was one of the last land masses to be populated by humans, and its isolation has contributed to the large number of endemic species that the country is known for. With increased global movement of people and goods this historic advantage no longer exists. In the last several decades numerous legislative, policy and operational tools have been used to protect New Zealand's special areas and biota from invasive alien species. With the benefit of 25 years of dedicated protection efforts by the Department of Conservation, best practice alien plant control techniques have been developed, building on lessons from animal pest eradications, trophic relationships, and on-the ground pragmatism and experience. Increasingly, an essential tool to achieving greater success will be working with other agencies, businesses and communities to harness resources. Three case studies illustrate the approaches and lessons learnt from alien plant management in New Zealand in the last 25 years: Raoul Island in the far north of New Zealand, Hen and Chicken Islands to the east of North Auckland Peninsula, and Fiordland National Park in south-western South Island.

Keywords Eradication • Fiordland National Park • Hen and Chicken Islands • Raoul Island • Seed bank • Seed dormancy • Zero-density

14.1 Introduction: New Zealand No Longer Isolated

New Zealand (named Aotearoa by indigenous Māori people) lies approximately 1,600 km east of Australia in the southern Pacific Ocean, between latitudes 29°S (the Kermadec Islands) and 52°S (Campbell Island). It is an archipelago of approximately 700 islands more than 1 ha in size upon the mostly submerged continent

C.J. West (✉) • A.M. Thompson
Department of Conservation, P.O Box 10-420, Wellington 6143, New Zealand
e-mail: cwest@doc.govt.nz; amthompson@doc.govt.nz

L.C. Foxcroft et al. (eds.), *Plant Invasions in Protected Areas: Patterns, Problems and Challenges*, Invading Nature - Springer Series in Invasion Ecology 7, DOI 10.1007/978-94-007-7750-7_14, © Springer Science+Business Media Dordrecht 2013

285

Zealandia (Campbell and Hutching 2007), and extends approximately 15,000 km in length. The New Zealand biota evolved in geographical isolation, and the unique flora and fauna that is characteristic of New Zealand is particularly vulnerable to the impacts of alien species generally (Diamond 1990).

Historically, New Zealand had no land mammals, other than three species of bat, one now presumed extinct. Instead, the fauna was laden with endemic birds (many of them flightless), lizards and invertebrates, set against an equally high endemic flora of conifer/broad-leaved forests, tussock grasslands, and subalpine communities (Williams and West 2000). Many endemic plant species, including approximately half of New Zealand's threatened plant species, are found in historically rare ecosystems that occupy just a fraction of the land surface (Williams et al. 2007).

The first introduced plant species resulting from human habitation came with Māori settlement approximately 800–1,000 years ago, mostly as food sources. However, the number of plant species introduced by Māori was substantially less than the number of plants introduced by European settlers in the early nineteenth century. Both groups brought food plants, but Europeans also replicated their homelands with ornamental garden flowers and shrubs. Introductions from Europe flourished, whereas plants introduced by the Polynesians largely died out (some surviving only on Raoul Island in the far north of New Zealand), primarily because the climate is more similar to Europe than Polynesia. Today, New Zealand has 2,418 species of native plants, and over 80 % of these are endemic (NZPCN 2012). In contrast, there are 25,049 species of introduced plants (Diez et al. 2009) and 2,536 of these have naturalised (NZPCN 2012). The Department of Conservation (DOC) recognises 328 environmental invasive alien plants (IAPs; Howell 2008).

This chapter focuses on IAP management on public conservation land; that is, land managed or administered by the Government's Department of Conservation. Public conservation land makes up about 8.5 million ha of land, about one-third of New Zealand. Intensive management of IAPs is carried out only within about 500,000 ha in areas of high value ecosystems and species (DOC 2012), in recognition that alien plant invasions can eliminate some native species (Williams and Timmins 1990).

14.2 Legislation and Policy Tools

14.2.1 Legislation to Support Protected Area Management

Legislation has been used for several decades to help prevent new IAPs entering the country or a region, and for limiting the spread of a new introduced species. The Biosecurity Act 1993, and the Hazardous Substances and New Organisms Act 1996, provide legislative and regulatory tools to prevent the unwanted importation of new pests into and throughout New Zealand.

The Conservation Act 1987 established the Department of Conservation and is the umbrella legislation for the protection of public conservation lands, and natural and historic resources. The Act also restricts the transfer and release of live aquatic life into any freshwater environment. Three further Acts: The National Parks Act 1980, the Reserves Act 1977 and the Marine Reserves Act 1971 are administered by DOC and establish the purpose, principles, and powers for managing national parks, reserves and marine reserves, respectively. This legislation prohibits alien species from being introduced without authorisation.

14.2.2 Policy Platforms for Invasive Alien Plant Management

In 1995, DOC developed a strategic plan and supporting tools to identify, prioritise and manage IAPs. This framework (Owen 1998) contains five objectives, namely to: (i) minimise the risk of introductions of new plant taxa that are potentially invasive, (ii) minimise the numbers, or contain the distribution of significant new IAPs where feasible ('Weed-led' programmes), (iii) protect land, freshwater and marine sites that are important to New Zealand's natural heritage from the impacts of IAPs ('Site-led' programmes), (iv) sustain and improve skills, control techniques, information and relationships to support DOC's management of IAPs, and (v) maintain and improve the quality of DOC's invasive alien plant management systems.

The New Zealand Biodiversity Strategy (Department of Conservation and Ministry for the Environment 2000) was developed in response to the Convention on Biological Diversity held in Rio de Janeiro in 1993, and with the recognition that New Zealand's indigenous biodiversity was declining. Later, a national Biosecurity Strategy (The Biosecurity Council 2003) was developed to meet the increased challenges associated with excluding, eradicating and managing risks to New Zealand's economy, environment, and the health of its citizens. The Biosecurity Strategy included institutional arrangements, Māori capacity, improved science, border protection, incursion response and pest management.

14.2.3 Managing for Outcomes

DOC undertakes work according to an 'Outcomes Model', comprising five intermediate outcomes that state the high level results that DOC aims to achieve, and the steps to be taken to achieve those results. Invasive plant management contributes primarily to the first Outcome, namely, protecting biodiversity: "The diversity of our natural heritage is maintained and restored". Supporting the outcomes model are the scientific tools that enable improved prioritisation of where and how work is done, and measures to assess the outcomes of that conservation work (see Lee et al. 2005).

DOC's 2012 annual report was, for the first time, informed by the data collected and analysed for 14 biodiversity indicators measured throughout New Zealand. Invasive alien plants occurred in 33 % of the locations sampled (fewer than previously recorded from the same plots), but the distribution and frequency appears similar to the situation in about 2000. Invasive plants are most commonly recorded in grassland areas and near human habitation, with most IAPs being non-woody and shade intolerant (MacLeod et al. 2012). The data also supported DOC's IAP management priorities, for example, the current focus on forest margins close to grasslands and habitation.

14.3 Impacts on Protected Areas in New Zealand

14.3.1 Grasslands

Before human habitation, grasslands dominated by native tussocks grew in alpine or dry climates with limited soil fertility and high light. After extensive burning by both Māori and European settlers the quantity of the grasslands increased (McGlone 2004). Northern hemisphere conifers, particularly *Pinus contorta* (lodgepole pine), were sown above the natural treeline and have since spread considerably (Brockerhoff et al. 2004; Craine et al. 2006), affecting native grasslands at all altitudes within range of any plantations. The high altitude plantations were sown by the New Zealand government in the (erroneous) belief that they would combat the high levels of erosion caused by overgrazing by stock, and wild introduced animals such as deer (primarily red deer, *Cervus elaphus*) and rabbits (common rabbit, *Oryctolagus cuniculus*; Bellingham and Lee 2006). Wilding conifers alter not only landscape values but also the ecosystem services such as the water supply provided by grasslands (Mark and Dickinson 2008).

Hieracium and *Pilosella* species, particularly *Pilosella officinarum* (mouse-ear hawkweed), *P. piloselloides* subsp. *praealta* (king devil hawkweed), *H. caespitosum* (field hawkweed) and *Hieracium lepidulum* (tussock hawkweed) have invaded short tussock montane and alpine grasslands in recent decades (Duncan et al. 1997). The extent of their cover is now hundreds of thousands of hectares, resulting in displaced native species and possible higher nutrient deposition than from tussock vegetation (Wiser and Allen 2000).

14.3.2   Forests and Shrublands

New Zealand's forests covered 80 % of the land before human settlement, but today less than 25 % of the original forest remains. There are two main types of indigenous forest in New Zealand, the southern beech forests and the conifer/broad-leaved forests, and each type is invaded by a different suite of IAPs.

Nothofagus spp. (beech) forests provide a range of ecosystem services, including the production of honeydew from native scale insects (*Ultracoelostoma* spp.), which in turn provide a rich food source for birds. These forests have been found to be susceptible to colonisation by *H. lepidulum*, with species-rich sites more likely to be invaded (Wiser et al. 1998; Wiser and Allen 2000). A different suite of IAPs invade conifer/broad-leaved forests. These species often originate as alien ornamental garden plants, and the smothering combination of woody vines (*Clematis vitalba*, old man's beard; *Asparagus scandens*, climbing asparagus; *Hedera helix*, ivy) and herbaceous groundcovers (*Tradescantia fluminensis*, wandering Jew; *Plectranthus ciliatus*, plectranthus; *Vinca major*, periwinkle) can have major impacts. Shrublands are similarly vulnerable to the spread of garden plants, with frequent invasion of *Cotoneaster glaucophyllus* (cotoneaster), *Lonicera japonica* (Japanese honeysuckle) and *Chrysanthemoides monilifera* (boneseed), as well as many others that combine to out-compete native plant communities.

14.3.3   Wetlands and Estuaries

The effects of IAPs are equally apparent in wetlands. Only 10 % of the original extent of New Zealand wetlands remain (Peters and Clarkson 2010). Wetland IAPs can be particularly difficult to manage because of the restrictions placed on the use of herbicides near and on waterways, combined with the volume and mass of the types of alien plants often present, such as *Salix* spp. (willows), *Glyceria maxima* (floating sweetgrass) and *Osmunda regalis* (royal fern), with *Spartina* spp. (cordgrass) particularly impacting estuaries. *Spartina* spp. have been controlled in all estuaries in the South Island that had been invaded either from deliberate planting (for land reclamation) or long-distance spread by sea. This work has been very successful and eradication from the South Island is now being considered. However, finding *Spartina* spp. plants in South Island estuaries is difficult as they typically occur amongst grasses in drains and wet pasture. To improve the likelihood of finding these last individuals, DOC is about to train a '*Spartina* detection dog' (K Vincent pers. comm.).

14.3.4   Sand Dunes

Invasive alien plants such as *Ammophila arenaria* (marram), *Lupinus arboreus* (tree lupin), *Stenotaphrum secundatum* (buffalo grass) and *Carpobrotus aequilaterus* (iceplant) dominate many sand dunes. These weed species invade dunes but, with the exception of *A. arenaria*, are not as effective at binding the sand as the native species *Spinifex sericeus* (silvery sand grass) and *Ficinia spiralis* (pingao). The relatively open nature of the cover of the native sand binding species readily enables invasion by alien species. In about 2012 the alien plant *Euphorbia paralias* (sea spurge) has naturalised at one known site on the west coast of the North Island in the Waikato region. As forewarned by Hilton (2001, 2003) this species has dispersed from Australia where it has rapidly invaded sand dunes along southern and eastern coastlines. The relevant agencies have responded with a control programme, including public communications and alerts, with the goal of eradication at this known site.

14.3.5   Lakes and Rivers

Few water bodies are free of introduced plants, although the abundance and impact of the IAPs varies between water bodies. Freshwater bodies on Stewart Island and the New Zealand sub-Antarctic Islands are not known to contain any IAPs.

The predominant introduced alien plants in lakes and low-gradient rivers are *Elodea canadensis* (Canadian pondweed), *Egeria densa* (egeria), *Lagarosiphon major* (Lagarosiphon), *Hydrilla verticillata* (Hydrilla) and *Ceratophyllum demersum* (hornwort). These species are spread by vegetative fragments, which is an important consideration when attempting to reduce their spread through public education. As many lakes on the South Island's West Coast, including Fiordland, have not yet been invaded by many aquatic IAPs, precautionary sanitation measures between water bodies are essential to prevent alien plant spread into unaffected lakes.

Perhaps the most widely known invasive alien species that has impacted New Zealand's freshwater resources and values is *Didymosphenia geminata* (didymo). This diatomaceous alga, descriptively named 'rock snot' and capable of producing substantive algal blooms, was discovered in the South Island in 2004 and has spread widely throughout the Island (Ministry for Primary Industries 2012). An intensive publicity campaign ('Check, Clean, and Dry' between waterways) has helped restrict its spread in the South Island, and so far it is not known to have spread to the North Island or Stewart Island.

14.3.6 Marine Coastlines

A number of marine alga species have been introduced to New Zealand via international shipping, usually in ballast water. Most are not invasive but *Undaria pinnatifida* (wakame), first recorded as naturalised in 1987 in Wellington Harbour, strongly modifies rocky sub-tidal and intertidal communities (Russell et al. 2008). This species has spread rapidly, primarily via coastal shipping and subsequently by natural dispersal from all foci. Eradication has been successful in one location to date (a fouled vessel that ran aground in the Chatham Islands, Wotton et al. 2004), and is currently underway in Dusky Sound, Fiordland.

14.4 Case Studies

We present two case studies of IAP control on New Zealand islands, and a third case study from the mainland. These regions are undergoing extensive IAP management because of their intrinsic natural values, including endemic species and landscape features, and the likelihood that sustained intervention would succeed in maintaining or improving the conservation values. Further, the islands selected are mostly free of mammalian pests, resulting in the IAPs being the primary inhibiting factor to achieving ecological integrity.

14.4.1 Case Study 1: Raoul Island: Rangitahua

Raoul Island is the largest island in the Kermadec Group and constitutes the northernmost region of New Zealand, lying about 1,000 km north-east of Auckland city; it is the only subtropical environment in New Zealand. The island, 2,943 ha in extent and rising to 516 m a.s.l., is the rugged, emergent summit of a large, active volcano.

Raoul Island is forested, with beach strand and rocky headland plant communities and a central, volcanically active crater. The dominant species are the hardwood *Metrosideros kermadecensis* (Kermadec pohutukawa) and *Rhopalostylis baueriana* (Kermadec nikau palm) with associated, primarily endemic, subcanopy trees and shrubs (e.g. *Myrsine kermadecensis,* Kermadec mapou; *Coprosma acutifolia; Homalanthus polyandrus,* Kermadec poplar; *Cyathea kermadecensis,* Kermadec tree fern and *C. milnei,* Milne's tree fern) (Sykes et al. 2000).

A high degree of natural disturbance is normal for Raoul Island, and the vegetation has evolved in response. Until the early twentieth century, Raoul Island was home to immense numbers of burrowing wedge-tailed shearwaters (*Puffinus pacificus*) as well as thousands of sooty terns (*Onychoprion fuscatus*) and many other seabirds of tropical and subtropical distribution (Gaskin 2011; Veitch et al. 2011b). Cyclones are frequent and the island occasionally experiences more than one during the cyclone season (December to March). Extensive patches of forest are blown down and coastal vegetation is defoliated by salt-spray. In addition, volcanic eruptions have been intermittent, the most recent being in 2006 (West 2011). Forest within the blast zone is felled or defoliated and may be buried in ejecta (pers. obs.). Even though the native vegetation is adapted to recover from these disturbances, many of the IAPs on the island also benefit.

Multiple human-derived disturbances originated with Polynesian voyagers about 960 AD (Anderson 1980), who introduced Pacific rats/kiore (*Rattus exulans*) and plants (e.g. *Aleurites moluccana*, candlenut; *Colocasia esculenta*, taro and *Cordyline fruticosa*, ti pore). This continued with European explorers and whalers in the late 1700s to early 1800s who introduced goats (*Capra hircus*), pigs (*Sus scrofa*) and possibly cats (*Felis catus*) (West 2002). The island was then settled intermittently by Europeans from 1836 to 1914 and the number of introduced plant species quickly exceeded the native plant species, many of which are endemic (Sykes et al. 2000). In 1934, Raoul Island was gazetted as a Flora and Fauna Reserve and subsequently transferred to nature reserve status under the Reserves Act 1977. A meteorological station was staffed from 1937 to about 1992 when occupation passed to DOC, as the primary work on the island was IAP eradication (West 2002). An eradication programme was started in 1972 by the Department of Lands and Survey.

Like most outlying islands of New Zealand, Raoul Island is a priority for the restoration of ecosystems and threatened species. The target for restoration is that Raoul Island once again becomes a seabird-dominated island, specifically, to "restore the Raoul Island ecosystem to a high level of ecological integrity by assisting its recovery from multiple disturbances" (unpublished Draft Kermadec Islands Restoration Plan 2009–2019). Following the successful eradication of all introduced mammals (Broome 2009), preventative biosecurity measures and IAP eradication are the key focus for management.

14.4.1.1   Raoul Island Restoration: The Story So Far

Raoul Island is free of all introduced mammals, with goats being eradicated in 1984 (Sykes and West 1996), and rats and cats eradicated in 2002 and 2004, respectively (Broome 2009). As a consequence seabirds are now returning to breed on Raoul Island and, each year, are recorded in greater numbers. Also, red-crowned parakeets (*Cyanoramphus novaezelandiae cyanurus*) and spotless crakes (*Porzana tabuensis*) have re-colonised from the nearby Meyer Islets and are plentiful (Gaskin 2011; Veitch et al. 2011b).

The response of the vegetation to mammal eradication has been similarly striking. After goats were eradicated canopy cover increased, resulting in a decline of light-demanding IAPs like *Alocasia brisbanensis*. Many preferentially browsed native species recovered, some from near extinction, e.g. *Veronica breviracemosa* (Kermadec koromiko; West and Havell 2011), and *Homalanthus polyandrus*, which is now widely distributed and relatively common. There was no noticeable increase of IAP species, which is attributed to the lower light levels within the forest and the effectiveness of the IAP eradication programme. Also the IAPs targeted for eradication appear to have been unpalatable as they were not recorded in the diet of goats on Raoul Island (Parkes 1984).

Eradication of rats enabled greater recruitment of many native plant species, as it did for a number of IAPs. Indeed, many IAPs that did not fruit in the presence of rats began to fruit and recruit seedlings for the first time, e.g. *Hibiscus tiliaceus* (fou), *Catharanthus roseus* (rosy periwinkle) and *Bryophyllum pinnatum* (airplant; West and Havell 2011). However, this outcome was anticipated and *Vitis vinifera* (grape), the species most likely to spread, was targeted for eradication before the rat eradication was undertaken (West 2011). Understanding species interactions such as this has contributed to an efficient IAP eradication programme.

14.4.1.2 Invasive Alien Plant Species: Eradication Successes and Remaining Challenges

The eradication programme for the range of IAPs on Raoul Island was described by West (1996) and progress in achieving eradication was subsequently reported (West 2002). With the eradication of rats in 2002 shown to be a significant factor in recovery of native sea and land birds on Raoul Island, eradication of IAPs is now essential for complete ecosystem restoration (West 1996, 2002). To date 11 IAPs have been eradicated (Table 14.1). For some of the historic species (listed in Table 14.1), some adult specimens are retained (the original planted individuals or, in the case of *Aleurites moluccana*, their adult offspring), but the progeny are eradicated or controlled to zero-density. *Aleurites moluccana* and *Araucaria heterophylla* (Norfolk pine) seed freely and seedlings are common but easily located and removed, and time to maturity is several years. Mature *A. heterophylla* specimens are now confined to a small grove within a historic site on the northern terraces. *Aleurites moluccana* has no dispersers and the large seeds fall beneath the parent plants which are relatively localised at easily managed and confined sites. *Araucaria heterophylla*, on the other hand is wind-dispersed from tall, historic individuals but the seed shadow distance is known and predictable. *Ficus cairica* (fig), although it fruits prolifically now that rats have been eradicated, has no pollinator present so there is no viable seed production and control is limited to removing vegetative spread. *Phoenix dactylifera* (date) is dioecious with one gender assumed to be present and no fruit has ever been observed on the mature palms despite the absence of rats (West 2011). The few plants removed since 1995 are likely to have grown from discarded date stones as they were located by the roadside.

Determining when a plant species has been eradicated is difficult for two reasons. First, plants can be very difficult to detect and, in a forested environment such as on Raoul Island where the transformer IAPs are vines, trees and shrubs, the IAPs blend in well with the native species. *Olea europaea* subsp. *cuspidata* (African olive) is an example of such a species, and therefore how attenuated the eradication time can be. In the last nine years only seven individuals have been

Table 14.1 Species listed in the Raoul Island alien plant database, indicating when eradication began, the last time a species was recorded (and removed) and whether they might have been eradicated

Species	Common name	Family	Eradication began	Last record	Eradicated?
*Aleurites moluccana	candlenut	Euphorbiaceae	1993	2013	No
Anredera cordifolia	Madeira vine	Basellaceae	1995	2013	No
*Araucaria heterophylla	Norfolk pine	Araucariaceae	1974	2013	No
Bryophyllum pinnatum	airplant	Crassulaceae	1998	2013	No
Caesalpinia decapetala	Mysore thorn	Fabaceae	1974	2013	No
Cortaderia selloana	pampas grass	Poaceae	1984	1993	Yes
*Ficus cairica	fig	Moraceae	1996	2012	No
Ficus macrophylla	Moreton Bay fig	Moraceae	1996	1999	Yes
Foeniculum vulgare	fennel	Apiaceae	1969	1999	Yes
Furcraea foetida	Mauritius hemp	Asparagaceae	1974	2002	Yes
Gomphocarpus fruticosus	swan plant	Asclepiadaceae	1979	2002	Yes
Macadamia tetraphylla	macadamia	Proteaceae	1996	2003	Yes
Olea europaea subsp. cuspidata	African olive	Oleaceae	1973	2011	No
Passiflora edulis	black passion fruit	Passifloraceae	1980	2013	No
*Phoenix dactylifera	date palm	Arecaceae	1995	1999	Yes
Phyllostachys aurea	bamboo	Poaceae	1996	2001	Yes
Populus nigra	poplar	Salicaceae	1995	2003	Yes
Prunus persica	peach	Rosaceae	1994	2013	No
Psidium cattleianum	purple guava	Myrtaceae	1973	2013	No
Psidium guajava	yellow guava	Myrtaceae	1972	2013	No
Ricinus communis	castor oil plant	Euphorbiaceae	1990	2012	No
Selaginella kraussiana	selaginella	Selaginellaceae	1998	2013	No
Senecio jacobaea	ragwort	Asteraceae	1980	1980	Yes
Senna septemtrionalis	Brazilian buttercup	Fabaceae	1978	2013	No
Tropaeolum majus	nasturtium	Tropaeolaceae	1999	2013	No
Urochloa mutica	para grass	Poaceae	1996	2009	No
Vicia sativa	vetch	Fabaceae	1996	2013	No
Vitex lucens	puriri	Verbenaceae	1997	1997	Yes
Vitis vinifera	grape	Vitaceae	1995	2012	No

* indicates species that have some mature plants retained because of their historic significance but all progeny are removed

found (just one mature plant), whereas 700 mature trees were removed from one location in a single year in an earlier phase of the eradication programme (West 1996). Second, the longevity of the seed bank for all species is unknown and can only be inferred from data from other members of the same families or genera, acknowledging that there is substantial inter- and intra-specific variability in recorded longevity (Thompson et al. 1997). In evaluating progress toward environmental weed eradication in New Zealand, Howell (2012) suggests that infestations should be checked annually for at least 3 years after the last plant has been removed and that this time frame should be significantly longer for species with long-lived seed banks. Data from Raoul Island indicate that 3 years is insufficient for some species as suckers may develop after that time from large individuals, e.g. *Vitis vinifera*, or individuals may persist in a seedling bank, e.g. *A. heterophylla* seedlings found more than 6 years after parent trees were felled (D Havell pers. comm.).

All seven species that were tentatively described as eradicated in 2002 (West 2002) are confirmed to be eradicated (Table 14.1) as no individuals have been detected for at least 10 years. In addition, a further four species (*Furcraea foetida*, Mauritius hemp; *Gomphocarpus fruticosus*, swan plant; *Phoenix dactylifera* and *Phyllostachys aurea*, walking stick bamboo) are also now confirmed as eradicated.

For seven of the targeted species still present on Raoul Island the challenges for eradication are based primarily on their biology, but also on the difficulty of accessing the terrain (West 2002). For *Senna septemtrionalis* (Brazilian buttercup) and *Caesalpinia decapetala* (Mysore thorn) the persistent seed bank (possibly decades, Thompson et al. 1997) is the largest problem, though the highly disturbed environment on Raoul Island is potentially an advantage, in that soil movement can bring seed to the surface and increased light at ground level induces germination. A large population of *S. septemtrionalis* was detected during aerial surveillance (May 2009), with approximately 1,500 mature individuals, 4,500 adolescents and tens of thousands of seedlings. The spread was however limited by dense stands of *Imperata cheesemanii* (imperata), an endemic grass (C Ardell pers. comm.). The detection of this infestation further proves the value of aerial surveillance whenever it can be achieved.

The challenges to eradicating *Anredera cordifolia* (Madeira vine) are the herbicide-resistant tubers and the terrain (the main population is situated at the top of 50 m bluffs above the sea). In 2003 more than 3.5 tonnes of tubers were removed and since then some sites have remained free of plants after multiple surveys. However, new populations are occasionally discovered downhill of known sites. A total of almost 17 tonnes of tubers has been removed since 1999, with a total of over 5,000 h of effort. It was hypothesised that this infestation arose via sea dispersal of tubers from the original plant dumped in Bell's Ravine (West 1996). Therefore, a goal of this programme is to avoid tubers falling into the sea to minimise the risk of distant infestations establishing, as experiments have shown that some *A. cordifolia* tubers will float for at least 30 days in fresh water (Vivian-Smith et al. 2007).

It is often possible to anticipate a species' behaviour based on biological traits, however, *Prunus persica* (peach) proved an exception. This species generally requires considerable winter chilling for strong foliage growth and fruit crops (Lyle 2006) and it is likely that chilling would be required to break seed dormancy (Martínez-Gómez and Dicenta 2001). However, the climate on Raoul Island is humid and warm temperate, substantially different from the optimal conditions described for cultivation. Therefore, it was anticipated that seeds would rot and

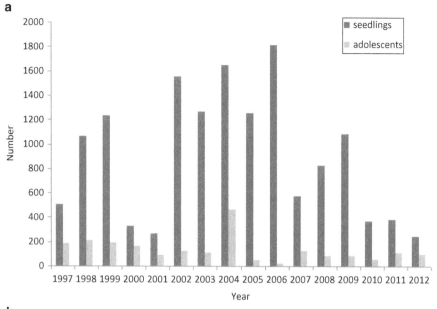

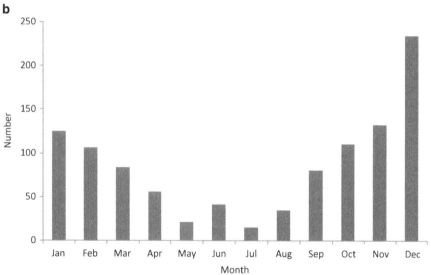

Fig. 14.1 (a) *Prunus persica* seedlings and adolescents removed from Raoul Island (**b**) The average number of *P. persica* seedlings on Raoul Island removed each month (1997–2008)

viability rapidly reduce. Despite this, *P. persica* naturalised away from planted individuals, most likely to have been inadvertently spread by staff on the island. Because of the amount of naturalisation and the tendency for felled, poisoned trees to resprout, *P. persica* was added to the eradication programme in 1994.

Approximately 300 mature trees were felled in the first 4 years (since 1994) and about 30 have been detected since. The longevity of seeds was unexpected, and seedlings are still germinating more than 12 years after the adult trees were removed (Fig. 14.1a).

It would appear that *P. persica* on Raoul Island has physiological seed dormancy (sensu Finch-Savage and Leubner-Metzger 2006) and that the difference between winter and summer temperatures is sufficient to break seed dormancy for a proportion of the seed bank each year. Some seedlings are recorded during winter months, although the majority are found in spring and summer (Fig. 14.1b). Spraying gibberellic acid on the ground in the infested sites could potentially break the seed dormancy and extinguish the seed bank more quickly. However, although there are many laboratory tests that demonstrate the effectiveness of gibberellic acid at breaking seed dormancy (e.g. Evans et al. 1996), it appears that this technique has not been used in the field.

The reason that *P. persica* behaved differently than expected is most likely because, with the possible exception of *Vicia sativa* (vetch) and *Foeniculum vulgare* (fennel), it is the only species targeted for eradication (see Table 14.1) that produces physiologically dormant seeds. All others, if they seed on Raoul Island, would appear to have non-dormant seed that may or may not form a seed bank.

The highly disturbed environment on Raoul Island presents challenges as well as the potential advantages described above. After the 2006 eruption, in which a staff member was killed, staff were not permitted to enter the crater (for safety reasons) for more than two years. This meant that several target species were able to reproduce in the crater and add new seeds to the seed bank (e.g. *Passiflora edulis*, black passion fruit; *Psidium cattleianum*, purple guava and *Senna septemtrionalis*). Cyclones that cause widespread but patchy treefall and intense rainfall events that create slips make access more difficult and result in time being spent on clearing tracks and roadways and slow the rate of progress when grid-searching in weed plots. The frequency of these events is very variable but can reduce weeding time significantly in some years.

14.4.2   Case Study 2: The Hen (Taranga) and Chicken (Marotere) Islands

Taranga and the Marotere Islands were originally settled by the indigenous Ngātiwai people but were named the Hen and Chicken Islands by Captain James Cook, who sighted the island group in 1769. The group lies approximately 12 km east of the Bream Head Scenic Reserve on the east coast of Northland, at latitude 35°S. The islands vary in size, from 2–3 to 489 ha, with a range of native vegetation across the lands, emerging from eroded volcanic remnants. The islands were designated as a scenic reserve in 1925, prior to becoming a nature reserve in

1977. Before protection of these islands was implemented, various, but not extensive, human activities ensued, including the gathering of seabirds and *Phormium* spp. (New Zealand flax) and the brief introduction of cattle to Mauimua (Towns and Parrish 2003). Mauipae (Coppermine Island) underwent several attempts at mining despite its protected status (Moore 1984).

There has been a long history of scientific interest in the islands, from the late nineteenth century when noted botanists Kirk and Cheeseman visited, followed by Cockayne and numerous others (Atkinson 1973), resulting in a wide range of records for flora and fauna. Many plant and animal species exist on these islands that are rare or absent on the mainland. Special fauna include the endemic ancient reptile the tuatara (*Sphenodon punctatus*), various lizards (*Oligosoma townsi, Oligosoma ornatum*) and birds such as saddleback (*Philesturnus carunculatus*) and kākā (*Nestor meridionalis*). Numerous endemic plants are found on these islands, many of which are rare, declining, or at risk. These range from annual and perennial herbs such as *Euphorbia glauca* (shore spurge), *Lepidium oleraceum* (Cook's scurvy grass) and *Rorippa divaricata* (New Zealand water cress), to coastal shrubs like *Senecio scaberulus* (fireweed) and trees including *Meryta sinclairii* (puka) and *Streblus banksii* (turepo).

Hen Island (Taranga) is the largest of the group (489 ha), with a steep coastline giving way to undulating valleys. Vegetation varies from *Kunzea ericioides* (kānuka) shrubland and *Beilschmiedia tarairi* (taraire) and *B. tawa* (tawa) forest, with 235 native and 43 adventive species recorded in 1978 (Wright 1978). The three main Chicken Islands are Lady Alice (Mauimua, 151 ha), Middle Chicken (Whatupuke, 99 ha), and Coppermine (Mauipae, 77 ha). In 1984, 245 indigenous vascular plants and 73 introduced species were recorded (Cameron 1984). Since kiore were eradicated during the 1990s, the health of the native plant communities, particularly the fruiting species, is expected to improve (Towns and Parrish 2003).

14.4.2.1 A Partnership at Work

The Department of Conservation and the Ngātiwai Trust Board jointly manage these islands, guided by a 10-year restoration plan (Towns and Parrish 2003). This plan addresses all biodiversity aspects of the three largest islands. Significantly, Mauitaha and Araara Islands are managed as kiore refuges. Kiore were introduced by Māori approximately 800 years ago, and are regarded as a taonga, or treasure. Despite this, DOC and Ngātiwai worked together to successfully eradicate the kiore from Hen Island in 2011. Now, all but the two islands containing kiore are free of mammalian pests resulting in improved conservation outcomes for both the birds and the invertebrates.

14.4.2.2   Implementing the Management Plan

The islands have been ranked as a priority ecosystem under the DOC's Outcomes Model. In practice, this integrates the site-led weed programme into a holistic 'prescription' that aims to mitigate all threats to the islands as well as using best practice species management techniques.

One of the goals of the 2003 restoration plan is to eradicate or control plant and animal pests that have the potential to compromise other restoration goals (Towns and Parrish 2003). To support this goal a Weed Strategy and Operational Plan was developed to identify the priority IAPs and their management objectives, as well as the operational methodology for the islands (M. Valdes pers. comm.) The IAPs include several escaped ornamental garden plants, as well as plants derived of seed from wind and bird dispersal from the mainland.

Three classes of IAPs (Classes 1, 2 and 3) were determined based on the invasiveness of the plant and their likely competition with desirable plants (Table 14.2).

Class 1: IAPs that have the potential to spread quickly and result in the highest impact on the surrounding ecosystems. The objective of these for Hen Island is sustained control by 2018, i.e. limiting each species to its present distribution and, where possible, reducing the abundance of the species (M. Valdes pers. comm.). For the Chicken Islands, eradication is the objective.

Class 2: IAPs in Class 2 represent the next level of invasiveness and impact after Class 1 IAPs. The objective for these is eradication from all of the islands.

Class 3: IAPs that have been judged to have less of an environmental impact, although their spread may be rapid. The objective of Class 3 IAPs is also eradication from all of the islands.

The known IAP sites are displayed on a map and GPS unit, so the sites can be easily found and thoroughly searched. This is especially important as the terrain is difficult to work on, with some cliff sites accessed by abseiling (Fig. 14.2). Management is by hand, in order to limit collateral damage to desirable plants and landscapes. Seed heads are bagged and removed from the island (M. Valdes pers. comm.).

14.4.2.3   Assessing Progress to Date

In the 20 years since IAP management was strategically considered and resourced, progress has been achieved in the management objectives. This success has been analysed using the Total Count method (Holloran 2006). Individuals are counted when they are removed/killed and recorded in one of three size classes: seedlings, adolescents and matures. Progress towards eradication is

Table 14.2 The Hen and Chicken Islands: Priority invasive alien plants are classified according to their environmental impact and management objective. Class 1 are the most environmentally damaging

Class 1	Class 2	Class 3
Hen: Sustained control	Hen: Eradication	Hen: Eradication
Chicken: Eradication	Chicken: Eradication	Chicken: Eradication
Ageratina adenophora	*Lycium ferocissimum*	*Cirsium* spp.
Ageratina riparia	*Asparagus asparagoides*	*Phytolacca octandra*
Araujia hortorum	*Senecio elegans*	*Senecio cineraria*
Cortaderia jubata	*Physalis peruviana*	*Senecio bipinnatisectus*
Cortaderia selloana	*Pennisetum clandestinum*	*Cannabis* sp.
	Paraserianthes lophantha	
	Erigeron karvinskianus	
	Myosotis sylvatica	
	Gladiolus spp.	
	Senecio jacobaea	
	Hakea sericea	

Fig. 14.2 Sites invaded primarily by *Ageratina adenophora*, accessed from the 'Don't be silly' track on Taranga. Each site measures between 15 and 150 m^2 (Photo Toby Shanley, Department of Conservation)

shown as a reduced number in each size-class or a reduced number of adults or juveniles over time.

On Hen Island, which has the most challenging terrain of the islands, there have been intensive search efforts in recent years, with a focus on large, active, or difficult IAP sites (Shanley 2010). *Araujia hortorum* (moth plant), with its tuberous roots, and shade tolerant seedlings wind-blown seed poses a particular challenge. There are nine existing sites, eight of which are located on the windward western side of the island. The ninth site on the eastern side is reportedly clear of the plants. Since the early 2000s, the number counted has steadily decreased, but the numbers of adults did increase as a result of intensive searching and control on a particularly steep site.

The results of *Ageratina adenophora* (Mexican devil) control have been less consistent, with a large number of adult plants removed in 2009–2010. Records show that 17 new or rediscovered plant sites required control. These sites appear to have been neglected due to their inaccessibility, and point to the need to have appropriately trained and competent staff on the island (Shanley 2010). The combination of the high seed production of the Asteraceae, and the rapid growth of *A. adenophora* seedlings, means that this short-lived, historical lack of control at certain sites is likely to have increased the work required over time to achieve sustained control on the island. Similarly, *Ageratina riparia* (mist flower) is presently known at two sites on Hen Island. However, one site has needed repetitive work due to inadequate previous management.

Currently there are approximately 100 sites with records of *Cortaderia* spp. (pampas), although some sites are found to be clean when examined, and there are increasing numbers of archived sites. To counter the ability of *Cortaderia* spp. to colonise cliffs, detection and surveillance of new plants is achieved by using a boat to patrol the coast.

The objective for the class 1 IAPs is sustained control, and this appears to be succeeding for four of the five species. Both *Araujia hortorum* and *A. riparia* plant numbers show general trends and numbers of *Cortaderia* spp. tally at dozens, with some sites now reported as clear of this species. However, *A. adenophora* numbers have increased markedly in the last few years (Unpublished data, DOC 2012).

There are some likely near-eradications for two species from class 2 (*Physalis peruviana*, Cape gooseberry and *Hakea sericea*, prickly hakea), and one eradication of a class 3 species, *Cannabis* (marijuana). For *P. peruviana*, all known sites were found clean in 2012. No *H. sericea* has been recorded on Hen Island since 2009, and no *Cannabis* plants have been found since 1997 (unpublished data, DOC 2012). These three IAPs all produce abundant amounts of seed, but the factor that is likely to have helped these potential eradications succeed is the very limited distribution that the plants seem to have had. Both *Cannabis* and *H. sericea* had been recorded at single sites, and *P. peruviana* was recorded at only three sites (unpublished data, DOC 2012). Further, hygiene on the island preventing seed spread has been vital, as have the strict biosecurity procedures that are part of any excursion to the islands.

The Chicken islands have had less consistent weeding efforts but overall have more invasive plant sites than Hen Island. However, similar progress appears to be occurring, with some sites on these islands also found to be clean of previously recorded IAPs, including *Araujia hortorum, Pennisetum clandestinum* (kikuyu grass) and *Gladiolus* spp. (gladiolus) on Lady Alice and *A. riparia* on Coppermine (Shanley 2010).

14.4.3   Case Study 3: Fiordland National Park

Fiordland National Park (FNP), gazetted in 1952, is New Zealand's largest national park (1,260,740 ha) and is part of the Te Wahi Pounamu South-West New Zealand World Heritage Area designated in 1990 (Fig. 14.3). The Park is mountainous and the myriad U-shaped valleys reveal past glaciation from the Pleistocene era. Forests dominate the slopes to treeline (800–1,000 m a.s.l.), with shrubland, tussock grassland and permanent snow above. Inland valleys have tussock grass flats. Very little of the original vegetation has been cleared although it is subject to high rates of disturbance through tectonic activity and heavy precipitation (rain and snow). Fiordland National Park lies in the belt of Southern Hemisphere westerly winds known as the 'Roaring 40s', so the combination of strong, moist, onshore winds and steep topography leads to the high rainfall. There are just three roads in FNP (Fig. 14.3). Access to the park is by road, water (sea, lakes and rivers) and air (helicopters and light aircraft).

14.4.3.1   Documentation of Alien Plant Invasion

The major impact on biodiversity in FNP is due to the establishment of alien mammal species, and IAPs, though the invasion of IAPs has been slower. Red deer (*Cervus elaphus*), stoats (*Mustela erminea*) and rodents (*Rattus* spp. and *Mus musculus*) occur throughout the park. Australian brush-tail possums (*Trichosurus vulpecula*) have become abundant in the drier and warmer eastern and northern parts of the park. Stoats are the main driver behind the reduced populations of many seed-dispersing native bird species (Dilks et al. 2003). Pigs and chamois (*Rupicarpa rupicarpa*) are more confined in their distribution (DOC 2002) and goats have been eradicated (M. Willans pers. comm.).

Captain James Cook spent 6 weeks in Dusky Sound in 1773, and during that time created a vegetable garden (Thomson 1922), but by 1791 no traces of the vegetables could be found (McNab 1907). The first record of naturalised plants (sensu Richardson et al. 2000) comes from Poole (1951) who, from February to May 1949, recorded ten herbaceous species within the area bordered by George Sound and Caswell Sound. In 1962, Bryony Macmillan recorded four naturalised plant species at Deep Cove, Doubtful Sound (Given 1973). At Puysegur Point, the site of a lighthouse that was permanently staffed from 1879 until 1980, G.I. Collett recorded 37 naturalised plant

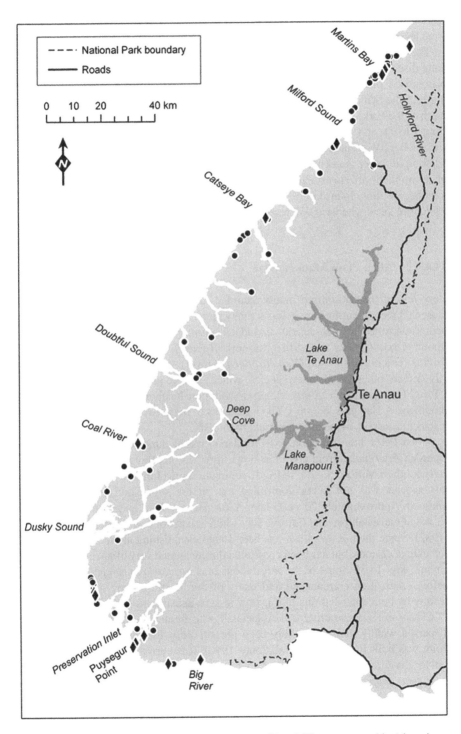

Fig. 14.3 Locations that *Ammophila arenaria* (*diamonds*) and *Ulex europaeus* (*dots*) have been recorded from in coastal Fiordland National Park

species in 1963 (Given 1973, Johnson 1982). In 1972, four naturalised plant species were recorded in Dusky Sound and Wet Jacket Arm (Given 1973).

The most comprehensive assessment of the distribution and abundance of naturalised plants was undertaken between 1969 and 1979 by Johnson (1982). Naturalised plants from 51 locations throughout FNP, from Martins Bay in the north to the Wairaurahiri River mouth in the south (as well as coastal areas to the north of the Park) were recorded. The number of species recorded at each site ranged from 1 to 71, and 136 species were recorded in total (Johnson 1982). Between 1996 and 2000, CJW resurveyed all 27 of the coastal sites surveyed by Johnson (1982). In addition all other areas where people may have come ashore, or where naturalised plants might be able to establish between Milford and Puysegur Point, were surveyed, totalling 99 sites. Johnson (1982) recorded a total of 100 naturalised species at the 27 coastal sites, whereas CJW recorded 93 species.

14.4.3.2 Alien Plant Management

Very few of the naturalised plants recorded in FNP are transformer species, but those that were identified are being actively managed. The top priority IAPs that are being controlled to zero-density include *A. arenaria*, *Crocosmia×crocosmiiflora* (montbretia), *Cytisus scoparius* (broom), *Rubus fruticosus* (blackberry), *Salix fragilis* (crack willow) and *Ulex europaeus* (gorse). Active surveillance occasionally detects other transformer species, which have required on-going eradication efforts, for example, *Calluna vulgaris* (common heather), and *Buddleia davidii* (buddleia, A Hay pers. comm.). The incursions of these two species represent long-distance transport by people, most likely tourists.

When Peter Johnson surveyed Fiordland coastal dunes in the mid-1970s, he recorded *A. arenaria* from eight locations from Martins Bay in the north (immediately south of Milford Sound) to Big River in the south (Johnson 1982). At that time this species was not being controlled but, at his suggestion, eradication was initiated. Approximately 20 years later *A. arenaria* was no longer present at Neck Cove but had established at Catseye Bay where it had not been recorded by Johnson (1982). Since then, *A. arenaria* has been found establishing at Neck Cove on two separate occasions, but has been controlled during annual surveillance (A Hay pers. comm.; Fig. 14.3). Some of the infestations of *A. arenaria* expanded substantially before eradication commenced and many of the locations can only be accessed readily by helicopter but this IAP is now at zero-density.

Crocosmia×crocosmiiflora is associated with human settlements and has not dispersed widely. It has probably been present since the lighthouse at Puysegur Point was built in the 1870s (Hall-Jones 1990). It is currently at zero density and active surveillance continues.

Cytisus scoparius has never been abundant in FNP, but small populations have been detected and controlled at several sites, possibly introduced with road gravel. Johnson (1982) recorded *C. scoparius* in nine locations and these are the places that are actively managed today. This species is at zero-density and subject to active searches of known locations as well as broader surveillance.

Rubus fruticosus agg. was recorded from four locations in coastal Fiordland: and also at Deep Cove in Doubtful Sound. The infestation at Deep Cove is likely to have arisen during the construction of the Manapouri power scheme and possibly after the road over the Wilmott Pass was built in the mid-1960s (Peat 1995) since it was not recorded in 1962 (Given 1973). The other locations are likely to date from the late 1800s. Some of the infestation sites are large and control is on-going.

Eradication of *Salix fragilis* has been achieved in coastal Fiordland. This species was known only from Cromarty, the site of a town that sprang up in 1892 to support a gold rush in Preservation Inlet (Hall-Jones 1990). However, in eastern Fiordland National Park, *S. fragilis* is being controlled to zero-density. Active surveillance is required because this species is well-established immediately adjacent to the east of the National Park in a stream flowing through private land that flows into the Eglinton River and thence into Lake Te Anau. It is also along the Waiau River from the outlet of Lake Manapouri at Pearl Harbour because two major rivers lined with *S. fragilis* enter from the east and lake-level manipulations as part of the hydro power generation scheme result in back eddies of water that contain stem fragments. Aside from this location on Lake Manapouri, *S. fragilis* has been controlled to zero-density elsewhere on the lake edge, a programme that was begun by National Park staff in the 1970s (Johnson 1982). Surveillance of the previously invaded areas as well as areas downstream within the National Park is undertaken regularly and any regeneration of this species, which only reproduces vegetatively in New Zealand, is controlled.

Ulex europaeus has established in a multitude of spots along the Fiordland coast (Fig. 14.3) but given that it is easy to detect when flowering and is a very well-known but not well liked plant by many, most people who encounter it in Fiordland pull it out or report the location to National Park staff so they can control the plants as soon as possible (Johnson 1982). Fishermen observing the bright yellow flowers from sea often report infestations to Park staff (pers. obs.). Johnson (1982) recorded 15 locations of *U. europaeus* within the National Park whereas CJW recorded 22 locations in her survey and was aware of additional sites. However, all known sites have been controlled to zero-density and any newly reported sites are added to the inventory of sites for annual helicopter-based surveillance and control. Every effort is made to kill *U. europaeus* before it has seeded for the first time as the seed is known to be viable for at least 40 years under normal seed bank conditions (Hill et al. 2001).

There are three other species that are controlled whenever they are detected within Fiordland National Park: *Berberis darwinii* (Darwin's barberry), *Hypericum androsaemum* (tutsan), and *Leycesteria formosa* (Himalayan honeysuckle). All three species are bird dispersed and have populations too large to control effectively outside the National Park, often on private land or public land not managed by DOC.

Berberis darwinii grew densely on the foreshore of Lake Manapouri in the township of Manapouri and for a number of years was controlled by a community "Weedbusters" group. Now this infestation is being managed by Southland District Council but it has given rise to new populations immediately adjacent to the

National Park that are not being controlled. Around Lake Te Anau all known incursions of *B. darwinii* are controlled. These are derived from a large population on private land east of the lake. This species was not recorded naturalised anywhere by Johnson (1982) and has invaded the eastern edge of Fiordland National Park from hedgerows and plants in private gardens within the last decade.

Hypericum androsaemum has entered the National Park as a garden escape through bird dispersal from Milford Sound village. The fruit-eating New Zealand pigeon (*Hemiphaga novaeseelandiae*) is a strong flier, well capable of flying the distances involved to spread the seed from Milford Sound to Anita Bay and Bligh Sound where it has been found (Powlesland et al. 2011). *Hypericum androsaemum* is widespread, though not abundant. However, given the rugged terrain of Fiordland and the ability of *H. androsaemum* to persist under a forest canopy (Johnson 1982), this species is controlled wherever it is found but is not actively searched for because birds can disperse it anywhere within forest over a vast area, making it very difficult to find.

Also invading from the east is *Leycesteria formosa*: it is relatively common on the eastern side of Lake Te Anau on private land and within the National Park is controlled whenever it is encountered.

14.4.3.3   Coping with the Current

All of the *Ulex europaeus* in FNP originates from the West Coast of the South Island, where it is abundant. The seeds do not float but the wood does (Johnson 1982) and when rivers on the West Coast are in flood entire plants can be uprooted and discharged to the sea where they are swept along by the Southland Current. If there are strong onshore winds along the Fiordland coast, surface water drift will transport the *U. europaeus* and other flood debris onto the rocky coast above the normal strand line or push debris into the fiords where, again, it will strand on downwind shores or in river deltas. The natural vegetation in these locations is typically low shrubland, often windshorn, which is ideal habitat for *U. europaeus*.

Ammophila arenaria is also dispersed to Fiordland from the West Coast on the Southland Current and was recorded by Johnson (1982) at Big Bay and Cascade Bay, both increasingly further north of FNP. It was apparently planted at Cascade Bay and is abundant on some beaches further north. Konlechner and Hilton (2009) have demonstrated that rhizome fragments of *A. arenaria* can be dispersed more than 600 km and remain viable in seawater for up to 70 days. This is ample time, given the rate of movement of the Southland Current and any associated wind-assisted surface movement (Stanton 1976) for *A. arenaria* to be dispersed to FNP from points north.

In order to reduce the rate of reinvasion of *A. arenaria*, Southland DOC staff (who manage FNP) requested West Coast DOC staff to eradicate *A. arenaria* from Cascade Bay since they were doing the same at Big Bay, north of the National Park. West Coast staff agreed and Cascade Bay is now free of *A. arenaria*. The ideal situation for Fiordland National Park regarding *U. europaeus* is that all rivers on the West Coast, south of latitude 42°S would have this species cleared from the maximum flood zone. This could be achieved via the West Coast Regional Council whose responsibility is to consider IAP management and then consult with the public about it.

Given the constant pressure of propagules (higher for *U. europaeus*, lower for *A. arenaria*) dispersing from the West Coast it is important that the surveillance and control programme for both species continues on an annual basis. After the eradication of *A. arenaria* commenced in the 1980s there was a significant lapse of commitment and populations at most locations expanded so that, in some cases, significant knock-down work was required to achieve zero-density. The effort to maintain zero-density status, however, is slight in comparison. Meanwhile, longer term strategies, as outlined above, can be implemented.

In addition to surveillance for these two known invaders from the West Coast we know that dispersal of *Euphorbia paralias* from Australia to FNP sand dunes is highly likely (see Sect. 14.3.4). This species will need to be included in the *A. arenaria* surveillance programme.

14.4.3.4 Contributors to Success

Sites or regions within FNP with the greatest and most prolonged human contact have more IAPs (Timmins and Williams 1991). Success has depended, firstly, upon having clear goals in relation to FNP as an iconic natural landscape and in recognition that the ecosystems and species within the park are of high value. Second, a collaborative approach with the communities and agencies who live and work alongside and in FNP, has allowed for early control of some species (e.g. fishermen reporting *Ulex europaeus*; Fiordland Marine Guardians prioritising biosecurity and supporting the eradication of *Undaria pinnatifida* from Dusky Sound: or minimised reinvasion (e.g. Manapouri Weedbusters controlling *Berberis darwinii*; West Coast DOC staff controlling *Ammophila arenaria* up-current). However, possums are still spreading uncontrolled within the park. The potential impact they could have in modifying the habitat to the advantage of IAPs, or if they are likely to disperse seeds that so far have not been dispersed by native and alien fauna within the park, is unknown. Part of the reason for the low invasion rates of IAPs in FNP is the intact forest cover, lack of roads and tracks, and the small number of huts. Possums will have an impact on this at canopy level whereas deer have been minimal promoters of IAPs in forest communities (CJW pers. obs.).

14.5 Invasive Alien Plant Management in New Zealand: Adapting to a Different World

The ecology of alien plants and on-the-ground experience has reinforced to weed managers that achieving successful control and eradication of IAPs is not easy or achievable in the short-term. In the past 25 years DOC has improved management techniques and knowledge, chastened by the length of time that has been required to eradicate some IAPs.

Despite New Zealand, like Australia, having one of the best border biosecurity systems in the world and very strong legislation internally, IAPs continue to arrive from overseas, e.g. *Euphorbia paralias*, or establish from cultivated plants (Esler 1988; Williams and Cameron 2006). The scale of the problem means that a strategic approach is essential, as is a stringent surveillance and monitoring regime that can be used to react to new introductions. Techniques need to be adaptable to the specific situation.

Eradicating IAPs or controlling them to zero-density is difficult and can be expensive, and most work shows that eradication is usually only effective when the population is very small (occupying <1 ha, Howell 2012). Innovative approaches are going to be needed if the control is to be successful, particularly with species that have rapid growth rates to maturity, persistent seed banks, bird dispersed seeds, and are hard to locate. In the case of Raoul Island, for example, it is particularly difficult to estimate the time it might take to achieve eradication of any species. A compromise in ecological integrity may need to be accepted, for example, where some IAPs are not eradicated in order to achieve the restoration of a seabird-driven ecosystem. Genetic markers could in future also be used to identify sources of invasion to then more effectively manage these sources or their pathways, as is done with some mammal control programmes (Russell et al. 2005).

From some of the mammal eradication work on islands (Veitch et al. 2011a), and changes in land use on the mainland, insight into how IAPs respond to the removal of invasive alien browsers and seed predators or dispersers is being improved. As invasive alien mammal eradication becomes a reality, the understanding of the biology and responses of IAPs to altered trophic relationships needs to be improved.

In tandem with the broader scale approach, the 'managing for outcomes' framework that DOC has developed (see Sect. 14.2.3) relies on 'prescriptions' that describe the management actions required to ensure ecosystem integrity and/or species persistence. Many of the actions require the control of IAPs to specific levels, and monitoring to understand the effectiveness of these actions, and whether the outcomes are being achieved. This presents further opportunities to understand trophic interactions and adapt management practices to achieve the desired outcomes.

Acknowledgements We thank Monica Valdes, David Havell, Michelle Gutsell, Alastair Hay, and Verity Forbes for providing data and reviewing sections of this chapter. We are grateful to Sarah Fell, Cindy Smith Yvonne Sprey and Tamra Gibson, on Raoul Island, for sending recent data on the number of individuals of some target species that have been killed. Paul Hughes was extremely helpful in providing the map of Fiordland National Park. Kevin O'Connor, Susan-Jane Owen and an anonymous reviewer also provided helpful reviews of this chapter for which we offer our thanks. Finally, we'd like to thank all those DOC staff and volunteers plus others elsewhere who do the hard work of protecting natural areas from the impacts of IAPs.

References

Anderson A (1980) The archaeology of Raoul Island (Kermadecs) and its place in the settlement history of Polynesia. Archaeol Phys Anthropol Ocean 15:131–141

Atkinson IAE (1973) Protection and use of the islands in Hauraki Gulf Maritime Park. Proc N Z Ecol Soc 20:103–114

Bellingham PJ, Lee WG (2006) Distinguishing natural processes from impacts of invasive mammalian herbivores. In: Allen RB, Lee WG (eds) Biological invasions in New Zealand, Ecological studies 186. Springer, Berlin, pp 323–336

Brockerhoff EG, Hoffmann JH, Roques A (2004) Is biological control an option for the management of wilding pines (*Pinus* spp.) in New Zealand? In: Hill RL, Zydenbos SM, Bezar CM (eds) Managing wilding conifers in New Zealand: present and future. New Zealand Plant Protection Society, Christchurch, pp 65–78

Broome KG (2009) Beyond Kapiti – a decade of invasive rodent eradications from New Zealand islands. Biodiversity 10:7–17

Cameron EK (1984) Vascular plants of the three largest Chickens (Marotere) Islands: Lady Alice, Whatupuke, Coppermine: North-east New Zealand. Tane 30:53–75. University of Auckland, Auckland

Campbell H, Hutching G (2007) In search of ancient New Zealand. Penguin Books, New Zealand

Craine JM, Lee WG, Walker S (2006) The context of plant invasions in New Zealand: evolutionary history and novel niches. In: Allen RB, Lee WG (eds) Biological invasions in New Zealand, Ecological studies 186. Springer, Berlin, pp 167–177

Department of Conservation and Ministry for the Environment (2000) The New Zealand biodiversity strategy: Our chance to turn the tide – Whakakōhukihukitia te tai roroku ki te tai oranga. Department of Conservation and Ministry for the Environment, Wellington

Diamond JM (1990) New Zealand as an archipelago: an international perspective. In: Towns DR, Daugherty CH, Atkinson IAE (eds) Ecological restoration of New Zealand islands, Conservation sciences publication no 2. Department of Conservation, Wellington, pp 3–8

Diez JM, Williams PA, Randall RP et al (2009) Learning from failures: testing broad taxonomic hypotheses about plant naturalization. Ecol Lett 12:1174–1183

Dilks P, Willans M, Pryde M et al (2003) Large scale stoat control to protect mohua (*Mohoua ochrocephala*) and kaka (*Nestor meridionalis*) in the Eglinton Valley, Fiordland, New Zealand. N Z J Ecol 27:1–9

DOC (2002) Fiordland National Park management plan. Department of Conservation, Invercargill

DOC (2012) Department of Conservation annual report for the year ended 30 June 2012. Department of Conservation, Wellington

Duncan RP, Coulhan KM, Foran BD (1997) The distribution and abundance of *Hieracium* species (hawkweeds) in the dry grasslands of Canterbury and Otago. N Z J Ecol 21:51–62

Esler AE (1988) Naturalisation of plants in urban Auckland. DSIR, Wellington

Evans AS, Mitchell RJ, Cabin RJ (1996) Morphological side effects of using gibberellic acid to induce germination: consequences for the study of seed dormancy. Am J Bot 83:543–549

Finch-Savage WE, Leubner-Metzger G (2006) Seed dormancy and the control of germination. New Phytol 171:501–523

Gaskin CP (2011) Seabirds of the Kermadec region: their natural history and conservation, Science for conservation 316. Department of Conservation, Wellington

Given DR (1973) Naturalised flowering plants in south-west Fiordland. N Z J Bot 11:247–250

Hall-Jones J (1990) Fiordland explored: an illustrated history. Craigs, Invercargill

Hill RL, Gourlay AH, Barker RJ (2001) Survival of *Ulex europaeus* seeds in the soil at three sites in New Zealand. N Z J Bot 39:235–244

Hilton M (2001) Sea spurge (*Euphorbia paralias*): floating New Zealand's way. Coast News 18:9–10

Hilton M (2003) Potential new invasive plants of coastal dunes: bad news from Australia. Protect, Summer 2002–2003:26–29

Holloran P (2006) Measuring performance of invasive eradication efforts in New Zealand. N Z Plant Protect 59:1–7

Howell CJ (2008) Consolidated list of environmental weeds in New Zealand, DOC research & development series 292. Department of Conservation, Wellington

Howell CJ (2012) Progress towards environmental weed eradication in New Zealand. Invasions Plant Sci Manag 5:249–258

Johnson PN (1982) Naturalised plants in south-west South Island, New Zealand. N Z J Bot 20:131–142

Konlechner TM, Hilton MJ (2009) The potential for dispersal of *Ammophila arenaria* (marram grass) rhizome in New Zealand. J Coast Res, SI 56 (Proceedings of the 10th International Coastal Symposium), Lisbon, Portugal, pp 434–437

Lee W, McGlone M, Wright E (2005) Biodiversity inventory and monitoring: a review of national and international systems and a proposed framework for future biodiversity monitoring by the Department of Conservation. Landcare Research Contract Report: LC0405/122. Landcare Research New Zealand Ltd, Wellington

Lyle S (2006) Discovering fruit and nuts. David Bateman Ltd, Auckland

MacLeod CJ, Affeld K, Allen RB et al (2012) Department of Conservation biodiversity indicators: 2012 assessment. Landcare Research, Wellington

Mark AF, Dickinson KJM (2008) Maximizing water yield with indigenous non-forest vegetation: a New Zealand perspective. Front Ecol Environ 6:25–34

Martínez-Gómez P, Dicenta F (2001) Mechanisms of dormancy in seeds of peach (*Prunus persica* (L.) Batsch) cv. GF305. Sci Hortic 91:51–58

McGlone M (2004) Vegetation history of the South Island high country. Landcare Research Contract Report: LC0304/065. Landcare Research, Wellington

McNab R (1907) Murihiku and the Southern Islands: a history of the West Coast Sounds, Foveaux Strait, Stewart Island, the Snares, Bounty, Antipodes, Auckland, Campbell and Macquarie Islands, from 1770 to 1829. W. Smith, Invercargill

Ministry for Primary Industries (2012) Didymo, *Didymosphenia geminata*. http://www.biosecurity.govt.nz/didymo Accessed 20 Dec 2012

Moore PR (1984) Mineral exploration on Coppermine Island 1849–1969: an historical review. Tane 30:165–174

New Zealand Plant Conservation Network (2012) Frequently asked questions http://www.nzpcn.org.nz/page.asp?help_faqs_NZ_plants Accessed 20 Nov 2012

Owen S-J (1998) Department of Conservation strategic plan for managing invasive weeds. Department of Conservation, Wellington

Parkes JP (1984) Feral goats on Raoul Island II. Diet and notes on the flora. N Z J Ecol 7:95–101

Peat N (1995) Manapouri saved! New Zealand's first great conservation success story. Longacre Press, Dunedin

Peters M, Clarkson B (eds) (2010) Wetland restoration: a handbook for New Zealand freshwater systems. Manaaki Whenua Press, Lincoln

Poole AL (1951) Flora and vegetation of the Caswell and George Sounds district. Transact R Soc N Z 79:62–83

Powlesland RG, Moran LR, Wotton DM (2011) Satellite tracking of kereru (*Hemiphaga novaeseelandiae*) in Southland, New Zealand: impacts, movements and home range. N Z J Ecol 35:229–235

Richardson DM, Pyšek P, Rejmánek M et al (2000) Naturalization and invasion of alien plants: concepts and definitions. Divers Distrib 6:93–107

Russell JC, Towns DR, Anderson SH et al (2005) Intercepting the first rat ashore. Nature 437:1107

Russell LK, Hepburn CD, Hurd CL et al (2008) The expanding range of *Undaria pinnatifida* in southern New Zealand: distribution, dispersal mechanisms and the invasion of wave-exposed environments. Biol Invasions 10:103–115

Shanley T (2010) Hen and Chicken and Poor Knights Islands weed report 2009–10. Unpublished report, Department of Conservation, New Zealand

Stanton BR (1976) Circulation and hydrology off the west coast of the South Island, New Zealand. N Z J Mar Freshw Res 10:445–467

Sykes WR, West CJ (1996) New records and other information on the vascular flora of the Kermadec Islands. N Z J Bot 34:447–462

Sykes WR, West CJ, Beever JE et al (2000) Kermadec Islands flora: a compilation of modern materials about the flora of the Kermadec Islands, Special edn. Manaaki Whenua Press, Lincoln

The Biosecurity Council (2003) Tiakina Aotearoa – protect New Zealand: the biosecurity strategy for New Zealand. Biosecurity Council, Wellington

Thompson K, Bakker JP, Bekker RM (1997) The soil seedbanks of North-West Europe: methodology, density and longevity. Cambridge University Press, Cambridge

Thomson GM (1922) The naturalisation of animals and plants in New Zealand. Cambridge University Press, Cambridge

Timmins SM, Williams PA (1991) Weed numbers in New Zealand's forest and scrub reserves. N Z J Ecol 15:153–162

Towns D, Parrish R (2003) Restoration of the principal Marotere Islands. Department of Conservation, Wellington

Veitch CR, Clout MN, Towns DR (eds) (2011a) Island invasives: eradication and management. IUCN, Gland

Veitch CR, Gaskin C, Baird K et al (2011b) Changes in bird numbers on Raoul Island, Kermadec Islands, New Zealand, following the eradication of goats, rats and cats. In: Veitch CR, Clout MN, Towns DR (eds) Island invasives: eradication and management. IUCN, Gland, pp 372–376

Vivian-Smith G, Lawson BE, Turnbull I et al (2007) The biology of Australian weeds 46. Anredera cordifolia (Ten.) Steenis. Plant Prot Q 22:2–10

West CJ (1996) Assessment of the weed control programme on Raoul Island, Kermadec Group, Science & research series no 98. Department of Conservation, Wellington

West CJ (2002) Eradication of alien plants on Raoul Island, Kermadec Islands, New Zealand. In: Veitch CR, Clout MN (eds) Turning the tide: the eradication of invasive species. IUCN, Gland, pp 365–373

West CJ (2011) Consideration of rat impacts on weeds prior to rat and cat eradication on Raoul Island, Kermadecs. In: Veitch CR, Clout MN, Towns DR (eds) Island invasives: eradication and management. IUCN, Gland, pp 244–247

West CJ, Havell D (2011) Plant responses following eradication of goats and rats from Raoul Island, Kermadecs. In: Veitch CR, Clout MN, Towns DR (eds) Island invasives: eradication and management. IUCN, Gland, p 535

Williams JA, West CJ (2000) Environmental weeds in Australia and New Zealand: issues and approaches to management. Austral Ecol 25:425–444

Williams PA, Cameron EK (2006) Creating gardens: the diversity and progression of European plant introductions. In: Allen RB, Lee WG (eds) Biological invasions in New Zealand, Ecological studies 186. Springer, Berlin, pp 33–47

Williams PA, Timmins SM (1990) Weeds in New Zealand protected natural areas: a review for the Department of Conservation, Science and research series no 14. Department of Conservation, Wellington

Williams PA, Wiser S, Clarkson B (2007) New Zealand's historically rare terrestrial ecosystems set in a physical and physiognomic framework. N Z J Ecol 31:119–128

Wiser SK, Allen RB (2000) Hieracium lepidulum invasion of indigenous ecosystems, Conservation advisory science notes no 278. Department of Conservation, Wellington

Wiser SK, Allen RB, Clinton PW et al (1998) Community structure and forest invasion by an exotic herb over 23 years. Ecology 79:2071–2081

Wotton DM, O'Brien C, Stuart MD et al (2004) Eradication success down under: heat treatment of a sunken trawler to kill the invasive seaweed Undaria pinnatifida. Mar Poll Bull 49:844–849

Wright AE (1978) Vascular plants of Hen Island (Taranga) North-Eastern New Zealand. Tane 24:77–102

Chapter 15
Plant Invasions in Protected Areas of Tropical Pacific Islands, with Special Reference to Hawaii

Lloyd L. Loope, R. Flint Hughes, and Jean-Yves Meyer

Abstract Isolated tropical islands are notoriously vulnerable to plant invasions. Serious management for protection of native biodiversity in Hawaii began in the 1970s, arguably at Hawaii Volcanoes National Park. Concerted alien plant management began there in the 1980s and has in a sense become a model for protected areas throughout Hawaii and Pacific Island countries and territories. We review the relative successes of their strategies and touch upon how their experience has been applied elsewhere. Protected areas in Hawaii are fortunate in having relatively good resources for addressing plant invasions, but many invasions remain intractable, and invasions from outside the boundaries continue from a highly globalised society with a penchant for horticultural novelty. There are likely few efforts in most Pacific Islands to combat alien plant invasions in protected areas, but such areas may often have fewer plant invasions as a result of their relative remoteness and/or socio-economic development status. The greatest current needs for protected areas in this region may be for establishment of yet more protected areas, for better resources to combat invasions in Pacific Island countries and territories, for more effective control methods including biological control programme to contain

L.L. Loope (retired) (✉)
Formerly: USGS Pacific Island Ecosystems Research Center, Haleakala Field Station, Makawao (Maui), HI 96768, USA

Current: 751 Pelenaka Place, Makawao, HI 96768, USA
e-mail: lll@aloha.net

R.F. Hughes
Institute of Pacific Islands Forestry, USDA Forest Service, 60 Nowelo Street, Hilo, HI 96720, USA
e-mail: fhughes@fs.fed.us

J.-Y. Meyer
Délégation à la Recherche, Government of French Polynesia, B.P. 20981, Papeete, Tahiti, French Polynesia
e-mail: jean-yves.meyer@recherche.gov.pf

L.C. Foxcroft et al. (eds.), *Plant Invasions in Protected Areas: Patterns, Problems and Challenges*, Invading Nature - Springer Series in Invasion Ecology 7, DOI 10.1007/978-94-007-7750-7_15, © Springer Science+Business Media Dordrecht 2013

intractable species, and for meaningful efforts to address prevention and early detection of potential new invaders.

Keywords Feral ungulates • Grass-fire cycle • Haleakala National Park • Hawaii Volcanoes National Park • National Park of American Samoa • South-eastern Polynesia

15.1 Introduction

The Pacific Ocean is enormous 20,000 km across from Singapore in the west to Panama in the east. It contains approximately 25,000 islands; more than the rest of the world's oceans combined (Gillespie et al. 2008); fewer than 800 of those are considered habitable by humans (Douglas 1969). Its isolated islands are known for high levels of endemism, collectively contributing a very significant portion of the earth's biodiversity (Myers et al. 2000). Its islands are notable for large numbers of endangered species and high rates of extinction. Although less famously documented than predatory and herbivorous alien animals (rats, cats, ungulates and others), plant invasions significantly contribute to undermining biodiversity on islands (e.g. Meyer and Florence 1996; Kueffer et al. 2010a).

An analysis of 25 of the world's 'biodiversity hotspots' by Myers et al. (2000) found that, collectively, the Pacific islands of Polynesia/Micronesia (excluding New Zealand), with slightly over half their flora endemic to the region, have an endemic flora comprising 1.1 % (3,334 species) of the world's flora; the remaining primary vegetation was estimated to cover 10,024 ha (21.8 % of its original extent), with 49 % in protected areas. The large continental island of New Caledonia (part of Melanesia), with 5,200 ha (28 % of the original) of remaining primary vegetation but only 10 % of it in protected areas, was noted as an exceptionally rich hotspot, with an astounding 1,865 endemic plant species.

The inherent vulnerability of Pacific islands to biological invasions was recognised as early as Darwin's observations in the Galapagos archipelago and elsewhere in the 1830s (Darwin 1859), and re-emphasised by Charles Elton in 1958. Pacific island ecosystems are recognised as typically having higher representation of alien species than mainland systems, and the severity of the impact of invasions (i.e. detrimental effects on native species) on islands generally increases with isolation of the islands, though rigorous explanation of these phenomena is elusive (D'Antonio and Dudley 1995). Isolated islands, and particularly their 'protected areas', provide extraordinary living museums of speciation and evolutionary radiation; the limitation in our ability to fully protect such areas from degradation has at least provided dynamic laboratories for better understanding invasions and their interactions with ecological processes (Vitousek et al. 1987).

Here we present and evaluate efforts by managers and researchers to address plant invasions in relatively well-studied protected areas among Pacific Islands and to highlight associated contributions to invasion biology and management.

We focus largely on several notable protected areas where substantial effort to combat invasions has been possible, both to illuminate the various approaches to the problems posed by, and the solutions applied to address the impacts of alien plant invasions on native floras of the Pacific Islands.

15.2 Evolution of the Protected Areas Network and Strategies for Addressing Plant Invasions

Hawaii Volcanoes National Park (NP), island of Hawaii (often called the "Big Island" with its one million ha of terrestrial surface), Hawaii, USA, has the longest history of management of any terrestrial protected area in the tropical Pacific Islands. It was originally established as the largest unit of Hawaii NP in 1916, primarily to protect its volcanic scenery and for geologic study. It currently covers 1,293 km^2, following the Kahuku addition in 2004 (see below), which added 60 % to the park's previous area. This represents about 12.5 % of the area of Hawaii Island or about 7.8 % of the land area of the state of Hawaii; this area is larger than most Pacific Islands. About 80 % of the total protected area is now comprised of fire derived, degraded grasslands dominated by alien species, or sparsely to unvegetated, volcanic terrain (including the upper slopes of 4,169 m a.s.l. Mauna Loa, a shield volcano and the second – to nearby 4,205 m a.s.l. Mauna Kea - highest peak of the tropical Pacific Islands). The remaining area of about 250–300 km^2 contains diverse plant communities. Hawaii Volcanoes NP has been on a trajectory of management for protection of native/endemic biodiversity since the early 1970s when Park managers set out to eliminate the entire population of about 15,000 feral goats (*Capra hircus*) with the aid of fencing. This single event was a 'game changer', initially opposed by almost everyone, including (in 1970–1971) the Director of the U.S. National Park Service (Sellars 1997), as well as by State government agencies in Hawaii.

Herbivory and associated disturbance by feral goats had been rampant in the area of Hawaii Volcanoes NP for nearly two centuries since their introduction to the islands in the 1780s, and more than 70,000 goats had been removed from the Park area since its establishment in 1916, with no long-term population control (Sellars 1997). Conventional wisdom at the time was that any serious effort toward biological conservation in Hawaii was impractical, if not impossible. The impetus for eliminating goat populations came from a strong movement (described in Sellars 1997), mostly within the U.S. National Park Service, to steer the national parks (nationwide) toward biological preservation. The goat eradication effort gradually received increasing and sustained local support, with feral goats largely eliminated in the Park by the end of the 1970s. Arguably, actions and rationale used by the National Park Service in the 1970s at Hawaii Volcanoes NP provided the critical momentum for the rise of 'active' conservation in the State of Hawaii by federal, state, and private entities, with meaningful public support. A generally environment-friendly climate in the USA in the 1970s and the federal Endangered

Species Act of 1973 added impetus to active conservation in Hawaii. By the late 1990s, there was broad buy-in to the concept of biological preservation in Hawaii and over 25 % of the state's land area had been incorporated under varying degrees of protected area management (Loope and Juvik 1998). Nevertheless, the national parks set the standard for what may be possible, and for developing reasonable and thoughtful strategies to try to achieve identified goals.

In Hawaii, alien plant management in protected areas started on a significant scale at Hawaii Volcanoes NP in the 1980s, at which time a sophisticated strategy was developed and articulated. A 1986 symposium entitled "Control of Introduced Plants in Hawaii's Native Ecosystems" was held in conjunction with Hawaii Volcanoes NP's Sixth Conference in Natural Sciences. A book edited by Stone, Smith, and Tunison, "Alien Plant Invasions in Native Ecosystems of Hawaii: Management and Research" (Stone et al. 1992), produced 6 years later, provides remarkable documentation of the development of Hawaii Volcanoes NP's strategy for addressing alien plant issues, as well as of Hawaii's state-wide situation. It was stated that the "The severity of the alien plant problem and the fact that it is so widespread in the Islands make a rigorously organised approach based on relevant information especially necessary. Moreover, development of a variety of approaches to weed control to deal with different situations... are necessary components of weed management programs." (Tunison et al. 1992a).

Tunison et al. (1992a) described the key features of the Hawaii Volcanoes NP alien plant control strategies "to protect native species assemblages." In summary: (i) controlling feral pigs and goats; (ii) excluding fire; (iii) mapping the distribution of important alien plants; (iv) controlling localised alien plants throughout the Park; (v) controlling all disruptive alien plants in Special Ecological Areas (the most diverse and intact areas in the Park); (vi) confining one widespread species, *Cenchrus setaceus* (= *Pennisetum setaceum*, fountain grass), to the area it currently infests; (vii) developing herbicidal control methods for target species; (viii) developing biological controls for some widespread species; (ix) researching the ecology, seed biology, and phenology of important alien plant pest species; (x) educating the public to the importance of alien plant control; and (xi) working with other agencies and groups in alien plant management.

Haleakala NP, island of Maui, Hawaii, was at first a disjunct part of Hawaii NP, established in 1916 and protecting the 3,055 m Haleakala volcano above about 2,000 m elevation. It was designated as Haleakala NP in 1961. The important addition of Kipahulu Valley, a highly pristine rainforest watershed stretching from sea level to above 2,000 m, and other adjacent lands, have resulted in a current Park area of 121 km^2. In the 1980s, this smaller, but more topographically diverse Park, eliminated goats with the aid of fencing and initiated alien plant management, building on the experience of Hawaii Volcanoes NP.

Points that deserve emphasis are the importance and difficulty of creating protected areas in the first place and the challenges most Pacific island countries have in funding management of such areas despite the widely acknowledged need to do so. The addition of the 48,245 ha Kahuku lands to Hawaii Volcanoes NP was enormously important for biodiversity protection in Hawaii, and the Park has been able to devote sufficient resources to take steps to initiate management on those added lands. The National Park of American Samoa (NPSA) was established in 1988 through an innovative and effective concept involving a 50-year lease to the U.S. National Park Service for the park land by the local Samoan village councils (Cox and Elmqvist 1991); the agreements were finalised on the island of Tutuila in 1993, with expansion to other islands in 2002. As such, NPS efforts in both Hawaii and American Samoa can be viewed as useful and distinct models for creating protective areas in other island nations across the Pacific.

South-eastern Polynesia is likely representative of most of the Pacific island regions regarding potential obstacles in establishing and managing protected areas. An important reason why SE Polynesia has few and small protected areas (e.g. less than 2 % of French Polynesia's land area) is that in the main inhabited islands (e.g. Tahiti, Pitcairn, Rarotonga) the land tenure situation is problematic. Most of the land is privately owned by families or clans with multiple beneficiaries that may not be capable of reaching consensus regarding establishment of protected areas. In French Polynesia for instance, the small atolls of Manuae (Scilly) and Motu One (Bellinghausen) in the Society Islands and the islets of Mohotani (Motane), Eiao and Hatutaa (Hatutu) in the Marquesas were more easily declared natural reserves in 1971 because they were uninhabited and public lands. This was also the case for the TeFaaiti Natural Park (750 ha) on Tahiti in 1989, the Vaikivi Natural Park and Reserve (240 ha) on the island of Ua Huka in 1997, and the Temehani Ute Ute plateau (69 ha) on the island of Raiatea in 2010 (Table 15.1). The lack of local capacity (long-term funding support to manage these protected areas, trained conservation managers and scientists, influential nature protection NGOs, etc.), but also the weak local political will and public support are important constraints in SE Polynesian countries and territories (pers. comm. from many people working in the South Pacific, to J.-Y. Meyer). Unfortunately, there are few or no management efforts or programmes to combat invasive alien plants in protected areas in SE Polynesia, although many of these islands provide extraordinary case-studies for illustrating both the impacts of invasive alien plants on biodiversity and cultural assets (e.g. the monumental stone statues in Easter Island or Rapa Nui), and for potential habitat restoration projects. Paradoxically, there are many fencing and weeding projects in French Polynesia recently conducted by local authorities, communities and NGOs in Tahiti, Raiatea (Society Is.) and Rapa Iti (Australs), but primarily in unprotected areas (J.-Y. Meyer, unpub. data).

15.3 How Successful Have Strategies for Alien Plant Management Been in Hawaii Volcanoes NP and Other Protected Areas?

In this section, we report progress in implementation of a generalised version of the 11 items of Tunison's (1992b) visionary strategy 20 years later.

Table 15.1 Characteristics of protected areas in South-eastern Polynesian islands, their dominant invasive plants, and other major ecological threats

Island(s)	Protected area and date of creation	Area (ha)	Main habitat and vegetation type(s)	Dominant invasive plant(s)	Other dominant threats	Source reference
Rapa Nui (Easter Island, Chile)	Rapa Nui National Park (1935), UNESCO World Heritage Site (1995)	6,650	Coastal vegetation and forest, Low and Mid-elevation grassland, savannas and shrubland (0–511 m)	*Cirsium vulgare, Crotalaria grahamiana, Melinis minutiflora, Psidium guajava*	Horses, cattle, human presence and frequentation	Meyer (2008, 2012) and unpub. data
Henderson (Pitcairn Is., UK)	UNESCO World Heritage Site (1988)	3,700	Coastal and raised limestone vegetation and forest (0–30 m)	None identified	Pacific rats	Florence et al. (1995) and Brooke et al. (2004)
Rarotonga (Cook Islands)	Takitumu Conservation Area (1996)	155	Low and Mid-elevation valley forest (50–250 m)	*Ardisia elliptica, Cestrum nocturnum, Psidium cattleianum, Syzygium jambos, Lantana camara, Spathodea campanulata, Merremia peltata, Mikania micrantha, Cardiospermum halicacabum*	Black and Pacific rats, human frequentation	Meyer (pers. obs. 1997, 2002); E. Saul (pers. com. 2011)
French Polynesia Tahiti (Society)	TeFaaiti Natural Park (1989)	750	Mid-elevation valley rainforest (70–500 m) and Mid-elevation plateau (500–700 m), High elev. cloud forest and subalpine vegetation on steep slopes (up to 2110 m)	*Miconia calvescens, Rubus rosifolius, Spathodea campanulata, Tecoma stans*	Black and Pacific rats, human frequentation	Meyer (pers. obs. 1998–2004)

Raiatea (Society)	Temehani Ute Ute Management Area (2010)	69	High-elevation plateau (415–817 m)	*Rhodomyrtus tomentosa, Miconia calvescens, Psidium cattleianum, Cecropia peltata, Rubus rosifolius*	Feral pigs, black and Pacific rats	Meyer (1996a)
Manuae (Scilly) and Motu One (Bellinghausen) (Society)	Natural reserves (1971)	1180	Atoll coastal vegetation and forest (0–2 m)	*Stachytarpheta cayennensis, Cenchrus echinatus*	Coconut plantations, human presence	Sachet (1983)
Ua Huka (Marquesas)	Vaikivi Natural Park and Natural Reserve (1997)	240	Mid-to high elevation rainforest (−880 m)	*Coffea arabica, Stachytarpheta cayennensis, Psidium guajava*	Feral goats, horses, Pacific rats, human frequentation	Meyer (1996a, 2005)
Eiao (Marquesas)	Natural reserve (1971), Management Area (2004)	3,920	Coastal vegetation and forest, and semi-dry and mesic forest (0–577 m)	*Acacia farnesiana, Leucaena leucocephala, Annona squamosa*	Feral sheep and pig, black and Pacific rats	Meyer (unpubl. data 2010)
Mohotani (Marquesas, French Polynesia)	Natural Reserve (1971), Management Area (2004)	1,280	Coastal forest, and semi-dry and mesic forest (0–531 m)	*Senna occidentalis, Pityrogramma calomelanos*	Feral sheep, Pacific rats	Meyer (1996a, 2000)
Hatutaa (Marquesas)	Natural reserve (1971), Management Area (2004)	660	Coastal vegetation and forest, and semi-dry forest (0–428 m)	*Passiflora foetida, Senna occidentalis*	Pacific rats	Meyer (unpubl. data 2010)
7 atolls: Fakarava, Aratika, Niau, Raraka, Taiaro, Kauehi, Tou (Tuamotu)	UNESCO Biosphere Reserve (Taiaro in 1972 and 6 other atolls in 2006)		Atoll and raised limestone coastal forests (0–6 m)	*Stachytarpheta cayennensis*	Black and Pacific rats, human presence, coconut plantations (fires)	Meyer (unpubl. data 2007)

15.3.1 Feral Ungulates and Implications for Managing Plant Invasions

Hawaii and other isolated islands lack an evolutionary history of ungulate presence, though large flightless birds may have filled similar ecological niches. Ungulates are still absent in wildland areas of many Pacific islands (Merlin and Juvik 1992), but many others have them. Feral goats – as well as deer (*Axis axis*), sheep (*Ovis aries*), mouflon (*Ovis musimon*), and other ungulates – continue to deplete biodiversity outside fenced areas in Hawaii but protected areas have become increasingly fenced (though at great cost) over the past three decades. Feral pigs (*Sus scrofa*) are currently considered primary modifiers of remaining Hawaiian rainforest and have substantial effects on other ecosystems. Although pigs were brought to the Hawaiian Islands by Polynesians roughly a millennium ago, the current severe environmental damage inflicted by pigs apparently began much more recently and seems to have resulted entirely from release of domestic, non-Polynesian genotypes (Diong 1982). Polynesian pigs were much smaller, more docile, and less prone to taking up a feral existence than those introduced in historical times (Tomich 1986). Much of the damage to plants by pigs is direct, involving physical rooting and feeding. Much damage also occurs from invasion of opportunistic plant species, often alien, that contribute to further displacement of native species. Seeds of alien plants are carried on pigs' coats or in their digestive tracts, and they thrive upon germination on the forest floor where pigs have exposed mineral soil (Diong 1982; Medeiros 2004).

Feral pigs have proven much harder to eliminate than goats. Hawaii Volcanoes NP has established 13 pig-free fenced units (often corresponding with SEAs – see below) with a combined area of approximately 16,000 ha plus. At Haleakala NP, feral pigs were eliminated in remote Kipahulu Valley in the late 1980s with fencing and snaring (Anderson and Stone 1993), and the entire Park has since been largely pig-free.

Once aggressive plant invaders have obtained a new foothold in the forest – often as a result of feral ungulate disturbance and dispersal – they spread opportunistically, aided by pigs and alien birds. Removal of pigs stops the mechanical damage and some of the seed dispersal, and is essential for halting direct degradation of biodiversity, but experience has showed that it does not stop plant invasions (e.g. Huenneke and Vitousek 1990; Medeiros 2004). Frequently, after pigs are removed from an area, native species may undergo various degrees of recovery, but plant invasions occupy the sites that had been kept bare by pig-digging. This trend was recently documented by Cole et al. (2012) who measured a substantial increase in cover (51 %) of common native species within a large (1024 ha) fenced exclosure in Hawaii Volcanoes NP from which pigs had been excluded for 16 years; within the same exclosure, the invasive tree *Psidium cattleianum* (strawberry guava) underwent a fivefold increase. Similar effects have been noted with response of alien vegetation to feral goat removal (e.g. Kellner et al. 2011).

Haleakala NP currently faces serious invasive plant problems in the remote Kipahulu Valley, especially with *Clidemia hirta* (Koster's curse), *Hedychium gardnerianum* (Kahili Ginger), and *P. cattleianum*. The original expectation was that removal of feral pigs from Kipahulu Valley would not only reduce direct pig impact on native vegetation but would also reduce expansion of these invasions to some degree, ideally facilitating effective management by mechanical/chemical means (Loope et al. 1992); in retrospect, removal of pigs has allowed substantial recovery of native vegetation, but *H. gardnerianum* and *C. hirta* have also expanded substantially even with concerted control effort and are currently posing severe problems, with biological control urgently needed (Medeiros 2004; Arthur Medeiros, U.S. Geological Survey, pers. comm.).

Feral pigs have recently become a serious problem in the National Park of American Samoa (NPSA), creating disturbance to native vegetation; substantial control effort has been made, but the problem persists (Tavita Togia, NPSA, pers. comm.). Ungulates are also present and create serious disturbance, facilitating alien plant invasions in some protected areas of SE Polynesia (Meyer, pers. obs.). Feral sheep infest Eiao and Mohotani, as do feral goats and horses (*Equus caballus*) in the Vaikivi Natural Park of Ua Huka in the Marquesas; feral pigs thrive on the Temehani Ute Ute plateau on Raiatea in the Society Islands, threatening rare endemic plants (Jacq and Meyer 2012); and cattle (*Bos taurus*) and horses are causing forest destruction and facilitating invasion by light-demanding weeds on Rapa Nui (Meyer, pers. obs.). The reserves on the dry, uninhabited islands of Eiao (with long-standing serious ungulate problems, Fig. 15.1) and Hatutaa (without ungulates) in the Marquesas provide a dramatic comparison of the persistence of native vegetation where ungulates are absent and the degradation of native vegetation with replacement by alien plant species in the presence of ungulates (Merlin and Juvik 1992; Meyer, pers. obs.).

15.3.2   Fire Management and the Intractable Grass-Fire Cycle at Hawaii Volcanoes NP

From the inception of Hawaii Volcanoes NP in 1916 until the 1960s, fire had a small and infrequent footprint across park lands. Although ignition sources were present (e.g. lava flows, lightning, human activity), vegetation was either too discontinuous or of too high a moisture content to provide adequate fuels to burn. In the subsequent decades following 1960, however, fire frequency increased tenfold and the extent of fires increased an astounding 60-fold (Tunison et al. 1995) in the extensive mid-elevation seasonal fire management unit (also known as the seasonal submontane zone). What caused the fire regime to change so dramatically? It was the local establishment and proliferation during the 1960s of alien C_4 grass species such as *Andropogon virginicus* (broomsedge) from the south-eastern US, *Schizachyrium condensatum* (beardgrass) from South America, and the spread of the *Melinis minutiflora* (African molasses grass) during the 1980s. Each of these species exhibit attributes that make them very prone to burning and very adept at re-establishing following fire; all create a continuous fuel bed,

322

L.L. Loope et al.

Fig. 15.1 Isolated native tree *Pisonia grandis*, in severely eroded landscape overgrazed by feral sheep, island of Eiao, Marquesas Islands, French Polynesia (Photo J-Y Meyer, November 2010)

maintain high dead-to-live biomass ratios throughout the year, and exhibit high extinction moisture content, allowing them to burn at high relative humidity (Hughes et al. 1991). These alien grasses readily invaded the interstices of what had been woodlands and shrublands dominated by native woody plants such as *Metrosideros polymorpha* ('Ohi'a lehua) and *Leptecophylla tameiameiae* (Puhatikiei) among others. Once invaded by alien grass species, these systems became exceedingly prone to fire, and when they did almost inevitably burn, the grasses rapidly recolonised the burned areas (Hughes et al. 1991); in general alien grass cover increased 33 %, and grass biomass increased 2- to 3-fold following fire (Tunison et al. 1995; D'Antonio et al. 2000). *Melinis minutiflora* in particular increased dramatically from pre-fire cover and biomass values (Hughes et al. 1991).

In stark contrast, an average of 55 % of *M. polymorpha* individuals suffered mortality following fire, and this is likely an inflated survivorship given that many of the surviving trees were located on rocky outcrops and thus experienced little in the way of fire effects (D'Antonio et al. 2000). Post-fire *M. polymorpha* seedling recruitment and establishment is non-existent (Tunison et al. 1995), and most of the common native shrub species were sharply reduced with respect to both cover and stem density (Fig. 15.2; Hughes et al. 1991) immediately following fire as well as after two decades of post-fire succession (D'Antonio et al. 2011). Successive fires lead to increased dominance of grasses and further diminution of native woody and herbaceous species populations. Collectively, alien grasses now dominate extensive areas of dry and seasonally dry habitats in Hawaii. They have been demonstrated to

Fig. 15.2 Fire-degraded grass-shrubland, elevation ∼ 900 m, Hawaii Volcanoes National Park, Hawaii. Shrubs in foreground are native *Dodonaea viscosa*, surrounded by a matrix of alien C4 grass species. Note large, dead *Metrosideros polymorpha* tree in middle ground (Photo RF Hughes, January 2013)

effectively compete with native species (D'Antonio et al. 1998) and alter both light regimes and soil nutrient dynamics (Hughes and Vitousek 1993; D'Antonio and Mack 2006). A more recent study (D'Antonio et al. 2011) documenting long-term patterns of post-fire succession in the absence of subsequent fire events demonstrated that replacement of native woody species by alien grasses, even in the absence of fire, persists over the long-term; results indicated that in spite of multiple 'fire-free' decades of post-fire succession, grasses maintained their dominance and native species failed to recover. As such, fire suppression by itself is inadequate to restore these systems to any sort of a native-dominated state.

As a consequence, the Hawaii Volcanoes management rule regarding wildfire – particularly in the mid-elevation seasonal fire management unit – has been one of active and concerted fire suppression (Hawaii Volcanoes National Park 2005). This is primarily in order to limit disturbance to, and mortality of, non-fire adapted native species and limit further proliferation of pyrophytic alien grasses. An exception to blanket suppression is in the dry coastal lowlands where fire effects studies and prescribed burns have demonstrated the positive effect of fire on the native grass *Heteropogon contortus* (Spear Grass; Tunison et al. 1994; D'Antonio et al. 2000). In these areas fire may be allowed to occur with minimal interference as a way to enhance cover of this native grass. Recently, large-scale efforts have been undertaken to plant a suite of native species that exhibit fire tolerant characteristics into burned areas. These planting efforts have met with success in terms of the establishment and survival of meaningful population sizes, and it is hoped that these relatively fire-tolerant native plant populations will persist and sustain themselves in the event that such areas experience successive fires in the future (Rhonda Loh, Hawaii Volcanoes NP, pers. comm.).

15.3.3   Mapping of Important Alien Plant Species

Adequate knowledge regarding the abundance, extent (i.e. hectares invaded) and distribution of invasive species is critical for developing effective control strategies and establishing workload requirements. Hawaii Volcanoes NP undertook a systematic programme to map the distribution of 38 widespread alien plant species in 1983–1985 (Tunison et al. 1992b). Results were successful in determining locations of many untreated populations, helping to assess feasibility of possible local eradication, shifted priorities, and showing that eight species were too widespread for control with the resources available – so that efforts for these species were shifted from a parkwide emphasis to control in selected areas with high biological value. Mapping and monitoring of an expanding suite of alien plant species continues at Hawaii Volcanoes NP. An important recent report (Benitez et al. 2012) reviews the expanding survey/mapping work and control history since 2000, reporting on results for 134 species surveyed by foot, vehicle and helicopter; 33 of those are widespread species and beyond park-wide control.

15.3.4   Park-Wide Control and Eradication of Localised but Potentially Problematic Alien Plant Species

In the 1980s, Hawaii Volcanoes NP adopted a strategy of controlling certain localised alien plant species on a park-wide basis while controlling widespread alien species in Special Ecological Areas (SEAs, see below). The purpose of the former effort has been to prevent the spread of potentially disruptive alien species while they are still manageable. Of the 41 species that were initially targeted, mapped, treated with appropriate herbicides and monitored, control of 21 was considered highly effective, with partial control for 17 additional species; three species were recalcitrant to control (Tunison and Zimmer 1992). By 2004, at least 15 of the initial species were considered eradicated and workloads reduced for most of the others (Timothy Tunison, Hawaii Volcanoes NP (retired), pers. comm.). The local eradication strategy has been extended opportunistically to prevent encroachment of new invaders, especially high impact species such as *Falcataria moluccana* (batai wood). A current analysis (Benitez et al. 2012) details progress/setbacks for 134 alien species (101 of them localised), including 16 species newly established in the past decade; the most problematic new species for attempted containment may be the shrub *C. hirta*, first detected in 2003, and *Cyathea* (= *Sphaeropteris*) *cooperi* (Australian tree fern), first detected in 2000. The Park's alien plant programme (including eradication/control of localised species and control in SEAs) has expanded significantly in scope and complexity over the past 3 decades. Since the early 1980s, the annual number of worker days spent in the field searching for and removing weeds has increased from <200 to >500 by the early 1990s and exceeds 1,200 worker days currently (Benitez et al. 2012).

15.3.5 Controlling All Disruptive Alien Plant Species in Selected High-Value Areas

In Hawaii Volcanoes NP, management units called Special Ecological Areas (SEAs) (Tunison and Stone 1992) were first established in 1985 to control 20+ highly disruptive invasive plant species recognised as too widespread for park-wide eradication to be feasible. SEAs are prioritised for intensive weed management based on their (i) ecological representativeness or rarity, (ii) manageability (accessible and with high recovery potential for native species), (iii) species diversity and rare species, and (iv) value for research and interpretation. Control methods varied from manual uprooting to chemical treatments depending on species. Typically, initial search and knockdown of weeds (knockdown phase) by control crews is followed by subsequent revisits (normally at 1–5 year intervals) as needed to keep infestations at low or manageable levels (maintenance phase) in SEAs.

The SEA concept has proved remarkably effective and flexible to date – whereas initial weed control may focus on only a few prime areas, the number and size of units can be expanded with time as opportunities become available. It has provided Hawaii Volcanoes NP with the ability to protect key biodiversity sites, even when alien plant species are uncontrollable on a large scale. The SEA programme started in 1985 with six SEAs and a total of 5,000 ha; by 2007, it had been expanded to 27 SEAs and 27,500 ha. Meanwhile, costs per ha declined from roughly $28/ha to $8.15/ha (Tunison and Stone 1992; Loh and Tunison 2009).

While we are not aware that any other protected area in the Pacific islands has formally adopted an SEA approach, many areas focus invasive plant control effort in areas where rare/endangered species are being threatened by plant invasions.

15.3.6 Local Eradication or Containment of Certain High-Impact Species

Fountain grass has for decades been considered one of the most disruptive alien plant species in Hawaii. It is believed capable of invading all Hawaii Volcanoes NP's plant communities, except closed rainforest, from sea level to 2,500 m elevation. It increases fuel loadings, thus increasing fire potential and most notably invades and reaches high densities on largely barren lava flows, which normally are pristine sites with few aliens, potentially making extensive new areas vulnerable to fire (Tunison 1992a). Hawaii Volcanoes NP's strategy in general has targeted problematic widespread species for biological control (believed to be unlikely for grasses) and/or for conventional control in Special Ecological Areas. However, a single exception was made, beginning in 1976 and intensified in 1983, for an 8,000 ha infestation of fountain grass, largely localised in the south-western corner of the Park, with outliers, especially along roads. The strategy for fountain grass has involved controlling all outlying populations, scouting the areas between these populations by helicopter, and controlling the periphery of the main infestation; the grass's pattern of spread suggested that such strategy could succeed (Tunison 1992a). Workloads were initially large and expensive but have gradually declined substantially in effort and cost over more than three decades of sustained effort (Benitez et al. 2012). At this point in time, the considerable investment to contain fountain grass seems justifiable and sustainable.

The only other case of such an ambitious strategy being applied in a Pacific island protected area may be the example of *F. moluccana* in the National Park of American Samoa (Case study 1).

15.3.6.1 Case Study 1: *Falcataria moluccana* at the National Park of American Samoa

Falcataria moluccana (= *Paraserianthes falcataria*, *Albizia falcataria*) is a very large, nitrogen-fixing tree of the legume family (Wagner et al. 1999). As an invasive species, it is daunting as the fastest growing tree species in the world, capable of 2.5 cm of growth per day (Walters 1971; Footman 2001). Individuals reach reproductive maturity by the age of four and subsequently produce copious amounts of wind dispersed seed (Parrota 1990). Canopies of single mature trees extend over 0.5 ha, and canopies of multiple trees commonly coalesce across multiple hectares up to square kilometres (Hughes and Denslow 2005).

Although valued by some in the Pacific, *F. moluccana* has become invasive in forests and developed landscapes across many Pacific islands. An archetypical early successional (i.e. pioneer) species, *F. moluccana* is generally found in mesic to wet forest environments and favours open, high light environments such as disturbed areas; its capacity to readily acquire nitrogen via its symbiotic association with *Rhizobium* bacteria makes it able to colonise even very young, highly N-limited lava flows such as those found on Hawaii Island (Hughes and Denslow 2005).

Previous research on the impacts of *F. moluccana* invasion on native Hawaiian forests demonstrated that wherever it invades, it utterly transforms the entire ecosystem by substantially increasing inputs of nitrogen, facilitating invasion by other weeds, while simultaneously suppressing native species. Hughes and

Denslow (2005) described the impacts of *F. moluccana* invasion on some of the last intact remnants of native wet lowland forest ecosystems undergoing primary succession in Hawaii. Nitrogen inputs via litterfall were 55 times greater in *F. moluccana* stands compared to native-dominated forests (Hughes and Denslow 2005), and at 240 kg N ha^{-1} year^{-1}, were commensurate to typical fertilizer N inputs of industrialised corn cropping systems of the US Midwest (Jaynes et al. 2001). Changes in nutrient status coincided with dramatic compositional and structural changes as well; *Falcataria moluccana* facilitated an explosive increase in densities of understory alien plant species, particularly *Psidium cattleianum*. In contrast, native species, especially the keystone over-story tree *Metrosideros polymorpha* suffered widespread mortality to the point of effective elimination from forests that they formerly dominated. Even where *F. moluccana* populations are killed and/or removed, native species are typically challenged to grow quickly enough to outpace the rapid recruitment of abundant *F. moluccana* seedlings that promptly germinate in response to increased understory light availability. Based on these findings, Hughes and Denslow (2005) concluded that the continued existence of native-dominated lowland wet forests in Hawaii largely will be determined by the future distribution of *F. moluccana*. As such, detection and control of individuals or small numbers of *F. moluccana* has become the default approach in protected areas such as Hawaii Volcanoes NP.

In American Samoa *F. moluccana* has invaded large areas of the forests of the National Park (NPSA) and neighbouring areas on Tutuila Island. The species likely was first introduced to Samoa on the island of Upolu perhaps as early as the 1830s and is thought to have spread to Tutuila Island in the early 1900s. By the 1980s *F. moluccana* was noted as locally common within NPSA boundaries (Whistler 1980, 1994) and by 2000, approximately 35 % of Tutuila Island (~6,725 ha) including much of NPSA, was infested with *F. moluccana*. This prompted NPSA efforts to begin aggressive measures to control this species (Fig. 15.3; Hughes et al. 2012). Research results indicate that *F. moluccana* displaces native Samoan trees; although aboveground biomass of intact native forests did not differ from those invaded by *F. moluccana*, greater than 60 % of the biomass of invaded forest plots was accounted for by *F. moluccana*, and biomass of native species was significantly greater in intact native forests. Following removal of *F. moluccana* (i.e. killing of mature individuals), a number of native Samoan trees grew rapidly, filling the resulting light gaps, achieving secondary succession without a reinvasion of *F. moluccana*. The presence of successional native tree species appeared to be the most important reason why *F. moluccana* removal is likely a successful management strategy. Once *F. moluccana* is removed, native tree species grow rapidly, exploiting the legacy of increased available soil N – left from *F. moluccana* litter inputs – and available sunlight. In addition, recruitment by shade intolerant *F. moluccana* seedlings was severely constrained to the point of being non-existent, likely a result of the shade cast by re-establishing native trees in management areas (Hughes et al. 2012). Like many Pacific islands, American Samoa commonly experiences large-scale, cataclysmic disturbances from cyclones (Mueller-Dombois and Fosberg 1998) as

Fig. 15.3 The large, non-native, N-fixing tree *Falcataria moluccana* has spread over extensive areas in and around the National Park of American Samoa. The Park has responded with a remarkably successful campaign of killing trees in place and relying on reproduction and growth of pioneer-type native tree species to quickly fill resulting light gaps. Photo from June 2006, 8 months after girdling of the *F. moluccana* trees (Photo Tavita Togia, NPSA)

well as more frequent but less cataclysmic disturbances in the form of cyclone 'near misses', tropical storms, and tropical depressions. Collectively these create what has been termed a 'chronic disturbance' regime (Webb et al. 2011), which has played a potent evolutionary role in shaping the composition, structure, and function of Samoa's native forests (Webb et al. 2006). Indeed, nearly 40 % of the common native trees in forests of American Samoa could be classified as successional (i.e. regenerating readily in disturbed forest) (Hughes et al. 2012). This is a critical evolutionary feature for the forest species of American Samoa, and one that makes large-scale control of *F. moluccana* feasible. Further, this scenario stands in stark contrast to results from experimental removal of the alien, N2-fixing tree, *Morella faya* (faya tree) from forests of Hawaii Volcanoes NP, where successful reestablishment and recovery of native forest species following control of the *M. faya* is much less certain given the presence of highly competitive alien species and the relatively slow-growing character of the native species – species that have not evolved in such a frequent storm disturbance environment (Loh and Daehler 2008).

Yet, if native Samoan tree species are so well adapted to small, forest gap forming disturbances, as well as large-scale disturbances such as cyclones, how did *F. moluccana* attain recent dominance in forest stands in the first place? Moreover, and what is to keep it from returning to dominance following disturbance in the future? Answers can be found in the growth characteristics of

F. moluccana. Because this species becomes very tall, very quickly (Walters 1971; Parrota 1990), and its seedlings accompany recruitment of native species, *F. moluccana* will likely outpace other species in the race to canopy dominance. In addition, since mature *F. moluccana* individuals attain heights well above those exhibited by most of the native Samoan tree species (Whistler 2004), *F. moluccana* will maintain overstory canopy dominance. As long as cyclones occur at sufficient frequencies, *F. moluccana* populations will likely persist and expand in the absence of on-going management practices. However, removal of mature *F. moluccana* individuals, re-establishment of native Samoan tree species, and exhaustion of the *F. moluccana* seed bank prior to subsequent large-scale disturbances may suffice to break the cycle of *F. moluccana* establishment and proliferation.

15.3.7   Develop Optimally Effective Herbicidal Control Methods

Any protected area in the Pacific that seriously addresses plant invasions must use herbicides effectively as part of the overall strategy. Hawaii Volcanoes NP conducted formal experiments in the 1980s to develop a safe, effective and efficient arsenal of treatments appropriate for a large number of target species for which there were no standard treatments (e.g. Santos et al. 1992). In some cases standard methods may not be sanctioned by National Park Service or other protected area guidelines. Over the past three decades, standard treatments have become increasingly available (e.g. Langeland and Stocker 1997; Motooka et al. 2002). Common active ingredients used in natural area weed management have included 2,4-D, triclopyr, glyphosate, metsulfuron methyl, imazapic and imazapyr; picloram and hexazinone were proven to be effective in Hawaii as well, but have less utility now due to restrictions in their use (James Leary, University of Hawaii, pers. comm.). The new herbicide active ingredients aminopyralid and aminocyclopyrachlor are proving to be highly effective on many target weed species in Hawaii, particularly those in the Fabaceae or legume family (J. Leary, pers. comm.). Individual plant treatment techniques include directed foliar applications, basal bark applications and basal injections. Additionally, on-going research in Hawaii has identified effective injection techniques for 16 invasive woody species so far (e.g. *F. moluccana*), where effective doses are less than 0.5 g active ingredient for large mature specimens (J. Leary, unpubl. data). Furthermore, an herbicide injection is a clean, safe, and efficient technique for delivering very small aliquots of concentrated formulations directly to the target. Besides proven target efficacy, this provides a practical use pattern in remote natural areas where a total payload that weighs a fraction of one kilogram is enough to treat hundreds of individuals in an all-day effort.

Benitez et al. (2012) summarise current herbicide treatments used for targeted alien plant species at Hawaii Volcanoes NP.

15.3.8 Facilitate Biological Control for Otherwise Intractable Species

Biological control was recognised in the early 1980s as a potentially major component of effort to address the most severe plant invasions in national parks and other protected areas in Hawaii. Hawaii already had a long history (dating back to about 1900) of biological control to assist against insect pests to protect agriculture, and some plants had been successful targets. The National Park Service (NPS) entered an agreement in the early 1980s with the U.S. Forest Service, the Hawaii Departments of Land and Natural Resources and of Agriculture, and University of Hawaii to "intensify biological control efforts on forest pests in the State" (Tunison 1992b). A quarantine (containment) facility was constructed at 1,200 m elevation at Hawaii Volcanoes NP for plant biocontrol using insects; the facility was completed and became operational in 1984 (Markin et al. 1992). Hawaii Department of Agriculture (HDOA) had the only other biocontrol containment facility in the state in Honolulu, which was capable of accommodating insects and fungal agents for testing (Markin et al. 1992).

Some details of progress (and lack thereof) with biocontrol are given for *Morella* (= *Myrica*) *faya* and *Hedychium gardnerianum* in Case study 2 and 3, respectively. The challenges of conducting a biocontrol programme in the public arena are raised below. Experience with *Clidemia hirta*, another high-priority target as one of Hawaii's most aggressive invaders, has been particularly discouraging; of 17 agents tested for host specificity in Hawaii, six arthropods and one fungal agent were field released: a thrips (1953), a fungus (1986), a beetle (1988), and four moths (one in 1970, three in 1995) (Conant 2002). Five established, but none were truly successful. Biotic interference (by alien ants and/or parasitoids) has been demonstrated for the thrips and strongly suspected for the moths (Conant 2002). DeWalt et al. (2004) reported observing promising potential biocontrol agents in *C. hirta*'s native range in Costa Rica. A nematode (*Ditylenchus gallaeformans*), under consideration for biocontrol of closely-related *M. calvescens* (miconia), may be the most promising agent for *C. hirta* in the long run (Johnson 2010, Tracy Johnson, U.S. Forest Service, pers. comm.); Hawaii has no native melastomes.

Experience with the recent release of an extremely important biocontrol agent for strawberry guava provided meaningful insights into the importance (and limitations) of public outreach. After 15 years of exploration and testing in Brazil (Wikler and Smith 2002), *Tectococcus ovatus* (Homoptera: Eriococcidae), a leaf-galling scale insect, was brought to Hawaii for intensive experimental testing to ensure its safety as a biocontrol agent; Tracy Johnson of the U.S. Forest Service conducted Hawaii-specific laboratory testing in the Hawaii Volcanoes NP biocontrol facility, beginning in 1999. In 2005, a release permit application was issued. The Hawaii Board of Agriculture held public meetings in 2005–2007, and federal and state permits for release were obtained in early 2008. However, Johnson then applied for additional permission for release of *T. ovatus* on state land (in order to facilitate intensive post-release monitoring), which triggered the need for a state

Environmental Assessment (Warner and Kinslow 2011). After an unexpected additional 3 ½ years of contentious, 'high-profile' interaction, primarily with a local (Hawaii-island) critic and his supporters (described in some detail by Warner and Kinslow 2011), *T. ovatus* was finally released on state land near Hawaii Volcanoes NP in December 2011. Whereas there was organised opposition on Hawaii island, public opinion on Maui was generally positive toward the release.

Although biological control of plant invaders has proved very effective in some countries of the world (see Van Driesche and Center 2014, this volume for a synthesis), it has had few successes for helping protected areas in Hawaii during nearly three decades of good intention and very significant effort. In general, after initial enthusiasm for biocontrol was tarnished by some early failures and dead ends, there has been an apparent tendency among key agencies to fund on-the-ground mechanical/chemical plant control more generously than biological control. The two existing containment facilities are far less than adequate to accommodate state-wide needs, especially given that HDOA biocontrol efforts are primarily targeted at agricultural pests. The regulatory process for biological control has become more rigorous, and some would say unnecessarily slow, perhaps especially for Hawaii (Messing and Wright 2006). Most recognise the importance of much increased efforts to monitor the fate of biocontrol releases (e.g. Denslow and D'Antonio 2005). Biological control continues to have much promise for addressing Hawaii's most serious invasive plant issues in protected areas and elsewhere, but an injection of major resources (not easily obtained, especially in harsh economic times) will be required for sustained success.

15.3.8.1 Case Study 2: *Morella faya*

Morella (= *Myrica*) *faya* is a small evergreen tree (5–10 m tall), an actinorrhizal nitrogen-fixer, native to the Canary Islands, Azores, and Madeira in the North Atlantic. It was brought to Hawaii by Portuguese immigrants in the 1880s, probably as an ornamental, and later was planted on multiple islands by the Territorial Department of Forestry for watershed reclamation in the 1920s and 1930s. Its aggressive invasiveness was recognised by the 1930s, by which time it was present on five of the six major Hawaii islands. The worst invasion is in Hawaii Volcanoes NP, where it colonises early successional open-canopied forests, on young volcanic substrates, achieving drastic alteration of N-levels and eventually forming nearly monospecific, closed canopy stands (Vitousek and Walker 1989). Feral pigs may have assisted its spread (before localised pig eradication), but ample bird-dispersal is highly efficient (Woodward et al. 1990). *Morella faya* increased from one tree found in 1961 to an estimated 12,200 ha by 1985, to 15,800 ha by 1992, and 30,495 ha at present, with about half of that total comprised of dense infestations (Benitez et al. 2012). Analysis of airborne imaging spectroscopy, focusing on a 1,360 ha forested area (originally endemic *Metrosideros polymorpha*) in the heart of Hawaii Volcanoes NP at 1200 m elevation, suggests that about 28 % of the landscape is dominated by *M. faya*, with an additional 23 % undergoing transformation as *M. faya* grows into the canopy (Asner and Vitousek 2005). In these sites, the rate of diameter growth is 15-fold more rapid than that of *M. polymorpha*; the rate of nitrogen-fixation in dense *M. faya* stands was measured at 18 kg N ha^{-1} year^{-1}, resulting in soil N-levels of about five times that of uninvaded *M. polymorpha* stands (Vitousek and Walker 1989).

Morella faya is apparently seriously invasive only in Hawaii, though it has been introduced to Australia, New Zealand and elsewhere. It has been a target of biological control, one of the first under Hawaii's interagency biocontrol agreement. An expedition to the *M. faya* home range in 1984 was relatively unsuccessful in finding promising agents. However, two moth species were brought back to Hawaii and tested, and one was released but found ineffective (Smith 2002).

An invasive leafhopper (*Sophonia rufofascia*) from Asia that was first recorded in Hawaii in 1987 has a very broad host range and attacks both *M. faya* and (to a lesser extent) *M. polymorpha*. Negative effects of the leafhopper on *M. polymorpha* are damaging and have been shown to be more severe where there is adjacent *M. faya* (Lenz and Taylor 2001); *M. faya* has shown significant mortality. Hawaii Volcanoes NP has explored options for optimal control of *M. faya* and concluded that girdling *M. faya* trees and leaving the dead trees in place is the most promising methodology for restoring native species (e.g. Loh and Daehler 2007).

15.3.8.2   Case Study 3: *Hedychium gardnerianum*

Hedychium gardnerianum (Fig. 15.4) is a serious invader of rainforests at both Hawaii Volcanoes and Haleakala NPs as well as elsewhere in the Hawaiian Islands. Its case is unusual in that the first known collection in the state was made in the Hawaii Volcanoes NP employees' housing area in 1943; it apparently became well-established during the 1960s–1980s (Linda Pratt, U.S. Geological Survey, pers. comm.). It is a large herb (1–3 m tall) that occupies about 3,000 ha (Benitez et al. 2012) at 750–1,300 m elevation in Hawaii Volcanoes NP and tends to establish a monospecific understory, gradually smothering native understory species and generally preventing native tree seedling recruitment (Minden et al. 2010a, b). The invasive tree *Psidium cattleianum* is the only species that appears able to reproduce successfully through a dense thicket of *H. gardnerianum* (Minden et al. 2010b). Experience at Haleakala NP (where it was discovered in the mid-1980s and is rapidly spreading in spite of control efforts) shows that *H. gardnerianum* is fully capable of smothering rare Lobeliaceae (Stephen Anderson, Haleakala NP, pers. comm.). Analysis of airborne imaging spectroscopy, combined with ground-based analyses, indicated that *M. polymorpha* forest over-stories had significantly lower leaf N concentrations in areas with *H. gardnerianum* understory – likely a competitive effect (Asner and Vitousek 2005).

Fig. 15.4 *Hedychium gardnerianum*, a large herbaceous ginger from the Himalayan region, is becoming an increasingly serious invader of Hawaiian middle and high-elevation rain forests, capable of smothering and preventing reproduction of most other species in invaded stands (Photo RF Hughes, January 2013)

Hedychium gardnerianum, native to high elevations of Nepal and India, is a serious invader in many rainforest areas worldwide, especially on islands (Kueffer et al. 2010a). Exploration and preliminary testing for biocontrol agents is underway through a multi-country collaborative project with CAB International (Djeddour and Shaw 2011).

15.3.9   Support/Encourage Research on Biology and Control Strategies of Alien Plant Species

Scientific research regarding the ecology, impact and control of alien plant species is critical for intelligent, efficient, and effective management of protected areas. It can be instrumental in helping to understand and document the nature and relative threat posed by respective alien species; in its absence, managers risk basing management decisions on inaccurate conventional wisdom, assumptions, and hearsay. A relatively rich collection of scientific literature has helped inform and prioritise management efforts in Hawaii Volcanoes NP. Vitousek and Walker (1989) and Asner and Vitousek (2005) documented ecosystem-scale influences of *Morella faya* invasions, and Loh and Daehler (2007, 2008) addressed successional

trajectories following *M. faya* removal. Huenneke and Vitousek (1990) documented
mechanisms of *Psidium cattleianum* invasion and their impact for native forest
management. Asner and Vitousek (2005) and Minden et al. (2010a, b) elucidated
the community and ecosystem impacts of *Hedychium gardnerianum* invasion
into Hawaii's native forest based on remote sensing and plot-level investigations.
La Rosa (1992) described the characteristics of the alien vine, *Passiflora tarminiana*
(= *P. mollissima*, banana passion flower/banana poker) as well as the mechanisms of
its invasion into mesic to wet native forests. The grass fire cycle has been the topic of
numerous publications that have helped managers determine the most appropriate
approaches to protect native woodland/shrubland ecosystems, initially by Hughes
et al (1991) and most recently D'Antonio et al. (2011). In Haleakala NP, Medeiros
(2004) addressed patterns and mechanisms of *P. cattleianum*, *H. gardnerianum*,
and *Clidemia hirta* invasion. Diong (1982) addressed the synergy of feral pigs with
increased dominance of *P. cattleianum* in Haleakala NP. Elsewhere, research in
Tahiti (Meyer 1996a, 2010 and many other papers) has been very important in
alerting Hawaii and for developing strategies against *M. calvescens*. Research by
Hughes and Denslow (2005) concerning Hawaii and Hughes et al. (2012) concerning
American Samoa – discussed in Case study 1 – illustrates particularly well how the
importance of understanding certain plant invasions in depth can illuminate strategies
for managing those invasions when circumstances are right.

Despite a rich literature documenting the threats posed by invasive alien species,
recent research has addressed the supposition that alien species may prove unavoidable components of island ecosystems and should be "embraced" where appropriate
(Hobbs et al. 2006, 2009). Lugo (2004) and Kueffer et al. (2010b) demonstrated in
forests of Puerto Rico and the Seychelles, respectively, the potential usefulness of
alien trees for providing suitable conditions for native plant recruitment following
major anthropogenic disturbances such as deforestation associated with mining or
conversion to agriculture. Clearly, any potential benefits incurred from alien plant
species will be determined by the particular characteristics of the species as well as
the characteristics of the ecosystems they happen to inhabit, and costs and/or benefits
should be evaluated on a case by case basis with an underpinning of ecological
understanding of that alien plant/ecosystem interaction. It appears very unlikely that
such aggressive invaders in the Pacific as *F. moluccana*, *H. gardnerianum*, and
M. calvescens will ever prove beneficial for native biodiversity, but keeping open
minds to conservation possibilities for 'novel ecosystems' may be warranted.

15.3.10   Education Within Agencies and Outreach
to the Public

Resource managers at Hawaii Volcanoes and Haleakala NPs have tried to thoroughly educate co-workers in the National Park Service so that they will continue to

support alien plant control programmes. Alien plant problems are a minor theme of some park interpretive programmes that emphasise geological, cultural, and other biological messages to the public. The Resources Management Division at Hawaii Volcanoes NP has an active volunteer programme but has been more successful at generating interest in assignments with wildlife (especially sea turtles) than with alien plant control activities (Tunison 1992b, R. Loh, pers. comm.). On the whole, conservation groups and scientists are well informed about alien plant problems, but it has been difficult to interest the local public about damage caused by alien plants and the need to control them. One reason may be that Hawaii Island (although it has native ecosystems more accessible to the public than any other island) has had a largely rural and small-town culture with little interest in conservation (Tunison 1992b). Fortunately, that situation has been improving as schools at every level have begun including Hawaiian natural history topics in their syllabi. There is much greater awareness of invasive species issues for communities in and around the Park than in more distant parts of Hawaii Island (R. Loh, pers. comm.). Establishment and progress of Watershed Partnerships and the Big Island Invasive Species Committee over the past two decades have also stimulated considerable interest. On the island of Maui, public support for combating alien plants may be more developed, at least partly because of perceived need for watershed protection. (Hawaii Island has relatively few watersheds in relation to its size, because of the relatively gentle topography and porous substrates.) Maui County funding has been very important for the campaign to combat *M. calvescens* on East Maui, for example, where the species threatens not only biodiversity but ecosystem services in watersheds. At a state-wide level, the population is largely urban (mostly on Oahu island); many citizens of the state may have little contact with or knowledge of Hawaii's endemic biota.

The control of *F. moluccana* in the NPSA of Tutuila Island (Case study 1) provides an instructive example of how to effectively involve the public in a meaningful manner that both builds substantive support and accomplishes stated objectives to protect native biodiversity. From the outset, managers addressed the need for action, for example, control of *F. moluccana* populations in national park boundaries - with the surrounding village chiefs who are the relevant bodies of authority (T. Togia, NPSA, pers. comm.). Second, widespread public knowledge of, and support for, the control effort was cultivated through the use of local media outlets on a consistent basis. Lastly, and perhaps most importantly, the majority of funds acquired for *F. moluccana* control were dedicated to the employment of large numbers of young people from the respective surrounding villages to actually carry out the control efforts. This approach created a truly meaningful connection between the villages at large and execution of control efforts, galvanizing strong and tangible support in a manner that would be difficult to engender by other means. To date, NPSA field crews have killed over 6,000 mature trees and restored approximately 1,500 ha of native Samoan forest in the process. This is a model that bears consideration when contemplating invasive species control efforts elsewhere in the Pacific.

15.3.11   Collaborative Work with Other Agencies to Try to Address the Alien Plant Problem at Its Roots

15.3.11.1   Early Detection and Rapid Response Outside Protected Area Boundaries

Individuals in Hawaii became aware in the late 1970s and early 1980s that an aggressively invasive tree, *M. calvescens*, with likely serious implications for extinguishing biodiversity, was undergoing a rampant invasion in Tahiti and that the species was already present and spreading on the island of Hawaii. They were unfortunately unable to stimulate sustained agency or collective action until about 2 years after the species was discovered by a Haleakala NP employee on East Maui, about 8 km from the Park's remote Kipahulu Valley rainforest, in 1988. By the time a sustained alarm was raised, increasing information was becoming available from Tahiti's *M. calvescens* invasion (culminating in the accounts of Meyer 1996b and Meyer and Florence 1996). Based on the behaviour of the species in Tahiti, it was concluded that remote Kipahulu Valley's rainforest would not be defendable from a wave of *M. calvescens* invasion, so that a more proactive strategy was urgently needed. Haleakala NP employees first conducted surveys and removal efforts in 1991, other agencies helped, and a small interagency organization, the "Melastome Action Committee" (later the Maui Invasive Species Committee, MISC) coalesced, to publicise the problem and try to marshal resources to address the local and state-wide invasion of *M. calvescens* and other serious weeds. It became evident that partnerships are the only opportunity to have a chance of dealing effectively with such enormous shared threats. From the beginning, containment of *M. calvescens* was regarded as a holding action until biological control, first investigated in 1993, could become available (Medeiros et al 1997). Two decades later, *M. calvescens* is under containment by MISC on windward East Maui (only a few small plants have been found and removed in Haleakala's lower Kipahulu Valley over the years), but at considerable sustained cost (Meyer et al. 2011). Haleakala NP has been a major funding source and provider of expertise and manpower for *M. calvescens* containment. Unfortunately, though *M. calvescens* biocontrol efforts are progressing in recent years and can soon put forward multiple agents (Johnson 2010), biocontrol has not received nearly the level of resources that on-the-ground control has received. The best hope is for continued *M. calvescens* containment as biocontrol agents are released and monitored on Maui to assure that Maui's *M. calvescens* invasion is not allowed to 'explode' before the biocontrol agents can take over the job.

The concept of "stopping the next miconia" (i.e. locating and targeting potentially serious plant invasions early, while eradication is still a possibility, wherever on an island they arise) has gained traction in Hawaii in the past decade, especially on Maui, with development of pragmatic early detection methodology and actual eradications by MISC (Loope et al. 2004; Kueffer and Loope 2009; Penniman et al. 2011). Kraus and Duffy (2010) have described Hawaii's current relatively effective "existing functional management model for the eradication of incipient populations of invasive species that avoids reliance on official governmental response". The individual island-based model "involves formation of informal multi-partner committees that utilise outside funding to achieve pest-management goals". The model has evolved from the efforts to address the *M. calvescens* problem, first on Maui and later on other islands.

French Polynesia has devoted substantial education and regulatory effort to prevent *M. calvescens* from spreading (seeds can be easily spread via contaminated construction equipment, hiking boots, etc.) from the hub of Tahiti to other high islands. As a result, no other island has been invaded since 1997 (Meyer 2010). The National Park of American Samoa has conducted public outreach with the aim of reducing the spread of well-known plant invaders to its islands as well as eliminating small populations of barely established species such as *P. cattleianum* (T. Togia, NPSA, pers. comm.) The Pacific Invasives Learning Network (PILN, www.sprep.org) and collaborating organizations have encouraged similar efforts throughout the Pacific.

15.3.11.2 State-Wide and Countrywide Efforts at Reducing New Invasions

Given Hawaii's high vulnerability to invasions, there is an obvious need to continually reassess possibilities to improve Hawaii's network for prevention of new invasions of all taxa. Hawaii's Coordinating Group on Alien Pest Species (CGAPS, www.hawaiiinvasivespecies.org/cgaps/) was launched in 1995 (Holt 1996) with hopes of "an alliance of biodiversity, health, agriculture, and business interests for improved alien species management in Hawaii". The CGAPS has evolved into an important voluntary forum of 14 State, Federal, and private organizations directly involved in or with a major stake in invasive species prevention and/or management in Hawaii. It is well integrated with related efforts such as the relatively new and governmental Hawaii Invasive Species Council (HISC, www. hawaiiinvasivespecies.org/hisc/). Interagency communication has been greatly facilitated CGAPS, and progress is continually taking place. However, it must be said that its task is Herculean, given the forces of globalization, Hawaii's vulnerability to invasions, and perhaps inevitably fragmented government response (CGAPS 2009; Kraus and Duffy 2010). Notably, there has been exceptionally little progress in the past two decades in implementation of regulations to restrict high-risk plant imports into Hawaii, in spite of such efforts as the Hawaii-Pacific weed risk assessment (Daehler et al. 2004; Denslow et al. 2009), good public information on the subject, support from an influential segment of the plant industry (Kueffer and Loope 2009), and even seemingly good support in the state legislature (Mark Fox, The Nature Conservancy of Hawaii, pers. comm.).

The Pacific Invasives Initiative (PII, www.pacificinvasivesinitiative.org) and associated programmes are working to facilitate improvement of biosecurity capacity in Pacific island countries.

15.4 Overview of Plant Invasions of Protected Areas in South-Eastern Polynesia

We use SE Polynesia as a region that may be somewhat representative of the vast complexity of Pacific islands and their protected areas, though obviously this is an oversimplification. Meyer (2004) has reviewed the status of the invasive plant threat to native flora and vegetation of the region. The islands of SE Polynesia include French Polynesia (a French overseas territory formed by five archipelagos, namely the Australs, Marquesas, Society, Tuamotu and Gambier, and comprising 120 islands), Cook Islands (an independent country in free association with New Zealand, 16 islands), Pitcairn Islands (a UK overseas territory with four islands, namely Pitcairn, Henderson, and Oeno and Ducie atolls), and Easter Island or Rapa Nui (a Chilean territory). These archipelagoes are comprised of relatively small tropical islands, the largest being Tahiti (1,045 km^2) in the Society Islands. Except for the National Park of Rapa Nui (6,660 ha) created in 1930 and a World Heritage Cultural Site since 1995, and the uninhabited raised atoll of Henderson (3,700 ha) declared a World Heritage Natural Site in 1988, there are few protected areas in SE Polynesia (Table 15.1). Only one is found in the main island of Rarotonga (Cook Islands), the "Takitumu Conservation Area," which is a community-based management area of about 155 ha. There are eight in French Polynesia with different protection status (natural parks, reserves, 'management areas', and one UNESCO Biosphere Reserve comprising seven atolls in the Tuamotu), but their size is relatively small (total area of about 10,000 ha, less than 2 % of French Polynesia's terrestrial surface).

Unfortunately, the number of invasive alien plants is high in many islands; some of them are dominant such as the small tree *M. calvescens* in low to mid-elevation rainforest in Te Faaiti Natural Park; the thorny shrub *Acacia farnesiana* (klu bush, kolu) in dry coastal areas in Eiao Management Area, a remote islet in the northern Marquesas; the small tree *Rhodomyrtus tomentosa* (downy rose myrtle) on the high elevation wet plateau (400–800 m) of Temehani Ute Ute Management Area (Fig. 15.5); the shrubs *Ardisia elliptica* (shoebutton) and *Cestrum nocturnum* (night-blooming jasmine) in the lowland rainforests of Takitumu Conservation Area. Most of Rapa Nui National Park is covered by grassland, which has become invaded by *M. minutiflora*, (Fig. 15.6), *Cirsium vulgare* (bull thistle), *Crotalaria grahamiana* (rattle-pod) and *Psidium guajava* (common guava). In addition to the presence of grazing and browsing alien ungulates in some reserves, natural disturbances such as cyclones may also facilitate plant invasions such as the vines *Merremia peltata* (merremia), *Mikania scandens* (climbing hempweed) and *Cardiospermum grandiflorum* (balloon vine) in the Takitumu Conservation Area (E. Saul, pers. comm., Fig. 15.7), or the spiny shrub *Rubus rosifolius* (thimbleberry) and the African tulip tree *Spathodea campanulata* (African tulip tree) in Te Faaiti Natural Park.

There are very few programmes to manage invasive alien plants in protected areas in SE Polynesia, perhaps because the proportion of protected areas occurring across the areas is commensurately small. Manual and mechanical control of the

Fig. 15.5 Temehani Plateau, Raiatea, French Polynesia. Invasion of native shrubland by the alien shrubs *Chrysobalanus icaco* and *Rhodomyrtus tomentosa* (Photo J-Y Meyer, April 2009)

Fig. 15.6 Landscape dominated by the alien grass *Melinis minutiflora* near Anakena, Rapa Nui National Park, Rapa Nui (Easter Island) (Photo J-Y Meyer, February 2012)

Fig. 15.7 Invasion of native rain forest near Te Manga, island of Rarotonga (Cook Islands), South-eastern Polynesia, by the alien vines *Merremia peltata*, *Mikania micrantha*, and *Cardiospermum halicacabum* (Photo J-Y Meyer, October 2009)

weeds *Asclepias curassavica* (Mexican butterfly weed), *Cenchrus clandestinus* (= *Pennisetum clandestinum*, kikuyu grass), *Crotalaria grahamiana* (bushy rattlepod), and *Melinis minutiflora* has started in Rano Raraku in 2011 by Rapa Nui National Park (CONAF) in collaboration with the international branch of the French National Forestry Office (ONF-International). A pilot project to eradicate the thorny tree *Robinia pseudoacacia* (black locust) in the Rano Kau crater (Fig. 15.8) has been launched in 2012 (Meyer, unpub. data), and habitat restoration by reintroducing native and endemic plants (including the famous *Sophora toromiro* (toromiro) which was extinct in the wild) is planned. A biological control programme to contain *M. calvescens* with a fungal pathogen *Colletotrichum gloeosporioides* forma specialis *miconiae*, successfully released (Fig. 15.9) in Tahiti in 2000, has resulted in the recruitment of native and endemic plants in the understory of heavily invaded upland rainforests, such as in Te Faaiti Natural Park (Meyer et al. 2011).

Fig. 15.8 Dense forest of invasive alien *Robinia pseudoacacia*, RanoKau crater, Rapa Nui National Park, Rapa Nui (Easter Island) (Photo J-Y Meyer, February 2012)

Fig. 15.9 Sapling of the highly invasive alien tree *Miconia calvescens* showing effects of the biocontrol fungal pathogen *Colletotrichum gloeosporioides* forma specialis *miconiae*, Fare Mato, Tahiti, French Polynesia (Photo J-Y Meyer, November 2009)

15.5   Conclusions

From an invasive plants standpoint, protected areas in the Hawaiian Islands, compared with those of other Pacific island governments, have had the advantage of a relatively favourable economic situation. They also have the disadvantages of a highly globalised economy, a history of purposeful alien plant introductions with an astounding 10,000 species introduced to the islands between 1906 and 1960 (Woodcock 2003), and a society that values horticultural novelty (Meyer and Lavergne 2004; Denslow et al. 2009; Kueffer et al. 2010a). Most protected areas in Pacific islands outside Hawaii have little invasive plant management, but may often have fewer plant invasions as a result of being 'off the beaten path', though that may not remain the case in the future. Increased regulation pertaining to the importation of alien plant and animal material is called for in addition to increased enforcement of existing importation regulations.

Hawaii has provided, and continues to provide, a useful laboratory for experiencing, understanding, and combating invasions. Hawaii Volcanoes NP has led conceptually and by example, and the strategies adopted there 2 or 3 decades ago serve as credible approaches for slowing invasions though not stopping them. However, there are largely intractable situations of the grass-fire cycle and certain invasive plant species. These problems are enormous, and the inability of the conservation community to work successfully through the political process to find a practical strategy for substantially reducing continued plant invasions has been a major disappointment, though this shortcoming is by no means unique to Hawaii and Pacific islands. Fortunately, improved herbicide technology offers promising tools for eradications and control as well for resurrection of native biodiversity after catastrophic invasions (e.g. Baider and Florens 2011).

Accelerated efforts involving biocontrol are warranted, along with more restrictive plant material importation rules and enforcement to help stem the tide of the most serious invasions and reduce the probability of new ones. If Hawaii's biocontrol efforts can thrive in coming decades, the prospect of enhanced international collaboration with Pacific island biocontrol projects is appealing.

Historically, the common conservation strategy has been to set up protected areas in sites with high ecological values, often in pristine habitats or remnants of natural ecosystems where native and endemic species are dominant. In some cases in the Pacific Islands, however, protected areas have been set in historically human-disturbed areas (e.g. in South-eastern Polynesia, lowland rainforests have been modified by Polynesians during the past 1,000 years) or are currently found in alien-dominated forests as a result of plant succession after anthropogenic or even natural disturbances (such as fires, floods or cyclones). Innovative conservation strategies in these 'novel' ecosystems (sensu Hobbs et al. 2006, 2009) should be explored where possible and appropriate.

A major question for the future, however, is whether plant invasions will become even less manageable and more problematic in protected areas of Pacific Islands with increasing manifestations of climate change. Oceanic islands may benefit to some degree from moderation of the rate of change by maritime influences, in comparison with continental areas. Nevertheless, some, and perhaps many, invasive plant species on islands will likely have an extra edge in competitive ability due to increased CO_2 availability, disturbance from extreme climate events, and ability to invade higher elevation habitats as climates warm (Bradley et al 2010).

Acknowledgements We thank Rhonda Loh for advice on the grass-fire section and several other sections, David Benitez for advice on the status of mapping of invaders, James Leary for advice for the herbicides section, and Christoph Kueffer for numerous suggestions for improving the manuscript. We thank our respective agencies for supporting our participation in writing this chapter.

References

Anderson SJ, Stone CP (1993) Snaring to control feral pigs *Sus scrofa* in a remote Hawaiian rainforest. Biol Conserv 63:195–201
Asner GP, Vitousek PM (2005) Remote analysis of biological invasion and biogeochemical change. Proc Natl Acad Sci U S A 102:4383–4386
Baider C, Florens FBV (2011) Control of invasive alien weeds averts imminent plant extinction. Biol Invasions 13:2641–2646
Benitez DM, Loh R, Tunison T et al (2012) The distribution of invasive plant species of concern in the Kilauea and Mauna Loa strip areas of Hawaii Volcanoes National Park, 2000–2010. Technical report 179. Pacific Cooperative Studies Unit, University of Hawaii, Honolulu
Bradley BA, Blumenthal DM, Wilcove DS et al (2010) Predicting plant invasions in an era of global change. Trends Ecol Evol 25:310–318
Brooke de M L, Hepburn L, Trevelyan RJ (2004) Henderson Island world heritage site management plan 2004–2009. Foreign and Commonwealth Office, London, 40 pp
CGAPS (Coordinating Group on Alien Pest Species) (2009) CGAPS vision and action plan. http://www.hawaiiinvasivespecies.org/cgaps/pdfs/20100204cgapsactionsbyagency.pdf
Cole RJ, Litton CM, Koontz MJ et al (2012) Vegetation recovery 16 years after feral pig removal from a wet Hawaiian forest. Biotropica 44:463–471
Conant P (2002) Classical biological control of *Clidemia hirta* (Melastomataceae) in Hawaii using multiple strategies. In: Smith CW, Denslow J, Hight S (eds) Proceedings of workshop on biological control of native ecosystems in Hawaii. Technical report 129. Pacific Cooperative Studies Unit, University of Hawaii, Honolulu. pp 13–20
Cox PA, Elmqvist T (1991) Indigenous control of tropical rain-forest reserves: an alternative strategy for conservation. Ambio 20:317–321
Daehler CC, Denslow JS, Ansari S et al (2004) A risk assessment system for screening out invasive pest plants from Hawaii and other Pacific Islands. Conserv Biol 18:360–368
D'Antonio CM, Dudley TL (1995) Biological invasions as agents of change on islands versus mainlands. In: Vitousek P, Loope L, Adsersen H (eds) Biological diversity and ecosystem function on islands. Ecological studies 115. Springer, New York, pp 103–119
D'Antonio CM, Mack MC (2006) Nutrient limitation in a fire-derived, nitrogen-rich Hawaiian grassland. Biotropica 38:458–467
D'Antonio CM, Hughes FR, Mack M et al (1998) The response of native species to removal of invasive exotic grasses in a seasonally dry Hawaiian woodland. J Veg Sci 9:699–712
D'Antonio CM, Tunison JT, Loh RK (2000) Variations in impact of exotic grasses and fire on native plant communities in Hawaii. Austral Ecol 25:507–522
D'Antonio CM, Hughes RF, Tunison JT (2011) Long-term impacts of invasive grasses and subsequent fire in seasonally dry Hawaiian woodlands. Ecol Appl 21:1617–1628

Darwin CR (1859) On the origin of species by means of natural selection, or the preservation of favoured races in the struggle for life. John Murray, London

Denslow JS, D'Antonio CM (2005) After biocontrol: assessing indirect effects of insect releases. Biol Control 35:307–318

Denslow JS, Space JC, Thomas PA (2009) Invasive exotic plants in the tropical Pacific islands: patterns of diversity. Biotropica 41:162–170

DeWalt SJ, Denslow JS, Ickes K (2004) Natural enemy release facilitates habitat expansion of the invasive tropical shrub *Clidemia hirta*. Ecology 85:471–483

Diong CH (1982) Population biology and management of the feral pig (*Sus scrofa* L.) in Kipahulu Valley, Maui. Ph.D. dissertation, Department of Zoology, University of Hawaii, Honolulu. http://www.hear.org/articles/diong1982/

Djeddour D, Shaw R (2011) Biological control of kahili ginger, *Hedychium gardnerianum*. CABI annual report, Europe UK 2010. CABI, Egham. http://www.cabi.org/uploads/file/Centrereports/UK_Centre_Report_2010.pdf

Douglas G (1969) Checklist of Pacific oceanic islands. Micronesica 5:327–464

Florence J, Waldren S, Chepstow-Lusty AJ (1995) The flora of the Pitcairn Islands: a review. Biol J Linn Soc 56:79–119

Footman T (2001) Guinness world records 2001. Bantam Books, New York

Gillespie RG, Claridge EM, Roderick GK (2008) Biodiversity dynamics in isolated island communities: interaction between natural and human-mediated processes. Mol Ecol 17:45–57

Hawaii Volcanoes National Park (2005) Fire management plan. Division of Resources Management, National Park Service, Volcano

Hobbs RJ, Arico S, Aronson J et al (2006) Novel ecosystems: theoretical and management aspects of the new ecological world order. Glob Ecol Biogeogr 15:1–7

Hobbs RJ, Higgs E, Harris JA (2009) Novel ecosystems: implications for conservation and restoration. Trends Ecol Evol 24:599–605

Holt A (1996) An alliance of biodiversity, health, agriculture, and business interests for improved alien species management in Hawaii. In: Sandlund OT, Schei PJ, Viken A (eds) Proceedings of the Norway/UN conference on alien species. Directorate for Nature Management and Norwegian Institute for Nature Research, Trondheim, pp 155–160. http://www.hear.org/AlienSpeciesInHawaii/articles/norway.htm

Huenneke LF, Vitousek PM (1990) Seedling and clonal recruitment of the invasive tree *Psidium cattleianum*: implications for management of native Hawaiian forests. Biol Conserv 53:199–211

Hughes RF, Denslow JS (2005) Invasion by a N2-fixing tree alters function and structure in wet lowland forests of Hawaii. Ecol Appl 15:1615–1628

Hughes RF, Vitousek PM (1993) Barriers to shrub reestablishment following fire in the seasonal submontane zone of Hawaii. Oecologia 93:557–563

Hughes F, Vitousek PM, Tunison T (1991) Alien grass invasion and fire in the seasonal submontane zone of Hawaii. Ecology 72:743–746

Hughes RF, Uowolo AL, Togia TP (2012) Effective control of *Falcataria moluccana* in forests of American Samoa: managing invasive species in concert with ecological processes. Biol Invasions 14:1393–1413

Jacq F, Meyer J-Y (2012) Taux de mortalité et causes de disparition de *Apetahia raiateensis* (Campanulaceae), une plante endémique de l'île de Raiatea (Polynésie française). Revue d'Ecologie (Terre et Vie) 67:57–72

Jaynes DB, Colvin TS, Karlen DL et al (2001) Nitrate loss in subsurface drainage as affected by nitrogen fertilizer rate. J Environ Qual 30:1305–1314

Johnson MT (2010) Miconia biocontrol: where are we going and when will we get there? In: Loope LL, Meyer J-Y, Hardesty BD et al (eds) Proceedings of the international Miconia conference, Keanae, Maui, Hawaii, May 4–7, 2009. Maui Invasive Species Committee and Pacific Cooperative Studies Unit, University of Hawaii, Honolulu, 11pp. http://www.hear.org/miconia2009/pdfs/johnson.pdf

Kellner JR, Asner GP, Kinney KM et al (2011) Remote analysis of biological invasion and the impact of enemy release. Ecol Appl 21:2094–2104

Kraus F, Duffy DC (2010) A successful model from Hawaii for rapid response to invasive species. J Nat Conserv 18:135–141

Kueffer C, Loope L (eds) (2009) Prevention, early detection and containment of invasive, non-native plants in the Hawaiian Islands: current efforts and needs. Pacific Cooperative Studies Unit, University of Hawaii, Honolulu. Technical report no. 166. http://www.botany. hawaii.edu/faculty/duffy/techr/166

Kueffer C, Daehler CC, Torres-Santana W et al (2010a) A global comparison of plant invasions on oceanic islands. Perspect Plant Ecol Evol Syst 12:145–161

Kueffer C, Schumacher E, Dietz H et al (2010b) Managing successional trajectories in alien-dominated, novel ecosystems by facilitating seedling regeneration: a case study. Biol Conserv 143:1792–1802

Langeland KA, Stocker RK (1997) Control of non-native plants in natural areas of Florida. Institute of Food and Agricultural Sciences publication SP242. University of Florida, Gainesville

LaRosa AM (1992) The status of banana poka in Hawaii. In: Stone CP, Smith CW, Tunison JT (eds) Alien plant invasions in native ecosystems of Hawaii: management and research. University of Hawaii Press for Cooperative National Park Resources Studies Unit, University of Hawaii, Honolulu, pp 271–299

Lenz L, Taylor JA (2001) The influence of an invasive tree species (*Myrica faya*) on the abundance of an alien insect (*Sophonia rufofascia*) in Hawaii Volcanoes National Park. Biol Conserv 102:301–307

Loh R, Daehler CC (2007) Influence of invasive tree kill rates on native and invasive plant establishment in a Hawaiian wet forest. Restor Ecol 15:199–211

Loh RK, Daehler CC (2008) Influence of woody invader control methods and seed availability on native and invasive species establishment in a Hawaiian forest. Biol Invasions 10:805–819

Loh RK, Tunison JT (2009) Long term management of invasive plant species in Special Ecological Areas at Hawaii Volcanoes National Park: a review of the last 20 years, or where do we go from here? In: Kueffer C, Loope L (eds) Prevention, early detection and containment of invasive, non-native plants in the Hawaiian Islands: current efforts and needs. Technical report no. 166. Pacific Cooperative Studies Unit, University of Hawaii, Honolulu, pp 33–35

Loope LL, Juvik SP (1998) Protected areas. In: Juvik SP, Juvik JO (eds) Atlas of Hawaii, 3rd edn. University of Hawaii Press, Honolulu, pp 154–157

Loope LL, Nagata RJ, Medeiros AC (1992) Introduced plants in Haleakala National Park. The status of banana poka in Hawaii. In: Stone CP, Smith CW, Tunison JT (eds) Alien plant invasions in native ecosystems of Hawaii: management and research. University of Hawaii Press for Cooperative National Park Resources Studies Unit, University of Hawaii, Honolulu, pp 551–576

Loope LL, Starr F, Starr KM (2004) Protecting endangered plant species from displacement by invasive plants on Maui, Hawaii. Weed Technol 18:1472–1474

Lugo AE (2004) The outcome of alien tree invasions in Puerto Rico. Front Ecol Environ 2:265–273

Markin GP, Lai P-Y, Funasaki GY (1992) Status of biological control of weeds in Hawaii and implications for managing native ecosystems. The status of banana poka in Hawaii. In: Stone CP, Smith CW, Tunison JT (eds) Alien plant invasions in native ecosystems of Hawaii: management and research. University of Hawaii Press for Cooperative National Park Resources Studies Unit, University of Hawaii, Honolulu, pp 466–482

Medeiros AC (2004) Phenology, reproductive potential, seed dispersal and predation, and seedling establishment of three invasive plant species in a Hawaiian rain forest. Ph.D. Dissertation, Department of Botany, University of Hawaii, Honolulu. http://www.hear.org/ articles/medeiros2004dissertation

Medeiros AC, Loope LL, Conant P et al (1997) Status, ecology, and management of the invasive tree *Miconia calvescens* DC (Melastomataceae) in the Hawaiian Islands. In: Evenhuis NL,

Miller SE (eds) Records of the Hawaii biological survey for 1996. Bishop Mus occasional paper 48:23–35

Merlin MD, Juvik JO (1992) Relationships among native and alien plants on Pacific islands with and without significant human disturbance and feral ungulates. The status of banana poka in Hawaii. In: Stone CP, Smith CW, Tunison JT (eds) Alien plant invasions in native ecosystems of Hawaii: management and research. University of Hawaii Press for Cooperative National Park Resources Studies Unit, University of Hawaii, Honolulu, pp 597–624

Messing RH, Wright MG (2006) Biological control of invasive species: solution or pollution? Front Ecol Environ 4:132–140

Meyer J-Y (1996a) Espèces et espaces menacés de la Société et des Marquises. Contribution à la biodiversité de la Polynésie française N°1-5 [unpublished report]. Délégation à l'Environnement/ Délégation à la Recherche, Papeete, 245 pp

Meyer J-Y (1996b) Status of *Miconia calvescens* (Melastomataceae), a dominant invasive tree in the Society Islands (French Polynesia). Pacific Sci 50:66–76

Meyer J-Y (2000) Rapport de mission aux Marquises Sud (Hiva Oa, Fatu Hiva, Mohotani) du 6 au 20 février 2000 [unpublished report]. Délégation à la Recherche, Papeete, 19 pp

Meyer J-Y (2004) Threat of invasive alien plants to native flora and forest vegetation of Eastern. Polynesia. Pacific Sci 58:357–375

Meyer J-Y (2005) Rapport de mission botanique à Ua Huka (Marquises, Groupe Nord) du 24 au 31 juillet 2005 et description de la végétation et de la flore de la réserve naturelle de Vaikivi [unpublished report]. Délégation à la Recherche, Papeete, 29 pp

Meyer J-Y (2008) Rapport de mission d'expertise à Rapa Nui du 02 au 11 juin 2008: plan d'action stratégique pour lutter contre les plantes introduites envahissantes sur Rapa Nui (Île de Pâques) [Strategic action plan to control invasive alien plants on Rapa Nui (Easter Island) [unpublished report]. Délégation à la Recherche, Papeete, 62 pp

Meyer J-Y (2010) The miconia saga: 20 years of study and control in French Polynesia (1988–2008). In: Loope LL, Meyer J-Y, Hardesty BD et al (eds) Proceedings of the international Miconia conference, Keanae, Maui, Hawaii, May 4–7, 2009. Maui Invasive Species Committee and Pacific Cooperative Studies Unit, University of Hawaii at Manoa. http://www.hear.org/miconia2009/pdfs/meyer.pdf

Meyer J-Y (2012) Rapport de mission d'expertise à Rapa Nui du 14 au 20 février 2012: projet de restauration écologique dans des parcelles de démonstration sur Rapa Nui (île de Pâques) [unpublished report]. Délégation à la Recherche, Papeete, 50 pp

Meyer J-Y, Florence J (1996) Tahiti's native flora endangered by the invasion of *Miconia calvescens* DC (Melastomataceae). J Biogeogr 23:775–781

Meyer J-Y, Lavergne C (2004) Beautés fatales: Acanthaceae species as invasive alien plants on tropical Indo-Pacific Islands. Divers Distrib 10:333–347

Meyer J-Y, Loope L, Goarant A-C (2011) Strategy to control the invasive alien tree *Miconia calvescens* in Pacific islands: eradication, containment or something else? In: Veitch CR, Clout MN, Towns DR (eds) Island invasives: eradication and management. IUCN, Gland, pp 91–96. http://www.issg.org/pdf/publications/Island_Invasives/pdfwebview/1Meyer.pdf

Minden V, Hennenberg KJ, Porembski S et al (2010a) Invasion and management of alien *Hedychium gardnerianum* (kahili ginger, Zingiberaceae) alter plant species composition of a montane rainforest on the island of Hawaii. Plant Ecol 206:321–333

Minden V, Jacobi JD, Porembski S et al (2010b) Effects of invasive alien *Hedychium gardnerianum* on native plant species regeneration in a Hawaiian rainforest. Appl Veg Sci 13:5–14

Motooka PL, Ching G, Nagai G (2002) Herbicidal weed control methods for pastures and natural areas of Hawaii. CTAHR, Cooperative Extension Service, University of Hawaii. Technical bulletin WC-8. http://www2.ctahr.hawaii.edu/oc/freepubs/pdf/wc-8.pdf

Mueller-Dombois D, Fosberg FR (1998) Vegetation of the tropical Pacific islands. Springer, New York

Myers N, Mittermeier RA, Mittermeier CG et al (2000) Biodiversity hotspots for conservation priorities. Nature 403:853–858

Parrotta JA (1990) *Paraserianthes falcataria* (L.) Nielsen. Batai, Moluccan sau. Leguminosae (Mimosoideae) Legume family. Report SO-ITF-SM-31. US Department of Agriculture, Forest Service, Southern Forest Experiment Station, Institute of Tropical Forestry, New Orleans, 5 pp

Penniman TM, Buchanan L, Loope LL (2011) Recent plant eradications on the islands of Maui County, Hawaii. In: Veitch CR, Clout MN, Towns DR (eds) Island invasives: eradication and management. IUCN, Gland, pp 325–331. http://www.issg.org/pdf/publications/Island_Inva sives/pdfHQprint/3Penniman.pdf

Sachet M-H (1983) Végétation et flore terrestre de l'atoll de Scilly (Fenua Ura). J Soc Océanistes 77:29–34

Santos GL, Kageler D, Gardner DE et al (1992) Herbicidal control of selected alien plants in Hawaii Volcanoes National Park. The status of banana poka in Hawaii. In: Stone CP, Smith CW, Tunison JT (eds) Alien plant invasions in native ecosystems of Hawaii: management and research. University of Hawaii Press for Cooperative National Park Resources Studies Unit, University of Hawaii, Hawaii, pp 341–375

Sellars RW (1997) Preserving nature in the National Parks: a history. Yale University Press, New Haven

Smith CW (2002) Forest pest biological control program in Hawaii. In: Smith CW, Denslow J, Hight S (eds) Proceedings of workshop on biological control of native ecosystems in Hawaii. Technical report 129. Pacific Cooperative Studies Unit, University of Hawaii, Honolulu, pp 91–102. http://www.botany.hawaii.edu/faculty/duffy/techr/129.pdf

Stone CP, Smith CW, Tunison JT (eds) (1992) Alien plant invasions in native ecosystems of Hawaii: Management and research. University of Hawaii Press for Cooperative National Park Resources Studies Unit, University of Hawaii, Honolulu. http://www.hawaii.edu/hpicesu/ book/1992_chap/default.htm

Tomich QC (1986) Mammals in Hawaii: a synopsis and notational bibliography, 2nd edn. B.P. Bishop Museum special publication 76, Honolulu

Tunison JT (1992a) Fountain grass control in Hawaii Volcanoes National Park: management considerations and strategies. In: Stone CP, Smith CW, Tunison JT (eds) Alien plant invasions in native ecosystems of Hawaii: management and research. University of Hawaii Press for Cooperative National Park Resources Studies Unit, University of Hawaii, Honolulu, pp 376–393

Tunison JT (1992b) Alien plant control strategies in Hawaii Volcanoes National Park. In: Stone CP, Smith CW, Tunison JT (eds) Alien plant invasions in native ecosystems of Hawaii: management and research. University of Hawaii Press for Cooperative National Park Resources Studies Unit, University of Hawaii, Honolulu, pp 485–505

Tunison JT, Stone CP (1992) Special Ecological Areas: an approach to alien plant control in Hawaii Volcanoes National Park. In: Stone CP, Smith CW, Tunison JT (eds) Alien plant invasions in native ecosystems of Hawaii: management and research. University of Hawaii Press for Cooperative National Park Resources Studies Unit, University of Hawaii, Honolulu, pp 781–798

Tunison JT, Zimmer NG (1992) Success in controlling localized alien plants in Hawaii Volcanoes National Park. In: Stone CP, Smith CW, Tunison JT (eds) Alien plant invasions in native ecosystems of Hawaii: management and research. University of Hawaii Press for Cooperative National Park Resources Studies Unit, University of Hawaii, Honolulu, pp 506–524

Tunison JT, Smith CW, Stone CP (1992a) Alien plant management in Hawaii: conclusions. In: Stone CP, Smith CW, Tunison JT (eds) Alien plant invasions in native ecosystems of Hawaii: management and research. University of Hawaii Press for Cooperative National Park Resources Studies Unit, University of Hawaii, Honolulu, pp 821–833

Tunison JT, Whiteaker LD, Cuddihy LW et al (1992b) The distribution of selected localized alien plant species in Hawaii Volcanoes National Park. Technical report 84. Cooperative National Park Resources Studies Unit, University of Hawaii, Honolulu

Tunison JT, Leialoha JAK, Loh RK et al (1994) Fire effects in the coastal lowlands, Hawaii Volcanoes National Park. Technical report 88. Cooperative National Park Resources Studies Unit, University of Hawaii, Honolulu

Tunison JT, Loh RK, Leialoha JAK (1995) Fire effects in the submontane seasonal zone, Hawaii Volcanoes National Park. Technical report 97. Cooperative National Park Resources Studies Unit, University of Hawaii, Honolulu

Van Driesche R, Center T (2014) Chapter 26: Biological control of invasive plants in protected areas. In: Foxcroft LC, Pyšek P, Richardson DM, Genovesi P (eds) Plant invasions in protected areas: patterns, problems and challenges. Springer, Dordrecht, pp 561–597

Vitousek PM, Walker LR (1989) Biological invasion by *Myrica faya* in Hawaii – plant demography, nitrogen-fixation, ecosystem effects. Ecol Monogr 59:247–265

Vitousek PM, Loope LL, Stone CP (1987) Introduced species in Hawaii: biological effects and opportunities for ecological research. Trends Ecol Evol 2:224–227

Wagner WL, Herbst DR, Sohmer SH (1999) Manual of the flowering plants of Hawai'i. University of Hawaii Press, Honolulu

Walters GA (1971) A species that grew too fast – *Albizia falcataria*. J For 69:168

Warner KD, Kinslow F (2011) Manipulating risk communication: value predispositions shape public understandings of invasive species science in Hawaii. Public Underst Sci. doi:10.1177/0963662511403983

Webb EL, van de Bult M, Chutipong W et al (2006) Composition and structure of lowland rain-forest tree communities on Ta'u, American Samoa. Pacific Sci 60:333–354

Webb EL, Seamon JO, Fa'aumu S (2011) Frequent, low-amplitude disturbances drive high tree turnover rates on a remote, cyclone-prone Polynesian island. J Biogeogr 38:1240–1252

Whistler WA (1980) The vegetation of Eastern Samoa. Allertonia 2:45–190

Whistler WA (1994) Botanical inventory of the proposed Tutuila and Ofu units of the NPSA. CPSU technical report 87. University of Hawaii, Honolulu

Whistler WA (2004) Rainforest trees of Samoa. Isle Botanica, Honolulu

Wikler C, Smith CW (2002) Strawberry guava (*Psidium cattleianum*) prospects for biological control. In: Smith CW, Denslow J, Hight S (eds) Proceedings of workshop on biological control of native ecosystems in Hawaii. Technical report 129. Pacific Cooperative Studies Unit, University of Hawaii, Honolulu, pp 108–116

Woodcock D (2003) To restore the watersheds: early twentieth-century tree planting in Hawai'i. Ann Assoc Am Geogr 93:624–635

Woodward SA, Vitousek PM, Matson K et al (1990) Use of the exotic tree *Myrica faya* by native and exotic birds in Hawai'i Volcanoes National Park. Pacific Sci 44:88–93

Chapter 16
A Pragmatic Approach to the Management of Plant Invasions in Galapagos

Mark R. Gardener, Mandy Trueman, Chris Buddenhagen, Ruben Heleno, Heinke Jäger, Rachel Atkinson, and Alan Tye

Abstract This chapter presents an overview of the process undertaken to understand alien plant invasions and work towards their effective management in the Galapagos Islands. Galapagos is a unique case study for the management of alien plants in protected areas because much the archipelago has few alien plants and the original ecosystems are relatively intact. We discuss a pragmatic approach developed over 15 years to help prioritise management of 871 plant species introduced to the islands. This approach includes understanding invasion pathways; identifying which species are present and their distribution; determining invasive species impact on biodiversity, ecosystem function and mutualisms;

M.R. Gardener (✉)
Charles Darwin Foundation, Santa Cruz, Galapagos Islands, Ecuador

School of Plant Biology, University of Western Australia, Crawley, WA 6009, Australia
e-mail: mark.r.gardener@gmail.com

M. Trueman
School of Plant Biology, University of Western Australia, Crawley, WA 6009, Australia

Charles Darwin Foundation, Santa Cruz, Galapagos Islands, Ecuador
e-mail: truemandy@gmail.com

C. Buddenhagen
Department of Biological Science, Florida State University, Tallahassee, FL 32306, USA

Charles Darwin Foundation, Santa Cruz, Galapagos Islands, Ecuador
e-mail: cbuddenhagen@gmail.com

R. Heleno
Department of Life Sciences, Centre for Functional Ecology, University of Coimbra, Coimbra 3001-401, Portugal

Charles Darwin Foundation, Santa Cruz, Galapagos Islands, Ecuador
e-mail: rheleno@uc.pt

L.C. Foxcroft et al. (eds.), *Plant Invasions in Protected Areas: Patterns, Problems and Challenges*, Invading Nature - Springer Series in Invasion Ecology 7, DOI 10.1007/978-94-007-7750-7_16, © Springer Science+Business Media Dordrecht 2013

prioritising management using weed risk assessment; guidelines to prevent further introduction through quarantine and early intervention; and developing methods to control or eradicate priority species. Principal barriers to application of the approach are limited capacity and coordination among managers and inherent difficulties arising from invasive species traits such as seed banks and dispersal and their interactions with ecosystems. We also discuss the approach of managing invasive species individually and suggest it may be more appropriate, when feasible, for the relatively intact uninhabited islands and dry regions of Galapagos. The more degraded highlands of the inhabited islands need a more complex approach that balances costs with prioritised outcomes for biodiversity and ecosystem functionality.

Keywords Impacts • Islands • Ecosystem function • Mutualisms • Weed risk assessment • Priorities • Quarantine • Eradication

16.1 Introduction

The Galapagos Islands are a special case for the management of alien plants in protected areas. The majority of the land area – on the uninhabited islands – contains very few alien plant species and is in relatively pristine condition (Bensted-Smith et al. 2002). However, the core areas of human impact – on the four islands inhabited by humans – contain more introduced than native plant species and ecosystems are highly altered from their historical state (Snell et al. 2002a). Thus most of this chapter focuses on the 4 % of the Galapagos archipelago which has been colonised by humans and the parts of the protected area immediately surrounding these nuclei.

H. Jäger
Department of Ecology, Technische Universität Berlin, Berlin 12165, Germany

Charles Darwin Foundation, Santa Cruz, Galapagos Islands, Ecuador
e-mail: heinke.jaeger@tu-berlin.de

R. Atkinson
Charles Darwin Foundation, Santa Cruz, Galapagos Islands, Ecuador
e-mail: ratkinson27@gmail.com

A. Tye
Charles Darwin Foundation, Santa Cruz, Galapagos Islands, Ecuador
e-mail: alantye@gmail.com

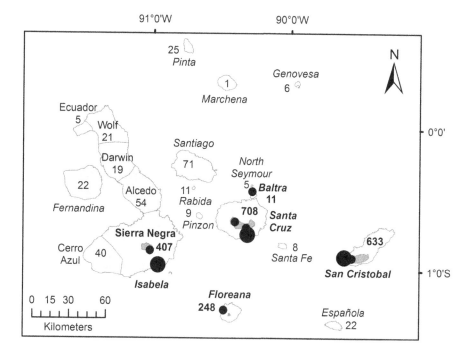

Fig. 16.1 The number of alien plant species on each island larger than 150 ha (with the six Isabela volcanoes shown separately because they are mostly ecologically isolated by bare lava on the lower flanks). Islands with a humid highland zone are *shaded light grey*. Currently inhabited islands are labelled in *bold*, with the inhabited area *shaded dark grey*; all remaining land areas constitute the Galapagos National Park. Towns and villages are shown as .... *black dots*. (Plant data from Charles Darwin Foundation Collections Database (CDF 2009) and Trueman et al. (2010); island names from Snell et al. (1996))

Eighty of the 871 non-native vascular plant species[1] recorded in Galapagos (up to year 2012) are found on the 46 islands and islets that have never been colonised. Figure 16.1 shows that the inhabited islands/volcanoes have an order of magnitude more alien plant species compared to the uninhabited ones. Most of the species on the uninhabited islands are herbaceous, annual or short lived perennials, probably accidentally introduced and unlikely to cause great impact (e.g. *Porophyllum ruderale*, ruda gallinazo). In contrast, all of the 871 species are found on at least one of the four main islands inhabited by people: Santa Cruz, San Cristóbal, Isabela and Floreana. Between 21 and 100 % of the humid highlands on these islands have been degraded through agriculture and alien plant and animal invasion (Watson et al. 2009), which effectively forms degraded nuclei surrounded

---

[1] We have used species as the taxonomic unit throughout this chapter for consistency, but the 871 'species' in Galapagos actually include subspecies. In this chapter taxonomy and common names follows Jaramillo and Guézou (2012).

by less degraded dry peripheries. Baltra, the other colonised island, is an exception by having few alien species, due to being a small, low island without a humid zone. Many plant species introduced to the areas used by humans have spread into the neighbouring Galapagos National Park (GNP), including some that have been transmitted to uninhabited islands.

In the mid-1990s the Charles Darwin Research Station started a systematic process to catalogue, understand and prioritise invasive plant management, and over the last 15 years the authors of this chapter, and many others, have worked together to develop a pragmatic approach to managing plant invasions in Galapagos. The goal has always been to facilitate realistic management decisions that produce achievable and valuable outcomes. The development of this process was accelerated between 2002 and 2007 by the Global Environment Facility (GEF) funded project entitled "Control of Invasive Species in the Galapagos Archipelago". This ambitious programme had a number of elements including: baseline inventories, quarantine development, research on invasions, experimental eradications, awareness and participation programmes, capacity building, and development of a Galapagos-wide planning and policy framework. Whilst the consolidation of the results continues to this day, the legacy from this project is significant.

This chapter aims to outline the pragmatic management process as applied in Galapagos under three main umbrellas: (i) Understanding the problem (identify introduction and invasion pathways; identify what species are present and their spatial/temporal extent; determine their impact on biodiversity, ecosystem function and mutualisms); (ii) Developing management tools (prioritise species for management and border biosecurity; develop methods to control or eradicate priority species); and (iii) Addressing the challenges of applying our approach to attain achievable goals (the relative benefits of managing single species versus whole ecosystems for different parts of Galapagos; the importance of engaging people and the differing visions of various stakeholders). We use examples to illustrate each step of the management process and discuss limitations. We finish with a summary of lessons learned and an outline of management opportunities that are specific to the unique case of Galapagos.

16.2 Understanding the Problem

16.2.1 Plant Introductions and Invasion Pathways

Alien plants have likely been introduced to Galapagos ever since its discovery in 1535 when the archipelago was first opened up to humans. Early visitors included buccaneers, whalers and fishermen. Few introductions would have occurred prior to human settlement of the islands which began in 1807 and led to permanent human presence since 1879 (Grenier 2007). On his visit to Galapagos in 1835, Charles Darwin recorded 17 alien species in Floreana (a colony that was repeatedly

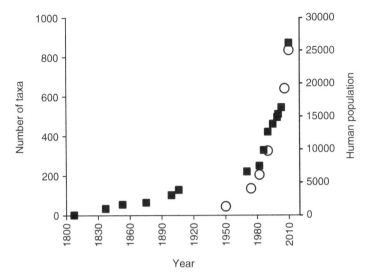

Fig. 16.2 Cumulative number of alien plant taxa recorded (*black squares*) and human population (*open circles*) in Galapagos since settlement in 1807 (Sources: alien species (Jaramillo and Guézou 2012; Tye 2006); human population (Grenier 2007; INEC 2007, 2011))

abandoned), all of which had been deliberately introduced for agriculture (Hooker 1851). Tortoise hunters had vegetable gardens on Santa Cruz in the nineteenth century, before that island was permanently colonised in the early 1900s (Lundh 2006).

In 2011, the alien flora of Galapagos amounted to 871 species (Jaramillo and Guézou 2012). Most of these were introduced in the past 30–50 years (Fig. 16.2; Tye 2006; Trueman et al. 2010). The recordings of alien plant species in Galapagos has increased exponentially or multi-stage linearly since records began in 1807, partly reflecting the introduction rate, but more importantly illustrating the changes in species recording method and effort (Tye 2006). Many species may have been present for years, even decades, before they were first recorded hence biasing rate of increase (Tye 2006). There is further uncertainty about the total number of aliens, because approximately 60 species are classified as questionably native, which pending further evidence could be native or introduced. Using fossil pollen and other plant remains, van Leeuwen et al. (2008) and Coffey et al. (2011) showed that nine species classified as questionably native or definitely alien had actually been present in Galapagos before humans. Nevertheless, anthropogenic introductions far outweigh the natural introduction rate by a factor of at least 13,000 (Tye 2006).

The impact of invasive plants on the Galapagos biota has been emphasised only since the 1970s (Schofield 1973), though the first invasions probably began prior to permanent human settlement. Examples include *Citrus* spp. that were introduced as anti-scorbutics by visiting pirates or whalers (Lundh 2006) and still persist in the wild today, and *Furcraea hexapetala* (Cuban hemp, cabuya) which was planted to mark a trail and had naturalised on Santa Cruz Island by 1905 (Lundh 2006).

354   M.R. Gardener et al.

Fig. 16.3 Many parts of the highland grass and fern zone within the Galapagos National Park on Santa Cruz Island (**a** – 1970) have been invaded by the tree *Cinchona pubescens* (**b** – 2004) (Photographs by Frank J. Sulloway, taken looking east from Cerro Crocker, reproduced with permission from the photographer)

Widespread plant invasions and ecosystem transformation did not begin until well after settlement, and were facilitated by agricultural clearing and seed dispersal. One example is *Syzygium jambos* (rose-apple) that started to form thickets in San Cristóbal in the 1950s (Lundh 2006). On Santa Cruz the tree *Cinchona pubescens* (red quinine) began to spread widely in the 1970s (Hamann 1974; Fig. 16.3) and several other plant invasions accelerated after the 1983 El Niño

Fig. 16.4 Forest of the native *Scalesia pedunculata* (lechoso; young stand) and *Psidium galapageium* (guayabillo; with the brown epiphytic liverworts) (**a** – in 1975) has been replaced by taller forest of *Cedrela odorata* with a mid-storey of *Psidium guajava* (**b** – in 2004). Both photographs are from a permanent quadrat in the 'caseta' area in the Tortoise Reserve, Santa Cruz Island (Photographs by Ole Hamann, reproduced with permission from the photographer)

event (e.g. *Psidium guajava*, guava; *Rubus niveus*, blackberry; *Cedrela odorata*, Cuban cedar; *Cestrum auriculatum*, sauco; Fig. 16.4). In the south of Isabela, *P. guajava* now forms a monoculture of cover over 40,000 ha (Laura Brewington, personal communication, 2009) and was probably dispersed by feral cattle and established after a big fire in the 1980s.

Table 16.1 Current transformer species (those that change the character, condition, form or nature of ecosystems over a substantial area) and their biological characteristics

Family	Species	Habit	Dispersal	Evidence of seedbank in Galapagos/vegetative reproduction
Agavaceae	*Furcraea hexapetala*	Herb	Wind	Yes/yes
Boraginaceae	*Cordia alliodora*	Tree	Wind	Unknown/no
Commelinaceae	*Tradescantia fluminensis*	Herb	Animal[a]	Yes/yes
Crassulaceae	*Bryophyllum pinnatum*	Herb	Wind/vegetative fragments	Unknown/yes
Lauraceae	*Persea americana*	Hree	Animal[a]	No/no
Meliaceae	*Cedrela odorata*	Tree	Wind	No/no
Mimosaceae	*Leucaena leucocephala*	Small tree	Water, soil	Yes/no
Myrtaceae	*Syzygium jambos*	Tree	Water/animal	No/yes
Myrtaceae	*Psidium guajava*	Small tree	Animal[a]	Unknown/no
Passifloraceae	*Passiflora edulis*	Climber	Animal[a]	No/no
Poaceae	*Melinis minutiflora*	Grass	Animal[a]/wind	Yes/yes
Poaceae	*Panicum maximum*	Grass	Animal[a]	Unknown/yes
Poaceae	*Pennisetum purpureum*	Grass	Animal[a]	Yes/yes
Poaceae	*Urochloa decumbens*	Grass	Animal[a]/wind	Yes/yes
Rosaceae	*Rubus niveus*	Shrub	Animal[a], soil	Yes[b]/yes
Rubiaceae	*Cinchona pubescens*	Tree	Wind[a]	Yes[b]/yes
Solanaceae	*Cestrum auriculatum*	Small tree	Animal[a]	Yes/no
Verbenaceae	*Lantana camara*	Shrub	Animal[a]	Yes/yes

[a]Animal dispersal data from Blake et al. (2012), Connett et al. (in press), Guerrero and Tye (2011), Heleno et al. (2011, 2013b); and references therein)
[b]Seedbank data from Landazuri (2002) and Rentería (2002); All other data from author observations

To help understand the problem of alien plants in any place it is important to know how far the flora has progressed along the invasion continuum (Richardson et al. 2000); this can be done by determining which introduced species fall into each of the categories of naturalised, invasive and transforming. Of the alien flora in Galapagos, 332 species (38 %) are naturalised (Jaramillo and Guézou 2012) and 32 of these species (4 % of total) are considered invasive (Tye 2001). Table 16.1 shows a subset of 18 of these invasive species that are widespread and considered to be problematic for native ecosystems; these fit the bill of transformer species (sensu Richardson et al. 2000; Pyšek et al. 2004). They are all perennial, have effective dispersal mechanisms and mostly have a seedbank and/or vegetative reproduction (Table 16.1). This overall measure of progression along the invasion continuum is difficult to compare with other locations because of differences in sampling effort and methods, and differences in the definition of 'invasive' (Guézou et al. 2010). However the outnumbering of native flora by alien flora is similar to other oceanic islands, while the proportion naturalised and invasive is lower than on other oceanic islands (Trueman et al. 2010). Thus Galapagos is at an early stage of the invasion process (Tye 2006; Trueman et al. 2010). Even if no more introductions occur, some of the species already present will become naturalised and turn into transformers that impact native ecosystems.

Different types of plants have been introduced over various periods of human colonisation and this has influenced patterns of invasion and impact. Initially, intentional introductions were focused on useful, cultivated plants to be grown in the humid agricultural regions, while the focus has shifted more recently towards ornamental plants, grown mainly in gardens in the lowland urban areas (Tye 2006), which now form the majority of alien plant species in Galapagos (Atkinson et al. 2010). Many species have recently been recorded for the first time and although they may have been introduced much earlier, most of them still have a very limited distribution on private properties (Guézou et al. 2010; Trueman et al. 2010). A similar number of alien species are now grown in the urban areas of the dry lowlands compared with the rural humid highlands areas of Galapagos, although the most common species in each area are different (Trueman 2008; Guézou et al. 2010). Due to their mesic climate and earlier introduction history, the humid highland areas have suffered more invasions so far (Snell et al. 2002a; Watson et al. 2009). Invasions of ornamental species in the dry coastal towns have only recently begun. For example, *Kalanchoe tubiflora* (= *Bryophyllum delagoense*, chandelier plant) is now naturalising and dispersing into the adjacent GNP. The variability of the Galapagos climate also affects invasions. In particular, wetter periods associated with El Niño assist the establishment and spread of alien species in the islands (Hamann 1985; Luong and Toro 1985; Aldaz and Tye 1999; Tye and Aldaz 1999). Also, *Leucaena trichodes* (wild tamarind) persisted in gardens for many years, until a recent run of wet years enabled it to naturalise and establish a seed bank (R. Atkinson, personal observation, 2009). There is concern that a future warmer, wetter climate would favour further invasions in the dry lowland areas which occupy the majority of Galapagos land area and contain most of the endemic species (Trueman and d'Ozouville 2010).

16.2.2   Inventories: Baseline Data for Early Detection

To understand and manage alien plants we need to know what they are, where they are, whether they pose a problem, and which management options might be suitable. Exclusion and rapid response to incursions are widely considered to be the most viable, cost effective means of managing invasive species (Panetta and Timmins 2004), yet detailed distribution information is needed in order to carry these out. This information can be obtained by carrying out surveys of alien species (Hosking et al. 2004). A particular concern in Galapagos is the spread of alien species from the inhabited zones (private properties) into the GNP, so the inhabited areas have been the target for such inventories.

Between 2002 and 2007 a team of botanists carried out one of the most extensive surveys of alien species in any Pacific archipelago, visiting 6,031 private properties (97 % of the total in urban and rural zones) in the islands of Floreana, Santa Cruz, San Cristóbal and Southern Isabela (Guézou et al. 2010). The survey of 253 km^2 cost US$300,000 and 17 person-years of trained botanists' time (Guézou et al. 2010).

The survey results constitute a useful tool for management. Almost all species are represented by specimens in the Charles Darwin Research Station herbarium (CDS). With few exceptions each species record is assigned to a particular property. Most species were uncommon, with 252 species recorded in five properties or less (Trueman et al. 2010). If a decision were made to attempt to manage some uncommon but potentially invasive species in the future, the inventory data could be used to select potential targets and guide more detailed surveys.

This inventory was also an excellent community outreach tool, and was well received as evidenced by the small percentage of properties where entry was refused (<1 %). The team met with landowners and discussed invasive plant problems, thus increasing community awareness. All species detected were subsequently evaluated by the Galapagos Weed Risk Assessment which is discussed further below.

16.2.3   Impacts on Biodiversity and Ecosystem Function

There is little hard evidence that invasive plant species cause extinctions (Gurevitch and Padilla 2004). However, there is much evidence that they cause changes in ecosystem structure and function, and an understanding of such impacts can be used to determine whether changes are reversible and to inform management.

Cinchona pubescens and *R. niveus* are the two best studied transformer species (sensu Pyšek et al. 2004) in Galapagos; there is quantitative information on their biotic and abiotic impacts. Both species were introduced to the highlands of Santa Cruz Island, *C. pubescens* in the 1940s, now covering over 11,000 ha (Buddenhagen et al. 2004), and *R. niveus* in the 1960s (Lawesson and Ortiz 1990), now covering more than 30,000 ha on five islands. *Cinchona pubescens* is converting formerly treeless vegetation zones into forests, and *R. niveus* has formed dense impenetrable thickets up to 4 m high.

A 7-year study showed that *C. pubescens* significantly reduced species diversity and the cover of most species by at least 50 % in the invaded area. Endemic herbaceous species were more adversely affected than non-endemic native species (Jäger et al. 2009). Despite the fact that no species went locally extinct throughout this period where *C. pubescens* cover averaged 20 %, species did disappear when *C. pubescens* cover reached 100 % (Jäger et al. 2007). Similarly *R. niveus* had adverse impacts on the native plant community had it invaded (Rentería 2011;

Rentería et al. 2012a, b). Species richness was reduced by 56 % when *R. niveus* cover exceeded 60 %, with herbs being more affected than ferns. In addition, abundance of almost all species was significantly reduced in heavily invaded sites (>60 % *R. niveus* cover), compared with medium to low invaded sites (<60 % cover). Such studies provide a scientific basis for management intervention, although in these particular cases more information, such as social values of different elements of the invaded ecosystem, would further inform management.

Tradescantia fluminensis (wandering jew) is an example of a species that has fast become a transformer. It was first recorded in 1985, though apparently introduced in 1972 (Patricia Jaramillo, personal communication, 2012) from Loja in continental Ecuador as a ground-cover under coffee plantations (Anne Guézou, 2012). It was first observed to be invasive in 2005 around Los Gemelos (Rentería and Buddenhagen 2006), the most important remnant of *Scalesia pedunculata* forest, and is now widespread in the humid and transition zones on Santa Cruz Island (M. Gardener, M. Trueman, R. Atkinson, H. Jäger, A. Guézou, 2011). It spreads vegetatively and its seeds are dispersed long distances in the guts of giant tortoises (Blake et al. 2012) and birds (Heleno et al. 2013b). It forms a thick mat on the forest floor which inhibits recruitment of native herbaceous species (M. Gardener, personal observation, 2011). Although these observations and impacts have not been quantified in Galapagos, it is likely that this species is already causing reduced diversity and abundance in Galapagos as it has done in New Zealand (Standish et al. 2002).

In a meta-analysis, Vilà et al. (2011) showed that apart from changing biodiversity patterns, alien plant species often change ecosystem functions, including changed physical conditions which can act as irreversible barriers to restoration. In Galapagos, photosynthetically active radiation was reduced by 87 % under the *C. pubescens* canopy, while precipitation increased by 42 % because of enhanced fog interception (Jäger et al. 2009). In very dense *R. niveus* stands (>80 % cover), sunlight reaching the understory (0.5 m height) was reduced by 94 % whereas in medium to low invaded sites (<20 % cover), sunlight was reduced by 45 % (Rentería 2011). These altered abiotic conditions are expected to lead to alterations in both alien and native species composition and abundance, although in both these cases there is no reason to believe the changes are irreversible, as removal of the invader could eventually restore the original physical conditions.

Some of these impacts can be attributed to specific features of the invader. *Rubus niveus* showed faster growth rates and biomass production than four co-occurring native woody species and had a larger seed bank than native species in the invaded areas (Rentería 2011). Invasive species typically have faster growth than native species, and often also have higher specific leaf area (SLA; Daehler 2003). The latter is true for *C. pubescens*, which has a significantly higher SLA than the endemic dominant *Miconia robinsoniana* (Galapagos miconia) and the native *Pteridium arachnoideum* (bracken fern) that naturally occur in the invaded area, as well as a higher leaf turnover rate (Jäger et al. 2013). Total nitrogen, ammonium and phosphorus concentrations in soil were significantly higher in invaded compared to non-invaded areas in the Miconia zone (Jäger et al. 2013). Leaf litter from invaded areas also contained more phosphorus (Jäger et al. 2013). These results suggest that a greater cover of *C. pubescens* over *M. robinsoniana* and *P. arachnoideum* means more faster-decomposing leaves, accelerated nutrient cycling and thus increased nutrient availability in the soil (Jäger et al. 2013).

16.2.4 Impacts on Plant Reproductive Mutualisms

Plant animal mutualisms can be used as a measure of ecosystem functionality in invaded systems. Recent research in Galapagos has deepened our understanding of pollination and seed-dispersal and the role of alien plants (Buddenhagen and Jewell 2006; McMullen et al. 2008; Guerrero and Tye 2011; Heleno et al. 2011, 2013b; Chamorro et al. 2012). Many frugivorous animals, particularly birds but also reptiles, are trophic generalists which readily incorporate invasive plants into their diets and thus facilitate their spread (Bartuszevige and Gorchov 2006; Williams 2006). This makes containment difficult. In Galapagos, 12 bird and three reptile species were found to disperse seeds of alien plants (Heleno et al. 2013b). Whilst only 5 % of the total number of seeds where alien, they were present in 17 % of nearly 3,000 animal droppings and represented 24 % of all seed species. Although this suggests high usage of alien plants by frugivores, these values are still moderate when compared to relative abundance of alien plants in the Galapagos and values from other oceanic archipelagos such as Hawaii or the Azores (Chimera and Drake 2010; Heleno et al. 2013a). As in other oceanic islands, reptiles are important seed dispersers in Galapagos. The widespread lava lizards (*Microlophus* spp.) are known to disperse seeds of at least 27 species, including eight aliens (Heleno et al. 2013b) while the giant tortoises (*Chelonoidis nigra*) are known to disperse at least 48 species, including 16 aliens, over large distances (Blake et al. 2012).

Galapagos has few invasive birds, which may have limited the spread of invasive plants; however other alien vertebrates such as cattle, donkeys, rats and horses, and also native vertebrates, are contributing to their spread (Clark 1981; Fowler and Johnson 1985; Rentería and Buddenhagen 2006; Chimera and Drake 2010; Heleno et al. 2011, 2013b). The potential role of the alien bird, smooth billed ani (*Crotophaga ani*), in plant invasions is hard to evaluate, however Guerrero and Tye (2011) and Connett et al. (in press) show that it might be an important disperser of at least four invasive species including *R. niveus* and *Lantana camara* (lantana).

A community level assessment of the importance of native and introduced pollinators for both native and invasive plants is still in progress, but preliminary results suggest that while the invasive plants are predominantly affecting pollination networks on inhabited islands, introduced insects are changing species interactions patterns even on the most remote islands (Chamorro et al. 2012; Traveset et al. 2013; Traveset and Chamorro, unpublished data). The proportion of self-compatible species in the Galapagos flora, and evolutionary trends in this trait, are unclear (Tye and Francisco-Ortega 2011). The Galapagos native flora is characterised by small, drab–coloured flowers, with poor rewards and associated with a depauperate pollinator fauna (McMullen 2009; Chamorro et al. 2012). Tight co-evolution between plants and pollinators is likely to be rare in Galapagos, compared with older archipelagos, especially those with specialist bird pollinators, like Hawaii. Given the relatively recent human presence in Galapagos and the few documented extinctions, there is no reason to suspect that native plants are already threatened by limited pollination. However, this could change abruptly with extinction of a keystone pollinator species like the endemic carpenter bee (*Xylocopa darwini*) which is known to visit at least 84 plant species (McMullen 1989; Chamorro et al. 2012).

In conclusion, it is still too early to say if shifts in the assemblages of mutualists, either by direct introductions or as a consequence of vegetation shifts, will disrupt patterns of seed production and dispersal of insular plants in Galapagos. However accumulating evidence suggests that some of these disruptive processes might have already begun to occur.

16.3   Developing Management Tools

16.3.1   Weed Risk Assessment to Prioritise Management

A precautionary approach to alien plant species is especially warranted in Galapagos because 96 % of the land area is a national park which supports a world-renowned unique biodiversity. Correctly distinguishing potential invaders from non-invaders has been the main focus of weed risk assessment worldwide particularly for pre-border screening of potentially useful species (Gordon et al. 2008; Weber et al. 2009; Koop et al. 2012). While there is less pressure from new introductions in Galapagos at present, due to quarantine laws, some introductions do still occur.

With 871 alien plants already present in Galapagos and limited resources available, there is a need for a tool to prioritise management based on the risk of each species. An objective risk assessment system, using a modified Australian Weed Risk Assessment protocol (AWRA) (Pheloung et al. 1999; Daehler and Carino 2000; Gordon et al. 2008; Weber et al. 2009; Koop et al. 2012) was therefore developed, with a focus on assessment of already-introduced species. Of the 49 questions in the AWRA, a small subset are known to be most useful to predict invasiveness, i.e. questions that relate to climate match in home range, dispersal, the presence of a congeneric invader, evidence for invasiveness elsewhere, a positive response to disturbance, short maturation and ability to propagate vegetatively (Caley and Kuhnert 2006; Weber et al. 2009).

The Galapagos WRA (GWRA) has two parts (Tye, Buddenhagen and Mader, unpublished data): one to assess the potential invasiveness of a species that may be introduced to Galapagos (a screening function similar to the AWRA) and the other, which we focus on here, to describe or predict the invasiveness of plant species already present in Galapagos. Literature and internet resources provide answers to questions regarding behaviour of a species outside Galapagos, and local expert opinion provides information on behaviour in Galapagos. The answers contribute to a scoring system that allocates species into mutually exclusive categories based on definitions by Richardson et al. (2000) and Pyšek et al. (2004): (1). Transformer (e.g. *R. niveus*); (2). Naturalised likely transformer (e.g. *Piper peltatum*, Santa María); (3). Naturalised but no evidence for future invasive behaviour (i.e. it could become a transformer or integrator (e.g. *Kalanchoe tubiflora*); (4). Integrator - long established in Galapagos but believed to have low impact (e.g. *Plantago major*, greater plantain, or *Pseudelephantopus spicatus*, dogs tongue); (5). Not naturalised potential transformer (e.g. *Pueraria phaseoloides*, kudzu, or *Acacia nilotica*, Nile acacia); and (6). Not naturalised in Galapagos and not a known invader elsewhere (e.g. *Plumeria rubra*, frangipani).

Management options can then be considered for each species. One option is to do nothing: many alien plants appear to be largely harmless wherever they occur (Category 6). 'Integrators' (Category 4) would also not normally be a focus of management for conservation purposes. These are typically plants that have been naturalised in Galapagos for decades and have spread to, or had the opportunity to spread to, all suitable habitats but are still either clearly confined to areas of human disturbance or have low impact in natural areas (e.g. ephemeral, low density, small stature or low growth habit, and without other negative effects). Category 5 species are obvious targets for eradication to prevent potential future impacts. Most populations of transformers (Category 1) might be regarded as impossible to eradicate (if they are too widespread) although some rarer ones, or species likely to hybridise with endemic species, might be susceptible to eradication or worth subjecting to site-based control (to protect localised biodiversity of high value), or biocontrol or inter-island quarantine measures. Some preliminary results from the GWRA have been used to inform pilot eradication projects, though the full assessment has yet to be formally implemented in Galapagos. Once the results are communicated to management agencies, then appropriate management decisions can be made for each species.

16.3.2   Management Options

For a conservation manager, the ultimate goal of any action against invasive species should be the conservation or restoration of native species or ecosystems (Genovesi 2007). The hierarchical approach promoted by the Convention on Biological Diversity (CBD 2002) is a useful framework for invasive species management, based on the rationale that financial investment early on in the invasion process is more cost effective, and puts less strain on the natural system than controlling already established invasive species (McNeely 2001). Exclusion is the first goal, but if an incursion occurs, then early detection and timely eradication are crucial to prevent establishment. If these methods fail and the alien has established in the wild, options include containment, control, biocontrol and no action.

Prevention is by far the most important tool for managing invasion. In recognition of this the Quarantine Inspection System for Galapagos (QISG) began in 1999. In 2009 its function was incorporated into Agrocalidad – a national agency responsible for animal and plant health. The principle that all species are potentially invasive until proven otherwise, combined with a permitted species list, is the basis of the QISG. Inspectors are based at sea and air ports in continental Ecuador and in Galapagos. There are also strict quarantine rules and inspections for travel between islands, especially to uninhabited islands, which, for the most part are free of transformer species. When species that are not permitted are detected, they are identified and either lodged in reference collections or destroyed. Although the QISG has created a high level of consciousness about invasive species and may have slowed the importation of alien plants, an evaluation in 2007 found it to be largely ineffective because it was under-funded and under-staffed (Zapata 2007).

Eradication is the best alternative after prevention fails, before spread is significant. The GWRA was used to choose plant species for a pilot eradication programme in Galapagos. Thirty populations (where a population = one species on one island) were chosen for evaluation for eradication feasibility, based on three criteria: (i) limited distribution as known from inventories, field surveys, and interviews with landowners (e.g. Soria et al. 2002); (ii) proven behaviour as invasive elsewhere in the world with a climate similar to Galapagos (e.g. Buddenhagen 2006); or (iii) already invasive somewhere in Galapagos. Information was collected about the factors affecting feasibility of eradication and its cost, including among other things discussions with landowners. Twenty-one populations (18 species) reached management stage, the earliest in 1996. Four of the eradication operations were successful, all in Santa Cruz; *Rubus adenotrichus* (mora silvestre) and *R. megalococcus* (Sarsa mora; Buddenhagen 2006), *P. phaseoloides* (Tye 2007), and *Cenchrus pilosus* (abrojo). Each of these species covered less than 1 ha in net area and was on land with a single owner (Gardener et al. 2010b). It is likely that two more species, *Persea americana* (avocado) and *Sapindus saponaria* (soapberry), have been eradicated in Santiago: both these are trees with slow maturation times, no long-lived seed bank, and limited distributions. Most of the remaining projects were abandoned after 1 or 2 years of data collection, mainly because their extent was greater than initially thought, and because search and control costs were too high. In some cases, preventing long distance seed dispersal and managing a long-lived seed bank further hampered success, for example *R. niveus* (Rentería et al. 2012a). Another barrier to eradicating species with small distributions was a lack of permission from land-owners to remove plants (e.g. *A. nilotica* and *Cryptostegia grandiflora*, rubber vine). The Special Law for Galapagos (Congreso Nacional 1998) actually has provisions that allow government officials to enter private property for managing invasive species; however, unfortunately this provision, like many others, has not been enforced.

Containment is a component of eradication projects and also a management option in its own right. The requirements for containment are the reduction of long distance dispersal and the timely detection of new foci (Panetta and Cacho 2012). Because most invasive species in Galapagos are either wind or animal dispersed, the vectors of dispersal are difficult to manage, as is plant fecundity. As we can see from the *R. niveus* example above, these conditions may be impossible to meet when a plant is widely naturalised.

Control or maintenance management is controlling an invader at a density sufficiently low to produce the desired conservation outcome. Typical maintenance options include mechanical, chemical, and biological control (Simberloff 2003). In Galapagos, both manual and chemical controls are used and have been developed for various species (Gardener et al. 1999; Buddenhagen et al. 2004). Control, unlike eradication, is indefinite and, although it may not have a large initial cost, can become accumulatively expensive in the long-term. It is thus important to undertake control only where the conservation objective is achievable and where potential biodiversity or socio-economic losses are considered to be unacceptable if no management action is taken. This means that the target species should only be reduced to below a threshold of impact.

Local control is carried out by the Galapagos National Park Service (GNPS) in sites of high biodiversity value or importance for tourism. A recent evaluation of alien plant control projects in the GNP between 2005 and 2010 included 17 projects on five islands covering 11 species (García and Gardener 2012). Approximately US $3,350,000 were spent on these projects to manage a total area of nearly 2,000 ha at a cost of US$280 per ha per year. Control outcomes were varied: some, such as the control of the tree species *C. odorata* and *C. pubescens*, reduced densities to below a threshold of impact whereas others, such as control of *R. niveus*, were ineffective and the disturbance may have facilitated further invasion; a major constraint was rapid recolonisation from the seed bank or vegetative shoots (García and Gardener 2012).

Biological control of widespread invasive species has been used effectively for the last century, sometimes providing exceptional value for money. For example, a recent economic analysis has shown an overall benefit–cost ratio of 23:1 for biological weed control programmes in Australia (Paige and Lacey 2006). Van Driesche et al. (2010) reviewed weed biological control projects for protection of natural ecosystems worldwide and found that 60 % achieved useful levels of control. There are also inherent risks associated with biological control that can lead to unintended consequences, thus informed decision-making is critically important (Simberloff 2012). Biological control has been successfully implemented in Galapagos for the management of the invasive *Icerya purchasi* (cottony cushion scale) using a cardinal ladybird (*Rodolia cardinalis*; Calderón-Alvarez et al. 2012). This success led to interest for other projects including the development of biological control agents for *R. niveus* and *L. camara* (Rentería and Ellison 2007; Atkinson et al. 2009). If implemented, biological control for *R. niveus* would require a significant up-front investment and could take up to 10 years – the development of a control agent was quoted at USD$660,000 by the Centre for Agricultural Bioscience International (Mark Gardener, 2011) and does not include management and coordination costs. To put this in context, the cost of control action for *R. niveus* on a single island (Santiago) has been approximately USD $582,000 over 6 years (Rentería et al. 2012a).

16.4   Challenges of Applying Management

16.4.1   Managing Species or Ecosystems

All of the above methods are aimed at managing alien species individually. Legislation is also often prescriptive at species level (e.g. declared weeds), and management of single species is operationally easier to achieve. However, single weed management strategies can result in unwanted or unexpected negative outcomes at the community or ecosystem levels (Zavaleta et al. 2001). The most common of these unwanted outcomes is where an invasive species is reduced in density only to be replaced by another. For example, in Galapagos the disturbance created by control of *C. pubescens* may have facilitated invasion by *R. niveus* (Jäger and Kowarik 2010). The recognition that interactions among species are crucial to maintain ecosystem functioning (Duffy et al. 2007) has recently highlighted the importance of framing conservation efforts at the community level. This means including information on how species interact (Millennium Ecosystem Assessment 2005).

In Galapagos, where invasions have not drastically altered ecosystem processes, such as on uninhabited islands and to a lesser extent dry lowlands of inhabited islands, managing alien species individually is still a worthwhile approach. There is also a moral imperative to keep these areas as close to their pre-human state as possible. Galapagos still has 95 % of its original, pre-human biodiversity (Bensted-Smith et al. 2002) and relative to most other oceanic archipelagos is in good ecological order. In these areas, a suitable goal is "the restoration of the populations and distributions of all extant native biodiversity and of natural ecological/evolutionary processes to the conditions prior to human settlement" (Snell et al. 2002b). In these areas we therefore support a species-led approach focusing on eradication and containment of priority species.

However, on the inhabited islands, particularly in the humid highlands, ecosystems are highly degraded. There are so many different invasive species and ecological impacts that the removal of one species is likely to result in replacement by another invasive. Here, the pre-human state is not fully attainable given realistically available resources. Novel ecosystems, those that have new species combinations arising through either species invasions or environmental change, are widespread on continents and islands and often objects of conservation for their own sake (Chapin and Starfield 1997; Vitousek et al. 1997; Jackson and Hobbs 2009). Owen (1998) describes a site-led approach which is based on the biodiversity value of a site, level of disturbance, the risk of aliens, and the cost of management. A more sophisticated tool for the same approach is scenario planning, which uses social and biological data to model outcomes of different management strategies, assisting stakeholders to make informed decisions (Roura-Pascual et al. 2011; Hulme 2012). Such approaches may be more suitable for planning conservation in the more degraded ecosystems in Galapagos. The goal would be to maintain as much native biodiversity as possible, together with original functionality, and undertake management interventions that maximise benefits over the total area of intervention (Gardener et al. 2010a). Transformer species that directly impact biodiversity will still require drastic intervention to prevent biodiversity loss, and biocontrol may be an option for such species.

16.4.2   Involving People

Management of alien plants is a matter of societal choice: perceptions, values, motivations, desires and needs of people will determine if and how management goals are determined and attained. In Galapagos, various stakeholders are responsible for this management. The GNPS (national government) manages alien plants within the GNP. However, alien plant invasions in the GNP spread from the inhabited areas. The municipalities (local governments) are responsible for managing urban and agricultural lands but the provincial Governing Council and the Ministry of Agriculture (national government) also have important roles on these private lands. The prevention of new introductions comes under the remit of the QISG, which is part of Agrocalidad (another national government department). It is a challenge to have all these agencies working together for overall coordinated management of alien plants in the archipelago. Additionally, all of this top-down management can only be effective with community support. We expand on some of the challenges below, relating them to the various aspects of alien plant management discussed above.

Within the GNP, major challenges to successful management are limited technical capacity and scientific knowledge. The lack of scientific knowledge is compounded by the long history of poor information transfer from researchers to managers, including language barriers. In the past this has resulted in some conflict within and between stakeholder groups, or distrust from the community. As a result, an inter-institutional committee supported by a trust fund for the management of invasive species was established in 2011 to mitigate some of these problems. This has helped in some ways. For example, GNPS provided free herbicides and equipment for the control of invasive species on private land. Another issue is that private landowners rarely work together, so that even if one farmer controls invasions on his or her own land, reinvasion is highly likely from surrounding farms and the GNP.

The limitation in technical capacity is perhaps more concerning. Within GNP, there is no long-term institutional commitment to an adaptive, well designed invasive plant management strategy. For example, management goals (including eradication, control and biocontrol) are not chosen strategically and appropriate techniques are not employed. Working toward achievable objectives, tracking costs, documenting plans, results, and failures for any project that is implemented requires much discipline. This long-term vision needs to be part of the organisational culture if effective management is to be achieved. There is also a need for regular training and for monitoring, evaluation, and adaptive management.

Differences in values held by various sectors of the community can lead to conflicts when species that produce benefits to the community are also invasive. Examples are *C. odorata* and *C. pubescens* that are both valuable timber trees and also highly invasive in the GNP and the agricultural zone. Extraction of both trees is allowed under permit from the GNPS. Ironically, both species (which are native to Ecuador) are considered threatened by over harvesting in their native range. *Cedrela odorata* is listed by the IUCN as Vulnerable (IUCN 2011). There is pressure to develop a sustainable harvest of *C. odorata* in Galapagos, despite a GNPS campaign to control it in some parts of Santa Cruz.

Another challenge is changing values over time. Approximately 80 % of the alien plants species in Galapagos were introduced on purpose, for medicine, food, timber, forage, or ornamentation. Many of them are already invasive. Thus some species that were earlier regarded as successful cultivations are now considered serious pests. Other species may become invasive in the future and suffer the same change in value. An example is *Aristolochia elegans* (Dutchman's pipe) which is said to cure stomach problems; there are currently less than 20 plants on a single farm in Galapagos. However it is known to be invasive and problematic in Australia (Skoien and Csurhes 2009) and could potentially cause future problems in Galapagos. This would constitute an easy eradication target but permission was not given by the landholder because it is considered a useful (yet harmless) species. In this and other examples in Galapagos, it has sometimes proven difficult to explain the precautionary principle to the public.

One of the big challenges in Galapagos is the continual influx of people from the mainland, many of whom are not aware of the vulnerability of Galapagos ecosystems to their actions. Thus a continual awareness-raising and education approach is needed for all stakeholders to produce a community based vision and integrated action plan for invasive species management. The best example of a win-win solution so far in Galapagos has been with a native garden project – providing native species as alternatives to introduced ornamentals, thus reducing the future threat to the GNP and to agriculture on the islands (Atkinson et al. 2010). There has been a high level of community and local government support for the project which has raised awareness about the potential harm of alien plants, whilst also satisfying a social need for ornamental gardens.

16.5 Conclusion: Core Lessons Learnt

Research into alien plant issues in Galapagos has contributed greatly to our understanding of introductions, invasions, impacts and management options. Nine core lessons have emerged from this work over the last few decades and can provide valuable advice to other regions:

(i)   Most plants were introduced to Galapagos in the last 50 years and appear to be still in the early stage of invasion. Even if no new introductions occur, the number of naturalised, invasive and transforming species will increase.

(ii)   Detailed surveys focused on inhabited areas (the nuclei of plant introductions), and freely available results, form the cornerstone of effective weed management. Continuous monitoring is required to detect new introduced species and help to understand how naturalised species might be spreading.

(iii) Impacts of invasive plants are not fully understood. Where quantified, increasing invasive species cover was correlated with decreases in native diversity and abundance. Impacts were also observed on physical parameters such as light, water and nutrients. There is little understanding of ecosystem-level impacts and the dynamic nature of the invasions.

(iv)   Our understanding of mutualistic networks in Galapagos and the impact of invasive plants on them is still limited. To date it appears that seed-dispersal and pollination networks are still not severely impacted but are changing.

(v)   The Galapagos Weed Risk Assessment system can help prioritise the management of alien plant species.

(vi)   A number of tools for prevention, eradication and control have been developed in Galapagos. Disciplined application of prevention and early intervention strategies should lead to successes with fewer resources. However, it is still common practice to try to manage species that are too widespread for cost-effective control.

(vii) In the more pristine areas of Galapagos, a species-led approach to invasive plant management is appropriate. In highly degraded areas, a more complex approach is needed to prioritise management spending at a site level to achieve optimal outcomes for biodiversity and ecosystem function.

(viii) Within institutions responsible for alien plant management, the main challenges to successful outcomes are the limitation in technical capacity and scientific knowledge, and lack of information transfer.

(ix)   Local communities are central to weed management and must be fully involved in management planning and implementation.

As discussed throughout the chapter, most alien plants and their impacts are concentrated on the four inhabited islands, especially in the humid highlands. The more pristine uninhabited islands have been protected from plant introductions due to their remoteness. This remoteness has two forms. First, human visits to these uninhabited islands are very limited and highly regulated, so direct impacts are minimal. Second, the physical (insular) distance from the inhabited islands limits the dispersal of introduced plants. However, increased human traffic from tourism, conservation activities and illegal camps reduce remoteness and increase the probability of chance introductions. Furthermore, there are early signs that the insularity is decreasing. Research shows that introduced plants are dispersed by frugivores (Heleno et al. 2013a) and some of these may be capable of inter-island movement. For these reasons the management of the inhabited and fully protected islands must be integrated.

Galapagos is iconic in the conservation world and many people feel that if it is impossible to manage plant invasions there, then there is not much hope for the rest of the world. A number of opportunities exist in Galapagos that are not available in other protected areas: (i) Insular geography with limited pathways for further introductions; (ii) 96 % of the land area is protected; (iii) 95 % of known native species are still extant; (iv) Few invasive plants are established or widespread in the dry lowlands: preventative action here would be beneficial to biodiversity protection; (v) There is a relatively high level of awareness of invasive species, with supporting legislation; (vi) The small size of the human population presents the opportunity to further raise awareness among the majority of residents; (vii) Labour costs are relatively low, so management costs are not prohibitive; (viii) Few introduced vertebrates are established, resulting in limited seed dispersal; and (ix) Pollinator and disperser failure has not occurred in native plants yet.

Scientists, managers, governing agencies and the community need to work together to overcome barriers and make the most of these opportunities for effectively managing plant invasions. As in Hawaii, both conservation science and management must "focus on explicit planning that adequately reflects biological and fiscal realities, rather than impractical, unfocused, and unachievable wish lists" (Duffy and Kraus 2006). A shared, realistic vision and a pragmatic approach to achieve that vision will be essential for success in Galapagos.

Acknowledgements   We thank Dane Panetta and Anne Guézou for providing helpful comments on a draft of this chapter. Thanks also to Ole Hamann and Frank Sulloway for contributing photographs. This chapter is contribution number 2054 of the Charles Darwin Foundation for the Galapagos Islands.

References

Aldaz I, Tye A (1999) Effects of the 1997–98 El Niño event on the vegetation of Alcedo Volcano, Isabela Island. Noticias de Galápagos 60:25–28

Atkinson RJ, Guézou A, Rentería J et al (2009) Diagnostico y planificacion para el desarrollo de un agente de control biológico para *Rubus niveus* en las islas Galápagos. Fundacion Charles Darwin y Servicio Parque Nacional Galápagos, Puerto Ayora, Galapagos

Atkinson R, Trueman M, Guézou A et al (2010) Native gardens for Galapagos – can community action prevent future plant invasions? In: Toral-Granda MV, Cayot L, Marin-Luna A (eds) Galapagos report 2009–2010. Charles Darwin Foundation, Galapagos National Park and Governing Council of Galapagos, Puerto Ayora, pp 159–163

Bartuszevige AM, Gorchov DL (2006) Avian seed dispersal of an invasive shrub. Biol Invasion 8:1013–1022

Bensted-Smith R, Powell G, Dinerstein E (2002) Planning for the ecoregion. In: Bensted-Smith R (ed) A biodiversity vision for the Galapagos Islands. Charles Darwin Foundation and World Wildlife Fund, Puerto Ayora, pp 11–16

Blake S, Wikelski M, Cabrera F et al (2012) Seed dispersal by Galápagos tortoises. J Biogeogr 39:1961–1972

Buddenhagen CE (2006) The successful eradication of two blackberry species *Rubus megalococcus* and *R. adenotrichos* (Rosaceae) from Santa Cruz Island, Galapagos, Ecuador. Pac Conserv Biol 12:272–278

Buddenhagen CE, Jewell KJ (2006) Invasive plant seed viability after processing by some endemic Galapagos birds. Ornit Neotrop 17:73–80

Buddenhagen CE, Rentería JL, Gardener M et al (2004) The control of a highly invasive tree *Cinchona pubescens* in Galapagos. Weed Technol 18:1194–1202

Calderón-Alvarez C, Causton CE, Hoddle MS et al (2012) Monitoring the effects of *Rodolia cardinalis* on *Icerya purchasi* populations on the Galapagos Islands. Biocontrol 57:167–179

Caley P, Kuhnert PM (2006) Application and evaluation of classification trees for screening unwanted plants. Aust Ecol 31:647–655

CBD (2002) Decision VI/23: alien species that threaten ecosystems, habitats or species to which is annexed. Guiding principles for the prevention, introduction and mitigation of impacts of alien species that threaten ecosystems, habitats or species. Sixth Conference of the Parties, 7–19 Apr 2002, The Hague

CDF (2009) Charles Darwin Foundation collections database. Charles Darwin Foundation, Puerto Ayora, Galapagos. http://www.darwinfoundation.org/datazone/collections. Accessed 13 July 2009

Chamorro S, Heleno R, Oslesen JM et al (2012) Pollination patterns and plant breeding systems in the Galápagos: a review. Ann Bot 110:1489–1501

Chapin FS, Starfield AM (1997) Time lags and novel ecosystems in response to transient climatic change in Alaska. Clim Change 35:449–461

Chimera C, Drake D (2010) Patterns of seed dispersal and dispersal failure in a Hawaiian dry forest having only introduced birds. Biotropica 42:493–502

Clark DA (1981) Foraging patterns of black rats across a desert-montane forest gradient in the Galapagos Islands. Biotropica 13:182–194

Coffey EED, Froyd CA, Willis KJ (2011) When is an invasive not an invasive? Macrofossil evidence of doubtful native plant species in the Galapagos Islands. Ecology 92:805–812

Congreso Nacional (1998) Ley de regimen especial para la conservacion y desarrollo sustentable de la provincia de Galápagos. Quito, Registro Oficial No 278, 18 de marzo, de, p 31

Connett L, Guézou A, Herrera HW et al (in press) Gizzard contents of the Smooth-billed Ani, *Crotophaga ani* (L.), In: The Galapagos Islands Ecuador. Galap Res

Daehler CC (2003) Performance comparisons of co-occuring native and alien invasive plants: implications for conservation and restoration. Annu Rev Ecol Evol Syst 34:183–211

Daehler CC, Carino DA (2000) Predicting invasive plants: prospects for a general screening system based on current regional models. Biol Invasion 2:93–102

Duffy D, Kraus F (2006) Science and the art of the solvable in Hawai'i's extinction crisis. Environ Hawaii 16:3–6 http://manoa.hawaii.edu/hpicesu/papers/2006_Science_and_the_Art.pdf. Accessed 15 Dec 2012

Duffy JE, Carinale BJ, France KE et al (2007) The functional role of biodiversity in ecosystems: incorporating trophic complexity. Ecol Lett 10:522–538

Fowler LE, Johnson MK (1985) Diets of giant tortoises and feral burros on Volcan Alcedo, Galapagos. J Wildl Manage 49:165–169

García G, Gardener MR (2012) Evaluación de proyectos de control de plantas transformadoras y reforestación de sitios de alta valor en Galápagos. Galapagos National Park and Charles Darwin Foundation, Puerto Ayora, Galapagos, Ecuador

Gardener MR, Tye A, Wilkinson SR (1999) Control of introduced plants in the Galapagos Islands. In: Bishop AC, Boersma M, Barnes CD (eds) Proceedings from the twelfth Australian weeds conference, Hobart, pp 396–400

Gardener M, Atkinson R, Rueda D et al (2010a) Optimizing restoration of the degraded highlands of Galapagos: a conceptual framework. In: Toral-Granda MV, Cayot L, Marin-Luna A (eds) Galapagos report 2009–2010. Charles Darwin Foundation, Galapagos National Park and Governing Council of Galapagos, Puerto Ayora, Galapagos, pp 164–169

Gardener MR, Atkinson R, Rentería JL (2010b) Eradications and people: lessons from the plant eradication program in Galapagos. Restor Ecol 18:20–29

Genovesi P (2007) Limits and potentialities of eradication as a tool for addressing biological invasions. In: Nentwig W (ed) Biological invasions. Springer, Berlin, pp 385–402

Gordon DR, Onderdonk DA, Fox AM et al (2008) Consistent accuracy of the Australian weed risk assessment system across varied geographies. Divers Distrib 14:234–242

Grenier C (2007) Conservación contra natura, Las Islas Galápagos, 2nd edn, Travaux de l'Institut Francais d'Etudes Andines. Instituto Francés de Estudios Andinos (IFEA), Lima

Guerrero AM, Tye A (2011) Native and introduced birds of Galapagos as dispersers of native and introduced plants. Ornit Neotrop 22:207–217

Guézou A, Trueman M, Buddenhagen CE et al (2010) An extensive alien plant inventory from the inhabited areas of Galapagos. PLoS ONE 5:e10276

Gurevitch J, Padilla DK (2004) Are invasive species a major cause of extinctions? Trends Ecol Evol 19:470–474

Hamann O (1974) Contribution to the flora and vegetation of the Galapagos Islands III. Five new floristic records. Bot Notiser 127:309–316

Hamann O (1985) The El Niño influence on the Galápagos vegetation. In: Robinson G, del Pino E (eds) El Niño in the Galápagos Islands: the 1982–1983 Event. Charles Darwin Foundation, Quito, pp 299–330

Heleno R, Blake S, Jaramillo P et al (2011) Frugivory and seed dispersal in the Galápagos: what is the state of the art? Integr Zool 6:110–128

Heleno R, Ramos JA, Memmott J (2013a) Integration of exotic seeds into an Azorean seed dispersal network. Biol Invasion 15:1143–1154

Heleno R, Olesen JM, Nogales M et al (2013b) Seed-dispersal networks in the Galápagos and the consequences of plant invasions. Proc R Soc B Biol Sci 280:20122112

Hooker JD (1851) On the vegetation of the Galapagos Archipelago. Trans Linn Soc Lond 20:235–262

Hosking JR, Waterhouse BM, Williams PA (2004) Are we doing enough about early detection of weed species naturalising in Australia. In: 14th Australian weeds conference: papers and proceedings. Weed management – balancing people, planet, profit. Weed Society of NSW, Wagga Wagga, New South Wales, p 718

Hulme PE (2012) Weed risk assessment: a way forward or a waste of time? J Appl Ecol 49:10–19

INEC (2007) Difusión de resultados definitivos del censo de Población y Vivienda 2006. Instituto Nacional de Estadistica y Censos, Quito

INEC (2011) Instituto Nacional de Estadística y Censos 2010. http://www.inec.gov.ec/estadisticas. Accessed 2 Apr 2012

IUCN (2011) The IUCN red list of threatened species. Version 2011.2. http://www.iucnredlist.org. Accessed 2 Apr 2012

Jackson ST, Hobbs RJ (2009) Ecological restoration in the light of ecological history. Science 325:567–569

Jäger H, Kowarik I (2010) Resilience of native plant community following manual control of invasive Cinchona pubescens in Galapagos. Restor Ecol 18:103–112

Jäger H, Tye A, Kowarik I (2007) Tree invasion in naturally treeless environments: impacts of quinine (Cinchona pubescens) trees on native vegetation in Galapagos. Biol Conserv 140:297–307

Jäger H, Kowarik I, Tye A (2009) Destruction without extinction: long-term impacts of an invasive tree species on Galápagos highland vegetation. J Ecol 97:1252–1263

Jäger H, Alencastro MJ, Kaupenjohann M et al (2013) Ecosystem changes in Galápagos highlands by the invasive tree Cinchona pubescens. Plant Soil 371:629–640

Jaramillo P, Guézou A (2012) CDF Checklist of Galapagos vascular plants. Charles Darwin Foundation. http://www.darwinfoundation.org/datazone/collections/. Accessed 9 Dec 2012

Koop A, Fowler L, Newton L et al (2012) Development and validation of a weed screening tool for the United States. Biol Invasion 14:273–294

Landazuri O (2002) Distribución, fenología reproductiva y dinámica del banco de semillas de mora (*Rubus niveus* Thunb.) en la parte alta de la isla Santa Cruz, Galápagos. Universidad Central de Ecuador, Quito, Ecuador

Lawesson JE, Ortiz L (1990) Plantas introducidas en las Islas Galapagos. In: Lawesson JE, Hamann O, Rogers G et al (eds) Botanical research and management in the Galapagos Islands. Monogr Syst Bot Missouri Bot Garden 32:201–211

Lundh J (2006) The farm area and cultivated plants on Santa Cruz, 1932–1965, with remarks on other parts of Galapagos. Galapagos Res 64:12–25

Luong TT, Toro B (1985) Cambios en la vegetación de las islas Galápagos durante "El Niño" 1982–1983. In: Robinson G, del Pino E (eds) El Niño in the Galápagos Islands: the 1982–1983 event. Charles Darwin Foundation, Quito, pp 331–342

McMullen C (1989) The Galápagos carpenter bee, just how important is it? Noticias Galápagos 48:16–18

McMullen CK (2009) Insular flora: more than "wretched-looking little weeds". In: De Roy T (ed) Galápagos: preserving Darwin's legacy. David Bateman Ltd, Albany, pp 60–66

McMullen CK, Tye A, Hamann O (2008) Botanical research in the Galápagos Islands: the last fifty years and the next fifty. Galapagos Res 65:43–45

McNeely JA (2001) The great reshuffling: human dimensions of invasive alien species. IUCN, Gland

Millennium Ecosystem Assessment (2005) Ecosystems and human well-being: a synthesis. Island Press, Washington, DC

Owen SJ (1998) Department of conservation strategic plan for managing invasive weeds. Department of Conservation, Wellington

Paige AR, Lacey KL (2006) Economic impact assessment of Australian weed biological control. CRC for Australian Weed Management Technical Series No. 10, Glen Osmond, Australia

Panetta FD, Cacho OJ (2012) Beyond fecundity control: which weeds are most containable? J Appl Ecol 49:311–321

Panetta FD, Timmins SM (2004) Evaluating the feasibility of eradication for terrestrial weed incursions. Plant Protect Q 19:5–11

Pheloung PC, Williams PA, Halloy SR (1999) A weed-risk assessment model for use as a biosecurity tool evaluating plant introductions. J Environ Manage 57:239–251

Pyšek P, Richardson DM, Rejmánek M et al (2004) Alien plants in checklists and floras: towards better communication between taxonomists and ecologists. Taxon 53:131–143

Rentería JL (2002) Ecología y manejo de la cascarilla (*Cinchona pubescens* Vahl), en Santa Cruz, Galápagos. Área Agropecuaria y de Recursos Naturales Renovables. Universidad Nacional de Loja, Loja, Ecuador

Rentería JL (2011) Towards an optimal management of the invasive plant *Rubus niveus* in the Galapagos Islands. Imperial College London, London

Rentería JL, Buddenhagen CE (2006) Invasive plants in the *Scalesia pedunculata* forest at Los Gemelos, Santa Cruz, Galapagos. Galapag Res 64:31–35

Rentería JL, Ellison C (2007) Potential biological control of *Lantana camara* in the Galapagos using the rust *Puccinia lantanae*. In: Julien MH, Sforza R, Bon MC (eds) Proceedings of the XII international symposium on biological control of weeds, La Grande Motte, France, 22–27 Apr 2007, p 361

Rentería JL, Gardener MR, Panetta FD et al (2012a) Management of the invasive hill raspberry (*Rubus niveus*) on Santiago Island, Galapagos: eradication or indefinite control? Invasive Plant Sci Manage 5:37–46

Rentería JL, Gardener MR, Panetta FD et al (2012b) Possible impacts of the invasive plant *Rubus niveus* on the native vegetation of the Scalesia forest in the Galapagos Islands. PLoS ONE 7: e48106

Richardson DM, Pyšek P, Rejmanek M et al (2000) Naturalization and invasion of alien plants: concepts and definitions. Divers Distrib 6:93–107

Roura-Pascual N, Richardson DM, Chapman RA et al (2011) Managing biological invasions: charting courses to desirable futures in the Cape Floristic Region. Reg Environ Change 11:311–320

Schofield EK (1973) Galápagos flora: the threat of introduced plants. Biol Conserv 5:48–51

Simberloff D (2003) Eradication- preventing invasions at the onset. Weed Sci 51:247–253

Simberloff D (2012) Risks of biological control for conservation purposes. Biocontrol 57:263–276

Skoien P, Csurhes S (2009) Weed risk assessment Dutchman's pipe *Aristolochia elegans*. Queensland Primary Industries and Fisheries, Brisbane

Snell HM, Stone PA, Snell HL (1996) A summary of geographical characteristics of the Galápagos Islands. J Biogeogr 23:619–624

Snell HL, Tye A, Causton CE et al (2002a) Current status of and threats to the terrestrial biodiversity of Galapagos. In: Bensted-Smith R (ed) A biodiversity vision for the Galapagos Islands. Charles Darwin Foundation and World Wildlife Fund, Puerto Ayora, Galapagos, pp 30–47

Snell HL, Tye A, Causton C et al (2002b) Projections for the future: a terrestrial biodiversity vision. In: Bensted-Smith R (ed) A biodiversity vision for the Galapágos Islands. Charles Darwin Foundation and World Wildlife Fund, Puerto Ayora, pp 48–59

Soria M, Gardener MR, Tye A (2002) Eradication of potentially invasive plants with limited distributions in the Galapagos islands. In: Veitch D, Clout M (eds) Turning the tide: the eradication of invasive species. Invasive Species Specialty Group of the World Conservation Union (IUCN), Auckland, pp 287–292

Standish RJ, Robertson AW, Williams PA (2002) The impact of an invasive weed *Tradescantia fluminensis* on native forest regeneration. J Appl Ecol 38:1253–1263

Traveset A, Heleno R, Chamorro S et al. (2013) Invaders of pollination networks in the Galápagos Islands: Emergence of novel communities. Proc R Soc - B 280:2012–3040.

Trueman M (2008) Minimising the risk of invasion into the Galapagos National Park by introduced plants from the inhabited areas of the Galapagos Islands. Masters, Charles Darwin University, Darwin

Trueman M, d'Ozouville N (2010) Characterizing the Galapagos terrestrial climate in the face of climate change. Galapagos Res 67:27–37

Trueman M, Atkinson R, Guézou AP et al (2010) Residence time and human-mediated propagule pressure at work in the alien flora of Galapagos. Biol Invasion 12:3949–3960

Tye A (2001) Invasive plant problems and requirements for weed risk assessment in the Galápagos islands. In: Groves RH, Panetta FD, Virtue JD (eds) Weed risk assessment. CSIRO Publishing, Melbourne, pp 153–175

Tye A (2006) Can we infer island introduction and naturalization rates from inventory data? Evidence from introduced plants in Galápagos. Biol Invasion 8:201–215

Tye A (2007) Cost of rapid-response eradication of a recently introduced plant, tropical kudzu (*Pueraria phaseoloides*), from Santa Cruz Island, Galapagos. Plant Protect Q 22:33–34

Tye A, Aldaz I (1999) Effects of the 1997–98 El Niño event on the vegetation of Galápagos. Noticias Galápagos 60:22–24

Tye A, Francisco-Ortega J (2011) Origins and evolution of Galapagos endemic vascular plants. In: Bramwell D, Caujapé-Castells J (eds) The biology of island floras. Cambridge University Press, Cambridge, pp 89–153

Van Driesche RG, Carruthers RI, Center T et al (2010) Classical biological control for the protection of natural ecosystems. Biol Control 54:S2–S33

van Leeuwen JFN, Froyd CA, van der Knaap WO et al (2008) Fossil pollen as a guide to conservation in the Galápagos. Science 322:1206

Vilà M, Espinar JL, Hejda M et al (2011) Ecological impacts of invasive alien plants: a meta-analysis of their effects on species, communities and ecosystems. Ecol Lett 14:702–708

Vitousek PM, D'Antonio OM, Loope LL et al (1997) Introduced species: a significant component of human-caused global change. NZ J Ecol 21:1–16

Watson J, Trueman M, Tufet M et al (2009) Mapping terrestrial anthropogenic degradation on the inhabited islands of the Galápagos archipelago. Oryx 44:79–82

Weber J, Panetta FD, Virtue J et al (2009) An analysis of assessment outcomes from eight years' operation of the Australian border weed risk assessment system. J Environ Manage 90:798–807

Williams PA (2006) The role of blackbirds (*Turdus merula*) in weed invasion in New Zealand. NZ J Ecol 30:285–291

Zapata CE (2007) Evaluation of the quarantine and inspection system for Galapagos (SICGAL) after seven years. In: Cayot L (ed) Galapagos report 2006–2007. Charles Darwin Foundation, Galapagos National Park & INGALA, Puerto Ayora

Zavaleta ES, Hobbs RJ, Mooney HA (2001) Viewing invasive species removal in a whole-ecosystem context. Trends Ecol Evol 16:454–459

Chapter 17
Invasive Alien Plants in the Azorean Protected Areas: Invasion Status and Mitigation Actions

Hugo Costa, Maria José Bettencourt, Carlos M.N. Silva,
Joaquim Teodósio, Artur Gil, and Luís Silva

Abstract This chapter addresses plant invasions in the protected areas of the Azores (Northern Atlantic), whose flora encompasses a considerable proportion of alien species (about 70 %). The chapter includes (i) a general characterization of the Azores, with particular reference to their Island Natural Parks covering 24 % of the inland surface; (ii) an assessment of the plant invasion status of the Island Natural Parks (based on distribution data and expert evaluation of potential impacts and possibility of control of invasive alien plants); and (iii) a report about on-going and recent management initiatives embracing the control of invasive alien plants.

H. Costa
CIBIO, Centro de Investigação em Biodiversidade e Recursos Genéticos, InBIO Laboratório Associado, Pólo dos Açores Departamento de Biologia, Universidade dos Açores, Apartado 1422, Ponta Delgada 9501-801, Portugal

School of Geography, University of Nottingham, NG7 2RD, Nottingham, UK
e-mail: lgxhag@nottingham.ac.uk

M.J. Bettencourt
Direção Regional do Ambiente dos Açores, Rua Cônsul Dabney –Colónia Alemã, Horta 9901-014, Portugal
e-mail: maria.jv.bettencourt@azores.gov.pt

C.M.N. Silva • J. Teodósio
Sociedade Portuguesa para o Estudo das Aves SPEA, Avenida João Crisóstomo 18-4D, Lisboa 1000-179, Portugal
e-mail: carlos.silva.spea@gmail.com; joaquim.teodosio@spea.pt

A. Gil
Azorean Biodiversity Group, CITA-A, Departamento de Biologia, Universidade dos Açores, Rua da Mãe de Deus 13A, Ponta Delgada 9501-801, Portugal
e-mail: arturgil@uac.pt

L. Silva (✉)
CIBIO, Centro de Investigação em Biodiversidade e Recursos Genéticos, InBIO Laboratório Associado, Pólo dos Açores Departamento de Biologia, Universidade dos Açores, Apartado 1422, Ponta Delgada 9501-801, Portugal
e-mail: lsilva@uac.pt

L.C. Foxcroft et al. (eds.), *Plant Invasions in Protected Areas: Patterns, Problems and Challenges*, Invading Nature - Springer Series in Invasion Ecology 7, DOI 10.1007/978-94-007-7750-7_17, © Springer Science+Business Media Dordrecht 2013

The results show that the Island Natural Parks of Santa Maria Island is potentially the most threatened by invasive alien plants, followed in decreasing order by the Island Natural Parks of the islands of Graciosa, São Jorge, Corvo, Faial, São Miguel, Terceira, Flores and Pico. Some of the most threatening species are highlighted. Due to the innovative assessment methodology, the results do not fully corroborate previous studies, showing that just species listing may not provide a full understanding of the potential effects of invasive alien plants on native biodiversity, thus bringing new insights that may assist management initiatives. Several invasive alien plants management projects run by the Azorean Government and the Portuguese Society for the Study of Birds are described, with reference to those supported by the LIFE programme and PRECEFIAS (an Azorean project devoted to control of invasive alien plants in protected areas). Finally, a holistic discussion is provided stressing strengths and weaknesses of all topics covered in the chapter so that more effective invasive alien plant management strategies can be achieved in the future.

Keywords Azores • Invasion status • Island Natural Parks • Plant invasions • Protected areas

17.1 Introduction

Biological invasions are a major threat to native biodiversity (Mack et al. 2000; Millennium Ecosystem Assessment 2005), particularly on oceanic islands (Reaser et al. 2007; Caujapé-Castells et al. 2010) because their biota have evolved in isolation and their environment has suffered extensive anthropogenic changes (Loope and Mueller-Dombois 1989). Also, islands generally correspond to relatively small areas with large proportions of endemic species with little chance of escaping from sudden threats, including habitat destruction and invasion by alien species (Martín et al. 2010). With reference to plants, often alien species form dominant stands and can modify the three-dimensional structure of native plant communities or even replace them entirely (Asner et al. 2008).

Many studies have addressed the issue of alien species impacts over oceanic islands (e.g. Castro et al. 2010; Kueffer et al. 2010) and much attention has been given to the presence of alien plants in either the Azores as a whole or in particular islands or particular protected areas (e.g. Ramos 1996; Silva and Smith 2004, 2006; Schaefer et al. 2011; Arosa et al. 2012). One protected area located in São Miguel Island has received much attention, because of the presence of the endangered Azores bullfinch (*Pyrrhula murina*) (Heleno et al. 2009, 2010; Ceia et al. 2011b). However, the only archipelago-wide study carried out so far addressing the whole protected areas, was that of Silva et al. (2008). The authors analysed biological invasions in the European Macaronesia, including fauna and flora, and defined the top 100 invasive alien species (IAS) with management priority. They concluded that: (i) priority habitats or habitats listed on the Habitats Directive are affected by a large majority of the top IAS (54 and 33 %, respectively); (ii) 95 % of the top IAS affect legally protected areas; and (iii) 83 % of the IAS are potentially able to expand their distribution range (i.e. they are not at equilibrium). From this top 100 IAS, 83 % are vascular plants. These numbers reinforce the conservation concerns about the presence and impacts of IAS in the Azores, particularly regarding protected areas.

In the present chapter, we address plant invasions in protected areas of the Azores. The following sections provide a global overview, specific case studies, and on-going or recent management actions. Specifically, we present (i) a general characterization of the Azores, with particular reference to their Island Natural Parks (INP); (ii) an assessment of the invasion status of the INP; and (iii) management initiatives embracing invasive alien plant (IAP) control run by the Azorean Government and the Society for the Study of Birds (SPEA). We finish with a holistic discussion stressing strengths and weaknesses of all topics covered in the chapter.

17.2 The Azores

17.2.1 Biophysical Background

The Azores are located in the North Atlantic Ocean approximately 1,500 km from mainland Portugal (Fig. 17.1). The archipelago has a total surface of 2,322 km^2 and consists of nine volcanic islands, spanning 615 km and aligned on a west/northwest–east/south-east trend. It is divided into three groups: the western group of Corvo and Flores; the central group of Faial, Pico, Graciosa, São Jorge, and Terceira; and the eastern group of São Miguel and Santa Maria. The largest island is São Miguel (745 km^2), and the smallest is Corvo (17 km^2). Pico has the highest maximum elevation (2,351 m), and Graciosa the lowest maximum elevation (402 m a.s.l.). Five other islands have elevations near 1,000 m (Borges et al. 2010).

The climate is temperate oceanic. The mean annual temperature is 17° C at sea level, the relative humidity is high and the rainfall ranges from 1,500 to more than 3,000 mm per year, increasing with altitude and from east to west (Silva and Smith 2006). The abiotic conditions of the Azores, namely the extreme isolation, from both mainland and neighbouring islands within the archipelago, and the relatively young age of most of the Azorean islands, are responsible for the unique character of the Azorean biota. The Azores present unusually high numbers of widespread native species and low raw numbers of diversification events (and thus small numbers of single island endemics) when compared to similar Macaronesian archipelagos (Carine and Schaefer 2009; Borges et al. 2011).

As regards native plant communities, the Azores are covered by coastal vegetation, wetland vegetation (lakeshore and seashore communities and a variety of bogs), several types of meadows, different types of native scrubland (e.g. coastal,

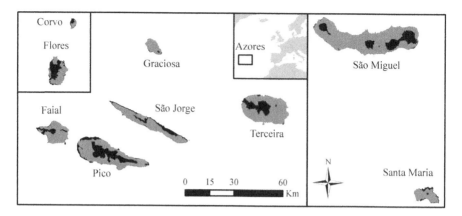

Fig. 17.1 The Azores and its inland protected areas, the Island Natural Parks (*black area*)

pioneer, mountain scrubland) and forest (e.g. *Morella, Erica, Laurus, Juniperus* dominated forests). The Azorean *Laurus* dominated forests differ from those found on the Macaronesian archipelagos of Madeira and Canaries, as it includes a single species of Lauraceae, several species of sclerophyllous and microphyllous trees and shrubs (Borges et al. 2010). Nevertheless, the Azores present a large proportion of native and endemic plant species occurring in most islands, which is associated to the high compositional uniformity of the Azorean native forest (Sjögren 1973a). The absence of some endemic species in some Azorean islands is likely due to recent anthropogenic land-use changes and local extinctions (Cardoso et al. 2010; Triantis et al. 2010).

After human settlement began in the fifteenth century, other types of vegetation cover have become progressively dominant while many alien species been introduced. Presently, the Azorean landscape includes pastureland (65 %), *Cryptomeria japonica* (Japanese cedar) dominated production forest (about 20 % of the forested area), mixed woodland (dominated by alien taxa, more than 30 % of the forested area), field crops, orchards, vineyards, hedgerows and gardens (Silva and Smith 2006).

17.2.2 Alien Flora

The Azores present a very high proportion of alien flora. The introduction of alien species is clearly associated with human colonization. Nowadays, around 70 % of the approximately 1,000 vascular plants that make up the Azorean flora have been purposefully or accidentally introduced. Introductions have been made with different purposes, such as ornamental, agricultural and forestry. Also, a considerable percentage of plants have been probably introduced accidentally, since they are considered as weeds. Indeed, the positive relationship between human population density and the diversity of alien taxa has been statistically demonstrated as well as a decline in the percentage of alien species with altitude (Silva and Smith 2004). The current very high proportion of alien flora in the Azores is not a result of a gradual introduction of alien species over time. According to Sjögren (1973a, b), the number of plant taxa increased by almost 100 % in the previous 150 years, with alien plants affecting plant community composition at all the altitudes and in different biotopes. The percentage of alien taxa increased from 57 % (Trelease 1897) to 62 % in 1966 (Palhinha 1966) and to 69 % in 2004 from a total of 1,002 taxa (Silva and Smith 2004), which represents about 205 introduced plants per 10 km^2. In the Azores, the number of alien species per square km varies between one or two times the number of native species, depending on the island, which is more than in any other Macaronesian archipelago (Silva et al. 2008). At a local scale, some differences are found among the Azorean islands. The percentage of alien taxa is lower for Corvo, Flores and São Jorge, and relatively higher for Graciosa, with intermediate values for the remaining islands (Silva and Smith 2006).

Many vascular plant species introduced in the Azores are now considered as either naturalised or frequently escaped (Silva et al. 2010), occurring not only in marginal habitats but also in a variety of different systems, such as crops, stone walls and coastal and wooded areas (Silva and Smith 2006). A negative impact on the native community of phytophagous insects, lichens and molluscs is expected, as well as changes in vegetation structure, difficulties in the regeneration of endemic species, and competition for dispersal agents, leading to a reduction in the frequency and abundance of native plant taxa (Silva et al. 2008). For one of the best studied protected areas in the Azores, at Serra da Tronqueira (São Miguel), Moniz and Silva (2003) found changes in vegetation structure in areas invaded by *Clethra arborea* (lily of the Valley Tree). Other species invading protected areas in São Miguel were targeted with research, including *Gunnera tinctoria* (Chilean rhubarb; Silva et al. 1996), *Hedychium gardnerianum* (Kahili ginger lily; Cordeiro and Silva 2003), *Dicksonia antarctica* (soft tree fern; Arosa et al. 2012) and *Leycesteria formosa* (Himalayan honeysuckle; Silva et al. 2009), the latter being found in Terceira.

Other studies revealed the effect of alien plants on animal communities. For instance, Silva and Tavares (1995) showed significant differences in phytophagous insect composition in *Pittosporum undulatum* (sweet Pittosporum) as compared to native tree species. Heleno et al. (2009), in a more comprehensive work, found that plant and insect species richness, along with plant species evenness, declined as the level of plant invasion increased. Overall, insect abundance was not significantly affected by alien plants, but insect biomass significantly decreased at increasing numbers of alien plants. Furthermore, the impact of alien plants was sufficiently severe to invert the otherwise expected pattern of species-richness decline with increasing elevation. Ceia et al. (2009) found that birds showed a higher relative abundance in native than in exotic forests. Also, the distribution of the Azores bullfinch was heavily restricted by the spread of exotic forest, as previously reported by Ramos (1996). Other interesting studies include the use or remote sensing to detect invasion by *P. undulatum and C. arborea* at Serra da Tronqueira (Gil et al. 2011, 2013). Globally, the effect of the introduction of alien plants has led to an homogenization of island floras around the world, making the present Azorean flora more similar to that of very distant archipelagos (e.g. Hawaii) than before human settlement had begun (Castro et al. 2010).

17.2.3 Protected Areas

The Macaronesian region is part of the Mediterranean biodiversity hotspot as one of the most important areas for conservation worldwide (Myers et al. 2000). In the specific case of the Azores, its relatively high level of endemism is of great relevance for conservation. In this regard, protected areas in the archipelago are strategically important for effective biodiversity conservation. Progress in the conservation of the Azorean biodiversity also depends on long-term studies about the distribution and abundance of threatened species, the implementation of species recovery programmes and the control of invasive species coupled to habitat restoration.

The first protected areas in the Azores were officially created in 1972 (integral reserves of "Caldeira do Faial" and "Montanha do Pico"). Two important European directives stimulated the establishment of protected areas: Birds Directive (Directive 2009/147/EC) and Habitats Directive (Directive 92/43/EEC). The latter is the means by which the European Union meets its obligations under the Bern Convention. Since the early 1980s, several types of protected areas have been created following the implementation of those directives in the Azores, including 23 Sites of Community Importance (SCI) and 15 Special Protected Areas (SPA) that integrate the Natura 2000 network. In 2007 the regional network of protected areas was reclassified to establish a model based on management criteria, according to the International Union for the Conservation of Nature (IUCN) Category System. As a result, two basic units (the Island Natural Parks and the Azores Marine Park) and five categories of protected areas (Natural Reserve; Natural Monument; Habitats or Species Management Protected Area; Protected Landscape; Resource Management Protected Area) were defined. The terrestrial areas classified under the system in each island compose the Island Natural Park, covering 24.1 % of the land surface of the Azores (INP, Fig. 17.1). For more details, the reader is referred to Calado et al. (2009).

17.3 Invasion Status of the Azorean Protected Areas

As previously mentioned, most of the current Azorean flora is alien, including many invasive plants. Much attention has been given to the presence of alien plants in the Azores as a whole, and several studies have addressed plant invasions in particular protected areas. However, although the protected areas encompass the most important species and habitats of the archipelago, there has only been one assessment of the protected area network as a whole, while in the meantime the Azorean protected areas have been redefined. For those reasons, assessing the invasion status of all the Azorean INP is appropriate and needed.

17.3.1 Data

For this analysis, we defined the study area, the target IAPs and used species occurrence data. To identify the Azorean protected areas we used the GIS layer available at the Azores Government website at http://www.azores.gov.pt in vector format (ESRI geodatabase). Given the scope of the assessment, the Azores Marine Park was excluded from the analysis, and we did not distinguish between the different categories of protected areas defined inside each INP. Therefore, in each of the nine islands, all the inland protected areas of different categories belonging to the INP were deemed as a single protected area. Hereafter we use the terms INP and protected areas interchangeably.

As target IAP, we used those included in the top 100 IAS described in Silva et al. (2008). This rank of 100 species was built in order to identify management priorities for IAS in European Macaronesia. The species of the top 100 were identified by considering their effects on native biodiversity and habitats and also the possibility to successfully control or eradicate them. The system was built upon previous invasive species assessment protocols, particularly the works of Hiebert (1997) and Morse et al. (2004), and also the work of Marsh et al. (2007) regarding the allocation of management resources for wildlife conservation.

Species occurrence data were obtained from ATLANTIS, a regional species database with information on approximately 5,000 species, reported on a 500 m grid. In more detail, most of the data rely on a comprehensive literature survey (dating back to the nineteenth century) as well as unpublished records from field surveys. In ATLANTIS it is possible to store detailed information for each species, including their taxonomy, distribution data, species description, conservation and biogeographic status. More details are found in Borges et al. (2010) and at the Azorean Biodiversity Portal (http://www.azoresbioportal.angra.uac.pt/).

17.3.2 Analysis

We considered all the ATLANTIS 500 m cells overlapping the protected areas that present plant records (hereafter sample). From this sample, we held the records of the target IAP (hereafter IAP records). By dividing the IAP records by the sample for each island, we estimated the relative frequency, fr, of the IAP as an estimate of their invasion extent inside the islands' protected areas. Then, we used the outcomes of Silva et al. (2008) upon which the Top 100 was built. We briefly explain them hereafter.

A set of two criteria was used to evaluate and score (i) known and potential effect of the IAP on native biodiversity and on natural and semi-natural habitats; and (ii) the probability of their successful control or eradication.

In the present analysis, we used the Silva et al. (2008) scores of criteria (i) and (ii) to assess the level of threat that each target IAP represents for the Azorean protected areas. We set a quotient between the scores of criteria (i) and (ii), meaning that the noxious effects of a given IAP may be faded if the possibility of its successful control or eradication is high. Such quotient was weighted according to the species relative frequency, *fr*. Thus, the overall threat to the Azorean protected areas was calculated through the following index (Eq. 17.1):

$$IT_i = \sum_{s=1}^{n} \frac{score_s 1}{score_s 2} \times fr_{si} \tag{17.1}$$

where IT_i is the index of threat in the protected areas of island i ($i = 1,\ldots, 9$), n is the number of target IAP falling inside the protected areas of island i, *score 1* and *score 2* are the outcomes of Silva et al. (2008) for each target species s, as explained above, and *fr* is the relative frequency of species s in the protected areas of island i.

In addition, in order to better understand what the most threatening IAP in the whole Azores protected areas are, we calculated a homologous index by summing up the contributions from all islands for each species (Eq. 17.2):

$$IT_s = \sum_{i=1}^{9} \frac{score_s 1}{score_s 2} \times fr_{si} \tag{17.2}$$

where IT_s is the index of threat of species s in the Azorean protected areas and the remaining features are as above.

In order to statistically compare the index of threat among islands, a bootstrap analysis was used. For each island, we randomly sampled 500 INP species compositions, based on the initial frequency of each IAP. We then used the resulting 500 estimates to calculate the mean and the 95 % confidence interval for the index. Calculations were performed using R (R Development Core Team 2013).

17.3.3   Results

All islands' protected areas included IAP within their bounds, ranging from 15 to 34 species. In total, records of 49 IAP from the Top 100 IAS were found in the protected areas of the Azores. Table 17.1 summarises the results for the nine islands.

As shown in Table 17.1, the protected areas with the highest index of threat were those of Santa Maria, followed in decreasing order by Graciosa, São Jorge, Corvo, Faial, São Miguel, Terceira, Flores and Pico. Bootstrap results show that the values among islands were significantly different (no 95 % CI overlap) except between São Miguel and Terceira. In general, this result is explained by (i) the invaded area

Table 17.1 Impact of Invasive Alien Plants (*IAP*) on the Azorean protected areas

Island	Total area (km^2)	Island Natural Park area			IAP[a]	Sample with IAP (%)[b]	Mean IAP frequency fr (%)[b,c]	IT$_i$[b]	95 % CI
		(km^2)	(%)	Sample (%)					
Corvo	17.16	7.79	45.4	91.8	15	35.8 (8)	11.7 (1)	2.19 (4)	2.134–2.256
Faial	173.06	30.19	17.4	87.7	26	54.5 (3)	6.0 (5)	1.82 (5)	1.779–1.866
Flores	141.33	60.74	43.0	77.1	28	53.6 (4)	4.6 (7)	1.47 (8)	1.429–1.509
Graciosa	60.66	3.31	5.5	84.2	20	66.7 (2)	10.1 (3)	2.60 (2)	2.513–2.683
Pico	444.80	157.10	35.3	75.1	32	22.2 (9)	2.7 (9)	1.07 (9)	1.049–1.089
São Jorge	243.65	56.51	23.2	67.7	30	44.9 (6)	6.0 (4)	2.33 (3)	2.297–2.369
Santa Maria	96.89	16.80	17.3	73.1	26	69.9 (1)	10.3 (2)	3.39 (1)	3.323–3.458
São Miguel	744.58	142.37	19.1	86.4	34	41.1 (7)	4.0 (8)	1.62 (6)	1.596–1.647
Terceira	400.27	85.36	21.3	94.7	22	51.1 (5)	5.2 (6)	1.58 (7)	1.555–1.605
Azores	2322.37	560.18	24.1	80.7	49	34.1	6.7		

Islands surface area, Island Natural Park area (total, as a percentage of island area, and sampled portion), number of Top IAP found in islands' INP, proportion of the ATLANTIS sample with IAP, mean IAP relative frequencies, and index of threat (IT$_i$). The 95 % confidence interval was obtained by re-sampling 500 times the IAP composition of each INP

[a]Number of alien invasive plants included in the Top 100 (Silva et al. 2008)

[b]Rank position in brackets

[c]Mean of all relative frequencies, fr, of the IAP falling inside the island's protected areas

extent, which is expressed by the proportion of the sample with IAP; and (ii) by the relative frequency, *fr*, of the IAP inside the islands' protected areas. In the first case, the two islands whose protected areas showed the highest index values (Santa Maria and Graciosa) are those presenting higher proportion of the sample with IAP, whereas Pico's protected areas stand on the opposite position. In the second case, the larger the frequencies, *fr*, the greater the index of threat. However, two islands are out of this trend: Corvo and São Miguel. The former presents, on average, the largest frequencies, *fr*, for the corresponding IAP, but just the fourth index of threat. This means Corvo's protected areas are mostly invaded by IAP which are relatively innocuous and/or have a high probability of successful control or eradication. In contrast, São Miguel presents, on average, the eighth largest frequencies, *fr*, for the corresponding IAP but the sixth index of threat. This means São Miguel's protected areas are mostly invaded by IAP which are relatively noxious and/or have a low probability of successful control or eradication. This is in agreement with several of the reports cited above, where several noxious invaders in São Miguel are addressed. The rank of this island is not higher, only because several of its IAPs are present only on this island, thus scoring relatively lower in the Top 100 score.

Regarding the species' index of threat in the Azorean protected areas, the species with highest index values were *P. undulatum*, *H. gardnerianum*, *Conyza bonariensis* (fleabane), *Arundo donax* (giant reed) and *Acacia melanoxylon* (Australian blackwood), to cite but the first five. This result corroborates previous publications on invasive plants in the Azores. *P. undulatum* has been pointed as the most important woody invasive plant in the Azores (Lourenço et al. 2011). *Hedychium gardnerianum*, *A. donax* and *A. melanoxylon* have been mentioned as well as some of the most invasive plants in the archipelago (e.g. Silva and Smith 2006; Silva et al. 2011). Instead, the ruderal species *C. bonariensis* has received less attention, perhaps because it is smaller (terophyte of 10–150 cm) or less conspicuous than other IAP, or because it is particularly present at dry pastures and disturbed locations near the coast where its impact on native plant communities is considered reduced (although the species is very abundant in such communities).

Species such as *C. arborea*, *G. tinctoria* and *L. formosa*, were expected to have a lower score, since they are only present in São Miguel, although they have been reported to pose a real threat to the preservation of the Azorean biodiversity (see several references cited above).

17.4   Invasive Alien Plant Control

17.4.1   *Regional Plan of Eradication and Control of IAP: PRECEFIAS*

In 2004 a Regional Plan of Eradication and Control of Invasive Plant Species in Sensitive Areas (PRECEFIAS) was approved. This plan focus on more than

20 species and sensitive areas in all islands of the Azores, and aims to (i) improve the conservation status of natural habitats and populations of priority species; (ii) reduce the effects of invasive plants; (iii) produce a list of invasive or potentially invasive species; and (iv) raise awareness of the problems created by current invasive species and by eventual new introductions to the flora in the Azores. Therefore, PRECEFIAS encompasses four measures: (i) inventory; (ii) removal and recovery; (iii) promotion and awareness; and (iv) monitoring.

During inventory, vulnerable areas were identified in all islands (areas with high natural value –with natural habitats and populations of priority species). Most of these areas coincided with the regional network of protected areas defined later in 2007. Then, a list of target invasive species was produced (Table 17.2). Most are included in the Top 100 (Silva et al. 2008) but it also includes other species. For example, *Pteridium aquilinum* (bracken), although not considered as invasive, is reported as a weed in pastures, causing acute poisoning as well as haematuria in cattle (Pinto et al. 2007); some species tend to expand vegetatively after long periods of cultivation (e.g. *Aloe arborescens*, krantz aloe; *Phyllostachys* sp.); other species were recently recognised as potential invaders, such as *Tetrapanax papyriferus* (rice paper plant), mostly around human settlements and along water courses, or in the margins of the laurel forest (*Scrophularia scorodonia,* balm-leaved figwort). Table 17.2 and Fig. 17.2 show a list of interventions undertaken in the archipelago. Moreover, some endemic species have been promoted as an alternative to alien species used for ornamental purposes (e.g. using the endemic *Viburnum treleasei* (folhado), instead of the alien *H. macrophylla*), although with reduced adhesion from stakeholders and demanding extreme care to avoid negative impacts on the endemic species (changes in population genetic structure).

The biomass resulting from control actions undergoes several kinds of valorisation to increase the sustainability of the plan. For instance, trunks with economic value are used whereas valueless wood is provided to local populations for home heating. In the particular case of *P. undulatum*, secondary branches and the foliage are powdered and used as a source of compost for pineapple plantations in greenhouses in São Miguel. In this context, much could still be done to make IAP control more sustainable, by using biomass for energy production (Lourenço et al. 2011).

For promotion and awareness, the environment services of the Azores, jointly with schools and non-governmental organizations, have developed several campaigns to draw attention to the issue of the impacts caused by invasive plants in island systems and, therefore, to the need for their control. After the recent implementation of the regional network of protected areas, around €500,000 are spent annually on these activities.

An interesting case study was the eradication of *H. gardnerianum* and of *Canna indica* (canna), two introduced herbaceous perennial species expanding in the eastern part of Corvo. Work was carried out from 2004 to 2006 and comprised four phases: (i) detection of invaded areas in the entire island; (ii) application of

Table 17.2 Interventions performed in the Azores, within the framework of the Regional Plan of Eradication and Control of Invasive Plant Species in Sensitive Areas (*PRECEFIAS*), for removing alien plant species

Island	Location	Area (ha)	Target species
Corvo	Biosphere reserve	16.5	*Canna indica; Hedychium gardnerianum; Hydrangea macrophylla; Leycesteria formosa; Tetrapanax papyferus*
Faial	Monte da Guia, Varadouro, Morro de Castelo Branco and Capelinhos volcano	6.2	*Arundo donax; Carpobrotus edulis; Hedychium gardnerianum; Hydrangea macrophylla; Lantana camara; Pittosporum undulatum*
Flores	Burrinha, Caldeira Branca and Morro Alto	372.0	*Acacia melanoxylon; Arundo donax; Hedychium gardnerianum; Lantana camara; Pittosporum undulatum*
Graciosa	Ponta Branca, Ponta do Carapacho, Caldeira da Graciosa and Caldeirinha de Pero Botelho	3.6	*Agave americana; Aloe arborescens; Arundo donax; Hedychium gardnerianum; Hydrangea macrophylla; Phormium tenax; Pittosporum undulatum; Pteridium aquilinum; Rubus ulmifolius*
Pico	Mistério da Praínha, Caveiro, Ponta da Ilha, Caldeirão da Ribeirinha and Landscape of Vineyard Culture	17.0	*Acacia melanoxylon; Ageratina adenophora; Ailanthus altissima; Carpobrotus edulis; Cryptomeria japonica; Hedychium gardnerianum; Hydrangea macrophylla; Metrosideros excelsa; Persicaria capitata; Pittosporum undulatum; Rubus ulmifolius*
São Jorge	Bocas do Fogo and Ponta dos Rosais	14.0	*Carpobrotus edulis; Hydrangea macrophylla; Pittosporum undulatum*
Santa Maria	Barreiro da Faneca and Pico Alto	7.2	*Acacia melanoxylon; Hedychium gardnerianum; Pinus pinaster; Pittosporum undulatum; Pteridium aquilinum; Ulex europaeus*
São Miguel	Sete Cidades and Furnas	121.5	*Acacia melanoxylon; Clethra arborea; Cortaderia selloana; Egeria densa; Eichhornia crassipes; Gunnera tinctoria; Hedychium gardnerianum; Hydrangea macrophylla; Leycesteria formosa; Pteridium aquilinum; Persicaria capitata;*

(continued)

Table 17.2 (continued)

Island	Location	Area (ha)	Target species
			Scrophularia scorodonia; Solanum mauritianum; Ulex europaeus
Terceira	Algar do Carvão, Lagoínha da Serreta and Pico Alto	13.7	*Hedychium gardnerianum; Hydrangea macrophylla; Pittosporum undulatum; Pteridium aquilinum; Rubus ulmifolius*

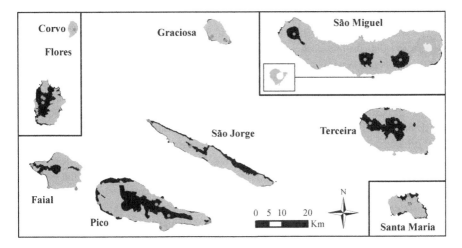

Fig. 17.2 Location of the main actions for invasion alien plant control in the Azores: PRECEFIAS actions (*red points*) and Island Natural Parks regarded in LIFE projects (*Yellow*: "PRIOLO –Azores bullfinch habitat recovery in Pico da Vara/Ribeira do Guilherme SPA" LIFE project; *Green*: "Sustainable Laurisilva-Recovery, conservation and sustainable management of Tronqueira/Planalto dos Graminhais" LIFE project; *Blue*: "Safe islands for seabirds" LIFE project; *Black*: remaining protected areas)

control actions; (iii) monitoring in order to detect/avoid re-infestation; and (iv) public-relations work in order to raise awareness about the importance of invasive plant eradication in the management of natural areas. The control actions included removal of the whole plants which was restricted to riparian zones. Chemical treatments with Ally® were used in other habitats (orchards, woodland, pastures and roadside slopes). The two species were detected and controlled in a total of 36.6 ha.

17.4.2   LIFE Projects

Birds are a good example of how IAS are threatening the Azorean biodiversity. The Azores bullfinch, an endemic bird of São Miguel adapted to the native laurel forest, saw its population reaching such low numbers that came to be judged extinct in the mid-twentieth century. One of the main reasons for this fact is that the native vegetation is today highly degraded due to the invasion by alien plants such as *P. undulatum, H. gardnerianum, C. arborea* and *A. melanoxylon* (Ramos 1996). Invasive alien plants provide neither quality food nor refuge for the bullfinch, which threatens its existence (Ceia et al. 2009, 2011b).

Also, the Azores were inhabited by millions of seabirds. However, these colonies have decreased greatly as a result of the introduction of IAS, including plant and mammal species, such as rats, mice and cats. Currently, the seabird populations other than Cory's Shearwater (*Calonectris diomedea*) are confined to tiny uninhabited islets and some remote and inaccessible cliffs.

The Portuguese Society for the Study of Birds (SPEA), a non-profit scientific association that promotes the study and conservation of birds and their habitat in Portugal, has run several projects in the Azores to protect birds by focusing on the issue of IAS. The LIFE project "PRIOLO –Azores bullfinch habitat recovery in Pico da Vara/Ribeira do Guilherme SPA" (LIFE03 NAT/P/000013) ran from 2003 to 2008 in partnership with the Royal Society for the Protection of Birds (RSPB), Nordeste Municipal Council, University of the Azores and the Azores Government. The project aimed to control major IAP, in particular *H. gardnerianum* and *C. arborea*, over nearly 230 ha of native forest in the Serra da Tronqueira (Fig. 17.2) and also to undertake restoration actions. Thus, more than 60,000 specimens of native and endemic species were planted (e.g. *Vaccinium cylindraceum*, Azores blueberry; *Morella faya*, Firetree; *Juniperus brevifolia*, Azorean juniper; *Ilex perado* subsp. *azorica*, Azorean holly; *Picconia azorica*, Azorean picconia; *Frangula azorica*, Buckthorne; *Laurus azorica*, Azorean laurel; and *Erica azorica*, Azorean heather).

The results achieved in this project showed a reduction by over 92 % of the coverage of *H. gardnerianum* and *C. arborea* (Silva 2007). This promoted, for instance, a substantial increase of flowering of native species, which is necessary for the survival of the Azores bullfinch. Heleno et al. (2010) assessed restoration success of the invaded forest and found that 2 years after removing alien plants there were increases in the abundance of native seeds, phytophagous insects, insect parasitoids, and birds in the experimental plot compared to an un-manipulated plot. Therefore, this work provides evidence of the positive effects of weeding cascading through the food web from native plants to phytophagous insects, insect parasitoids, and birds.

Clethra arborea however has a dual role in the diet of the Azores bullfinch. It is a crucial winter food resource but it lowers the availability of native laurel forest species that compose most of the bird's diet throughout the year (Ceia et al. 2011b). In order to evaluate the first responses of the Azores bullfinch to habitat restoration, these authors studied bird diet, foraging behaviour, food availability and habitat occupancy in managed (without *C. arborea*) and control areas. They found a significant increase in the availability of native food resources in managed areas. The project also had a positive impact in the local community with creation of jobs and the promotion of Nordeste and Tronqueira, and it was awarded with the "Best of the Best – Nature" prize in 2009. The main achievement of the programme was the downgrade of the Azores bullfinch status from Critically Endangered to Endangered in the IUCN Red list (Ceia et al. 2011a).

The "Sustainable *Laurisilva*-Recovery, conservation and sustainable management of Tronqueira/Planalto dos Graminhais" (LIFE 07 NAT/P/000630) started in 2009. This on-going 4 year LIFE project is being implemented by SPEA in partnership with the Azores Government and Povoação Municipal Council. This project aims mainly to extend the actions of previous projects to the raised bogs of Planalto dos Graminais (Fig. 17.2). The project included the establishment of a plant nursery dedicated to the production of native plants for conservation purposes. So far, more than 50,000 plants were produced to be planted in the spaces left open by the control of IAP (e.g. *G. tinctoria* and *P. undulatum*). Adding to the area accounted since 2003 within the first LIFE project previously described, approximately more 50 ha of laurel forest and 90 ha of raised bogs are undergoing interventions. In the latter case, additional measures, such as closure of drainage ditches and rehabilitation of water lines are being implemented. This project also aims to raise awareness among the local community and tourists on the benefits provided by the protected habitats in terms of regularization of the local water system, and provision of fresh water for human use. In addition, this project aims at promoting sustainable tourism through participatory processes, based on the European Charter for Sustainable Tourism. The ultimate goal is to find a model that allows for the conservation of an important natural heritage, at the same time improving the quality of life of the local populations.

The most recent LIFE project dealing with IAS is the 4 year programme "Safe islands for seabirds" (LIFE07 NAT/P/000649), implemented by SPEA since 2009 in partnership with RSPB, Corvo Municipal Council and the Azores Government. It is a demonstration project that arose from the need to study the feasibility of recovering terrestrial habitats for seabirds. The project aims to evaluate and plan for the feasibility of controlling and eradicating both invasive plants and alien predators in coastal areas that are important for seabirds. The seashore of Corvo (most of it established as protected area; Fig. 17.2) was chosen to develop great part of this project for much of its geographical location and habitat availability, which make it an ideal place for thousands of seabirds that nest there every year.

Two sites were selected to apply control techniques and restoration actions, namely a site of 3 ha covered by *Tamarix africana* (African tamarisk) and another one of 12 ha covered by *Hydrangea macrophylla* (bigleaf hydrangea). However, the control work has been difficult to implement and results are not available yet. In addition, the islet of Vila Franca do Campo (a protected area at south of São Miguel; Fig. 17.2) is another area of intervention to test control methods (cutting and herbicide application) mainly focused on *A. donax* (giant reed). This IAP blocks the entrance of Cory's shearwater nest burrows and out-competes with native flora. First results of control actions show a reduction of 92 % of *A. donax* stems after 8 months from first interventions (Silva et al. 2011).

17.5 Discussion

The history of biological invasions in the Azores started along with human colonisation in the fifteenth century. In the mid-nineteenth century, the alien flora was already identified as more numerous than the native Azorean flora (Trelease 1897). Nowadays, many alien species are considered as either naturalised or frequently escaped (Silva et al. 2010) and many are an important threat to native biodiversity (Ramos 1996; Silva and Smith 2004, 2006; Silva et al. 2008; Castro et al. 2010; Kueffer et al. 2010; Arosa et al. 2012).

The conservation, landscape and scientific values of the Azores were recognised under the recently redrawn regional network of protected areas, which covers 24.1 % on the inland surface. According to the assessment results here presented, the invasions status of the protected areas differ from island to island and do not necessarily follow previous studies that have addressed IAP in the whole Azores.

According to Silva and Smith (2006), Corvo, Flores and São Jorge present relatively low percentage of alien taxa as opposed to Graciosa, with intermediate values for the remaining islands. However, using a modified ranking methodology based on the index of threat focused on protected areas and taking into account spatial information and the potential for controlling the species, the results are somewhat different, with the protected areas of Corvo, Flores and São Jorge ranking fourth, eighth and third, respectively, and Graciosa's protected areas not emerging as the most threatened. The highest rank was attributed to the protected areas of Santa Maria, although the island does not present a particularly high percentage of alien taxa.

Some limitations of the method applied should be pointed out. The method focused only on protected areas, which cover a variable proportion of the islands (from only 5.5 to 45.4 % –Graciosa and Corvo, respectively). Thus, potential IAP occurring outside the protected areas bounds were not considered, although in some cases these may invade protected areas in the future. Furthermore, since ATLANTIS gathers data from different sources, the proportion of sampled cells also varies among islands and protected areas, thus introducing a possible bias in the index of threat. Moreover, the method tends to give a lower rank to species mainly affecting one island. This was the case of *C. arborea*, *G. tinctoria*, and *L. formosa*. Finally, new invasions are being reported, as is the case of *D. antarctica* (Arosa et al. 2012). Indeed, the establishment of new alien species in natural areas has not stopped. For instance, invasions by *Rhaphiolepis umbellata* (yeddo hawthorn), a garden plant, have been detected in coastal areas of Flores; two species of *Drosera* have been recently located in a São Miguel's protected area (2013) as well as the tree fern *Cyathea medullaris* (black tree fern) (Silva L and Jiménez S, unpublished data).

Despite these limitations, the assessment results provide new insights into the IAP potential impacts in the Azores, namely within protected areas, and shows that characterizing the flora of a region by listing native and alien species may not permit to fully understand the potential effects of IAP over native biodiversity. In addition, the assessment results may provide a basis for prioritization of management. The results may in fact provide guidelines for allocating resources for IAP control. Thus, more attention should be given to the highly ranked protected areas, such as those of Santa Maria and Graciosa. Nevertheless, these guidelines should be cross-analyzed with data regarding the conservation value of each site in order to increase the efficiency of the control/restoration actions.

Although the results of the assessment presented here as well as the publications mentioned throughout the chapter suggest that the serious situation of the Azores requires more work to protect its unique biota, remarkable projects have already been put into practice addressing some of the protected areas threatened by IAP in the archipelago. For instance, the control and eradication of IAP in Corvo within PRECEFIAS, especially *H. gardnerianum* and *C. indica*, was considered a successful case by the Azorean authorities and an example to follow. This is a very positive outcome of the plan as linking nature conservation and decision-making at political level is not always achieved. Encouraging results have also been achieved in the framework of LIFE projects, which have enabled native habitats to be recovered in favor of the Azores bullfinch and seabirds, which ultimately benefits other groups of fauna and even the local populations.

Nevertheless, more work is necessary to match the needs for IAP control identified herein and in other publications. Clearly, the establishment of an operational system devoted to early detection and control of new invaders, particularly in the Azorean protected areas, is one of the priorities in a strategy to better manage plant invasions. However, the limited available financial resources and the high cost associated to long-term operational projects are a major constraint for ensuring effective IAP control and eradication policies in the Azorean protected areas. In order to promote a consistent, integrated and cost-effective IAP assessment and monitoring in the Azores, the establishment of an *Invasion Biology Regional Observatory* (ORBI) was proposed in 2011, during the first *Regional Meeting on Invasive Alien Species* (http://www.invasoras.uac.pt/). ORBI would allow for a better coordination of actions undertaken by researchers, NGOs and administration bodies, including managers of protected areas. ORBI prior missions would be to gather, manage, update and spread information on IAS in the Azores, in order to foster and support more cost-effective control and eradication policies.

It is also important to stress that more efforts should be devoted to objectively assess the outcomes of management/conservation efforts undertaken in the Azores. This will be important to evaluate the efficacy of the control actions and the efficiency in the use of the resources allocated to the regional programmes.

Moreover, the results obtained in the Azores could then be more objectively transferable to other regions and to the scientific community as a whole. For this, a closer connection between management and research should be established in the future.

Considering the strict link between biological invasions and human activities, the preservation of native biota in the protected areas of the Azores will depend on the support of the local society, as a whole, to the implementation of a strategic approach to the issue, crucial to reverse or reduce the on-going homogenization of the Azorean flora.

Acknowledgements The manuscript was improved following the suggestions of two anonymous reviewers. Hugo Costa was supported by a scholarship from the project "Woody Biomass" included in "MIT –Green Islands Project" funded by the Azorean Government. Part of the work was also supported by a scholarship from CIBIO.

References

Arosa ML, Ceia RS, Quintanilla LG et al (2012) The tree fern *Dicksonia antarctica* invades two habitats of European conservation priority in São Miguel Island, Azores. Biol Invasions 14:1317–1323

Asner GP, Hughes RF, Vitousek PM et al (2008) Invasive plants transform the three-dimensional structure of rain forests. Proc Natl Acad Sci U S A 105:4519–4523

Borges PAV, Gabriel R, Arroz AM et al (2010) The Azorean biodiversity portal: an internet database for regional biodiversity outreach. Syst Biodivers 8:423–434

Borges PAV, Cardoso P, Cunha R et al (2011) Macroecological patterns of species distribution, composition and richness of the Azorean terrestrial biota. Ecologia 1:22–35

Calado H, Lopes C, Porteiro J et al (2009) Legal and technical framework of Azorean protected areas. J Coast Res SI 56:1179–1183

Cardoso P, Arnedo MA, Triantis KA et al (2010) Drivers of diversity in Macaronesian spiders and the role of species extinctions. J Biogeogr 37:1034–1046

Carine MA, Schaefer H (2009) The Azores diversity enigma: why are there so few Azorean endemic flowering plants and why are they so widespread? J Biogeogr 37:77–89

Castro SA, Daehler CC, Silva L et al (2010) Floristic homogenization as a teleconnected trend in oceanic islands. Divers Distrib 16:902–910

Caujapé-Castells J, Tye A, Crawford DJ et al (2010) Conservation of oceanic island floras: present and future global challenges. Perspect Plant Ecol Evol Syst 12:107–129

Ceia RS, Heleno RH, Ramos J (2009) Summer abundance and ecological distribution of passerines in native and exotic forests in São Miguel, Azores. Ardeola 56:25–39

Ceia RS, Ramos J, Heleno RH et al (2011a) Status assessment of the critically endangered Azores bullfinch. Bird Conserv Int 21:477–489

Ceia RS, Sampaio H, Parejo S et al (2011b) Throwing the baby out with the bathwater: does laurel forest restoration remove a critical winter food supply for the critically endangered Azores bullfinch? Biol Invasions 13:93–104

Cordeiro N, Silva L (2003) Seed production and vegetative growth of *Hedychium gardnerianum* Ker-Gawler (Zingiberaceae) in São Miguel Island (Azores). Arquipélago. Life Mar Sci 20A:31–36

Gil A, Yu Q, Lobo A et al (2011) Assessing the effectiveness of high resolution satellite imagery for vegetation mapping in small islands protected areas. J Coast Res SI 64:1663–1667

Gil A, Lobo A, Abadi M et al (2013) Mapping invasive woody plants in Azores Protected Areas by using very high-resolution multispectral imagery. Eur J Remote Sens 46:289–304

Heleno RH, Ceia RS, Ramos J et al (2009) The effect of alien plants on insect abundance and biomass: a food web approach. Conserv Biol 23:410–419

Heleno RH, Lacerda, Ramos J et al (2010) Evaluation of restoration effectiveness: community response to the removal of alien plants. Ecol Appl 20:1191–1203

Hiebert RD (1997) Prioritizing invasive plants and planning for management. In: Luken JO, Thieret JW (eds) Assessment and management of plant invasions. Springer, New York, pp 195–212

Kueffer C, Daehler CC, Torres-Santana CW et al (2010) A global comparison of plant invasions on oceanic islands. Perspect Plant Ecol Evol Syst 12:145–161

Loope LL, Mueller-Dombois D (1989) Characteristics of invaded islands, with special reference to Hawaii. In: Drake JA, Mooney HA, Di Castri F et al (eds) Biological invasions –a global perspective. Wiley, Chichester, pp 257–280

Lourenço P, Medeiros V, Gil A et al (2011) Distribution, habitat and biomass of *Pittosporum undulatum*, the most important woody plant invader in the Azores Archipelago. For Ecol Manage 262:178–187

Mack RN, Simberloff D, Lonsdale WM et al (2000) Biotic invasions: causes, epidemiology, global consequences and control. Ecol Appl 10:689–710

Marsh H, Dennis A, Hines H et al (2007) Optimizing allocation of management resources for wildlife. Conserv Biol 21:387–399

Martín J, Cardoso P, Arechavaleta M et al (2010) Using taxonomically unbiased criteria to prioritize resource allocation for oceanic island species conservation. Biodivers Conserv 19:1659–1682

Millennium Ecosystem Assessment (2005) Ecosystems and human well-being: biodiversity synthesis. World Resources Institute, Washington, DC

Moniz J, Silva L (2003) Impact of *Clethra arborea* Aiton (Clethraceae) in a special protection area of São Miguel island, Azores. Arquipélago. Life Mar Sci 20A:37–46

Morse LE, Randall JM, Benton N et al (2004) An invasive species assessment protocol: evaluating non-native plants for their impact on biodiversity. Version 1. NatureServe, Arlington

Myers N, Mittermeier RA, Mittermeier CG et al (2000) Biodiversity hotspots for conservation priorities. Nature 403:853–858

Palhinha RT (1966) Catálogo das plantas vasculares dos Açores. Sociedade de Estudos Açorianos Afonso. Chaves, Lisbon

Pinto CA, Peleteiro MC, Lobo MA et al (2007) Intoxicação aguda pelo feto comum (*Pteridium aquilinum* (L.) Kühn) em bovinos. Rev Port Ciências Vet 102:289–298

R Development Core Team (2013) R: A language and environment for statistical computing. R Foundation for Statistical Computing, Vienna

Ramos J (1996) Introduction of exotic tree species as a threat to the Azores bullfinch population. J Appl Ecol 33:710–722

Reaser JK, Meyerson LA, Cronk Q et al (2007) Ecological and socioeconomic impacts of invasive alien species in island ecosystems. Environ Conserv 34:1–14

Schaefer H, Hardy OJ, Silva L et al (2011) Testing Darwin's naturalization hypothesis in the Azores. Ecol Lett 14:389–396

Silva CMN (2007) Utilização do herbicida ALLY® no controlo da invasão de laurissilva dos Açores. Dissertation, Escola Superior Agrária do Instituto Politécnico de Castelo Branco

Silva L, Smith C (2004) A characterization of the non-indigenous flora of the Azores Archipelago. Biol Invasions 6:193–204

Silva L, Smith C (2006) A quantitative approach to the study of non-indigenous plants: an example from the Azores Archipelago. Biodivers Conserv 15:1661–1679

Silva L, Tavares J (1995) Phytophagous insects associated with endemic, Macaronesian and exotic plants in the Azores. In: Editoriall C (ed) Avances en entomología Ibérica. Museo Nacional de Ciencias Naturales (CSIC) y Universidad Autónoma de Madrid, Madrid, pp 179–188

Silva L, Tavares J, Pena A (1996) Ecological basis for the control of *Gunnera tinctoria* (*Molina*) Mirbel (Gunneraceae) in São Miguel Island. In: Proceedings of 2nd international weed control congress, Copenhagen. Department of Weed Control and Pesticide Ecology, Flakkebjerg, pp 233–239

Silva L, Ojeda-Land E, Rodríguez-Luengo JL (eds) (2008) Invasive terrestrial flora and fauna of Macaronesia. Top 100 in Azores, Madeira and Canaries. Arena, Ponta Delgada

Silva L, Marcelino J, Resendes R et al (2009) First record of the top invasive plant *Leycesteria formosa* in Terceira island, Azores. Arquipélago. Life Mar Sci 26:69–72

Silva L, Moura M, Schaefer H et al (2010) List of vascular plants (Tracheobionta). In: Borges PAV, Costa A, Cunha R et al (eds) A list of the terrestrial and marine biota from the Azores. Princípia, Cascais

Silva CMN, Silva L, Oliveira N et al (2011) Control of giant reed *Arundo donax* on Vila Franca do Campo Islet, Azores, Portuga. Conserv Evid 8:93–99

Sjögren E (1973a) Recent changes in the vascular flora and vegetation of the Azores Islands. Mem Soc Broteriana 22:1–113

Sjögren E (1973b) Vascular plants new to the Azores and to individual islands in the Archipelago. Bol Mus Munic Funchal 124:94–120

Trelease W (1897) Botanical observations on the Azores. Annu Rep Mo Bot Gard 8:77–220

Triantis KA, Borges PAV, Ladle RJ et al (2010) Extinction debt on oceanic islands. Ecography 33:285–294

Chapter 18
Invasive Alien Plants in Protected Areas in Mediterranean Islands: Knowledge Gaps and Main Threats

Giuseppe Brundu

Abstract Protected areas in the Mediterranean islands (Pamis) hold a unique level of biodiversity and are very sensitive to all human-generated environmental changes, including their synergies. There is a general lack of information on the presence and abundance of invasive plants for the whole set of Pamis, which are more commonly considered at country level and very rarely as a whole. This is a serious hindrance for management at international levels. Available information on a selected set of Pamis, key invasive alien plants and their impacts are briefly reported in this chapter. Given the negative ecological and socioeconomic impacts of many invasive alien plants on Pamis, their management has become an important challenge and a high priority for the conservation of native species and natural areas in the Mediterranean. Specific policies and strategies, including a dedicated definition of priorities, are urgently needed.

Keywords *Carpobrotus* spp. • Invasive alien plants • Island endemic species • Mediterranean islands • Mediterranean endemic plants

18.1 Introduction

The word *island* creates an image of fantasy, from which to escape the normal, routine and stressful lives, and visit a *paradise* with an *exotic island lifestyle* (Baum 1997 cited in Lim and Cooper 2009). Insularity often becomes an attraction and motivates people to travel across political, social and emotional boundaries for the *island experience*. While promoting ecotourism may provide economic benefits (Kafyri et al. 2012), including fundraising for conservation projects (Lindsay et al. 2008), there are, however, also negative aspects of island tourism that threaten

G. Brundu (✉)
Department of Science for Nature and Environmental Resources (DIPNET),
University of Sassari, Via Piandanna 4, 07100 Sassari, Italy
e-mail: gbrundu@uniss.it

L.C. Foxcroft et al. (eds.), *Plant Invasions in Protected Areas: Patterns, Problems and Challenges*, Invading Nature - Springer Series in Invasion Ecology 7, DOI 10.1007/978-94-007-7750-7_18, © Springer Science+Business Media Dordrecht 2013

sustainability (Lim and Cooper 2009). Conservation strategies represent a crucial issue in the Mediterranean basin because this area, which represents less than the 2 % of the world's land surface, houses 10 % of the world's total floristic richness (Médail and Quézel 1997, 1999), including high levels of endemism (e.g. Cowling et al. 1996). In some areas, particularly the mountains and the islands, rates of endemism often exceed 10 % and sometimes 20 % of the local flora (Médail and Quézel 1997; Thompson 2005). A characteristic element of this endemism is that of all the species endemic to the Mediterranean basin, 60 % are narrow endemic species, in that they have a distribution which is restricted to a single well-defined area within a small part of the Mediterranean basin (Thompson 2005; Thompson et al. 2005).

The Mediterranean basin and its islands face serious short-term and long-term threats (Cuttelod et al. 2008). Increases in the tourism sector and population growth are two of the major causes (e.g. Chapin et al. 2000), placing increasing demands on already limited water and energy resources. Climate change and associated impacts, not yet fully understood, are expected to impact the Mediterranean basin (Schröter et al. 2005), of which an increased fire risk is likely (Pinol et al. 1998; Moriondo et al. 2006; Giannakopoulos et al. 2009).

Invasive alien species occur in all nature reserves, including those in the tropics (Usher 1988, 1991) and predictions are that the importance of invasive alien species in nature reserves will increase in the future, unless effective control measures are adopted (Macdonald et al. 1988, 1989; Pyšek et al. 2002). Protected areas (PAs) situated on islands are well known to be more vulnerable than those located on mainland's (Brockie et al. 1988; Holt 1992) and the degree to which a nature reserve is invaded is, often, closely related to the number of human visitors (Usher 1988; Lonsdale 1999). The purpose of this chapter is to highlight gaps in the strategy for nature conservation in PAs of Mediterranean basin islands (PAMIs), as they are effectively being jeopardised by the presence of invasive alien plant (IAP) species, the lack of coordinated actions, and information sharing at an international level.

18.2 The Mediterranean Islands: How Many Islands? How Many Native Species?

When assessing the risks posed by IAPs on PAMIs we immediately face the problem of a general lack of data (see also Hopkins 2002). First, the total number of Mediterranean islands is uncertain. This uncertainty is the result of their large number, the absence of a common definition and the absence of a database for Mediterranean islands. Information on the Med islands is more often collected at country level and less commonly at a geographic or biogeographic level.

For example, Delanoë et al. (1996) report the presence of nearly 5,000 islands, with 162 islands covering at least 10 km^2 and 4,000 islets under 10 km^2. Later,

Delanoë and de Montmollin (1999) describe the presence of nearly 10,000 islands in the Mediterranean basin, of all sizes and origin. Noteworthy, the Aegean archipelago alone includes more than 7,000 islands and islets, of which the vast majority do not exceed 1 km^2 (Triantis and Mylonas 2009). As many as 10,000 islands have been listed for Greece (European Commission 2011).

The eastern Croatian coast is, other than the Greek coast, one of the most diverse in the Mediterranean basin. There are 1,151 islands, islets and reefs, and, depending on tides, 80 additional reefs periodically appear above sea level (Stražičić 1989). Sixty Croatian islands have coast lines of longer than 10 km, while 653 islets have their coastal lines shorter than 10 km, but with developed soil and vegetation. Further, 438 islets and reefs have also their coastal lines shorter than 10 km but with neither soil nor vegetation (Nikolić et al. 2008). However, as with other countries, the number of islands listed for Croatia also differs between sources (e.g. according to http://www.europeanislands.net/ there are 1,185 islands, including 718 islands, 389 sounds and 78 reefs, with a total area of 3,300 km^2).

A national assessment on the total number of Italian islands has never been published, although the islands and islets surrounding Sicily and Sardinia have been studied in more detail. For example, according to Arrigoni and Bocchieri (1995) there are 399 islets (with a surface greater than 300 m) around the island of Sardinia, giving a total surface area of 279 km^2, with the flora of 71 of them having been surveyed. According to Damery (2008) and Serrano (2008), there are 157 islets surrounding Corsica, with 130 of them clearly demarcated and named.

According to the Millennium Ecosystem Assessment (MEA) categories, islands are defined as "lands isolated by surrounding water and with a high proportion of coast to hinterland"; it stipulates that they must be populated, separated from the mainland by a distance of at least 2 km, and measure between 0.15 km^2 and the size of Greenland (2.2 million km^2). For mapping and statistical purposes, the MEA uses the ESRI ArcWorld Country Boundary dataset, which contains nearly 12,000 islands, including islands belonging to the Association of Small Island States, and the Small Island Developing States Network (Hassan et al. 2005). However, this definition obviously does not suit many Mediterranean islands and especially PAs in the Mediterranean Islands (e.g. Greuter 2001). Thus, although a total of 10,000–15,000 islands for the Mediterranean basin may be a reasonable estimate, this figure still needs confirmation. It is especially important if, for example, it is to be used for the purpose of developing a common database of plant genetic resources and invasion threats.

It is generally reported that the Mediterranean basin flora includes 24,000–25,000 species of seed-plants, i.e. 10 % of the world's plant species, with 13,000 endemic taxa (e.g. López and Correas 2003). The total number rises to 29,000–30,500 if subspecies are considered (Heywood 1995 and references cited therein). In general, a considerable portion of all plant species worldwide are restricted to islands (Kreft et al. 2008). However, the total number of species of the Mediterranean flora on islands, and their conservation status, is mainly based on estimates (e.g. Greuter 2001), although some of the islands have been more intensively assessed and have more precise numbers. The Med-Checklist taxonomic database provides a species list (presence/absence data) for 27 geographical areas. Their borders coincide largely with the political borders of countries around the Mediterranean basin. It is however not possible to retrieve information directly for all islands, but only for large clusters, for example, 1,410 species have been listed in the 'East Aegean Island' region.

18.3   Protected Areas on Mediterranean Islands and Their Alien Floras

At the global scale, PAs have been designated by national governments for their value in protecting particular species, habitats or landscapes, and cover about 21 million km^2. This includes 12.2 % of the global terrestrial area and 5.9 % of the world's oceans (Price et al. 2010; UNEP-WCMC 2010). A number of international networks exist, largely comprising land and sea belonging to PAMIs that are already designated at the national level. Four of these networks are global, two derive from international conventions that emphasise conservation: the Convention on Wetlands (Ramsar Convention), signed in 1971, and the World Heritage Convention, signed in 1972, respectively (Price et al. 2010). The other two networks are the World Network of Biosphere Reserves, including biosphere reserves designated since 1976 under the Man and the Biosphere (MAB) programme of the United Nations Educational, Scientific, and Cultural Organisation and the Global Geoparks Network, established in 2004 (Price et al. 2010).

At the European and Mediterranean basin level, 'Natura 2000' is the cornerstone of the regions nature conservation policy, and is regulated mainly by two directives: the 1979 Bird Directive and the 1992 Habitat Directive 92/43/EEC (Maiorano et al. 2007). The bird directive identified 193 endangered species and subspecies, for which the member states are required to designate Special Protection Areas (SPAs). The habitat directive aims to protect animals (other than birds), plants, and habitats, for which each member state is required to identify Sites of Community Importance (SCIs). The SCIs and SPAs make up the Natura 2000 network, whose aim is to conserve an extensive range of habitat types and wildlife species throughout Europe, maintaining listed habitat and species at "favourable conservation status" (European Commission 2000a, b; Maiorano et al. 2007).

The list of Natura 2000 sites in the Mediterranean region was first adopted in July 2006, and updated in March and December 2008. Within the Mediterranean region there are 2,928 SCIs under the Habitats Directive and further 999 SPAs under the Birds Directive. There is often considerable overlap between some SCIs and SPAs which means that the figures are not cumulative. Nevertheless, it is estimated that together they cover around 20 % (174,930 km^2) of the total land area in this region (European Commission 2009a). Within the EU, the Macaronesian region consists of three archipelagos: the Azores (see Costa et al. 2013 for a more detailed discussion), Madeira (both belonging to Portugal) and the Canaries (Spain). Altogether, within the Macaronesian Region there are 211 SCIs and further 65 SPAs. It is estimated that together they cover more than a third of the total land area in this region, i.e. about 3,516 km^2 (European Commission 2009b).

The majority of conservation programmes are applied at a national level (Mace et al. 2000; Halpern et al. 2006; Karka et al. 2009), but with increasing internationalisation of conservation efforts, global and regional coordination is becoming more common. Within these programmes, government, and non-governmental organisations in particular, spend some of their resources abroad (Rodríguez et al. 2007). However, compared to local programmes, collaboration between countries can be costly, complicated, and often requires additional logistics and resources (Karka et al. 2009). This results in the current situation where this type of data is more commonly collated at the regional or global level, for example, in the World Database on PAs and there is no common database for PAMIs. There is, however, probably more concern and general interest in networking across marine PAs than terrestrial PAMIs (UNEP-WCMC 2008).

The European inventory of nationally designated PAs began under the CORINE programme. It is now maintained for European Environment Agency (EEA) by the European Topic Centre on Biological Diversity and annually updated through Eionet. EEA provides the European inventory of nationally designated areas to the World Database of Protected Areas and to Eurostat. The database can be downloaded from the EEA Data Service. The nationally designated areas data can also be queried online in the European Nature Information System (EUNIS), yet it is not possible to make any direct query on PAMIs.

To try to assess the total number and size of PAMIs according to Natura 2000, the "European database on Natura 2000" (available at the EEA data center) was downloaded and examined. It consists of data submitted by Member States to the European Commission. These data are subject to regular validation and updating processes. After validation, a new EU-wide Natura 2000 database is released. The spatial data (borders of sites) submitted by each Member State is validated by the EEA and linked to the descriptive data. An analysis of the data set with GIS software to calculate the data presented in Table 18.1, was performed excluding those SCIs and SPAs that exclusively cover water surfaces or *Poseidonia oceanica* (Neptune grass) beds.

For assessing the status of knowledge on the presence of alien plants in PAMIs, one useful starting reference is the work by Delanoë et al. (1996) and their list of "major islands and archipelagos in the Mediterranean" and of "principal protected areas in the Mediterranean islands" listing a total of 39 PAs distributed as follows: Croatia (7), Cyprus (2), Egypt (1), Spain (3), France (6), Greece (6), Italy (7), Lebanon (1), Malta (2), Tunisia (3) and Turkey (1). This list has been checked and updated in Table 18.2.

The analysis on the number of PAMIs and their alien flora (Tables 18.1 and 18.2) highlight the general lack of information. At a national level, not all countries clearly distinguish the PAs present in the mainland from those located on islands. All Croatian islands have an area of about 3,267 km^2, and total PA on them is slightly more than 285 km^2, i.e. 8.7 % of Croatian island area undergoes some kind of protection (Boršić, *pers. comm.*). However, a complete inventory of the alien

Table 18.1 Number and total surface of Natura 2000 sites within protected areas in the Mediterranean and Macaronesian islands, with total values for each country

Country//Islands	No	km^2
Cyprus	**61**	**2,272.4**
French islands (total)	**91**	**4,205.1**
Corsica	83	2,196.5
Other French islands	8	2,008.6
Greek islands (total)	**177**	**13,604.4**
Crete	53	3,783.9
Other Greek islands	124	9,820.6
Italian islands (total)	**402**	**14,794.4**
Sardinia	120	6,289.2
Sicily	210	6,177.5
Other Italian islands	72	2,327.7
Malta	**39**	**241.8**
Portugal islands (total)	**48**	**992.1**
Azores	36	453.1
Madeira	12	539.0
Spanish islands (total)	**371**	**8,322.7**
Balearics	158	2,399.2
Canary islands	203	5,793.3
Other Spanish islands	10	130.2
Total	**1,189**	**44,432.9**

The Macaronesian islands are included here as an example. Data source: EEA Data Centre (After GIS analysis and specific elaboration for islands and PAMIs)

flora of Croatia is not available; there are inventories for some islands (Dobrović et al. 2005) but not for all PAMIs. The same knowledge gap exists for archipelagos with fewer islands. Although reasonably well studied and having full inventories for the total alien flora (e.g. in the Balearics; see Moragues and Rita 2005), information on single PAMIs may be difficult to obtain or is not available at all. Larger islands like Sicily, Sardinia and Corsica have very precise and updated inventories (e.g. Celesti-Grapow et al. 2009; Jeanmonod et al. 2011) but information on alien flora in PAMIs is limited. Furthermore, there are also knowledge gaps in defining and prioritising invasive aliens in the PAMIs. The research done by Pretto et al. (2012) on 37 small Mediterranean islands revealed that the main determinants of both total and established alien species richness are to tourism development and the sprawl of artificial surfaces. The variation in total alien flora composition is mainly driven by environmental variables, whereas when established taxa are considered, human-mediated factors account for most of the explained variation. We can assume that these drivers are also significant for PAMIs. The EU funded project EPIDEMIE provided a large amount of information on plant invasions in Mediterranean islands (e.g. Lloret et al. 2005; Lambdon and Hulme 2006a, b; Vilà et al. 2006; Lambdon 2008; Lambdon et al. 2008a, b), yet the presence and impacts of invasive plants in PAMIs were not specifically addressed within that project.

Table 18.2 Total numbers of alien taxa for a selected set of protected areas in the Mediterranean islands (List modified from Delanoë et al. 1996)

Country	PAMI	Alien flora	Main references
Algeria	îles Habibas (Oran)	4[a]	Véla et al. (2011)
Croatia	Brijuni Islands	N/A	
	Kornati Archipelago	1[a]	Henkens et al. (2010)
	Limski Zajlev (including the island of Lim)	N/A	
	Lokrum Island	6[a]	Bogdanovic et al. (2006)
	Malotonski Bay (including the island of Vepar)	N/A	
	Island of Mljet	>10	Boršić et al. (2009)
	Dundo Forest of the island of Rab	N/A	
Cyprus	Larnaka Lake	>2	Hand (2003)
	Limassol Lake (Akrotiri) and Limassol Forest-Kyparissia Area	>3	Hadjisterkotis (2004) and Hand (2004)
Egypt	Tanees Island (Ashtum El Gamil, Lake Manzala)	N/A	
Spain	Chafarinas Islands (Congreso, Isabel II and Rey Francisco)	1[a]	Calderon (1894)
	Cabrera NP (Majorca)	5[a]	Ministerio de Medio Ambiente y Medio Rural y Marino (2011)
	Columbretes Islands	5[a]	Juan and Crespo (2001) and Fabregat et al. (2007)
France	Cerbicale Islands (Corsica)	5[a]	Paradis et al. (2006)
	Sanguinares Islands (Corsica)	14	Paradis and Piazza (2003) and Paradis and Appietto (2005)
	Fango biosphere reserve (Corsica)	5	Muséum national d'Histoire naturelle (2003–2012)
	Scandola (Corsica)	11	Muséum national d'Histoire naturelle (2003–2012)
	Lavezzu Islands (Corsica)	7[a]	Gamisan and Paradis (1992), Paradis et al. (1994), and Paradis (2009)
	Port-Cros and Hyères Islands	8–174	Bossu (2010) and Crouzet (2009)
Greece	Kithira island	80	Yannitsaros (1998)
	Samaria Gorges (Crete)	2[a]	Brundu, personal observation
	Palm grove in Vaï (Crete)	2[a]	Brundu, personal observation
	Mt Ainos NP (Cephalonia, Ionian Islands)	N/A	
	Sigri petrified forest (Lesbos)	3[a]	Brundu, personal observation
	Yioura (Γιούρα) Island (Northern Sporades)	4[a]	Kamari et al. (1988)
Italy	Alicudi (Nature reserve, Sicily)	29	Domina and Mazzola (2011)
	Asinara Island (Sardinia)	65	Bocchieri (1988)
	La Maddalena archipelago (Sardinia)	49[a]	Pretto et al. (2012)
	Lampedusa Island (Pelagie Islands, Sicily)	20–23	Pretto et al. (2012) and Domina and Mazzola (2011)

(continued)

Table 18.2 (continued)

Country	PAMI	Alien flora	Main references
	Montecristo Island (Archipelago Toscano NP)	59	Pretto et al. (2012)
	Zannone Island (Circeo NP)	6	Pretto et al. (2012)
	Pantelleria (Sicily)	33	Domina and Mazzola (2011)
	Elba Island NP	53	Domina and Mazzola (2011)
	Ustica Island (Lipari Archipelago, Sicily)	32–62	Hammer et al. (1999) and Domina and Mazzola (2011)
	Isola delle Femmine Nature reserve	16	Caldarella et al. (2010)
	Lachea – Faraglioni dei ciclopi	12	Siracusa (2000)
	Tavolara and Molara (Sardinia)	11, 6	Pretto et al. (2012)
	Vulcano (Sicily)	40	Domina and Mazzola (2011)
Lebanon	Palm Islands (Archipelago of Tripoli)	3[a]	Ministry of Environment of Lebanon (2004)
Lybia	Geziret Ghara and Geziret Al Elba (Ghara islands)	N/A	
Malta	Ghadira (Malta)	N/A	
	Fungus rock (Il Gebla tal-General, Dwejra, Gozo)	3[a]	Cassar et al. (2004)
	Selmunett Islands (St Paul's islands)	1[a]	MEPA (2012)
	Filfla Island	N/A	
	Comino and Cominotto	8[a]	Pavon (2008)
Tunisia	Island of Galiton (La Galite Archipelago)	1[a]	Pavon and Véla (2011)
	Zembra and Zembretta Islands	1[a]	Pavon and Véla (2011)
	Djeziret Bessila (Kneiss Islands)	N/A	
Turkey	Islets included in Olympos-Beydağları	N/A	

N/A indicates that no information on the alien flora was found following standard practices for conducting meta-analysis on Journal databases and contacting local experts. However information may be present in grey literature and investigations may be in progress in several of these PAMIs, but results are not currently available
[a]Indicates that the total number of alien species is likely to be underestimated as it is not derived from a specific inventory for that PAMI. In these cases the values may be derived from a generic flora list, not specifically dedicated to alien species or still in progress, from a generic report or document on the PAMI, or from expert communications and observations

18.4   Key Invasive Alien Plants in Protected Areas on Mediterranean Islands and Their Impacts

There are significant differences in the abundance and presence of invasive alien plants in Mediterranean islands in general (Lloret et al. 2005) and, as a result, in PAMIs. Nevertheless, some key species need particular attention for prevention (if they are not yet arrived in the single PAMIs), for early detection (if they are already present e.g. in nearby islands on the same archipelago), for eradication and for control and management if already established in the territory.

When assessing the potential or actual risks posed by invasive alien plant species on PAMIs, land managers face an additional problem. This is due to difficulties in identifying species native to other parts of the world that might be scarcely studied or might be introduced as cultivated varieties or hybrids.

Marsilea azorica (clover fern) was described in 1983 from the isolated Azores archipelago in the northern Atlantic, where it is restricted to a single roadside pond. Thought to be an extremely local endemic, it was subsequently listed as a conservation priority species for the Azores, Macaronesia, and Europe, included as 'critically endangered' on the IUCN Red list, and as 'strictly protected' species by the Bern convention and the European Union's habitats directive. However, Schaefer et al. (2011) demonstrated that *M. azorica* is conspecific with *M. hirsuta* (rough water clover), a species native to Australia, but widely cultivated and locally invasive in the southern U.S.A. Based on DNA data, they conclude that these plants are most likely a recent introduction to the Azores from Florida.

The South African *Carpobrotus edulis* (ice plant) and *C. acinaciformis* (Hottentot fig) are perennial, clonal, trailing succulents characterised by a mat-forming habit and thick, three dimensional, elongated leaves with a triangular cross-section. Within their native range of South Africa, the two taxa can be visually distinguished by flower colour: *C. edulis* is the only member of its genus having distinctly yellow flowers fading to light pink, while C. acinaciformis has vivid magenta flowers (Wisura and Glen 1993; Suehs et al. 2002). In contradiction to several decades of recording, the presence of two or more species of *Carpobrotus*, Akeroyd and Preston (1990, 1993), basing their observations on foliar characteristics, concluded that *C. edulis* was the only species present in the Mediterranean basin, although in the form of several varieties with different flower colours (and that citations *for C. acinaciformis* in the Mediterranean basin were misidentifications of a magenta-flowered *C. edulis* var. *rubescens*). Further complicating the situation is the possibility of hybridisation between *Carpobrotus* species, which has been clearly demonstrated between *C. edulis* and *C. chilensis* (sea fig) in California (Albert et al. 1997; Gallagher et al. 1997), and noted between *C. edulis* and three other species, including *C. acinaciformis* in South Africa (Wisura and Glen 1993) and *C. virescens* (coastal pigface) in Australia (Blake 1969). However, recent taxonomic investigation has brought into question just exactly which *Carpobrotus* taxon is present in the Mediterranean basin and is the most aggressive towards the indigenous flora (Suehs et al. 2002). The study from Suehs et al. (2002), based on morphological and isozyme analysis, clearly discriminates two invasive *Carpobrotus* taxa, *C. edulis* and *C. acinaciformis* (or *C. aff. acinaciformis*) in the Hyères archipelago off the south-eastern coast of France (French National Park of Port-Cros). Many studies have demonstrated that despite their very similar appearance and habit, these two taxa differ dramatically in their reproductive strategies (e.g. Suehs et al. 2005).

Carpobrotus edulis, C. acinaciformis and their hybrids are generally considered highly invasive species in Mediterranean coastal areas, on islands and PAMIs. Most studies have focused on the effects of *Carpobrotus* spp. on the invaded community (Weber and D'Antonio 1999; Suehs et al. 2001; Moragues and Traveset 2005; Bartomeus et al. 2008; Jakobsson et al. 2008; Vilà et al. 2008) as well as clonality and growth (Sintes et al. 2007; Traveset et al. 2008a, b; de la Peña et al. 2010). Based on these studies, it appears that *Carpobrotus* spp. have a significant impact on the diversity, structure and dynamics of the native vegetation. Moreover, the invasion has a dramatic impact on the characteristics of the invaded soil, leading to an increase in organic matter content and a decrease of pH (Vilà et al. 2006; Conser and Connor 2009; de la Peña et al. 2010) and modification of the lichen and bryophyte soil community (Zedda et al. 2010; Cogoni et al. 2011). *Carpobrotus* spp. may have also negative impacts on several terrestrial animal species (Palmer et al. 2004; Orgeas et al. 2007; Galán 2008). The fleshy fruits bear a large number, often over a thousand, of small seeds (Bartomeus and Vilà 2009) that are eaten and widely dispersed by several mammals such as rabbits (D'Antonio 1990) and rats (Bourgeois et al. 2005). Several studies have shown that *Carpobrotus* spp. fruits and gull-derived resources are significant resources for introduced black rats on Mediterranean islands and that these enriched resources alters individual growth rates, reproductive output and rat population densities at a local scale (Bourgeois et al. 2005; Ruffino et al. 2011).

Several endemic plant species with a very narrow distribution are present on PAMIs (Table 18.3) and are threatened by *Carpobrotus* spp. For example, the critically endangered *Apium bermejoi* (api d'en Bermejo), occurs in the eastern part of Minorca island where it is only found in two small areas, separated by a rocky zone about 200 m wide. The total population numbers less than 100 individuals and covers an area of just a few dozen square metres (de Montmollin and Strahm 2005). Additionally, *Carpobrotus* species are particularly problematic as they are very attractive ornamental species, very easily propagated by cuttings, rarely perceived as a problem by local communities (e.g. Bardsley and Edwards-Jones 2006, 2007) and, on the contrary, considered very useful for soil stabilisation and for providing a green covering of the soil also during the arid season, on mostly every type of soils and with minimal water requirements. Therefore all planned interventions to control the species must include awareness raising, education, promotion of alternative species, to avoid the possible re-introduction in the restored sites.

De Montmollin and Strahm (2005) report a list of 'top 50' endangered Mediterranean plants, most of which occur in PAMIs. The main factor raising the risk of extinction for these 50 species is linked to the size of their population and their distribution. In almost every case, due to the small number of individuals or tiny area of distribution, any major disturbance (for example, fire or construction work, competition with other species) might just push the species to extinction or seriously reduce its chances of survival (de Montmollin 2011). The island experts that contributed to reporting the conservation status of the 'top 50' indicated invasive alien plants as a major threat for conservation in the 12 cases (Table 18.3).

Indeed, most of PAMIs' threatened species face more than one threat. Many of these species have a very limited range, often less than 1 ha, and might be present on a single islet, thus any habitat modification or space occupation by alien taxa might represent an additional threat. Habitat modification might also drive increases in the local abundance of invaders (Didham et al. 2007) and invasive

Table 18.3 Twelve endemic and critically endangered Mediterranean plants from the IUCN 'Top 50 Mediterranean Island Plant' list, which are threatened by alien plant invasions (according to de Montmollin and Strahm 2005)

Endemic species	Island//PAMIs	Invasive alien plants
Anchusa crispa	Sardinia (Italy), Corsica (France) (SCIs)	*Carpobrotus* spp.
Abies nebrodensis	Sicily NP (Italy)	Alien fir
Apium bermejoi	Minorca SCI (Spain)	*Carpobrotus edulis*
Calendula maritima	Sicily SCI (Italy)	*Carpobrotus edulis*
Centaurea gymnocarpa	Capraia – Tuscan Archipelago NP (Italy)	*Carpobrotus acinaciformis, Senecio angulatus*
Centranthus trinervis	Corsica (France)	*Centranthus ruber, Cortaderia selloana*
Cheirolophus crassifolius	Gozo and Malta	*Agave americana, Carpobrotus edulis, Opuntia ficus-indica*
Cremnophyton lanfrancoi	Gozo and Malta	*Agave americana, Carpobrotus edulis, Opuntia ficus-indica*
Helichrysum melitense	Gozo	*Agave americana, Carpobrotus edulis, Opuntia ficus-indica*
Medicago citrina	Columbretes (Spain)	*Cuscuta sp., Opuntia maxima*
Silene hicesiae	Aeolian islands (Italy)	*Ailanthus altissima*
Viola ucriana	Sicily (Italy)	Alien conifers (plantations)

SCI Site of Community Interest according to Directive 92/43/EEC, *NP* National Park

plant species generally have a strong, albeit variable, influence on patterns of biodiversity at relatively small spatial scales (Gaertner et al. 2009; Powell et al. 2011). Quite often the rare species on PAMIs are, like the Hawaiian endemics, imperilled by the "vicious triumvirate" of feral pigs, goats and alien plants or similar consortiums (e.g. Pisanu et al. 2012). This is a particular problem when these threatened plants occur in areas that are protected, or are otherwise not currently subject to habitat loss or direct habitat destruction by humans, however fires and herbivores might arrive anyway (see Gurevitch and Padilla 2004). All known populations of the endangered *Medicago citrina* (alfalfa arborea) are found on islands and small islets near the eastern Iberian Peninsula (Aizpuru et al. 2000; Juan et al. 2004), more precisely only ten isolated subpopulations are currently known (four from Ibiza, three from Cabrera, two from Columbretes islands and one from an offshore islet in northern Alicante province), constituting a severely fragmented genetic system. The alien scale insect *Icerya purchasi* was reported to attack *M. citrina* at Columbretes islands (Juan 2002) which suffer from the combined effect of multiple alien herbivores (European rabbits, *Oryctolagus cuniculus*; black rats, *Rattus rattus*; and house mouse, *Mus musculus*) in all four different life stages (Traveset et al. 2009; Latorre et al. 2013).

Many invasive alien species are present since a long time in most of the PAMIs (Table 18.4) such as, for example, *Acacia* spp., *Agave* spp., *Ailanthus altissima* (tree of heaven), *Amaranthus* spp., *Cotula coronopifolia* (brass buttons), *Nicotiana glauca* (tree tobacco), *Opuntia* spp., *Oxalis pes-caprae* (Bermuda buttercup), *Ricinus*

Table 18.4 Key invasive alien plant species in the Mediterranean island protected areas and their main impacts on biodiversity (I1), agriculture and forestry (I2) and other sectors (I3), according to the definitions of the EPPO prioritisation method (Brunel et al. 2010a, b)

Species	I1	I2	I3	References for impacts on PAMIs
Acacia spp.	H	L	Na	Hadjikyriakou and Hadjisterkoti (2002), Cardinale et al. (2008), Le Maitre et al. (2011), and Wilson et al. 2011
Agave spp.	H	Ns	Na	Camarda et al. (2004) and Lambdon et al. (2008a, b)
Akebia quinata	H	Na	Na	Brunel et al. (2010b)
Ailanthus altissima	H	M	Y	Vilà et al. (2006), Traveset et al. (2008a), and Jeanmonod et al. (2011)
Alternanthera philoxeriodes	H	M	Y	Brunel et al. (2010b)
Amaranthus spp.	L	H	Y	Brundu et al. (2003, 2004) and Camarda et al. (2004)
Ambrosia artemisiifolia	L	H	Y	Brunel et al. (2010b)
Aptenia cordifolia	M	Ns	Na	Brundu et al. (2003, 2004) and Camarda et al. (2004)
Araujia sericifera	M	M	Y	Brunel et al. (2010b)
Baccharis halimifolia	H	L	Y	Brunel et al. (2010b)
Carpobrotus acinaciformis	H	L	Ns	Vilà et al. (2006, 2010)
Carpobrotus edulis	H	L	Ns	de Montmollin and Strahm (2005), Vilà et al. (2006), and Jeanmonod et al. (2011)
Cabomba caroliniana	H	L	Na	Brunel et al. (2010b)
Cortaderia selloana	H	L	Y	Brunel et al. (2010b) and Jeanmonod et al. (2011)
Cotula coronopifolia	M	L	Ns	Biondi and Bagella (2005) and Paradis et al. (2006)
Delairea odorata	M	Na	Y	Brunel et al. (2010b)
Eichhornia crassipes	H	H	Y	Brunel et al. (2010b)
Eucalyptus spp.	H	Ns	Ns	Brundu et al. (2003, 2004) and Díez (2005)
Fallopia baldschuanica	M	Na	Na	Brunel et al. (2010b)
Gomphocarpus fruticosus	H	L	Y	Brundu et al. (2003, 2004) and Jeanmonod et al. (2011)
Hakea sericea	H	Na	Y	Brunel et al. (2010b)
Humulus japonicus	H	Na	Y	Brunel et al. (2010b)
Hydrilla verticillata	H	L	Y	Brunel et al. (2010b)
Ludwigia grandiflora	H	H	Y	Brunel et al. (2010b)
Ludwigia peploides	H	H	Y	Brunel et al. (2010b) and Jeanmonod and Schlüssel (2010)
Microstegium vimineum	H	Na	Na	Brunel et al. (2010b)
Myriophyllum heterophyllum	H	H	Na	Brunel et al. (2010b)

(continued)

Table 18.4 (continued)

Species	I1	I2	I3	References for impacts on PAMIs
Nassella spp.	M	M	Na	Brunel et al. (2010b) and Bourdôt et al. (2012)
Nicotiana glauca	H	L	Y	Bogdanovic et al. (2006)
Opuntia spp.	H	L	Y	Vilà et al. (2003), Camarda et al. (2004), Monteiro et al. (2005), and Bartomeus et al. (2008)
Oxalis pes-caprae	M	H	Y	Moragues and Rita (2005), Vilà et al. (2006), and Caldarella et al. (2010)
Paspalum paspaloides	H	M	Na	Camarda et al. (2004) and Fraga i Arguimbau (2008)
Paspalum dilatatum	H	M	Na	Brundu et al. (2003, 2004)
Pennisetum setaceum	H	L	Na	Moragues and Rita (2005), Brunel et al. (2010b), Caldarella et al. (2010), and Pasta et al. (2010)
Pistia stratiotes	H	M	Y	Brunel et al. (2010b)
Salvinia molesta	H	L	Y	Paradis and Miniconi (2011)
Senecio inaequidens	M	M	Y	Brundu et al. (2003, 2004)
Sesbania punicea	M	L	Y	Brunel et al. (2010b) and Jeanmonod et al. (2011)
Solanum elaeagnifolium	M	H	Y	Pavletić et al. (1978), Tscheulin et al. (2009), and Brunel et al. (2010b)
Solanum spp.	H	H	Y	Camarda et al. (2004)
Verbesina encelioides	M	M	Na	Brunel et al. (2010b)

L low impact, *M* medium impact, *H* high impact, *Y* impacts recorded also for other sectors, *Na* no available information on impacts, *Ns* not significant impact

communis (castor oil plant), *Senecio* spp., *Solanum* spp. Their distribution and impacts can be found in specific literature and floras, e.g. in Camarda et al. (2004) for Sardinian national parks, in Pretto et al. (2012), for a set of small Italian islands, including several PAMIs. In Table 18.4 the value for I1 (impact on biodiversity) addresses the potential for that alien to induce long term population loss affecting rare and threatened species and to cause serious habitat or ecosystem effects that are difficult to reverse and it is ranked as high, medium, low according to the definitions in Brunel et al. (2010a, b). For example, a high impact on biodiversity is reported for those invasive aliens that colonise habitats that have a value for nature conservation where it forms large, dense and persistent populations. For example, *Eichhornia crassipes* (water hyacinth) and *Ludwigia grandiflora* (water primrose) in water bodies and *Carpobrotus* spp. in dune ecosystems.

Acacia species are among the most serious plant invaders worldwide (e.g. Kull et al. 2011; Le Maitre et al. 2011; Wilson et al. 2011). They have been initially introduced in Mediterranean islands to stabilise sand dunes and as ornamental and are now very common in PAMIs. In the invaded areas they decrease native plant diversity (Marchante et al. 2003), significantly alter soil properties and impact water cycling. Nevertheless, there are also indirect unexpected effects of the introduction of *Acacia* spp. in PAMIs. Symbiotic rhizobia isolated from seven

wild legume shrubs native of Sicily (*Genista* spp.) were identified as being of
tropical and Australian origin and their presence in Sicily and some neighbouring
islands could be explained by the introduction of Australian native *Acacia* spp. into
the Mediterranean basin (Cardinale et al. 2008). A similar case is reported
for *Eucalyptus* spp., very common tree species in many PAMIs, even if the effects
of the introduction of alien root-symbiotic fungi together with *Eucalypts* have
received little attention but have been demonstrated by Díez (2005) for *Eucalyptus
camaldulensis* (river red gum) and *E. globulus* (Tasmanian blue gum), i.e. the most
common species in PAMIs.

Prickly pear cacti (*Opuntia* spp.) are one of the most well-known examples of a
plant genus that has invaded a variety of habitats around the world (Cronk and
Fuller 1995). In Mediterranean islands and PAMIs, *Opuntia* spp., notably *Opuntia
ficus-indica* (prickly pear), *Opuntia stricta* var. *dillenii* (erect prickly pear),
Austrocylindropuntia cylindrica (= *O. cylindrica*) (cane cactus), and *O. maxima*
have been used for fruit consumption, livestock foraging, fencing and as ornamentals. In the past, *Opuntia* spp. were also used for the production of a red dye that was
obtained from the cochineal insect which infested the plants. Vegetative reproduction is common, with cladodes breaking off and rooting, usually near the parental
ramet, often forming conspicuous patches (Vilà et al. 2003). Sexual reproduction is
also very important for mid- to long-distance dispersal, which is facilitated by
animals that consume their fruits. In some cases, disperser may also be alien rodents
introduced into the PAMIs (Traveset et al. 2009; Padrón et al. 2011). Invasion by
Opuntia spp. is also considered an example in which a single plant species poses an
opportunity for a novel ecosystem to develop because the invader has functional
traits qualitatively different from the other colonising native species (Vilà
et al. 2003; Monteiro et al. 2005). Furthermore *Opuntia* spp. might also compete
for pollinators with native species (Bartomeus et al. 2008).

Nicotiana glauca is a woody invader generally found in open and disturbed
areas, including wastelands, roadsides and archaeological sites. Extensive stands
may be found for lengthy periods on stream floodplains and temporary river beds.
Studies have demonstrated that *N. glauca* is highly toxic to humans and animals,
and is usually avoided as it is unpalatable. However, during severe drought, when
food resources are scarce, livestock may consume the plant and die as the plant
contains the alkaloid anabasine, which is considerably more toxic than nicotine
(Florentine and Westbrooke 2005).

Eichhornia crassipes, *Salvinia molesta* (floating water fern) and *Pistia stratiotes*
(water lettuce) have spread throughout the world's waterways as a result of anthropogenic activities. With the potential to double in biomass in a few days or weeks
and the ability to spread easily due to their free-floating vegetative form, these
species can successfully colonise new habitats, form dense mats along shorelines
and displace native vegetation. In doing so, they affect ecological processes within
rivers and lakes, often decreasing biodiversity. Many native hydrophyte and
helophyte species are found in permanent and temporary freshwater ecosystems
in PAMIs and are threatened with extinction at a regional level (Dudgeon
et al. 2006; Villamagna and Murphy 2010; Hussner 2012). Temporary ponds in

the Mediterranean are priority conservation habitats under the European Union Habitats Directive, but those of natural origin are scarce as many of them have been destroyed or transformed into permanent waters. An annual plant, *Cotula coronopifolia*, can build up dense populations that crowds out native vegetation in these habitats (Biondi and Bagella 2005; Paradis et al. 2006).

A major step in tackling invasive alien plants consists of identifying those species that represent a future threat to managed and unmanaged habitats in PAMIs. The European and Mediterranean Plant Protection Organisation (EPPO) reviews and organises data on alien plants in order to build an early warning system (Brunel et al. 2010a, b). During a series of dedicated workshops, the EPPO prioritisation system (Brunel et al. 2010a) has been applied to the Mediterranean basin. Surveys and rapid assessments of spread and impact have allowed identification of the following emerging invasive alien plants: *Alternanthera philoxeroides* (alligator weed), *Ambrosia artemisiifolia* (common ragweed), *Baccharis halimifolia* (saltbush), *Cortaderia selloana* (pampas grass), *Eichhornia crassipes*, *Fallopia baldschuanica* (Russian vine), *Hakea sericea* (silky hakea), *Humulus japonicus* (Japanese hop), *Ludwigia grandiflora* and *L. peploides* (creeping water primrose), *Hydrilla verticillata* (hydrilla), *Microstegium vimineum* (Japanese stiltgrass), *Myriophyllum heterophyllum* (twoleaf watermilfoil), *Pennisetum setaceum* (fountain grass), *Pistia stratiotes*, *Salvinia molesta* and *Solanum elaeagnifolium* (silverleaf nightshade) (Table 18.4). These species should represent priorities for action. Some other species are placed on the observation list, as available information does not allow them to be included among the worst threats: *Akebia quinata* (five-leaf akebia), *Araujia sericifera* (moth plant), *Delairea odorata* (cape ivy), *Cabomba caroliniana* (Carolina fanwort), *Nassella neesiana* (Chilean needle grass), *N. tenuissima* (Mexican feather grass) and *N. trichotoma* (serrated tussock), *Sesbania punicea* (rattlebox) and *Verbesina encelioides* (golden crownbeard).

Researches and botanical surveys conducted as part of the "Mediterranean Small Islands Initiative" (PIM, http://www.initiative-pim.org/) are providing a useful source for the presence of alien plant species on PAMIs that were poorly studied in the past. For example, Pavon and Véla (2011) recently recorded *Conyza floribunda* (tropical horseweed) on Tunisian islands, while the taxon was not recorded in previous expeditions (Bocchieri and Mossa 1985). In the framework of the PIM initiative, the flora of the nature reserve of Habibas (NW Algeria) was studied by Véla et al. (2001). The Authors highlight the presence of several non-native species including *Carpobrotus* sp., *Oxalis pes-caprae* and *Opuntia* sp.

18.5   Pathways for Invasive Plant Introduction in Protected Areas on Mediterranean Islands

Mediterranean islands often have a high human population density, a dense road network, high ports/harbours and airports per capita (or per area), large dependence on imports and a high flux of humans across their borders, especially through tourism (Hulme et al. 2008b). These attributes are also generally present in PAMIs, in fact, many of them are leading tourist destinations, especially during the summer season. All these attributes strongly facilitate the introduction of alien species as contaminants of trade and/or hitchhikers on transport vectors, from the mainland to the islands, from the island to the PAMI, between different islands and PAMIs. Yet even with the increased opportunities for accidental introductions, the majority of naturalised species arise from intentional introductions that have subsequently escaped from gardens, agriculture or forestry and botanical gardens (Hulme et al. 2008b and references cited therein). The use of alien species in farming, forestry and for recreational purposes has increased in the Mediterranean since the middle of the twentieth century (e.g. Naveh 1975). In many cases, especially in the past, PAMIs have not received dedicated forestry policies, so that large areas have been planted with alien *Acacia*, *Eucalyptus* and *Pinus* species, in some case even with the purpose of promoting the recovery of native vegetation. *Carpobrotus* spp. have often been used to stabilise sand dunes, and associated with the planting of *Acacia* and *Pinus* species. Escapes of ornamental plants represent the largest single source of naturalised alien species. Almost half of all plant introductions to Mediterranean islands stem from the increasing popularity of gardens and landscaping associated with tourist developments and housing and gardens estate. It therefore follows that this is likely to be a major source of naturalised species (de Montmollin and Strahm 2005; Hulme et al. 2008a, b and references cited therein).

Bird activities have several direct and indirect effects on vegetation. First, there are a number of physical effects, such as trampling, sitting, digging, and uprooting, and second, the chemical effects due to guano manure and increasing nitrogen, phosphorus and potassium compounds. Moreover, the degradation of the vegetation and the creation of stripped areas favour erosive phenomena on nesting sites, particularly under a Mediterranean bioclimate with violent rainfalls (Vidal et al. 1998 and references cited therein). Therefore species such as the yellow-legged gull, which is extremely abundant in many PAMIs (e.g. more than 15,000 pairs in the Tuscany archipelago), can promote the spread of *Carpobrotus* spp. and other alien taxa.

A great variety of taxa, including reptiles, birds and mammals actively participate in the seed dispersal of *Opuntia* spp. (*O. maxima*, *O. stricta* var *dilleni*). Phenology of *Opuntia* fruits in Minorca and Tenerife overlaps with only a few native fleshy-fruited plants present in the area, providing an advantage for the invader. *Opuntia* spp. are further integrated into native communities by means of mutualistic interactions, with both native and alien dispersers (Padrón et al. 2011).

Exogenous drivers, such as the European Union Common Agricultural Policy (CAP) and tourism (Ioannides et al. 2001) have also considerably impacted the composition and dynamics of many of the Mediterranean island landscapes, forestry and agriculture activities, especially on larger islands (Tzanopoulos and Vogiatzakis 2011), often directly promoting specific land-use changes and introduction or re-introduction of plant species. The Mediterranean region as a whole has one of the lowest levels of protection of the five Mediterranean regions of the world (Karka et al. 2009) and land conversion exceeds protection by a factor of 8:2 (Underwood et al. 2009; Cox and Underwood 2011). On larger islands such as

Sicily, Sardinia and Crete, investments in irrigation were made and intensive agriculture is now present (Vogiatzakis et al. 2008), thus potentially representing an important pathway that may affect PAs on these islands. On the smaller islands however, agricultural decline is widespread (Petanidou et al. 2008) and this may also promote the spread of certain taxa (e.g. *A. altissima* in abandoned olive groves) to PAMIs. The small island of Ustica, a nature reserve in the north of Sicily (Italy) has been investigated for its plant genetic resources in 1997. A checklist of the agri- and horticultural crop plants comprises more than 110 species, including several alien taxa, e.g. *Acacia saligna* (Port Jackson willow), (Hammer et al. 1999).

In larger islands some pathways may be similar to those already well described for the mainland (e.g. see Hulme et al. 2008a, b; Hulme 2009), including the presence of Botanical Gardens, arboreta, acclimatisation gardens (Hulme 2011). For example, the Sóller Botanic Garden (Serra de Tramuntana, Mallorca, Balearic Islands) has an important collection of species of the Balearic Islands, consisting of about 308 endemic (rare or endangered taxa) but also species of the flora of other Mediterranean islands, i.e. Corsica, Sardinia, Sicily, Malta, Crete and Cyprus. Another important botanical garden is located in the island of Lokrum (Croatia). The purpose of the garden was to investigate the introduction and adaptation of tropical and sub-tropical plants, especially those important for forestry, horticulture and pharmaceutical purposes. Seeds were obtained from other botanical gardens around the world on an exchange basis. Attention focused mostly on trees and shrubs from similar world climates, such as: central Chile, southern and eastern Australia, central and southern coastal California and the south-western Cape region of South Africa (www.imp-du.com).

18.6 Eradication and Management of Invasive Plants in PAMIs: A Sisyphean Task?

Ideally, each PAMIs' management plan should have an invasive alien plant strategy as a part of their general management plan. In order to abate the threat of invasive alien plants, the major strategies that should be included are to (i) inventory the islands alien flora, list the species and the pathways; (ii) assess invasive species threats (existing and potential); (iii) prevent new introductions and further spread of established invaders (for example in other islands of the same archipelago); (iv) control high priority species and pathways in priority habitats or ecosystems; (v) restore/rehabilitate native species and communities in high priority sites (e.g. Tu 2009); (vi) implement an early detection and rapid response approach and contingency plans for possible outbreaks.

To avoid negative impacts from invasive alien species in PAMIs it is best to prevent their entry. Yet it is unrealistic to expect exclusion measures to be 100 % effective and policies for preventing invasions must include monitoring and control measures. Furthermore, many alien plant species have been deliberately introduced in the PAMIs before they were declared, and in some cases also after being declared, for example, as a result of unwise forestry policies.

Recognising that resources for managing invasive alien plants threats on PAMIs are limited, it is essential to set priorities, identifying those species or populations and those PAMIs that are of highest priority for active management. PAMIs could be ranked according to their conservation value in several ways, e.g. identifying critically endangered species (birds, reptiles, amphibians and mammals breeding on islands, native and endemic plant species, etc.) according to IUCN or other methods. Other important elements for setting priorities for invaders and PAMIs is the feasibility of the control actions, the 'conservation gain', the possibility to disrupt processes of 'invasional meltdown', legislative obligations, the possibility of minimising disturbance to other species, the cost of eradication, the possibility to control pathways responsible for unwanted re-introduction (including the risk related to the presence of the visitors), the island size and topography, the distance from the mainland and/or form other islands in the same archipelago, the possible synergies with climate change that will likely radically change both land use and threats from alien plants in the coming decades in PAMIs. On larger Mediterranean islands where several PAMIs are present (see Table 18.2) it is also necessary to set priorities between PAs and invasive species within the same island. There is no standard method for prioritising alien plants, or control actions and eradication on islands (or PAMIs), as to date most of the studies on island prioritisation have been done on alien vertebrates, both at regional and global scale (e.g. Robertson and Gemmell 2004; Capizzi et al. 2010; Harris et al. 2011).

The management of plant invasions in general requires a level of public awareness and support, in particular when there are conflicts of interest (Andreu et al. 2009). For example, *Acacia* spp., were largely introduced into many PAMIs, have many beneficial uses (Griffin et al. 2011). They are often seen as a precious resource that is central to many rural livelihoods by providing fuel wood, food, fodder, and shelter (Kull et al. 2011; Wilson et al. 2011).

In 1992, the European Union approved the LIFE programme and many *LIFE Nature* projects were funded with the aim of implementing actions to preserve natural habitats, flora, and wildlife protected under the Birds Directive and the Habitat Directive. These actions are mostly carried out within the Natura 2000 Network (Table 18.1), which in many cases protects entire islands or significant portions of islands and guarantees the conservation of many threatened and declining species and habitats. Out of a total of 715 LIFE Nature projects financed from 1992 to 2002, 14 % included actions addressed at alien species (European Commission 2004). Nevertheless, as already indicated by Genovesi (2005) for Europe in general, only localised removal of alien plants have been achieved in PAMIs to date.

In 2001 the eradication of *Carpobrotus* spp. started in Minorca island (Spain), funded by a *LIFE Nature* project, with the purpose of removing *Carpobrotus* spp. from all Sites of Community Interest. The intervention also included the control and removal of *Carpobrotus* spp. in private and public gardens. Action to monitor and or locally control *Carpobrotus* spp. have been achieved, or are in progress, in PAMIs located in Italy (e.g. Sardinia and Montecristo island in Tuscany), France (e.g. Port-Cros National Park; Archipel de Riou Nature Reserve; île Lavezzu and île Mezzu Mare in the archipelago îles Sanguinaires in Corsica, see Paradis et al. 2008) and Malta. According to the 4th National Report to the CBD (MEPA 2004) *Carpobrotus* was successfully eradicated from *Ir-Ramla tat-Torri* (northern coast of the island of Malta) and *Ir-Ramla l-Ħamra* (along the northern coast of the island of Gozo). *LIFE Nature* projects are also funding the eradication of *Ailanthus altissima* from Montecristo island in Italy (2010–2014) and control of *Acacia saligna* in Cyprus (Conservation Management in Natura 2000 sites in Cyprus). *A. saligna* is considered a serious threat to the habitat of the salt lake of Larnaca, so that it was planned to remove a number of such plants from the area.

On the Italian island Isola delle Femmine (Nature reserve) *Opuntia stricta* and *Solanum sodomaeum* (apple of Sodom) have been locally eradicated by hand removal, while *Ailanthus altissima* was removed by hand removal and herbicide applications (Genovesi and Carnevali 2011). Nevertheless the flora of Isola delle Femmine it is still rich of alien taxa, including *Oxalis pes-caprae* and *Pennisetum setaceum* that are considered highly invasive on the island (Caldarella et al. 2010) and undoubtfully of more difficult control.

In Malta, according to a national report (MEPA 2004) a number of invasive species are being earmarked from removal from a number of PAs. Preliminary efforts have been undertaken, or are on-going, and are aimed at controlling the spread in the Maltese Islands of *Carpobrotus edulis* from sand dunes, *Arundo donax* and *Vitis vinifera* (common grape vine) from Ir-Ramla l-Ħamra, *A. saligna* from Għajn Tuffieħa (western coast of the island of Malta), *Agave* spp. from Rdum tal-Madonna. Other species being active managed on Malta are *Ricinus communis* and various *Opuntia* spp.

Unfortunately however, many PAMIs belongs to countries that do not yet have a national strategy on biological invasions, thus control efforts against invasive plants are not always accompanied by relevant measures to prevent further introductions.

18.7   Conclusions

Most of the PAMIs hold a unique level of biodiversity and are very sensitive to all human-generated environmental changes, including their synergies. Many invasive alien plants represent a serious threat for nature conservation in PAMIs, as they out-compete several endangered native plants. Nevertheless, information on PAMIs is included in country level statistics and is not specific to Mediterranean islands and PAMIs. In addition, international databases, European and international legislation do not often distinguish PAMIs from other PAs. There are also gaps in the knowledge on the presence of invasive alien plants in PAMIs as not all of them have been investigated at the same level of detail. This is a serious hindrance for management at an international level and specific policies and strategies, including a dedicated definition of priorities, are urgently needed.

G. Brundu

Acknowledgments Several people provided very useful information for the preparation of this chapter, as many activities and data on plant invasions in PAs on Mediterranean islands are available only in 'grey literature' and local reports or not published yet. I therefore gratefully acknowledge for fruitful discussions and for the information provided: Margarita Arianoutsou-Faragitaki, Yiannis Bazos, Fabrice Bernard, Igor Boršić, Sarah Brunel, Ignazio Camarda, Lucilla Carnevalli, Laura Celesti-Grapow, Katija Dolina, Gianniantonio Domina, Bruno Foggi, Piero Genovesi, Costas Kadis, Luka Katušić, Daniel Jeanmonod, Isabel Lorenzo Iñigo, Isabelle Mandon, Pietro Mazzola, Frederic Mèdail, Eva Moragues, Daniel Pavon, Francesca Pretto, Bernard Recorbet, Carl Smith, Paolo Sposimo, Anna Traveset, Errol Vela. Two anonymous reviewers greatly helped in improving the paper.

References

Aizpuru I, Ballester G, Bañares A et al (2000) Lista roja de la flora vascular española (valoración según categorías UICN). Conserv Veg 6:11–38

Akeroyd JR, Preston CD (1990) Notes on some Aizoaceae naturalized in Europe. Bot J Linn Soc 103:197–200

Akeroyd JR, Preston CD (1993) *Carpobrotus* N.E. Br. In: Tutin TG, Burges NA, Chater AO et al (eds) Flora Europaea, vol 1: Psilotaceae to Platanaceae. Cambridge University Press, Cambridge

Albert ME, D'Antonio CM, Schierenbeck KA (1997) Hybridization and introgression in *Carpobrotus* spp. (Aizoaceae) in California. I. Morphological evidence. Am J Bot 84:896–904

Andreu J, Vilà M, Hulme PE (2009) An assessment of stakeholder perceptions and management of noxious alien plants in Spain. Environ Manage 43:1244–1255

Arrigoni PV, Bocchieri E (1995) Caratteri fitogeografici della flora delle piccole isole circumsarde. Biogeographia 18:63–90

Bardsley D, Edwards-Jones G (2006) Stakeholders' perceptions of the impacts of invasive exotic plant species in the Mediterranean region. GeoJournal 65:199–210

Bardsley D, Edwards-Jones G (2007) Invasive species policy and climate change: social perceptions of environmental change in the Mediterranean. Environ Sci Policy 10:230–242

Bartomeus I, Vilà M (2009) Breeding system and pollen limitation of two supergeneralist alien plants invading Mediterranean shrublands. Aust J Bot 57:1–8

Bartomeus I, Vilà M, Santamaria L (2008) Contrasting effects of invasive plants in plant-pollinator networks. Oecologia 155:761–770

Baum T (1997) The fascination of islands: a touristic prospective. In: Lockhart D, Drakakis Smith DW (eds) Island tourism: trends and prospects. Pinter, London, pp 21–36

Biondi E, Bagella S (2005) Vegetazione e paesaggio vegetale dell'arcipelago di La Maddalena (Sardegna nord-orientale). Fitosociologia 42:3–99

Blake ST (1969) A revision of *Carpobrotus* and *Sarcozona* in Australia, genera allied to *Mesembryanthemum* (Aizoaceae). Contr Qld Herb 7:1–65

Bocchieri E (1988) L'Isola Asinara (Sardegna Nord-occidentale) e la sua flora. Webbia 42:227–268

Bocchieri E, Mossa L (1985) Risultati di una escursione geobotanica a La Galite (Tunisia Settentrionale). Boll Soc Sarda Sci Nat 24:207–225

Bogdanovic S, Božena M, Ruščic M et al (2006) *Nicotiana glauca* Graham (Solanaceae), a new invasive plant in Croatia. Acta Bot Croat 65:203–209

Boršić I, Jasprica N, Dolina K (2009) New records of vascular plants for the Island of Mljet (Southern Dalmatia, Croatia). Nat Croat 18:295–307

Bossu E (2010) Quinze ans d'expérience dans la lutte contre les plantes exotiques envahissantes forestières de l'île de Porquerolles (Provence, France). Sci Rep Port-Cros Natl Park Fr 24:199–204

Bourdôt GW, Lamoureaux SL, Watt MS et al (2012) The potential global distribution of the invasive weed *Nassella neesiana* under current and future climates. Biol Invasions 14:1545–1556

Bourgeois K, Suehs CM, Vidal E et al (2005) Invasional meltdown potential: facilitation between introduced plants and mammals on French Mediterranean islands. Ecoscience 12:248–256

Brockie RE, Loope LL, Usher MB et al (1988) Biological invasions of island nature reserves. Biol Conserv 44:9–36

Brundu G, Camarda I, Satta V (2003) A methodological approach for mapping alien plants in Sardinia (Italy). In: Child LE, Brock JH, Brundu G et al (eds) Plant invasions: ecological threats and management solutions. Backhuys Publishers, Leiden, pp 41–62

Brundu G, Camarda I, Hulme PE et al (2004) Comparative analysis of the abundance and distribution of alien plants on Mediterranean islands. In: Arianoutsou M, Papanastasis V (eds), Ecology, conservation and management of Mediterranean climate ecosystems. Proceedings 10th MEDECOS Conference, 25 April – 1 May 2004, Rhodes, Greece. Millipress, Rotterdam, pp 1–9

Brunel S, Branquart E, Fried G et al (2010a) The EPPO prioritization process for invasive alien plants. EPPO Bull 40:407–422

Brunel S, Schrader G, Brundu G et al (2010b) Emerging invasive alien plants for the Mediterranean Basin. EPPO Bull 40:219–238

Caldarella O, La Rosa A, Pasta S et al (2010) La flora vascolare della Riserva Naturale Orientata Isola Delle Femmine (Sicilia Nord-Occidentale): aggiornamento della check-list e analisi del turnover. Naturalista Sicil 34:421–476

Calderon S (1894) Las Chafarinas. Anales de la Sociedad Española de Historia Natural 23, I. Bolívar, Tesorero, Madrid

Camarda I, Brundu G, Carta L et al (2004) Invasive alien plants in the national parks of Sardinia. In: Camarda I, Manfredo MJ, Mulas F et al (eds), Global challenges of parks and protected area management. Procceedings 9th ISSRM, 10–13 October, 2002, La Maddalena, Carlo Delfino Editore, Sassari, Italy, pp 111–123

Capizzi D, Baccetti N, Sposimo P (2010) Prioritizing rat eradication on islands by cost and effectiveness to protect nesting seabirds. Biol Conserv 143:1716–1727

Cardinale M, Lanza A, Bonnì ML et al (2008) Diversity of rhizobia nodulating wild shrubs of Sicily and some neighbouring islands. Arch Microbiol 190:461–470

Cassar LF, Lanfranco S, Schembri PJ (2004) Report on a survey of the terrestrial ecological resources of the Qawra/Dwejra area, Western Gozo, commissioned by Nature Trust Malta. Ecoserv, Malta

Celesti-Grapow L, Alessandrini A, Arrigoni PV et al (2009) Inventory of the non-native flora of Italy. Plant Biosyst 143:386–430

Chapin FS, Zavaleta ES, Eviner VT et al (2000) Consequences of changing biodiversity. Nature 405:234–242

Cogoni A, Brundu G, Zedda L (2011) Diversity and ecology of terricolous bryophyte and lichen communities in coastal areas of Sardinia (Italy). Nova Hedwigia 92:159–175

Conser C, Connor EF (2009) Assessing the residual effect of *Carpobrotus edulis* invasion, implication for restoration. Biol Invasions 11:349–358

Costa H, Bettencourt MJ, Silva CMN et al (2013) Chapter 17: Invasive alien plants in the Azorean protected areas: invasion status and mitigation actions. In: Foxcroft LC, Pyšek P, Richardson DM, Genovesi P (eds) Plant invasions in protected areas: patterns, problems and challenges. Springer, Dordrecht, pp 375–394

Cowling RM, Rundel PW, Lamont BB et al (1996) Plant diversity in Mediterranean-climate regions. Trends Ecol Evol 11:362–366

Cox RL, Underwood EC (2011) The importance of conserving biodiversity outside of protected areas in Mediterranean ecosystems. PLoS ONE 6:e14508

Cronk CB, Fuller JL (1995) Plant invaders. Chapman & Hall, London

Crouzet N (2009) La flore vasculaire de Porquerolles et de ses îlots: mise à jour critique des inventaires. (Hyères, Var, France). Sci Rep Port-Cros Natl Park Fr 23:47–87

Cuttelod A, García N, Abdul Malak D et al (2008) The Mediterranean: a biodiversity hotspot under threat. In: Vié J-C, Hilton-Taylor C, Stuart SN (eds) The 2008 review of the IUCN Red list of threatened species. IUCN, Gland, pp 1–13

D'Antonio CM (1990) Seed production and dispersal in the non-native, invasive succulent *Carpobrotus edulis* (Aizoaceae) in coastal strand communities of central California. J Appl Ecol 27:693–702

Damery C (2008) Développement des actions et politiques de protection du Littoral au travers de l'Initiative PIM: élaboration de la base de données PIM. Rapport de stage de Master 2 professionnel, Universités de Montpellier I, II et III

De la Peña E, de Clercq N, Bonte D et al (2010) Plant-soil feedback as a mechanism of invasion by *Carpobrotus edulis*. Biol Invasions 12:3637–3648

de Montmollin B (2011) Conservation of threatened plants in the Mediterranean islands. Naturalista Sicil 35:73–78

de Montmollin B, Strahm W (eds) (2005) The top 50 Mediterranean island plants: wild plants at the brink of extinction, and what is needed to save them. IUCN/SSC Mediterranean Islands Plant Specialist Group. IUCN, Gland/Cambridge

Delanoë O, de Montmollin B (1999) MIPSG-Mediterranean Islands plant specialist group conservation of Mediterranean Island plants strategy for action. In: Heywood VH, Skoula M (eds) Wild food and non-food plants: information networking. Chania: CIHEAM-IAMC, 1999. (Cahiers Options Méditerranéennes; v. 38). Regional Workshop of the MEDUSA Network "Wild Food and Non-food Plants: Information Networking". 2, 1997/05/01–03, Port El Kantaoui (Tunisia), pp 99–103

Delanoë O, de Montmollin B, Olivier L et al (1996) Conservation of Mediterranean island plants. 1. Strategy for action. IUCN, Gland/Cambridge

Didham RK, Tylianakis JM, Gemmell NJ et al (2007) Interactive effects of habitat modification and species invasion on native species decline. Trends Ecol Evol 22:489–496

Díez J (2005) Invasion biology of Australian ectomycorrhizal fungi introduced with eucalypt plantations into the Iberian Peninsula. Biol Invasions 7:3–15

Dobrović I, Bogdanović C, Boršić I et al (2005) Analisi delle specie esotiche della flora croata. Inf Bot Ital 37:330–331

Domina G, Mazzola P (2011) Considerazioni biogeografiche sulla presenza di specie aliene nella flora vascolare del Mediterraneo. Biogeographia 30:269–276

Dudgeon D, Arthington AH, Gessner MO et al (2006) Freshwater biodiversity: importance, threats, status and conservation. Biol Rev 81:163–182

European Commission (2000a) Natura 2000. Managing our heritage. EU, Luxembourg

European Commission (2000b) Managing Natura 2000 sites. The provisions of article 6 of the 'Habitats' Directive 92/43/CEE. EU, Luxembourg

European Commission (2004) LIFE Focus/Alien species and nature conservation in the EU. The role of the LIFE program. Office for Official Publications of the European Communities, Luxembourg

European Commission (2009a) Natura 2000 in the Mediterranean region. Office for Official Publications of the European Communities, Luxembourg

European Commission (2009b) Natura 2000 in the Macaronesian region. Office for Official Publications of the European Communities, Luxembourg

European Commission (2011) EU Maritime police, facts and figures. Greece. Office for Official Publications of the European Communities, 2011, Luxembourg http://ec.europa.eu/maritimeaffairs/documentation/studies/documents/greece_01_en.pdf. Accessed November.

Fabregat C, Pitarch J, Guaita S et al. (2007) Propuesta para la realización de investigación aplicada a la evaluación del estado de conservación de la vegetación y propuestas para mejorar su gestión en la reserva natural de las islas columbretes. Generalitat Valenciana

Florentine SK, Westbrooke ME (2005) Invasion of the noxious weed *Nicotiana glauca* R. Graham after an episodic flooding event in the arid zone of Australia. J Arid Environ 60:531–545

Fraga i Arguimbau P (2008) Vascular flora associated to Mediterranean temporary ponds on the island of Minorca. Anales Jard Bot Madrid 65:393–414

Gaertner M, Breeyen AD, Hui C et al (2009) Impacts of alien plant invasions on species richness in Mediterranean-type ecosystems: a meta-analysis. Prog Phys Geogr 33:319–338

Galán P (2008) Efecto de la planta invasora *Carpobrotus edulis* sobre la densidad del eslizón tridáctilo (*Chalcides striatus*) en una localidad costera de Galicia. Bol Asoc Herpetol Esp 19:117–121

Gallagher KG, Schierenbeck KA, D'Antonio CM (1997) Hybridization and introgression in *Carpobrotus* spp. (Aizoaceae) in California II. Allozyme evidence. Am J Bot 84:905–911

Gamisan J, Paradis G (1992) Flore et végétation de l'île Lavezzu (Corse du sud). Travaux scientifiques du Parc naturel regional et des riserves naturelles des Corse

Genovesi P (2005) Eradications of invasive alien species in Europe: a review. Biol Invasions 7:127–133

Genovesi P, Carnevali L (2011) Invasive alien species on European islands: eradications and priorities for future work. In: Veitch CR, Clout MN, Towns DR (eds) Island invasives: eradication and management. IUCN, Gland, pp 56–62

Giannakopoulos C, Le Sager P, Bindi M et al (2009) Climatic changes and associated impacts in the Mediterranean resulting from a 2 °C global warming. Glob Planet Change 68:209–224

Greuter W (2001) Diversity of Mediterranean island floras. Bocconea 13:55–64

Griffin AR, Midgley S, Bush D et al (2011) Global plantings and utilisation of Australian acacias – past, present and future. Divers Distrib 17:837–847

Gurevitch J, Padilla DK (2004) Are invasive species a major cause of extinctions? Trends Ecol Evol 19:470–474

Hadjikyriakou G, Hadjisterkotis E (2002) The adventive plants of Cyprus with new records of invasive species. Zeitschrift für Jagdwissenschaft 48:59–71

Hadjisterkotis E (2004) Alien species on Cyprus. Aliens 19–20:43

Halpern BS, Pyke CR, Fox HE et al (2006) Gaps and mismatches between global conservation priorities and spending. Conserv Biol 20:56–64

Hammer K, Laghetti G, Perrino P (1999) A checklist of the cultivated plants of Ustica (Italy). Genet Resour Crop Evol 46:95–106

Hand R (ed) (2003) Supplementary notes to the flora of Cyprus III. Willdenowia 33:305–325

Hand R (ed) (2004) Supplementary notes to the flora of Cyprus IV. Willdenowia 34:427–456

Harris DB, Gregory SD, Bull LS et al (2011) Island prioritization for invasive rodent eradications with an emphasis on reinvasion risk. Biol Invasions 14:1251–1263

Hassan R, Scholes R, Ash N (eds) (2005) Ecosystems and human well-being: current state and trends, vol 1. Findings of the condition and trends working group. The millennium ecosystem assessment series, Island Press, Washington, DC

Henkens RJHG, Ottburg FGWA, der Sluis TV et al (eds) (2010) Biodiversity monitoring in the Kornati Archipelago. Protocols for the monitoring of Natura 2000 and Croatian Red list habitats and species. Wageningen, Alterra, Alterra-report 1963

Heywood VH (1995) The Mediterranean flora in the context of world diversity. Ecol Mediterr 21:11–18

Holt RA (1992) Control of alien plants on nature conservancy preserves. In: Stone CP, Smith CW, Tunison JT (eds) Alien plants invasions in native ecosystems of Hawaii: management and research. University of Hawaii Press, Honolulu, pp 525–535

Hopkins L (2002) IUCN and Mediterranean Islands: opportunities for biodiversity conservation and sustainable use. IUCN Centre for Mediterranean Cooperation. http://www.uicnmed.org/web2007/documentos/Island_book_hopkins.pdf

Hulme PE (2009) Trade, transport and trouble: managing invasive species pathways in an era of globalization. J Appl Ecol 46:10–18

Hulme PE (2011) Addressing the threat to biodiversity from botanic gardens. Trends Ecol Evol 26:168–174

Hulme PE, Bacher S, Kenis M et al (2008a) Grasping at the routes of biological invasions: a framework for integrating pathways into policy. J Appl Ecol 45:403–414

Hulme PE, Brundu G, Camarda I et al (2008b) Assessing the risks to Mediterranean islands ecosystems from non-native plant introductions. In: Tokarska-Guzik B, Brock JH, Brundu G et al (eds) Plant invasions: human perception, ecological impacts and management. Buckhuys Publishers, Leiden, pp 39–56

Hussner A (2012) Alien aquatic plant species in European countries. Weed Res 52:297–306

Ioannides D, Apostolopoulos Y, Sonmez S (2001) Mediterranean islands and sustainable tourism development: practices, management, and policies. Continuum Publishers, London

Jakobsson A, Padron B, Traveset A (2008) Pollen transfer from invasive *Carpobrotus* spp. to natives – a study of pollinator behaviour and reproduction success. Biol Conserv 141:136–145

Jeanmonod D, Schlüssel A (2010) Notes and contributions on Corsican flora, XXIII. Candollea 65:267–290

Jeanmonod D, Schlüssel A, Gamisans J (2011) Status and trends in the alien flora of Corsica. EPPO Bull 41:85–99

Juan A (2002) Estudio sobre la morfología, variabilidad molecular y bio-logía reproductiva de *Medicago citrina* (Font Quer) Greuter (Leguminosae). Bases para su conservación. PhD thesis, Universidad de Alicante, Spain

Juan A, Crespo MB (2001) Anotaciones sobre la vegetación nitrófila del Archipiélago de Columbretes (Castellón) [Notes on the nitrophilous plant communities of the Columbretes Archipelago (Castellón)]. Acta Bot Malacitana 26:219–224

Juan A, Crespo MB, Cowan RS et al (2004) Patterns of variability and gene flow in *Medicago citrina*, an endangered endemic of islands in the western Mediterranean, as revealed by amplified fragment length polymorphism (AFLP). Mol Ecol 13:2679–2690

Kafyri A, Hovardas H, Poirazidis K (2012) Determinants of visitor pro-environmental intentions on two small Greek islands: is ecotourism possible at coastal protected areas? Environ Manage 50:64–76

Kamari G, Phitos D, Snogerup B et al (1988) Flora and vegetation of Yioura, N Sporades, Greece. Willdenowia 17:59–85

Karka S, Levinc N, Granthamb HS et al (2009) Between-country collaboration and consideration of costs increase conservation planning efficiency in the Mediterranean Basin. Proc Natl Acad Sci 106:15368–15373

Kreft H, Jetz W, Mutke J et al (2008) Global diversity of island floras from a macroecological perspective. Ecol Lett 11:116–127

Kull CA, Shackleton CM, Cunningham PJ et al (2011) Adoption, use and perception of Australian acacias around the world. Divers Distrib 17:822–836

Lambdon PW (2008) Why is habitat breadth correlated strongly with range size? Trends amongst the alien and native floras of Mediterranean islands. J Biogeogr 35:1095–1105

Lambdon PW, Hulme PE (2006a) How strongly do interactions with closely-related native species influence plant invasions? Darwin's naturalization hypothesis assessed on Mediterranean islands. J Biogeogr 33:1116–1125

Lambdon PW, Hulme PE (2006b) Predicting the invasion success of Mediterranean alien plants from their introduction characteristics. Ecography 29:853–865

Lambdon PW, Lloret F, Hulme PE (2008a) Do alien plants on Mediterranean islands tend to invade different niches from native species? Biol Invasions 10:703–716

Lambdon PW, Lloret F, Hulme PE (2008b) Do non-native species invasions lead to biotic homogenization at small scales? The similarity and functional diversity of habitats compared for alien and native components of Mediterranean floras. Divers Distrib 14:774–785

Latorre L, Larrinaga AR, Santamaría L (2013) Combined impact of multiple exotic herbivores on different life stages of an endangered plant endemism, *Medicago citrina*. J Ecol 101:107–117

Le Maitre DC, Gaertner M, Marchante E et al (2011) Impacts of invasive Australian acacias: implications for management and restoration. Divers Distrib 17:1015–1029

Lim CC, Cooper C (2009) Beyond sustainability: optimising island tourism development. Int J Tour Res 11:89–103

Lindsay K, Craig J, Low M (2008) Tourism and conservation: the effects of track proximity on avian reproductive success and nest selection in an open sanctuary. Tour Manage 29:730–739

Lloret F, Médail F, Brundu G et al (2005) Species attributes and invasion success by alien plants on Mediterranean islands. J Ecol 93:512–520

Lonsdale WM (1999) Global patterns of plant invasions and the concept of invasibility. Ecology 80:1522–1536

López A, Correas E (2003) Assessment and opportunities of Mediterranean networks and action plans for the management of protected areas. IUCN, Gland

Macdonald IAW, Graber DM, DeBenedetti S et al (1988) Introduced species in nature reserves in Mediterranean type climatic regions of the world. Biol Conserv 44:37–66

Macdonald IAW, Loope LL, Usher MB et al (1989) Wildlife conservation and the invasion of nature reserves by introduced species: a global perspective. In: Drake JA, Mooney HA, di Castri F et al (eds) Biological invasions: a global perspective. Wiley, Chichester, pp 215–255

Mace GM, Balmford A, Boitani L et al (2000) It's time to work together and stop duplicating conservation efforts. Nature 405:393

Maiorano L, Falcucci A, Garton EO et al (2007) Contribution of the Natura 2000 network to biodiversity conservation in Italy. Conserv Biol 21:1433–1444

Marchante H, Marchante E, Freitas H (2003) Invasion of the Portuguese dune ecosystems by the exotic species *Acacia longifolia* (Andrews) Willd.: effects at the community level. In: Child LE, Brock JH, Brundu G et al (eds) Plant invasions: ecological threats and management solutions. Backhuys Publishers, Leiden, pp 75–85

Ministerio de Medio Ambiente y Medio Rural y Marino (2011) Plan de control y eliminación de especies vegetales invasoras de sistemas dunares. REF: 28/5101. http://www.magrama.gob.es/es/costas/temas/el-litoral-zonas-costeras/TODO_tcm7-187197.pdf

Médail F, Quézel P (1997) Hot-spots analysis for conservation of plant biodiversity in the Mediterranean Basin. Ann Mo Bot Gard 84:112–127

Médail F, Quézel P (1999) Biodiversity hotspots in the Mediterranean Basin: setting global conservation priorities. Conserv Biol 13:1510–1513

MEPA (2004) Malta Environment and Planning Authority. Fourth National Report to the CBD – Malta

MEPA (2012) Malta Environment and Planning Authority. Protecting the diversity of species. http://www.mepa.org.mt/outlook9-article6. Accessed 15 Nov 2012

Ministry of Environment of Lebanon (2004) Final Report. Biodiversity assessment and monitoring in the protected areas//Lebanon LEB/95/G31. Palm Island Nature Reserve. http://93.185.92.38/MOEAPP/ProtectedAreas/publications/FinalReportPalm.pdf

Monteiro A, Cheia VM, Vasconcelos T et al (2005) Management of the invasive species *Opuntia stricta* in a botanical reserve in Portugal. Weed Res 45:193–201

Moragues E, Rita J (2005) Els vegetals introduïts a les Illes Balears. Documents tècnics de conservació. II època, núm.11. Govern de les Illes Balears

Moragues E, Traveset A (2005) Effect of *Carpobrotus* spp. on the pollination success of native plant species of the Balearic Islands. Biol Conserv 122:611–619

Moriondo M, Good P, Durao R et al (2006) Potential impact of climate change on fire risk in the Mediterranean area. Clim Res 31:85–95

Muséum national d'Histoire naturelle (ed) (2003–2012) National inventory of natural heritage. http://inpn.mnhn.fr. Accessed Nov 2012

Naveh Z (1975) Degradation and rehabilitation of Mediterranean landscapes. Neotechnological degradation of Mediterranean landscapes and their restoration with drought resistant plants. Landsc Plan 2:133–146

Nikolić T, Antonić O, Alegro AL et al (2008) Plant species diversity of Adriatic islands: an introductory survey. Plant Biosyst 142:435–445

Orgeas J, Ponel P, Fadda S et al (2007) Conséquences écologiques de l'envahissement des griffes de sorcière (*Carpobrotus* spp.) sur les communautés d'insectes d'un îlot du Parc national de Port-Cros (Var). Sci Rep Port-Cros Natl Park Fr 22:233–257

Padrón B, Nogales M, Traveset A et al (2011) Integration of invasive *Opuntia* spp. by native and alien seed dispersers in the Mediterranean area and the Canary Islands. Biol Invasions 13:831–844

Palmer M, Lindea M, Pons GX (2004) Correlational patterns between invertebrate species composition and the presence of an invasive plant. Acta Oecol 26:219–226

Paradis G (2009) Biodiversité végétale des îlots satellites. Stantari 16:37–44

Paradis G, Appietto A (2005) Compléments à l'inventaire floristique de l'archipel des Îles Sanguinaires (Ajaccio, Corse). Le Monde des Plantes 487:1–6

Paradis G, Miniconi R (2011) Une nouvelle espèce aquatique invasive découverte en Corse, au sud du golfe d'Ajaccio : *Salvinia molesta* D.S.Mitch. (Salviniaceae, Pteridophyta). J Bot Soc Bot France 54:45–48

Paradis G, Piazza C (2003) Flore et végétation de l'archipel des Sanguinaires et de la presqu'île de la Parata (Ajaccio, Corse). Bull Soc Bot Centre-Ouest 34:65–136

Paradis G, Lorenzoni C, Piazza C (1994) Flore et végétation de l'île Piana (Réserve des Lavezzi, Corse du Sud). Trav Sci Parc Natl Rég Rés Nat Corse Fr 50:1–87

Paradis G, Piazza C, Pozzo di Borgo M-L (2006) Contribution à l'étude de la flore et de la végétation des îlots satellites de la Corse. 12e note: île Pietricaggiosa (archipel des îles Cerbicale). Bull Soc Bot Centre-Ouest 37:223–250

Paradis G, Hugot L, Spinosi P (2008) Les plantes envahissantes: une menace pour la biodiversité. Stantari 13:18–26

Pasta S, Badalamenti E, La Mantia T (2010) Tempi e modi di un'invasione incontrastata: *Pennisetum setaceum* (Forssk.) Chiov. (Poaceae) in Sicilia. Naturalista Sicil 34:487–525

Pavletić Z, Devetak Z, Trinajstić I (1978) Novo značajno nalazište neotofita u flori hrvatskog primorja. Fragm Herbol Jugosl 6:69–72

Pavon D (2008) Report of naturalists prospections realized in Malta. Petites îles de Méditerranée 08. IMEP Marseille

Pavon D, Vela E (2011) Espèces nouvelles pour la Tunisie observées sur les petites îles de la côte septentrionale (archipels de la Galite et de Zembra, îlots de Bizerte). Fl Medit 21:273–286

Petanidou T, Kizos T, Soulakellis N (2008) Socioeconomic dimensions of changes in the agricultural landscape of the Mediterranean basin: a case study of the abandonment of cultivation terraces on Nisyros Island, Greece. Environ Manage 41:250–266

Pinol J, Terradas J, Lloret F (1998) Climate warming, wildfire hazard, and wildfire occurrence in coastal eastern Spain. Climate Change 38:345–357

Pisanu S, Farris E, Filigheddu R et al (2012) Demographic effects of large, introduced herbivores on a long-lived endemic plant. Plant Ecol 213:1543–1553

Powell KI, Chase JM, Knight TM (2011) A synthesis of plant invasion effects on biodiversity across spatial scales. Am J Bot 98:539–548

Pretto F, Celesti-Grapow L, Carli E et al (2012) Determinants of non-native plant species richness and composition across small Mediterranean islands. Biol Invasions 14:2559–2572

Price MF, Park JJ, Bouamrane M (2010) Reporting progress on internationally designated sites: the periodic review of biosphere reserves. Environ Sci Pol 13:549–557

Pyšek P, Jarošík V, Kučera T (2002) Patterns of invasion in temperate nature reserves. Biol Conserv 104:13–24

Robertson BC, Gemmell NJ (2004) Defining eradication units to control invasive pests. J Appl Ecol 41:1042–1048

Rodríguez JP, Taber AB, Daszak P et al (2007) Globalization of conservation: a view from the south. Science 317:755–756

Ruffino L, Russell JC, Pisanu B et al (2011) Low individual-level diet plasticity in an island-invasive generalist forager. Popul Ecol 53:535–548

Schaefer H, Carine MA, Rumsey FJ (2011) From European priority species to invasive weed: *Marsilea azorica* (Marsileaceae) is a misidentified alien. Syst Bot 36:845–853

Schröter D, Cramer W, Leemans R et al (2005) Ecosystem service supply and vulnerability to global change in Europe. Science 310:1333–1337

Serrano M (2008) Les Petites Iles de Méditerranée (Initiative PIM): elaboration d'une base de données et premiers éléments de gestion. Master 2 Professionnel Expertise Ecologique et Gestion de la Biodiversité, Université de Marseille

Sintes T, Moragues E, Traveset A et al (2007) Clonal growth dynamics of the invasive *Carpobrotus* aff. *acinaciformis* in Mediterranean coastal systems: a non-linear model. Ecol Modell 206:110–118

Siracusa G (2000) Riserva naturale integrale isola Lachea e faraglioni dei Ciclopi. Flora dell'isola Lachea. G. Maimone, Catania

Stražičić N (1989) Pomorska geografija Jugoslavije. Školska knjiga, Zagreb

Suehs CM, Médail F, Affre L (2001) Ecological and genetic features of the invasion by the alien *Carpobrotus* (Aizoaceae) plants in Mediterranean island habitats. In: Brundu G, Brock J, Camarda I et al (eds) Plants invasions: species ecology and ecosystem management. Backhuys Publishers, Leiden, pp 145–158

Suehs CM, Affre L, Médail F (2002) Invasion dynamics of two alien *Carpobrotus* (Aizoaceae) taxa on a Mediterranean island: I. Genetic diversity and introgression. Heredity 92:31–40

Suehs CM, Affre L, Médail F (2005) Unexpected insularity effects in invasive plant mating systems: the case of *Carpobrotus* (Aizoaceae) taxa in the Mediterranean Basin. Biol J Linn Soc 85:65–79

Thompson JD (2005) Plant evolution in the Mediterranean. Oxford University Press, Oxford

Thompson JD, Lavergne S, Affre L et al (2005) Ecological differentiation of Mediterranean endemic plants. Taxon 54:967–976

Traveset A, Brundu G, Carta L et al (2008a) Consistent performance of invasive plant species within and among islands of the Mediterranean basin. Biol Invasions 10:847–858

Traveset A, Moragues E, Valladares F (2008b) Spreading of the invasive *Carpobrotus* aff. *acinaciformis* in Mediterranean ecosystems: the advantage of performing in different light environments. Appl Veg Sci 11:45–54

Traveset A, Nogales M, Alcover JA et al (2009) A review on the effects of alien rodents in the Balearic (Western Mediterranean Sea) and Canary Islands (Eastern Atlantic Ocean). Biol Invasions 11:1653–1670

Triantis KA, Mylonas M (2009) Greek islands, biology. In: Gillespie R, Glague DA (eds) Encyclopedia of islands. University of California Press, Berkeley, pp 388–392

Tscheulin T, Petanidou T, Potts SG et al (2009) The impact of *Solanum elaeagnifolium*, an invasive plant in the Mediterranean, on the flower visitation and seed set of the native co-flowering species *Glaucium flavum*. Plant Ecol 205:77–85

Tu M (2009) Assessing and managing invasive species within protected areas. In: Ervin J (ed) Protected area quick guide series. The Nature Conservancy, Arlington, pp 1–40

Tzanopoulos J, Vogiatzakis NV (2011) Processes and patterns of landscape change on a small Aegean island: the case of Sifnos, Greece. Landsc Urban Plan 99:58–64

Underwood EC, Klausmeyer KR, Cox RL et al (2009) Expanding the global network of protected areas to save the imperiled Mediterranean biome. Conserv Biol 23:43–52

UNEP-WCMC (2008) National and regional networks of marine protected areas: a review of progress. UNEP-WCMC, Cambridge. http://www.unep.org/regionalseas/publications/otherpubs/pdfs/MPA_Network_report.pdf. Accessed May 2012

UNEP-WCMC (2010) Biodiversity indicators partnership: coverage of protected areas. Indicator Factsheet 1.3.1. UNEP-WCMC, Cambridge

Usher MB (1988) Biological invasions of nature reserves: a search for generalisation. Biol Conserv 44:119–135

Usher MB (1991) Biological invasions into tropical nature reserves. In: Ramakrishnan PS (ed) Ecology of biological invasions in the tropics. International Scientific Publishers, New Delhi, pp 21–34

Vèla E, Saatkamp A, Pavon D (2011) Flora of Habibas Islands (N-W Algeria). Book of abstracts, II Jornades de Botànica a Menorca. Illes i plantes: conservació i coneixement de la flora a les illes de la Mediterrànea. Es Mercadal, 26–30 April 2011

Vidal E, Médail F, Tatoni T et al (1998) Functional analysis of the newly established plants induced by nesting gulls on Riou archipelago (Marseille, France). Acta Oecol 19:241–250

Vilà M, Burriel JA, Pino J et al (2003) Association between *Opuntia* species invasion and changes in land-cover in the Mediterranean region. Glob Chang Biol 9:1234–1239

Vilà M, Tessier M, Suehs CM et al (2006) Regional assessment of the impacts of plant invaders on vegetation structure and soil properties of Mediterranean islands. J Biogeogr 33:853–861

Vilà M, Siamantziouras ASD, Brundu G et al (2008) Widespread resistance of Mediterranean island ecosystems to the establishment of three alien species. Divers Distrib 14:839–851

Vilà M, Basnou C, Pyšek P et al (2010) How well do we understand the impacts of alien species on ecosystem services? A pan-European cross-taxa assessment. Front Ecol Environ 8:135–144

Villamagna AM, Murphy BR (2010) Ecological and socio-economic impacts of invasive water hyacinth (*Eichhornia crassipes*): a review. Freshw Biol 55:282–298

Vogiatzakis IN, Pungetti G, Mannion A (2008) Mediterranean island landscapes: natural and cultural approaches, vol 9, Landscape Series. Springer, Dordrecht

Weber E, D'Antonio CM (1999) Phenotypic plasticity in hybridizing *Carpobrotus* spp. (Aizoaceae) from coastal California and its role in plant invasion. Can J Bot 77:1411–1418

Wilson JRU, Gairifo C, Gibson MR et al (2011) Risk assessment, eradication, and biological control: global efforts to limit Australian acacia invasions. Divers Distrib 17:1030–1046

Wisura W, Glen HF (1993) The South African species of *Carpobrotus* (Mesembryanthema–Aizoaceae). Contr Bolus Herb 15:76–107

Yannitsaros A (1998) Additions to the flora of Kithira (Greece) I. Willdenowia 28:77–94

Zedda L, Cogoni A, Flore F et al (2010) Impacts of alien plants and man-made disturbance on soil-growing bryophyte and lichen diversity in coastal areas of Sardinia (Italy). Plant Biosyst 144:547–562

Chapter 19
Threats to Paradise? Plant Invasions in Protected Areas of the Western Indian Ocean Islands

Stéphane Baret, Cláudia Baider, Christoph Kueffer, Llewellyn C. Foxcroft, and Erwann Lagabrielle

Abstract The islands of the Western Indian Ocean are well known for their unique biodiversity. However, much of the native habitat has been destroyed and the remainder is threatened by invasive alien species. In this review we assessed the different protected area systems, synthesised the history of invasive alien plants and actions against them, and compared contrasting approaches in habitat management across the different island groups. Of the total terrestrial area of the Western Indian Ocean Islands, a third is under formal protection (defined as all six IUCN categories

S. Baret (✉)
Parc national de La Réunion, 258 rue de la République,
97431 La Plaine des Palmistes, La Réunion, France
e-mail: stephane.baret@reunion-parcnational.fr

C. Baider
The Mauritius Herbarium, Agricultural Services, (ex MSIRI-MCIA),
Ministry of Agro-Industry and Food Security, Réduit, Mauritius
e-mail: clbaider@gmail.com

C. Kueffer
Institute of Integrative Biology – Plant Ecology, ETH Zurich, CH-8092 Zürich, Switzerland
e-mail: christoph.kueffer@env.ethz.ch

L.C. Foxcroft
Conservation Services, South African National Parks, Private Bag X 402,
Skukuza 1350, South Africa

Centre for Invasion Biology, Department of Botany and Zoology,
Stellenbosch University, Private Bag X1, Stellenbosch 7602, South Africa
e-mail: Llewellyn.foxcroft@sanparks.org

E. Lagabrielle
Université de La Réunion et Institut de Recherche pour le Développement - UMR
228 ESPACE-DEV, 2 rue Joseph Wetzell, 97492 Sainte-Clotilde Cedex, La Réunion, France

Parc Technologique Universitaire, 2 rue Joseph Wetzell, Sainte-Clotilde Cedex,
La Réunion, France
e-mail: erwann.lagabrielle@ird.fr

L.C. Foxcroft et al. (eds.), *Plant Invasions in Protected Areas: Patterns, Problems and Challenges*, Invading Nature - Springer Series in Invasion Ecology 7, DOI 10.1007/978-94-007-7750-7_19, © Springer Science+Business Media Dordrecht 2013

of protected areas), with the proportion of protected areas and conservation status differing substantially between the islands. The awareness of the problems related to protected areas and specific invasive alien plant control actions, and which are supported by official government strategic documents, are further developed in Mauritius-Rodrigues, La Réunion, and Seychelles, but are still to be developed for the Comoros archipelago. We discuss the different approaches to management across the islands, the varying habitat types, fragmentation and degree of invasion. Invaded habitats are being managed by a range of approaches, including restoration, re-creation or inclusion as a novel ecosystem. We conclude by suggesting improvements in the protected area system in the Western Indian Ocean Islands, including priority actions that are necessary to prevent further invasion and control of invasive alien species already in the region.

Keywords Biodiversity hotspot • Endangered species • Novel ecosystem • Oceanic island • Tropical forest

19.1 Introduction

Together with Madagascar, the Western Indian Ocean Islands (WIOI) are regarded as one of 34 global biodiversity hotspots (Myers et al. 2000; Mittermeier et al. 2004) and are well known for their unique biodiversity and high endemic species richness (Kier et al. 2009; Thébaud et al. 2009; Baider et al. 2010; Strijk 2010). Unfortunately, the islands are also known for their widespread loss of native habitat (Vaughan and Wiehe 1937; Strasberg et al. 2005; Caujapé-Castells et al. 2010), and the high extinction rates of both animal and plant species (Cheke and Hume 2008). One of the most important threats to the biodiversity of the region's islands is the invasion and widespread habitat transformation by invasive alien plants (IAPs) (Kueffer et al. 2004; Caujapé-Castells et al. 2010; Kueffer et al. 2010a). Further habitat destruction has been halted on most islands, with, unfortunately, the exception of the Comoros archipelago (Louette et al. 2004). In order to halt further biodiversity loss in the region, it is critical to protect the remaining areas of native habitat, control IAPs, and where possible, restore habitat.

To safeguard global biodiversity and thereby work towards fulfilling the obligations of the Convention on Biological Diversity (CBD 1983), numerous countries have set up different systems of protected areas (PAs), with various level of protection and different types of management. However, on most oceanic islands especially, the creation of protected areas alone is not sufficient to maintain the diversity of uniquely evolved species, as even within protected areas active conservation management is required against the threats posed by IAS (Kueffer et al. 2004; Baret et al. 2006; Macdonald 2010). Because of their uniqueness – islands of different geological origin and ages, with rich biodiversity and with different political systems, land use history and historical dynamics of species invasion – the WIOI provide an excellent opportunity to (i) analyse the different terrestrial protected areas systems in place, (ii) compare the history and status of invasive alien plant management and (iii) learn from the unique approaches and effectiveness of habitat management and restoration across the islands. We discuss six case studies: (i) La Réunion, (ii) Seychelles, (iii) Mauritius, (iv) Rodrigues, (v) Mayotte and (vi) Comoros.

19.2   The Western Indian Ocean Islands

The Western Indian Ocean Islands include six islands or groups of islands (Fig. 19.1, Table 19.1). The Comoros archipelago is comprised of Grande Comore ($1,148$ km^2), Anjouan (424 km^2) and Mohéli (290 km^2), forming the Union of the Comoros. The fourth island of the Comoros archipelago is Mayotte, which is French territory. Because of their difference in sovereignty, which affects the PA system and their management, the Comoros ($1,862$ km^2) and Mayotte (363 km^2) are discussed further as separate island groups.

The Mascarenes are formed by the islands of La Réunion ($2,512$ km^2), Mauritius ($1,865$ km^2) and Rodrigues (109 km^2). Similar to the Comoros archipelago, La Réunion is also French territory, while Mauritius and Rodrigues are part of the same independent country. However, Mauritius and Rodrigues are also discussed separately because of their different colonisation histories, which resulted in varied approaches to PAs and management of invasive alien plants.

The Republic of Seychelles (455 km^2) encompasses two types of islands with different geological origin, comprising about 70 coralline and 40 granitic islands (continental fragments that formed part of the Gondwana supercontinent and became fully isolated about 65 million years ago). Where not specified otherwise, we restrict our discussions to the inner (granitic and main populated) islands.

19.3   Protected Area and Conservation Status of the Western Indian Ocean Islands

Islands are the ecosystems that have been most severely affected by invasions of alien species globally (Reaser et al. 2007). In order to provide context within which to discuss the invasion of alien species and current management approaches, a description of the status of PAs and conservation in the WIOI is necessary. As many different PA systems are used across the islands, we used the IUCN categories (see Dudley 2008 for the full definitions) in our discussion. We only included formally declared terrestrial PAs, including the terrestrial sections of marine protected areas, in the analysis.

When classified according to the six IUCN categories, a large proportion (32.9 %, Table 19.2) of the total terrestrial surface area of the WIOI is currently under protection (Table 19.3). About 20.3 % of the total land mass has been given

426 S. Baret et al.

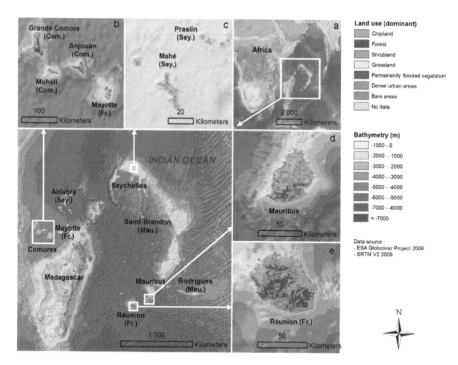

Fig. 19.1 Distribution of the Western Indian Ocean Islands. (**a**) the WIOI in relation to Africa, (**b**) the main islands of the Comoros archipelago, (**c**) Seychelles, (**d**) Mauritius and (**e**) Réunion

Table 19.1 Salient features of the Western Indian Ocean Islands

	Land Area (km²)	Inhabitants	Summit (m)	Native remnant vegetation cover Land area (km²)	Land area (%)
Comoros[a]	1,862	584,400	2361	335,2$^{/372,4}$	18,0$^{/20,0}$
Réunion	2,512	820,000	3070	999,8$^{/1313,8}$	39,8$^{/52,3}$
Mauritius	1,865	1,288,684	828	29,8$^{/93,3}$	1,6$^{/5}$
Mayotte	363	252,425	660	18,2	5,0
Rodrigues	109	38,039	398	0,6$^{/1,1}$	0,5$^{/1}$
Seychelles[b]	455^{240}	82,247^{82,000}	906	4,6$^{/45,5}$	1,0$^{/10}$
TOTAL	7,167	3,065,795	–	1388,0$^{/1844,2}$	19,4$^{/25,7}$

Native vegetation cover (forest with at least 50 % canopy cover of natives: intact and lightly or moderately invaded, in index including: highly invaded or novel ecosystems, when it is known)
[a]Data for the Comoros was estimated because different sources have confirmed that native habitat continues to be exploited
[b]Including outer coralline and granitic inner islands. In the columns 'land area' and 'inhabitants' index gives separate data for granitic inner islands

high protection status (falling within the upper four IUCN PA categories). These areas are either currently undergoing invasive alien plant management, or are in the planning phases (Table 19.2). However, the size of the PAs across the islands varies considerably. The largest formally protected areas are on La Réunion (1,039.6 km²

Table 19.2 Sum of land areas (km² and %) according to IUCN categories I to IV (heavily protected or with IAP control on-going); and from category I to VI, including indigenous plantations and novel ecosystems	I + II + III + IV		I to VI	
	(km²)	%	(km²)	%
Comoros	16,4	0,9	16,4	0,9
Réunion	1039,6	41,4	1937,2	77,1
Mauritius	87,9	4,7	87,9	4,7
Mayotte	41,8	11,5	41,8	11,5
Rodrigues	0,7	0,6	0,7	0,6
Seychelles[a]	271,5	59,7	272,5	59,9
WIOI total	**1457,9**	**20,3**	**2356,5**	**32,9**

[a]Including coralline and granitic islands

or 41.4 % of the island), and in the Seychelles (>59 % or 271.5 km², for both coralline and granitic islands). In La Réunion, most of the intact habitats and threatened plant species are found within PAs (about 41 % of the land area of the island). Only the Seychelles include PAs with the highest level of protection (IUCN category Ia; 209 km², 45.9 %). There are three national parks in the inner (granitic and main inhabited) Seychelles island group. If the Aldabra atoll UNESCO World Heritage Site is added to the above PA system, about 75 % of the land area of the Republic of Seychelles is formally protected. Significantly lower, PAs in Mauritius and Mayotte total only 87.9 km² (4.7 %) and 41.8 km² (11.5 %) respectively. Comoros (16.4 km², 0.9 %) and Rodrigues (0.7 km², 0.6 %) have the least area protected.

19.4   Invasion History and Management of Invasive Alien Plants

19.4.1   La Réunion

La Réunion was colonised by the French in 1642 and has remained under French administration. Awareness of the impact of the invasive species dates back to the end of the nineteenth century, with the botanist Cordemoy already concerned about the proliferation of *Rubus alceifolius* (giant bramble; Cordemoy 1895). Driven by scientists at the beginning of the 1980s (Cadet 1977; Lavergne 1978), and later by decision-makers and administrators, a strong political desire to control invasive alien plants began to emerge. Due to this commitment, and with the support of many stakeholders, for example scientists, consultants, the Regional and General Councils, the Forest Department (Office National des Forêts, ONF and Conservatoire d'Espaces Naturels, CEN-GCEIP) and the local ministry of environment (DIREN, now DEAL), both invasive alien species control and awareness campaigns were launched. The first comprehensive assessment of the extent of the invasive alien plant problem in La Réunion was conducted in 1989, forming the

S. Baret et al.

Table 19.3 Protected areas (in km^2) and % of land area per the six IUCN categories

IUCN cat	Ia km^2	%	Ib km^2	%	II km^2	%	III km^2	%	IV km^2	%	V km^2	%	VI km^2	%	RAM km^2	%	UNES/BIO km^2	%
Comoros					16,4	0,9									160,3	8,6	0,0	
Réunion					1008,8	40,2			30,8	1,2	897,6	35,7			0,0		1058,4	42,1
Mauritius	0,8	0,04	0		74,5	4,0	5,0	0,3	7,6	0,4					0,5		35,9	1,9
Mayotte									41,8	11,5					0,0		0,0	
Rodrigues									0,4	0,3					0,0		0,0	
Seychelles[a]	209,1	45,9			62,4	13,7	0,2	0,2	0,1	0,1			1,0	0,2	0,3	0,1	208,2	45,8
TOTAL	209,8	2,9	0	0	1162,2	16,2	5,2	0,1	80,7	1,1	897,6	12,5	1,0	0,0	161,1	2,3	1302,5	18,2

When it overlaps with other protection categories, as Ramsar sites and World Heritage-UNESCO, MAB biosphere these were distinguished. The areas of PAs were calculated with geographic information system (ArcGis 10) based on data from www.protectedareas.org (IUCN and UNEP-WCMC 2012) as follows: For Réunion, we used the classification system of Lefèbvre and Moncorps (2010), who under the French IUCN committee, analysed and classified PAs in France and its overseas territories according to the IUCN system. The adoption of the IUCN categories for PAs for the Republic of Mauritius was proposed in the Mauritius National Biodiversity Strategic and Action Plan (MNBSAP Mauritius 2006–2015), but it has not yet been adopted. Thus, in the case of Mauritius and Rodrigues, the IUCN categories were assessed by one of the authors (CB). For Mayotte and Comoros, the IUCN categories were assessed using information extracted from various sources: the reserve's websites, NGOS reports, scientific articles and protected area creation decrees

[a]Including coralline and granitic islands

basis for the first official management strategy (Macdonald 1989; Macdonald et al. 1991).

Invasive alien plant control programmes were, and continue to be, supported by collaborative research programmes undertaken by the University of La Réunion, the Agronomical Research Centre (CIRAD) and the National Botanical Garden (Conservatoire Botanique National de Mascarin, CBNM). These programmes contributed through providing information on the scale of invasion, species involved and their specific attributes (e.g. colonisation rates and reproductive strategies, and other biological and ecological attributes), and their impact on ecosystems or native species. In-depth scientific studies were conducted on various invasive alien species, including *R. alceifolius* (Amsellem 2000; Amsellem et al. 2000, 2001, 2002; Baret 2002; Baret et al. 2003, 2004, 2007, 2008; Baret and Strasberg 2005; Le Bourgeois et al 2013), *Ligustrum robustum* subsp. *walkeri* (Sri Lankan privet; Lavergne et al. 1999; Radjassegarane 1999; Lavergne 2000), *Hedychium flavescens* (yellow ginger; Radjassegarane 1999) and *Acacia mearnsii* (black wattle; Tassin 2002; Tassin and Balent 2004; Tassin et al. 2009).

At the same time, by ONF demonstrating the benefits and use of mechanical and/or chemical control sites, invasive species removal was initiated and became a major component of forest management. However, the lack of IAP control in all the PA (areas too wide according to the limited number of employees and the too slooping areas) but also on the surrounding private areas reduces the efficiency of the management interventions.

Various local strategies, which partially integrate the management of IAP in La Réunion have been produced such as ORF: (Orientations Régionales Forestières 2002) and ORGFH (Orientations Régionales de Gestion de la Faune Sauvage et de ses Habitats; Salamolard 2002). Ecological restoration guidelines (Triolo 2005) and the La Réunion Biodiversity Strategy (DIREN and ONCFS 2005) have also been developed. A review and assessment of the management practices used by the ONF (Hivert 2003), and the information on invasive plant management (Kueffer and Lavergne 2004), were compiled nearly a decade ago, resulting in many ground-level and in-depth syntheses. As from 2003, procedures for early detection and rapid response to plant invasions were initiated by the ONF. In 2006, the local environmental ministry created a technical group on biological invasions. The French branch of the IUCN published an assessment on the current status of invasive plant species, with recommendations for their management for La Réunion and all other French overseas territories (Soubeyran 2008). A detailed management plan for their control was created by a 'watch unit' (Salamolard et al. 2008).

The advancements of invasive plant control in La Réunion is partially a result of increased global awareness of the problems caused by IAS, leading to increasing on-the-ground efforts to control the spread of alien plant species within the French State and its territories. The Grenelle law, adopted in 2009, declares the State responsible for implementing management plans against both terrestrial and marine IAS, in order to limit their establishment and spread and to reducing their negative impacts. Thus, the local environmental ministry was instructed to produce an invasive plants strategy for La Réunion, which was finalised in 2010 (Baret et al. 2010).

On La Réunion, most PAs are threatened by alien plant species (Baret et al. 2006; Macdonald 2010). Therefore, control of invasive plants is essential, and in some cases, habitat restoration is also necessary. These activities are mainly implemented by the Forest Department. Even, in highly degraded ecosystems, some native threatened plant species continue to persist, however in extreme cases, only a few individual of endemic plants are known (which is also true for other WIOI). Although at first these areas would be considered of low conservation value, their restoration can serve as corridors between fragmented native habitats, and they are essential for conservation of some species on the brink of extinction. For example, in order to connect lowland to upland, a project called LIFE+ COREXERUN was initiated to conserve semi-xerophilous forest by restoring 30 ha. Unfortunately, these areas are often very steep and human access for management impossible, and thus the preservation of linear corridors is difficult. Restoration of unused areas (frequently abandoned land and now invaded), could connect natural areas through the arrangement of patches in the shape of Japanese stepping stones. These areas could be set up mainly outside PAs in order to limit further alien plant plantations, but also to facilitate connectivity between natural habitat fragments. As about 30 % of the land area of La Réunion is still relatively intact (Strasberg et al. 2005), control of invasive plants is considered the most important conservation management action. In some small areas, eradication of invasive plants is complemented with active habitat restoration. In other small areas (outside PAs), maintaining plantations of indigenous plants is also important. These have the dual objectives of preserving both the natural and cultural heritage of the region. Indeed, the people of Réunion have utilised the local flora for a range of needs (for example as medicinal plants, for bee-keeping and wood harvesting). The sustainable use of plants under controlled conditions will also limit the unsustainable, unmanaged harvest of plants in PAs. This is already in place in La Réunion and the National Parks continue working towards this aim.

19.4.2 The Seychelles

The Seychelles islands were not permanently settled until 1770, initially as a French colony, but it was officially turned over to Britain in 1815. Since 1976 the Seychelles have been an independent country. In the following, we summarise the major historic phases of alien species introductions and management responses based on Kueffer and Vos (2004), Beaver and Mougal (2009), and Kueffer et al. (2013).

The early settlement phase was characterised by agriculture and forestry based on a wide range of commodities, and almost complete deforestation of the islands. As early as the 1820s, most of Mahé was cleared of its original forest, and by the 1870s the only native vegetation that remained was in small patches in inaccessible mountain areas. Alien species formed the basis of subsistence and the economy. Major invasive species introduced in this period include the spice and fruit trees

Cinnamomum verum (cinnamon), *Psidium cattleianum* (strawberry guava) and *Syzygium jambos* (rose apple), cats *(Felis catus),* black rat *(Rattus rattus),* the common myna bird *(Acridotheres tristis),* and the giant African land snail *(Lissachatina fulica).* *Cinnamomum verum* became the most widespread alien plant species after colonising large parts of the deforested areas, and the black rat became a major threat to the native fauna. Individual naturalists noted the problem of widespread IAS in the nineteenth century already (cf. Gerlach 1993).

During the early British colonial era, in the first half of the twentieth century, the presence of British consultants for agriculture and forestry shaped environmental management. The colonial government promoted the re-afforestation of degraded land with alien tree species, some of which later became invasive; especially *Falcataria moluccana* (Moluccan albizia), and to some degree *Chrysobalanus icaco* (cocoplum). With the opening of the international airport in 1971 and the subsequent development of the tourism industry, which was strongly promoted by the Seychelles government after independence in 1976, nature conservation, already a concern among experts since the 1960s, rapidly increased in importance. The government prepared the first National Environmental Management Plan in the 1980s, and by 1997, the country had produced its first National Biodiversity Strategy and Action Plan, in response to the Convention on Biological Diversity (Dogley 2010). The 1990s were also the time when invasive alien plant species were recognised as a major conservation concern in Seychelles (Stoddart 1984; Gerlach 1993; Carlstroem 1996; Fleischmann 1997). However, although awareness of the problem increased, the increase in travel and transportation also led to many new introductions of alien species. These include *Clidemia hirta* (Koster's curse), the ring-necked parakeet *(Psittacula krameri),* crested tree lizard *(Calotes versicolor),* the spiralling whitefly *(Aleurodicus dispersus),* and a vascular wilt pathogen *(Verticillium calophylli),* which affects the native coastal tree *Calophyllum inophyllum* (takamaka).

Until very recently the planting of alien *Acacia* species including *A. mangium* (black or hickory wattle), for erosion control after forest fires was still promoted. Since the mid-1990s, the Ministry of Environment, together with the Swiss Federal Institute of Technology (ETH) have collaborated through a long-term project to investigate the biology and management options of plant invasions in Seychelles (Kueffer 2006). The project documented baseline information (Kueffer and Vos 2004), distribution of invasive plants (Fleischmann 1997), functional ecology of invasive trees (Kueffer et al. 2007, 2009; Schumacher et al. 2008, 2009), and impacts on ecosystem processes of these trees (Kueffer et al. 2008; Kueffer 2010). This information was used to propose a habitat restoration strategy for widespread *C. verum* dominated forests (Kueffer et al. 2010a, b). So far only small scale pilot projects have been implemented, but two larger scale projects aimed at IAP and habitat restoration have been funded for Inselberg (rocky outcrop) vegetation on Mahé and the World Heritage Site Vallée de Mai and its surroundings on Praslin (C. Kaiser-Bunbury, personal communication, 2012).

While awareness of the invasive species problem is high within the Seychelles Government and the local population, capacity is lacking for implementing an effective biosecurity system and large-scale invasive plant control and habitat restoration projects. There are however three main aspects that support, and provide opportunities for invasive plant management: (i) only relatively few alien plant species became widespread invaders, possibly because one alien plant, *C. verum*, dominates most habitats and hinders invasion of other alien plants, or due to the relatively recent increase in travel and transport in the 1970s, (ii) large alien animals such as pigs (*Sus scrofa*), goats (*Capra hircus*), or deer (*Cervus* sp.) did not establish in the wild on these small islands with a rugged topography, and (iii) on small offshore islands eradication of invasive species, and habitat restoration is feasible.

The most successful IAS control programmes in the Seychelles to date come from work on small offshore islands of some 20–200 ha. These areas are managed primarily for, or with high priority given to, nature conservation purposes (e.g. Aride, Conception, Cousin, Cousine, Frégate, North). While some of these islands are strict nature reserves, others support a luxury hotel, surrounded by land which is managed as a natural area. Once the alien vegetation is replaced by restored native vegetation, invasive plants are generally a relatively minor problem on these islands, in contrast to invasive animals (especially rigorous prevention of the reinvasion by alien mammals such as rats). Some species, including *Cocos nucifera* (coconut) saplings, *Carica papaya* (papaya), and some creepers require management (Kueffer and Vos 2004). On the populated main islands different pilot trials of chemical or mechanical control of different alien plant and animal species have been initiated, but no large-scale control or habitat rehabilitation effort has been implemented yet. The implementation of an effective biosecurity system on a national scale is currently being investigated through a project entitled 'Mainstreaming prevention and control measures for invasive alien species into trade, transport and travel across the production landscape' and funded by GEF through UNDP.

Strategies for invasive plant management are currently being developed for PAs (Kueffer et al. 2013), including widespread species such as *C. verum*. On the main island of Mahé, in the Morne Seychellois National Park, a novel forest ecosystem dominated by alien *C. verum* has formed at mid elevations from ~400 to 600 m a.s.l. Although these mid-elevation forests are dominated by alien trees, many threatened native plant and animal species still occur in this habitat. Typically, 70–90 % of adult trees are *C. verum*, and consequently the eradication across large areas is not a feasible management option. As an alternative, habitat management strategies that allow for maintaining native biodiversity in *C. verum*-dominated forests are being developed (Kueffer 2003; Kueffer et al. 2010a, b, 2013). It is hoped that some characteristics of the novel forest ecosystem can be used as opportunities for biodiversity conservation, while negative impacts are being mitigated. One novel factor of these forests that strongly shapes plant regeneration is root competition by adult *C. verum* trees, which produce a dense root mat just below the soil surface (Kueffer et al. 2007). While this can negatively affect native species regeneration, it also appears that such strong belowground competition functions as an effective barrier against the invasion by other alien plants such as *P. cattleianum*.

Invasion by *P. cattleianum* is likely to be more problematic because they form very dense aboveground thickets, thereby strongly suppressing native plants. Fruit characteristics of *C. verum* are also novel (Kueffer et al. 2009). *Cinnamomum verum* produces fruits that are high in protein and total energy content than those of native species (Kueffer et al. 2009), and in combination with alien fruit trees from nearby gardens and plantations, has probably contributed to the current high population densities of endemic frugivorous birds and fruit-bats. This novelty may mean that *C. verum* contributes to a high habitat quality for native frugivorous animals, but also that *C. verum* competes with native plant species for seed dispersal services and thereby negatively impacts native plant regeneration.

One current idea is that given these potential positive functions of *C. verum* it might be possible to establish small patches of completely native vegetation stands of a few meters in diameter in a matrix of *C. verum* forest (Kueffer et al. 2010a, b). These stands can be established by either weeding remnant stands of mostly native trees or by forming artificial gaps through the felling of a few *C. verum* trees and replanting them with native plants. If such a management strategy, which aims to establish and maintain small native-dominated vegetation patches in a matrix dominated by *C. verum*, is successful, then this may help to overcome some of the problems associated with a *C. verum* dominated landscape. These native vegetation patches could for instance function as sources of native seed, which may promote a scattered establishment of native species in the *C. verum* matrix forest.

19.4.3   Mauritius

The island was intermittently colonised by small Dutch groups from 1638 to exploit its natural wealth capital, especially black ebony (*Diospyros tesselaria*). At the same time alien species of animal and plants were introduced, although a permanent settlement did not start until 1722, when France took possession of the island. Mauritius has a very well documented history, including precise dates of introduction of many species that would later become invasive; many of which are also invasive across the other WIOI. For example, many of the worst invasive plant species in Mauritius (e.g. *P. cattleianum* and *Hiptage benghalensis*, liane cerf), were introduced through the Pamplemousses Botanical Garden, the first botanical garden in the southern hemisphere (Cheke and Hume 2008). Within 370 years of human presence on Mauritius, only 5 % of the original habitat has survived as fragments in areas unsuitable for agriculture. Extinction rates of endemic species varies from a 'low' of 10.9 % among angiosperms, to 44.4 and 58.8 % among land snails and land birds respectively, while proportions of threatened endemics ranged from 80 to 100 % (Griffiths and Florens 2006; Cheke and Hume 2008; Baider et al. 2010). In 1766, probably some of the world's earliest legislation for IAP control was promulgated, followed in 1767 by legislation on re-afforestation using native species, as well as creation of forest, mountain and river reserves.

Similar laws, regulations and ordinances were issued during the next 200 years, however due to the lack of enforcement Mauritius was to become one of the most ecologically devastated areas globally. This resulted in Mauritius having amongst the highest extinction rates and endangered species per area; earning it the very symbol of human-induced species extinction, the dodo (*Raphus cucullatus*).

When Mauritius became a British colony in 1810, the rate of deforestation for sugar cane plantations rose to unprecedented levels (Vaughan and Wiehe 1937). The negative effect of this rapid deforestation coupled with the rapid spread of invasive plants was an issue of substantial concern (Bouton 1838). The post-introduction spread of invasive plants was documented as very rapid, taking as little as 30 years for *Rubus alceifolius* to be considered a serious problem (Brouard 1963). Most of the country was under alien forest and scrub by 1870s (Thompson 1880). As from the 1900s plantations of *Eucalyptus* spp., *Pinus* spp., *Cryptomeria japonica* (Japanese cedar), *Casuarina equisetifolia* (coast sheoak), and *Ligustrum robustum* var. *walkeri* were established (Brouard 1963), with the latter species becoming an important invader. In the 1950s, *Camellia sinensis* (tea) was introduced as a cash crop (Brouard 1963), with some large areas of remaining native forest destroyed. It was to persist for only a few decades as an important economic activity, thereafter spreading into the native wet forests. At this time, a series of small nature reserves representing different vegetation types, were created (Brouard 1963). The control of invasive plants in native forest remnants was pioneered in the 1930s (Vaughan and Wiehe 1937), when a small patch of wet forest was manually weeded and fenced against large alien mammals such as Java deer (*Cervus timorensis*) and feral pigs (*Sus scrofa*). These areas later became known as Conservation Management Areas in Mauritius. A few successful biological control agents were introduced for controlling *Cordia curassavica* (black sage), *Opuntia vulgaris* (prickly pear) and to a lesser extent also *Lantana camara* (lantana; Fowler et al. 2000).

On the mainland, mostly in 1990s, new Conservation Management Areas were created after the proclamation of the Black River Gorges National Park. In the mid-2000s, use of chemical control and ring barking was shown to be less damaging to native saplings and seedlings (Seepaul 2006) and much cheaper than manual control (Larose 2005). It was quickly adopted by private land owners, and more recently by Government agencies as well. Aldabra tortoises (*Aldabrachelys gigantea*) are in the process of being released in Round Island to control invasive alien plants, but also as an analogue to the endemic, now extinct *Cylyndrapsis* species, and it is expect that they should help disseminating seeds of native species (Griffiths et al. 2010).

Recognising the deleterious effect of invasive plants, especially in health and agriculture and native biodiversity for the Republic of Mauritius (including Rodrigues), national multi-sectored strategy and action plan against IAP was approved in early 2010. This document is considered as first step to promote a long-term integrated approach to IAP, but little of its implementation has formally started. Management of invasive alien plants is an important component of the GEF funded project to increase the PA network on Mauritius, that also should includes creation of incentives for private landowner to manage, restore and protect native forest remnants.

Invasion of the native mainland habitats by alien plants is such that the vegetation canopy consists of more alien than native plants over about two-thirds (or 60 km^2) of the surviving remnants (Page and D'Argent 1997; Safford 1997). For example, a study at five of the best preserved wet native forest areas showed all to be extremely invaded in the understory, with alien plants making up 78.6 % of all woody stems of ≥ 1 cm diameter at breast height (Florens et al. 2012). All five sites are dominated, in terms of their relative importance, by at least one or more IAP, which in most cases is *P. cattleianum* (Florens et al. 2008). Thus due to the need to save species and with little to lose if nothing was to be done, Mauritius has become a laboratory to test various restoration approaches, like alien mammal eradication, invasive plants' control, reintroduction of locally extinct species (Thébaud et al. 2009), species translocation (Christinacce et al. 2008) or introduction of analogous species to restore ecosystem function (Griffiths et al. 2010). In contrast to some *C. verum* forest in the Seychelles, control of alien plants is feasible and research shows that it has led to an improved survival rate of native plant species (Florens 2008). These native species also grow faster and have a higher reproductive output, increasing the regeneration of previously invaded areas (Baider and Florens 2006, 2011; Florens 2008; Monty et al. 2013). Invasive alien plant management has also been shown to be beneficial to other groups, with the diversity and abundance of native butterflies much higher in these areas than in invaded forests (Florens et al. 2010). In some cases, endemic plants (Baider and Florens 2011) and animals (Florens and Baider 2007) that were considered extinct have 'reappeared' in forests cleared of invasive plants. It has unfortunately also been shown that some native groups, for example molluscs, do experience negative side effects of IAP control (Florens et al. 1998). Additionally, some interactions, like increased herbivory on endangered plant species have resulted in lower fruit set (Kaiser et al. 2008). Some of the negative effects are transient, or could be reduced or eliminated with gradual removing instead. The long-term survival of Mauritian species and ecosystems calls for control of alien plants over large areas. It has been recognised that the area of native habitats requiring control needs to increase tenfold, or to an area of 1,000 ha (MNBSAP 2006); the areas currently being managed still total less than 100 ha. The success of Mauritius in saving species and restoring ecosystem has in most cases been supported by sound research. Research has played an important role in native forest restoration and in saving species on the brink of extinction, by pioneering the development of improved and cheaper techniques. Many of these concepts and techniques have been used in the other WIOI. In contrast to La Réunion, there is usually good communication between researchers and managers, although sometimes incorrect but ingrained ecological concepts persist (Florens and Baider 2013). However, the Mauritian Government is currently imposing stringent new conditions for conducting ecological and conservation research in the National Parks (Florens 2012). In the long-term this will risk the leading role that Mauritius is playing in the control of invasive alien plants in the region. Moreover, it will also impact on the training of new a generation of local scientists to take over the daunting task of saving the unique species, ecosystems and their functions, at a time where there is still hope to do so.

19.4.4 Rodrigues

Rodrigues, a previously completely forested island (Leguat 1708), was not colonised until the nineteenth century. The first small group of men stayed in Rodrigues from 1671 to 1673 (Leguat 1708), followed by very occasional visitors up to 1735, when an outpost to control the collection of tortoises was created and maintained until 1771. During this period goats and cats were introduced, and fire was recorded to have destroyed part of the island in 1761, reoccurring in the ensuing years. Although there were few people, overexploitation of native tortoises and spread of invasive animals (e.g. rats, cats) led many Rodriguan endemic animal species to extinction in a very short time. A small permanent settlement was established in 1792 and by 1809, when the British took possession of the island, Rodrigues was already covered in shrubby vegetation, citrus trees, crop plantations, with few patches of native *Latania* (latans) and *Pandanus* (screw-pines) remaining. By 1845, invasive alien grasses turned coastal areas into fire prone grasslands, abandoned croplands were covered by spiny weeds, large areas of land were left bare, and little native forest remained. Balfour (1879), who described the flora of the island, reported that goats and cattle (*Bos primigenius*) had eaten everything within their reach, fire regularly burned the island, and invasive plants could be found in large numbers, preventing any regeneration of native species; many of which were already rare. For example, Balfour (1879) mentioned that *Leucaena leucocephala* (white popinac), introduced in about 1850, covered large areas of Rodrigues, forming dense and impenetrable stands with no other species growing in-between. Regulations to prevent cutting of trees were issued, with the result that some areas later turned into secondary alien forest, mainly by the expansion of *Syzygium jambos*, and to a lesser extent *Psidium cattleianum, Litsea glutinosa* (Indian laurel) and *Ravenala madagascariensis* (traveller's palm; Strahm 1989; Cheke and Hume 2008). Although some native vegetation was also recorded to have expanded by 1914, some areas remained as bare land because of the heavy use by cattle, goats and pigs (Cheke and Hulme 2008). Plantations of alien plants for food, for example *S. jambos* and *Artocarpus heterophyllous* (bread fruit), and for forestry, started in 1915 (Brouard 1963). Other species, like *Lantana camara*, were introduced before 1938, to become one of the worst invaders in the coastal areas. The few pockets of native trees that had survived were substantially reduced during the 1950–1960s and eventually nature reserves were declared in 1980s. However, at the same time *Acacia nilotica* (prickly acacia), which was later to become invasive, was introduced for soil protection.

Due the almost complete transformation of the island by invasive plants and fire, Rodrigues has virtually no original habitat remaining, with only some small patches in the outer islets, on mountain slopes or in one very dry limestone area. As a result the extinction rates of endemic species are very high; from 21 % for plants (around 77 % of the endemic species are considered threatened), to 100 % for reptiles (Cheke and Hume 2008; Baider et al. 2010). The main reference on the forest structure and association of species that can be used for restoration are from brief descriptions in 1879 (Balfour 1879), although by this time most of the primary forests had disappeared. With improved information available, Strahm (1989) also briefly described the main habitats and associated species. Therefore, restoration in Rodrigues mainly consists of IAP removal followed by planting native species. Since 1986, restoration has being carried out in two of the mainland nature reserves, Grande Montagne (13 ha) and Anse Quitor (8 ha) (Payandee 2003). There is a strong community participation component in the restoration programmes in Rodrigues, but little scientific guidance, monitoring and dissemination of techniques and results that could be used as a model in other islands. An area in Anse Quitor, at the François Leguat Giant Tortoise Reserve, is being replanted with native plants and giant Aldabra tortoises (*Aldabrachelys gigantea*) and a smaller Madagascan species (*Astrochelys radiata*, radiated tortoise) were introduced to restore ecological function of the extinct giant tortoises, with "encouraging results" (Burney 2011). Although most of the native plant species in Rodrigues were known to be endangered (Strahm 1989), after nearly two decades of restoration efforts, the populations of many of these most threatened species have not recovered, and still remain on the brink of extinction (A. Waterstone, personal communication, 2010; Strahm, Baider & Florens, unpl. data). However, the restoration programmes have allowed for the populations of some groups of insects (Orthoptera) to recover in these areas (Hugel 2012). This result shows the importance of the restoration work, as also shown in Mauritius, where some species surviving as tiny populations (e.g. with two adults of *Ixora vaughanii*, Baider and Florens 2011), are able to recover after removal of invasive plants.

19.4.5   Mayotte

Only about 5 % of the native vegetation remains in Mayotte (DAAF 2012) with intensive deforestation starting around the mid-nineteenth century for the production of sugar, vanilla and ylang-ylang. When most of these fields were abandoned, secondary native forest started reclaiming those areas, a process rarely or not seen, on any other of the Mascarene Islands or the Seychelles. Around 48 % (or 625 species) of all plant species on the island are alien, and some of them are invasive, for example: *Litsea glutinosa, L. camara, R. alceifolius and Antigonon leptopus* (coral vine), *Spathodea campanulata* (African tulip tree), *C. verum* and *Mangifera indica* (mango). *Leucaena leucocephala* and *Albizia lebbeck* (lebbeck) are invasive in the lowlands (Vos 2004; DAAF 2012). Also, some silvicultural species, such as *Acacia mangium*, used for re-afforestation and rehabilitation of eroded steep slopes, are becoming invasive (Vos 2004; DAAF 2012). The problem of alien plant invasions has only recently become an important issue on Mayotte, with some invasive plant control programmes now emerging. A programme for the control of *L. camara* in natural forests that cost 240,000 Euros was initiated in the late 1990s, however, with questionable results (DAAF 2012). A more recent survey of invasive plants has been done in the Majimbini reserve, resulting in the development of a management proposal (Cathala 2011). An awareness campaign only started in 2010 through a workshop organised by the National Botanical Garden (CBNM), which has an office on Mayotte. This workshop was aimed at stakeholders managing natural areas such as the ONF and local NGOs. The recent outbreak of *Salvinia molesta* (giant salvinia or kariba weed), in one of the very few reservoirs for potable water in the island, illustrated the sanitary and economic problems associated with invasive plants. The need for an official list of invasive plants has been recognised by the local authorities to allow for strict entry control measures on the island, as well as to implement management plans in the field (DAAF 2012).

Except for localised invasive alien plant control, no important management actions against invasive plants are on-going. After a few attempts over the last few decades, a new project to control the alien liana *Merremia peltata* (merremia), which invades forested areas, has started (DAAF 2012; Lesur 2012). Management actions that have been suggested for Mayotte include the restoration of the some of the most well preserved habitats, but also restoration of some highly disturbed habitats in an attempt to link forest fragments (such as corridors along rivers or specific managed areas), improving plants dissemination and animal movement (DAAF 2012).

19.4.6 Comoros

The Grande Comore, the largest of the three islands of the Comoros, rises to 2,361 m a.s.l., and includes the largest area of native vegetation in the region, mostly on the slopes of the active volcano Mount Karthala. Anjouan, the second largest of the Comorian islands, has almost no native vegetation left. In just 10 years (1973–1983), forested areas decreased by 36 % on Grand Comore, 53 % on Mohéli and 73 % on Anjouan (Louette et al. 2004). The rate of deforestation seems to have decreased, but is still on-going. The Comoros was considered the country with highest annual net loss of forest in the world for the period 2000–2010 (Chakravarty et al. 2012). Other than the native baobab (*Adansonia digitata*), the large vegetated patches on these islands are plantations of *Cocos nucifera, Cananga odorata* (ylang-ylang). Invasive alien plant species on Comoros differ across altitude and precipitation gradients: for example on the Grande Comore, *Pteridium aquilinum* (bracken fern) is found in lowland areas, while *P. cattleianum* is found at the mid-elevation of the scarps of Mount Karthala. The latter was not present on Mohéli in 2004, but *L. camara* and *S. jambos* were considered the most problematic species (Louette et al. 2004). On the Comoros, the negative impacts of invasive alien plants are not fully understood by the government or most stakeholders (Hamidou 2012).

For example, an official list of invasive plants has not yet been produced, although 16 alien plant species are considered invasive. Since 1994, laws and regulations relative to environmental protection have been created, including the prohibition of the introduction of some alien plant and animals. To date, control of invasive plants is sporadic (Hamidou 2012).

19.5 Lessons Learnt from Invasive Alien Plant Management in the Western Indian Ocean Islands

Invasive species management has been a conservation concern for many years in most of the WIOI, and some type of invasive plant management is now in place on all island groups. However, the intensity and type of activities varies substantially across the region. Managers and scientists in different island groups have to some extent followed different paths, and combining experiences from the different islands may provide improved opportunities for effective management. In La Réunion much emphasis has been placed on developing a comprehensive strategy for managing invasive species that links many different actors and stakeholders. The strategy encompasses all stages of IAS management, from border control and early detection, to control and habitat rehabilitation (Baret et al. 2010). Control of invasive plants in La Réunion is limited by the high cost of labour and other priorities that the field managers are required to deal with. Unfortunately IAS pressure is increasing, and with new invaders establishing and limited management capacity, maintenance of native habitats will become increasingly difficult. In La Réunion for example, following a fire that covered 850 ha of native upland habitat (Payet 2012), *Ulex europaeus* (gorse) was detected over half of the burned area, and as the seeds germinate quickly following fire, control was urgently required. Due to the lack of coordination, management was delayed and the plants were able to establish over a large area. The control is now expected to be very costly. Fortunately, public awareness is growing, with volunteers from NGOs and the military collaborating on invasive plant control programmes. This further emphasises the importance of a coordinated effort between volunteers and professionals, to ensure long-term success. Coordinating control actions, generating sufficient resources for invasive plant management and identifying priorities is therefore important and urgently needed in most of the WIOI (Baret et al. 2006, 2010; Lagabrielle et al. 2009, 2011; Macdonald 2010). Such a comprehensive approach to IAS and PA is not yet integrated in most WIOI, although some activities and strategies in this direction are under way (e.g. through a national biosecurity project in the Seychelles and the IAS strategy and expansion of PA in Mauritius). An important aspect of a multi-stakeholder strategy is to consider the local uses and values of native plants (e.g. medicinal plants, bee-keeping, wood). For instance, sustainable plantations of native tree species outside of PA can serve as a buffer against alien plant invasions and ecological corridors (or Japanese style stepping stones) for native species (Baret et al. 2012).

Interestingly, some of the Western Indian Ocean Islands invasive plant management approaches, and also PA system improvements, serve as models for other islands within the region. On Seychelles and Mauritius, for instance restoration of small offshore islands has been going on for decades with successful results. On Mauritius, restoration of larger inland habitats can be used and replicated on other WIOI like Seychelles and the Union of the Comoros, where labour is still relatively affordable for undertaking restoration projects over larger areas. Of course, the status of the native vegetation and its potential for recovery will influence the management techniques of invasive plants in PAs. Recovery of native vegetation is possible in areas where removal of alien species has been carried out, as the experience of Conservation Management Areas in Mauritius has shown (Baider and Florens 2011). In other areas like on Rodrigues, restoration of native vegetation from completely degraded vegetation has been achieved within just a few years, with the support from local communities an important factor in its success. Finally, Seychelles is leading the conceptual thinking and research on how to manage some types of alien-dominated novel forest ecosystems (Kueffer et al. 2013); and based on the results of these, new ideas could also be adopted in the other islands. Unfortunately in some cases, for example the Union of the Comoros, IAS management and habitat restoration is less advanced than on the other islands (Hamidou 2012), mainly because of the low-income and low level of development of the country, which directs priorities away from conservation.

Therefore, building on the existing experience in the region, it should be possible to achieve rapid progress for the control of IAS and improvement of PA system for the region as a whole. Strong collaboration across the Western Indian Ocean Islands, supported by implementation on the ground, is essential to conserve its rich and unique biodiversity. Invasive alien plant strategies should underline the conservation needs in the short, medium and long-term. We suggest some priorities for each island group (Table 19.4), based on the current status of invasive plant control and PAs of the islands. Such priorities range from setting-up local strategies, to increasing awareness of the local population. For each of the WIOI, except the Seychelles, one of the most important priorities is to consolidate or increase the PA system, in order to reach at least 17 % of their area. Additionally, all types of native habitats, including novel ecosystems if necessary, need to be included. In a few cases some of the priorities that are important in some islands have already been achieved on others, for example in La Réunion where early detection of invasive plants and rapid response management programmes are in place.

Table 19.4 Priorities to improve the PA system, IAP detection and control, and habitat restoration on the different WIOI

Priorities/WIOI	COM	MAY	RUN	MAU	ROD	SEY
Protected areas						
Create new protected areas and revisit classification system	2	2		1		
Elevated IUCN categories of PA						
Aim at obtain ≥17 % of PA in all different types native habitats with connectivity				2	1	
Create PA in novel ecosystems when original habitat do not exist anymore to achieve the ≥17 % of land mass under the network					2	
Methods						
Improved and cheaper IAP management				2		3
Promote expert exchanges between the WIO islands	4	5	5			5
Perspectives						
Set up a local strategy	1	1		3	3	
Improve legislation and enforcement	5	3				2
Develop research to improve IAP control and habitat restoration				4	4	4
To create local and regional programmes of early detection, and rapid response, restoration	2	4				1
Demonstrate to decision-makers the need to set up new PA and/or to elevate IUCN categories of existing PA				4	5	
Increase local population awareness about negative effect of IAP and techniques of control	1	2				
Implicate local population in IAP or/and in plantation of indigenous species (in lieu of IAP)				1		
Increase coordination of stakeholders and awareness about necessary resources to prevent and control IAP				3	5	

Only five priorities are provided for each island/country

19.6 Conclusion

As Margules and Pressey (2000) have argued, PAs have two main roles, (i) they should sample or represent the biodiversity of each region and (ii) they should separate this biodiversity from processes that threaten its persistence. On highly disturbed and invaded oceanic islands in particular, it would be naïve to believe that the establishment of PAs is enough to safeguard threatened native biodiversity. Management for biodiversity conservation is essential, and this involves in particular the control of IAS. As the restoration of degraded habitat is often more cost-efficient than species-focused recovery programmes (e.g. Balmford et al. 2002), conservation and sustainable utilisation of the rich and unique biodiversity of the WIOI is not an impossible task, but requires rapid and concerted action. Although there are a number of potential solutions to protect natural habitat and associated

442

S. Baret et al.

biodiversity, in order to identify priorities decision-making tools are still necessary. One element that can support effective conservation measures is networking and knowledge sharing at local to global scales (e.g. Kueffer 2012). At a global scale, the recently established Global Island Plant Conservation Network (GIPCN, www. bgci.org/ourwork/islands) may become a useful institution for collaborative activities among island conservationists. At a regional scale, the creation of an invasive alien species network in the Western Indian Ocean was identified as a priority at a recent workshop organised by IUCN in Mayotte in January 2012 (http://www. especes-envahissantes-outremer.fr/actualites.php). Such a network is currently being developed with support from the Indian Ocean Commission (IOC – COI) and the IUCN, with a discussion list already functional (Western Indian Ocean – Invasive Alien Species, WIO-IAS, https://list.auckland.ac.nz/sympa/info/wio-ias).

References

Amsellem L (2000) Comparaison entre aire d'origine et d'introduction de quelques traits biologiques chez Rubus alceifolius Poir. (Rosaceae), plante envahissante dans les îles de l'Océan Indien. PhD thesis, Université de Montpellier II, France

Amsellem L, Noyer J-L, Le Bourgeois T et al (2000) Comparison of genetic diversity of the invasive weed Rubus alceifolius Poir. (Rosaceae) in its native range and in areas of introduction, using amplified fragment length polymorphism (AFLP) markers. Mol Ecol 9:433–455

Amsellem L, Noyer J-L, Hossaert-McKey M (2001) Evidences of a switch in the reproductive biology of Rubus alceifolius (Rosaceae) towards apomixis, between its native range and its area of introduction. Am J Bot 88:2243–2251

Amsellem L, Noyer J-L, Pailler T et al (2002) Characterisation of pseudogamous apospory in the reproductive biology of the invasive weed Rubus alceifolius (Rosaceae), in its area of introduction. Acta Bot Gallica 149:217–224

Baider C, Florens FBV (2006) Current decline of the 'Dodo–tree': a case of broken–down interactions with extinct species or the result of new interactions with alien invaders? In: Laurance W, Peres CA (eds) Emerging threats to tropical forest. Chicago University Press, Chicago, pp 99–107

Baider C, Florens FBV (2011) Control of invasive alien weeds averts imminent plant extinction. Biol Invasion 13:2641–2646

Baider C, Florens FBV, Baret S et al (2010) Status of plant conservation in oceanic islands of the Western Indian Ocean. Proceedings of the 4th Global Botanic Gardens Congress, Dublin. Botanic Gardens Conservation International (BGCI). http://www.bgci.org/files/Dublin2010/papers/Baider-Claudia.pdf. Accessed 10 July 2012

Balfour B (1879) An account of the petrological, botanical, and zoological collections made in Kerguelen's Land and Rodriguez during the Transit of Venus Expeditions. Philos Trans R Soc B 168:302–387

Balmford A, Bruner A, Cooper P et al (2002) Economic reasons for conserving wild nature. Science 297:950–953

Baret (2002) Mécanismes d'invasion de Rubus alceifolius à l'île de La Réunion: interaction entre facteurs écologiques et perturbations naturelles et anthropiques dans la dynamique d'invasion. PhD thesis, Université de la Réunion, France

Baret S, Strasberg D (2005) The effects of opening trails on exotic plant invasion in protected areas on La Réunion island (Mascarene archipelago, Indian Ocean). Rev Ecol (Terre et Vie) 60:325–332

Baret S, Nicolini E, Le Bourgeois T et al (2003) Developmental patterns of the invasive bramble (*Rubus alceifolius* Poiret, Rosaceae) in Réunion Island: an architectural and morphometric analysis. Ann Bot 91:39–48

Baret S, Maurice S, Le Bourgeois T et al (2004) Altitudinal variation in fertility and vegetative growth in the invasive plant *Rubus alceifolius* Poiret (Rosaceae), on Réunion Island. Plant Ecol 172:265–273

Baret S, Rouget M, Richardson DM et al (2006) Current distribution and potential extent of the most invasive alien plant species on La Réunion (Indian Ocean, Mascarene Islands). Aust Ecol 31:747–758

Baret S, Le Bourgeois T, Rivière J-N et al (2007) Can species richness be maintained in logged endemic *Acacia heterophylla* forests (Réunion island, Indian ocean)? Rev Ecol (Terre et Vie) 62:273–284

Baret S, Cournac L, Thébaud C et al (2008) Effects of canopy gap size on recruitment and invasion of the non-indigenous *Rubus alceifolius* in lowland tropical rain forest on Réunion. J Trop Ecol 24:337–345

Baret S, Julliot C, Radjassegarane S (2010) Stratégie de lutte contre les espèces invasives à la Réunion. Rapport DIREN, Conseil Général de la Réunion, Région Réunion, Office National des Forêts, Parc national de La Réunion. Graphica, Studionaut', Parc national de La Réunion, La Réunion

Baret S, Lavergne C, Fontaine C et al (2012) Une méthodologie concertée pour la sauvegarde des plantes menacées de l'île de La Réunion. Rev Ecol (Terre et Vie) 11(Suppl):85–100

Beaver K, Mougal J (2009) Review and evaluation of invasive alien species (IAS) control and eradication activities in Seychelles and development of a field guide on IAS management. Consultancy report. Ministry of Environment-UNDP-GEF Project, Victoria, Seychelles

Bouton L (1838) Sur le décroissement de forêt à l'Ile Maurice. Port Louis, Mauritius

Brouard NR (1963) A history of woods and forest in Mauritius. Government Printer, Port Louis, Mauritius

Burney DA (2011) Rodrigues Island: hope thrives at the François Leguat Giant Tortoise and Cave Reserve. Madag Conserv Dev 6:3–4

Cadet T (1977) La végétation de l'île de La Réunion. PhD thesis, University of Marseille, France

Carlstroem A (1996) Endemic and threatened plant species on the granitic Seychelles. Conservation & National Parks Section. Division of Environment, Ministry of Foreign Affairs, Planning and Environment, Mahé, Seychelles

Cathala C (2011) Un plan de gestion pour l'espèces exotiques envahissantes végétales dans la réserve forestière de Majimbini (Mayotte). MSc dissertation, Université Catholique de l'Ouest, Angers, France

Caujapé-Castells J, Tye A, Crawford DJ et al (2010) Conservation of oceanic island floras: present and future global challenges. Perspect Plant Ecol Evol Syst 12:107–130

CBD (1983) Convention on biological diversity. http://www.cbd.int/convention/text/. Accessed 3 Mar 2013

Chakravarty S, Ghosh SK, Suresh CP et al (2012) Deforestation: causes, effects and control strategies. In: Okia CA (ed) Global perspectives on sustainable forest management. InTech. http://www.intechopen.com/books/global-perspectives-on-sustainable-forest-management/deforestation-causes-effects-and-control-strategies. Accessed 12 Feb 2013

Cheke A, Hume J (2008) Lost land of the dodo: an ecological history of Mauritius, Réunion & Rodrigues. T & AD Poyser, London / A&C Black Publishers Ltd, London, UK

Cordemoy EJ (1895) Flore de l'île de La Réunion (Phanérogames, Cryptogames vasculaires, Muscinées). Paul Klincksieck, Paris, France

Cristinacce A, Ladkoo A, Switzer R et al (2008) Captive breeding and rearing of Critically Endangered Mauritius fodies *Foudia rubra* for reintroduction. Zoo Biol 27:255–268

DAAF (2012) Analyse de la forêt de Mayotte. http://daf976.agriculture.gouv.fr/IMG/pdf/Diagnostic_foret_Mayotte_rapport_final_cle825916.pdf. Accessed 12 Feb 2013

DIREN, ONCFS (2005) Stratégie Réunionnaise pour la Biodiversité. DIREN, ONCFS, Préfecture de La Réunion, La Réunion

Dogley D (2010) A government's perspective on safeguarding biodiversity: the Seychelles experience. Biotropica 42:572–575

Dudley N (2008) Guidelines for applying protected area management categories. IUCN, Gland

Fleischmann K (1997) Invasion of alien woody plants on the islands of Mahé and Silhouette Switzerland, Seychelles. J Veg Sci 8:5–12

Florens FBV (2008) Écologie des forêts tropicales de l'île Maurice et impact des espèces introduites envahissantes. PhD thesis, Université de La Réunion, Réunion

Florens FBV (2012) National parks: Mauritius is putting conservation at risk. Nature 489:29

Florens FBV, Baider C (2007) Relocation of *Omphalotropis plicosa* (Pfeiffer, 1852), a Mauritius endemic landsnail believed extinct. J Mollusc Stud 73:205–206

Florens FBV, Baider C (2013) Ecological restoration in a developing island nation: how useful is the science. Rest Ecol 21:1–5

Florens FBV, Daby D, Jones CG (1998) The impact of controlling alien plants and animals on the snail fauna of forests of Mauritius. J Conchol S2:87–88

Florens FBV, Mauremootoo JR, Fowler SV et al (2010) Recovery of indigenous butterfly community following control of invasive alien plants in a tropical island's wet forests. Biodiver Conserv 19:3835–3848

Florens FBV, Baider C, Martin GMN et al (2012) Surviving 370 years of human impact: what remains of tree diversity and structure of the lowland wet forests of oceanic island Mauritius? Biodiver Conserv 21:1239–1267

Fowler SV, Ganeshan S, Mauremootoo J et al (2000) Biological control of weeds in Mauritius: past successes revisited and present challenges. In: Spencer NR (ed) Proceedings of the X International Symposium on Biological Control of Weeds, 4–14 July 1999. Montana State University, Bozeman, pp 43–50

Gerlach J (1993) Invasive Melastomataceae in Seychelles. Oryx 27:22–26

Griffiths OL, Florens FBV (2006) A field guide to the non-marine molluscs of the Mascarene Islands (Mauritius, Rodrigues, Réunion) and the northern dependencies of Mauritius. Bioculture Press, Mauritius

Griffiths CJ, Jones CG, Hansen DM et al (2010) The use of extant non-indigenous tortoises as a restoration tool to replace extinct ecosystem engineers. Restor Ecol 18:1–7

Hamidou S (2012) Espèces exotiques envahissantes, union des comores. Atelier de travail "océan Indien" sur les espèces exotiques envahissantes. Mamoudzou – Mayotte 23–26 janvier 2012. http://www.especes-envahissantes-outremer.fr/initiative-article-81.html

Hivert J (2003) Plantes exotiques envahissantes: état des méthodes de lutte mises en œuvre par l'Office National des Forêts à La Réunion. ONF, La Réunion

Hugel S (2012) Impact of native forest restoration on endemic crickets and katydids density in Rodrigues Island. J Insect Conserv 16:473–477

IUCN, UNEP-WCMC (2012) The World Database on Protected Areas (WDPA): February 2012 [On-line]. UNEP-WCMC, Cambridge, UK. www.protectedplanet.net. Accessed 1 Mar 2012

Kaiser CN, Hansen DM, Müller CB (2008) Habitat structure affects reproductive success of the rare endemic tree *Syzygium mamillatum* (Myrtaceae) in restored and unrestored sites in Mauritius. Biotropica 40:86–94

Kier G, Kreft H, Lee TM et al (2009) A global assessment of endemism and species richness across island and mainland regions. Proc Natl Acad Sci USA 106:9322–9327

Kueffer C (2003) Habitat restoration of mid-altitude secondary cinnamon forests in the Seychelles. In: Mauremootoo JR (ed) Proceedings of the regional workshop on invasive alien species and terrestrial ecosystem rehabilitation in Western Indian Ocean Island States. Sharing experience, identifying priorities and defining joint action. Indian Ocean Commission, Quatre Bornes, Mauritius, pp 147–155

Kueffer C (2006) Integrative ecological research: case-specific validation of ecological knowledge for environmental problem solving. GAIA 15:115–120

Kueffer C (2010) Reduced risk for positive soil-feedback on seedling regeneration by invasive trees on a very nutrient-poor soil in Seychelles. Biol Invasions 12:97–102

Kueffer C (2012) The importance of collaborative learning and research among conservationists from different oceanic islands. Rev Ecol (Terre et Vie) 11:125–135

Kueffer C, Lavergne C (2004) Case studies on the status of invasive woody plant species in the Western Indian Ocean. 4. Réunion. Forestry Department, Food and Agriculture Organization of the United Nations, Rome, Italy

Kueffer C, Vos P (2004) Case studies on the status of invasive woody plant species in the Western Indian Ocean. 5. Seychelles. Forest Health & Biosecurity Working Papers FBS/4-5E. Forestry Department, Food and Agriculture Organization of the United Nations, Rome, Italy

Kueffer C, Vos P, Lavergne C et al (2004) Case studies on the status of invasive woody plant species in the Western Indian Ocean. 1. Synthesis. Forestry Department, Food and Agriculture Organization of the United Nations, Rome, Italy

Kueffer C, Schumacher E, Fleischmann K et al (2007) Strong belowground competition shapes tree regeneration in invasive *Cinnamomum verum* forests. J Ecol 95:273–282

Kueffer C, Klingler G, Zirfass K et al (2008) Invasive trees show only weak potential to impact nutrient dynamics in phosphorus-poor tropical forests in the Seychelles. Funct Ecol 22:359–366

Kueffer C, Kronauer L, Edwards PJ (2009) Wider spectrum of fruit traits in invasive than native floras may increase the vulnerability of oceanic islands to plant invasions. Oikos 118:1327–1334

Kueffer C, Daehler CC, Torres-Santana CW et al (2010a) A global comparison of plant invasions on oceanic islands. Perspect Plant Ecol Evol Syst 12:145–161

Kueffer C, Schumacher E, Dietz H et al (2010b) Managing successional trajectories in alien-dominated, novel ecosystems by facilitating seedling regeneration: a case study. Biol Conserv 143:1792–1802

Kueffer C, Beaver K, Mougal J (2013) Management of novel ecosystems in the Seychelles. In: Hobbs RJ, Higgs E, Hall C (eds) Novel ecosystems. Intervening in the new ecological world order. Wiley-Blackwell, Oxford, pp 228–238

Lagabrielle E, Rouget M, Durieux L et al (2009) Identifying and mapping biodiversity processes for systematic conservation planning in insular regions. Case study of Réunion Island (Indian Ocean). Biol Conserv 142:1523–1535

Lagabrielle E, Rouget M, Le Bourgeois T et al (2011) Integrating conservation, restoration and land-use planning in islands: an illustrative case study in Réunion Island (Western Indian Ocean). Landsc Urban Plan 101:120–130

Larose A (2005) Investigating techniques for improved control of selected invasive alien weeds in native forests of Mauritius. BSc thesis, University of Mauritius, Mauritius

Lavergne R (1978) Les pestes végétales de l'île de La Réunion. Info Nat 16:9–59

Lavergne C (2000) Étude de la stratégie d'invasion du Troène de Ceylan, *Ligustrum robustum* subsp. *walkeri*, à la Réunion et des caractéristiques du milieu envahi. PhD thesis, Université de la Réunion, France

Lavergne C, Rameau J-C, Figier J (1999) The invasive woody weed *Ligustrum robustum* subsp. *walkeri* threatens native forests on La Réunion. Biol Invasion 1:1–15

Le Bourgeois T, Baret S, Desmier de Chenon R (2013) Biological control of *Rubus alceifolius* (Rosaceae) in La Réunion Island (Indian Ocean): from investigations on the plant to the release of the biocontrol agent Cibdela janthina (Argidae). XIII International Symposium on Biological Control of Weeds. Wu Y, Johnson T, Sing S et al (eds) Waikoloa, Big Island, Hawaii, USA, 11–16 September, 2011. FHTET: 153–158.

Lefebvre T, Moncorps S (2010) Les espaces protégés français : une pluralité d'outils au service de la conservation de la biodiversité. Comité français de l'UICN, Paris, France

Leguat F (1708) Voyage et avantures de François Leguat et de ses compagnons en deux isles desertes des Indes Orientales. J.-L. de Lorme, Amsterdam, Netherlands

Lesur D (2012) La gestion des réserves forestières à Mayotte. Atelier de travail « océan Indien » sur les espèces exotiques envahissantes. Mamoudzou – Mayotte 23–26 janvier 2012. http://www.especes-envahissantes-outremer.fr/pdf/atelier_ocean_Indien_2012/Gestion_forestiere_Mayotte.pdf

Louette M, Meirte D, Jocqué R (2004) La faune terrestre de l'archipel des Comores. MRAC, Tervuren

Macdonald IAW (1989) Stratégie de recherche et de gestion pour le contrôle à long terme des pestes végétales à La Réunion. Rapport de mission du 19 au 26 février 1989. Rapport University of Cape Town – Office National des Forêts – Région Réunion, La Réunion

Macdonald IAW (2010) Final report on the 2010 resurvey of alien plant invaders on the island of Reunion. Stellenbosch University of Cape Town, Parc national de La Réunion, Université de La Réunion, La Réunion

Macdonald IAW, Thébaud C, Strahm WA et al (1991) Effects of alien plant invasions on native vegetation remnants on La Reunion (Mascarene Islands, Indian Ocean). Environ Conserv 18:51–61

Margules CR, Pressey RL (2000) Systematic conservation planning. Nature 405:243–253

Mittermeier RA, Gil PR, Hoffman M et al (2004) Hotspots revisited. CEMEX, Mexico City

MNBSAP (2006) Mauritius national biodiversity strategy and action plan (2006–2015). Port Louis, Mauritius. www.cbd.int/doc/world/mu/mu-nbsap-01-en.pdf

Monty MLF, Florens FBV, Baider C (2013) The impact of invasive alien plants on the reproductive output of natives trees in the wet lowland forest of an oceanic island (Mauritius, Mascarenes). Trop Conserv Sci 6:35–49

Myers N, Mittermeier RA, Mittermeier CG et al (2000) Biodiversity hotspots for conservation priorities. Nature 403:853–858

Orientations Régionales Forestières (2002) Volume 1: État des lieux et travaux préparatoires aux orientations. Volume 2: Gestion durable des milieux naturels forestiers et développement stratégique des entreprises locales du bois. ONF, La Réunion

Page W, D'Argent GA (1997) A vegetation survey of Mauritius (Indian Ocean) to identify priority rainforest areas for conservation management. IUCN/MWF report, Mauritius

Payendee JR (2003) Restoration Projects in Rodrigues carried out by the Mauritian Wildlife Foundation. In: Mauremootoo JR (ed) Proceedings of the regional workshop on invasive alien species and terrestrial ecosystem rehabilitation for Western Indian Ocean Island States – sharing experience, identifying priorities and defining joint action, Seychelles 13–17 Oct 2003. Indian Ocean Commission, Quatre Bornes, Mauritius, pp 95–98

Payet G (2012) Impact d'un incendie sur un habitat naturel en cœur de Parc national: le cas du Maïdo, île de La Réunion (Océan Indien): État des lieux et suggestions pour une gestion conservatoire à court, moyen et long terme. Mastère spécialisé Forêt, Nature et SociétéMSc dissertation. AgroParisTech-Engref, Montpellier – Parc national de La Réunion

Radjassegarane S (1999) Les plantes envahissantes de l'île de la Réunion – Étude de deux exemples: Hedychium flavescens (Zingiberaceae) et Ligustrum robustum subsp. walkeri (Oleaceae) – Recherche préliminaire pour une lutte biologique. PhD thesis, Université Paul Sabatier, Toulouse, France

Reaser JK, Meyerson LA, Cronk Q et al (2007) Ecological and socioeconomic impacts of invasive alien species in island ecosystems. Environ Conserv 34:98–111

Safford RJ (1997) A survey of the occurrence of native vegetation remnants on Mauritius in 1993. Biol Conserv 80:181–188

Salamolard M (2002) Orientations Régionales de Gestion de la Faune sauvage et d'amélioration de la qualité de ses Habitats de La Réunion: Annexe 1. État des lieux. Rapport SEOR-DIREN, La Réunion

Salamolard M, Lavergne C, Cambert H et al (2008) Mise en place d'un dispositif de veille et d'intervention pour la prévention des invasions biologiques à la Réunion –cahier des charges. ARDA –ARVAM –CBNM –ONF –SÉOR –DIREN, La Réunion

Schumacher E, Kueffer C, Tobler M et al (2008) Influence of drought and shade on seedling growth of native and invasive trees in the Seychelles. Biotropica 40:543–549

Schumacher E, Kueffer C, Edwards PJ et al (2009) Influence of light and nutrient conditions on seedling growth of native and invasive trees in the Seychelles. Biol Invasion 11:1941–1954

Seepaul PP (2006) Investigating herbicide use in controlling weeds in native forests. BSc thesis, University of Mauritius, Mauritius

Soubeyran Y (2008) Espèces exotiques envahissantes dans les collectivités françaises d'outre-mer. État des lieux et recommandations. Collection Planète Nature. Comité français de l'UICN, Paris

Stoddart DR (ed) (1984) Biogeography and ecology of the Seychelles Islands. Junk DR W Publishers, The Hague

Strahm W (1989) Plant red data book for Rodrigues. Koeltz Scientific Books, Konigstein

Strasberg D, Rouget M, Richardson DM et al (2005) An assessment of habitat diversity, transformation and threats to biodiversity on Reunion Island (Mascarene Islands, Indian Ocean) as a basis for conservation planning. Biodiv Conserv 14:3015–3032

Strijk J (2010) Species diversification and differentiation in the Madagascar and Indian Ocean Islands Biodiversity Hotspot. PhD thesis, Université de Toulouse, France

Tassin J (2002) Dynamiques et conséquences de l'invasion des paysages agricoles des hauts de La Réunion par *Acacia mearnsii* de Wild. PhD thesis, Université Paul Sabatier, Toulouse, France

Tassin J, Balent G (2004) Le diagnostic d'invasion d'une essence forestière en milieu rural: exemple d'*Acacia mearnsii* à La Réunion. Rev For Fr 56:132–142

Tassin J, Médoc JM, Kull CA et al (2009) Can invasion patches of *Acacia mearnsii* serve as colonizing sites for native plant species on Réunion (Mascarene archipelago)? Afr J Ecol 47:422–432

Thébaud C, Warren BH, Cheke AC et al (2009) Mascarene Islands, biology. In: Gillespie RG, Clague D (eds) Encyclopedia of islands. University of California Press, Berkeley

Thompson R (1880) Report on the forests of Mauritius – their present condition and future management. Mercantile Record Company, Port Louis, Mauritius

Triolo J (2005) Guide pour la restauration écologique de la végétation indigène. Office National des Forêts, Direction Régionale de La Réunion. ONF -Région Réunion -Europe, La Réunion, France

Vaughan RE, Wiehe PO (1937) Studies on the vegetation of Mauritius: I. A preliminary survey of the plant communities. J Ecol 25:289–343

Vos P (2004) Case studies on the status of invasive woody plant species in the Western Indian Ocean. 2. The Comoros Archipelago (The Union of the Comoros and Mayotte). Forestry Department, Food and Agriculture Organization of the United Nations, Rome, Italy

Chapter 20
Southern Ocean Islands Invaded: Conserving Biodiversity in the World's Last Wilderness

Justine D. Shaw

Abstract The Southern Ocean islands are some of the most isolated landmasses in the world. Few of the islands support permanent human settlements or land based industries and as such they remain as some of the most uninvaded landscapes globally. Over 250 non-native plants are currently established across the region. Most are grasses and small herbs, and as such have similar growth forms to the native vegetation. Many of the invasive plants present today arrived several hundred years ago with whaling and sealing gangs, others have been introduced more recently with cargo and building programmes associated with research stations. The Southern Ocean Islands provide a unique opportunity to study and manage invasive plants as the islands have low propagule pressure and very few confounding factors that drive invasion processes elsewhere, such as herbivores, agriculture and land clearance. Invasive plants vary in abundance and distribution on islands and residence time has been shown to significantly influence their area of occupancy. The high protection status of the islands has led to numerous management actions and restoration programmes, some involving invasive plant eradication attempts. Rigorous biosecurity measures are essential to stem future introductions and ensure the island ecosystems remain intact.

Keywords Sub-Antarctic • Cool temperate • Eradication • Herbivores • Residence time

J.D. Shaw (✉)
Environmental Decision Group, School of Biological Sciences, The University of Queensland, St Lucia, Australia

Australian Antarctic Division, Department of Environment, Hobart, Australia
e-mail: shaw.justine@gmail.com

L.C. Foxcroft et al. (eds.), *Plant Invasions in Protected Areas: Patterns, Problems and Challenges*, Invading Nature - Springer Series in Invasion Ecology 7, DOI 10.1007/978-94-007-7750-7_20, © Springer Science+Business Media Dordrecht 2013

449

20.1   Introduction

There are 25 islands and archipelagos scattered across the Southern Ocean between latitudes 37°S and 55°S (Fig. 20.1). These Southern Ocean Islands (SOI) represent some of the most remote landmasses in the world. Almost all of them are entirely protected areas and as such represent some of the most pristine environments remaining on earth today. All these islands are truly oceanic, and have been separated in time and space from continental landmasses. This physical isolation has benefited the conservation values of these protected areas by limiting propagule pressure and as such the threat of invasive alien species. With varied geological origins, either continental, volcanic or opholitic, the islands differ in age. Some islands were extensively glaciated during the last glacial maximum while others remained little affected (Bergstrom and Chown 1999). Latitude strongly influences island climate with northern islands generally being warmer; however, climate is also influenced by island position in relation to the Antarctic Polar Frontal Zone and its movement. In general, the oceanic climate is cool, wet and windy. Despite their variance in origins and positions the islands form a broad biogeographical unit (Chown et al. 2001) although differences between them are well noted (Greve et al. 2005; Shaw et al. 2010; Terauds et al. 2012). They are notable for their vast numbers and diversity of seabirds and marine mammals, but perhaps less well known are the distinct terrestrial ecosystems that occur nowhere else in the world, containing many endemic species of plants, invertebrates and even a few endemic land birds. These terrestrial island ecosystems often benefit from marine-derived nutrients deposited by the hundreds of thousands of seals and seabirds (Smith 1987).

Floristically the islands are dominated by bryophytes and lichens, with vascular plants being less speciose; however, vascular plants dominate the vegetation structure at least at low elevations. The southern islands (also known as sub-Antarctic) support vegetation described as tundra-like, with major growth forms being tall tussock grasses, megaherbs (Fig. 20.2a), cushions plants (Fig. 20.2b), small grasses and herbs (Bliss 1979). The more northern (cool-temperate) islands (e.g. Auckland, Campbell, Snares, Antipodes, Tristan and Amsterdam groups) support shrubs and short trees. The major vegetation communities on the islands are short grasslands, tussock grasslands, feldmark and herbfield communities. These islands represent the southern limit of diverse vascular plant communities; Antarctica to the south only supports two indigenous vascular species. Indigenous vascular plant species richness varies across the islands and can be best explained by the size and temperature of the islands (Chown et al. 1998), with bigger, warmer islands having more species. Chown et al. (1998) concluded that Southern Ocean Island indigenous species richness is determined by the severity of island climates and the effect of island area on heterogeneity of habitat and possibly population size. For example, small, cold McDonald Island has only five indigenous vascular plants while the larger, warmer Auckland Island supports over 180 species (Shaw et al. 2010). Islands that are close together have more similar floras (Greve et al. 2005; Shaw et al. 2010).

Tristan da Cunha and Gough Is.
AFRICA
40°S
30°W
40°E
50°S
Prince Edward Is.
South Georgia   60°S
60°E
Crozet Is.
Falkland Is.
Kerguelen Is.
Amsterdam
& St. Paul Is.
Heard & McDonald Is.
90°W   ANTARCTICA   90°E
110°W
110°E
140°W
Macquarie I.
Campbell Is.
Antipodes Is.   Auckland Is.
Bounty Is.   The Snares
160°W   140°E   AUSTRALIA

Fig. 20.1 Map of the Southern Ocean Islands

Due to the low intensity of human activity, lack of land-based industry, the low intensity of invasive species and the abundance of wildlife these islands are recognised as having high conservation and scientific value (Dingwall 1995; Chown et al. 2001). These islands also have high conservation value given that they support some of the few terrestrial ecosystems to occur in mid to high southern latitudes (Chown et al. 1998; Bergstrom and Chown 1999). The endemic landbird, invertebrate and plant species all depend on these terrestrial island ecosystems for their survival. The islands provide nesting grounds for many of the world's procellariiform species, with many islands listed as habitat critical for several IUCN listed species.

452   J.D. Shaw

Fig. 20.2 Intact, invaded and recovering ecosystems of the Southern Ocean Islands. (**a**) World Heritage Campbell Island was once grazed by sheep. After their eradication native vegetation expanded, in particular the native megaherbs *Pleurophyllum speciosum* (Campbell Island daisy), (**b**) Volcanic World Heritage Heard Island with the native cushion plants *Azorella selago* (cushion plant), megaherbs *Pringlea antiscorbutica* (Kerguelen cabbage) and prostrate herb *Acaena magellanica* (buzzy) present around a low altitude drainage line, (**c**) Heavy rabbit grazing on World Heritage Macquarie Island has reduced the native tussock grasses *Poa foliosa* (tussock grass) to fibrous pedestals, which have been invaded by alien *Poa annua* (annual bluegrass; Shaw et al. 2011) as shown at this nutrient enriched coastal site, (**d**) The red inflorescences of alien *Rumex acetosella* (sheep sorrel), growing in a small patch near the meteorological station on Marion Island, (**e**) Hundreds of thousands of King penguins (*Aptenodytes patagonicus*) breed on the islands and deposit vast amounts of guano. Invasive plants, such as *Sagina procumbens* (procumbent pearlwort) thrive under physically disturbance and nutrient enrichment (Frenot et al. 1998), (**f**) Once established invasive plant propagules may be dispersed by native animal movement; the picture shows moulting Southern Elephant seals (*Mirounga leonina*) lying on a lawn of the invasive grass *Poa annua* – propagules are often observed adhering to their hair. (Photos credits (**a**) and (**e**) Aleks Terauds, (**b**) Kate Kiefer, Australian Antarctic Division, (**c**), (**d**), and (**f**) Justine Shaw)

20.2   Reserve Status and Protected Areas

Most Southern Ocean islands are wholly protected and in a global context this makes the region unique. In other parts of the world protected areas are typically remnants or fragments of a once broader ecosystem or ecoregion (Piekielek and Hansen 2012) and are different from their surrounds (Seiferling et al. 2012). Only the Falkland Islands and Tristan da Cunha support resident populations of people with whom agriculture and private land tenure are associated and as such there is less protected area coverage on these islands and island groups.

Five countries have sovereignty over the 25 islands: France, United Kingdom, Australia, South Africa and New Zealand. All islands have high-order protection under these sovereign nations; nine have the highest World Conservation Union (IUCN) ranking and eight are World Heritage Sites (UNEP). Chown et al. (2001) examined the higher plant, insect and bird taxa in conjunction with alien species presence of all the Southern Ocean Islands, showing that the eight World Heritage islands together with seven others represent 90 % of terrestrial species across the region. Macquarie Island is the only UNESCO Biosphere Reserve. Marion, Prince Edward, Crozet, Kerguelen, Gough, Inaccessible, Amsterdam and St Paul Islands are all Ramsar sites either individually or as island groups (RAMSAR 2012). All nations governing these islands are signatories to the Convention on International Trade in Endangered Species of Wild Fauna and Flora (CITES), except France, which has agreed but is not yet a signatory (de Villiers et al. 2006). With the exception of the Falkland Islands, all governing authorities require that visitors obtain a permit before landing (de Villiers et al. 2006), and all are governed by management plans, which are administered by the sovereign nations. In addition to the high-level reserve status, many islands have zoning systems restricting human activity for conservation purposes (Prince Edward Islands Management Plan Working Group 1996; Australian Antarctic Division 2005; Parks and Wildlife Service 2006; Lebouvier and Frenot 2007).

The risk posed by alien species is a well-recognised threat to island biota (Bergstrom and Chown 1999; Chown et al. 2001; Frenot et al. 2005; Kueffer et al. 2010). The Convention on Biological Diversity's Conference of Parties calls upon signatories to "establish quarantine measures to ensure protection of islands within their nation states from alien species threats, which could damage island ecosystems and induce biodiversity loss" (Convention on Biological Diversity 2006). In all of the islands' management plans provisions are made for invasive species management. Some documents identify priority species for management on the islands, while others state more generally that invasive species should be managed. This chapter provides a summary of the state of alien plant invasions and their current management in protected areas (in this case, islands) restricted to the Southern Ocean. Established alien plants are described and the islands that are most invaded are identified. Pathways of plant introductions, from historic to present day and the mechanisms through which alien plants impact the island ecosystems are discussed, as are the implications of these introductions. The level of area protection in the region, and its implications for plant invasions and their management, both in the region and more broadly is also explored.

20.3   Invasive Alien Plants in the Southern Ocean Islands

Across the 25 islands examined here approximately 280 alien species have been recorded (excluding transients; see Table 20.1), including plants originally introduced for cultivation that have escaped and become established. However, most of the alien plants were not intentional introductions. The alien status of these species is based on the long history of botanical survey relative to human arrival (Hooker 1847; Eaton 1875; Moseley 1879) and the origin of these plants is typically relatively easy to determine. For example, the most commonly occurring (widespread) invasive alien plants (IAPs) across the region are native to Europe (Frenot et al. 2005; Shaw et al. 2010) rather than the closest Southern Hemisphere continents. Most alien plants are concentrated at sites of intense human activity, including research stations and historic sites such as whaling stations or sealing depots. Some alien species have also been found at sites of farming and gardening activity (Meurk 1975) Monitoring of recent arrivals is usually straightforward given the low human activity and small size of the islands (Bergstrom and Chown 1999).

Like the dominant native plants, most alien plant species are grasses and herbs, with no woody structure and low stature. A few alien shrubs occur on some lower latitude (i.e. typically warmer) islands (Meurk 1982; Lee et al. 1991) although a single individual shrub has established recently on sub-Antarctic Marion Island (Gremmen and Smith 2008). Most alien plants are persistent rather than invasive, and have restricted distributions on the islands they have colonised (Frenot et al. 2005). At Kerguelen only seven of the 69 alien plants are invasive and widespread, extending beyond the bounds of the research station; while at Crozet seven of the 58 introduced plants are considered invasive (Frenot et al. 2001). Seventeen alien plants are established on the Prince Edward Islands, six of which have spread rapidly to become widespread (le Roux et al. 2013). Across the region some recent arrivals have been documented as spreading rapidly (Gremmen and Smith 1999; Frenot et al. 2001; Ryan et al. 2003; Scott and Kirkpatrick 2005; le Roux et al. 2013) even in the absence of pollinators and natural dispersers, whereas others occur as single individuals or in small patches. However, the most invasive species occupy large areas, and displace native plant species and impact broadly on the ecosystems. Some of these species have colonised numerous islands (Frenot et al. 2005; Shaw et al. 2010; see Table 20.2) and have invaded other areas around the world (Holm et al. 1977; Chwedorzweska 2008; Pyšek et al. 2009).

Table 20.1 Alien plant richness of Southern Ocean Islands. Small islands and islets that have not been well surveyed have been excluded

	Island groups/archipelagos	Major individual islands	Area $(km^2)^a$	Number of alien plants[b]
1	South Georgia		3,755	37
2	Prince Edward Islands	Prince Edward	44	3
3		Marion	290	14
4	Crozet	Cochons	70	6
5		Apôtres	3	2
6		Pinguoins	3.2	1
7		Est	130	5
8		Possession	150	58
9	Kerguelen		7,200	67
10	Heard and McDonald	Heard	368	1
11		McDonald	2.6	0
12	Amsterdam		55	57
13	St Paul		8.1	10
14	Macquarie		128	3
15	Campbell		113	70
16	Auckland		626	33
17	Snares		3.3	2
18	Bounty		1.3	0
19	Antipodes		21	4
20	Tristan da Cunha Group	Gough	57	24
21		Inaccessible	12	28
22		Nightingale	4	8
23		Tristan	86	133
24	Falkland Islands	West Falkland	3,500	95
25		East Falkland	5,000	97

[a]Areas are taken from Chown et al. (1998)
[b]Represents the alien plants that have been recorded, but does not include plants that were deliberately introduced for cultivation or are known to be transient and have disappeared since discovery (Taken from Shaw et al. 2010)

Table 20.2 The most widespread alien plant species in the Southern Ocean Islands

Plant species	Family	Islands occupied
Poa annua	Poaceae	17
Stellaria media	Caryophyllaceae	13
Cerastium fontanum	Caryophyllaceae	12
Sagina procumbens	Caryophyllaceae	11
Rumex acetosella	Polygonaceae	10
Holcus lanatus	Poaceae	9
Poa pratensis	Poaceae	9
Agrostis stolonifera	Poaceae	8
Festuca rubra	Poaceae	8
Sonchus oleraceus	Asteraceae	7
Agrostis capillaris	Poaceae	6
Taraxacum officinale	Asteraceae	6
Trifolium repens	Fabaceae	6
Plantago lanceolata	Plantaginaceae	6
Poa trivialis	Poaceae	6

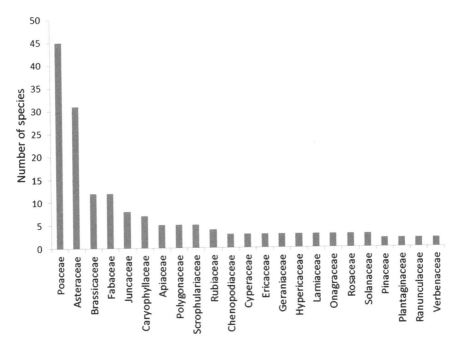

Fig. 20.3 The number of alien plant species per family (with two or more species) in the alien flora of the Southern Ocean Islands. Thirteen other families are represented by one species

The common invasive plants of the Southern Ocean Islands are from globally invasive taxa such as Poaceae, Asteraceae and Leguminosae (containing Fabaceae; Pyšek 1998), with each family having numerous invasive species across the region (see Fig. 20.3). There are several species which are common across many islands (see Table 20.2). The grass *Poa annua* (annual bluegrass) is the most widespread alien occurring on 23 of the 25 islands examined here (Fig. 20.2c). The only islands where this species has not been recorded lack any alien plants. *Poa annua* has a long history on most islands, being recorded on many early botanical surveys cross the region (although see Walton and Smith 1973). Heard Island is a notable exception as the grass was only first detected in two small patches in 1987 (Scott 1989). It has since spread but so far remains relatively restricted (Scott and Kirkpatrick 2005), which may be an artefact of its short residence time. On other islands it is widespread (e.g. Johnson and Campbell 1975; Meurk 1975; Copson 1984; Bergstrom and Smith 1990; Frenot et al. 2001; Gremmen and Smith 2008; le Roux et al. 2013). In terms of the intra-island distribution, some species are typically widespread, whereas others vary in their distributions. *Poa annua* and *Cerastium fontanum* (common mouse-eared chickweed) are generally widespread where they occur (Walton and Smith 1973; Gremmen and Smith 1999; Copson and Whinam 2001; Frenot et al. 2001; le Roux et al. 2013). *Sagina procumbens* (procumbent pearlwort) has undergone rapid expansion on several islands

(Gremmen and Smith 1999; Frenot et al. 2001; Ryan et al. 2003; Cooper et al. 2011; le Roux et al. 2013). By contrast *Agrostis stolonifera* (creeping bentgrass) has had a differing response across different islands, expanding rapidly on Marion (Gremmen and Smith 1999; le Roux et al. 2013) while remaining localised on Kerguelen, Possession (Crozet), Gough and the Tristan archipelago (Frenot et al. 2001; Ryan 2007).

The extent of a species distribution can differ markedly between islands. The probability of invasion increases with time since introduction (Rejmánek 2000). These islands offer a unique opportunity to study residence time, as unlike many other areas the timing of introduction is often known or can be determined, given the islands' long history of botanical research. By incorporating the date of first detection, current distribution and in some instances expansion rates the role of residence time (see Rouget and Richardson 2003) in driving plant invasions in protected areas can be investigated. A significant positive relationship was found between residence time and area of occupancy for all established alien plants on the Marion Island (le Roux et al. 2013). It is important to consider that some restricted species may still be in a lag phase (Richardson and Pyšek 2006) and as such should not preclude them from management actions.

It has been widely acknowledged that changing climates in the region (Smith and Steenkamp 1990; Pendlebury and Barnes-Keoghan 2007; le Roux 2009) increase the potential for new invasive species to establish if propagules continue to be introduced (Bergstrom and Chown 1999; Frenot et al. 2005; Convey et al. 2006; Chown et al. 2008). There is a relationship between the number of native plants and alien plants on Southern Ocean Islands (Shaw et al. 2010) similar to protected areas in the USA (McKinney 2002), other temperate regions (Timmins and Williams 1991; Pyšek et al. 2002), and global patterns more generally (Stohlgren et al. 1999; Sax and Gaines 2008). This suggests that the islands have low biotic resistance to invasion. Native plant species richness is a proxy variable for area and habitat heterogeneity, and in a comprehensive study Chown et al. (1998) showed that larger Southern Ocean Islands are the most susceptible to the establishment of alien plants because of their higher habitat heterogeneity and larger number of human occupants. Chown et al. (1998) concluded that climatically severe islands have a lower probability of propagules of invasive alien species surviving on them than more temperate islands. So while visitor frequency influences invasive plant species richness on Southern Ocean Island energy availability is also an important driver (Chown et al. 2005).

20.4 History and Pathways of Introduction

Due to their climate and isolation, the islands have a relatively short history of human activity compared with most other areas. The earliest recorded landings on the islands were in the mid-1600s. For example, South Georgia was sighted in 1675 by Antoine de la Roche (Burton 2005) and Ile Amsterdam was discovered in 1522 with the first landing recorded in 1696 (Micol and Jouvetin 1995). The advent of sealing in the late 1700s and early 1800s saw an expansion of shipping and land-based activity into the region. Fur seals were widely hunted for their skins across the Southern Ocean. Once this resource was depleted, elephant seals and to a lesser extent penguins were harvested at many locations for their oil. In later years, as whaling became industrialised, South Georgia and Kerguelen supported whaling stations, albeit with differing success and longevity (Tønnessen and Johnsen 1982). The advent of these activities saw an increase in activity more generally with whalers, tradesmen and fishermen using some islands for food or bait sources (see Micol and Jouventin 1995). These early human inhabitants directly harvested native wildlife and intentionally introduced other animals (Cumpston 1968; Jouventin and Roux 1983). Sheep (*Ovis aries*), cattle (*Bos primigenius*), goats (*Capra hircus*), rabbits (*Oryctolagus cuniculus*), pigs (*Sus scrofa*) and cats (*Felis sylvestris*) were all repeatedly introduced across the region to enhance food supply and livelihood (see Convey and Lebouvier 2009 for a comprehensive assessment). By the 1950s scientific research and weather stations were established on numerous islands by national agencies (Walton and Smith 1973; Frenot et al. 2001) resulting in an expansion in human presence and infrastructure, and in several cases an increase in alien plant (Frenot et al. 2001; Chown et al. 2005) and insect introductions (Gaston et al. 2003).

Several alien plants occur on the islands as a result of deliberate attempts to grow fruit trees or vegetable gardens (Walton and Smith 1973; Meurk 1975; Wace and Holdgate 1976) and in unattended gardens established by whalers on Auckland Island turnips, cabbage, potatoes and kale continued to grow successfully for many years after their propagators had departed (Johnson and Campbell 1975). Similar phenomena have been observed on Inaccessible Island (apples, potatoes) and Gough (potatoes) (P. G. Ryan, personal communication, 2012). However, over time few of these garden plants have become invasive. Some IAPs are directly associated with historic sites, such as whalers and sealers camps (Wace 1960; Huntley 1971; Walton 1975; Gremmen and Smith 1999). Seeds of several grass species were deliberately sown for sheep farming during the 1970s on Kerguelen and some have become invasive (Frenot et al. 2001). Invasive plant species have also become established across the islands through accidental introductions. Probably the greatest historic source of invasive plants came from livestock fodder taken to the islands (Walton 1975). Hay and grain from various sources such as the Falkland Islands, Australia, New Zealand, South Africa and possibly Norway provided a source of propagules.

Understanding the modern pathways of introduction to these islands has application to protected area management globally. In recent times plants have become established near research stations, probably after being introduced via building materials, in particular sand, and other cargo (Bergstrom and Smith 1990). Expansion of infrastructure on islands has been implicated as a driver of an increase in introductions. Recent studies have quantified propagule pressure to these islands, with clothing and cargo identified as sources (Whinam et al. 2005; Lee and Chown 2009). Chown et al. (2012) found that the scientists visiting these islands are more likely than tourists to transport plant propagules to the region (although it is important to acknowledge the greater number of tourists that visit the region than scientists). They found that at least half of the propagules reaching the region originate from environments with similar climates. It is important to consider the vectors and also the origins of propagules entering protected areas. Seabirds have been suggested as vectors of invasive plant introductions over short distances. For example birds may have been responsible for transferring two invasive plants to Prince Edward Island from nearby (approximately 22 km) Marion Island (Bergstrom and Smith 1990; Ryan et al. 2003). Dean et al. (1994) reported invasive plant species colonising albatross nests and at nest entrances of burrow nesting seabirds, and suggested that birds may act as vectors. Surface disturbance and nutrient enrichment associated with seabirds (Dean et al. 1994) and marine mammals (Frenot et al. 1999) also facilitate invasive plant. It is common to see IAPs such as, *Poa annua, Cerastium fontanum, Stellaria media* (common chickweed) and *Sagina procumbens* (among others) associated with colonies of burrowing seabirds, surface nesting seabirds (Fig. 20.2e) and marine mammals (Fig. 20.2f).

Human traffic has been implicated as a driver of some invasive species distributions within islands (Scott and Kirkpatrick 1994; Gremmen and Smith 1999; Convey et al. 2006). Human movement, via walking has been implicated as a driver of *P. annua* distribution on Macquarie Island (Scott and Kirkpatrick 1994). While on Marion Island, human intervention plays a role in local spread of species by increasing propagule pressure between areas (le Roux et al. 2013). However, it is important to note that species are able to spread in the absence of human movement such as on Prince Edward Island (le Roux et al. 2013) and Heard Island (Scott and Kirkpatrick 2005).

Different species have different rates of spread within the same island, for example, On Marion Island *S. procumbens* spread at 1.84 km^2 year^{-1}, while *Juncus* cf. *effusus* (common rush) spread at 0.3 km^2 year^{-1} (le Roux et al. 2013). This highlights the importance of removing alien plants in protected areas as soon as they are detected. On Prince Edward Island, a protected area with very low human visitation and strict biosecurity, *S. procumbens* spread 2.36 km^2 year^{-1}, highlighting that wind, water or animal dispersal act as drivers of dispersal in protected areas with low human activity.

20.5 Impacts on Sub-Antarctic Ecosystems

To date most research on invasion biology on the Southern Ocean Islands has focussed on the impacts of invasive vertebrates on island biotas (Bonner 1984; Convey et al. 2006). Cats and rats (*Rattus rattus*) have been shown to impact on seabirds (e.g. Jones et al. 2008; Medina et al. 2011) and as such these have been the main species targeted for management to date (Chapuis et al. 1994, 2011; Copson and Whinam 2001; Courchamp et al. 2003). Furthermore, reindeer (*Rangifer tarandus*), sheep, rabbits, cattle and goats have been introduced across the region, with rabbits being the most widespread. However, there have been few studies quantifying the impacts of invasive plants on ecosystems they invade.

Shaw et al. (2010) showed that plant introductions have resulted in differentiation of island floras. The addition of alien plants caused neighbouring islands to have more different floral assemblages. Several studies have shown that invasive plants expand their distribution with time and disturbance, thereby outcompeting native vegetation (Huntley 1971; Gremmen and Smith 1999; Frenot et al. 2001; Scott and Kirkpatrick 2005). *Sagina procumbens* is invasive on Marion Island and reduces native plant richness where it occurs (Gremmen 1997; Chown et al. 2010). Using historical photos from Marion Island, Chown et al. (2010) were able to demonstrate large scale displacement of native vegetation over 40 years. In one of the few studies to quantify the impacts of an invasive plant Gremmen et al. (1998) showed that *A. stolonifera* has modified the Marion Island terrestrial ecosystem. This invasive grass is thought to have been introduced with fodder brought to the island in the 1950s and was first detected in 1965 at the meteorological station at Transvaal Cove (see de Villiers and Cooper 2008). It occurs across a wide range of habitats, and is particularly abundant along drainage lines. When *A. stolonifera* invades a region it significantly reduces bryophyte biomass and alters vegetation structure (Gremmen et al. 1998). Vegetation dominated by *A. stolonifera* has low species richness, and significantly lower abundance of indigenous plants. In drainage line habitat, 30 % of native species were found to have disappeared (Gremmen et al. 1998). Invaded habitats were found to have significantly less biomass of enchytraeids (Annelida) and significantly more of one adult weevil (*Ectemnorhinus similis*) and the snail (*Notodiscus hookeri*) (Gremmen et al. 1998). More research and modelling are required to quantify how invasive plants impact on protected areas. Particularly as these questions are fundamental to decision makers and management authorities who fund management actions.

20.6   Invasive Species Interactions

The complexity of invasive species interactions has been documented for vertebrates on several Southern Ocean Islands (Huyser et al. 2000; Bergstrom et al. 2009a; Chapuis et al. 2011) and the impact of introduced herbivores on vegetation structure and dynamics is similarly well studied (Costin and Moore 1960; Meurk 1982; Leader-Williams 1985; Copson and Whinam 1998; Micol and Jouventin 1995; Scott and Kirkpatrick 2008). Vertebrate grazing can lead to a dominance of invasive plant species (Walton 1975; Leader Williams 1985; Micol and Jouventin 1995). For example on South Georgia, reindeer grazing influences the abundance and distribution of *P. annua* (Leader Williams et al. 1987). However, while this grass is tolerant of trampling, urine scorching, grazing and uprooting (Grime 1979) it is intolerant of competition from native vegetation on South Georgia. Within 4 years of reindeer exclusion *P. annua* became greatly reduced in mesic meadows, while under grazing it thrived, forming closed swards

(Leader Williams et al. 1987). Similar reductions in *Cirsium vulgare* (spear thistle) were observed following cattle eradication on Ile Amsterdam and grazing rabbits have been shown to cause an increase of *P. annua* on Macquarie Island through similar mechanisms (Copson and Whinam 1998). Few native species respond in a similar positive manner to grazing by introduced mammals, allowing *P. annua* to outcompete native species in response to grazing pressure.

The interactions between invasive species are complex with many different factors influencing their strength and direction (Bergstrom et al. 2009b; Dowding et al. 2009; Raymond et al. 2011). There are instances where managing one invasive species may assist in the management of others. For example thickets of the invasive plant *Ulex europaeus* (common gorse), on a small island of the Falkland Islands, supports higher numbers of invasive rodents than most native vegetation types, as it provides cover from predation (Quillfeldt et al. 2008). Clearing of these thickets may assist in the management of rodents.

Perhaps more complex management issues relate to the removal of one invasive, which can lead to an increase in another. The removal of rabbits on Ile Verte and Ile Gillou, small islands within the Kerguelen archipelago, led to a significant increase in abundance and cover of invasive *Taraxacum officinale* (dandelion) across all habitats (Chapuis et al. 2004). *Poa annua*, *C. fontanum* and *Senecio vulgaris* (common groundsel) also increased following rabbit removal, albeit on a smaller scale. The different response of these invasive plants to rabbit removal is attributed to the nature of rabbit grazing on each species and their responses. While rabbits grazed on *P. annua* and *C. fontanum* on both islands (Bousses et al. in Chapuis et al. 2004) they also actively removed inflorescences of *T. officinale* thereby reducing the plants reproductive output (Chapuis et al. 2004). *Senecio vulgaris*, however, was not grazed by rabbits and appeared to increase due to climate change (Chapuis et al. 2004).

These cascade effects present conundrums for land managers who are forced to make trade-offs for conservation outcomes. For example, rabbit management on Iles Kerguelen has led to increases in invasive plants. If the original motivation of rabbit removal was to restore native vegetation then it is not a straightforward management issue. Further complications arise when also considering the conservation of burrowing seabirds which are directly and indirectly affected by rabbits (Brothers and Bone 2008; Brodier et al. 2011). Based on the lessons learnt from Iles Kerguelen Chapuis et al. (2004) suggested that in some instances rabbit control via myxomatosis may be preferable to eradication in order to allow a small population of rabbits to continue grazing on the invasive *Taraxacum* species facilitating the re-establishment of native vegetation and thereby reducing erosion which will benefit seabirds (Chapuis et al. 2011).

Experience of alien species invasions on Southern Ocean Islands provides valuable lessons on effective management of protected areas, particularly with regard to trophic cascades associated with invasive vertebrate introduction (Crafford 1990; Chapuis et al. 1994; Courchamp et al. 2003; Quillfeldt et al. 2008), removal (Micol and Jouventin 1995; Huyser et al. 2000; Bergstrom et al. 2009a), control (Copson and Whinam 1998) and the associated responses of invasive plants (Leader Williams et al. 1987; Copson and Whinam 1998; Brodier et al. 2011; Chapuis et al. 2011). The traits of plant species need to be considered when trying to predict their responses to vertebrate control and eradication. For example, when predicting the ecosystem response to eradication of rabbits on Macquarie Islands it is relevant that *Taraxacum* species is absent from the island, and the three invasive species (*P. annua*, *C. fontanum* and *S. media*) present are likely to respond differently given their species traits and the associated native vegetation.

20.7 Mitigation and Management

While the threat of invasive species to the islands biodiversity has long been recognised (Carrick 1964), recent research has highlighted how biosecurity measures can reduce propagule pressure to the region (Chown et al. 2012). Furthermore Lee and Chown (2011) quantified how human movement intra-regionally, i.e. within an island or between islands facilitates invasive species expansion. Effective quarantine protocols are essential to reduce the threat of alien species introductions to protected areas. The high reserve status and associated governance of the Southern Ocean Islands have enabled stringent quarantine protocols to be adopted for all visitors and cargo to the islands. The permit system for accessing, building and undertaking research in these protected areas also facilitates the enforcement of strict quarantine protocols. While the rigour of this practise varies between island and operators, overall awareness has increased markedly in recent years. Tourist activities are heavily regulated through permits and observers where tourism currently occurs. Tour operators are required to ensure that tourists clean their boots prior to disembarking tourist vessels and no food products are permitted ashore. People travelling to most of the islands are not permitted to take outdoor clothing that has been utilised in other areas. Most programmes provide their personnel with new or clean boots and outer clothing. These biosecurity principles are shared between countries and applied to other areas, such as Antarctica (Chown et al. 2012) and other remote islands, and have broader applicability to protected areas more generally. Some national programmes have gone further than others, for example, the South African government has banned the importation of fresh fruit and vegetables to Marion and Prince Edward Island in an effort to reduce the risk of alien introductions. Heard Island has similar restrictions. As Nature Reserves, National Parks and World Heritage Areas there is a sense of responsibility to restore these islands where possible. In the past there has been a focus on vertebrate pests, such as cats and rats. As over time these restoration efforts have been successful, there has been a shift in focus to invasive plant eradications across the region.

The lack of resident human populations within or adjacent to these island protected areas affords a unique management scenario, one that is for the most part devoid of conflicting interests from stakeholders. In other protected areas, conflicts can arise over invasive species removal or restrictions on livestock grazing or numerous of other land use practises that drive plant invasions due to differing values systems, perceptions and land use by different user groups (Simberloff 2003; Ford-Thompson et al. 2012; van Wilgen 2012).

The provision for invasive plant removal is identified in most island management plans for the purpose of maintaining biodiversity. To date there have been 21 recorded management actions to control or eradicate alien plants on Southern Ocean Islands, focusing on small restricted populations. These projects have had varied success. Mechanical removal has been used in the majority of cases with herbicides also used in some instances. In most situations, the target species were selected due to their low area of occupancy, or restricted distribution. Indeed to date the few successful plant eradications involved small populations (or just a few individuals) and were detected early. The grasses *Anthoxanthum odoratum* (sweet vernal grass) and *Rumex crispus* (curled dock) were physically removed from Macquarie Island, both plants had very restricted distributions (Copson and Whinam 2001). *Rumex acetosella* (sheep sorrel) occurs in only two patches on Marion Island (Fig. 20.2d) and as such it is a candidate for management (Gremmen and Smith 2008; le Roux et al. 2013). A *Rumex* species was eradicated from Enderby Island, an islet of the Auckland group, through hand pulling over several years by researchers who regularly visited the island (Department of Conservation unpublished data). However, it is worth noting that repeated removal of *R. acetosella* over several years failed to remove this species from a <1 m^2 area on the plateau of Inaccessible Island (P. G. Ryan, personal communication, 2012). Eradication of several small patches of *Agrostis gigantea* (redtop) was attempted on Marion Island (Gremmen and Smith 2008), and in 2006 an attempt to eradicate a single, small patch of *Agropyron repens* (couch grass) commenced through mechanical removal and repeated herbicide application (de Villiers and Cooper 2008). In one of the largest campaigns to date, a sustained effort has been made to remove *S. procumbens* from Gough Island since it was initially detected in 1998. The site of occupancy has been treated with herbicide, hot water, gas burners and mechanical removable of plants and top soil, yet germinants still continued to emerge after several years of treatment (Cooper et al. 2011). Attempts to remove *P. annua* and *Stellaria media* on the Snares Island through chemical and manual removal have failed, due to the established seedbanks and difficulties associated with the logistics required for ongoing actions (Department of Conservation, New Zealand unpublished data). The control attempts of *Cardamine flexuosa* (wavy-leaved bittercress) on South Georgia have not yet been successful (Government South Georgia South Sandwich Islands 2010) due to seedbank persistence.

The eradication of single alien plants or a restricted small patch of a plant, have been proposed for several islands (Department of Conservation, New Zealand unpublished data; Parks and Wildlife Service 2006; Osbourne et al. 2009; le Roux et al. 2013). To date populations such as these are the only types that have been successfully eradicated on Southern Ocean Islands. These cover numerous species, for example, the eradication of 20 species is proposed on Possession

Island of the Crozet group (National Nature Reserve of French Southern Lands, unpublished data). In such proposals, reserve managers consider the resources required to undertake invasive plant management in the context of the likely outcomes and associated conservation benefits (de Villiers and Cooper 2008). These risk analyses must include the logistical constraints of physical isolation, limited personnel, weather, technological limitations and wildlife disturbance. For populations of invasive plants that cover large areas the cost cannot be justified given the low probability of success. With the current technologies available it is unlikely that widespread invasive plants will be completely removed in the near future. A complicating factor for managing invasive species in high conservation protected areas is there are often legislative limits on what management actions are permissible. Approval for strategies such as the introduction of biocontrol agents is often harder to achieve, the same too for broad scale herbicide application, given the off target impacts on other species of high conservation value. While eradication of invasive plants may not be feasible in other protected areas it is still possible to manage transport of invasive propagules within areas by restricting human movements to invaded areas, or conversely, protecting uninvaded areas. Designated intra-island pathways and wash down points have been used on scientific expeditions to manage *P. annua* spread within Heard Island.

Given their current reserve status, Southern Ocean Islands could be considered some of the most protected areas in the world. Unlike many other protected areas they also have the added benefit of being geographically isolated and have low human visitation and activity. Currently the two inhabited island groups the Falkland Islands and Tristan da Cunha have the highest numbers of established alien plant species than all other islands which are wholly protected areas (i.e. no resident human populations). Chown et al. (1998) discuss the relationship of human occupants and island area in driving this alien plant richness. What is perhaps most pertinent to the current management of invasive plants on these inhabited islands is their reduced protected area status (i.e. not whole island) resulting in conflict between invasive plant management and land use and tenure (United Kingdom Overseas Territories 2011). While efforts are currently underway to develop a Falkland Islands Protected Areas Strategy, aiming to build a network of protected areas it is undeniable that current agricultural practices, in general, complicate invasive plant management. Large scale (i.e. whole island) protected area status has alleviated such land management conflict on most other Southern Ocean Islands.

Changing climate, such as warming and drying, together with increased visitation have the potential to increase invasions to these islands (Bergstrom and Chown 1999; le Roux and McGeoch 2008; Davies et al. 2011); however, strict legislation and governance has ensured threats are minimised to date. Nevertheless, propagules continue to be introduced to the region (Whinam et al. 2005; Lee and Chown 2009) and invasive species already present are likely to spread under climatic change and in some instance already appear to be doing so (Frenot et al. 1998; Chown et al. 2013; Scott and Kirkpatrick 2013). Invasive species pose a considerable threat to the biodiversity of these protected areas, and for this reason it is critical that new introductions are prevented and where possible existing invasives are managed. International collaboration on invasive species research (Frenot et al. 2005; Chown et al. 2012) is critical for developing successful mitigation strategies, and will be required into the future to ensure the on-going protection of these islands from the establishment of invasive species.

References

Australian Antarctic Division (2005) Heard Island and McDonald Islands Marine Reserve. Department of National Parks, Canberra

Bergstrom DM, Chown SL (1999) Life at the front: history, ecology and change on southern ocean islands. Trends Ecol Evol 14:472–477

Bergstrom DM, Smith VR (1990) Alien vascular flora of Marion and Prince Edward Islands: new species, present distribution and status. Antarct Sci 2:301–308

Bergstrom DM, Lucieer A, Kiefer K et al (2009a) Management implications of the Macquarie Island trophic cascade revisited: a reply to Dowding et al. (2009). J Appl Ecol 46:1133–1136

Bergstrom DM, Lucieer A, Kiefer K et al (2009b) Indirect effects of invasive species removal devastate world Heritage Island. J Appl Ecol 46:73–81

Bliss LC (1979) Vascular plant vegetation of the Southern Circumpolar Region in relation to Antarctic, alpine and arctic vegetation. Can J Bot 57:2167–2178

Bonner WN (1984) Introduced mammals. In: Laws RM (ed) Antarctic ecology. Academic, London, pp 237–278

Brodier S, Pisanu B, Villers A et al (2011) Responses of seabirds to the rabbit eradication on Ile Verte, sub-Antarctic Kerguelen Archipelago. Anim Conserv 14:459–465

Brothers N, Bone C (2008) The response of burrow-nesting petrels and other vulnerable bird species to vertebrate pest management and climate change on sub-Antarctic Macquarie Island. Pap Proc R Soc Tasmania 142:123–148

Burton R (2005) The discovery of South Georgia. In: Poncet S, Crosbie K (eds) A visitors guide to South Georgia. Wildguides Ltd, Hampshire, p 32

Carrick R (1964) Problems of conservation in and around the Southern Ocean. In: Carrick R, Holdgate M, Prevost J (eds) Biologie antarctique. Herman, Paris, pp 589–598

Chapuis J-L, Bousses P, Barnaud G (1994) Alien mammals, impact and management in the French subantarctic islands. Biol Conserv 67:97–104

Chapuis J-L, Frenot Y, Lebouvier M (2004) Recovery of native plant communities after eradication of rabbits from the subantarctic Kerguelen Islands, and influence of climate change. Biol Conserv 117:167–179

Chapuis J-L, Pisanu B, Brodier S et al (2011) Eradication of invasive herbivores: usefulness and limits of biological conservation in a changing world. Anim Conserv 4:471–473

Chown SL, Gremmen NJM, Gaston KJ (1998) Ecological biogeography of southern ocean islands: species-area relationships, human impacts and conservation. Am Nat 152:562–575

Chown SL, Rodrigues ASL, Gremmen NJM et al (2001) World Heritage status and conservation of Southern Ocean Islands. Conserv Biol 15:550–557

Chown SL, Hull B, Gaston KJ (2005) Human impacts, energy availability and invasion across Southern Ocean Islands. Glob Ecol Biogeogr 14:521–528

Chown SL, Lee JE, Shaw JD (2008) Conservation of Southern Ocean Islands: invertebrates as exemplars. J Insect Conserv 12:277–291

Chown SL, Terauds A, Huntley BJ et al (2010) South Africa's southern sentinel: terrestrial environmental change at sub-Antarctic Marion Island. In: Zietsman L, Pauw J, Van Jaarseveld AS et al (eds) Earth observations of environmental changes in Southern Africa: causes, consequences and responses. SAEON, Pretoria, pp 139–146

Chown SL, Huiskes AHL, Gremmen NJM et al (2012) Continent-wide risk assessment for the establishment of nonindigenous species in Antarctica. Proc Natl Acad Sci USA 109:4938–4943

Chown SL, le Roux PC, Ramaswiela T et al (2013) Climate change and elevational diversity capacity: do weedy species take up the slack? Biol Lett 9:20120806

Chwedorzweska K (2008) *Poa annua* L. in Antarctica: searching for the source of introduction. Polar Biol 31:263–268

Convention on Biological Diversity (2006) Conference of the parties, Decision VIII/1 Island biodiversity, Convention on biological diversity, Secretariat, Brazil. http://www.cbd.int/deci sions/cop/?m=cop-08

Convey P, Lebouvier M (2009) Environmental change and human impacts on terrestrial ecosystems of the sub-Antarctic islands between their discovery and the mid-twentieth century. Pap Proc R Soc Tasmania 143:33–44

Convey P, Frenot Y, Gremmen N et al (2006) Biological invasions. In: Bergstrom DM, Convey P, Huiskes AHL (eds) Trends in Antarctic terrestrial and limnetic ecosystems. Springer, Dordrecht, pp 193–220

Cooper J, Cuthbert RJ, Gremmen NJM et al (2011) Earth, fire and water: applying novel techniques to eradicate the invasive plant, procumbent pearlwort *Sagina procumbens*, on Gough Island World Heritage Site in the South Atlantic. In: Veitch CR, Clout MN, Towns D (eds) Island invasives: eradiation and management. IUCN, Gland, pp 162–165

Copson GR (1984) Annotated atlas of the vascular flora of Macquarie Island. ANARE Res Not 18:71

Copson G, Whinam J (1998) Response of vegetation on subantarctic Macquarie Island to reduced rabbit grazing. Aust J Bot 46:15–24

Copson GR, Whinam J (2001) Review of ecological restoration programme on subantarctic Macquarie Island: pest management progress and future directions. Ecol Manag Restor 2:129–138

Costin AB, Moore DM (1960) The effects of rabbit grazing on the grasslands of Macquarie Islands. J Ecol 48:729–732

Courchamp F, Chapuis J-L, Pascal M (2003) Mammal invaders on islands, impact, control and control impact. Biol Rev 78:347–383

Crafford JE (1990) The role of feral house mice in ecosystem functioning on Marion Island. In: Kerry KR, Hempel G (eds) Antarctic ecosystems. Ecological change and conservation. Springer, Berlin, pp 359–364

Cumpston JS (1968) Macquarie Island, 1st edn. Antarctic Division, Department of External Affairs, Melbourne

Davies KF, Melbourne BA, McClenahan JL et al (2011) Statistical models for monitoring and predicting effects of climate change and invasion on the free-living insects and a spider from sub-Antarctic Heard Island. Polar Biol 34:119–125

Dean W, Milton SJ, Ryan PG et al (1994) The role of disturbance in the establishment of indigenous and alien plants at Inaccessible and Nightingale Islands in the South Atlantic Ocean. Vegetatio 113:13–23

de Villiers MS, Cooper J (2008) Conservation and management. In: Chown SL, Froneman PW (eds) The Prince Edward Islands: land-sea interactions in a changing ecosystem. Sun Press, Stellenbosch, pp 301–330

de Villiers MS, Coope J, Carmichael N et al (2006) Conservation management at Southern Ocean Islands: towards the development of best practise guidelines. Polarforschung 75:113–131

Dingwall PR (1995) Progress in conservation of the sub-Antarctic Islands. IUCN, Gland

Dowding JE, Murphy EC, Springer K et al (2009) Cats, rabbits, myxoma virus and vegetation on Macquarie Island: a comment on Bergstrom et al. (2009). J Appl Ecol 46:1129–1132

Eaton AE (1875) First report of the naturalist attached to the transit of venus expedition to the Kerguelen's Islands, December 1875. Proc R Soc Lond Ser B Biol Sci 33:351–356

Ford-Thompson AFS, Snell C, Saunders G et al (2012) Stakeholder participation in management of invasive vertebrates. Conserv Biol 26:345–356

Frenot Y, Gloaguen JC, Cannavacciuolo M et al (1998) Primary succession on glacier forelands in the subantarctic Kerguelen Islands. J Veg Sci 9:75–84

Frenot Y, Aubry M, Misset MT et al (1999) Phenotypic plasticity and genetic diversity in *Poa annua* L. (Poaceae) at Crozet and Kerguelen Islands, (subantarctic). Polar Biol 22:302–310

Frenot Y, Gloaguen J-C, Masse L et al (2001) Human activities, ecosystem disturbances and plant invasions in the French islands of the southern Indian Ocean (Crozet, Kerguelen and Amsterdam islands). Biol Conserv 101:33–50

Frenot Y, Chown SL, Whinam J et al (2005) Biological invasions in the Antarctic: extent, impacts and implications. Biol Rev 80:45–72

Gaston KJ, Jones AG, Hanel C et al (2003) Rate of species introductions to a remote oceanic island. Proc R Soc Lond Ser B Biol Sci 270:1091–1098

Gremmen NJM (1997) Changes in the vegetation of sub-Antarctic Marion Island resulting from introduced vascular plants. In: Battaglia B, Valencia J, Walton DWH (eds) Antarctic communities: species, structure and survival. Cambridge University Press, Cambridge, pp 417–423

Gremmen NJM, Smith VR (1999) New records of alien vascular plants from Marion and Prince Edward Islands, sub-Antarctic. Polar Biol 21:401–409

Gremmen NJM, Smith VR (2008) Terrestrial vegetation and dynamics. In: Chown SL, Froneman PW (eds) The Prince Edward Islands. Land sea interactions in a changing ecosystem. Sun Press, Stellenbosch, pp 215–241

Gremmen NJM, Chown SL, Marshall DJ (1998) Impact of the introduced grass *Agrostis stolonifera* L. on vegetation and soil fauna of drainage line communities at Marion Island, sub-Antarctic. Biol Conserv 85:223–231

Greve M, Gremmen NJM, Gaston KJ et al (2005) Nestedness of Southern Ocean island biotas: ecological perspectives on a biogeographical conundrum. J Biogeogr 32:155–168

Grime JP (1979) Plant strategies and vegetation processes. Wiley, Chichester

Government South Georgia South Sandwich Islands (2010) Government South Georgia South Sandwich Islands newsletter. http://www.sgisland.gs/index.php/%28h%29South_Georgia_Newsletter%2C_December_2010

Holm LG, Plucknett DL, Pancho JV et al (1977) The world's worst weeds: distribution and biology. University of Hawaii Press, Honolulu

Hooker JD (1847) The botany of the Antarctic Voyage of H.M. discovery ships Erebus and Terror in the years 1839–43 under the command of James Clarke Ross I. Reeve Brothers, London

Huntley B (1971) Vegetation. In: Van Zinderen Bakker EM, Winterbotttom JM, Dyer RA (eds) Marion and Prince Edward Islands: report on the South African biological and geological expedition 1965–1966. A.A. Baklema, Cape Town, pp 98–160

Huyser O, Ryan PG, Cooper J (2000) Changes in population size, habitat use and breeding biology of lesser sheathbills (*Chionis minor*) at Marion Island: impacts of cats, mice and climate change? Biol Conserv 92:299–310

Johnson PN, Campbell DJ (1975) Vascular plants of the Auckland Islands. N Z J Bot 13:665–720

Jones HP, Tershy BR, Zavaleta ES et al (2008) Severity of the effects of invasive rats on seabirds: a global review. Conserv Biol 22:16–26

Jouventin P, Roux J-P (1983) The discovery of a new albatross. Nature 305:181

Kueffer C, Daehler CC, Torres-Santana CW et al (2010) A global comparison of plant invasions on oceanic islands. Persp Plant Ecol Evol Syst 12:145–161

Leader Williams N (1985) The sub-Antarctic islands- introduced mammals. In: Bonner WN, Walton DWH (eds) Key environments- Antarctica. Pergamon Press, Oxford, pp 318–328

Leader Williams N, Smith RIL, Rothery P (1987) Influence of introduced reindeer on the vegetation of South Georgia: results from a long term experiment. J Appl Ecol 24:801–822

Lebouvier M, Frenot Y (2007) Conservation management in the French sub-Antarctic islands and surrounding seas. Pap Proc R Soc Tasmania 14:23–28

Lee JE, Chown SL (2009) Breaching the dispersal barrier to invasion: quantification and management. Ecol Appl 19:1944–1959

Lee JE, Chown SL (2011) Quantification of intra-regional propagule movements in the Antarctic. Antarct Sci 23:337–342

Lee WG, Bastow Wilson J, Meurk CD et al (1991) Invasion of subantarctic Auckland Islands, New Zealand by the Asteroid tree *Olearia lyallii* and its interaction with a resident Myrtaceous tree *Meterosideros umbellata*. J Biogeogr 18:493–508

le Roux PC (2009) Climate and climatic change. In: Chown SL, Froneman PW (eds) The Prince Edward Islands: land-sea interactions in a changing ecosystem. Sun Press, Stellenbosch, pp 39–64

le Roux PC, McGeoch MA (2008) Rapid range expansion and community reorganization in response to warming. Glob Chang Biol 14:2950–2962

le Roux PC, Ramaswiela T, Kalwij JM et al (2013) Human activities, propagule pressure and alien plants in the sub-Antarctic: tests of generalities and evidence in support of management. Biol Conserv 161:18–27

McKinney ML (2002) Influence of settlement time, human population, park shape and age, visitation and roads on the number of alien plant species in protected areas in the USA. Divers Distrib 8:311–318

Medina FM, Bonnaud E, Vidal E et al (2011) A global review of the impacts of invasive cats on island endangered vertebrates. Glob Change Biol 17:3503–3510

Meurk CD (1975) Contributions of the flora and plant ecology of Campbell Island. N Z J Bot 13:721–742

Meurk CD (1982) Regeneration of subantarctic plants on Campbell Island following the exclusion of sheep. N Z J Ecol 5:51–58

Micol T, Jouventin P (1995) Restoration of Amsterdam Island, South Indian Ocean, following control of feral cattle. Biol Conserv 73:199–206

Moseley HN (1879) Notes by a naturalist on the "Challenger". John Murray, London

Osbourne J, Borosova R, Briggs M et al (2009) Survey for baseline information on introduced vascular plants: South Georgia. South Atlantics invasive species project. Royal Society for the Protection of Birds. http://www.kew.org/ucm/groups/public/documents/document/kppcont_047462.pdf

Parks and Wildlife Service (2006) Macquarie Island Nature Reserve and world heritage area management plan. Department of Tourism, Heritage and the Arts, Hobart

Pendlebury SF, Barnes-Keoghan IP (2007) Climate and climate change in the sub-Antarctic. Pap Proc R Soc Tasmania 141:67–81

Piekielek NB, Hansen AJ (2012) Extent of fragmentation of coarse-scale habitats in and around U.S. National Parks. Biol Conserv 155:13–22

Prince Edward Islands Management Plan Working Group (1996) Prince Edward Islands management plan. Department of Environmental Affairs & Tourism, Pretoria

Pyšek P (1998) Is there a taxonomic pattern to plant invasions? Oikos 82:282–294

Pyšek P, Jarošík V, Kučera T (2002) Patterns of invasion in temperate nature reserves. Biol Conserv 104:13–24

Pyšek P, Lambdon P, Arianoutsou M et al (eds) (2009) Handbook of alien species of Europe. Springer, Berlin

Quillfeldt P, Schenk I, Mc Gill RA et al (2008) Introduced mammals coexist with seabirds at New Island, Falkland Islands: abundance, habitat preferences and stable isotope analysis of diet. Polar Biol 31:333–349

RAMSAR (2012). http://www.ramsar.org/cda/en/ramsar-documents-list/main/ramsar/1-31-218_4000_0__

Raymond B, McInnes J, Dambacher JM et al (2011) Qualitative modelling of invasive species eradication on subantarctic Macquarie Island. J Appl Ecol 48:181–191

Rejmánek M (2000) Invasive plants: approaches and predictions. Aust Ecol 25:497–506

Richardson DM, Pyšek P (2006) Plant invasions: merging the concepts of species invasiveness and community invasibility. Progr Phys Geogr 30:409–431

Rouget M, Richardson DM (2003) Inferring process from pattern in plant invasions: a semimechanistic model incorporating propagule pressure and environmental factors. Am Nat 162:713–724

Ryan PG (2007) Field guides to the animals and plants of Tristan da Cunha and Gough Island. Pisces Publications for the Tristan Island Government, Newbury

Ryan PG, Smith VR, Gremmen NJM (2003) The distribution and spread of alien vascular plants on Prince Edward Island. S Afr J Mar Sci 25:555–562

Sax DF, Gaines SD (2008) Species invasions and extinction: the future of native biodiversity on islands. Proc Natl Acad Sci U S A 105:11490–11497

Scott JJ (1989) New records of vascular plants from Heard Island. Polar Rec 25:37–42

Scott JJ, Kirkpatrick JB (1994) Effects of human trampling on the sub-Antarctic vegetation of Macquarie Island. Polar Rec 30:207–220

Scott JJ, Kirkpatrick JB (2005) Changes in subantarctic Heard Island vegetation at sites occupied by *Poa annua*, 1987–2000. Arct Antarct Alp Res 37:366–371

Scott JJ, Kirkpatrick JB (2008) Rabbits, landslips and vegetation change on coastal slopes of subantarctic Macquarie Island, 1980–2007: implications for management. Polar Biol 31:409–419

Scott JJ, Kirkpatrick JB (2013) Changes in the cover of plant species associated with climate change and grazing pressure on the Macquarie Island coastal slopes, 1980–2009. Polar Biol 36:127

Seiferling IS, Proulx R, Peres-Neto P et al (2012) Measuring protected-area isolation and correlations of isolation with land-use intensity and protection status. Conserv Biol 26:610–618

Shaw JD, Spear D, Greve M et al (2010) Taxonomic homogenization and differentiation across Southern Ocean islands differ among insects and vascular plants. J Biogeogr 37:217–228

Shaw JD, Terauds A, Bergstrom DM (2011) Rapid commencement of ecosystem recovery following aerial baiting on sub-Antarctic Macquarie Island. Ecol Manag Restor 12:241–243

Simberloff D (2003) Confronting introduced species: a form of xenophobia? Biol Invasions 5:179–192

Smith VR (1987) Production and nutrient dynamics of plant communities on a sub-Antarctic island. II. Standing crop and primary production of fjaelmark and fernbrakes. Polar Biol 7:125–144

Smith VR, Steenkamp M (1990) Climatic change and its ecological implications at a subantarctic island. Oecologia 85:14–24

Stohlgren TJ, Binkley D, Chong GW et al (1999) Exotic plant species invade hot spots of native plant diversity. Ecol Monogr 69:25–46

Terauds A, Chown SL, Morgan F et al (2012) Conservation biogeography of the Antarctic. Divers Distrib 18:726–741

Timmins SM, Williams PA (1991) Weed numbers in New Zealand's forest and scrub reserves. N Z J Ecol 15:153–162

Tønnessen JN, Johnsen AO (1982) The history of modern whaling. Australian National University Press, Canberra

United Kingdom Overseas Territories (2011) Falkland Islands protected areas strategy: cooperative management of biological diversity (OTEP XOT 803). United Kingdom Overseas Territories Conservation Forum. http://www.ukotcf.org/infoDB/infoSourcesDetail2.cfm?refID=314

van Wilgen BW (2012) Evidence, perceptions, and trade-offs associated with invasive alien plant control in the Table Mountain National Park, South Africa. Ecol Soc 17:23

Wace NM (1960) The botany of the Southern Ocean Islands. Proc R Soc Lond Ser B Biol Sci 152:475–490

Wace NM, Holdgate MW (1976) Man and nature on the Tristan da Cunha Islands. International Union for Conservation of Nature and Natural Resources, Morges

Walton DWH (1975) European weeds and other alien species in the sub-Antarctic. Weed Res 15:271–282

Walton DWH, Smith RIL (1973) Status of the alien vascular flora of South Georgia. Br Antarct Surv Bull 36:79–97

Whinam J, Chilcott N, Bergstrom DM (2005) Subantarctic hitchhikers: expeditioners as vectors of introductions of alien organisms. Biol Conserv 121:207–219

Part III
Managing Invasions in Protected Areas: From Prevention to Restoration

Chapter 21
Manipulating Alien Plant Species Propagule Pressure as a Prevention Strategy for Protected Areas

Laura A. Meyerson and Petr Pyšek

Abstract In this chapter we argue that preventing the introduction and spread of alien species in protected areas is still a highly relevant and critically important management strategy despite current and future global change. There has been a provocative and attention grabbing call of late in the conservation literature to accept alien species invasions as inevitable and perhaps even desirable. Such 'novel ecosystems', it has been argued, may function equivalently or better under future conditions. However, we suggest that it is the very uncertainty that global change and its associated impacts bring that makes prevention more necessary than ever for protected areas. Here we focus on the variables affecting protected areas that can and cannot be manipulated to strengthen prevention efforts. Because so much has been learned about alien species prevention, we also outline different approaches for existing protected areas and those that are planned in the future, with particular emphasis on the management of invasive alien plant pathways and propagule pressure.

Keywords Invasive plants • Protected areas • Prevention • Management • Climate change

L.A. Meyerson (✉)
Department of Natural Resources Science, University of Rhode Island, 1 Greenhouse Road, Kingston, RI 02881, USA

Department of Invasion Ecology, Institute of Botany, Academy of Sciences of the Czech Republic, Průhonice CZ 252 43, Czech Republic
e-mail: Laura_Meyerson@uri.edu

P. Pyšek
Department of Invasion Ecology, Institute of Botany, Academy of Sciences of the Czech Republic, Průhonice CZ 252 43, Czech Republic

Department of Ecology, Faculty of Science, Charles University in Prague, CZ 128 44 Viničná 7, Prague 2, Czech Republic
e-mail: pysek@ibot.cas.cz

L.C. Foxcroft et al. (eds.), *Plant Invasions in Protected Areas: Patterns, Problems and Challenges*, Invading Nature - Springer Series in Invasion Ecology 7, DOI 10.1007/978-94-007-7750-7_21, © Springer Science+Business Media Dordrecht 2013

21.1 Introduction

Preventing the introduction and colonization of invasive alien plants is an integral component of the management plans of protected areas (PAs) around the world. However, absolute exclusion of unwanted species is recognised as an unrealistic goal in most cases (Rejmánek and Pitcairn 2002; Pluess et al. 2012a, b) and the porosity of PA borders has been well documented since the first global effort to survey invasions in PAs (Usher 1988). Global climate change, atmospheric N deposition, human population growth, land conversion and associated disturbances, and higher levels of trade are all factors contributing to accelerating rates of invasions in most ecosystems (e.g. Meyerson and Reaser 2002; Meyerson and Mooney 2007; Hulme et al. 2009; Chytrý et al. 2012), and present further challenges to the successful prevention of invasive plants incursion into PAs. Other challenges, such as the activities that foster plant introductions in the surrounding matrix (e.g. horticulture, erosion control) and global economic fluctuations that affect the ability to staff and manage PAs for prevention of invasive plants, further complicate management.

Given this rather grim view, it is appropriate to ask whether prevention attempts are still relevant, worthwhile, or even possible. Prevention is a key management option, generally considered as more effective than mitigation and restoration after invasion has taken place (Pyšek and Richardson 2010). Prevention has been identified by the Convention on Biological Diversity in 2002 as the priority management action, with early detection, rapid response, and possible eradication only to follow when prevention fails, and long-term management being the last option (Simberloff et al. 2013). Generally, prevention is applicable at different stages (e.g. Pyšek and Richardson 2010; Blackburn et al. 2011), starting with screening and constricting pathways and vectors, intercepting movements at national borders and assessing risks resulting from international trade; these approaches repeatedly proved successful in reducing the propagule pressures of potential invaders and saving substantial amounts of money to economies (see Simberloff et al. 2013 for examples).

In this chapter we focus on invasive alien plant species prevention as a critical component of PA management. Here, we define 'prevention' as the protection of a defined reserve from invasive alien plants which includes both (i) trying to prevent the arrival of new invasive species and (ii) eradicating or controlling those already present. Therefore, this definition includes limiting the spread of an invasive species already established in a PA, as prevention from on-going and potentially increasing impacts within the reserve. This perspective takes a page from the successful biosecurity approaches that have been employed by New Zealand (Hulme 2011a) and other countries at their borders to manage incursions of unwanted organisms (www.biosecurity.govt.nz/biosec).

Specifically, we highlight the variables controlling plant invasions in PAs with regard to whether or not they can be manipulated. This strategy may be both more effective for preventing the introduction and spread of invasive plants, and more efficient in terms of allocating scarce resources to invasive plants management in PAs.

21.2 Prevention from a Practical Perspective: The Relevance in the Face of Global Change

Over the last several decades, management approaches of PAs have necessarily evolved to recognise and incorporate global, regional and local changes and to consider uncertainty in future trajectories. This evolution holds also for management of invasive species in PAs where absolute exclusion may not only be impractical from an economic perspective but also simply impossible given new and evolving introduction pathways that facilitate high levels of propagule pressure.

In the face of inevitable global change, why is prevention of plant invasions into PAs still relevant? Much literature in the last decade has focused on changing temperature and precipitation patterns that could favour invasive species and negatively impact habitats for native species (e.g. Thuiller 2007; Lambdon et al. 2008; Palmer et al. 2008; Potts et al. 2010). This, of course, includes PAs where many unique biotic and abiotic features and interactions are regulated by temperature and precipitation (Baron et al. 2009). Recent work by Diez et al. (2012) analysed three regional North American datasets of flowering phenology over time. They found that predicting general patterns of phenological response to climate change may be possible at the community level once regional climate drivers (i.e. in addition to temperature) are accounted for (Diez et al. 2012). Such predictions may help to inform the timing and types of management efforts, particularly when interbreeding and hybridization are of concern.

Research has also predicted that both native and introduced species will migrate towards the poles as the climate changes (Walther et al. 2002; Parmesan 2006; Hellmann et al. 2008; Baron et al. 2009), as is already occurring for several species (Parmesan and Yohe 2003). This implies that at least some of the native species that reside within PA boundaries will migrate out of the reserve and new species (both native and alien) will expand their range, migrate in, and colonise the PA resulting in novel biotic assemblages (Baron et al. 2009). Some plants already present in the PA will likely increase in abundance and perhaps also in their distributional range. How individual PA management will respond to the immigration of natives not previously present and to the proliferation of other species already within the reserve should depend on the particular management goals of each site, including how those native immigrants interact with important extant fauna and flora. A further challenge to the relevance of preventing plant invasions is the recent high-profile species translocation conservation strategy (assisted migration) and the controversies that this has generated (Hoegh-Guldberg et al. 2008; Ricciardi and Simberloff 2009; Richardson et al. 2009; IUCN 2012; Schwartz et al. 2012).

However, as has been well documented, invasive species cause significant and lasting changes in natural landscapes over time through ecosystem engineering (e.g. hydrology, accretion, erosion), allelopathy and nutrient cycling (e.g. N-fixation, resource acquisition), and altering light and temperature regimes, that result in profound changes in faunal and floral communities (habitat loss, extirpation, decrease in species diversity, community structure and trophic relationships, etc.); these various types of impacts have received much attention recently (Farnsworth and Meyerson 2003; Liao et al. 2008; Gaertner et al. 2009; Pyšek and Richardson 2010; Vilà et al. 2011; Pyšek et al. 2012; Strayer 2012; Simberloff et al. 2013). Many PAs contain endangered species or rare ecosystem types that could be permanently altered or driven to local extinction by an unchecked invasion. In many countries, PAs are not only considered national treasures because of their historical significance and the species they harbour (e.g. Yellowstone National Park, USA), but are also critical for providing ecosystem goods and services, such as water (van Wilgen et al. 2011, 2012) and an income source for those who live nearby. The potential loss of some or all of these native species, ecosystems and eco-services due to biological invasions keeps the prevention strategy both relevant and crucial.

21.3   Proactive Prevention: To Manipulate or Not

In all PAs, there are variables that can be changed and other variables that are beyond the control of PA management (Table 21.1). While each PA is unique in terms of its location and placement in the larger landscape matrix, some general factors can be applied across most PAs. Detailing these factors for each PA is an important starting point for a PA management plan to better understand where effort and resources can result in positive change and where they cannot. However, the first and perhaps most important distinction in proactive prevention in PAs is whether manipulations to prevent plant invasions are intended for established reserves or for those still in the planning stages.

Globally, the percentage of terrestrial area dedicated to protection of nature has grown over the last two decades. Figure 21.1 presents data on the percentage change in PAs for different regions of the world. Developed regions are grouped while developing regions, arguably hot spots of biodiversity, are distinguished. All regions show at least some increase in the percentage of protected area since 1990 and some (both developed and developing) show additional gains since 2000. If this increasing trend in protected area creation continues, a bifurcated approach for prevention in established PAs versus prevention planned PAs is needed. This dual prevention protocol would allow best practices to be implemented at the earliest stages in newly established PAs, and lessons to be learned from existing PAs where prevention has been successfully applied. However, major knowledge gaps persist. We still do not have satisfactory answers to questions such as:

- Do prevention opportunities differ between established and planned protected areas?
- Is it possible to design new parks that aid in prevention?
- Are prevention strategies 'built in' to reserve design more effective than retroactive prevention?

Below we discuss the major factors affecting invasions in nature reserves and begin to address these questions.

Table 21.1 Overview of factors aimed at preventing or minimizing alien species incursions into extant and future PAs

Factor type	Established PAs	Planned PAs
Manipulatable	Establish buffer zones	Location in landscape matrix (with respect to IAS propagule drivers and socioeconomic fabric of region)
	Restrict/regulate access to fragile areas	Planning buffer zones
	Input of propagules (intensity and timing of park visits, boundary control, biosecurity in PAs)	Education and implementation of codes of conduct
	Human behaviour, education, and codes of conduct	Location of visitor centres, entrance gates, roads and hiking trails
	Water regime	
	Regulated fires	
	Managing/eradicating invasive species already present	
	Allocation of funds into preventive measures of IAS management	
Non-manipulatable	Location in landscape matrix	Climate change
	Climate change	Natural fires
	N-deposition	N-deposition
	Natural fires	Human population development and activities
	Human population development and activities	Stochastic events
	Stochastic events	

Factors are divided according to whether or not they can be manipulated, completely or partially, with the aim to reduce invasive plant species and their impacts, and presented separately for PAs already existing and those to be established in the future. See text for discussion and examples of how the management can be accomplished or mitigation implemented to address those factors that could not be changed or managed

21.4 Vulnerability to Invasion

21.4.1 Location in the Extant Matrix: Managing Natural and Human-Induced Drivers of Propagule Pressure

For planned reserves, drivers of propagule pressure can, to a certain extent, be managed according to the PA location in the landscape matrix. In the Czech Republic and elsewhere in Europe, there are different types of PAs that differ in their levels of protection. In large-scale PAs that include extensive sections of inhabited landscapes such as national parks or so-called 'protected landscape areas', human activities, including commercial activities, are regulated but not excluded. Within these PAs, the most biologically valuable parts are declared as nature reserves and strict protection measures imposed, with the aim to completely

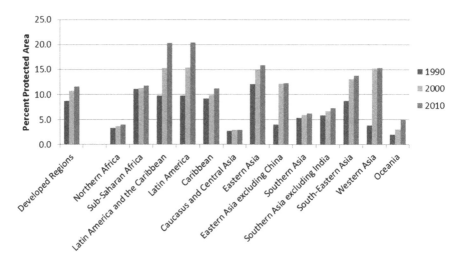

Fig. 21.1 Proportion of terrestrial area protected in developed and developing regions in 1990, 2000, 2010. The differently *shaded bars* represent pooled data for both categories for each of three decades. Pooled data are shown for developed regions in the *left* set of bars, developing regions are shown by individual regions. Data Source: IUCN and UNEP-WCMC (2012) The World Database on Protected Areas (WDPA): Accessed March 2012. Cambridge, UK: UNEP-WCMC

remove them from human influence and allow only natural processes to act. Data from the Czech Republic indicate that small-scale nature reserves located within larger PAs resulted in lower levels of invasion than in reserves established in the non-protected area landscapes.

This finding resulted in the formulation of the "several small inside single large" principle (SSISL) and was suggested as a biodiversity maintenance tool to help prevent plant invasions in nature reserves (Pyšek et al. 2002). This is because of the higher propagule pressure resulting from more intense human activities in landscapes that are not subject to any protection (Chytrý et al. 2008). Similarly, small dry rainforest reserves in Queensland, Australia that do not contain extensive areas of surrounding habitat are unlikely to be secure in the long term (Fensham 1996), pointing to the importance of the surrounding zones in management of small-scale reserves (see also Chown et al. 2003; Jarošík et al. 2011a). Establishment of future small-scale PAs, where this option is available or it is being decided among several possible areas of comparable quality to preserve, should take into account the character of surrounding landscape and connectivity of its more natural parts in the matrix. The above examples suggest that the most convenient strategy from the prevention point of view is to place, whenever possible, new PAs in regions of low human activity, ideally in partially protected landscapes.

Generally, managing pathways and propagule pressure is the key concept behind prevention. A recent study of alien plants incursion into Kruger National Park (KNP), South Africa (Foxcroft et al. 2011), indicated that understanding how drivers of propagule pressure operate can be potentially applied to the management of not only planned but also existing PAs. In this study, the number of alien invasive plants colonizing the park by crossing its border was primarily determined by the amount of water runoff and density of roads outside the park in its close surroundings. Of these two major vectors of propagules, the water runoff had a stronger effect, but both were complementary – where there was no river outside the park, roads acted as an effective driver for propagule pressure. Although the specific effects of factors likely differ among individual PAs, the role of the two major propagule drivers, representing natural and human-induced dispersal, seems to be generally applicable to PAs (Foxcroft et al. 2011). Interestingly, if the general KNP model is applied to individual invasive species, predictors from inside the park, such as the presence of main rivers and species-specific effect of vegetation types, also become important. Landscape characteristics outside the park, such as location of rivers, may serve as guidelines for management to enact proactive interventions to manipulate landscape features near the KNP to prevent further incursions. Predictors from the inside the KNP can be used to identify high-risk areas to improve the cost-effectiveness of management, to locate invasive plants and target them for eradication (Jarošík et al. 2011b).

However, the fact that rivers are the most powerful drivers of alien species propagules (Richardson et al. 2007; Pyšek et al. 2010) creates a conflict of interest because rivers increase habitat heterogeneity and maintain biodiversity and are therefore important landscape elements of many PAs. At present, park managers have little control over the upper reaches of the rivers that flow through KNP. Nevertheless, quantifying the threshold value of water runoff from surrounding areas below which invasion is less likely opens the way to prioritise control measures such as targeting particular riparian areas outside the PA for removal of invasive plants more urgently than others. Or, further spread of alien plants into KNP may be limited by relocating entrance gates to areas where water runoff is low, or follow the same principle if there is a need to create additional entrance gates (Foxcroft et al. 2011).

21.4.2   Buffer Zones: Making Use of Natural Vegetation Resistance to Invasions

Globally, there are twice as many alien species outside of nature reserves than are present within PAs (Lonsdale 1999), suggesting that there are some mechanisms conferring resistance to invasion on PAs. There is some rigorous evidence in the literature that natural vegetation in PAs acts as a buffer against invasion by aliens species. Vegetation of temperate reserves in the Czech Republic was shown to act as an effective barrier against the establishment of alien plants; old reserves had initially fewer aliens than young reserves, and over time it was more difficult for an alien species to invade a nature reserve than a corresponding section of non-protected landscape (Pyšek et al. 2003). The Kruger National park study revealed that in addition to the drivers of propagule pressure, the presence of natural vegetation decreased the abundance of invasive species inside the park. The number of records of invasive plants declined rapidly beyond 1,500 m inside the park, indicating that the park boundary limited their spread (Foxcroft et al. 2011). This phenomenon, of natural vegetation creating a buffer zone, can be potentially used as a prevention tool, if such zones are planned for future reserves, or established around existing PAs. Using vegetation buffer zones as a long term prevention strategy could help minimise the costs, disturbances, and risks associated with on-going active management in core protected areas. For existing parks, such as the KNP example, by focusing only on a subset of vegetation types identified as high-risk for invasion along the park boundary, and fine-tuning the target areas by using information on the presence of rivers and vegetative buffers, management can be made more cost effective (Jarošík et al. 2011b).

21.4.3   Visitors: A Goldmine for Prevention?

It has been repeatedly documented that the number of alien species that occur in a PA is closely related to that of human visitors, this pathway being one of the commonly used surrogates for propagule pressure in invasion studies (Macdonald et al. 1988; Usher 1988; Lonsdale 1999; McKinney 2002; Pyšek et al. 2002; Lee and Chown 2009a, b). In general, horticulture is considered as the most important pathway of introduction of invasive plants (e.g. Mack 2000; Hulme 2011b), but unintentional introductions, such as by visitors to PAs, also generate potentially invasive species. In the Czech flora, unintentionally introduced plant species are less likely to become invasive, but those that do represent a threat to natural areas because they invade a wider range of semi-natural habitats than species escaped from horticulture (Pyšek et al. 2011). The accidental introduction of invasive plant propagules by visitors interacts with the different land-uses within the PA. This can increase the spread of alien plants from areas within the PA such as tourist camps and staff villages, which serve as source areas of propagules (Foxcroft 2001; Foxcroft and Downey 2008).

In many parts of the world, measures to prevent the introduction and spread of invasive plant propagules by visitors to established reserves are already in place. For example, some parks regulate pathways by controlling the number of visitors and the seasons during which people can visit the park, or the accessible areas (e.g. Galapagos, Ecuador [www.galapagospark.org]). Many parks ask visitors to clean their shoes, clothes, tires and vehicles, equipment and pets before entering (e.g. Olympic National Park, Washington, USA and Yosemite National Park, California [www.nps.gov/yose/naturescience/invasive-plants.htm], USA) and some even provide cleaning stations for sanitizing hiking and hunting gear prior to entering the PA (e.g. Fiordland National Park, NZ, www.doc.govt.nz). However, additional prevention measures are possible and desirable, particularly for PAs that are still in the planning stages. For example, informed by scientific knowledge, strategic placements of entrance gates, visitor centres, or staff villages could be considered to reduce threats from plant invasions (Jarošík et al. 2011b; Foxcroft et al. 2011). Finally, an important part of preventive measures aimed at reducing propagule pressure and disturbances caused by visitors is education. Human behaviour is key to preventing biological invasions and it can be assumed that the majority of people visiting PAs are open to being educated about how to contribute to solving problems rather than creating them. Educated visitors can become a part of the on-going monitoring system for new introductions to PAs and all visitors should be educated on appropriate biosecurity precautions and their importance for the PA's preservation. In addition to directly changing behaviour, educated visitors can create culture change in which peer pressure helps to maintain biosecurity standards. In Europe, 'codes of conduct' relating to invasive plants exist for both the horticulture industry (e.g. Heywood and Brunel 2009) and for botanical gardens (Heywood and Sharrock 2012). They are aimed at educating those industries and the people that work in them. Developing and implementing a similar code of conduct for protected areas might be an effective way to formalise and encourage behaviour that strengthens prevention and results in a culture change by PA visitors. This is important because many PAs in Europe, such as large-scale national parks and protected landscape areas host gardening-related and other commercial activities, and as shown recently by Hulme (2011b), botanical gardens can represent serious threat in terms of alien plant invasions.

21.4.4 Factors that Cannot Be Manipulated

Some factors cannot be manipulated for a single PA, regardless of whether PAs are established or still being planned (Table 21.1). These include global climatic factors that change over time (e.g. 2 °C temperature increase per decade is projected by the IPCC 2007), the amount of N-deposition, natural fires beyond control and other stochastic events such as hurricanes or earthquakes, and the multiple pressures associated with human population development and growth. The United Nations is projecting that by 2100 the global population will exceed 10 billion with Africa and Asia as the most populous regions globally (www.un.org/esa/population). Population growth includes demographic and socioeconomic factors such as trade, but also political turmoil, changes in human life style, increases in consumption, behaviour potentially associated with increasingly scarce resources. International treaties and regulations to manage these global changes are notoriously difficult to enforce, particularly when they negatively affect trade and when they involve both developed and developing nations. While these larger global forces paint a somewhat gloomy picture, the take away message is that in terms of managing invasive plants in PAs, it is critical to dedicate energy and resources to the factors that can be controlled, but also to be prepared to adapt when outside forces influence prevention efforts.

21.5   Never Too Late: Eradication Can Reduce
Propagule Pressure

Eradicating existing invasions is another dimension of prevention where the aim is
to prevent propagules from spreading further within the target area, particularly
when eradication is part of an early detection and rapid response strategy to manage
'offensively' rather than 'defensively' (sensu Rejmánek and Pitcairn 2002). As
such, it is grouped with factors that can and should be manipulated to strengthen
prevention. Both inside reserves and in the landscape outside of reserves the goal is
to remove invasive populations. However, within reserves, eliminating or reducing
propagules from which other invasive populations could establish is particularly
crucial because most habitats in a PA are valuable and subject to protection. Indeed,
most eradication campaigns against invasive plants have been conducted in some
kind of PAs (23 of the 27 analysed in Pluess et al. 2012b; see also Genovesi 2011;
Simberloff 2014 for an overview of successful eradications). The analysis in Pluess
et al. (2012b) of factors affecting whether eradication will be successful or not
indicated that event-specific factors, such as the extent of the infested area, reaction
time and measures of sanitary control can be taken into account and even be
manipulated to some degree by authorities dealing with invasive species management. Specifically, initiating the campaign before the extent of infestation reaches a
critical threshold, starting to eradicate within the first 4 years since the problem was
detected, paying special attention to plant invaders escaped from cultivation, and
applying sanitary measures can substantially increase the probability of eradication
success (Pluess et al. 2012b).

21.6   Game Over or Game on?

Is prevention of plant invasions in PAs still a relevant strategy? Our conclusion is
emphatically yes. A policy shift away from invasive species prevention could in
fact be disastrous, perhaps most especially in highly diverse developing countries
where the stakes are especially high (Nuñez and Pauchard 2010; Lovei and
Lewinsohn 2012). In addition to managing PAs under the uncertainty of climate
change, an increased uncertainty would be added in terms of the effects and
interactions of invasive plants with resident flora and fauna and abiotic processes,
further complicating management and likely increasing costs. For these and other
reasons, we strongly argue that prevention remains relevant and is perhaps more
important than ever as the ability of ecosystems and native species to adapt to a
changing future comes to the forefront of management and research agendas.

Despite the recent flurry of articles in high impact journals touting the inevita-
bility of novel ecosystems and the benefits of invasive species, there have been
spectacular successes associated with the eradication of invasive species and
preventing their spread to new areas (e.g. Simberloff et al. 2011). We concede

that in the face of overwhelming global change and dynamic international trade many of the old ecological rules no longer apply and those hard and fast solutions for completely preventing biological invasions are not feasible. However, we argue that we need to develop innovative approaches for preventing new introductions and invasions and must continue working to eradicate existing invasions, particularly in PAs that serve as the repositories for our global biological wealth. These approaches would include different strategies for existing versus planned reserves, managing pathways both spatially and temporally, managing propagule pressure through the creation and management of buffer zones to reduce propagule pressure, and redoubling our efforts to educate the millions of annual visitors to PAs worldwide and harness their enthusiasm to reinforce prevention efforts.

Given the uncertainty of how the future climate will affect the introduction and spread of invasive plants and the ecosystems where they are introduced, we would be wise to invest our energies in scenario-based planning which allows for an array of alternative futures (Baron et al. 2009). By employing strategies informed by adaptive management and research that distinguishes management approaches that do and do not accomplish prevention, we can begin to close the gap between theory and practice in preventing invasions in protected areas.

Acknowledgements The study was supported by long-term research development project no. RVO 67985939 (Academy of Sciences of the Czech Republic). LAM also acknowledges support from the United States and Czech Fulbright Commissions, and PP from institutional resources of Ministry of Education, Youth and Sports of the Czech Republic, and Praemium Academiae award from the Academy of Sciences of the Czech Republic.

References

Baron J, Gunderson L, Allen C et al (2009) National parks and reserves for adapting to climate change. Environ Manage 44:1033–1042

Blackburn TM, Pyšek P, Bacher S et al (2011) A proposed unified framework for biological invasions. Trends Ecol Evol 26:333–339

Chown SL, van Rensburg BJ, Gaston KJ et al (2003) Energy, species richness, and human population size: conservation implications at a national scale. Ecol Appl 13:1233–1241

Chytrý M, Jarošík V, Pyšek P et al (2008) Separating habitat invasibility by alien plants from the actual level of invasion. Ecology 89:1541–1553

Chytrý M, Wild J, Pyšek P et al (2012) Projecting trends in plant invasions in Europe under different scenarios of future land-use change. Glob Ecol Biogeogr 21:75–87

Diez JM, Ibanez I, Miller-Rushing AJ et al (2012) Forecasting phenology: from species variability to community patterns. Ecol Lett 15:545–553

Farnsworth EJ, Meyerson LA (2003) Comparative ecophysiology of four wetland plant species along a continuum of invasiveness. Wetlands 23:750–762

Fensham RJ (1996) Land clearance and conservation of inland dry rainforest in north Queensland, Australia. Biol Conserv 75:289–298

Foxcroft LC (2001) A case study of human dimensions in invasion and control of alien plants in the personnel villages of Kruger National Park. In: McNeely JA (ed) The great reshuffling: human dimensions of invasive alien species. IUCN, Gland/Cambridge, pp 127–134

Foxcroft LC, Downey PO (2008) Protecting biodiversity by managing alien plants in national parks: perspectives from South Africa and Australia. In: Tokarska-Guzik B, Brock JH, Brundu G et al (eds) Plant invasions: human perception, ecological impacts and management. Backhuys Publishers, Leiden, pp 387–403

Foxcroft LC, Jarošík V, Pyšek P et al (2011) Protected-area boundaries as filters of plant invasions. Conserv Biol 25:400–405

Gaertner M, Breeyen AD, Hui C et al (2009) Impacts of alien plant invasions on species richness in Mediterranean-type ecosystems: a meta-analysis. Prog Phys Geogr 33:319–338

Genovesi P (2011) Are we turning the tide? Eradications in times of crisis: how the global community is responding to biological invasions. In: Veitch CR, Clout MN, Towns DR (eds) Island invasives: eradication and management. IUCN, Gland

Hellmann JJ, Byers JE, Bierwagen BG et al (2008) Five potential consequences of climate change for invasive species. Conserv Biol 22:534–543

Heywood V, Brunel S (2009) Code of conduct on horticulture and invasive alien plants, Nature and environment No. 155. Council of Europe Publishing, Strasbourg

Heywood V, Sharrock S (2012) European code of conduct for botanic gardens on invasive alien plants. Standing Committee 32nd meeting Strasbourg, 27–30 Nov 2012. https://wcd.coe.int/com.instranet.InstraServlet?command=com.instranet.CmdBlobGet&InstranetImage=2169478&SecMode=1&DocId=1943644&Usage=2. Accessed 31 Dec 2012

Hoegh-Guldberg O, Hughes L, McIntyre S et al (2008) Assisted colonization and rapid climate change. Science 321:345–346

Hulme PE (2011a) Biosecurity: the changing face of invasion biology. In: Richardson DM (ed) Fifty years of invasion ecology: the legacy of Charles Elton. Blackwell Publishing, Oxford, pp 301–314

Hulme PE (2011b) Addressing the threat to biodiversity from botanic gardens. Trends Ecol Evol 26:168–174

Hulme PE, Pyšek P, Nentwig W et al (2009) Will threat of biological invasions unite the European Union? Science 324:40–41

IPCC (2007) Fourth assessment report: climate change 2007. http://www.ipcc.ch/publications_and_data/ar4. Accessed 13 Nov 2012

IUCN (2012) IUCN Guidelines for reintroductions and other conservation translocations. IUCN Species Survival Commission (SSC). http://www.issg.org/pdf/publications/Translocation-Guidelines-2012.pdf. Accessed 31 Dec 2012

Jarošík V, Konvička M, Pyšek P et al (2011a) Conservation in a city: do the same principles apply to different taxa? Biol Conserv 144:490–499

Jarošík V, Pyšek P, Foxcroft LC et al (2011b) Predicting incursion of plant invaders into Kruger National Park, South Africa: the interplay of general drivers and species-specific factors. PLoS ONE 6:e28711

Lambdon PW, Pyšek P, Basnou C et al (2008) Alien flora of Europe: species diversity, temporal trends, geographical patterns and research needs. Preslia 80:101–149

Lee JE, Chown SL (2009a) Quantifying the propagule load associated with the construction of an Antarctic research station. Antarct Sci 5:471–475

Lee JE, Chown SL (2009b) Breaching the dispersal barrier to invasion: quantification and management. Ecol Appl 19:1944–1959

Liao C, Peng R, Luo Y et al (2008) Altered ecosystem carbon and nitrogen cycles by plant invasion: a meta-analysis. New Phytol 177:706–714

Lonsdale WM (1999) Global patterns of plant invasions and the concept of invasibility. Ecology 80:1522–1536

Lovei GL, Lewinsohn TM (2012) Megadiverse developing countries face huge risks from invasives. Trends Ecol Evol 27:2–3

Macdonald IAW, Graber DM, DeBenedetti S et al (1988) Introduced species in nature reserves in Mediterranean type climatic regions of the world. Biol Conserv 44:37–66

Mack RN (2000) Cultivation fosters plant naturalization by reducing environmental stochasticity. Biol Invest 2:111–122

McKinney ML (2002) Influence of settlement time, human population, park shape and age, visitation and roads on the number of alien plant species in protected areas in the USA. Divers Distrib 8:311–318

Meyerson LA, Mooney HA (2007) Invasive alien species in an era of globalization. Front Ecol Environ 5:199–208

Meyerson LA, Reaser JK (2002) Biosecurity: moving toward a comprehensive approach. BioScience 52:593–600

Nuñez MA, Pauchard A (2010) Biological invasions in developed and developing countries: does one model fit all? Biol Inviron 12:707–714

Palmer MA, Reidy Liermann CA, Nilsson C et al (2008) Climate change and the world's river basins: anticipating management options. Front Ecol Environ 6:81–89

Parmesan C (2006) Ecological and evolutionary responses to recent climate change. Ann Rev Ecol Evol Syst 37:637–669

Parmesan C, Yohe G (2003) A globally coherent fingerprint of climate change impacts across natural systems. Nature 421:37–42

Pluess T, Cannon R, Jarošík V et al (2012a) When are eradication campaigns successful? A test of common assumptions. Biol Inv 14:1365–1378

Pluess T, Jarošík V, Pyšek P et al (2012b) Which factors affect the success or failure of eradication campaigns against alien species? PLoS One 7:e48157

Potts SG, Biesmeijer JC, Kremen C et al (2010) Global pollinator declines: trends, impacts, drivers. Trends Ecol Evol 25:345–353

Pyšek P, Richardson DM (2010) Invasive species, environmental change and management, and health. Ann Rev Env Res 35:25–55

Pyšek P, Jarošík V, Kučera T (2002) Patterns of invasion in temperate nature reserves. Biol Conserv 104:13–24

Pyšek P, Jarošík V, Kučera T (2003) Inclusion of native and alien species in temperate nature reserves: an historical study from Central Europe. Conserv Biol 17:1414–1424

Pyšek P, Bacher S, Chytrý M et al (2010) Contrasting patterns in the invasions of European terrestrial and freshwater habitats by alien plants, insects and vertebrates. Glob Ecol Biogeogr 19:317–331

Pyšek P, Jarošík V, Pergl J (2011) Alien plants introduced by different pathways differ in invasion success: unintentional introductions as greater threat to natural areas? PLoS ONE 6:e24890

Pyšek P, Jarošík V, Hulme PE et al (2012) A global assessment of invasive plant impacts on resident species, communities and ecosystems: the interaction of impact measures, invading species' traits and environment. Global Change Biol 18:1725–1737

Rejmánek M, Pitcairn MJ (2002) When is eradication of exotic pest plants a realistic goal? In: Veitch CR, Clout MN (eds) Turning the tide: the eradication of invasive species. IUCN, Gland, pp 249–253

Ricciardi A, Simberloff D (2009) Assisted colonization is not a viable conservation strategy. Trends Ecol Evol 24:248–253

Richardson DM, Holmes PM, Esler KJ et al (2007) Riparian vegetation: degradation, alien plant invasions, and restoration prospects. Divers Distrib 13:126–139

Richardson DM, Hellmann JJ, McLachlan J et al (2009) Multidimensional evaluation of managed relocation. Proc Natl Acad Sci U S A 106:9721–9724

Schwartz MW, Hellmann JJ, McLachlan JM et al (2012) Managed relocation: integrating the scientific, regulatory, and ethical challenges. Bioscience 8:732–743

Simberloff D (2014) Chapter 25: Eradication: pipe dream or real option? In: Foxcroft LC, Pyšek P, Richardson DM, Genovesi P (eds) Plant invasions in protected areas: patterns, problems and challenges. Springer, Dordrecht, pp 549–559

Simberloff D, Genovesi P, Pyšek P et al (2011) Recognizing conservation success. Science 332:419

Simberloff D, Martin JL, Genovesi P et al (2013) Impacts of biological invasions: what's what and the way forward. Trends Ecol Evol 28:56–66

Strayer DL (2012) Eight questions about invasions and ecosystem functioning. Ecol Lett 15:1199–1210

Thuiller W (2007) Climate change and the ecologist. Nature 448:550–552

Usher MB (1988) Biological invasions of nature reserves: a search for generalisation. Biol Conserv 44:119–135

van Wilgen BW, Khan A, Marais C (2011) Changing perspectives on managing biological invasions: insights from South Africa and the working for water programme. In: Richardson DM (ed) Fifty years of invasion ecology: the legacy of Charles Elton. Blackwell Publishing, Oxford, pp 377–393

van Wilgen BW, Forsyth GG, Le Maitre DC et al (2012) An assessment of the effectiveness of a large, national-scale invasive alien plant control strategy in South Africa. Biol Conserv 148:28–38

Vilà M, Espinar JL, Hejda M et al (2011) Ecological impacts of invasive alien plants: a meta-analysis of their effects on species, communities and ecosystems. Ecol Lett 14:702–708

Walther G-R, Post E, Convey P et al (2002) Ecological responses to recent climate change. Nature 416:389–395

Chapter 22
Guidelines for Addressing Invasive Species in Protected Areas

Piero Genovesi and Andrea Monaco

Abstract Biological invasions pose a severe and increasing threat to the world's protected areas, and protected areas are being called upon to improve their management efficacy. Based on a review of best practice cases presented in this book and other sources, a set of guidelines to deal with invasive alien species is proposed, discussing the challenges and opportunities that protected areas face. These guidelines have a broader scope than the management of invasive alien plants within protected areas only, and include recommendations on all aspects related to invasive species, from raising awareness within the public and decision makers, to developing staff capacity, encouraging responsible behaviour, implementing prevention actions, improving the ability to react promptly to new incursions, developing surveillance and monitoring frameworks, integrating invasive species into management plans, and also for encouraging action at a broader scale than that of the protected area. These guidelines can help protected areas play a key role in preventing and mitigating the global effects of biological invasions, also catalysing more stringent action and policies at all scales.

Keywords Invasive alien species • Awareness raising • Prevention • Management • Information exchange

P. Genovesi (✉)
ISPRA, Institute for Environmental Protection and Research,
Via V. Brancati 48, I-00144 Rome, Italy

Chair IUCN SSC Invasive Species Specialist Group, Rome, Italy
e-mail: piero.genovesi@isprambiente.it

A. Monaco
ARP, Regional Parks Agency – Lazio Region, Via del Pescaccio 96,
I-00166 Rome, Italy
e-mail: amonaco@regione.lazio.it

L.C. Foxcroft et al. (eds.), *Plant Invasions in Protected Areas: Patterns, Problems and Challenges,* Invading Nature - Springer Series in Invasion Ecology 7, DOI 10.1007/978-94-007-7750-7_22, © Springer Science+Business Media Dordrecht 2013

22.1 Introduction

Invasive alien species (IAS) are of one the most important direct drivers of biodiversity loss and ecosystem service change (Brunel et al. 2013), globally increasing at an unprecedented pace (Butchart et al. 2010). Furthermore, the challenges related to this threat are expected to grow, because of the strong links between invasions and other factors of change such as global warming, growing human populations, and habitat loss (Simberloff et al. 2013; Spear et al. 2013). In particular, the potential synergic effects of invasions and climate change appear alarming (Willis et al. 2010), because global warming can exacerbate the rate of invasions (Dudley et al. 2010). Additionally, efforts to reduce climate change impacts, if not carefully planned, may introduce further IAS (Ricciardi and Simberloff 2009; IUCN 2013).

The impact of biological invasions can even be worse in protected areas (PAs) than elsewhere, because these areas preserve key elements of global biological diversity, ensuring the maintenance of essential services for the livelihood of many communities (Foxcroft et al. 2014). Addressing this issue requires reconsidering general PA policies, as well as overall priorities, posing complex challenges to PAs, for example, to find ways to ensure understanding and support by PA visitors and even staff. It is therefore urgent that PAs improve their strategies to address this, as well as other key threats, such as habitat loss and climate change.

The guidelines proposed here aim to guide PA managers on how to approach IAS, in order to prevent and mitigate the impacts of biological invasions within, as well as beyond, the borders of PAs. The guidelines are largely based on examples of invasive alien plants (IAP), which are the main focus of this volume, but the proposed principles apply to all other taxonomic groups. The guidelines are provided to enhance the pivotal role that PAs can play, making optimal use of the specific knowledge, skills and sensitivity of PAs in terms of, for example raising awareness, surveillance and monitoring, which are all key elements for a more effective response to invasions.

These guidelines are aimed mainly at PA managers and staff, practitioners and local communities, but also authorities, NGOs and funders. All these stakeholders are central to more effective management of invasive species, and at the same time to promote a more effective role of PAs on this issue at all levels.

22.2 Challenges and Opportunities for Protected Areas in Addressing Invasive Alien Species

The urgent need to address the threats being posed to PAs by biological invasions more effectively has been highlighted by several authors, who have tried to identify the obstacles that limit implementation (e.g. Laurance et al. 2012; Tu and Robison 2014). Based on a survey of PA managers, De Poorter (2007) highlighted the main impediments to more effective IAP management as (i) the lack of capacity for mainstreaming IAS management into overall PA management, (ii) the limited capacity of staff at site level, (iii) the low level of awareness, (iv) the gaps in information on IAS available to PA managers, (v) the lack of funding, (vi) legal or institutional impediments, (vii) and the clashes of interests between stakeholders. A recent survey in European PAs (in 2012; Pysek et al. 2014) highlighted that IAS are perceived as the second most serious threat to PAs after habitat loss and fragmentation, and even more important than tourism. The survey also highlighted the main impediments to action in Europe, largely confirming the findings by De Poorter (2007), with (i) limited resources indicated as the main problem, followed by (ii) the lack of capacity, (iii) lack of awareness, (iv) gaps in information, (v) little support by the public or stakeholders and (vi) the institutional and legal impediments to action. Apart from these constraints, the complexity of the issue and the need to implement measures that are specifically targeted at IAS, pose additional challenges to park mangers. For example, the interactions between IAS, which can show synergic patterns and cause surprising cascade effects, require the responses to be very careful planned (Shaw 2014). Further, the measures usually adopted by PAs, such as enforcing a protection regime not necessarily coupled by active management, are clearly not enough to reduce the impact of IAS. For example, many islands are protected, but still highly impacted by invasions (Bergstrom and Chown 1999; Frenot et al. 2005; Kueffer et al. 2010; Baret et al. 2014). This is because of the inherent vulnerability of islands, as well as of all isolated ecosystems, to the impacts of IAS (Brundu 2014; Loope et al. 2014; Shaw 2014), and the need to implement measures specifically tailored to these situations. Also, the unintended effects of the establishment of PAs may facilitate the introduction of IAS, for example in the Mediterranean islands, which are characterised by high tourism pressure (Brundu 2014).

The urgent need to specifically address this threat in PAs is also linked to the 'environmentalist's paradox' (Raudsepp-Hearne et al. 2010). Despite constant improvement of human wellbeing in many areas of the world, and the expansion of PAs, the state of the environment often continues to worsen, and invasions are becoming epidemic in scale (Cox and Underwood 2011; McNeely 2014; Mora and Sale 2011), challenging the global community to improve the efficacy of conservation measures.

The ability to maintain the ecological integrity of PAs depends extensively on the efficacy of management outside their borders, and therefore PAs are also called to catalyse a more effective approach to IAS management beyond their borders (Laurance et al. 2012; Spear et al. 2013). In this regard, PAs can play a key role in catalysing the participation of interest groups and communities, promoting more active support by society and of the measures needed to address invasions. Raising awareness on invasions at all levels is indeed one of the most important roles that PAs can play. Protected areas generally have high credibility in society, and could therefore be particularly effective in communicating and educating visitors, local communities and the general public on invasions, an issue that is particularly difficult to approach (Boshoff et al. 2008).

The broad strategic approach needed to address IAS is indeed well known. Article 8(h) of the Convention on Biological Diversity calls parties "as far as possible and as appropriate, (to) prevent the introduction of, control or eradicate those alien species which threaten ecosystems, habitats or species". Further details were provided in 2002 at the Conference of the Parties of the Convention on Biological Diversity, with decision VI/23 providing guiding principles for invasive alien species management, based on a 'hierarchical approach'. This approach calls for prevention as the first line of defence, early detection and rapid response when prevention fails, eradication as the best option to manage established species, and permanent management when the other options are not applicable (Wittenberg and Cock 2001).

All these measures need to be applied at the appropriate scale, from species-specific approaches to ecosystem management responses, and considering action at multiple scales, from local to regional and even global (Foxcroft et al. 2009; Seipel et al. 2012). Building on this concept, PAs should ideally address the problem of invasions at the earliest possible stage of their planning, possibly starting from the design of any new protected area itself (Meyerson and Pyšek 2014). The landscape configuration of the geographic context in which a PA is established, and the natural corridors connecting the PA with surrounding areas, affects the permeability of the PA and is important in determining the future patterns of invasions (Foxcroft et al. 2011; Meiners and Pickett 2014).

Measures addressing IAS not only are important to reduce their impacts on biodiversity, but can also be beneficial for other aspects, for example by reducing patterns of erosion or the risk of fires (Foxcroft et al. 2014), as well as for human safety. Several IAS have biological characteristics that pose a danger to the safety of park employees and visitors, as in the case of the lionfish (*Pterois volitans* and *P. miles*). Lion fish, which have poisonous spines that can be hazardous to people snorkelling and scuba diving, have invaded many of the south-eastern ocean and coastal parks of the USA (McCreedy et al. 2012; Whitfield et al. 2002). Often the danger presented by invasive species is unexpected by park employees and visitors, and improving awareness of these dangers can help reduce further harm.

Invasions are also relevant to the perception of PAs by the public. The appeal of PAs is linked to the natural scenery and biodiversity of these areas, and the reduction of native species or the extensive habitat alteration that IAS can cause, can affect the appreciation of PAs visitors. Also, the implementation of management actions in several cases have raised concerns and criticisms by the PA's visitors that need to be addressed carefully (van Wilgen 2012). For example pine trees (*Pinus* spp.) in the Cape peninsula (South Africa) have been grown for plantation forestry since the seventeenth century, are particularly damaging to the endemic *fynbos* biome (van Wilgen and Richardson 2012), but at the same time are regarded by people as attractive and ecologically beneficial (van Wilgen and Richardson 2012).

22.3 Guidelines: Eight Components to Improving Invasive Species Management in Protected Areas

22.3.1 Raise Awareness on Biological Invasions at All Levels

The limited awareness and concern of the public is a major constraint to the efforts to prevent and mitigate the impacts of IAS (Pyšek et al. 2014). A key role of PAs is to be a focal point for the diffusion of information and knowledge on biological invasions at all levels, from the PA staff and managers to the visitors, to local communities and the general public. Protected areas can in fact play a pivotal role in this regard, because of the credibility that these institutions generally have. More specifically, PAs have direct contact with visitors, which should be used to inform them about the threat posed by IAS, while at the same time communicating the value of native biodiversity for the preservation of nature, but also of the ecosystem services we all rely upon.

The awareness of IAS can also be raised through the involvement of the public in the different activities related to the monitoring and management of IAS. There are very valuable examples of the involvement of scuba divers in the detection of seaweeds, as in the case of the seaweed *Caulerpa webbiana* in the marine PAs of Azores (Amat et al. 2008), for which a specific webpage has been created to report observations of this invasive seaweed (http://www.horta.uac.pt/caulerpa/httpdocs/english.html). In the Adirondack Park (New York State, USA) The Nature Conservancy involved volunteers in a monitoring campaign that delineated the distribution of 13 IAP along major roadways, allowing for prioritization of actions (Brown et al. 2001). These two examples highlight the potential of local communities' involvement for monitoring and detecting IAP. Also, and perhaps more importantly, to mainstream conservation and the need to combat IAS, thereby profoundly influencing the perception of the public to impacts of IAP, and the severe effects of biological invasions more generally. There are also several examples show the efficacy of communities and volunteers for eradication and management of IAP. The "balsam blitzes" is an initiative aimed at controlling the *Impatiens glandulifera* (Himalayan balsam), in the Pembrokeshire Coast National Park (Wales, UK), involving volunteers mostly from local NGOs. The on-going eradication of *Lysichiton americanus* (American skunk cabbage), in the Taunus Nature Park (Germany), is carried on with the involvement of over 100 volunteers (Pyšek et al. 2014).

The involvement of local communities not only can support management, but also raise general awareness on the issue, or even have significant social and economic benefits. An example is the Tutuila Island in the National Park of American Samoa (Loope et al. 2014), where park managers have worked with chiefs of the villages surrounding the park to control the invasive tree *Falcataria moluccana* (batai wood). They have also circulated information through local media outlets. The campaign managed to raise funds that were used to employ

large numbers of young people from the villages, and galvanised strong support
by the local communities in the recovery of the native ecosystems. The best
known programme involving local communities for controlling IAP, at a much
larger scale and not restricted to PAs, is without any doubt the 'Working for
Water' programme, launched in South Africa in 1995. The programme has a dual
aim of controlling IAP and fighting poverty, and since its inception has cleared
more than one million hectares, providing jobs and training to approximately
around 20,000 people, in majority women (http://www.dwaf.gov.za/wfw/;
van Wilgen et al. 2012).

Several successful campaigns aimed at raising awareness on the issue of
invasive species in non-protected land can provide examples for PAs. The
'Weedbuster' (http://www.daff.qld.gov.au/4790_7012.htm) is an awareness and
education programme launched in Australia in 1994 (thereafter also in New
Zealand and South Africa) aimed at protecting the environment from weeds, by
active initiatives such as the 'weedbuster weeks' or the 'weedbusters dirty week-
ends'. Gardeners are asked to identify any weedy ornamental species that might
be growing on their properties and replace them with non-weedy alternatives from
local garden centres. An example of the many human dimensions related to
invasive species, and of the possible ways to address them, is the "Operation
No Release" in Singapore (http://www.nparks.gov.sg/cms/docs/operation_no_
release.pdf), aimed at discouraging the release of living animals done in the
Vesak day (holy day), a Buddhist celebration where thousands of birds, insects
and animals are released in a 'symbolic act to liberation' (Shiu and Stokes 2008).
This successful programme is based on an active role of National Parks rangers,
stationing at popular release sites and discouraging the public from releasing
animals.

Raising the awareness of the public requires effective communication strategies
and sensitive arguments, such as the already mentioned example of the direct
danger to the safety of people posed by the lionfish. In some cases even the PA
employees can be unaware of the threat of biological invasions, and require specific
communication efforts. For example, the staff of the Kruger National Park (South
Africa), in particular the longer-standing personnel, strongly opposed the parks
efforts to clear well-known invasive ornamental plants that had been in their
gardens for a long time, and supported the programme only after specific education
and communication efforts by the PA authorities (Foxcroft 2001).

22.3.2 Integrate Invasive Species and Protected Area Management

Addressing biological invasions raises serious technical challenges, often calling
for complex solutions. Different to other drivers of biodiversity loss, combating
IAS requires coordinated measures ranging from prevention to control, and

especially in the case of long, well established species, the interactions among species (native and other alien species) need to be taken into account. Furthermore, biological invasions interact in complex and non-additive ways with other drivers, such as climate change. This can alter the pathways of arrival of IAS, influence the probability of establishment, and modify the competitive and predatory effects on native species, also affecting the prevention and control strategies (Rahel and Olden 2008).

As also highlighted by Tu and Robison (2014), PAs are required to develop and enforce well planned, coordinated and effective strategies to address IAS, integrating all the elements from awareness raising and communication efforts, regulatory measures, prevention aspects, as well as eradication and management programmes into a single programme. There are indeed examples of coordinated and effective approaches to IAS in PAs. In North America, the National Park Service manages IAS on park lands at different scales, through an integrated approach of cooperation and collaboration, inventory and monitoring, prevention, early detection and rapid response, treatment and control, and restoration (http://www.nature.nps.gov/biol ogy/invasivespecies/). Most US parks have incorporated IAS management into long-term planning and routine PA management. For example, Curecanti and Glen Canyon National Recreation Areas have implemented 'boat checks' to help visitors make sure their boats are free of zebra and quagga mussels (*Dreissena polymorpha* and *D. bugensis*) prior to entering the park (http://www.nps.gov/cure/ planyourvisit/mussel_free_certification.htm).

Unfortunately, in many cases the approaches adopted by PAs tend to be limited in focus. There is a tendency to concentrate efforts on the reaction to invasions, often neglecting more proactive approaches. For example, South African National Parks, which are acknowledged as being among the best managed in Africa, often focus more on the control of widespread IAP and of some mammals, but have to a large extent not focused sufficient attention to possible prevention, early warning and rapid response programmes (e.g. see Foxcroft and Freitag-Ronaldson 2007).

The dynamic basis of biological invasions also calls for an adaptive management approach, although there are many obstacles to adopting this method for IAS, including the lack of frameworks for decision making and feedback mechanisms, and the inadequacy of the governance structures (Foxcroft and McGeoch 2011). However, there are interesting examples where adaptive management approaches have been successfully applied, for example in the Kruger National Park (Foxcroft and McGeoch 2011). In particular, it would be important that PAs base their activities on IAS on a priority setting exercise, in order to sustainably manage the available resources, and to direct them in the most effective way for minimizing the impacts of IAS (Randall 2011). There are examples of tools to support objective priority setting, such as the Alien Plants Ranking System developed in the USA (APRS; http://www.npwrc.usgs.gov/ resource/literatr/aprs/index.htm). The computer-based programme helps decision making by taking actual and potential impacts, as well as feasibility of control, into account.

22.3.3   Implementing Site-Based Prevention Actions as a Priority

Prevention includes screening and addressing pathways and vectors, intercepting movements at borders, and taking action based on risk assessment. These activities have been identified as a global priority by the Aichi Target 9 and adopted by the Convention on Biological Diversity, calling for the identification of key pathways of invasions and implementing measures to address them. Meeting this target requires action at multiple spatial scales, from global, to regional, and down to an individual PA or site-specific efforts, and linking the processes and responses operating at the different scales (Kueffer et al. 2014). Protected areas could do more than routine management of IAP, by encouraging responsible behaviour by private individuals and industries by, for example, promoting the adoption of agreed standards, best-practice guidelines or codes of conduct. An example, implemented at the level of an individual PA, is the code of conduct implemented by the Kruger National Park, which includes a list of IAP not to be planted, and to be immediately removed if observed (Foxcroft et al. 2008). Another example is the environmental code of conduct for terrestrial scientific field research in Antarctica (SCAR 2009). This includes provisions for all visitors, especially scientists, to the Antarctic and Sub-Antarctic by, for example, cleaning or sterilising equipment to remove propagules. At even larger scales, PAs could support and encourage the implementation of the codes on IAS and horticulture, and IAS and botanical gardens (Heywood and Brunel 2009; Heywood 2012). Other actions at the scale of a PA could include the on-going assessment of site-specific activities and vectors responsible of IAS introductions, and developing measures to reduce the risk of further invasions. In this regard PAs should identify potential new invaders, forecasting which IAS are expected to enter their borders, in order to intercept them when feasible. This approach has proved successful at larger scale contexts (see Simberloff et al. 2013 for examples), and should therefore also be adopted at site specific scales.

Although in general prevention is acknowledged by far the most cost effective way to address invasions, PAs often tend to focus more on management of IAS, than on the sources of invasions, or on addressing new invasions in their early stages. Prevention in PAs should also include the eradication or control of newly arrived IAS, before they become widespread (discussed more in detail in Sect. 22.3.5, and Table 22.1, Guideline v). As discussed by Meyerson and Pyšek (2014), reducing the rate of introductions into PAs is in fact a particularly important strategy, because it would significantly reduce the probability of establishment of IAP. There are however many examples of effective prevention efforts in PAs of all regions of the world (Meyerson and Pyšek 2014). Some parks, such as the Galapagos National Park and Marine Reserve, regulate the number of visitors and the periods of access. Many PAs in the USA and New Zealand impose cleaning of shoes, clothes, vehicles or equipment before entering, in some cases providing cleaning stations. The 'Check, clean, dry: didymo controls' programme of the

Table 22.1 A summary of guidelines for addressing invasive alien species in protected areas

Guideline	Rationale
(i) Raise Awareness on Biological Invasions at all Levels	Limited awareness and concern of the public is a major constraint to prevention and mitigation of impact of IAS, and PAs should thus give priority to informing on this issue. In some cases also PA employees are not fully aware of the issue
(ii) Integrate Invasive Species and Protected Area Management	Addressing IAS requires strategic approaches, based on coordinated prevention as well as management measures. Dynamic nature of invasions calls for more proactive rather than reactive approaches to the issue, and to adaptive management
(iii) Implement Site-Based Prevention Actions as a Priority	Prevention should be the first line of defence from invasions. Protected areas can do much in this respect, encouraging responsible behaviours by privates as well as enterprises, identifying most relevant vectors and pathways of invasion, or IAS expected to arrive to their territories, and developing focused measures to reduce risks. Prevention should also be linked to early warning and rapid response
(iv) Develop Staff Capacities for all Aspects of Invasive Species Management	Capacity and awareness of PA officials and staff are crucial for applying most of the guidelines. Trained staff are key to effective management, and can contribute to communicate to the visitors as well as to the general public
(v) Set up Rapid Detection and Prompt Response Framework	Early warning and rapid response is a key element of any strategic approach to invasions, as it is much more effective and cost effective than controlling invaders once they have established. It requires a coordinated framework for surveillance and monitoring activities, identification of invading species, assessment of risks, sharing of information, development of alarm lists and selection and enforcement of appropriate responses. Support by the public, and contingency action and funding are also very important
(vi) Manage Invasive Species Beyond the Protected Area Boundaries	The invasion of PAs often originates from the surrounding areas and this calls for a landscape perspective to planning. Establishment of buffer zones should be explored. To enhance prevention, PAs should cooperate with surroundings landowners and institutions, and lobby with competent authorities for implementing regulatory or voluntary measures to address activities such as forestry, horticulture, hunting, or botanical gardens
(vii) Implement Surveillance, Monitoring and Information Exchange Networks	Effective prevention and response to invasions – but also awareness – largely depend on knowledge basis. Information on the spread of invasive species, biological traits of the species, impacts, and available management alternatives are essential. Early warning and rapid response

(continued)

Table 22.1 (continued)

Guideline	Rationale
	require effective surveillance and access to information to identify new invaders and screen the associated risks. PAs should thus give priority to collection, sharing and access to information, also exploring the involvement of visitors and volunteers in data collection
(viii) Lobby with Institutions and Decision-Makers to Support Stringent Policy	Addressing biological invasions requires action by PAs at all levels, from the local to the global levels, including cooperation with institutions and all competent authorities for adopting regulatory or voluntary measures to address key pathways, and for identifying priorities. Protected areas should support the adoption of more stringent policies at the national as well as global scale, and influence donors and funding agencies policies. Protected areas can also document impacts, circulate best practices, and catalyse coordination amongst relevant institutions and stakeholders

Fiordland National Park (New Zealand) is aimed at preventing the establishment of the invasive freshwater algae *Didymosphenia geminata* (didymo) in the park, by encouraging visitors to check, clean and dry all gear before leaving the lake edge and moving into lake tributaries or other waterways (http://www.biosecurity.govt. nz/biosec/camp-acts/check-clean-dry). The Southern Ocean Islands have implemented stringent quarantine protocols for all visitors and cargo, including tourist vessels; these efforts have also much increased the awareness on biological invasions in recent years (Shaw 2014). Prevention efforts could be based on voluntary approaches such as the codes of conduct mentioned above, but should consider regulatory approaches, for example addressing the activities carried on within the PA's borders, or in the surrounding areas, that could cause a risk of introductions (forestry, livestock breeding, horticulture, etc.).

Despite the positive examples reported above, it is evident that much more could be done in terms of information and education of visitors, as the behaviour of people is essential to increase biosecurity of PAs. In French Polynesia intense education efforts as well as regulatory approaches, have successfully prevented *Miconia calvescens* (miconia) spreading from the hub of Tahiti to other islands, preventing further invasion since 1997 (Meyer et al. 2010; Loope et al. 2014). Indeed a major limit for adopting a more comprehensive and effective strategy to address this threat is the scarcity of resources. However, this constraint highlights the importance of addressing the causes of invasions instead of the symptoms, calling for better planning, and for prioritising actions such as prevention, instead of concentrating the staff and funding in managing to the most visible IAS, often with limited effects in terms of impact mitigation.

22.3.4 Develop Staff Capacity for All Aspects of Invasive Species Management

The capacity and awareness of PA officials are crucial for applying most of the guidelines presented. For example, PA managers have a key role in preventing further invasions, and streamlining employees' knowledge, experiences and skills, and would indeed significantly improve the ability of the PA to manage IAS (Tu and Robison 2014). In general the capacity of PA staff has been highlighted as essential for fulfilling the need for visitor's education on biological invasions and the value of biodiversity in PAs (Boshoff et al. 2008). One example of a programme aimed at improving skills and share experiences and ideas, is the Pacific Invasives Learning Network (http://www.sprep.org/Pacific-Invasives-Learning-Network-PILN/piln-welcome), launched in an area of the world with particular problems of isolation and access to knowledge (Micronesia, Polynesia, Melanesia and Hawaii). The programme builds on multi-agency teams, and is aimed at empowering effective invasive species management through a participant-driven network rapidly sharing skills and resources, and providing links to technical expertise and information. The capacity of the staff, both in terms of technical skills and of general awareness on the problems, is particularly important for enabling the rapid detection of new incursions, and the prompt reaction to these (see also Sect. 22.3.5, Table 22.1, Guideline v). One example in this regard are the SANParks 'honorary rangers', who volunteer to assist in a variety of activities in the organisation, as well as in the management of IAS. Improved public opinion is crucial to support PAs to be able to address the real causes of invasions, for example, by supporting the development of policies based on prevention, instead of only focusing on the 'symptoms' that affect their territories, such as widely established IAS.

Park rangers are often the front interface with the public, and informed staff can thus significantly help raising the awareness of the park visitors, and to ensure the public support to the control activities carried out by the PA. Once again an interesting example comes from the US National Park Service, where the park officials are trained to communicate the implications of the lionfish invasion, thereby improving the understanding of the need for lionfish removal (McCreedy et al. 2012).

22.3.5 Set Up Rapid Detection and Prompt Response Framework

Early warning and rapid response to new invasions is a key pillar of an effective strategy and PAs can indeed play a particularly important role in this respect, acting as 'miners' canaries' of incursions (Loope 2004). In this regard PAs need to improve their ability to rapidly enforce effective management of newly arrived

IAP, at the earliest possible stage after their introduction into the PA's territory. Prompt detection and rapid response can still be successful in eradication efforts that are likely to be challenging, such as for marine species. For example, in the case of the highly invasive Pacific alga, *Caulerpa taxifolia*, an incursion in California was quickly detected and successfully eradicated within 6 months of discovery, while procrastination in the Mediterranean allowed the species to invade thousands of hectares off the coasts of Spain, France, Monaco, Italy, Croatia, and Tunisia, making it ineradicable with current technologies (Simberloff et al. 2013). Prompt reaction not only is much more effective, but also more economically viable. A review of successful or attempted plant eradication programmes carried out in New Zealand revealed that early removal of plants costs on average 40 times less than removal carried out after an invasive plant has widely established (Harris and Timmins 2009).

To enable more effective early detection and rapid response requires a coordinated framework for surveillance and monitoring activities, species identification, risk assessment, information sharing, and selection and enforcement of appropriate responses (Genovesi et al. 2010). Developing alarm lists of possible new invaders can also enable more rapid reaction. An effective large scale approach to early detection and rapid response is the California Weed Action Plan (Schoenig 2005), which although generally enforced at a large scale, can provide valuable suggestions for PAs. The action plan, which is supported by a budget of about US$2.5 million/year, is based on an official list of noxious weeds for which prompt action is mandatory. A network of biologists, and trained farmers and volunteers enable early detection of new incursions, and grants are provided to implement weed control activities. The action plan has allowed the successful removal of over 2,000 infestations and the complete eradication of 17 weeds. The Californian example highlights the importance of coordinated and comprehensive frameworks for enabling prompt reaction to invasions. A questionnaire circulated to experts, decision makers, and practitioners in Europe identified the gaps for establishing an early warning and rapid response framework for IAS. These included (i) the limited funds available, (ii) the lack of early detection mechanisms, (iii) the absence of legal tools to regulate IAS introductions, (iv) the need for competent authorities to be able to carry out the appropriate responses, (v) the lack of legal tools to regulate possession of IAS, (vi) the limited ability to detect new invasions, (vii) the unclear assignment of roles and responsibilities, (viii) the technical constraints to management, and (ix) the legal obstacles to implementing control or eradication programmes (Genovesi et al. 2010). Many of these constraints also affect PAs and a coordinated approach needs to address all these aspects. Additionally, PAs need to identify priorities based on a rigorous risk assessment process, make best use of their resources, including the involvement of communities and volunteers (Pyšek et al. 2014), and enforcing effective responses once a new invader is detected (Simberloff 2014).

For the enforcement of early warning and rapid response systems to invasions it is essential to have adequate support from the public, and PAs should thus give particular attention to the communication thereof. It is also important to have methods in place to monitor the effects of the system in terms of outcomes, to allow improving the overall framework (Tu and Robison 2014). To improve their ability to respond promptly to new incursions, PAs could establish contingency plans, designed for species or broader taxonomic groups, as identified on the basis of an assessment of the most probable new invaders (see also Sect. 22.3.7, Table 22.1, Guideline vii). Contingency plans should include training on management alternatives, and possibly the establishment of dedicated task forces, which could be created for an individual PA or at a larger scale (see also Sect. 22.3.4, Table 22.1, Guideline iv). For example, the US National Park Service has developed an invasive plant management programme, creating 16 Exotic Plant Management Teams, which provide highly trained mobile assistance to parks throughout the National Park System (http://www.nature.nps.gov/biology/invasivespecies/EPMT_teams.cfm). Protected areas should also procure and maintain the basic equipment needed for managing different taxonomic groups, thereby assisting in improving the time taken to implement rapid response actions. The identification of contingency funding sources is crucially important to enable effective response to new invasions. For example, the successful eradication of *Caulerpa taxifolia* in California was made possible by the rapid procurement of substantial resources.

22.3.6 Manage Invasive Species Beyond the Protected Area Boundaries

Land use outside PA boundaries provides propagules for colonization (Meiners and Pickett 2014), with features such as river networks facilitating the spread of IAS (Foxcroft et al. 2011; Vardien et al. 2013). This is also the case of weeds entering PAs through agricultural practices adopted outside their borders (Bazzaz 1986; Hulme et al. 2014), and areas with high human population density (Spear et al. 2014). The IAP that are present in the adjacent areas are thus a key factor affecting the composition and number of individuals colonizing a PA (Rose and Hermanutz 2004; Dawson et al. 2011). This effect is particularly evident in the case of small PAs occurring in modified landscapes, where it is therefore particularly important to adopt a landscape perspective to planning (Meiners and Pickett 2014), and also consider the establishment of buffer zones where promoting lower-impact land uses and involving local communities (Laurance et al. 2012). Cooperation with surroundings landowners and institutions is thus an important element for enhancing prevention, and this can also be done at a much larger scale than the immediate surroundings of the PA, discussing and lobbying with the competent authorities at all levels the adoption of regulatory or voluntary measures to address activities potentially at risk of causing invasions, such as forestry, horticulture, hunting, or botanical gardens (see also Sect. 22.3.3, Table 22.1, Guideline iii). Also the establishment of buffer zones of land managed not to facilitate invasions can be an effective way to reduce risks of invasions in PAs (Foxcroft et al. 2011).

22.3.7 Implement Surveillance, Monitoring and Information Exchange Networks

The efficacy of any strategy to address IAS strictly depends on the available information, and on the sharing of data, knowledge and experience. For example, inventories of invasive species in PAs, based on rigorous scientific criteria, are an essential tool for PAs to prevent and control invasions (Pyšek et al. 2009). Furthermore, the effective management of IAS requires good quality data on the spread of invasive species, as well as access to information on the biological traits of the species, its impacts, and on the available management alternatives. In addition to the elements highlighted above (see also Sect. 22.3.3, Table 22.1, Guidelines iii and v), early warning and rapid response requires effective surveillance to detect emerging incursions, and access to information to correctly identify the new invaders and to screen the associated risks to implement responses (Genovesi et al. 2010). Also, meta-analyses of the available data can permit to prioritise pathways of introduction, as well as species, for example on the basis of the impacts they cause and the vulnerability to the control actions (Hulme et al. 2008).

Protected areas should implement surveillance and monitoring schemes, enabling the standardised collection of data on the distribution and abundance of IAS (Pyšek et al. 2014). Citizen science could significantly improve efficacy of surveillance and monitoring of IAS, and PAs should explore possible ways to involve visitors and volunteers in the collection of data (Gallo and Wait 2011; see also Sect. 22.3.1, Table 22.1, Guideline i). Information not only are important for the effective management of IAS, but also – as already stressed – to raise awareness on the issue, by providing to the public examples on the causes and consequences of invasions, including in particular the impacts on biological diversity as well as on ecosystem services.

Monitoring should not be limited to IAS, but also address the efficacy of management actions, collecting information on the effects of control activities, on the costs of management, and on the public perceptions of the issue. All this information is essential to avoid waste of resources, especially in the case of permanent management, that should always be based on an assessment of the cost/benefits, and to an evaluation of the sustainability of the required actions in the medium-long term. The importance to increase the sharing of information on IAS has also been stressed by the Convention on Biological Diversity, that with Decision X/38 started an initiative aimed at increasing the interoperability of databases on IAS, that has then led to the launch of the Global Invasive Alien Species Information Partnership (GIASIP; http://www.cbd.int/doc/meetings/sbstta/ sbstta-15/information/sbstta-15-inf-14-en.pdf); all major existing global information systems such as the Global Invasive Species Information Database of the IUCN SSC Invasive Species Specialist Group (http://www.issg.org/database/welcome/) and the Invasive Species Compendium of CABI (http://www.cabi.org/ISC/) have agreed to cooperate at improving the exchange of information within the GIASIP (http://giasipartnership.myspecies.info).

The importance of data sharing for PAs is twofold. On the one hand PAs need to access tools for identifying species, prioritising action, and enabling prompt reaction. This requires access to information on the management alternatives, as well as contacts of experts at the global scale. Access to information is particularly important in developing countries, or in remote areas such as oceanic islands, where local expertise is often limited (see the Pacific Invasive Learning Network in Sect. 22.3.4). On the other hand, PAs can provide data and information that can guide action, including examples of best practice, which can enable improved management in other contexts. For this reason PAs should implement web based information platforms to report information and data, and at the same time to work with national, regional and larger scale information services to improve the global sharing of information. Databases providing information about alien species are an important tool for building management capacity at a global level. An example of an initiative aimed at bridging gaps between IAP experts and managers working on IAS in mountain areas is the Mountain Invasion Research Network (MIREN, www.miren.ethz.ch), encompassing 11 regions across the globe (Dietz et al. 2006; Kueffer et al. 2014). The implementation of data sharing platforms could also permit the involvement of the public for the monitoring and management of IAS, for example through the use of applications developed for mobile phones, tablets, etc. (e.g. "PlantTracker" http://planttracker.naturelocator.org/; "Aliens Among Us app"; http://www.royalbcmuseum.bc.ca/TravellingExhibitions/default.aspx; "iAs_sess", http://ias-ess.org). Trained volunteers can indeed support monitoring, but could be particularly helpful for detecting new incursions; the "Eye on Earth" initiative provides an interesting example in this regard (Pyšek et al. 2014). One example of an effective information system for invasive species in PAs is the Marine Invasive Species Database, compiled by the US National Park Service using reports of invasive species in National Parks from several agencies and NGOs. The list permitted to identify marine invasive species documented within each park boundary, as well as a list of potential marine invasive species that are present within the ecoregion, but not yet documented in a park. The Great Lakes Invasive Species Database, also implemented by the US National Park Service, includes data on invasive species for five Great Lakes National Park units, both covering species recorded for the parks, as well as invasive species established in the region but not yet reported for a national park, the latter aimed at enabling early detection of new incursions.

22.3.8 Lobby with Institutions and Decision-Makers to Support Stringent Policies

Addressing biological invasions requires action at all levels, from the local to the global level. Trade regulations, which are important for preventing invasions (e.g. horticulture), can only be enforced at the national, regional or even global scale. Furthermore, legal frameworks can facilitate, but also constraint the efficacy of action, as highlighted by the results of the survey carried on in Europe on the issue, which reported the inadequacy of legal systems among the key constraints for combating IAS (Genovesi et al. 2010). Therefore, as also stressed in Guideline vi (Sect. 22.3.6, Table 22.1), PAs should cooperate with institutions and all competent authorities for adopting regulatory or voluntary measures to address key pathways such as forestry, horticulture, hunting, or botanical gardens (Hulme et al. 2008). Another area where it is important to cooperate at a larger scale than that of PAs, is the identification of priorities in terms of management of IAS. In order to make best use of the available resources these priorities should in fact be identified at the national scale and across all protected areas, basing the decision on a rigorous assessment of risks.

Protected areas, through synergic actions, can indeed do much more to promote the adoption of more stringent policies at the national as well as global scale, and to convince donors and funding agencies to secure budget and funding for IAS. Protected areas can document impacts and project future effects of IAS, and can provide information on the resources spent to address this threat. They can catalyse coordination amongst relevant institutions and stakeholders, so promoting more effective actions also beyond their territories (Tu 2009). Regional or national networks of PAs (e.g. IUCN World Commission on Protected Areas, Europarc for Europe, etc.) should encourage national and global institutions, such as the Convention on Biological Diversity, to adopt and enforce more effective policies, and to address the legal constraints to management of IAS, that in some cases have been shown to limit the effective response to invasions. Furthermore, interacting with relevant national or even supranational institutions can facilitate the access to available resources, as in the case of the European Union LIFE funding instrument. The LIFE funding programme has been particularly effective in promoting management of IAS in European PAs and incorporating control efforts in ecological restoration of protected land (Scalera and Zaghi 2004; Pyšek et al. 2014).

An example of a regional attempt to develop such guidelines, are the European Guidelines on Protected Areas and IAS, which is promoted by the Council of Europe and supported by the IUCN SSC Invasive Species Specialist Group (Monaco and Genovesi 2013). The guidelines are based on European legislation and international conventions on the mitigation of impacts caused by IAS to PAs, and the need for more effective management of this threat in PAs to preserve biodiversity. The European guidelines take into account the best practices in the regional PAs, and will provide non-binding recommendations to PAs to improve their ability to respond to this threat. Especially in the case of PAs, self-regulation can in fact be much more appropriate, effective and successful than any legally binding scheme.

22.4 Conclusions

Biological invasions affect protected areas all over the world. The effects of this threat to the biodiversity of PAs are dramatic and are expected to grow in the future, especially as they increasingly interlink with other factors of change such as climate change, habitat loss and human pressure. The impact of IAS on PAs has long been underestimated, and the concerns of scientists that this threat was going to increase (Usher 1988; Macdonald et al. 1989) was similarly ignored by many national and supranational institutions.

It is therefore urgent that PAs improve their management of IAS, if they want to fully play their role as champions of the protection of the global diversity and of the ecosystem services we all rely upon for our very existence. Letting nature take its course is not a strategy that can be used for IAS (Meiners and Pickett 2014; Meyerson and Pyšek 2014) and active management of this issue is therefore fundamental. However, only evidence-based policy and management, developed through rigorous science, will allow PAs to respond appropriately to the growing environmental crisis at all scales.

Protected areas can and should play a major role in the struggle against invasions, not only by improving the efficacy of IAS management within their territories, but also monitoring the patterns of invasions, raising awareness at all levels, improving the capacity of practitioners to deal with invaders, implementing site-based prevention efforts, enforcing early detection and rapid response frameworks, and catalysing action also beyond the park boundaries.

Protected areas cannot stop invasions, but can indeed be important in preventing and mitigating the global effects of this threat by being reservoirs of the heritage of native species and ecosystems. They can also be used as sentinels of incursions to speed up response at all levels, champions for increasing information and awareness with the different sectors of the society, as well as catalysts for action at all scales.

References

Amat JN, Cardigos F, Santos RS (2008) The recent northern introduction of the seaweed *Caulerpa webbiana* (Caulerpales, Chlorophyta) in Faial, Azores Islands (North-Eastern Atlantic). Aquat Invasions 3:417–422

Baret S, Baider C, Kueffer C et al (2014) Chapter 19: Threats to paradise? Plant invasion in protected areas of Western Indian Ocean islands. In: Foxcroft LC, Pyšek P, Richardson DM, Genovesi P (eds) Plant invasions in protected areas: patterns, problems and challenges. Springer, Dordrecht, pp 423–447

Bazzaz FA (1986) Life history of colonizing plants: some demographic, genetic, and physiological features. In: Mooney HA, Drake J (eds) Ecology of biological invasions of North America and Hawaii. Springer, New York, pp 96–110

Bergstrom DM, Chown SL (1999) Life at the front: history, ecology and change on southern ocean islands. Trends Ecol Evol 14:472–477

Boshoff AF, Landman M, Kerley GIH et al (2008) Visitors' views on alien animal species in national parks: a case study from South Africa. S Afr J Sci 104:326–328

Brown WT, Krasny ME, Schoch N (2001) Volunteer monitoring of non-indigenous, invasive species. Nat Areas J 21:189–196

Brundu G (2014) Chapter 18: Invasive alien plants in protected areas in Mediterranean islands: knowledge gaps and main threats. In: Foxcroft LC, Pyšek P, Richardson DM, Genovesi P (eds) Plant invasions in protected areas: patterns, problems and challenges. Springer, Dordrecht, pp 395–422

Brunel S, Fernández-Galiano E, Genovesi P et al (2013) Invasive alien species: a growing but neglected threat? In: Late lessons from early warning: science, precaution, innovation. Lessons for preventing harm. EEA report 1/2013, Copenhagen, pp 518–540

Butchart SHM, Walpole M, Collen B et al (2010) Global biodiversity: indicators of recent declines. Science 328:1164–1168

Cox R, Underwood C (2011) The importance of conserving biodiversity outside of protected areas in Mediterranean ecosystems. PLos One 6(1):e14508

Dawson W, Burslem DFRP, Hulme PE (2011) The comparative importance of species traits and introduction characteristics in tropical plant invasions. Divers Distrib 17:1111–1121

De Poorter M (2007) Invasive alien species and protected areas: a scoping report. Part 1. Scoping the scale and nature of invasive alien species threats to protected areas, impediments to invasive alien species management and means to address those impediments. Global Invasive Species Programme, Invasive Species Specialist Group. http://www.issg.org/gisp_publications_reports.htm

Dietz H, Kueffer C, Parks CG (2006) MIREN: a new research network concerned with plant invasion into mountain areas. Mt Res Dev 26:80–81

Dudley N, Stolton S, Belokurov A et al (2010) Natural solutions: protected areas helping people cope with climate change. WWF International, Gland

Foxcroft LC (2001) A case study of human dimensions in invasion and control of alien plants in the personnel villages of Kruger National Park. In: McNeely JA (ed) The great reshuffling: human dimensions of invasive alien species. IUCN, Gland/Cambridge, pp 127–134

Foxcroft LC, Freitag-Ronaldson S (2007) Seven decades of institutional learning: managing alien plant invasions in the Kruger National Park, South Africa. Oryx 41:160–167

Foxcroft LC, McGeoch MA (2011) Implementing invasive species management in an adaptive management framework. Koedoe 53:111–121

Foxcroft LC, Richardson DM, Wilson JRU (2008) Ornamental plants as invasive aliens: problems and solutions in Kruger National Park, South Africa. Environ Manage 41:32–51

Foxcroft LC, Richardson DM, Rouget M et al (2009) Patterns of alien plant distribution at multiple spatial scales in a large national park: implications for ecology, management and monitoring. Divers Distrib 15:367–378

Foxcroft LC, Jarošík V, Pyšek P et al (2011) Protected-area boundaries as filters of plant invasions. Conserv Biol 25:400–405

Foxcroft LC, Pyšek P, Richardson DM et al (2014) Chapter 2: The bottom line: impacts of alien plant invasions in protected areas. In: Foxcroft LC, Pyšek P, Richardson DM, Genovesi P (eds) Plant invasions in protected areas: patterns, problems and challenges. Springer, Dordrecht, pp 19–41

Frenot Y, Chown SL, Whinam J et al (2005) Biological invasions in the Antarctic: extent, impacts and implications. Biol Rev 80:45–72

Gallo T, Wait D (2011) Creating a successful citizen science model to detect and report invasive species. BioScience 61:459–465

Genovesi P, Scalera R, Brunel S et al (2010) Towards an early warning and information system for invasive alien species (IAS) threatening biodiversity in Europe. EEA technical report n.5/2010. European Environment Agency, Copenhagen

Harris A, Timmins SM (2009) Estimating the benefit of early control of all newly naturalised plants, Science for conservation No. 292. New Zealand Department of Conservation, Wellington

Heywood V (2012) European code of conduct for botanic gardens on invasive alien species. Council of Europe Document T-PVS/Inf (2012)1. Council of Europe, Strasbourg

Heywood V, Brunel S (2009) Code of conduct on horticulture and invasive alien plants. Nat Environ 155:1–35

Hulme PE, Bacher S, Kenis M et al (2008) Grasping at the routes of biological invasions: a framework for integrating pathways into policy. J Appl Ecol 45:403–414

Hulme PE, Burslem DFRP, Dawson W et al (2014) Chapter 8: Aliens in the arc: are invasive trees a threat to the Montane forests of East Africa? In: Foxcroft LC, Pyšek P, Richardson DM, Genovesi P (eds) Plant invasions in protected areas: patterns, problems and challenges. Springer, Dordrecht, pp 145–165

IUCN (2013) Guidelines for reintroductions and other conservation translocations. Adopted by SSC Steering Committee, 5th September 2012. IUCN SSC Reintroduction Specialist Group and Invasive Species Specialist Group

Kueffer C, Daehler CC, Torres-Santana CW et al (2010) A global comparison of plant invasions on oceanic islands. Perspect Plant Ecol Evol Syst 12:145–161

Kueffer C, McDougall K, Alexander J et al (2014) Chapter 21: Plant invasions into mountain protected areas: assessment, prevention and control at multiple spatial scales. In: Foxcroft LC, Pyšek P, Richardson DM, Genovesi P (eds) Plant invasions in protected areas: patterns, problems and challenges. Springer, Dordrecht, pp 473–486

Laurance WF, Useche DC, Rendeiro J et al (2012) Averting biodiversity collapse in tropical forest protected areas. Nature 489:290–294

Loope LL (2004) The challenge of effectively addressing the threat of invasive species to the National Park System. Park Sci 22(2):14–20

Loope LL, Flint Hughes R, Meyer J-Y (2014) Chapter 18: Plant invasions in protected areas of tropical pacific islands, with special reference to Hawaii. In: Foxcroft LC, Pyšek P, Richardson DM, Genovesi P (eds) Plant invasions in protected areas: patterns, problems and challenges. Springer, Dordrecht, pp 395–422

Macdonald IAW, Loope LL, Usher MB et al (1989) Wildlife conservation and the invasion of nature reserves by introduced species: a global perspective. In: Drake JA, Mooney HA, di Castri F et al (eds) Biological invasions: a global perspective. Wiley, Chichester, pp 215–255

McCreedy C, Toline CA, McDonough V (2012) Lionfish response plan: a systematic approach to managing impacts from the lionfish, an invasive species, in units of the National Park System. Natural Resource Report NPS/NRSS/WRD/NRR—2012/497. National Park Service, Fort Collins

McNeely J (2014) Chapter 6: Global efforts to address the wicked problem of invasive alien species. In: Foxcroft LC, Pyšek P, Richardson DM, Genovesi P (eds) Plant invasions in protected areas: patterns, problems and challenges. Springer, Dordrecht, pp 89–113

Meiners SJ, Pickett STA (2014) Chapter 3: Plant invasion in protected landscapes: exception or expectation? In: Foxcroft LC, Pyšek P, Richardson DM, Genovesi P (eds) Plant invasions in protected areas: patterns, problems and challenges. Springer, Dordrecht, pp 43–60

Meyer JY, Fourdrigniez M, Taputuarai R (2010) The recovery of the native and endemic flora after the introduction of a fungal pathogen to control the invasive tree *Miconia calvescens* in Tahiti, French Polynesia. Biol Control Nat 3:1–21

Meyerson LA, Pyšek P (2014) Chapter 21: Manipulating alien species propagule pressure as a prevention strategy in protected areas. In: Foxcroft LC, Pyšek P, Richardson DM, Genovesi P (eds) Plant invasions in protected areas: patterns, problems and challenges. Springer, Dordrecht, pp 473–486

Monaco A, Genovesi P (2013) European guidelines on protected areas and invasive alien species. Council of Europe Document T-PVS/Inf (2013) 22. Council of Europe, Strasbourg

Mora C, Sale P (2011) Ongoing global biodiversity loss and the need to move beyond protected areas: a review of the technical and practical shortcoming of protected areas on land and sea. Mar Ecol Prog Ser 434:251–266

Pyšek P, Hulme PE, Nentwig W (2009) Glossary of the main technical terms used in the handbook. In: DAISIE (ed) Handbook of alien species in Europe. Springer, Berlin, pp 375–379

Pyšek P, Genovesi P, Pergl J et al (2014) Chapter 11: Invasion of protected areas in Europe: an old continent facing new problems. In: Foxcroft LC, Pyšek P, Richardson DM, Genovesi P (eds) Plant invasions in protected areas: patterns, problems and challenges. Springer, Dordrecht, pp 209–240

Rahel FJ, Olden JD (2008) Assessing the effects of climate change on aquatic invasive species. Conserv Biol 22:521–533

Randall J (2011) Protected areas. In: Simberloff D, Rejmánek M (eds) Encyclopedia of biological invasions. University of California Press, Berkeley/Los Angeles, pp 563–567

Raudsepp-Hearne C, Peterson GD, Tengö M et al (2010) Untangling the environmentalist's paradox: why is human well-being increasing as ecosystem services degrade? BioScience 60:576–589

Ricciardi A, Simberloff D (2009) Assisted colonization is not a viable conservation strategy. Trends Ecol Evol 24:248–253

Rose M, Hermanutz L (2004) Are boreal ecosystems susceptible to alien plant invasion? Evidence from protected areas. Oecologia 139:467–477

Scalera R, Zaghi D (2004) Life focus/alien species and nature conservation in the EU: the role of the life program. European Commission, Office for Official Publications of the European Communities, Luxembourg

SCAR (2009) SCAR's environmental code of conduct for terrestrial scientific field research in Antarctica. In: Antarctic Treaty Consultative Meeting XXXII. Committee on Environmental Protection XII. Information Paper 004, 6–17 Apr 2009, Baltimore

Schoenig S (ed) (2005) California noxious and invasive weed action plan. California Department of Food and Agriculture (CDFA), California Invasive Weed Awareness Coalition (CALIWAC), Sacramento

Seipel T, Kueffer C, Rew LJ et al (2012) Processes at multiple scales affect richness and similarity of non-native plant species in mountains around the world. Glob Ecol Biogeogr 21:236–246

Shaw J (2014) Chapter 19: Invasion of Southern Ocean Islands: implications for isolated protected areas. In: Foxcroft LC, Pyšek P, Richardson DM, Genovesi P (eds) Plant invasions in protected areas: patterns, problems and challenges. Springer, Dordrecht, pp 423–447

Shiu H, Stokes L (2008) Buddhist animal release practices: historic, environmental, public health and economic concerns. Contemp Buddhism 9:181–196

Simberloff D (2014) Chapter 26: Eradication – pipe dream or real option? In: Foxcroft LC, Pyšek P, Richardson DM, Genovesi P (eds) Plant invasions in protected areas: patterns, problems and challenges. Springer, Dordrecht, pp 561–597

Simberloff D, Martin JL, Genovesi P et al (2013) Impacts of biological invasions: what's what and the way forward. Trends Ecol Evol 28:58–66

Spear D, Foxcroft LC, Bezuidenhout H et al (2013) Human population density explains alien species richness in protected areas. Biol Conserv 159:137–147

Tu M (2009) Assessing and managing invasive species within protected areas. Protected area quick guide series. In: J. Ervin (ed) The Nature Conservancy, Arlington

Tu M, Robison MA (2014) Chapter 24: Overcoming barriers to the prevention and management of alien plant invasions in protected areas: a practical approach. In: Foxcroft LC, Pyšek P, Richardson DM, Genovesi P (eds) Plant invasions in protected areas: patterns, problems and challenges. Springer, Dordrecht, pp 529–547

Usher MB (1988) Invasions of nature reserves: a search for generalizations. Biol Conserv 44:119–135

van Wilgen BW (2012) Evidence, perceptions, and trade-offs associated with invasive alien plant control in the Table Mountain National Park, South Africa. Ecol Soc 17:23

van Wilgen BW, Richardson DM (2012) Three centuries of managing introduced conifers in South Africa: benefits, impacts, changing perceptions and conflict resolution. J Environ Manage 106:56–68

van Wilgen BW, Forsyth GG, Le Maitre DC et al (2012) An assessment of the effectiveness of a large, national-scale invasive alien plant control strategy in South Africa. Biol Conserv 148:28–38

Vardien W, Richardson DM, Foxcroft LC et al (2013) Management history determines gene flow in a prominent invader. Ecography 36:1–10

Whitfield PE, Gardner T, Vives SP et al (2002) Biological invasion of the indo-pacific lionfish *Pterois volitans* along the Atlantic Coast of North America. Mar Ecol Prog Ser 235:289–297

Willis CG, Ruhfel BR, Primack RB et al (2010) Favorable climate change response explains non-native species' success in Thoreau's woods. PLoS One 5(1):e8878

Wittenberg R, Cock MJW (eds) (2001) Invasive alien species: a toolkit of best prevention and management practices. CAB International, Wallingford

Chapter 23
Protecting Biodiversity Through Strategic Alien Plant Management: An Approach for Increasing Conservation Outcomes in Protected Areas

Paul O. Downey

Abstract Despite wide acknowledgement of the significant threat posed by invasive alien plants to biodiversity, management strategies have not yet adequately addressed the problem. Among the reasons for this are the lack of knowledge of the biodiversity at risk from invasive alien plants, an emphasis on control rather than the outcome of such control actions, ineffective monitoring programmes, a lack of resources, institutional barriers to change, and mismatches between policy and management. To resolve this situation, strategies for managing invasive alien plants need to focus on specific biodiversity conservation outcomes and put in place a range of measures to ensure that the aims are achieved. One area where such a change would have significant conservation outcomes is the management of invasive alien plants in protected areas, given the threat posed to high-value biodiversity. Such a system would also enable conservation outcomes to be reported on. The lack of outcome reporting has been highlighted as a significant problem in numerous studies, including the recent assessment of progress towards the Convention of Biological Diversity targets. Here I present an overview of one approach that has been developed to ensure that invasive alien plant management delivers desired conservation outcomes. To achieve this, each step in the planning and management process was evaluated and modified to ensure that it could deliver the desired outcome. The potential application of this approach within protected areas to improve the management of invasive alien plants and increase the protection of biodiversity is discussed. Adoption of these processes by managers of protected areas will have long lasting benefits for both invasive alien plant control and biodiversity conservation as it prioritises management to areas where control is likely to have the greatest outcomes; something that is critical given the lack of

P.O. Downey (✉)
Parks and Wildlife Group, Office of Environment and Heritage, PO Box 1967, Hurstville, NSW 1481, Australia

Institute for Applied Ecology, University of Canberra, Canberra, ACT 2601, Australia
e-mail: paul.downey@canberra.edu.au

L.C. Foxcroft et al. (eds.), *Plant Invasions in Protected Areas: Patterns, Problems and Challenges*, Invading Nature - Springer Series in Invasion Ecology 7, DOI 10.1007/978-94-007-7750-7_23, © Springer Science+Business Media Dordrecht 2013

resources currently available to manage invasive alien plants in many protected areas across the globe.

Keywords Invasive alien plants • Threat • Conservation outcomes • Biodiversity • Protection • Management • Monitoring • Prioritisation • Triage

23.1 Introduction

One of the main mechanisms used to abate biodiversity declines globally has been the creation of Protected Areas (PA) and a PA network or reserve system, starting with the formal recognition of Bogd Khan Mountain in Mongolia as a PA in 1778 (UNESCO 2012) and then more famously the declaration of Yellowstone National Park in the USA in 1872 (through the *Yellowstone Park Act 1872*) (McNeely 1994). A century later there were more than 2,500 PAs globally, covering some 3.97 million km^2 (Macdonald et al. 1989), which had expanded to over 8,500 PAs and 5 % of the earth's land surface by 1994 (McNeely 1994) and almost 158,000 PAs by 2011 (IUCN and UNEP-WCMC 2012), with an estimated area of between 10.1 and 15.5 % of the earth's land surface (Soutullo 2010).

Individual PAs were initially administered independently because there were so few of them. As the number of PAs grew, however, governments formed agencies to administer PAs (e.g. the US National Parks Service and later through international oversight by the International Union for the Conservation of Nature; IUCN). Most agencies that administer PAs have some kind of reserve acquisition programme that identifies and assesses un-reserved land for potential inclusion in their reserve system. Such assessments are based on the notion of building a comprehensive, adequate and representative reserve system, and conserving the highest quality biodiversity. The process for selecting new PAs has been criticised as being *ad hoc* because explicit criteria have not always been used (for further discussion see Pressey 1994; Margules and Pressey 2000), especially with respect to biodiversity conservation (Cabeza and Moilanen 2001). Researchers now agree that both biodiversity persistence and initial conservation values need to be considered when designing reserves (Margules and Pressey 2000). The success of using biodiversity attributes in reserve design, however, is heavily reliant on the availability of reliable information about species distributions, which is not always the case (Cabeza and Moilanen 2001; Ervin 2003); although the use of data interpolation techniques have been used to account for these incomplete data sets (Polasky et al. 2000). In addition, broader analyses have been undertaken to determine the effectiveness of the reserve network with respect to the degree of species diversity represented (Rodrigues et al. 2004). Whilst such assessments have helped to influence reserve design, much less rigor has been applied to the ability of PA agencies to conserve the biodiversity within their existing reserve system, or as newly declared PAs are added; for example, the extent to which PA management fulfils its goal of protecting biodiversity (Rodrigues et al. 2004).

Despite these problems, there is no doubt that PAs are essential for the conservation of biodiversity (e.g. Bruner et al. 2001). However, it wasn't until 1994 that specific reference to biodiversity conservation was included by the IUCN in their classification system for PAs, when the phrase "the protection and mainte-nance of biological diversity" was added. However, as Boitani et al. (2008) argue the IUCN PA categories don't reflect the role of PAs in biodiversity conservation, despite mentioning conservation objectives, because they don't explicitly identify the biodiversity that needs protection, or the processes required to preserve such biodiversity and the outcomes to be measured. Others have also argued along similar lines, for example Hockings (2003) who added that it was imperative to include information on how conservation actions will be maintained over time, monitored and evaluated, and Mace and Baillie (2008) suggested that threat abate-ment actions should be used to measure conservation outcomes in PAs. To resolve these issues Boitani et al. (2008) suggest switching the IUCN PA categories from those based on management objectives (or statements of intent) to conservation outcomes.

International treaties such as the Convention on Biological Diversity (CBD; UNEP 1999) recognised the importance of the existing PA network for the conser-vation of biodiversity and the need for its future expansion. As a result the CBD requires signatory nations to establish PA systems as a response to the decline of biodiversity. However, as outlined above, because the specific details were not outlined, achieving the CBD targets would be difficult. This is illustrated by a recent assessment of progress towards the CBD target which showed that irrespective of (i) some local successes, (ii) an increase in extent and coverage of biodiversity within PAs, (iii) the policy responses to invasive alien species, and (iv) a range of other response variables measured, the rate of biodiversity loss does not appear to be slowing (Butchart et al. 2010; SCBD 2010).

23.2   Threats to Biodiversity Within Protected Areas

The process of identifying, assessing and declaring a parcel of land as a PA (e.g. National Park or Nature Reserve) effectively addresses the main driver of biodiversity decline for that individual PA, as any further destruction of habitat is (idealistically: see Ervin 2003) usually halted. However, such declarations alone do not halt the decline of biodiversity because many of the other drivers of biodiversity decline are still active, for example invasive alien plants (IAPs; Foxcroft and Downey 2008), land clearing and modification caused by competing interests (Satchell 1997), and poaching (Ervin 2003). Therefore active management is required for both existing and new PAs, especially those that previously had different land tenures or uses, such as the conversion of forestry or agricultural lands to PAs, or those with high visitation/usage.

In a recent assessment of PA management in four countries (Bhutan, China, Russia and South Africa), five threats to the biodiversity within each country were identified (Ervin 2003). The threats demonstrate the need for increased management action, which included: (i) poaching, (ii) invasive alien plants, (iii) tourism, (iv) logging, and (v) encroachment (e.g. by competing interests like agriculture, roads and railways). The assessment (Ervin 2003) also identified five management challenges that influenced the ability of PAs to deliver biodiversity conservation outcomes, being: funding, staffing, research and monitoring, resource inventories (e.g. of biodiversity and threats), and community relations. The issues identified by Ervin (2003) highlight the complexity of managing land (PAs) for biodiversity conservation and the need for active long-term management. Without active management within PAs some species may still go extinct, despite the broader protection offered by the designation of a PA (Serrouya and Wittmer 2010).

23.3   The Invasive Alien Plant Threat and Biodiversity Conservation in Protected Areas

Based on analyses of the major drivers of biodiversity decline (e.g. Wilcove et al. 1998) it could be argued that once a parcel of land is declared a PA the next major threat to the biodiversity contained within that parcel of land can be attributed to the invasion and subsequent impact of alien species. The problem of invasive alien plants within PAs has been widely documented (Macdonald 1983; Loope et al. 1988; Usher 1988; Usher et al. 1988; Macdonald et al. 1989; Macdonald 1990; Timmins and Williams 1991; Pyšek et al. 2002; De Poorter 2007a, b; Foxcroft and Freitag-Ronaldson 2007; Foxcroft and Downey 2008; Allen et al. 2009; Foxcroft et al. 2014), as has their invasion patterns (Pyšek et al. 2002; Pauchard and Alaback 2004) and history within specific PAs in terms of the number and diversity of alien species that have invaded (Usher 1988). For example, concerns over alien plants in South Africa's Kruger National Park were first raised over 70 years ago with the identification of six 'troublesome weeds' (Obermeijer 1937). Unfortunately, little was done to manage these alien plants in the Park for the next 20 years and even then it was insufficient to adequately address the problem (Foxcroft and Freitag-Ronaldson 2007). There are now over 350 alien plants in Kruger National Park (Spear et al. 2011), although not all of which are invasive. An assessment of the alien species problem within PAs led Macdonald (1990) to conclude that for most of the world's PAs there are more invasive species problems than there are resources available to abate the threat.

Managing alien plants in PAs to improve conservation outcomes requires active control of, at least, the invasive species, which is specifically targeted to the recovery of the biodiversity they threaten (Downey et al. 2010c). Despite numerous assessments of the alien plant problem within PAs, based on Downey et al.'s (2010c) definition of impact or threat, the actual impact to native biodiversity contained within PAs has been poorly documented. Compounding this problem is that when the threat from invasive alien plants to native species was assessed in detail, the number of species at risk was at least an order of magnitude greater than what was previously known (Downey 2008a). Although most PA managers and researchers acknowledge the threat (e.g. Usher 1988; Macdonald et al. 1989), the management practices and policies put in place to abate the threat and impact are rarely capable of delivering biodiversity conservation outcomes. One reason is that such strategies seldom include information on the species at risk or prioritise control based on achieving the desired outcome (Downey 2010a). Another reason is that many of the early assessments were focused on the invader rather than the biodiversity that needed protecting, with the exception of biodiversity in a broad sense (Macdonald 1990).

In a report on the world's PAs, Chape et al. (2008) outlined the major threats PAs face, which include invasive alien species. In their superficial treatment of the threat from invasive alien species, the authors identify four major management options for alien species in PAs, being: (i) prevention, (ii) early detection, (iii) eradication, and (iv) control. The authors provide no guidance other than the use of chemical and biological control, along with the precaution that using such measures might lead to off-target impacts. The authors also mention the presence of groups (e.g. the IUCN Invasive Species Specialist Group, ISSG) and conventions (e.g. Convention on Biological Diversity) which address alien species issues. Ironically, the authors don't mention management options around protection or the reduction of impacts on biodiversity, which must be fundamental objectives of any alien species management programme within PAs. The authors also provide a regional assessment of the PA network, which is almost exclusively on the status of the reserve system in terms of numbers of reserves in each region. As Grice (2004) and Downey (2008a) argue, such a lack of knowledge of the biodiversity at risk from alien species or specific management actions required, has greatly hindered our ability to adequately address the problem.

The lack of information on biodiversity at risk from alien plants, and how this has hampered conservation outcomes, is highlighted in the foreword to a recent report on why the rate of biodiversity loss has not declined, entitled *Global Biodiversity Outlook 3* (SCBD 2010). In the report the Executive Director of UNEP states, "Governments also need to rise to the challenge of Alien Invasive Species". This is followed, paradoxically, by the use of agricultural economic impacts as examples to highlight the urgency for action to protect biodiversity from alien species, rather than specific reference to the known impacts to biodiversity. The report showed that Goal 6 (control threats from invasive alien species) has not been achieved. The reasons for which are specifically related to the continued increase in the number of introductions, and that most countries lack effective management programmes for managing major alien species that threaten ecosystems, habitats or species (SCBD 2010).

In this chapter I examine the reasons for this mismatch between the conservation challenge and the management focus, and then outline a series of processes developed to rectify this situation for managing alien plants within PAs. These processes are neither species nor reserve specific, as they were developed specifically to ensure alien plant management targeted biodiversity conservation independent of land tenure. Thus, they can be applied to all PAs as a way of increasing the level of protection of biodiversity that is achieved through invasive alien plant management in a measurable, consistent and robust manner.

23.4 Managing Invasive Alien Plants for Biodiversity Conservation Outcomes

23.4.1 Mismatch Between Management Actions and Conservation Outcomes

The need to manage 'weedy' plants arose out of agriculture (mainly associated with growing crops) and as such the field of weed science was initially focused on the biology and control of alien plants in agricultural systems. As a result there is good information available on the control of alien plants. Over the past several decades, however, research has become increasingly focused on alien plants that invade non-agricultural lands, especially after the SCOPE (Scientific Committee on Problems in the Environment) programme of the 1980s (Drake et al. 1989), and specifically, what makes them invade such areas (Levine et al. 2003). Whilst researchers have focused on alien plants in native vegetation, many land managers have clung onto historical approaches to alien plant or weed management, seeing alien plant management as a war on weeds, with killing being the main objective. This approach has actually hampered alien plant management by maintaining a deep-seated emphasis on the act of killing all alien plants, rather than on the outcome of the control programmes (Downey 2011). Contributing to this situation is a widely held assumption amongst many managers that control of alien plant will lead to a positive response to any biodiversity present, because the target alien species has been removed. However, there is growing documentation that alien species management alone does not always produce positive biodiversity benefits (e.g. Humphries et al. 1993; Luken 1997; Turner and Virtue 2006; Beater et al. 2008; Reid et al. 2009; Gaertner et al. 2012). Thus management of invasive alien plants for conservation needs to address these serious mismatches in order to achieve the desired outcomes.

Several barriers that have hampered the transition to an outcome-orientated management system for alien plants have been identified. First, there is a lack of data and inventories on alien plants and/or biodiversity, let alone biodiversity at risk from alien plants or the alien plants posing the greatest impact, upon which to make decisions or influence management (Ervin 2003; Levine et al. 2003; Downey 2008a, 2010a, Downey et al. 2010c). Therefore many management decisions take place in the absence of robust data on impacts on biodiversity (Grice et al. 2004). Second, there is a lack of adequate monitoring programmes, despite the necessity of monitoring being raised more than a decade ago (Blossey 1999), and the increasing emphasis on the need to monitor the outcomes of alien plant management programmes (Grice 2004; Reid et al. 2009). Third, there is a lack of planning processes to ensure that actions result in conservation outcomes (including, for example, prioritisation and risk assessment, Downey et al. 2010c). Compounding these issues is the fact that many PAs have multiple alien plant problems (e.g. there are over 350 alien plants recorded in Kruger National Park, Spear et al. 2011), and there are generally insufficient funds to manage all alien plant problems (Macdonald 1990; Ervin 2003).

Whilst the administration of PAs has seen the advent of management systems for managing alien plants for biodiversity (e.g. Foxcroft and Downey 2008), there is often a failure to link management objectives with on-ground actions. As Downey (2010a) argues, this situation arises because such planning systems rarely address all the steps of the management process (i.e. from planning to implementation, monitoring and reporting). In addition, many of these actions require additional resources to those allocated specifically to controlling alien plants. Given the limited resources available, any actions, other than the physical control of alien plants, are rarely undertaken. To highlight this point Reid et al. (2009) found relatively little information on the tangible outcomes of invasive alien plant management, despite large numbers of programmes occurring. Along with many others (e.g. Blossey 1999), I argue that the lack of commitment to actions other than just physical control has greatly affected the outcomes of alien plant management, especially for conservation.

In addition, on-ground management of alien plants is not matched with alien plant policies. For example, Downey (2010b) found that many research and management studies focused on established and/or widespread alien plants, despite broad management policies focusing on early detection and eradication as the priority. There are several reasons for this mismatch. Firstly such policies fail to adequately address the problem of established and/or widespread alien species, which are the primary focus for many managers, especially within PAs. Second, the focus of these two areas are opposed, with early detection and eradication aimed at preventing future problems, and management of established species aimed at reducing or abating the current problem, which if left unchecked will increase. Compounding this problem is that some prioritisation processes have given less emphasis to widespread established species (e.g. Randall 2000), which has created problems for the management of these alien plants for conservation (Williams et al. 2009). Lastly, given limited resources and the scale of the problem, managers are typically making decisions on cost-benefit analyses, which show early intervention is cost-effective (e.g. van Wilgen et al. 2004). Because an evaluation, in monetary terms, of the impacts of alien plants on biodiversity is difficult, such assessments are often not attempted (van Wilgen et al. 2004). As a result comparisons between costs and benefits between early intervention and management of establish species for conservation outcomes have rarely been made, but are essential. This results in decisions being based on an understanding of only part of the problem and as such decisions on protecting biodiversity tend to receive lower priority.

As a way of addressing these issues, Downey (2011) argued that alien plant management needed to adopt an outcome-oriented approach that goes beyond the act of actual control, with specific reference to the protection of biodiversity. Such an approach would also help ensure that conservation outcomes were achieved through alien plant management, in which the greatest benefit would be to the management of invasive alien plants within PAs – given their conservation objectives and other challenges outlined above.

Macdonald (1990) [inadvertently] outlines the management problem succinctly with the statement that "setting priorities has two major components: targeting introduced species for control and targeting areas within reserves for control operations", by not including any mention of either the biodiversity at risk or the desired outcomes of such control operations. I argue that decisions similar to Macdonald (1990) are based on the pragmatics of removal of alien species rather than on the outcome of such activities, which is what is actually required to help PAs management meet their objectives or goals.

Other authors have also argued that consideration of the impacts from alien plants is critical when prioritising control efforts (e.g. Byers et al. 2002). In an attempt to better define the impacts of alien species Parker et al. (1999) outlined a measure of Impact (I); being the invader's Range (R), Abundance (A) and Effect (E) (or $I = R \times A \times E$). Whilst they highlight the challenges of defining 'impact', the problems are inadvertently compounded because their measure and the three variables used are not applicable in a non-theoretical context, which the authors themselves discuss, and thus cannot easily be used (Downey et al. 2010b). Such studies also highlight another major problem, in that many researchers have failed to deliver workable solutions for conservation managers.

23.4.2 Ensuring Alien Plant Management Delivers on Conservation Outcomes

Because species threatened by IAPs are not necessarily at risk in every known location (Downey 2010a), and since control cannot be achieved at every site, recovery of native species is not always possible (D'Antonio and Meyerson 2002). Decision support tools are therefore needed to help land managers focus on ensuring optimal conservation outcomes. To address many of the problems outlined above, Downey (2010a) developed a system to ensure that alien plant management delivered biodiversity conservation outcomes. This system was based on a significant body of work spanning almost a decade of managing *Chrysanthemoides monilifera* subsp. *rotundata* (bitou bush), a nationally significant invasive alien plant in New South Wales (NSW), Australia. This shrub was accidentally introduced to Australia from South Africa a century ago. After initially being recognised as a potentially invasive alien, its value in stabilising active sand dunes was realised and it was subsequently actively planted in the 1950s for this purpose. *Chrysanthemoides monilifera* now occupies over 80 % of the 1,100 km long NSW coastline (Weiss et al. 2008). Because of the threat it posed to coastal biodiversity, in 1999 *C. monilifera* was listed as a *Key Threatened Process* under the NSW threatened species legislation (DEC 2006), which subsequently initiated the programme developed by Downey (2010a). The underlying basis of the system

Table 23.1 The 11-step planning process developed by Downey (2010a) to ensure that alien plant management delivers conservation outcomes

Step	Management action	Description
1	Determine where the threat is active	Map the distribution and density of the target alien species
2	Determine the biodiversity at risk	Assess the native species that are threatened by the target alien species
3	Determine the distribution of the biodiversity at risk	Map the distribution of the native species threatened
4	Prioritise the species at risk	Establish a system to prioritise the threatened native species, based on the degree of threat posed to them
5	Prioritise sites for control	Establish a system to assess individual sites for control based on the actual threat present and the ability to deliver effective control
6	Provide best practice management guidance	Use or establish best practice guidelines that ensure control measures do not have adverse impacts on the native species being protected during control actions. These guidelines and any other supporting material must also be disseminated to all stakeholders
7	Identify stakeholders and ways to engage them	Most alien plant management programmes rely on the collaboration of a range of stakeholders which need to be engaged and remain committed to the programme
8	Prepare a formal strategy and coordinate its implementation	For any major programme an over-arching plan or strategy is needed to ensure that the host agency is committed to the on-ground programme/s. Many programmes fail without on-going coordination (including communication)
9	Establish site-specific management plans	Site-specific management plans are essential to ensure that local conditions, constraints and resources are accounted for. These plans need to have specific criteria to ensure that the outcomes are achieved. Such plans also enable local managers to have ownership, within a bigger programme which is critical for success
10	Monitor the effectiveness of on-ground actions	Establish a monitoring system that enables a wide range of end users to adopt it. Such a system must include instruction, datasheets, guidance on analysis and a repository for the data
11	Review and report on the outcomes of the first 10 steps	The last step is aimed at adaptive management in that the whole process is evaluated and revised, as needed, on a regular or strategic basis

is an 11-step planning process that aims to ensure that all alien plant management undertaken for biodiversity conservation is targeted to areas where the likelihood of a positive response to the native species at risk is greatest (Table 23.1).

Each of the 11-steps addresses a critical stage that is required to ensure alien plant management can deliver the desired conservation outcomes. To achieve this aim, new systems were needed for many of the steps. For example, Step 2 –

Determine the biodiversity at risk, required a system for rapidly determining the biodiversity at risk from alien plants. In response to this need, the Weed Impacts to Native Species (WINS) assessment tool was established by Downey (2006), involving four stages: (1) a review of the literature; (2) collection and assessment of the knowledge from land managers and botanists with specific involvement, either in managing the target alien species or the native species in invaded areas; (3) rigorous evaluation and examination of an interim list of species potentially at risk, and; (4) ranking the revised list using a model, which then forms Step 4 – *Prioritise the species at risk* in Downey's 11-step system. Although this process does not provide the same quality information as scientific assessments of the impact of an alien species (Downey 2006), the strength of the WINS approach lies in its ability to enable managers to make informed decisions in a short period of time, especially if they focus on the highest priorities. Previously many management decisions took place in the absence of such information (Grice et al. 2004), as illustrated by Turner and Downey (2010) who used the WINS system for *Lantana camara* (lantana). In addition, there are many challenges with data collection to be able to quantify the impact of invasive alien plants on biodiversity (Downey and Grice 2008). Several authors have, however, compiled information on native species at risk from alien plants. For example Coutts-Smith and Downey (2006) compiled a list of the native species listed under the NSW *Threatened Species Conservation Act 1995* at risk from alien plants. Such lists do not capture the entire range of species at risk, whereas the WINS system goes beyond such studies to ascertain a more comprehensive extent of species at risk, irrespective of whether they are listed as threatened or not. Evaluation of the native species compiled using the WINS system for the Threat Abatement Plan for *C. monilifera* (DEC 2006) showed that 65 % were not formally listed as threatened (i.e. under threatened species legislation). In addition, as the 11-step process includes an adaptive management component (i.e. Step 11), there are mechanisms to revise, refine or even replace any of the processes currently in place for each step.

Although there have been several meta-analyses of published studies on the impact of invasive alien plants on biodiversity (e.g. Gaertner et al. 2009; Vilà et al. 2011), these studies have perpetuated some of the problems outlined here. These studies focused on the alien species and the mechanisms for native species declines, rather than on the native species themselves, with the exception of broad-scale trends. Such studies, whilst useful scientifically, are of limited value from a management perspective. Downey (2010a) argued that it is critical for management to be based on information about the species at risk in order to protect them.

Another new system required was a method of assessing specific sites to determine the likelihood of management actions achieving conservation outcomes (Step 5 – *Prioritise sites for control*). Based on a site model outlined in the Threat Abatement Plan for *C. monilifera*, a process (Prioritisation of Impacts for Conservation of Sites, PIC-Sites) was developed (Downey 2008b), which accounts for several key factors that may influence the biodiversity outcome at a specific site. These include: (i) not all species threatened by invasive alien plants are at risk at every location; (ii) alien plant control cannot be achieved at all sites; (iii) native

Table 23.2 Triage matrix for the strategic management of alien plants for the protection of biodiversity (After Downey et al. 2010c)

		Probability of protecting biodiversity at specific sites		
		High	Medium	Low
Level of threat to biodiversity	High	A – Alien plant management is critical, immediate, targeted and long-term	B – Targeted management action needs to occur promptly and long-term	C – Broader management (i.e. of multiple threats simultaneously)
	Medium	D – Targeted management action needs to occur promptly and long-term	E – General management to reduce the impact of alien plant populations	F – General low level management to reduce the threat
	Low	G – Actions to minimise the threat and prevent further elevation of the problem	H – Low level of management only	I – No immediate action, management action required only after completion of higher priorities

species recovery is not possible at all sites; (iv) other threats to the native species at risk may also be active at a site and may not be abated easily; and (v) the density and health of the species at risk, and therefore their ability to recover following alien plant control programmes, will vary between sites. Downey et al. (2010c) added additional factors into a site model (again based on DEC 2006) including the (i) ability to achieve effective control at the site (i.e. is a control method available and can the site be accessed easily), (ii) the degree of impact posed by the alien species at each site; and (iii) the condition of the biodiversity present; including the value of the site to the species overall survival, and the physical condition of the site. By using and modifying these processes for Steps 4 and 5, Downey et al. (2010c) developed a triage matrix for strategically managing invasive alien plants, in which management is directed to the highest priority species (assets) at sites where control will have the greatest benefit in the first instance (Table 23.2).

Many existing systems needed to be modified to ensure that the aim was consistent with achieving conservation outcomes rather than control of alien species alone (e.g. Steps 6, 9 and 10). This is particularly why site-specific 'restoration plans' or management plans (Step 9) need to dovetail with alien control actions to improve conservation outcomes (Holmes et al. 2008). Step 10 – *Monitor the effectiveness of on-ground actions* is discussed in detail below. A summary of Downey's (2010a) 11-step programme is presented in Table 23.1.

Although several authors have outlined various components of the 11-steps (e.g. Macdonald 1990; Hiebert and Stubbendieck 1993; Goodall and Naudé 1998; Timmins and Owen 2001; Nel et al. 2004; Platt et al. 2005; Randall et al. 2008), none have put them into a comprehensive plan to ensure that conservation outcomes, in terms of specific species at risk, were achieved by examining every management stage (e.g. planning through to on-going action, monitoring and reporting). In addition, the emphasis on the actual biodiversity at risk is at best rudimentary, and often general in nature (e.g. Macdonald 1990). The basis for the development of these other processes is similar to those that underpin the processes outlined by Downey (2010a). For example, Goodall and Naudé (1998) argued that the current approach to invasive alien plant management was inadequate and that habitats needed to be ranked according to criteria that encompass the habitat's status, protection, and management, thereby enabling priorities to be established that balance urgent environmental needs with management budgets. Assessments of other major invasive alien plant management programmes have shown that management needs to prioritise both the species and areas for control, as control efforts may not decrease the problem, with invasions appearing to have increased, and continue to remain a serious threat in many biomes (van Wilgen et al. 2012). Others have suggested establishing priorities based on the likelihood of success (e.g. Fensham and Cowie 1998). The triage matrix (Downey et al. 2010c) uses a combination of the expert knowledge and models to establish alien plant and conservation priorities, something that Cowling et al. (2003) found has merit when they compared both expert-based and systematic, algorithm-based approaches to identifying priority areas for conservation. Downey (2010a) and Downey et al. (2010c) examined a range of existing processes when creating those that underpin the 11-steps. These authors go on to say that other processes could be substituted for those developed for *C. monilifera*, so long as they met the objectives of the specific step in question.

The 11-step process outlined in Table 23.1 has been adopted and modified for another nationally significant invasive alien plant in Australia, *L. camara* (Turner et al. 2010), which is a major problem in New South Wales (NSW) and Queensland. The modifications were mainly associated with differences in legislation and the classification of threatened species and ecological communities between the two States, and the inclusion of a significantly greater number of species identified as being at risk (Turner and Downey 2010). For example, these authors identified over 1,300 native plants and 150 native animals at risk from *L. camara* invasion, an order of magnitude greater than previously identified.

The 11-step process and the triage system are both based on (i) the target alien plant being identified at the outset, and (ii) a single species focus. When multiple alien plant species are present, or if there is a need to manage multiple alien species, a different process is needed to prioritise species for action in the first instance, which can then be used to feed into the processes outlined above. For example, Downey et al. (2010b) developed such a system to evaluate the potential threat posed by alien plants to biodiversity in NSW. Firstly, an initial assessment was used to screen the 1,650 naturalised plant species in that state by removing alien plants that were not known to threaten native species or weren't widely established (i.e. naturalised but not known to be invasive). This process reduced the list to 340 major alien plant species, which were then assessed with a model to determine those species that are likely to pose the greatest threat to biodiversity. Using another approach, Randall et al. (2008) developed an assessment protocol for ranking alien plants based in part on their biodiversity impacts, using a system of 20 multiple-choice questions. The authors are currently assessing alien plants in the USA using this system.

Additional justification for the processes outlined here comes from an assessment of the recovery plans developed under the US Endangered Species Act, which showed that the best outcome for managing multiple threatened species was to group them together based on similar threats (Clark and Harvey 2002). The Threat Abatement Plan for *C. monilifera* (DEC 2006), which forms the basis for the processes outlined here, is a threat specific management plan. It encompasses the protection of over 160 native species at risk from one common threat.

Establishing processes for managing the biodiversity at risk from multiple alien plants poses a range of other challenges. These relate particularly to individual native species threatened by multiple alien plants, and sites containing multiple alien plants as well as species at risk. These challenges were overcome when the processes outlined here were modified and adapted to address all alien plant threats to biodiversity within each of the 13 Natural Resource Management Regions (or Catchment Management Authorities) in NSW (DPI and OEH 2011).

Turning a strategy or strategic approach into on-ground action and outcomes is not easy. Examination of successful attempts to turn strategies into action reveals that several key factors are required (based on Gelderblom et al. 2003 and Downey 2010a), being: (i) management-based research, typically associated with a long history of such research; (ii) a history of knowledge and commitment to addressing the problem; (iii) scientists committed to both the planning and implementation phases, and; (iv) institutions committed to implementing the strategy long-term, through provision of resources (people and funding), institutional frameworks, policies and processes, and promotion of the strategy, its implementation and outcomes. As Downey (2010a) and others (e.g. Roura-Pascual et al. 2009) argue, coordination of management was a critical step in ensuring the delivery of conservation management outcomes. These successful attempts to turn a strategy into action have also strategically realigned existing resources to deliver improved conservation outcomes. Holmes et al. (2008) argue that specialists are needed to assist project managers when they develop site-specific restoration or management plans – something that Downey (2010a) also found.

Prioritisation of alien plants for management has also occurred for emerging species (e.g. Mgidi et al. 2007) and for eradication targets (Skurka Darin et al. 2011). Skurka Darin et al. (2011) used criteria that assess the relative impact, potential spread, and feasibility of eradication. The authors found that priorities based exclusively on species-level characteristics are less effective compared to priorities derived from a blended prioritization system that encompasses both species attributes and individual population and site parameters, because the level of impact may vary among species and populations; as also outlined by Downey (2010a). Thus the prioritisation system [WHIPPET] developed by Skurka Darin et al. (2011) aims to facilitate improved decision-making process by allocating limited resources to priority invasive plant infestations with the greatest predicted impacts. Whilst this approach has similarities to that outlined by Downey et al. (2010c), it has some major differences, which limits its application. First it

520   P.O. Downey

is aimed solely at determining priorities for eradication. Second it assesses relative impacts in terms of the surrounding habitat, rather than actual impact present at a site or the impact of the species (i.e. on biodiversity). Lastly it assesses invasiveness or rate of spread, which is not assessed by Downey et al. (2010c). Both systems assess feasibility of control – being feasibility of eradication in Skurka Darin et al.'s (2011) system. Because each system addresses different priorities there is significant benefit in using both systems to derive a complete range of priorities, from eradication to conservation outcomes.

Prioritisations systems are time-consuming to construct and test, and highly dependent on the availability of robust data, which is not always readily available (Downey et al. 2010c; Skurka Darin et al. 2011). Other attributes could also be included, but were specifically not included because this would make the system too cumbersome to use. Significant progress and investment has now been made towards developing prioritisation systems for managing alien plants. Such investment and progress should be used to establish priorities in other areas.

Some jurisdictions (especially in the developing world) may argue that they do not have the resources to collect and collate the data required to determine priorities using such systems (Khuroo et al. 2011). However, such arguments are rarely based on an assessment of the value of investing in robust prioritisation compared to the cost of their existing approaches. Instead these arguments are based simply around a lack of resources and capacity. There is a high initial cost, but the priorities established should lead to long-term saving by directing limited funds to priorities based on achievable outcomes. Alternatively, prioritising lists of invaders has serious limitations, which can compromise the overall effectiveness of any programme that includes spending limited resources on (i) low impact populations of alien plants, (ii) difficult to access populations of alien plants, or (iii) missing high impact populations of low priority alien species (Skurka Darin et al. 2011). Thus alien plant management for biodiversity conservation must move to a prioritised management system. Given that there are now several such systems available and a growing body of knowledge on invasive alien plants, such systems could be adopted in other locations without the significant inputs needed to develop them in the first instance.

A major problem with assessing the effectiveness of large-scale alien plant control programmes has been a lack of data on the cost-effectiveness. Whilst there have been some assessments (e.g. McConnachie et al. 2012), which are extremely useful, the benefits of such programmes to biodiversity are rarely encompassed. A notable exception is Sinden et al. (2008) who examined the C. monilifera programme and showed a return of AU$2.56 per AU$1 of investment. Forsyth et al. (2012) developed a prioritisation tool for invasive alien plant management in South Africa, which revealed that there are many high priority areas that are not receiving funding, and low priority areas which are receiving high levels of funding. The authors state that "clearly, there is a need for realigning priorities, including directing sufficient funds to the highest priority catchments to provide effective control", whilst limiting resources to low priorities. Perplexingly, neither Skurka Darin et al. (2011) nor Forsyth et al. (2012) make any reference to the work of Downey et al. (2010c), despite them both establishing prioritisation systems for invasive alien plant management, and the work of Downey et al. (2010c) being readily available (i.e. through internet searches). The work presented here has been based on a body of published studies over almost a decade (see Downey 2010a) and has been incorporated within broader weed risk assessment systems (Downey et al. 2010a).

23.4.3 Ensuring Alien Plant Management Delivers on Conservation Outcomes in Protected Areas

In an assessment of case studies on alien species management within 24 PAs, Usher (1988) found very few used prioritisation systems for management. Whilst several management frameworks have been developed (e.g. Hiebert and Stubbendieck 1993; Timmins and Owen 2001; Nel et al. 2004; Platt et al. 2005), they do not encompass the on-ground management or monitoring components under one holistic system.

The processes outlined here have been used to establish conservation priorities for alien plant management programmes across the 800+ PAs in NSW. In addition, the majority of sites outlined in DEC (2006), which formed the basis for the processes outlined above, are within PAs. Initial results to date have shown that this approach has been successful in ensuring the delivery of conservation outcomes (Hamilton et al. 2010). Whilst the input [cost] required to collect, collate and assess the data needed to determine the priorities can be high, the conservation benefit exceed such inputs as management programmes are then prioritised based on delivering the greatest conservation outcome from alien plant management. This is critical given management effectiveness in PAs is affected by limited resources (Ervin 2003) and that control of alien plants in PAs typically only occurs for a subset of the species present (Macdonald and Frame 1988). The processes outlined here give PA managers decision support tools to prioritise invasive alien plant control programmes to areas where they will deliver the greatest conservation outcome and alien plant control.

23.5 Monitoring the Conservation Outcome of Alien Plant Management

23.5.1 Historical Monitoring Shortfalls

Increasing emphasis is being placed on the importance of monitoring the effects of alien plant control programmes (Blossey 1999; Grice 2004; Martin and van Klinken 2006; Maxwell et al. 2009; Reid et al. 2009). Although many invasive alien plant control programmes underway in natural areas (including PAs) have conservation aims (i.e. the protection of native plant species), few have linked monitoring programmes for evaluating their success in terms of the recovery of native species. Land managers face a series of challenges regarding monitoring, including: what to monitor, which techniques to use, where to monitor, how to design a robust monitoring programme, how to analyse and report the data to their managers or others. There is also a shortage of resources dedicated towards monitoring, specifically time and money, and monitoring skills and guidance, and a lack of emphasis on monitoring from on-ground staff. Consequently, many invasive alien plant management programmes either (i) do not include any form of monitoring, (ii) use simplistic qualitative monitoring methods like photo points, or (iii) collect data that cannot be analysed for a number of reasons, mostly associated with the use of an inappropriate experimental design, inconsistent collection of data, or collection of data that can't answer the aim of the control programme or management, and (iv) do not contain mechanisms to report on the monitoring that has been undertaken (Downey and Hughes 2010). Long-term analysis of invasive alien plant control programmes are rare and those that exist show that the management of invasive alien plants in many instances does not always led to a return to a reference or pre-invasion state, but rather resetting the invasion clock, either with the initial or new invaders (Beater et al. 2008).

Lastly, assessment of monitoring efforts for the recovery of threatened species has shown that monitoring did not adequately address the specific threats affecting species (Campbell et al. 2002). Such assessment has also highlighted the need to carefully evaluate monitoring programmes to avoid unnecessary monitoring, either in terms of whole programmes or components, and the need to make monitoring more holistic. Given that the outcomes of many conservation programmes are difficult to predict, monitoring of their progress must be an integral component of any conservation initiative (Campbell et al. 2002).

23.5.2 Monitoring Conservation Outcomes

To overcome the shortcomings outlined above with respect to monitoring, Hughes et al. (2009) developed a manual aimed at ensuring that the response of native species is assessed following invasive alien plant control. The manual is separated into three tiers and each tier is aimed at a different set of stakeholders, depending on their skills and resources. To ensure greatest adoption the manual was field tested on a wide range of potential users (including volunteer groups) and the text was circulated to several experts and potential users for comment. In addition to instructions for all monitoring methods encompassed, the manual contains data sheets and guidance on simple analysis, and is freely available (www.environment. nsw.gov.au/bitouTAP/monitoring.htm).

Whilst this monitoring manual was initially developed for *C. monilifera*, it quickly became apparent during its development that it needed to be expanded to include other alien plants. Where possible the manual was developed to be applicable for other alien plants, although aquatic plants and vines are not adequately covered, given their unique life form.

One of the greatest advantages of this monitoring manual is that it ensures consistency across alien plant management programmes. A lack of consistency has greatly hampered the ability of conservation managers to make adaptive management decisions, or illustrate the problems and improve their own knowledge base. Given that the specific aim of this manual is to assess the response of native species to alien plant control, it has direct application for monitoring invasive alien plant programmes within PAs. For example, it has been adopted as standard for monitoring alien plant programmes within PAs in NSW. In addition, monitoring data is crucial to the successful implementation and continued use of any adaptive management strategy (e.g. Foxcroft and Downey 2008).

23.6 Conclusions

The management of invasive alien plants within PAs needs to be better aligned with ensuring the delivery of conservation outcomes, in terms of establishing robust priorities, management processes and monitoring protocols. Whilst the processes outlined here have not been specifically designed for managing invasive alien plants in PAs, they are directly applicable as they aim to ensure that alien plant management delivers the desired outcomes irrespective of where they occur; a key objective of PAs management. Adoption of these or other similar processes by organisations that administer PAs, will have significant long-term benefits to conservation of PAs, invasive alien plant management and the protection of biodiversity.

Acknowledgements I thank those who helped in the development of the processes outlined here. I also appreciate discussions with Peter Turner and Llewellyn Foxcroft, and thank Bill Sea for making comments on an earlier version. I thank the editors for the opportunity to submit a chapter in this landmark publication and the two referees for their suggested improvements.

References

Allen JA, Brown CS, Stohlgren TJ (2009) Non-native plant invasions of United States National Parks. Biol Invasions 11:2195–2207
Beater MMT, Garner RD, Witkowski ETF (2008) Impacts of clearing invasive alien plants from 1995 to 2005 on vegetation structure, invasion intensity and ground cover in a temperate to subtropical riparian ecosystem. S Afr J Bot 74:495–507
Blossey B (1999) Before, during and after: the need for long-term monitoring in invasive species management. Biol Invasions 1:301–311
Boitani L, Cowling RM, Dublin HT et al (2008) Change the IUCN protected area categories to reflect biodiversity outcomes. PLoS Biol 6:e66
Bruner AG, Gullison RE, Rice RE et al (2001) Effectiveness of parks in protecting tropical biodiversity. Science 291:125–128
Butchart SHM, Walpole M, Collen B et al (2010) Global biodiversity: indicators of recent declines. Science 328:1164–1168

Byers JE, Reichard S, Randall JM et al (2002) Directing research to reduce the impacts of nonindigenous species. Conserv Biol 16:630–640

Cabeza M, Moilanen A (2001) Design of reserve networks and the persistence of biodiversity. Trends Ecol Evol 16:242–248

Campbell SP, Clark JA, Crampton LH et al (2002) An assessment of monitoring efforts in endangered species recovery plans. Ecol Appl 12:674–681

Chape S, Spalding M, Jenkins MD (eds) (2008) The world's protected areas: status, values and prospects in the 21st century. University of California Press, Berkeley

Clark JA, Harvey E (2002) Assessing multi-species recovery plans under the Endangered Species Act. Ecol Appl 12:655–662

Coutts-Smith AJ, Downey PO (2006) Impact of weeds on threatened biodiversity in New South Wales, Technical series 11. CRC for Australian Weed Management, Adelaide

Cowling RM, Pressey RL, Sims-Castley R et al (2003) The expert or the algorithm? Comparison of priority conservation areas in the Cape Floristic Region identified by park managers and reserve selection software. Biol Conserv 112:147–167

D'Antonio C, Meyerson LA (2002) Exotic plant species as problems and solutions in ecological restoration: a synthesis. Restor Ecol 10:703–713

DEC (2006) NSW Threat Abatement Plan - invasion of native plant communities by *Chrysanthemoides monilifera* (bitou bush and boneseed). Department of Environment and Conservation NSW, Hurstville

de Poorter M (2007a) Invasive alien species and protected areas: a scoping report. Part 1. Scoping the scale and nature of invasive alien species threats to protected areas, impediments to invasive alien species management and means to address those impediments. Global Invasive Species Programme, Invasive Species Specialist Group. http://www.issg.org/gisp_publica tions_reports.htm

de Poorter M (2007b) Invasive alien species and protected areas: a scoping report. Part 2. Suggestions for an IUCN approach to addressing present and future threats from invasive alien species in protected areas. Global Invasive Species Programme, Invasive Species Specialist Group. http://www.issg.org/gisp_publications_reports.htm

Downey PO (2006) The weed impact to native species (WINS) assessment tool – results from a trial for bridal creeper (*Asparagus asparagoides* (L.) Druce) and ground asparagus (*Asparagus aethiopicus* L.) in southern New South Wales. Plant Prot Q 21:109–116

Downey PO (2008a) Determination and management of alien plant impacts on biodiversity: examples from New South Wales, Australia. In: Tokarska-Guzik B, Brock J, Brundu G et al (eds) Plant invasion: human perception, ecological impacts and management. Backhuys Publishers, Leiden, pp 369–385

Downey PO (2008b) Determining sites for weed control and biodiversity conservation. In: van Klinken RD, Osten VA, Panetta FD et al (eds) 16th Australian weeds conference, Cairns. Queensland Weeds Society, Brisbane, pp 387–388

Downey PO (2010a) Managing widespread alien plant species to ensure biodiversity conservation: a case study using an 11-step planning process. Invest Plan Sci Manage 3:451–461

Downey PO (2010b) Do the aims of weed management programs align with the objectives of weed policy? In: Zydenbos SM (ed) 17th Australasian weeds conference, Christchurch. NZ Plant Protection Society (Inc), Christchurch, pp 85–86

Downey PO (2011) Changing of the guard: moving from a war on weeds to an outcome-orientated weed management system. Plant Prot Q 26:86–91

Downey PO, Grice AC (2008) Determination and management of the impacts of weeds on biodiversity. In: van Klinken RD, Osten VA, Panetta FD et al (eds) 16th Australian weeds conference, Cairns. Queensland Weeds Society, Brisbane, pp 23–25

Downey PO, Hughes NK (2010) Monitoring protocols to assess the recovery of native plant species following the control of widespread weed species. In: Zydenbos SM (ed) 17th Australasian weeds conference, Christchurch. NZ Plant Protection Society (Inc), Christchurch, pp 445–448

Downey PO, Johnson SB, Virtue JG et al (2010a) Assessing risk across the spectrum of weed management. CAB Rev Perspect Agric Vet Sci Nutr Nat Resour 5(038):1–15

Downey PO, Scanlon TJ, Hosking JR (2010b) Prioritising alien plant species based on their ability to impact on biodiversity: a case study from New South Wales. Plant Prot Q 25:111–126

Downey PO, Williams MC, Whiffen LK et al (2010c) Managing alien plants for biodiversity outcomes – the need for triage. Invest Plan Sci Manage 3:1–11

DPI OEH (2011) Biodiversity priorities for widespread weeds. DPI (Department of Primary Industries NSW), OEH (Office of Environment and Heritage), Orange

Drake JA, Mooney HA, di Castri F et al (1989) Biological invasions: a global perspective. Wiley, New York

Ervin J (2003) Rapid assessment of protected area management effectiveness in four countries. BioScience 53:833–841

Fensham RJ, Cowie ID (1998) Alien plant invasions on the Tiwi Islands. Extent, implications and priorities for control. Biol Conserv 83:55–68

Forsyth GG, Le Maitre DC, O'Farrell PJ et al (2012) The prioritisation of invasive alien plant control projects using a multi-criteria decision model informed by stakeholder input and spatial data. J Environ Manage 103:51–57

Foxcroft LC, Downey PO (2008) Protecting biodiversity by managing alien plants in National Parks: perspectives from South Africa and Australia. In: Tokarska-Guzik B, Brock J, Brundu G et al (eds) Plant invasion: human perception, ecological impacts and management. Backhuys Publishers, Leiden, pp 387–403

Foxcroft LC, Freitag-Ronaldson S (2007) Seven decades of institutional learning: managing alien plant invasions in the Kruger National Park, South Africa. Oryx 41:160–167

Foxcroft LC, Richardson DM, Pyšek P et al (2014) Chapter 1: Plant invasions in protected areas: outlining the issues and creating the links. In: Foxcroft LC, Pyšek P, Richardson DM, Genovesi P (eds) Plant invasions in protected areas: patterns, problems and challenges. Springer, Dordrecht, pp 3–18

Gaertner M, Den Breeÿen A, Hui C et al (2009) Impacts of alien plant invasions on species richness in Mediterranean-type ecosystems: a meta-analysis. Program Phys Geogr 33:319–338

Gaertner M, Holmes PM, Richardson DM (2012) Biological invasions, resilience and restoration. In: van Andel J, Aronson J (eds) Restoration ecology: the new frontier. Wiley-Blackwell, Oxford, pp 265–280

Gelderblom CM, van Wilgen BW, Nel JL et al (2003) Turning strategy into action: implementing a conservation action plan in the Cape Floristic Region. Biol Conserv 112:291–297

Goodall JM, Naudé DC (1998) An ecosystem approach for planning sustainable management of environmental weeds in South Africa. Agric Ecosyst Environ 68:109–123

Grice AC (2004) Weeds and the monitoring of biodiversity in Australian rangelands. Aust Ecol 29:51–58

Grice AC, Field AR, McFadyen REC (2004) Quantifying the effects of weeds on biodiversity: beyond blind Freddy's test. In: Sindel BM, Johnson SB (eds) 14th Australian weeds conference, Wagga Wagga. Weeds Society of NSW, Sydney, pp 464–468

Hamilton MA, Turner PJ, Rendell N et al (2010) Reducing the threat of a nationally significant weed to biodiversity: four years of implementation of the Bitou Bush Threat Abatement Plan. In: Zydenbos SM (ed) 17th Australasian weeds conference, Christchurch. NZ Plant Protection Society (Inc), Christchurch, pp 166–169

Hiebert RD, Stubbendieck J (1993) Handbook for ranking exotic plants for management and control. Natural resources report NPS/NRMWRO/NRR-93/08. US Department of the Interior, National Parks Service, Denver

Hockings M (2003) Systems for assessing the effectiveness of management in protected areas. BioScience 53:823–832

Holmes PM, Esler KJ, Richardson DM et al (2008) Guidelines for improved management of riparian zones invaded by alien plants in South Africa. S Afr J Bot 74:538–552

Hughes NK, Burley AL, King SA et al (2009) Monitoring Manual: for bitou bush control and native plant recovery. Department of Environment and Climate Change, Sydney

Humphries SE, Groves RH, Mitchell DS (1993) Plant invasions: homogenizing Australian eco-systems. In: Moritz C, Kikkawa J (eds) Conservation Biology in Australia and Oceania. Surrey Beatty, Sydney, pp 149–170

IUCN, UNEP-WCMC (2012) The World Database on Protected Areas (WDPA): Feb 2012. Cambridge

Khuroo AA, Reshi ZA, Rashid I et al (2011) Towards an integrated research framework and policy agenda on biological invasions in the developing world: a case-study of India. Environ Res 111:999–1006

Levine JM, Vilá M, D'Antonio CM et al (2003) Mechanisms underlying the impacts of exotic plant invasions. Process R Soc Lond B 270:775–781

Loope LL, Sanchez PG, Tarr PW et al (1988) Biological invasions of arid land nature reserves. Biol Conserv 44:95–118

Luken JO (1997) Management of plant invasions: implicating ecological succession. In: Luken JO, Thieret JW (eds) Assessment and management of plant invasions. Springer, New York, pp 133–144

Macdonald IAW (1983) Alien trees, shrubs and creepers invading indigenous vegetation in the Hluhluwe-Umfolozi Game Reserve complex in Natal. Bothalia 14:949–959

Macdonald IAW (1990) Strategies for limiting the invasion of protected areas by introduced organisms. Monogr Syst Bot Miss Bot Gard 32:189–199

Macdonald IAW, Frame GW (1988) The invasion of introduced species into nature reserves in tropical savannas and dry woodlands. Biol Conserv 44:67–93

Macdonald IAW, Loope LL, Usher MB et al (1989) Wildlife conservation and the invasion of nature reserves by introduced species: a global perspective. In: Drake JA, Mooney HA, di Castri F et al (eds) Biological invasions: a global perspective. Wiley, New York, pp 215–255

Mace GM, Baillie JEM (2008) The 2010 biodiversity indicators: challenges for science and policy. Conserv Biol 21:1406–1413

Margules CR, Pressey RL (2000) Systematic conservation planning. Nature 405:243–253

Martin TG, van Klinken RD (2006) Value for money? Investment in weed management in Australian rangelands. Range J 28:63–75

Maxwell BD, Lehnhoff E, Rew LJ (2009) The rationale for monitoring invasive plant populations as a crucial step for management. Invest Plan Sci Manage 2:1–9

McConnachie MM, Cowling RM, van Wilgen BW et al (2012) Evaluating the cost-effectiveness of invasive alien plant clearing: a case study from South Africa. Biol Conserv 155:128–135

McNeely JA (1994) Protected areas for the 21st Century: working to provide benefits to society. Biodivers Conserv 3:390–405

Mgidi TN, Le Maitre DC, Schonegevel L et al (2007) Alien plant invasions - incorporating emerging invaders in regional prioritization: a pragmatic approach for Southern Africa. J Environ Manage 84:173–187

Nel JL, Richardson DM, Rouget M et al (2004) A proposed classification of invasive alien plant species in South Africa: towards prioritizing species and areas for management action. S Afr J Sci 100:53–64

Obermeijer AA (1937) A preliminary list of the plants found in the Kruger National Park. Ann Transv Mus 17(4):185–227

Parker IM, Simberloff D, Lonsdale WM et al (1999) Impact: towards a framework for under-standing the ecological effects of invaders. Biol Invasions 1:3–19

Pauchard A, Alaback PB (2004) Influence of elevation, land use, and landscape context on patterns of alien plant invasions along roadsides in protected areas of South-Central Chile. Conserv Biol 18:238–248

Platt S, Adair R, White M et al (2005) Regional priority setting for weed management on public land in Victoria. In: 2nd Victorian weeds conference, Bendigo. Weed Society of Victoria, Melbourne, pp 89-98

Polasky S, Cramm JD, Solow AR et al (2000) Choosing reserve networks with incomplete species information. Biol Conserv 94:1–10

Pressey RL (1994) Ad hoc reservations: forward or backward steps in developing representative reserve systems? Conserv Biol 8:662–668

Pyšek P, Jarošík V, Kučera T (2002) Patterns of invasion in temperate nature reserves. Biol Conserv 104:13–24

Randall RP (2000) 'Which are my worst weeds?' A simple ranking system for prioritizing weeds. Plant Prot Q 15:109–115

Randall JM, Morse LE, Benton N et al (2008) The invasive species assessment protocol: a tool for creating regional and national lists of invasive non-native plants that negatively impact biodiversity. Invest Plan Sci Manage 1:36–49

Reid AM, Morin L, Downey PO et al (2009) Does invasive plant management aid the restoration of natural ecosystems? Biol Conserv 142:2342–2349

Rodrigues ASL, Andelman SJ, Bakarr MI et al (2004) Effectiveness of the global protected area network in representing species diversity. Nature 428:640–643

Roura-Pascual N, Richardson DM, Krug RM et al (2009) Ecology and management of alien plant invasions in South African fynbos: accommodating key complexities in objective decision making. Biol Conserv 142:1595–1604

Satchell M (1997) Parks in peril. US Newsl World Rep 23:22–28

SCBD (Secretariat of the Convention on Biological Diversity) (2010) Global biodiversity outlook 3. Montréal, Canada

Serrouya R, Wittmer HU (2010) Imminent extinctions of woodland caribou from national parks. Conserv Biol 24:363–364

Sinden J, Downey PO, Hester SM et al (2008) Economic evaluation of the management of bitou bush (Chrysanthemoides monilifera subsp. rotundata) to conserve native plant communities in New South Wales. Plant Prot Q 23:34–37

Skurka Darin GM, Schoenig S, Barney JN et al (2011) WHIPPET: a novel tool for prioritizing invasive plant populations for regional eradication. J Environ Manage 92:131–139

Soutullo A (2010) Extent of the global network of terrestrial protected areas. Conserv Biol 24:362–363

Spear D, McGeoch MA, Foxcroft LC et al. (2011) Alien species in South Africa's National Parks (SANParks). Koedoe 53, Art. #1032

Timmins SM, Owen S-J (2001) Scary species, superlative sites: assessing weed risk in New Zealand's protected natural areas. In: Groves RH, Panetta FD, Virtue JG (eds) Weed risk assessment. CSIRO Publishers, Melbourne, pp 217–227

Timmins SM, Williams PA (1991) Weed numbers in New Zealand's forest and scrub reserves. NZ J Ecol 15:153–162

Turner PJ, Downey PO (2010) Ensuring invasive alien plant management delivers biodiversity conservation: insights from a new approach using Lantana camara. Plant Prot Q 25:102–110

Turner PJ, Virtue JG (2006) An eight year removal experiment measuring the impact of bridal creeper (Asparagus asparagoides (L.) Druce) and the potential benefit from its control. Plant Prot Q 21:79–84

Turner PJ, Hamilton MA, Downey PO (eds) (2010) Plan to protect environmental assets from Lantana. Report for Biosecurity Queensland on behalf of the National Lantana Management Group. Department of Employment, Economic Development and Innovation, Yeerongpilly

UNEP (1999) CBD (Convention on Biological Diversity) – In-situ conservation. Article 8. http://www.cbd.int/convention/articles/?a=cbd-08

UNESCO (2012) Mongolia sacred mountains: Bogd Khan, Burkhan Khaldun, Otgon Tenger. http://whc.unesco.org/en/tentativelists/936/

Usher MB (1988) Biological invasions of nature reserves: a search for generalisations. Biol Conserv 44:119–135

Usher MB, Kruger FJ, Macdonald IAW et al (1988) The ecology of biological invasions into nature reserves: an introduction. Biol Conserv 44:1–8

van Wilgen BW, de Wit MP, Anderson HJ et al (2004) Costs and benefits of biological control of invasive alien plants: case studies from South Africa. S Afr J Sci 100:113–122 van Wilgen BW, Forsyth GG, Le Maitre DC et al (2012) An assessment of the effectiveness of a large, national-scale invasive alien plant control strategy in South Africa. Biol Conserv 148:28–38

Vilà M, Espinar JL, Hejda M et al (2011) Ecological impacts of invasive alien plants: a meta-analysis of their effects on species, communities and ecosystems. Ecol Lett 14:702–708

Weiss PW, Adair RJ, Edwards PB et al (2008) *Chrysanthemoides monilifera* ssp. *monilifera* (L.) T. Norl. and ssp. *rotundata* (DC.) T. Norl. Plant Prot Q 23:3–14

Wilcove DS, Rothstein D, Dubow J et al (1998) Quantifying threats to imperilled species in the United States. BioScience 48:607–615

Williams MC, Auld BA, Whiffen LK et al (2009) Elephants in the room: widespread weeds and biodiversity. Plant Prot Q 24:120–122

Chapter 24
Overcoming Barriers to the Prevention and Management of Alien Plant Invasions in Protected Areas: A Practical Approach

Mandy Tu and Ramona A. Robison

Abstract This chapter familiarises protected areas managers with some of the basic background information and tools for preventing and managing invasive alien plants. The most efficient methods for protecting natural areas from the impacts of invasive alien plants include the prevention and early detection of new introductions, followed by rapid response to eliminate small populations, and to keep established populations at manageable levels. Barriers to achieving these outcomes are typically related to a lack of awareness of the problems, adequate knowledge, allocation of time from staff and volunteers, available technology, and continued funding for projects. It is critical to focus management actions on the most damaging top priority invasive alien plants and their pathways of introduction. An awareness of the problems and impacts caused, followed by an assessment of the biological and contextual situation, effective communication on the extent of the threats, and planning for and implementing dynamic interventions are essential. Simple steps can be taken to assess the situation and lessen the impacts. This chapter is designed to serve as a practical roadmap for addressing invasive alien plant invasions in protected areas.

Keywords Adaptive management • Assessment • Big basin state park • California • Early detection • Prioritization • USA

M. Tu (✉)
Independent Consulting Ecologist, 1608 NE 63rd Avenue, Hillsboro, OR 97124, USA
e-mail: mandytu.pdx@gmail.com

R.A. Robison
Botanist and Invasive Species Specialist, California State Parks, 1416 9th Street, Room 923, Sacramento, CA 95814, USA
e-mail: ramona.robison@parks.ca.gov

L.C. Foxcroft et al. (eds.), *Plant Invasions in Protected Areas: Patterns, Problems and Challenges*, Invading Nature - Springer Series in Invasion Ecology 7, DOI 10.1007/978-94-007-7750-7_24, © Springer Science+Business Media Dordrecht 2013

24.1   Introduction

Invasive alien plants (IAPs) can have a multitude of effects on protected area (PA) values. These can occur at the species, population and community levels, and the most damaging can significantly alter ecosystem functions and processes (Drake et al. 1989; Chornesky and Randall 2003). Invasive alien plants are, however, often not recognised as threats to PA values (De Poorter 2007). To our knowledge, the vast majority of PAs worldwide have not acknowledged or adequately addressed IAPs, and unfortunately, even where they are recognised, available funding and other resources for their management are often lacking (De Poorter 2007; Genovesi et al. 2010). Since there will never be adequate resources with which to manage all IAPs, it is essential to prioritise prevention and management interventions to achieve the highest possible benefit.

The best approach to ensuring that PA values are shielded from the impacts of IAPs originates from adequate knowledge and planning. This includes an awareness of threats and an adequate assessment of the biological and contextual situation. Effective communication on the extent of the threat to PA management and stakeholders is needed to facilitate capacity building, planning and implementing dynamic interventions. Although some resources are needed to effectively address these threats, simple steps can be taken to assess the situation and lessen the impacts of IAPs.

This chapter is designed to assist PA managers in overcoming some common obstacles encountered when addressing IAPs. Towards the end of the chapter, a list of specific questions is provided to serve as a roadmap for creating a realistic and practical IAP management plan.

24.2   Barriers to Invasive Alien Plant Prevention and Management

There may be many impediments to the adequate prevention and management of IAPs. In addition to the difficulty of sustaining active management interventions and monitoring their success, common obstructions to management may include a lack of: motivation and awareness of the seriousness of the threat to PA values; knowledge about what resources are available and how to implement prevention, early detection and management programmes; equipment, technology and other tools to map and control IAPs; available person-hours (allocation of staff and/or volunteers); and sustained funding for continued monitoring and control.

The above items are not discrete, and there are often overlaps in how to best address each barrier to action. For instance, lack of awareness of the IAP problem and knowledge of what resources are available and how best to proceed can be addressed through a range of avenues. These include outreach, education, background research, expert opinion from other land managers, and some preliminary data gathering. Here, we suggest that to overcome these common barriers, each PA should consider the following:

1. Motivation to get started and willingness to act. A lack of motivation to work on IAPs may initially stem from lack of awareness, or from feeling overwhelmed by the scope of the problem. Getting started by articulating clear goals and objectives in writing can provide needed motivation. Similarly, learning some of the potential impacts of IAPs to PA values, and communicating those threats to upper management, may result in having them included in overall PA management goals.

2. An assessment of the IAP threats. A basic assessment of the current situation is necessary to communicate the threats and potential impacts of IAPs to PA values. Once an initial assessment is complete, the PA manager can then decide how to proceed. This may determine if management actions will be taken, prevention practices put into place, or further assessments completed.

3. A written prevention and management plan for IAPs. When integrated into the overall PA management plan (Ervin 2003; Goodman 2003; Pomeroy et al. 2004), IAPs can start to gain priority in management decisions. There are unlikely to be sufficient resources to manage all IAPs, thus having a written plan that adequately assesses the situation, prioritises sites and species, and outlines possible management intervention options is needed.

4. Implementing prevention, early detection and control programmes. Only when an IAP plan is completed can funding and other resources be requested and allocated to projects. Top priorities for IAPs should include prevention and early detection programmes, focusing a portion of management on small infestations and/or the most damaging invaders. Optimally, each PA plan should integrate IAP threats and management options, coupled with recommendations for sufficient staff and funding for long-term prevention and management of IAPs.

5. Staff and funding dedicated to IAP monitoring and management. Dedicated and knowledgeable staff are essential to prevention and management success in the long-term. The IAP management plan should outline how many staff and volunteers are needed, and describe the equipment and other resources needed to prevent and manage top priority IAPs. Much can be accomplished with volunteers, but continuity of staff and written documentation are essential to sustain efforts. If funds are only available on an ad hoc basis, they might best be used towards assessing IAPs through mapping and monitoring, or for detecting and preventing the establishment of new invaders. It is crucial to have at least one staff member dedicated to overseeing IAP project progress.

The remainder of this chapter focuses on how a PA manager might complete a basic assessment of IAP threats, form of an early detection programme (with limited resources), and develop an adaptive IAP management plan.

24.3   Assessing the Ecological Situation and Available Resources

An assessment of the ecological context is needed before developing an IAP management plan. This assessment identifies the species and populations of IAPs that currently, or potentially, threaten PA values. Results from the assessment logically lead towards the development of management priorities and suitable interventions, and can also be used as a tool to communicate IAP threats with upper management, partners, funders, and other stakeholders.

A basic assessment should include a list of the most damaging IAPs within the PA and a description of the impacts that those IAPs may have on PA values. If further funds become available, a complete inventory of IAPs could be completed, including: (i) list of IAPs present within the PA, and their impacts; (ii) list of IAPs not yet present within the PA, but nearby in the region; (iii) map of IAP locations and their extents, as well as non-invaded areas; (iv) an investigation of site condition, in relation to IAP presence or abundance.

To identify IAP threats, the first thing a manager needs to know are which plant species are considered alien and could degrade PA values. After delineating PA boundaries, the manager should ideally compile a complete plant list, or at minimum a targeted IAP species list, and note their locations on maps. Mapping the location and extent of IAPs as well as their proximity to high value sites such as water, wetlands or rare communities will determine the most appropriate management tools and techniques. Mapping IAPs can also identify entry pathways and vectors. An additional purpose of mapping is to assist in monitoring management effectiveness.

Information on IAP distribution and abundance can be used to evaluate changes over time. Total numbers of IAP species, number of sites invaded, or hectares invaded are all examples of quantifiable measures. In addition to field surveys, information on what is invasive locally can be acquired through literature and internet database searches, collection and specimen records from local herbaria, interpretation of remotely-sensed data, and from local experts (Wittenberg and Cock 2001; Salafsky et al. 2003; GISD 2012).

To determine what resources are available and whether there are any limitations to the use of certain tools and techniques, the manager must be able to ascertain if the existing staff and volunteers have the capacity, ability, and technical knowledge to: (i) recognise and identify the IAP species that may cause damage to PA values; (ii) determine best management methods, including where to look for control information; (iii) recognise pathways of potential IAP entry; (iv) secure the labour needed to control and remove IAP; (v) identify gaps in practices, programmes, and policies to prevent IAP introduction; (v) identify who to educate on the issue to receive institutional and/or public support for IAP work; (vi) determine if there are limitations to the use of specific management tools or techniques; and (vii) know who to collaborate with to share equipment and resources, as well as how to develop community support for control projects.

24.3.1 Completing an Assessment

An IAP assessment may be accomplished at various levels of detail and complexity. At the most basic level, the IAPs with the greatest impacts in the PA should be identified, and those impacts understood and communicated. A few relatively easy methods for assessing and mapping IAPs when resources are limited may include:

- Find which species are alien locally by talking with a knowledgeable individual (from a nearby university, agency, or someone with expertise on local plants) or completing some online searches that may list IAPs in your region (GISD 2012)
- Walk through the PA with that knowledgeable individual to identify local IAPs
- Use hand-drawn maps to record locations and extents
- Rough estimates of infestations may be quantified through visual estimates of percent plant cover, photo-monitoring, or remote sensing technologies

24.3.2 Using Technology to Assist in Assessment and Management

Although not essential to IAP management, having access to and utilizing current technology can assist a PA manager in a variety of ways. While many PAs may not have access to the latest technologies in remote sensing, mapping software, or invasive plant control equipment, much can be accomplished simply by using a computer (or a smartphone) with internet access. The internet can serve as a portal, enabling a PA manager at a remote site to: discover what may be invasive in a country or region (e.g. through the Global Invasive Species Database, GISD 2012); communicate and ask questions to a large knowledgeable audience (such as through the Aliens-Listserve, ISSG 2012); find a regional IAP expert (Daisie 2012); become knowledgeable about IAPs and how to manage them (CIPM 2002/2007); identify possible IAP species, and search for control options and best management practices (i.e. search for 'species name' and 'control').

Additionally, there are now many free tools available online that can assist in mapping and assessing IAPs. Several open source GIS (geographic information system) software packages are available, there are smartphone applications that can assist in species identification and mapping, and services such as Google Earth may provide baseline maps. While funding for equipment and technologies will always be desired, today there are many technologies that can assist in IAP assessment and management that only a few years ago did not exist or were prohibitively expensive. If a PA manager only has a computer or smartphone with internet access, technology can be used as a tool to save time, increase accuracy, and assist in IAP management.

24.3.3   Case Study 1: Mapping Arundo donax at Cuatro Ciénegas, Coahuila, Mexico

The valley of Cuatro Ciénegas, Mexico is one of the world's most unique ecosystems, displaying a high rate of endemism that is associated with its hundreds of geothermal springs. These springs and nearby wetlands and rivers have recently been invaded by the alien grass *Arundo donax* (giant reed). To determine the extent of *A. donax* invasion in the valley, the non-profit conservation group ProNatura-Noreste took the lead in mapping and assessing the threat to the valley. They used a combination of GIS-produced maps and hand-drawn paper maps, which were then supplemented with GPS locations from handheld units, to create their initial assessment of the extent of *A. donax*. Later, enhancements to the maps were made using aerial photographs. While not complete, even the early maps showing the extent and damage incurred by *A. donax* to the geothermal springs were informative, and encouraged further mapping assessment and research.

24.4   Developing an Early Detection and Rapid Response Programme

After prevention, the early detection and identification of a new species or population at a given site, and a rapid response to eradicate or contain it before it spreads, is the next most worthwhile endeavour to minimise IAP damage (Hobbs and Humphries 1995; Simberloff 2014). Since prevention is covered by other chapters of this book (Genovesi and Monaco 2014; Meyerson and Pyšek 2014) it is not detailed here. Many IAPs are difficult or impossible to manage once they are well established, and therefore prompt management interventions are often the best use of limited management funds (Rejmánek and Pitcairn 2002; Harris and Timmins 2009).

Early detection and rapid response (EDRR) can be implemented with limited staff or as part of a comprehensive programme involving multiple staff and volunteers. When only one or a few staff or volunteers are available they can be given information on new species with potential to invade the area, and time to look for and report any they find. When additional staff or volunteers are available for surveillance, more thorough and comprehensive surveys are possible. Core elements of an EDRR programme include: cooperating with surrounding landowners, training staff and volunteers, detection and reporting, identification and voucher specimen collection, rapid assessment, planning, rapid response, and monitoring and evaluation (FICMNEW 2003; Genovesi et al. 2010).

Although an EDRR programme can be run entirely within the boundaries of a PA, a cooperative programme including surrounding landowners will provide additional value. All stakeholders should agree on the IAPs being targeted, and the survey and reporting responsibilities of each group. Regular communication between all landowners is required to indicate when new species are detected, or when surveillance confirms that specific areas are uninvaded. Additionally, partner involvement may be needed to develop a public outreach or communication plan to gain support for the problems and planned work.

Creating and maintaining an effective EDRR programme requires training for PA staff and volunteers, as well as researchers, visitors, and members of partner organisations that may be able to provide additional assistance. Those leading the programme should determine survey priorities, monitoring protocols, and mapping needs, and develop short surveillance or 'watch-lists' of IAPs nearby, but not yet in the PA. Also needed are clear reporting pathways to appropriate staff, agency or other entities, and direction on how and who will manage new infestations. Those who will be looking for new invaders and carrying out the EDRR plan should know how to identify IAP species on the watch-list, the most important places to look for new IAPs, survey protocols, and how and where to report new invaders.

Once the identification of a new IAP is confirmed, the manager should rapidly assess its distribution and evaluate its potential to spread further. Options for eradicating, containing or controlling it should also be determined. Ideally this information will already be in the IAP management plan. Deadlines should be short enough to ensure that the new invader does not have time to spread. Contingency plans and funding sources for the surveillance work and for emergency responses to new IAP introductions should be identified.

Eradication or containment of even a small population of a new IAP may need repeated treatments over many years. If managers decide not to take action on a new IAP, they should clearly document this decision and the reasons for it. Following management action, the fate of the population should be tracked with regular monitoring. This may require regular monitoring for many years to ensure that all individuals are eliminated and that no seeds or plant propagules remain. Monitoring data should be quickly evaluated and plans for further treatments modified as appropriate. The data and decisions based on monitoring should be recorded and reported.

An effective early detection programme can be accomplished in different ways, depending on the scope and scale of the project and available personnel and other resources. At its core, early detection is just finding or observing a new plant and reporting it. When approached holistically as part of the entire PA IAP programme, early detection may also employ training staff or volunteers to look for certain species, or to monitor a special site for new invaders. It also includes positively identifying the new observation, recording its location, which is then rapidly followed by efforts to control or eliminate the plant(s).

24.4.1 Case Study 2: A One Person EDRR Programme at Lassen Volcanic National Park, California, USA

Lassen Volcanic National Park is located in the southern Cascade Range in the western USA. The park has one botanist, who also manages the IAP programme, and thus there are limited resources available for IAP mapping, monitoring and EDRR. The park does, however, have several staff that travel around the park, completing maintenance work on trails, campgrounds, and other facilities. In the mid-2000s, the park botanist created a short-list of IAPs that are uncommon in the park, or are known from nearby, but were not yet on park property. She created a one-page colour hand-out that included photographs and identifying features of these IAPs, and gave this to maintenance workers throughout the park. This hand-out included information on who to contact if any of these plants were observed in the park. This way, even though the park only has one botanist with many responsibilities, she was able to get many people involved, looking for and reporting new plants and infestations.

24.4.2   Case Study 3: Outreach and Education: Encouraging the Reporting of New Invaders in Laojunshan National Park, Yunnan Province, China

Laojunshan Park is a relatively new national park in southwest China. Due to its isolation from major cities and highways, and its high biodiversity, it is relatively uninvaded by IAPs. In order to educate the local villagers, park visitors and park officials about challenges facing the park, the non-profit organisation *The Nature Conservancy*, worked with partners to develop outreach materials, including a very popular colour calendar depicting common IAPs. By asking locals to participate and report IAPs, they worked to engage and inform their stakeholders about invasive species problems.

24.5   Writing an Adaptive Prevention and Management Plan

In order to not become overwhelmed with IAP problems, it is necessary to have a written management plan. A written plan helps ensure continuity, especially as staff turnover is inevitable. The plan itself can be used to leverage additional funding and as a communication tool. There are numerous examples available of such management plans, ranging from single species plans to national-scale plans for all invasive species (ISSG 2012; NISC 2012).

An effective plan clearly states PA goals and objectives, and includes an assessment of the current situation and available resources. The management plan should not only identify those species that currently, or potentially, pose threats to goals and values, but should also list high value areas, and prioritise species or populations for management. It should also include details on treatment options, planned management actions, methods for monitoring, and evaluating and adapting

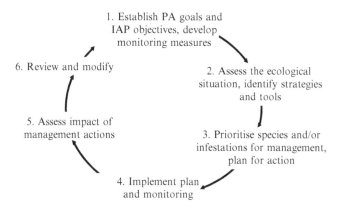

Fig. 24.1 Adaptive framework for invasive alien plant management (Modified from Tu and Meyers-Rice 2002; Tu et al. 2001)

results (Pomeroy et al. 2004). Comprehensive plans should also identify vectors and pathways of entry, data management needs, and opportunities for collaboration (Wittenberg and Cock 2001). The best management plans are those that can adapt to changing situations and conditions, and use a cyclical adaptive management framework that allows the PA manager to address problems and test solutions (Fig. 24.1), ensuring that certain measurable objectives are met (Tu et al. 2001).

Using this type of framework for IAPs, the PA manager should first (i) establish the overall goals of the PA, which identify the core values that the PA aims to maintain such as a rare species or community, overall biodiversity or ecosystem processes, open space, or cultural and historic values. The manager then identifies how IAPs could damage those values, developing specific IAP objectives to prevent, contain or reverse the damage caused. The process also assists in developing monitoring measures that can discern when objectives are met. (ii) The manager assesses the unique situation of the PA, including available resources and the ecological context. Suitable strategies and tools are identified that can be used to accomplish the objectives. (iii) Species and/or infestations are prioritised for management based on the value(s) of the site and the severity of impacts to PA values, and a plan of action is made. (iv) The plan is implemented and monitoring is completed. (v) The manager evaluates the impact of the management actions in terms of protecting PA goals and values, and of achieving objectives. Finally (vi) the cycle is repeated after re-evaluating conclusions, evaluating steps (i) through (iv) and modifying and improving objectives, priorities, methods and plans.

To assist PA managers in writing this type of IAP management plan, a template is provided by Tu and Meyers-Rice (2002). A short list of questions that can be used in the development of a plan is also provided in Table 24.1. Case study 4 briefly illustrates how a plan might be used to inspire action, and case study 5 provides an example from California, USA of their IAP situation.

Table 24.1 A roadmap for developing an invasive alien plant management programme

(i) Basic initial questions: setting the stage
1   What are the overall goals and mission of the PA?
2   How might IAPs interfere with these goals or mission(s)?
3   What are three IAP objectives that can be accomplished in the next 2 years?
4   Is there a recent plant inventory and map of IAP locations?
5   What personnel and other resources are available for IAP management within the PA?
(ii) Strategies and developing priorities
6   What are the main vectors and pathways by which IAPs could move into the PA?
7   Which IAPs are in the vicinity but not yet in the PA, and have the potential to severely impact management goals and objectives?
8   Which are the highest priority sites that are currently uninvaded or lightly invaded?
9   Which are the highest priority IAPs to prevent and control?
10 Are there small populations of IAPs that can realistically be eliminated?
(iii) Management intervention decisions and monitoring change
11 What are the top three IAP priorities (species or specific populations) for eradication, containment or control? What might be the best method or combination of methods to eradicate, contain or control those IAP?
12 Will current monitoring be able to determine if the objectives have been achieved?

The questions listed above aim to assist a PA manager in creating a realistic and practical adaptive IAP management plan. It should provide the basic outline for a plan, including objectives, assessment needs, suitable strategies, and priorities. Start with answering at least the questions in section (i). If additional resources are available, continue answering the remainder of the questions. If a question cannot be answered at this time, contemplate what steps are necessary in order to create an adequate response. Responses are intended to be used to develop specific action steps and direct future activities

24.5.1 Case Study 4: Using an Invasive Alien Plant Management Plan to Stimulate Action and Support at the Cosumnes River Preserve, California, USA

The Cosumnes River Preserve is a cooperative conservation project located in central California, USA, involving federal and state agencies and local non-profit groups. Even though IAPs are common in the region, they were not considered a high priority for management until one of the site managers wrote an IAP management plan, based on a template provided by The Nature Conservancy (Tu and Meyers-Rice 2002). With a written IAP plan in-hand, the scope of the IAP threat to the preserve could be described, and several workable solutions offered. In this way the PA managers were convinced to allocate some of the site managers' time, and that of several volunteers, to eliminate alien trees from the preserve. Additionally, the plan kept the staff focused on the top IAP priorities, and helped secure grants for equipment and IAP control work.

24.5.2 Case Study 5: Example of an Invasive Alien Plant Management Plan: Big Basin State Park, California, USA (Contributed by Ramona A. Robison and Tim Hyland)

Big Basin State Park (BBSP), located in the California central coast mountain range, is about 6,700 ha and is dominated by a variety of vegetation types, including old growth redwood forest, mixed conifer forest and chaparral. The following provides 12 aspects that are considered within BBSPs IAP management plan.

1. The overall mission of the California State Parks is "To provide for the health, inspiration and education of the people of California by helping to preserve the state's extraordinary biological diversity, protecting its most valued natural and cultural resources, and creating opportunities for high-quality outdoor recreation." Specific to BBSP, goals include managing vegetation to support high native species richness, maintaining or enhancing habitat for rare and threatened species, and detecting new populations of IAPs while still small, and eliminating them where feasible.
2. Invasive alien plants reduce biodiversity and restrict access for recreation.
3. Specific objectives for vegetation management within the old growth redwood forest at BBSP include:

   - Maintain the health of 1,862 ha of old growth redwood forest
   - Burn all prescribed burn plots every 7–15 years
   - Manage for rare and representative native plant species
   - Eliminate all reproductive plants of *Genista monspessulana* (French broom)
   - Prevent the establishment of *Ehrharta erecta* (panic veldtgrass) and *Brachypodium sylvaticum* (false brome)
   - Eliminate all populations of *Cortaderia jubata* (purple pampas grass)

4. BBSP has two on-going mapping programmes that include IAPs. The first is collected at the Park level and is conducted in conjunction with control efforts. This mapping is updated yearly and existing points are verified and new occurrence points are noted and added to the mapping system. The second mapping approach covers IAP occurrences within a sub-set of the entire California State Parks system (279 parks, comprising 607,000 ha). This mapping effort tracks a sub-set of selected species at designated parks, such as *G. monspessulana* and *Vinca major* (large periwinkle), which are mapped on 3 year intervals and the results are compared over time across the designated parks.
5. The Santa Cruz District of the California State Parks, which includes the BBSP, employs three full-time scientists, and two to three seasonal workers. In addition, there are seven full-time personnel who spend roughly 80 % of their time on vegetation management. As there are 29 parks in the district, totalling about 25,500 ha, all efforts to successfully manage IAPs must be prioritised.

6. Vectors of IAPs into BBSP include roads, trails, waterways, campground use and adjacent properties. Additionally, encroachment permits are requested for a variety of uses such as utility line maintenance, special events and movie filming. Currently the District prepares and approves encroachment permits, and conducts management unit inspections as part of an early detection rapid response system. District staff are also responsible for roadside and campground maintenance, although major highways are maintained by County or State departments.

7. The District is currently working to develop a 'watch list' of nearby IAPs that could be introduced into the Parks. California has several systems that could be integrated to contribute to this effort.

8. Within BBSP the highest priority habitats are old growth redwood forest, knobcone pine and chaparral communities. Fortunately, these three habitat types are relatively resistant to IAP invasion, but monitoring is needed to ensure new species are not introduced.

9. The most important IAPs to control in BBSP are those species which could invade the old growth redwood forest understory. *Ehrharta erecta* and *B. sylvaticum* have the potential to do significant damage and are therefore high priorities for early detection and elimination.

10. There are a few small populations of IAPs that can be eliminated. For example, there is a small patch of *Ammophila arenaria* (European beachgrass) of less than 0.5 ha, and three patches of *E. erecta* smaller than 0.1 ha, which are currently being treated.

11. The top three priorities for management in BBSP are

    (a) *Ehrharta erecta*, which is being managed with chemical control.
    (b) *Brachypodium sylvaticum* is on the surveillance list and is not currently widespread.
    (c) *Cortaderia jubata* is managed with a combination of mechanical removal and chemical control.

    Control of outliers and monitoring of perimeter areas are also high priorities. Plants are treated when they are most visible, but before they set seed. Efforts are made to ensure multiple visits are made to sites during one season, but unfortunately this is not always possible. Sites with the greatest ecological value are focused on first.

12. Monitoring consists of annual management unit inspections. Information is collected on presence and absence of certain species in designated locations. Additional monitoring is also carried out in areas with active management. In those cases additional information is collected and tracked to monitor progress.

All control efforts are documented, including number of plants, person hours, and the amount of herbicide used. These records can be used to determine if control efforts are succeeding. However, they do not quantify the effect on native plant communities.

24.5.3 Prioritizing Invasive Alien Plant Threats and Management Interventions

One large impediment to writing an IAP management plan is identifying the top priorities for management. There are often too many IAP species and populations to control, and never enough resources to manage all of them. Priorities help the manager direct funds and other resources to the most productive projects and minimise the workload needed to curtail long-term damage caused by IAPs. It is generally better to focus on a few IAPs that are not as widespread where there is a good chance of success, rather than on widespread IAPs which may be impossible to control. Preventing new infestations and assigning the highest priority to keeping clean areas uninvaded should always be the top priority. Guidelines for how to deal with IAPs are further addressed in detail in several chapters in this book (Larios and Suding 2014; Simberloff 2014; Van Driesche and Center 2014).

Decisions to prioritise IAPs should consider the environmental damage caused by the target species and the value of the threatened asset (Timmins and Owen 2001). Invasive alien plants that are fast-growing and disruptive to ecosystems, or occur within the most highly valued sites (such as communities of rare species), may be the highest priority species or areas considered for active management. The difficulty of control of a particular IAP species should also be considered, giving higher priority to infestations in high quality sites which are controllable with available technology and resources. Therefore, a high priority may be given to eliminating one or two IAP species over their entire invaded range, and to keeping several sites of high biodiversity or ecological importance free of a larger suite of damaging IAPs.

Once an IAP inventory has been completed for the PA, the manager should identify and determine those species and populations that currently or potentially threaten values. Many IAPs are able to move into and dominate natural communities or restoration projects. However, only a small proportion of IAPs may be true ecosystem engineers (i.e. those species that can completely change or transform the ecosystem), disrupting ecosystem processes such as fire regime, soil nitrogen levels, or hydrology, thereby shifting plant communities and altering landscapes (Chornesky and Randall 2003). It is these ecosystem engineers that are most likely to threaten the PA values and their management should be a high priority.

The priority-setting process can be difficult, partly because managers need to consider many factors. There are several prioritization methods available (Hiebert and Stubbendieck 1993; Timmins and Owen 2001; Tu and Meyers-Rice 2002; Skurka Darin et al. 2011) that work to rank the invasiveness of plants, invasibility of sites, and/or to assist in developing management priorities. There are many considerations to take into account when prioritizing sites and species for management. For example, public support or opposition to management intervention can determine if a project will be funded, or even be initiated (Genovesi 2007, 2011).

The following guidelines (modified from Owen 1998; Timmins and Owen 2001; Mazzu 2005) may help in developing IAP management priorities:

1. First priority. Prevent invasions by IAPs known from nearby PAs or from areas surrounding the PA (e.g. take action to prevent invasions by requiring use of weed-free hay, closing trailheads near known infestations, and urging all hikers to clean their shoes at stations installed at trailheads).
2. Second priority. New IAP populations in the PA or surrounding region; new IAP populations in sites of high conservation status or other specially designated sites, or in areas not yet invaded.
3. Third priority. Other IAP populations that are small, easy to control, or not yet widespread; IAP populations that threaten specific PA values, such as endangered or endemic species or communities, or fragile habitats; keeping large areas free of the most damaging IAPs; IAP populations that could be a source of propagules (e.g. upstream IAP populations, roadsides, trailheads, visitor centres).
4. Fourth priority. Containment of priority IAPs where they exist in large populations; control and suppression of existing large IAP populations to acceptable threshold levels.

24.5.4   Selecting the Most Appropriate Tools for Managing Invasive Alien Plants

One of the last barriers to writing and implementing an IAP management plan is deciding on the most appropriate tools and techniques. Fortunately, many invasions can be slowed, halted or reversed, and in certain situations even severely invaded areas can be restored to healthy ecosystems, but this often requires years of active management. The objectives of IAP management actions may be eradication, containment and/or control of the IAP (Wittenberg and Cock 2001; Hulme 2006), depending on the extent of the IAPs, condition of the surrounding environment, and available resources.

Eradication is the complete, long-term elimination of an IAP species within a defined area, while containment is the restriction of the distribution and spread of the IAP species in that area. Control is the suppression of IAP abundance within or between a given area, typically to below an acceptable threshold level that still allows the values of the PA to be maintained (García-Llorente et al. 2011). Eradication is covered in detail in (Simberloff 2014), thus is not further discussed here.

Often knowing what tools and methods are available for the control and management of IAPs, and selecting the best tool for a species or situation, can be daunting. There are advantages and disadvantages to each tool and technique. Further, control should be thought of as a long-term continual maintenance commitment, because if it is not continued indefinitely, biodiversity and other values will likely decline to pre-treatment levels. The long-term goals of management at the control level are to protect sites and values from becoming degraded, to keep IAPs from spreading and negatively impacting management targets, and to decrease the resources needed to maintain healthy ecosystems. Control costs may be high in the early stages of a project when the IAP species is abundant and widespread, but as its abundance and distribution in the PA decrease, control costs may decrease as well. Control actions are often combined with restoration efforts, which work to bring a project or area into a desired future state while simultaneously reducing IAP abundance and promoting native species and communities (Larios and Suding 2014).

Striving to increase the proportion of time and effort spent on prevention and EDRR practices, rather than on control and management, is an advisable objective. At the same time, it is necessary to keep in mind any potential off-target impacts of the planned management actions and any possible risks to human health and environmental safety. A decision to do nothing about an IAP species or invasion also has costs and benefits, and these should be evaluated and recorded. When selecting the method, choose those tools and techniques that will provide the most targeted and selective control, ensuring minimum non-target damage to the surrounding environment and native plant communities. Some tools are more selective or targeted than others, or can be made more selective depending on their application. Each tool, technique and methodology is different and has advantages and disadvantages depending on how it is applied (Table 24.2).

24.6 Conclusions

There can be many barriers to addressing and implementing a successful IAP prevention and management programme. Initiating a new programme for IAP management can seem daunting at first, especially if the PA manager is not already familiar with the problems associated with alien plant invasions and is already engaged in other land management activities. However, there are many sources of information available on IAP prevention and management to assist in the development of a simple, effective plan. The key is to get started by doing something, and to let the process evolve as more knowledge and management expertise is developed for solving the particular IAP management problem.

Obtaining sufficient resources will always be a challenge. The primary focus needs to be on the objectives and the aspects that can be achieved and will make a difference, such as prevention and EDRR, and the control of small populations of the most damaging invaders. When stymied, engage with willing and interested partners to gauge interest and share resources to raise awareness and initiate a course of action. Partnerships are critical to raise awareness and provide a venue to communicate how IAPs may harm and threaten PA values. Academics or experts from local institutions, or even from other regions of the world, may be able to assist in identifying and quantifying the IAP situation. Even if there are little to no resources immediately available, a manager can take steps to prevent new invasions by minimising the introduction and spread of IAPs. Many IAP infestations take several years of repeated treatments to produce tangible results, but with on-going monitoring, prevention and management efforts may show surprising successes.

Table 24.2 Brief descriptions of some plant control methods and techniques

Method description	Advantages	Disadvantages	References
Manual and mechanical			
Manual includes human labour using: shovels, picks, axes, saws or machetes. Mechanical includes use of machinery: mowers, brush-cutters, chainsaws or earth-moving equipment	Little training is needed for safe use of most tools and they can be used in a variety of situations; hand tools are relatively low cost and can provide very specific targeted control	Time and labour intensive; difficult to adequately control IAPs when populations are no longer small; often does not provide lasting control; tools may be sharp and dangerous	Cal-IPC (2004) and DiTomaso et al. (2013)
Land management practices			
Any method that modifies land practices (i.e. grazing, prescribed fire or irrigation/flooding) to the detriment of IAPs and benefits PA objectives	No chemical use; may be able to treat large areas; generally low intensity of human labour; native plants may respond positively; may return a site to a historic disturbance regime	May not be specific to IAPs; may result in more soil disturbance or damage to off-target native plants; generally does not eliminate the target IAPs from an area	Brooks and Lusk (2008) and Zouhar et al. (2008)
Biological			
Use of another species, such as insects, fungi, or microbes, to control an IAP species	Relatively low cost to acquire biological control agents; can keep IAPs at a low level across large landscapes; no chemical residues; no need to repeatedly treat IAP infestations once agents are established	Few IAP species have biological control agents available; may not work in all instances; risks of unintended consequences to native species and communities	Coombs et al. (2004) and Van Driesche and Center (2014)
Chemical			
Use of specific chemicals (herbicides) to kill undesirable plants	May be able to treat large populations at a lower cost than other methods; may be specific to the target IAP; may achieve good control; may not be too labour and/or time intensive	Chemical residues may remain in soil or water; may contaminate groundwater sources; may have off-target or unintended impacts on desirable native plants and communities; may have health exposure issues for applicators; may be expensive to obtain chemicals, application equipment, protective equipment, and applicator training	Bossard et al. (2000) and Tu et al. (2001)

(continued)

Table 24.2 (continued)

Method description	Advantages	Disadvantages	References
Restoration			
Process of returning an ecosystem to a close approximation of a historical or natural condition	Works to bring the project site to a desired and/or natural condition; establishes native plants to compete with IAPs; revegetation with native plants may keep IAPs from establishing	Restoring previously densely invaded sites is typically very expensive (i.e. for labour, cost of native plants and long-term monitoring and maintenance)	Falk et al. (2006) and Larios and Suding (2014)

Much information is available on techniques and management applications (Sheley and Petroff 1999; Bossard et al. 2000; Tu et al. 2001; Wittenberg and Cock 2001; Cal-IPC 2004; Hulme 2006; DiTomaso et al. 2013). Rather than repeating these techniques in detail we present the salient points, specifically within the context of PA management

References

Bossard CC, Randall JM, Hoshovsky MC (2000) Invasive plants of California's wildlands. University of California Press, Berkeley

Brooks M, Lusk M (2008) Fire management and invasive plants: a handbook. US Fish & Wildlife Service, Arlington

Cal-IPC (2004) The weed workers' handbook: a guide to techniques for removing Bay Area invasive plants. California Invasive Plant Council, Berkeley. http://www.cal-ipc.org/ip/man agement/wwh/index.php. Accessed 15 Mar 2012

Chornesky EA, Randall JM (2003) The threat of invasive alien species to biological diversity: setting a future course. Ann Miss Bot Gar 90:67–76

Center for Invasive Plant Management (2002/2007) Invasive plant management: CIPM online textbook. Center for Invasive Plant Management, Bozeman. http://www.weedcenter.org/text book/index.html. Accessed 12 Dec 2012

Coombs EM, Clark JK, Piper GL et al (2004) Biological control of invasive plants in the United States. Oregon State University Press, Corvallis

Daisie (2012) Delivering alien invasive species inventories for Europe website, able to search experts, species and regions. European Commission under the Sixth Framework Programme. http://www.europe-aliens.org. Accessed 12 Dec 2012

De Poorter M (2007) Invasive alien species and protected areas: a scoping report. Part 1. Scoping the scale and nature of invasive alien species threats to protected areas, impediments to invasive alien species management and means to address those impediments. Global Invasive Species Programme, Invasive Species Specialist Group. http://www.issg.org/gisp_publica tions_reports.htm

DiTomaso JM, Kyser GB, Oneto SR et al (2013) Weed control in natural areas in the Western United States. University of California Press, Berkeley

Drake JA, Mooney HA, di Castri E et al (1989) Biological invasions: a global perspective. Wiley, New York

Ervin J (2003) Rapid assessment of protected area management effectiveness in four countries. BioScience 53:833–841

Falk D, Palmer M, Zedler J (2006) Foundations of restoration ecology: the science and practice of ecological restoration. Island Press, Washington, DC

FICMNEW (2003) A national early detection and rapid response system for invasive plants in the United States: conceptual design. Federal Interagency Committee for the Management of

Noxious and Exotic Weeds, Washington, DC. http://www.fws.gov/ficmnew/FICMNEW_
EDRR_FINAL.pdf. Accessed 30 Mar 2012

García-Llorente M, Martín-López B, Nunes PALD et al (2011) Analyzing the social factors that
influence willingness to pay for invasive alien species management under two different
strategies: eradication and prevention. Environ Manage 48:418–435

Genovesi P (2007) Limits and potentialities of eradication as a tool for addressing biological
invasions. In: Nentwig W (ed) Biological invasions. Springer, Berlin, pp 385–401

Genovesi P (2011) Eradication. In: Simberloff D, Rejmánek M (eds) Encyclopedia of biological
invasions. University of California Press, Berkley/Los Angeles, pp 813–817

Genovesi P, Monaco A (2014) Chapter 22: Guidelines for addressing invasive species in protected
areas. In: Foxcroft LC, Pyšek P, Richardson DM, Genovesi P (eds) Plant invasions in protected
areas: patterns, problems and challenges. Springer, Dordrecht, pp 487–506

Genovesi P, Scalera R, Brunel S et al. (2010) Towards an early warning and information system
for invasive alien species (IAS) threatening biodiversity in Europe. European Environment
Agency Technical Report n.5/2010

GISD (2012) Global invasive species database website, able to search for species or by location.
Invasive Species Specialist Group (ISSG) of the SSC – Species Survival Commission of the
IUCN – International Union for Conservation of Nature. http://www.issg.org/database/wel
come/. Accessed 15 Mar 2012

Goodman PS (2003) Assessing management effectiveness and setting priorities in protected areas
in KwaZulu-Natal. BioScience 53:843–850

Harris S, Timmins SM (2009) Estimating the benefit of early control of all newly naturalised
plants. Science for Conservation 292. New Zealand Department of Conservation. http://www.
doc.govt.nz/Documents/science-and-technical/sfc292.pdf. Accessed 24 Jan 2013

Hiebert RD, Stubbendieck J (1993) Handbook for ranking exotic plants for management and
control. NPS/US DOI, Denver

Hobbs RJ, Humphries SE (1995) An integrated approach to the ecology and management of plant
invasions. Conserv Biol 9:761–770

Hulme PE (2006) Beyond control: wider implications for the management of biological invasions.
J Appl Ecol 43:835–847

ISSG (2012) Invasive species specialist group website, with tools, publications and networking.
Invasive Species Specialist Group, Species Survival Commission (SSC) of the International
Union for Conservation of Nature (IUCN). http://www.issg.org. Accessed 12 Dec 2012

Larios L, Suding KN (2014) Chapter 27: Restoration within protected areas: when and how to
intervene to manage plant invasions? In: Foxcroft LC, Pyšek P, Richardson DM, Genovesi P
(eds) Plant invasions in protected areas: patterns, problems and challenges. Springer, Dordrecht, pp 599–618

Mazzu L (2005) Common control measures, for invasive plants of the Pacific Northwest region.
Preventing and managing invasive plants. Final Environmental Impact Statement, Appendix.
USDA Forest Service, Washington, DC

Meyerson LA, Pyšek P (2014) Chapter 21: Manipulating alien species propagule pressure as a
prevention strategy in protected areas. In: Foxcroft LC, Pyšek P, Richardson DM, Genovesi
P (eds) Plant invasions in protected areas: patterns, problems and challenges. Springer,
Dordrecht, pp 473–486

NISC (2012) NISC website, with examples of management plans. National Invasive Species
Council. United States Department of Agriculture. http://www.invasivespecies.gov. Accessed
15 Mar 2012

Owen SJ (1998) Department of Conservation strategic plan for managing invasive weeds. Department of Conservation, Wellington

Pomeroy RS, Parks JE, Watson IM (2004) How is your MPA doing? A guidebook of natural and
social indicators for evaluating marine protected area management effectiveness. IUCN, Gland

Rejmánek M, Pitcairn MJ (2002) When is eradication of exotic pest plants a realistic goal? In:
Veitch CR, Clout MN (eds) Turning the tide: the eradication of invasive species. IUCN, Gland,
pp 249–253

Salafsky N, Salzer D, Ervin J et al (2003) Conventions for defining, naming, measuring, combining, and mapping threats in conservation: an initial proposal for a standard system. The Nature Conservancy, Arlington

Sheley RL, Petroff JK (1999) Biology and management of noxious rangeland weeds. Oregon State University Press, Corvallis

Simberloff D (2014) Chapter 25: Eradication – pipe dream or real option? In: Foxcroft LC, Pyšek P, Richardson DM, Genovesi P (eds) Plant invasions in protected areas: patterns, problems and challenges. Springer, Dordrecht, pp 549–559

Skurka Darin GM, Schoenig S, Barney JN et al (2011) WHIPPET: a novel tool for prioritizing invasive plant populations for regional eradication. J Environ Manage 92:131–139

Timmins SM, Owen SJ (2001) Scary species, superlative sites: assessing weed risk in New Zealand's protected natural areas. In: Groves RH, Panetta FD, Virtue JG (eds) Weed risk assessment. CSIRO, Canberra, pp 217–227

Tu M, Meyers-Rice B (2002) Site weed management plan template. The Nature Conservancy. Center of Invasive Species and Ecosystem Health, University of Georgia. http://www.invasive.org/gist/products.html. Accessed 15 Mar 2012

Tu M, Hurd C, Randall JM (2001) Weed control methods handbook: tools and techniques for use in natural areas. The Nature Conservancy. Center of Invasive Species and Ecosystem Health, University of Georgia. http://www.invasive.org/gist/handbook.html. Accessed 15 Mar 2012

Van Driesche R, Center T (2014) Chapter 26: Biological control of invasive plants in protected area. In: Foxcroft LC, Pyšek P, Richardson DM, Genovesi P (eds) Plant invasions in protected areas: patterns, problems and challenges. Springer, Dordrecht, pp 561–597

Wittenberg R, Cock MJW (2001) Invasive alien species: a toolkit of best prevention and management practices. CAB International, Wallingford

Zouhar K, Smith JK, Sutherland S et al (2008) Wildland fire in ecosystems: fire and non-native invasive plants. USDA Forest Service General technical report RMRS-GTR-42-vol.6, Ogden

Chapter 25
Eradication: Pipe Dream or Real Option?

Daniel Simberloff

Abstract Invasive alien plant populations have often been eradicated from very small areas, but pessimism about eradication of widely distributed plants pervades the management community. Contributing to this view are several legendary and expensive failed eradication campaigns, the inconspicuous nature of many plants, the existence of soil seed banks, and the perceived expense of eradication over large areas. However, if several years' worth of the cost of maintenance management campaigns could instead be devoted to a one-shot, well-funded eradication effort, projects that currently seem impossible might be brought within the range of feasibility. Factors in addition to cost that must be considered are whether adequate lines of authority can compel cooperation and prevent sabotage, whether there is sufficient knowledge of the target species to have identified a feasible approach to eradication that advances the goal of restoration, and the need for intensive monitoring and possible follow-up operations. Especially for PAs, the likelihood of reinvasion from nearby sites is a concern. If an eradication campaign would employ the same general methods as those that would have been used if the goal was maintenance control, there is likely little cost and much potential benefit to attempting eradication. Gradual improvement has occurred in plant eradication programmes through accumulated experience and incremental improvement of longstanding methods. However, the field of invasive plant management (including eradication) has not seen the advent of remarkably innovative new approaches and greatly improved records of eradication success that currently foster optimism and enthusiasm among managers dealing with invasive animals.

Keywords Extirpation • Monitoring • Small populations • Target species

D. Simberloff (✉)
Department of Ecology and Evolutionary Biology, University of Tennessee,
Knoxville, TN, USA
e-mail: dsimberloff@utk.edu

L.C. Foxcroft et al. (eds.), *Plant Invasions in Protected Areas: Patterns, Problems and Challenges*, Invading Nature - Springer Series in Invasion Ecology 7, DOI 10.1007/978-94-007-7750-7_25, © Springer Science+Business Media Dordrecht 2013

25.1 Introduction

By 'eradication', invasion scientists mean removing every individual of a discrete, more or less isolated population. This is distinct from 'extirpation', which means eliminating a segment of a population, but with conspecific individuals still present in contiguous or nearby populations. Unfortunately, the term 'eradication' is used quite colloquially, particularly in media reports and political statements advocating or announcing 'eradication' of some weed or pest species, when what is meant is really extirpation of the species in some defined area. Sometimes 'eradication' is used to mean simply killing a lot of individuals, not even all of them at the same site. Part of a general scepticism about the feasibility of eradication stems from the fact that campaigns colloquially announced as 'eradication' campaigns were never meant to be that, so of course they failed to eradicate the target population (Simberloff 2003a).

Another factor generating scepticism about the possibility of eradication is the history of several high-profile, expensive true eradication campaigns that not only failed to eradicate their target species but had enormous damaging non-target impacts. An example is the failed campaign in the United States to eradicate white pine blister rust, introduced in the early twentieth century on white pine seedlings from Germany (Maloy 1997). The campaign aimed to eliminate the fungus by eradicating both native and introduced species of *Ribes*, the alternate host. Labour costs alone were over $150 million and were particularly heavy during World War I. During World War II, prison inmates as well as German and Italian prisoners of war dug up *Ribes* and spread chemicals, including in wetlands and stream-sides. Non-target impacts were massive, and the campaign failed utterly. Another continent-wide campaign, this time to eliminate *Berberis vulgaris* (European barberry), was similarly motivated. In the United States, *B. vulgaris* is an alternate host of stem rust of cereals, which inflicted enormous losses on wheat growers. The campaign, detailed by Campbell and Long (2001) and Mack and Foster (2009), began in 1918, lasted 60 years, employed thousands, and rendered *B. vulgaris* a rare plant in much of the United States, even today. As the methods included use of rock salt, kerosene, and dynamite (Mack and Foster 2009), one can speculate about non-target impacts.

In a widely cited paper, Rejmánek and Pitcairn (2002) found a decade ago that eradication of agricultural weed populations smaller than a hectare is usually feasible, and that for infestations between one hectare and 1,000 ha, between a fourth and a third of attempts they surveyed had succeeded. However, in their survey the cost of eradications rose so rapidly with area that they felt it was unlikely that eradications of plant populations occupying more than 1,000 ha would be feasible. Panetta and Timmins (2004) agreed with this threshold and suggested that the prospects for eradication of plants from natural areas would tend to be much dimmer than those for agricultural weeds. Gardener et al. (2010) recently cast further doubt on the feasibility of most plant eradication projects other than very small ones, an assertion subsequently publicised in a high-visibility news report

(Vince 2011). However, the past decade has seen dramatic progress in eradication of invasive animal populations (see for example Genovesi 2011a, b) on ever-larger islands, plus further experience with invasive plant management. It is thus timely to reconsider Rejmánek and Pitcairn's pessimism regarding large-scale plant eradications and also ask under what circumstances should attempted eradication be the preferred response to a plant invasion of a PA as opposed to some sort of maintenance management, such as biological, chemical, mechanical, or physical control.

It is a commonplace that plants are generally harder to eradicate than animals, especially vertebrates. Seed banks may persist in the soil for many years (Panetta 2004), numbers of plant individuals may be enormous, individuals – even seedlings of trees – may be small and cryptic, and the attractive baits and traps that have so aided animal eradication are not applicable to plants. Thus, the degree of optimism that has begun to infuse the community of managers and policymakers dealing with invasive animals (e.g. Genovesi 2011a, b) has largely failed to engage those managing invasive plants. It is telling that, in a recent international conference on eradication of invasive species on islands (Veitch et al. 2011), of 94 papers, 89 were about eradicating animals and only 6 about eradicating plants; of 45 abstracts, 44 were on eradicating animals and none were on eradicating plants.

25.2 Successes and Failures

Many small plant invasions have been eradicated from sites other than PAs. Mack and Lonsdale (2002) describe eradication in Australia of small populations of North American *Eupatorium serotinum* (late boneset) in a cattle sales yard, as well as eradication of Old World *Centaurea trichocephala* (feather-head knapweed) in a degraded pasture in Washington state, USA. In the marine realm, the "killer alga" (*Caulerpa taxifolia*) was eradicated from two sites in California (Anderson 2005; Woodfield and Merkel 2006). At one site, about 0.13 ha of the alga was distributed widely among 42.3 ha of a 100.6 ha lagoon, and the other consisted of a group of shallow ponds totalling 1.1 ha connected to a harbour. In South Australia, *C. taxifolia* was eradicated from an artificial marine water body 7 km long by a few hundred meters wide (Walters 2009).

Some small populations of invasive plants have been eradicated in PAs. For instance, Rejmánek and Pitcairn (2002) cite two eradications of small populations in the Channel Islands National Park, California, while *Oryza rufipogon* (Asian common rice) was eradicated from an area of 0.1 ha in Everglades National Park (Westbrooks 1993). Macdonald (1988) reports ten invasive plant species as having been eliminated from Kruger National Park, South Africa. He identified only four of these species: *Opuntia aurantiaca* (jointed cactus), *Acacia dealbata* (silver wattle), *Bidens formosa* (cosmos), and *Nicotiana glauca* (tree tobacco). However, L. Foxcroft (personal communication, 2013) reports that the latter species is present cyclically. The Bermuda Department of Agriculture has eradicated *Livistona chinensis* (Chinese fan palm), *Pimenta dioica* (allspice), *Eugenia uniflora*

(Barbados cherry), and *Citharexylum spinosum* (fiddlewood) from Nonsuch Island (Bermuda), a wildlife sanctuary of 5.7 ha (Mack and Lonsdale 2002). In 1972, the New Zealand government targeted 29 non-native plant species for removal from 2,943 ha Raoul Island, a designated nature reserve. For seven species that occupied relatively small areas, including highly invasive ones such as *Cortaderia selloana* (pampas grass), success is believed to have been achieved, although continued monitoring is undertaken to ensure that resurgence does not occur from a soil seed bank (West 2002). However, for the seven main target species, all originally quite widespread on the island, progress toward eradication has been more gradual, with occasional setbacks as new infestations are detected (Holloran 2006). In the Galapagos, four non-native plant species have been eradicated from Santa Cruz Island (two of these are not found elsewhere in the archipelago), each from an area less than 0.1 ha (Gardener et al. 2010). Although the great majority of Santa Cruz is part of the Galapagos National Park, at least one of eradications took place on private land. *Cenchrus echinatus* (sandbur) was eradicated from 64 ha on Laysan Island (Hawaiian Islands; 411 ha), managed as a PA by the US Fish and Wildlife Service, in a 10-year campaign beginning in 1991 (Flint and Rehkemper 2002; E. Flint, personal communication, 2007).

A much larger success, although not in a PA, was the eradication of the pasture pest *Bassia scoparia* (burning bush) from several thousand ha distributed over a linear distance of 900 km in western Australia (Randall 2001; Dodd 2004), no doubt helped by the fact that locations of all plantings had been recorded. Perhaps the most ambitious current plant eradication programme rivals the *Ribes* and *Berberis* eradication campaigns of the early twentieth century. This is the attempt to eradicate a parasitic agricultural weed, *Striga asiatica* (witchweed), which is ongoing after over 50 years (Eplee 2001; Mack and Foster 2009) and has reduced the infested area from 162,000 ha to less than 1,000 ha in North and South Carolina. Success is likely within a decade (Mack and Foster 2009).

Many more attempted plant eradications have failed than have succeeded. Gardener et al. (2010), for example, cite failure to eradicate (so far) 26 targeted plant species in the Galapagos, comparing this record to the four successes cited above. For at least two of these failures, no campaign was actually implemented. For *Caulerpa taxifolia*, several eradication efforts have failed, as against the three successes noted above (Walters 2009).

25.3 Criteria for Success

Myers et al. (2000) and Simberloff (2002a, b, 2003a, b) have suggested several criteria that characterise successful eradications and that should be met before eradication is attempted. Of course the idiosyncrasies of each case will weigh heavily, but the following factors should always be borne in mind:

1. Economic resources. Are resources sufficient to complete the eradication as planned, and are those resources encumbered in such a way that they will be available for the duration of the project, even as the target population and its perceived impact are greatly reduced? Costs of removing the last few individuals may exceed those of removing all the rest, and funding agencies may be inclined to reduce support once the problem is lessened (Mack and Lonsdale 2002). Are resources needed to manage the species in areas near the target protected area to prevent reinvasion, and, if so, are they available?
2. Adequate lines of authority. Eradication is, by its nature, an all-or-none phenomenon. By contrast, in maintenance management by chemical or mechanical control, for instance, the refusal of a few landowners to permit the project to be carried out on their property would not necessarily prevent substantial reduction of the target species. However, an inviolable sanctuary for the target would prevent eradication by definition. Do such sanctuaries exist adjacent to or near the target protected area?
3. Enough must be known about the biology of the target species that a route to eradication can be identified that is feasible with available resources.
4. The eradication project, even if successful, must not produce an undesirable condition. For PAs, the ultimate goal would almost certainly be restoration of a semblance of the natural ecological community and the dynamic trajectory it was following before the invasion. Thus, for instance, high likelihood that an eradicated plant species would simply reinvade quickly or be replaced by another introduced species would weigh heavily against attempting eradication (although this would not necessarily be decisive; see an example below). It is also possible the method used in an attempted eradication would have a high risk of non-target impacts that would prevent the restoration goal from being achieved. Massive use of some persistent herbicide, for instance, or tremendous damage from machinery used in a scorched-earth operation, might so damage the prospects for restoration as to be untenable.

25.4 Monitoring: Determining Success and Detecting Reinvasion

The main issue concerning eradication of any invasive plant population is whether it is feasible and at what cost. If so, particularly if the campaign is costly, it then becomes important to consider whether reinvasion is likely, whether it would be detected quickly, and what could be done about it if it occurs. Protected areas adjacent to unprotected lands pose particular problems in this regards, whereas island refuges are obviously at an advantage. Of course to a great extent likelihood of reinvasion depends on the location of the site relative to extant populations and the means by which propagules of the eliminated species might arrive. Constant vigilance is needed first of all to ensure that an eradication effort really was successful, and secondly, to note and deal with any newly arrived individuals.

To know that every last individual of a plant species is gone is fraught with many difficulties, enumerated by Panetta and Timmins (2004). With plants, the existence of a soil seed bank poses particular problems (Panetta 2004) and, depending on seed longevity, can mean that many years must pass before one can ascertain that eradication had occurred. For animals, depending on the species, it is common practice to declare success (or concede failure) quite quickly, for example, often 1 year for rats and 4 years for the Asian longhorn beetle (*Anoplophora glabripennis*) in Chicago (Haack et al. 2010). For plants, sometimes 4 years of absence has been chosen as the criterion for success, for example, *C. taxifolia* in California (Woodfield and Merkel 2006), while for *Bassia scoparia* in Western Australia the criterion was 3 years (Randall 2001). However, several announcements of eradication have been premature. For instance, in Queensland, a 40-plus-year campaign to eradicate several small populations of the North American herb *Helenium amarum* (yellow sneezeweed), first detected in 1953, was declared successful in 2002 after annual searching for survivors failed to detect any (Mack and Lonsdale 2002; Csurhes and Zhou 2008). But in 2007 several individuals were found; these were removed and the area continues to be monitored (Csurhes and Zhou 2008).

25.5 When Should Eradication Be Attempted in Protected Areas?

Not all potential eradication projects that meet the above criteria for high likelihood of success can be undertaken, if only because resources would likely not suffice. However, in assessment of alternative management possibilities – in essence, (i) do nothing, at least for the present, (ii) attempt some sort of maintenance management, or (iii) attempt eradication – several factors suggest that eradication deserves more consideration than it often gets.

First, if an invasion is recent and the invaded area still small, it is likely that eradication is feasible, as suggested by the data in Rejmánek and Pitcairn (2002; see Pluess et al. 2012). Furthermore, the cost would be far less than if the effort were made after the invasion had spread. The likelihood of damaging non-target impacts would be less, both because the invader is unlikely to have established important interactions with native resident species and because whatever eradication method is attempted will not be employed over a large area. Finally, it is risky to wait to see if the species begins to spread or cause problems and, if it does, only then undertake an eradication campaign. Many introduced species, including plants such as *Schinus terebinthifolius* (Brazilian pepper) and *Arundo donax* (giant reed), have remained restricted for long periods, even decades, before rather suddenly spreading widely (see Crooks 2005). Because eradication campaigns typically take time to plan and implement, one could easily miss a window of opportunity by delaying an eradication attempt. Also, some invasive plants have major ecosystemic impacts that are nevertheless sufficiently subtle that they are not detected quickly; plants that fix nitrogen or concentrate phosphorus can fall in this category (Simberloff 2011). Waiting until such impacts become evident may allow a species to spread to a point at which eradication is vastly more expensive and perhaps not feasible. *Crupina vulgaris* (common crupina) in the American West and *Clidemia hirta* (Koster's curse) in the Hawaiian Islands, two non-native plants that were discovered soon after arrival and almost certainly could have been eradicated, without likely reinvasion, were allowed to spread while authorities questioned whether they would be very damaging. Both proved highly invasive and were well beyond the stage when they could have been eradicated by the time it was agreed that they should be controlled (Simberloff 2003b). The alga *Caulerpa taxifolia*, which has now spread throughout much of the near-shore western Mediterranean, could also almost certainly have been controlled had a campaign been undertaken soon after discovery (Meinesz 1999).

For PAs, the status of the target plant in neighbouring areas is a particular concern, as the funding for management, including attempted eradication, in the PA is unlikely to allow efforts beyond that area. Thus, for instance, the State of Florida and US federal agencies have mounted a promising programme using chemical and mechanical means to reduce or eliminate *Melaleuca quinquenervia* (broad-leaved paperbark tree) from state and federal lands (including PAs) in south Florida (F. Laroche, personal communication). However, by statute public funds cannot support such efforts on private lands adjacent to government properties. Three biological control insects have been released and, of course, do not respect property boundaries. These may contribute to an effective maintenance management programme in this case, but biological control in otherwise untreated areas would be unlikely to lead to eradication of this or other invasive plant species if this were the goal.

For widespread invasions, including longstanding ones, the expense of an eradication campaign can be forbidding even if the technology exists to suggest that success is possible. However, a comparison to ongoing costs of maintenance management in some cases leads to speculation about whether attempting eradication might be the truly most cost-effective approach (Simberloff 2003a). For instance, the United States spends $45 million annually on management of *Lythrum salicaria* (purple loosestrife) and $3 million to $6 million annually on control of *M. quinquenervia*). Thus, over a 10-year period, ongoing maintenance management costs tens or hundreds of millions of dollars. One can imagine that having such resources available over a much shorter period for an eradication attempt might make an eradication attempt feasible that would have been impossible with just a few million dollars. Another possible resource that has not been devoted to plant eradication is volunteer or prisoner labour. Such sources are now routinely used in a number of effective maintenance management programmes, especially in PAs (see Simberloff 2003a), and allow managers to marshal many more workers than could possibly have been paid. Use of vast amounts of manpower might make it possible to eradicate much more widespread invasions than would have been deemed feasible based on personnel costs alone.

If an eradication campaign uses the same method that would have been used had maintenance management been the goal, it may well be more cost-effective to invest added resources and attempt to eradicate the invader. This is because even failure to eradicate would be no great loss and would probably entail more complete maintenance management. One would have to tally the costs and potential benefits of the added effort. An excellent example is the project, begun in 1992, to eradicate *Ammophila arenaria* (European beachgrass) and hybrids of two African ice-plant species (*Carpobrotus edulis* × *C. chilensis*) from an 11 ha area of Lanphere Dunes in Humboldt Bay National Wildlife Refuge in California (Pickart 2013). The stated goal in terms of these invasive plants was eradication, and the goal for the system was restoration of the ecosystem to its trajectory before European modification, by restoring abiotic processes that maintain a dynamic dune ecosystem. Herbicides were precluded by local community objections, and the impact of heavy machinery on native vegetation, including two federally listed species, would have been too great, so the method chosen was digging and pulling by hand. *Ammophila arenaria* was almost wholly eliminated after 2 years, and ice-plant after 5. However, rare resprouts of both species are seen, and there is occasional reinvasion from nearby areas. Annual spot treatments control these at very low densities. Thus, complete eradication has not yet been achieved for either species (or else reinvasion quickly occurs), but the ultimate restoration goal has been met, and, if maintenance management rather than total eradication had been the stated goal, the method that was implemented would have been exactly the same. Further, the fact that quick reinvasion is likely does not invalidate the approach in this instance, as annual monitoring and spot treatments are feasible and inexpensive.

25.6   Further Advances?

Just within the last decade, animal eradication has advanced greatly, with projects that would have seemed impossible a decade or two ago now well within the realm of possibility (Genovesi 2011a, b), with better methods of avoiding non-target impacts (e.g. Caut et al. 2009) and with important conservation benefits (McGeoch et al. 2010). Some of these advances result from new technologies and others from incremental improvement of existing techniques, combined with ambition (Simberloff et al. 2013). Even if one grants the difficulties that are peculiar to invasive plant eradication, it seems as if greater successes are possible by the same routes that are forging progress in animal eradication. For instance, greater efforts using the same techniques that had previously failed to control invasive plant species on Motuopao Island (New Zealand) are leading towards successful eradication of several species (Beauchamp and Ward 2011). Assiduous application of longstanding techniques has led to eradication of small infestations of 12 non-native plant species from single islands in the Hawaiian archipelago and the imminent elimination of eight others (Penniman et al. 2011).

What do not seem to have arisen in invasive plant control generally, and eradication attempts in particular, are highly innovative new technologies. Meyer et al. (2011) suggest a new strategy, focused on preventing fruit production, which might permit eradication of small infestations of previously intractable *Miconia calvescens* (miconia). But absent are plant analogues to completely novel approaches such as the development of attractive pheromones that have greatly advanced sea lamprey management (Fine and Sorensen 2008), toxic micro-beads that have cleared some water facilities of zebra mussels (Aldridge et al. 2006), and the battery of genetic manipulations currently under way in attempts to eradicate populations of fishes (e.g. Thresher 2008) and insects (e.g. Pollack 2011). It seems unlikely that the biology of plants differs in characteristic ways from that of animals so as to inhibit the development of radically new control technologies.

References

Aldridge DC, Elliott P, Moggridge GD (2006) Microencapsulated BioBullets for the control of biofouling zebra mussels. Environ Sci Technol 40:975–979

Anderson LWJ (2005) California's reaction to *Caulerpa taxifolia*: a model for invasive species rapid response? Biol Invasion 7:1003–1016

Beauchamp AJ, Ward E (2011) A targeted approach to multi-species control and eradication of escaped garden and ecosystem-modifying weeds on Motuopao Island, Northland, New Zealand. In: Veitch CR, Clout MN, Towns DR (eds). Island invasives: eradication and management. Proceedings of the international conference on island invasives. IUCN, Gland, Switzerland and CBB, Auckland, New Zealand, pp 264–268

Campbell CL, Long DL (2001) The campaign to eradicate the common barberry in the United States. In: Peterson PD (ed) Stem rust of wheat: from ancient enemy to modern foe. APS Press, St. Paul, pp 16–43

Caut S, Angulo E, Courchamp F (2009) Avoiding surprise effects on Surprise Island: alien species control in a multi-trophic level perspective. Biol Invasion 11:1689–1703

Crooks JA (2005) Lag times and exotic species: the ecology and management of biological invasions in slow-motion. Écoscience 12:316–329

Csurhes S, Zhou Y (2008) Pest plant risk assessment. Bitter weed, *Helenium amarum*. Queensland Government, Department of Primary Industry and Fisheries, Brisbane

Dodd J (2004) *Kochia* (*Bassia scoparia* (L.) A.J. Scott) eradication in Western Australia: a review. In: Sindel BM, Johnson SB (eds) Proceedings of the fourteenth Australian weeds conference. Weed Science Society of New South Wales, Sydney, pp 496–499

Eplee RE (2001) Coordination of witchweed eradication in the USA. In: Wittenberg R, Cock MJW (eds) Invasive alien species: a toolkit of best prevention and management practices. CAB International, Wallingford, p 36

Fine JM, Sorensen PW (2008) Isolation and biological activity of the multi-component sea lamprey migratory pheromone and new information in its potency. J Chem Ecol 34:1259–1267

Flint EN, Rehkemper C (2002) Control and eradication of the introduced grass, *Cenchrus echinatus*, at Laysan Island, central Pacific Ocean. In: Veitch CR, Clout MN (eds) Turning the tide: the eradication of invasive species. IUCN Species Survival Commission, Gland, pp 110–115

Gardener MR, Atkinson R, Rentería JL (2010) Eradications and people: lessons from the plant eradication program in Galapagos. Restor Ecol 18:20–29

Genovesi P (2011a) Eradication. In: Simberloff D, Rejmánek M (eds) Encyclopedia of biological invasions. University of California Press, Berkeley, pp 198–203

Genovesi P (2011b) Are we turning the tide? Eradications in times of crisis: how the global community is responding to biological invasions. In: Veitch CR, Clout MN, Towns DR (eds) Island invasives: eradication and management. Proceedings of the International conference on island invasives. IUCN, Gland, Switzerland and CBB, Auckland, New Zealand, pp 5–8

Haack RA, Hérard F, Sun J et al (2010) Managing invasive populations of Asian longhorned beetle and citrus longhorned beetle: a worldwide perspective. Annu Rev Entomol 55:521–546

Holloran P (2006) Measuring performance of invasive plant eradication efforts in New Zealand. NZ Plant Protect 59:1–7

Macdonald IAW (1988) The history, impacts and control of introduced species in the Kruger National Park, South Africa. Trans R Soc S Afr 46:251–276

Mack RN, Foster SK (2009) Eradicating plant invaders: combining ecologically-based tactics and broad-sense strategy. In: Inderjit (ed) Management of invasive weeds. Springer, Dordrecht, pp 35–60

Mack RN, Lonsdale WM (2002) Eradicating invasive plants: hard-won lessons for islands. In: Veitch CR, Clout MN (eds) Turning the tide: the eradication of invasive species. IUCN SSC Invasive Species Specialist Group, IUCN, Gland, pp 164–172

Maloy OC (1997) White pine blister rust control in North America: a case history. Annu Rev Phytopathol 35:87–109

McGeoch MA, Butchart SHM, Spear D et al (2010) Global indicators of biological invasion: species numbers, biodiversity impact and policy responses. Divers Distrib 16:95–108

Meinesz A (1999) Killer algae. University of Chicago Press, Chicago

Meyer J-L, Loope L, Goarant A-C (2011) Strategy to control the invasive alien tree *Miconia calvescens* in Pacific islands: eradication, containment or something else? In: Veitch CR, Clout MN, Towns DR (eds) Island invasives: eradication and management. Proceedings of the international conference on island invasives. IUCN, Gland, Switzerland and CBB, Auckland, New Zealand, pp 91–96

Myers JH, Simberloff D, Kuris AM et al (2000) Eradication revisited: dealing with exotic species. Trends Ecol Evol 15:316–320

Panetta FD (2004) Seed banks: the bane of the weed eradicator. In: Sindel BM, Johnson SB (eds) Proceedings of the 14th Australian weeds conference. Weed Society of New South Wales, Wagga Wagga, NSW, Australia, pp 523–526

Panetta FD, Timmins SM (2004) Evaluating the feasibility of eradication for terrestrial weed incursions. Plant Protect Q 19:5–11

Penniman TM, Buchanan L, Loope L (2011) Recent plant eradications on the islands of Maui County, Hawai'i. In: Veitch CR, Clout MN, Towns DR (eds). Island invasives: eradication and management. Proceedings of the international conference on island invasives. IUCN, Gland, Switzerland and CBB, Auckland, New Zealand, pp 325–331

Pickart A (2013) Dune restoration over two decades at the Lanphere and Ma-le'l Dunes in northern California. In: Martinez ML, Gallego-Fernandez JB, Hesp PA (eds) Restoration of coastal dunes. Springer, New York, pp 159–171

Pluess T, Jarošík V, Pyšek P et al (2012) Which factors affect the success or failure of eradication campaigns against alien species? PLoS ONE 7:e48157

Pollack A (2011) Mosquito bred to fight dengue fever shows promise in a field test. New York Times, October 31, p B1

Randall R (2001) Eradication of a deliberately introduced plant found to be invasive. In: Wittenberg R, Cock MJW (eds) Invasive alien species: a toolkit of best prevention and management practices. CAB International, Wallingford, p 174

Rejmánek M, Pitcairn MJ (2002) When is eradication of exotic pest plants a realistic goal? In: Veitch CR, Clout MN (eds) Turning the tide: the eradication of invasive species. IUCN SSC Invasive Species Specialist Group, IUCN, Gland, pp 249–253

Simberloff D (2002a) Today Tiritiri Matangi, tomorrow the world! Are we aiming too low in invasives control? In: Veitch CR, Clout MN (eds) Turning the tide: the eradication of invasive species. IUCN SSC Invasive Species Specialist Group, IUCN, Gland, pp 4–12

Simberloff D (2002b) Why not eradication? In: Rapport DJ, Lasley WL, Ralston DE et al (eds) Managing for healthy ecosystems. CRC/Lewis Press, Boca Raton, pp 541–548

Simberloff D (2003a) Eradication – preventing invasions at the outset. Weed Sci 51:247–253

Simberloff D (2003b) How much information on population biology is needed to manage introduced species? Conserv Biol 17:83–92

Simberloff D (2011) How common are invasion-induced ecosystem impacts? Biol Invasion 13:1255–1268

Simberloff D, Martin J-L, Genovesi P et al (2013) Impacts of biological invasions – what's what and the way forward. Trends Ecol Evol 28:58–66

Thresher RE (2008) Autocidal technology for the control of invasive fish. Fisheries 33:114–121

Veitch CR, Clout MN, Towns DR (eds) (2011) Island invasives: eradication and management. Proceedings of the international conference on island invasives. IUCN, Gland, Switzerland and CBB, Auckland, New Zealand

Vince G (2011) Embracing invasives. Science 331:1383–1384

Walters L (2009) Ecology and management of the invasive marine macroalga *Caulerpa taxifolia*. In: Inderjit (ed) Management of invasive weeds. Springer, Dordrecht, pp 287–318

West CJ (2002) Eradication of alien plants on Raoul Island, Kermadec Islands, New Zealand. In: Veitch CR, Clout MN (eds) Turning the tide: the eradication of invasive species. IUCN SSC Invasive Species Specialist Group. IUCN, Gland, pp 365–373

Westbrooks RG (1993) Exclusion and eradication of foreign weeds from the United States by USDA APHIS. In: McKnight BN (ed) Biological pollution. Indiana Academy Science, Indianapolis, pp 225–241

Woodfield R, Merkel K (2006) Final report on the eradication of the invasive seaweed *Caulerpa taxifolia* from Agua Hedionda Lagoon and Huntington Harbour, California. Southern California Caulerpa Action Team, Long Beach, California

Chapter 26
Biological Control of Invasive Plants in Protected Areas

Roy Van Driesche and Ted Center

Abstract Classical weed biological control is widely used in natural areas. It is based on introduction of specialised natural enemies (herbivorous insects and fungal pathogens) from the weed's native range. It can be used safely if specialised natural enemies are selected and can be highly effective in suppressing weeds over large areas. Agents used in modern projects typically have genus or species level specificity and are safe when proper risk analysis and procedures are followed. Agents spread over large areas and can move into hard-to-reach areas. If correctly selected, agents are safe for use in areas too ecologically sensitive for chemical or mechanical control. Costs are independent of area to be treated because agents are self-reproducing, and results are self-sustaining. Biological control is most appropriate for use against widespread weeds, difficult to control with other methods that occur in critical habitats and damage biodiversity or ecosystem function. Finding suitable agents is easier against weeds distantly related to local native plants. Such targets reduce risk to native flora, facilitate agent screening, lower cost, and increase likelihood of success. Projects should be partnerships between biological control scientists and conservation biologists, and biological control activities should be done within a comprehensive restoration plan for the ecosystem. In some cases, suppression of the invasive weed may be sufficient, but sometimes additional actions, such as replanting native species or modifying ecosystem processes such as fire or flooding regimes may be essential.

R. Van Driesche (✉)
Department of Environmental Conservation, University of Massachusetts,
Amherst, MA 01003, USA
e-mail: vandries@cns.umass.edu

T. Center (✉)
USDA, ARS, Invasive Plant Research, 3225 College Avenue,
Fort Lauderdale, FL 33312, USA
e-mail: ted.center@ars.usda.gov

L.C. Foxcroft et al. (eds.), *Plant Invasions in Protected Areas: Patterns, Problems and Challenges*, Invading Nature - Springer Series in Invasion Ecology 7, DOI 10.1007/978-94-007-7750-7_26, © Springer Science+Business Media Dordrecht 2013

Keywords Benefits • Classical biological control • Ecological restoration • Natural areas • Natural enemy introduction

26.1 What Is Biological Control?

Biological is a tool that can be used in some cases to assist in the ecological restoration of areas affected by high density, damaging populations of invasive plants, reducing such plants to densities that pose less of a burden on native biodiversity and allowing ecosystems to avoid invasion-driven physical transformations (Van Driesche et al. 2008). This result is achieved by harnessing the power of herbivory or fungal infections to lower the fitness of targeted exotic plants, allowing the intrinsic competitive power of native plants to be more effective. If landscape-scale reduction of the target plant's density is acceptable (as in the case of weeds with no economic value), projects seek to lower plant fitness directly, resulting in smaller infestations, slower spread, and reduced weed biomass. All components of the target plant – seeds, foliage, stems, etc. – can then be attacked. In contrast, if plants have important social or economic uses (such as some introduced forestry species in South Africa), then the project's goals must be limited to the suppression of reproduction, selecting agents that attack only flowers or seeds. This can help limit the plant's spread from economic use sites into wild lands and can support manual clearance of existing stands in natural areas by minimizing regeneration. Weed biological control is an ecological restoration tactic whose risks are generally low (Pemberton 2000) and whose use is often effective. In South Africa, for example, 19 of 23 (83 %) projects were completely or partially successful (Hoffmann 1996; Clrruttwell McFadyen 1998) and in Hawaii 10 of 21 (50 %) targeted weeds were completely or partially suppressed (Markin et al. 1992; Gardner et al. 1995). Here we discuss the contribution of biological control of invasive plants to protection of natural areas, including legally protected preserves, and suggest steps for strategic integration of weed biological control into restoration ecology.

While the focus of this book is on controlling weeds in legally protected areas (parks, preserves), biological projects operate on a much wider scale. Unlike locally applied measures (chemical and mechanical control or replanting of native species), which are typically done inside preserves, biological control is applied to whole landscapes. As such, it is the case that preserves fall inside areas affected by biological control rather than the reverse. For example, control of *Euphorbia esula* (leafy spurge) in the northern prairie of North America was an areawide reduction of the pest over millions of ha, including a number of preserves as, for example, the Pine Butte Swamp Preserve of The Nature Conservancy in Montana (USA). Similarly, biological control of *Melaleuca quinquenervia* (melaleuca) over several hundred thousand ha of southern Florida (USA) overlapped with such preserves as the Everglades National Park (a World Heritage site) and the Big Cypress National Preserve. In the Northern Territories of Australia, biological control of *Mimosa pigra* (mimosa) certainly included areas within Kakadu National Park, while biological control of *Opuntia* spp. cacti and *Hakea sericea* (hakea) in South Africa affected Kruger National Park and various Protected Areas of the Cape Floral Region, respectively. Many other instances of overlap between legally protected areas and regions where biological control has reduced damaging invasive plants could be identified.

26.2 Advantages and Disadvantages of Biological Control Versus Other Methods

Invasive plants in natural areas may be locally suppressed by hand weeding, mechanical control (cutting, dredging, mechanical clearing of brush), and application of herbicides. Size of the area over which the weed must be controlled determines the practicality of these methods. Preserve managers may intend to suppress a weed in only a specific, often small, area (the preserve), but these patches are frequently part of a landscape-wide infestation. Weed reduction at the landscape level requires biological control, but doing so requires long-term commitment of resources and the expertise of specialised scientists. Consequently, biological control is not the first choice for weed control on a single preserve. In part this is because such projects cannot be initiated or carried out at the preserve level since they act over the whole landscape and require governmental approval, special skills, and years of effort before rewards are produced. The advantages of biological control, however, are especially important in cases where invasive plants are widespread and control is desired over large areas. Because biological control uses self-perpetuating living organisms, control spreads on its own after effective agents have been identified and established, until they reach their ecological limits. These features make biological control the only control method that is economically feasible for suppression of invasive plants over very large areas (millions of ha). Also, the method is free of both the disturbances characteristic of mechanical control and the pollution that may follow the widespread use of herbicides.

Disadvantages of biological control from the perspective of preserve stewards include the fact that the method is beyond their direct control to initiate against new target plants. Stewards can, however, participate in regional projects, releasing useful agents on their property after effective agents become available. The participation of The Nature Conservancy (TNC) in the control of *Euphorbia esula* on some prairie preserves in North America is an example of such participation. *Aphthona* beetles were introduced to preserves after being studied and proven effective at other sites (Cornett et al. 2006). In some cases, conservation groups may participate at earlier stages, as for example TNC participation in preparation for the release and evaluation of *Aphalara itadori*, a psyllid being studied in the United States and the U.K. (Shaw et al. 2009) for the control of *Fallopia japonica* (Japanese knotweed) along rivers (Gerber et al. 2008).

The most important disadvantage of biological control agents stems from their permanency. Once released, inappropriately selected agents can rarely be removed, although potentially they might themselves be amenable to suppression via biological control using insect parasitoids (e.g. Pemberton and Cordo 2001). The use of biological control against invasive plants requires a high level of certainty about the safety and desirability of each new herbivore before it is released. The track record of weed biological control insects released in the United States (including Hawaii) and the Caribbean since the 1960s provides strong evidence that modern agent selection processes provide this necessary level of certainly (Pemberton 2000). Finally, the ecological limits to spread for each newly released biological control agent must be predicted so that non-target flora in all potentially invaded regions can be considered. However, human-assisted accidental spread of agents may occasionally move agents to distant areas (Pratt and Center 2012), far beyond the area targeted for biological control. In this respect, biological control agents are no different than any other species, all being potentially subject to such chance events.

26.3   When Is Biological Control the Right Approach?

Biological control projects should not be under taken lightly as they require a long-term commitment of funds and scientific manpower to carry through to completion. Premature commitment to a project against a minor pest may consume resources better used against a more serious invader. Appropriate targets should be invasive non-native plants that are widespread (or potentially so) or are intransigent to other control methods in critical habitats and cause (or potentially cause) significant damage, usually to natural areas. Several factors further modify both feasibility and cost of projects. Securing agents with adequately narrow host ranges is more likely when targeted invasive plants are only distantly related to native plants. Targeting such species lowers the risk to native flora, expedites agent screening, is often less costly, and is more likely to succeed. *Melaleuca* and *Tamarix* spp. (saltcedar) in North America are both in subfamilies or tribes with few or no representatives in the native flora (saltcedar: subfamily, Baum 1967; Crins 1989, and melaleuca: tribe, Serbesoff-King 2003). In contrast, projects in North America directed against invasive thistles (*Carduus*, *Cirsium*, and *Silybum*) or knapweeds (*Centaurea*) are more complicated because there are many congeneric native species (e.g. for thistles; Schroder 1980), some of which may be endangered. While projects against targets with native congeners are not uncommon, screening of more species may be needed to find suitable agents.

While many projects begin with no known prospective agents, some have the advantage of being directed against species that have been controlled elsewhere, making them quicker and cheaper because potentially effective species are known. The principal cost in such cases is the screening of additional species from the flora of the new area. It is important to recognise that safety is contextual and a species safe to introduce in one country may be unsafe in another due to differences in the composition of resident plant communities.

Projects should be directed against plants that cause the most ecological damage to local ecosystems. Species that change the properties of the invaded community, such as increasing fire frequency or intensity (Brooks et al. 2004) or that have a structural form that allows them to overtop native species are likely to be highly damaging. Invasive floating aquatic plants, vines, and ground covers are likely to seriously damage native plant communities. For example, plants such as *Eichhornia crassipes* (water hyacinth) and *Salvinia molesta* (salvinia) severely alter the submersed aquatic communities that they blanket (Mitchell 1978; Thomas and Room 1986).

Availability of funding is also an important consideration in starting a classical biological control programme. Programmes without adequate, long-term funding (often 10 years or longer) and enough political support to see the work through the inevitable setbacks have little chance of success. It is also important to select species as targets for biological control that reflect local conservation priorities.

26.4   Mechanics of Biological Control

Steps and decision points involved in a classical weed biological control project are illustrated in Fig. 26.1. More detailed information on the actual mechanics involved is available in the literature (e.g. Harley and Forno 1992; Van Driesche et al. 2008). The flow chart in Fig. 26.1 is of a generalised nature and does not cover all aspects of a weed biological control project. Selection of a candidate agent often involves more than simple determination of host range. It may be warranted to attempt to predict the potential efficacy of the agent; however, how biological control agents actually perform in nature depends on a complex set of population processes that can't be tested in the laboratory, including the effects of parasitoids, predators, competitors, and climate. This step is often advocated, even though such predictions have never been proven to be possible. Any such predictions should therefore be validated after release to advance the science involved. Also, consideration must be given to the presence of pre-existing agents so that the plant parts attacked, the phenological timing of agent populations, and agent habitat preferences are all mutually complementary. Such considerations become important in projects requiring multiple agents. The order of introduction may also need to be considered, so that earlier introduced agents do not interfere with success of later introductions. Each project is unique so rules of thumb are often meaningless and projects must be adaptively managed. The overall purpose of this process is to balance the risk of introducing a novel herbivore or plant pathogen into the system against the risk imposed by the plant invader (i.e. the risk of doing nothing).

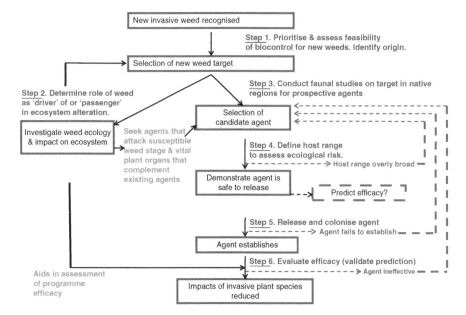

Fig. 26.1 The steps and decision points involved in biological control of an invasive plant species

26.5 Fitting Biological Control into a Holistic Approach to Ecological Restoration

Biological control of invasive plants in natural areas (including legally protected preserves) should be a partnership between biological control scientists and conservation biologists, and the biological control activities should be part of a larger, holistic programme for the restoration of the affected ecosystems or protected areas.

26.5.1 Partnerships and Goal Setting

Partnerships for such biological control efforts are critical because of the complicated nature of the problems being addressed, requiring the active participation of land managers and ecologists who are familiar with the systems in need of restoration. Skills needed for large scale ecological restoration projects are commonly spread over several agencies or universities, with distinct budgets and somewhat different perspectives on the problem. It is important to overcome this separation and form a team willing to work together to bring into play the full range of ideas and skills needed. Such partnerships also help avoid conflicts that might arise when actions are taken by single parties before all concerned have reached an accord.

Defining restoration goals and the role of biological control and other activities is a critical first step to be taken by the partners in a project. While biological control aims to lower the density of what is believed to be an invader damaging to the natural area, the goal per se is to restore the natural community, either in terms of biodiversity or ecological function. For example, *Lygodium microphyllum* (Old World climbing fern) has changed the fire cycle of invaded *Taxodium distichum* (cypress) forests by increasing fire frequency and intensity. Therefore, lowering these fire characteristics should be a goal of the project. Additionally, if certain native species have been displaced from a plant community, their return to pre-invasion levels is also a goal. Such native plants, however, may not recover spontaneously following the suppression of the responsible invasive species because propagules of desired plants may be lacking or other invasive weeds may increase quickly following biological control. For example, at some sites where the weevil *Rhinoncomimus latipes* reduced infestations of *Persicaria perfoliata* (mile-a-minute weed), the invasive plants *Microstegium vimineum* (Japanese stiltgrass) and *Rosa multiflora* (multiflora rose) increased in abundance (Lake 2011; Hough-Goldstein et al. 2012). In that case, planting native perennials, along with use of a pre-emergent herbicide to suppress the other invasive species, was used to restore such communities (Lake 2011). Similarly, Stephens et al. (2009) found that only non-native grasses showed consistent increases following reductions of *Centaurea diffusa* (diffuse knapweed) caused by the weevil *Larinus minuta*. Additional measures were, therefore, necessary to restore that community.

26.5.2 Determining Causality of Community Degradation

Invasive plants may be fundamentally responsible for ecosystem degradation or just symptoms of other processes. This has been described with the simile of 'drivers and passengers.' Also, if invasive plants are drivers, it is necessary to determine if they act alone or are facilitated by other factors. Because there is a long time lag between starting a biological control project and release of effective agents, surveys for prospective agents should start as soon as there is agreement that the targeted invasive plant is the driver of the observed habitat degradation. In some cases, biological control may not be advisable if the invader is merely a passenger, responding to some other disturbance (e.g. nutrification of an aquatic system stimulating growth of aquatic weeds).

Determining that an invasive plant is merely a passenger requires careful observation and knowledge of the community. For example, in the 1950s, several species of *Opuntia* cacti were seen as pests in pastures, including the two native species *O. stricta* and *O. triacantha* and one introduced species, *O. cochenillifera* (Pemberton and Liu 2007) on the Caribbean islands of Nevis and St. Kitts. However, these high density cactus stands were largely the consequence of pasture overgrazing, opening land to cactus invasion and reducing competition from other plants. However, the ability of the moth *Cactoblastis cactorum* to suppress *Opuntia* species was well known (Dodd 1940; DeBach 1974), and so it was released on Nevis, Montserrat, and Antigua (Simmonds and Bennett 1966), ignoring risks to native cacti. Given that the fundamental reason for damaging cacti levels was overgrazing by goats (Simmonds and Bennett 1966), the correct response would have been not biological control, but rather better livestock management. Another example in which the difference between a driver and passenger is not clear is that of *Alliaria petiolata* (garlic mustard), a European mustard that forms dense stands in deciduous forests of the north-eastern and north central United States, which are associated with low diversity of native forest herbaceous plants (Blossey et al. 2002). Based on those facts, biological control of *A. petiolata* seemed necessary to protect forest wildflower diversity. However, some have linked invasive earthworms (Maerz et al. 2009) and overgrazing by deer (Knight et al. 2009) as more fundamental drivers of change in these forests. However, it should be noted that things can change: species that were not originally drivers may become so once the community has been widely invaded if the invaders change fundamental aspects of the community. So, it is necessary to keep an open mind and continue to observe invaded systems in such cases.

That some invasive plants are drivers of change is well know from impacts in other locations and the invasion of new areas by such species should be seen as a cause for alarm and suggest the need for immediate plant suppression (Reichard and Hamilton 1997). Additionally, studies in the invaded area may reveal that invasive plants are drivers, having highly damaging effects either by suppressing native plants, e.g. *Miconia calvescens* (miconia) in Tahitian forests (Meyer and Florence 1996; Medeiros et al. 1997; Meyer 1998), or by changing a fundamental characteristic defining and creating a community, such loss of water depth due to soil accretion caused by *Melaleuca quinquenervia* in the Florida Everglades (Center et al. 2012). Such plants clearly merit being targeted for biological control.

Synergy among invaders is also possible in systems suffering from multiple invasions, such as the Florida Everglades (Simberloff and von Holle 1999). The ecological damage from a single invader, and hence the need for its biological control, may change due to its interactions with other invaders, requiring a community-based view to correctly assess risk and need for control. For example, a non-native plant group, such as the figs (*Ficus*), which require specific exotic pollinators, may persist at innocuous levels in the introduced range for decades but then rapidly become invasive after their pollinators invade (Nadel et al. 1992). Once able to reproduce, figs produce many fruits that are spread by exotic frugivorus birds (Kaufmann et al. 1991). Seed dispersal then greatly facilitates the spread of figs to new areas.

External factors may either enhance an important invader's impacts or make undoing its damage more difficult. For example, *Tamarix* spp. are invasive in riparian areas in deserts of the south-western United States. However, the invasion was strongly facilitated by altered river management, dammed rivers being more favourable for *Tamarix* and less favourable for reproduction of native cottonwoods and willows, thus reducing native competition to *Tamarix*. While biological control for the suppression of the species is necessary to restore invaded riparian communities (DeLoach et al. 2003), it may not be sufficient in some areas if native vegetation propagules are absent or conditions are unfavourable for their growth. In such cases, actions such as replanting of native species or resumption of natural flooding may be required (Shafroth et al. 2005). Similarly, eutrophication of waterways is well known to exacerbate infestation of invasive aquatic plants (Coetzee and Hill 2012).

26.5.3 Integrating Activities and Dealing with Complications

In some cases, biological control may need to be combined with other control tactics to suppress an invader. Also, invader suppression itself may need to be combined with other efforts to restore habitat conditions favourable for native species. If an invader is detected early, biological control may not be the right approach as it may be possible to simply eradicate small invader populations by chemical or mechanical means. However, if the infestation is spreading rapidly or is already widespread, biological control is likely to be needed. Biological, chemical, and mechanical controls, or replanting of native species are likely to be implemented by different restoration partners due to differences in expertise. In such cases, careful joint planning of the timing, placement, and degree of all such activities is critical to prevent delays or conflicts. In restoration projects, unforeseen complications are common and the restoration plan must adapt to new developments are they occur. Adaptive management relies on monitoring of the system as it changes in responses to control efforts, so that tactics and goals can be changed as needed if new insights are gained into how the system is currently functioning. While each case is different, complications (or indeed "surprises") are commonplace and should be anticipated, at least in general terms.

While releases of particular individual natural enemy species cannot be "undone" (and therefore must be carefully assessed for safety before release), in the aggregate over many projects and agents, post-release monitoring of outcomes is useful in determining if estimates of safety were well founded and, if not, monitoring data can be used to identify faulty assumptions or procedures that might need to be changed. More commonly, such monitoring is likely to validate predictions broadly but may detect small deviations from predictions that can be used to increase efficacy of pre-release safety procedures.

Invader replacement is one such complication. Suppression of a dominant invader by biological control agents frees space and resources for other plants. These plants may be native species or, in some cases, other invasive species formerly suppressed by the controlled invader. Invader replacement is particularly common in aquatic systems where anthropogenic nutrification has stimulated plant growth (Coetzee and Hill 2012). Rapid control of a problematic aquatic weed, particularly through the use of herbicides, often leads to an upsurge in the abundance of a different macrophyte or a massive algal bloom (Richard et al. 1984). Although spontaneous declines in invasive aquatic macrophytes are not well understood, nutrient depletion has been identified as one likely factor (Barko et al. 1994), and this relationship should be considered in planning the biological control of such species.

Unanticipated food web effects are another complication that may arise at some stages of a biological control project. When biological control agents establish and become common but do not by themselves lower the density of the pest plant they attack, then the biological control agent may supply a readily available food subsidy for resident predators. This is generally a temporary condition, subsiding as the target is controlled. One potential example is that of seed-head gall flies, *Urophora affinis* and *U. quadrifasciata*, introduced to control *Centaurea stoebe* subsp. *micranthos* (spotted knapweed). These flies became abundant and according to Pearson et al. (2000) failed to control *C. stoebe* subsp. *micranthos* (but see Story et al. 2008). During this period, galled seed-heads became a protein-rich food source for deer mice (*Peromyscus maniculatus*), enabling them to more readily survive winters and reproduce earlier in spring, leading to higher mouse populations in knapweed stands (Ortega et al. 2004). These authors further speculated that more mice would mean higher levels of Sin Nombre hantavirus, a human pathogen (Pearson and Callaway 2006). Although this example may seem cause for alarm, there is no evidence of increased incidence of this virus, and, in terms of the biological control project's goals, gall flies alone were never expected to entirely control the weed. Other agents have combined with them to reduce knapweed infestations (see Corn et al. 2006; Story et al. 2006, 2008). It does, however, serve as a reminder that such effects are possible when a new species is inserted into the trophic structure of a community.

Finally, society's view of the desirability of suppressing the target pest may change erratically over time as new facts emerge. For example, biological control of *Hydrilla verticillata* (hydrilla), a submerged aquatic weed, began in Florida in the 1980s. Two leaf-mining flies and two weevils were released as biological control agents (Center et al. 1997; Grodowitz et al. 1997; Wheeler and Center 2007), with partial suppression of plant density. Subsequently, a new herbicide called fluridone was developed that provided easier control of *H. verticillata* and biological control efforts stalled in the 1990s before the project could be completed. When *H. verticillata* became resistant to fluridone in the early 2000s (Michel et al. 2004), interest in biological control revived briefly, but stalled again when the endangered Florida snail kite (*Rostrhamus sociabilis plumbeus*), which normally feeds on the Florida apple snail (*Pomacea paludosa*), was discovered exploiting an exotic apple snail (*Pomacea insularum*) in mats of *H. verticillata*. However, this favourable view of *H. verticillata* reversed again when it was found that it supported growth of a toxic cyanobacterium (Wilde et al. 2005) that became linked to the deaths of thousands of American coots (*Fulica americana*) (Wilde et al. 2005) and possibly some bald eagles (*Haliaeetus leucocephalus*) and other predators via bioaccumulation of the toxin in food webs (Birrenkott et al. 2004; Fischer et al. 2006). These impacts on wildlife stimulated interest in the release of herbivorous fish to control *H. verticillata* in the south-eastern United States, but concerns arose that these fish might be sensitive to the toxin or might transfer it through the food web to piscivorous species (Wilde et al. 2005). Finally, to complicate matters further, in some locations manatees (*Trichechus manatus*) have been shown to benefit from *H. verticillata* infestations (Evans et al. 2008). While biological control of *H. verticillata* may or may not be resumed in the future, these events illustrate the rapidity with which societal views of the desirability of suppression of an invasive plant can change.

26.6   The Historical Record of Weed Biocontrol in Natural Areas

Biological weed control has targeted invasive species in many natural or semi-natural systems (Van Driesche et al. 2010) (Table 26.1) and these efforts have benefits to legally protected preserves within the affected regions or landscapes. Here we review many of these projects, to provide a sense of the magnitude of the benefits of biological control to ecosystem restoration. Projects are arranged by habitat to give an integrated sense of biological control's value in particular systems.

26.6.1   Fynbos Invaders

The South African fynbos supports 8,700 plant species, 68 % of which are endemic (Richardson et al. 1997; Holmes et al. 2000). Its infertile soils are readily invaded by nitrogen-fixing plants that raise soil fertility and depress native plant growth (Lamb and Klaussner 1988; Stock et al. 1995; Yelenik et al. 2004). Fynbos habitats have been invaded by various introduced woody plants in such genera as *Acacia*, *Pinus*, *Hakea*, and *Sesbania*. All of these but the pines have been targeted for biological control, with considerable success. Biological control, often in the form of seed reduction, together with manual clearance, has greatly reduced the threat of several invaders, including (i) *Acacia saligna*, controlled by the fungus, *Uromycladium tepperianum* (Wood and Morris 2007); (ii) *A. longifolia* (Dennill and Donnelly 1991); (iii) *A. pycnantha* (Moran et al. 2005); (iv) *A. cyclops*, which formed impenetrable stands in the lowland fynbos (Richardson et al. 1996) and threatened plant biodiversity in the Cape Peninsula (Higgins et al. 1999), was rendered less invasive by the seed weevil *Melanterius* cf. *servulus* (Impson et al. 2004) and the flower-galling midge *Dasineura dielsi* (Adair 2005; Impson et al. 2008, 2011); (v) *Hakea sericea*, controlled by five introduced insects and the pathogen *Colletotrichum gloeosporioides* in conjunction with manual removal (Gordon 1999; Esler et al. 2010; Gordon and Fourie 2011; Gordon, personal communication, 2012), and (vi) *Sesbania punicea* (Hoffmann and Moran 1998), a leguminous tree that formed dense bands 20–30 m wide along rivers until three

Table 26.1 Plants invasive in natural areas that have been targets of biological control, with notes on habitat/biome invaded, degree of project success, and location

Target species	Habitat/Biome	Success	Notes/Location	References
Acacia cyclops (rooikrans)	F, Co	C	Agents effective at reducing seed production but integrated control needed to reduce tree densities	Impson et al. (2004, 2008, 2011) and Adair (2005)
Acacia longifolia (long-leaved wattle)	F	C	South Africa, especially the fynbos region: bud galler and seed feeder effectively reduce seed crop	Dennill and Donnelly (1991), Dennill et al. (1999), and Moran et al. (2005)
Acacia nilotica subsp. *indica* (prickly acacia)	GD	IP	Western Queensland, Australia: transforms natural grassland into woody savannah	Palmer et al. (2012)
Acacia pycnantha (golden wattle)	F	C	South Africa, especially the fynbos region: extensive damage from bud galler	Moran et al. (2005) and Impson et al. (2011)
Acacia saligna (Port Jackson willow)	F	C	South Africa, especially the fynbos region: seed beetles destroy to 90 % of seeds; gall-forming rust fungus highly effective	Moran et al. (2005), Wood and Morris (2007), and Impson et al. (2011)
Ageratina adenophora & *A. riparia* (mistflowers)	WR, O	P	Africa, Asia, Australia, USA, New Zealand, Papua New Guinea, Philippines, Tahiti, Hawaii: invades stream banks and moist areas. Aggressive competitor; controlled in Hawaii, New Zealand, and possibly South Africa by a smut fungus	Heystek et al. (2011), Cruttwell McFadyen (2012), and Schooler et al. (2012a)
Alliaria petiolata (garlic mustard)	FW	IP	Northeast and north central USA: agents under evaluation	Blossey et al. (2001b)
Alternanthera philoxeriodes (alligator weed)	A, WR	C, P	USA, Australia, New Zealand, and China: aquatic infestations controlled by a chrysomelid leaf beetle but not terrestrial populations	Coulson (1977), Julien (1981), Julien and Griffiths (1998), Sainty et al. (1998), Buckingham (2002), and Julien et al. (2012a)
Anredera cordifolia (Madeira vine)	WR, FW	IP	Coastal eastern Australia, South Africa, New Zealand, Sri Lanka, Hawaii: agent surveys underway	van der Westhuizen (2006, 2011), Cagnotti et al. (2007), and Palmer and Senaratne (2012)

Species			Notes	References
Arundo donax (giant reed)	WR, GD	IP	South-western USA, especially Texas: adventive gall wasp found prior to release and armoured scale released	Tracy and DeLoach (1998), Racelis et al. (2009, 2010), and Goolsby et al. (2011)
Asparagus asparagoides (bridal creeper)	FW, Co	P, IP	Coastal areas of temperate Australia: little control to date	Turner et al. (2008b) and Morin and Scott (2012)
Azolla filiculoides (red water fern)	A	C	South Africa: weevil highly effective	McConnachie et al. (2004), Hill and McConnachie (2009), and Coetzee et al. (2011a)
Baccharis halimifolia (groundsel bush)	FW, Co	P	Australia: changes in climate and land use along with biological control have contributed to declines	Palmer and Sims-Chilton (2012)
Cabomba caroliniana (fanwort)	A	IP	Australia: agents under evaluation in quarantine	Schooler et al. (2012b)
Caesalpinia decapetala (Mauritius thorn)	FW	IP	New Zealand, Australia, USA, Kenya, Zimbabwe, South Africa. Transformer species increasing fire risk and causing trees to collapse in subtropical forests	Byrne et al. (2011)
Campuloclinium macrocephalum (pompom weed)	GD	IP	South Africa: disrupting grassland conservation efforts	McConnachie et al. (2011)
Cardiospermum grandiflorum (balloon vine)	Co	IP	Australia, the Cook Islands, Hawaii, New Zealand, South Africa	Simelane et al. (2011)
Centaurea diffusa & *C. stoebe* (diffuse and spotted knapweeds)	GD	IP	Western North America: plant densities not yet reduced at most locations, but declines observed in Oregon and California	Gutierrez et al. (2005), Pitcairn et al. (2005), and Smith (2007)
Cereus jamacaru (queen of the night cactus)	GD	C	South Africa: mealybug effective in most parts of country; cerambycid stem borer very effective when at high density	Paterson et al. (2011)
Cestrum laevigatum (inkberry)	FW, Co	IP	South Africa: forms dense stands in coastal forests and thickets	Fourie (2011)
Cestrum parqui (Chilean inkberry)	WR	IP	South Africa: along the Vaal river in the High Veld	Fourie (2011)

(continued)

Table 26.1 (continued)

Target species	Habitat/Biome	Success	Notes/Location	References
Chromolaena odorata (Siam weed)	GD, FW	P, IP	Significant control in Papua New Guinea and East Timor of one biotype; some control of second biotype in South Africa	Day and Bofeng (2007), Zachariades et al. (2009, 2011b), Day and Cruttwell McFadyen (2012), and Day, personal communication, 2012
Chrysanthemoides monilifera subsp. *rotundata* (bitou bush), *C. m.* subsp. *monilifera* (boneseed)	Co	P	Bitoubush along coastline of New South Wales, Australia: flowering and seed production widely suppressed. Boneseed in SE Australia. No reductions in density of either subspecies	Holtkamp (2002), Edwards et al. (2009), and Adair et al. (2012)
Cryptostegia grandiflora (rubber vine)	GD, FW	P	Dry tropics of Australia: excellent control achieved where rust fungus established	Evans and Tomley (1994), Mo et al. (2000), Vogler and Lindsay (2002), and Palmer and Vogler (2012)
Cylindropuntia spp. (jumping cholla)	GD	P, IP	Reported as injuring, even causing death of South African wildlife; effective biological control of some species but not others	Paterson et al. (2011) and Holtkamp (2012)
Cytisus scoparius (Scotch broom)	FW	IP	Australia, New Zealand, USA: causes loss of native plant species in Australia. Unsuccessful to date despite long history of effort (>50 years)	Hosking et al. (2012)
Dioscorea bulbifera (air potato)	FW	IP	Florida, USA: first agent released 2011	Pemberton (2009)
Dolichandra unguis-cati (= *Macfadyena unguis-cati*) (cat's claw)	WR, FW, O	IP	Invasive in Australia, South Africa, India, Mauritius, China, Hawaii and Florida in the USA, New Caledonia, St. Helena, and New Zealand	Dhileepan et al. (2007a, b), King et al. (2011), and Dhileepan (2012)
Eichhornia crassipes (water hyacinth)	A	C, P, IP	Southern USA, Mexico, East and West Africa, India, and other warm regions: complete control in many tropical areas; partial control in cooler regions	Beshir and Bennett (1985), Center et al. (2002), Coetzee et al. (2009, 2011a), and Julien (2012a)

Euphorbia esula (leafy spurge)	GD	P	Northern prairies of North America: complete control in many areas	Cornett et al. (2006), Cline et al. (2008), and Samuel et al. (2008)
Euphorbia paralias (sea spurge)	Co	IP	Coastal southern Australia: project in early stages	Scott (2012)
Fallopia japonica (Japanese knotweed)	WR	IP	United States and United Kingdom: first agent approved for release in UK	Shaw et al. (2009)
Genista monspessulana (Cape broom)	GD, FW	IP	Australia, USA: no agents released but one found adventive in Australia. International collaboration on-going	Sheppard and Henry (2012)
Hakea sericea & H. gibbosa (silky & rock hakea)	F	C	South Africa: control achieved in combination with manual clearing	Gordon (1999), Esler et al. (2010), Gordon and Fourie (2011), and Gordon (2012)
Hydrilla verticillata (hydrilla)	A	P, IP	Southern USA: partial control in a few areas	Balciunas et al. (2002), Coetzee et al. (2011a), and Grodowitz, personal communication, 2012
Hypericum perforatum (St. Johnswort)	GD, FW	C, P	Western USA: complete control; Australia: partial control	Huffaker and Kennett (1959), McCaffrey et al. (1995), Briese (1997), and Briese and Cullen (2012)
Jatropha gossypiifolia (bellyache bush)	WR	IP	Dry tropics of Australia: one agent released but not established	Heard et al. (2012)
Lantana camara hybrid complex (lantana)	GD, FW, Co	P, IP	Hawaii, Africa, Asia, Oceania, northern and eastern Australia: limited success in a few areas despite > 100 years history	Day and Zalucki (2009), Urban et al. (2011), and Day (2012a)
Leptospermum laevigatum (Australian myrtle)	F	IP	South Africa: negligible control by two biological control agents	Gordon (2011)
Leucaena leucocephala (lead tree)	FW, Co, O	IP	Hawaii, Taiwan, Fiji, Northern Australia, South America, Europe, India, SE Asia, USA: susceptible to psyllid infestations	Austin et al. (1996) and Olckers (2011a)
Lygodium microphyllum (Old World climbing fern)	WR, FW	IP	Southern Florida, USA: control is developing at release sites	Boughton and Pemberton (2009)

(continued)

Table 26.1 (continued)

Target species	Habitat/Biome	Success	Notes/Location	References
Lythrum salicaria (purple loosestrife)	WR	P	Northern USA and adjacent areas of Canada: control in some areas	Blossey et al. (2001a), Landis et al. (2003), Denoth and Myers (2005), and Grevstad (2006)
Marrubium vulgare (horehound)	GD, FW, O	P, IP	Southern Australia: weed suppression and seed reduction noted. Also invasive in North and South America and New Zealand	Weiss and Sagliocco (2012)
Melaleuca quinquenervia (melaleuca)	WR	C	Southern Florida, USA; control very effective in combination with mechanical and chemical control of mature plants	Pratt et al. (2005), Center et al. (2007), Rayamajhi et al. (2007, 2008, 2009), and Tipping et al. (2008b, 2009)
Miconia calvescens (miconia)	WR, FW, Co	P, IP	Partial control in Tahiti; no control yet in Hawaii	Seixas et al. (2004), Badenes-Perez et al. (2007), and Meyer et al. (2008, 2009)
Mikania micrantha (hemp vine)	WR, FW, Co	P, IP	Australia, Asia, Pacific Islands. A rust reducing infestations in some areas	Day (2012b)
Mimosa pigra (mimosa)	WR	P	Africa, Australia, & Asia. Seed banks reduced by about 90 % in northern Australian wetlands	Heard and Paynter (2009) and Heard (2012)
Moraea spp. (Cape tulips)	GD, FW, O	IP	Australia. Two South Africa species under investigation (M. flaccida & M. miniata)	Scott and Morin (2012)
Myriophyllum aquaticum (parrot's feather)	A	P	South Africa, control is considered satisfactory although not complete	Coetzee et al. (2011b)
Myriophyllum spicatum (Eurasian watermilfoil)	A	P, IP	United States and Canada: some control attained using a native weevil; South Africa: in progress	Newman (2004) and Coetzee et al. (2011a)
Opuntia spp. (prickly pear cacti)	GD	C	Australia, South Africa. Australia estimates benefit:cost at 147:1	Dodd (1940), Paterson et al. (2011), and Hosking (2012)

Opuntia robusta (wheel cactus)	GD, FW	IP	Australia: little to no control by agents from other *Opuntia* species. Also invasive in the Americas, New Zealand, South Africa, and Mediterranean Europe	Baker (2012)
Paraserianthes lophantha (stinkbean)	F	P	South African fynbos: transformer species, substantial control by seed weevil	Dennill and Donnelly (1991), Dennill et al. (1999) and Impson et al. (2009, 2011)
Parkinsonia aculeata (parkinsonia)	WR, GD	IP	Australia: no population level impacts yet realised	van Klinken and Heard (2012)
Parthenium hysterophorus (parthenium weed)	GD	P	Primarily a pasture weed; Queensland, Australia, control achieved in some areas; Also invasive in many parts of Asia & Africa	Dhileepan and Cruttwell McFadyen (2012)
Pereskia aculeata (Barbados gooseberry)	GD	IP	South Africa: Overtops and kills native flora, sometime collapsing large trees; chrysomelid beetle shows some promise	Paterson et al. (2011)
Persicaria perfoliata (mile-a-minute weed)	FW	P, IP	Eastern USA: Effective control at some locations	Hough-Goldstein et al. (2009, 2012)
Pistia stratiotes (water lettuce)	A	C, P	Papua New Guinea, Australia; several regions in Africa; and warm parts of North America. Complete control obtained in Queensland and some areas of South Africa	Harley et al. (1990), Dray and Center (1992), Ajuonu and Neuenschwander (2003), Mbati and Neuenschwander (2005), Neuenschwander et al. (2009), Coetzee et al. (2011a), and Day (2012c)
Phyla canescens (lippia)	WR	IP	Australia: Many prospective agents found but none yet released	Julien et al. (2012b)
Prosopis spp. (mesquite)	GD	P	Arid parts of Australia: control achieved in the Pilbara region; South Africa: seed beetles somewhat effective	van Klinken and Campbell (2009), Zachariades et al. (2011a), and van Klinken (2012)
Rubus spp. (blackberries)	FW, O	IP	Chile: reduction in size and competitiveness of plants; Australia, Hawaii, tropical Africa, West Indies, UK	Oehrens (1977), Oehrens and Gonzalez (1977), and Morin and Evans (2012)

(continued)

Table 26.1 (continued)

Target species	Habitat/Biome	Success	Notes/Location	References
Salsola spp. (tumble weeds)	GD	IP	Western USA, especially California	Smith (2005) and Smith et al. (2009)
Salvinia molesta (giant salvinia)	A	C	Australia, Papua New Guinea, parts of the USA, and parts of Africa, especially the Congo basin. Australia estimates benefit: cost up to 53:1	Room et al. (1981), Thomas and Room (1986), Mbati and Neuenschwander (2005), Diop and Hill (2009), Julien et al. (2009), Coetzee et al. (2011b), and Julien (2012b)
Schinus terebinthifolius (Brazilian peppertree)		IP	Florida, USA: agents under evaluation; Hawaii	Cuda et al. (2009)
Senecio jacobaea (tansy ragwort)	GD	C	Western USA: highly successful in northern California and western Oregan	Pemberton and Turner (1990), McEvoy et al. (1991), Turner and McEvoy (1995), and Coombs et al. (1996, 2004)
Sesbania punicea (sesbania)	F	C	South Africa, especially the fynbos region: three agents maintain plant at non-problematic levels	Hoffmann and Moran (1991, 1998)
Solanum mauritianum (bugwood, tree tobacco)	FW	IP	South Africa: colonises native forest margins overtopping and shading out native species	Olckers (2011b)
Solanum viarum (tropical soda apple)	GD	IP	Southeastern USA: control achieved at release sites, agent spreading	Medal et al. (2008) and Medal and Cuda (2010)
Tamarix ramosissima (saltcedar)	WR, GD	P, IP	Western USA; control developing around release sites	Hudgeons et al. (2007), Carruthers et al. (2008), DeLoach et al. (2008), and Dudley and Bean (2012)
Tecoma stans (yellow bells)	WR	IP	South Africa and neighbouring countries: a 'transformer' species, invades watercourses and rocky sites in tropical/subtropical areas including high-rainfall to semi-arid areas; rust fungus recently released but establishment unconfirmed	Madire et al. (2011)

Species	Habitat/Biome	Outcome	Notes	References
Triadica sebifera (Chinese tallow tree)	WR	IP	South-eastern USA: agents under evaluation; invades lake and pond margins; displaces native plants; reduces nesting habitat for birds	Wang et al. (2009)
Ulex europaeus (gorse)	WR, FW, Co	P, IP	Chile, Oregon (USA), Tasmania, Hawaii, New Zealand: some impact in Chile, Hawaii and Tasmania	Norambuena (1995), Norambuena and Piper (2000), Davies et al. (2007), Norambuena et al. (2007), Hill et al. (2008), and Ireson and Davies (2012)
Vincetoxicum nigrum & *V. rossicum* (swallow-worts)	WR, GD, FW	IP	North-eastern USA: surveys for agents in progress	Weed and Casagrande (2010)

Habitat/Biome symbols: *F* fynbos, *A* aquatic, *WR* wetland/riparian, *GD* grassland/desert, *FW* forest/woodlands, *Co* coastal, *O* other
Outcome symbols: *C* complete control, *P* partial control, *IP* in progress

beetles (*Trichapion lativentre*, *Rhyssomatus marginatus*, and *Neodiplogrammus quadrivittatus*) were introduced that destroyed its buds and seeds and bored in its stems (Hoffmann and Moran 1991), reducing its density >95 % (Hoffmann and Moran 1998) and returning rivers to pre-invasion conditions (Hoffmann 2011).

26.6.2 Floating Weeds

In warm regions, floating invasive plants may blanket water surfaces, e.g. *Eichhornia crassipes*, *Salvinia molesta*, *Azolla filiculoides* (red fern), *Pistia stratiotes* (water lettuce) (see Table 26.1), having profound effects on light penetration, changes in nutrients, oxygen, and pH (Toft et al. 2003) and affecting the whole aquatic community. Native benthic plants and associated invertebrates are strongly affected by these changes (Hansen et al. 1971). Biological control has been highly effective in some locations against *A. filiculoides* (Hill and McConnachie 2009; Coetzee et al. 2011a), *E. crassipes* (Center et al. 2002; Wilson et al. 2007; Coetzee et al. 2009, 2011a; Julien 2012a), *S. molesta* (Tipping et al 2008a; Julien et al. 2009), *P. stratiotes* (Neuenschwander et al. 2009; Coetzee et al. 2011a), and *Alternanthera philoxeriodes* (alligator weed; Buckingham 2002). Many water bodies have been relieved of burdening layers of these weeds by biological control and some, like Lake Victoria in East Africa, harbour globally important biota (here, cichlid fishes) (Anonymous 2000; Wilson et al. 2007).

26.6.3 Wetlands Invaders

Wetlands have been invaded by several non-aquatic plants, including the tree *Melaleuca quinquenervia*, the fern *Lygodium microphyllum*, the shrub *Mimosa pigra*, and the herbaceous perennials *Lythrum salicaria* (purple loosestrife) and *Fallopia japonica*. These plants have reduced native biodiversity through habitat change and competition with native plants. All five have been targeted with biological control and at least one (*M. quinquenervia*) has been successfully controlled, while *M. pigra*, *L. salicaria*, and *L. microphyllum* projects have had partial success or are in progress.

   Melaleuca quinquenervia formed dense monocultures in Florida, displacing native vegetation (Rayamajhi et al. 2002) and reducing biodiversity of freshwater marshes by 60–80 % (Austin 1978). The weevil *Oxyops vitiosa*, the psyllid *Boreioglycaspis melaleucae*, and the cecidomyiid *Lophodiplosis trifida* suppressed seeding and seedling survival (Center et al. 2007, 2012; Rayamajhi et al. 2007; Tipping et al. 2009), killing 85 % of seedlings, saplings, and suppressed understory trees, leading to a fourfold increase in plant biodiversity (Rayamajhi et al. 2009). This tree is now largely under biological control after an effective integrated control programme in which biological control contributed restraints on seed production, seedling survival, and stump regrowth, while cutting or application of herbicides removed mature trees (Center et al. 2012).

Lygodium microphyllum from Australia smothers trees in Everglades hammocks, cypress swamps, and pine flatwoods in Florida (Pemberton and Ferriter 1998) and increases fire intensity by forming flammable skirts on tree trunks (Pemberton and Ferriter 1998). The pyralid moth *Neomusotima conspurcatalis* has established and now is defoliating the fern at some release sites, allowing regrowth of native plants (Boughton and Pemberton 2009). It has been slow to disperse but is now found several miles from release sites (Center, personal observation).

Mimosa pigra invaded tropical wetlands in Australia, Asia, and Africa, particularly along margins of wetlands, lakes, and channels, but also in open plains and swamps (Cook et al. 1996). In Australia, *M. pigra* converts several vegetation types into homogeneous shrublands with little biodiversity (Braithwaite et al. 1989), threatening vulnerable plant and animal species (Walden et al. 2004). Among two fungi and nine insects established, two species have shown the most impact to date: the sesiid borer *Carmenta mimosa* and the leaf-mining gracillariid *Neurostrota gunniella*, which together have reduced seed set and seedling regeneration, causing *M. pigra* stands to shrink at the edges (Heard and Paynter 2009). Seed banks are now 90 % below pre-biological control levels (Heard 2012).

Lythrum salicaria is a Eurasian perennial that has extensively invaded wetlands in North America, damaging plants, birds, amphibians, and insects (Blossey et al. 2001a; Maerz et al. 2005; Brown et al. 2006; Schooler et al. 2009). The leaf feeding beetles *Galerucella calmariensis* and *Galerucella pusilla*, the root-mining weevil *Hylobius transversovittatus*, and the flower-feeding weevil *Nanophyes marmoratus* were released (Blossey et al. 2001a) and caused defoliation at many sites (Blossey et al. 2001a; Landis et al. 2003; Denoth and Myers 2005; Grevstad 2006). In Michigan, *G. calmariensis* reduced plant height by 61–95 % (Landis et al. 2003) and in many sites where loosestrife has been suppressed, native species have increased (Landis et al. 2003).

26.6.4 *Grassland and Desert Invaders*

Grasslands and deserts have been invaded by many plant groups, including toxic forbs, woody shrubs, cacti, and grasses (the latter, often introduced for grazing). Toxic forbs have been repeatedly targeted for biological control because of their harm to grazing, e.g. *Centaurea diffusa*, *C. maculosa* and *C. solstitialis* (yellow startistle), *Euphorbia esula*, *Hypericum perforatum* (St. John's wort), *Salsola* spp., and *Senecio jacobaea* (tansy ragwort). These comprise some of the earliest weed biological control projects. Projects against *S. jacobaea* (McEvoy et al. 1991; Turner and McEvoy 1995; Coombs et al. 1996) and *H. perforatum* (Huffaker and Kennett 1959; McCaffrey et al. 1995) are considered complete successes, at least in some countries. In coastal prairies in Oregon, biological control of *S. jacobaea* led to a 40 % increase of the rare hairy-stemmed checkered-mallow (*Sidalcea hirtipes*; Gruber and Whytemare 1997). In natural California grasslands dominated by St. John's wort, biological control allowed native grasses such as *Danthonia californica* (California oatgrass) and *Elymus glaucus* (blue wild rye) to increase (Huffaker and Kennett 1959).

Projects against invasive shrubs in these habitats include ones against *Lantana camara* (lantana), *Prosopis* spp. (mesquite), and *Tamarix* spp. Of these, little has yet been achieved against *L. camara* (Day and Zalucki 2009; Urban et al. 2011), but the project against *Prosopis* spp. has been partially successful (van Klinken and Campbell 2009; Zachariades et al. 2011a) and saltcedar is currently being repeatedly defoliated by introduced chrysomelids in the south-western United States. Vegetative change from reduction of saltcedar, however, has yet to occur (Dudley and Bean 2012).

Invasive cacti have been controlled by biocontrol agents several times. Targeted species include *O. stricta* (prickly pear cactus), *Cylindropuntia fulgida* var. *fulgida* (jumping cholla), *Pereskia aculeata* (Barbados "gooseberry"), and *Cereus jamacaru* (queen of night cactus). While little has been achieved against *P. aculeata* (Paterson et al. 2011), *O. stricta* has been completely controlled by *Cactoblastis cactorum* in several locations (Dodd 1940; Paterson et al. 2011) and partial control has been achieved against jumping cholla (Paterson et al. 2011). While not specifically documented, dense stands of cacti such as those that once dominated large regions in South Africa, certainly caused declines in abundance of native species (Hoffmann 2011).

Among invasive plants, grasses may be particularly damaging to biodiversity because of their effects on fire cycles (Brooks and Pyke 2001). However, few grasses have been targets for biological control because of concerns for the economic value of introduced grasses and the assumption that grass-feeding insects were not sufficiently specialised for introduction. Currently some grasses (e.g. *Arundo donax* in the United States) are targets of biocontrol projects (Goolsby and Moran 2009; Goolsby et al 2011) and pathogenic fungi as well as insects have been of particular interest (e.g. Palmer et al. 2008).

26.6.5   Forest Invaders

Invasive plants in forest communities that have been targeted for biological control include (i) forbs: *Alliaria petiolata*, (ii) vines: *Anredera cordifolia* (Madeira vine), *Cryptostegia grandiflora* (rubber vine), *Dioscorea bulbifera* (air potato), *Dolichandra unguis-cati* (= *Macfadyena unguis-cati*, cats claw), and *Persicaria perfoliata*, (iii) shrubs: *Solanum mauritianum* (tree tobacco), and (iv) trees: *Caesalpinia decapetala* (Mauritius thorn), *M. calvescens* (Table 26.1). Of these, projects against *M. calvescens*, *C. grandiflora*, and *P. perfoliata* have had some success.

Miconia calvescens is a small, broad-leaved tree from the Americas that invaded natural forests on Pacific islands, including Hawaii and Tahiti and formed dense monocultures that suppressed native vegetation (Meyer and Florence 1996; Medeiros et al. 1997; Meyer 1998). The fungus *Colletotrichum gloeosporioides* forma specialis *miconiae* from Brazil (Killgore et al. 1999) was released in Tahiti and caused partial defoliation (up to 47 %) in mesic and wet forests below 1,400 m, which allowed substantial recovery of native vegetation (Meyer et al. 2008, 2009).

Cryptostegia grandiflora invaded forested areas along rivers in the dry tropics of Queensland, Australia, and later spread into adjacent grasslands and savannas (Tomley 1995). Dense stands killed eucalyptus trees and reduced native biodiversity, with infested areas being avoided by native birds (Bengsen and Pearson 2006) and lizards (Valentine et al. 2007). In drought-prone areas, *C. grandiflora* has been controlled by the rust *Maravalia cryptostegiae* (Evans and Tomley 1994; Vogler and Lindsay 2002) and the pyralid moth *Euclasta whalleyi* (Mo et al. 2000), allowing increased growth of local grasses (Palmer and Vogler 2012).

Persicaria perfoliata, a spiny annual vine of Asian origin, invades forest edges and disturbed open areas within forests in the mid-Atlantic region of the United States (Hough-Goldstein et al. 2008), degrading wildlife habitat and out-competing native plants, due to its early germination, rapid growth, and ability to climb over other plants (Wu et al. 2002). *Rhinoncomimus latipes* established at release sites (Hough-Goldstein et al. 2009, 2012) and reduced spring plant densities by 75 % within 2–3 years.

26.6.6 Coastal Invasive Plants

Plants of several forms have invaded a variety of coastal habitats, including mudflats, sand dunes, littoral grasslands, and forests. Species targeted for biological control have included *Chrysanthemoides monilifera* subsp. *rotundata* (bitou bush), *Asparagus asparagoides* (bridal creeper), *Acacia cyclops* (rooikrans), *Spartina* spp. (cordgrasses), and *Ulex europaeus* (gorse). Of these, populations of *C. monilifera*, *A. asparagoides*, and *A. cyclops* (discussed above under fynbos invaders) have been partially suppressed.

Chrysanthemoides monilifera subsp. *rotundata* invaded over 80 % of the coastline of New South Wales, Australia (Thomas and Leys 2002), where it dominated sand dunes, coastal grasslands, heath, woodlands, and rainforests and drastically altered these communities, becoming the dominant threat to 150 native plants in 24 plant communities (DEC 2006). Four introduced insect species established (Adair et al. 2012) and reduced flowering and seed production (Holtkamp 2002; Edwards et al. 2009), making a contribution toward suppression. Plant density, however, has yet to decline (Adair et al. 2012).

Asparagus asparagoides invaded coastal shrublands, woodlands, and forests in Australia (Morin et al. 2006a), where it smothered natural vegetation. In Western Australia, areas infested with this species had only half as many native plant species as nearby non-invaded areas (Turner et al. 2008a). It also threatened four endangered ecological communities in New South Wales – littoral rainforest, river-flat eucalypt forest on coastal floodplains, swamp-oak floodplain forest, and subtropical coastal floodplain forest (Downey 2006), as well as threatening many native plants, including the orchid *Pterostylis arenicola* (Sorensen and Jusaitis 1995) and the shrub *Pimelea spicata* (Willis et al. 2003). An introduced rust fungus *Puccinia myrsiphylli*, a leaf beetle *Crioceris* sp., and an undescribed Erythroneurini leafhopper have established. The leafhopper has had some effect, but the rust fungus caused significant reduction in *A. asparagoides* densities (Morin and Edwards 2006; Morin et al. 2006b; Turner et al. 2008b; Morin and Scott 2012).

26.7 Conclusions

The affection of people for novel plants ensures that plants will continue to be moved into new biogeographical regions where some will become invasive, sometimes in protected nature reserves. Given that prospect, use of biological control to dampen the impacts of the most damaging of these species in protected areas and landscapes generally is and will likely remain an important restoration tool. For example, without biological control the Everglades, a World Heritage Site, may have been abandoned to become a biologically impoverished *Melaleuca quinquenervia* swamp forest, many tropical rivers around the world would be burdened with over capping layers of floating exotic weeds, and fynbos habitats would be converted to woodlands of exotic trees. Both the benefits of classical weed biological control to native plants (Van Driesche et al. 2010) and the limited nature of the entailed risks (Pemberton 2000) are now better recognised. Improved communication between biological control scientists and conservation biologists (Van Driesche 2012) and emerging mutual trust should allow the use of biological control to help resolve some of the worst cases of invasive plants in natural areas, including in legally protected reserves.

References

Adair RJ (2005) The biology of *Dasineura dielsi* Rübsaamen (Diptera: Cecidomyiidae) in relation to the biological control of *Acacia cyclops* (Mimosaceae) in South Africa. Aust J Entomol 44:446–456

Adair RJ, Morley T, Morin L (2012) *Chrysanthemoides monilifera* (L.) T. Norl. – bitou bush and boneseed. In: Julien M, Cruttwell McFadyen RE et al (eds) Biological control of weeds in Australia. CSIRO, Melbourne, pp 170–183

Ajuonu O, Neuenschwander P (2003) Release, establishment, spread and impact of the weevil *Neohydronomus affinis* (Coleoptera: Curculionidae) on water lettuce (*Pistia stratiotes*) in Bénin, West Africa. Afr Entomol 11:205–211

Anonymous (2000) Harvesters get that sinking feeling. Biocon News Inform 21:1N–8N

Austin DF (1978) Exotic plants and their effects in southeastern Florida. Environ Conserv 5:25–34

Austin MT, Williams MJ, Hammond AC et al (1996) Psyllid population dynamics and plant resistance of *Leucaena* selections in Florida. Trop Grasslands 30:223–228

Badenes-Perez FR, Alfaro-Alpizar MA, Castillo-Castillo A et al (2007) Biological control of *Miconia calvescens* with a suite of insect herbivores from Costa Rica and Brazil. In: Julien MH, Sforza R, Bon MC et al (eds) Proceedings of the XII international symposium of biological control of weeds, La Grande Motte, France, 22–27 Apr 2007. CAB International, Wallingford, UK, pp 129–132

Baker J (2012) *Opuntia robusta* H. L. Wendl.ex Pfeiff. – wheel cactus. In: Julien M, Cruttwell McFadyen RE, Cullen J (eds) Biological control of weeds in Australia. CSIRO, Melbourne, pp 425–430

Balciunas JK, Grodowitz MJ, Cofrancesco AF et al (2002) *Hydrilla*. In: Van Driesche R, Blossey B, Hoddle M et al (eds) Biological control of invasive plants in the eastern United States. FHTET-2002-04. USDA, Forest Service, Morgantown, pp 91–114

Barko JW, Smith CS, Chambers PA (1994) Perspectives on submersed macrophyte invasions and declines. Lake Reserv Manage 10:1–3

Baum BR (1967) Introduced and naturalized tamarisks in the United States and Canada (Tamaricaceae). Baileya 15:19–25

Bengsen AJ, Pearson RG (2006) Examination of factors potentially affecting riparian bird assemblages in a tropical Queensland Savanna. Ecol Manage Restor 7:141–144

Beshir MO, Bennett FD (1985) Biological control of waterhyacinth on the White Nile, Sudan. In: Delfosse ES (ed) Proceedings of the VI international symposium on biological control of weeds. Agriculture Canada, Ottawa, Canada, pp 491–496

Birrenkott AH, Wilde SB, Hains JJ et al (2004) Establishing a food-chain link between aquatic plant material and avian vacuolar myelinopathy in mallards (*Anas platyrhynchos*). J Wildlife Dis 40:485–492

Blossey B, Skinner LC, Taylor J (2001a) Impact and management of purple loosestrife (*Lythrum salicaria*) in North America. Biodiv Conserv 10:1787–1807

Blossey B, Nuzzo V, Hin H et al (2001b) Developing biological control of *Alliaria petiolata* (M. Bieb.) Cavara and Grande (garlic mustard). Nat Areas J 21:357–367

Blossey B, Nuzzo VA, Hinz HL et al (2002) Garlic mustard. In: Van Driesche R, Blossey B, Hoddle M et al (eds) Biological control of invasive plants in the eastern United States. FHTET-2002-04. USDA, Forest Service, Morgantown, pp 365–372

Boughton AJ, Pemberton RW (2009) Establishment of an imported natural enemy, *Neomusotima conspurcatalis* (Lepidoptera; Crambidae) against an invasive weed, Old World climbing fern, *Lygodium microphyllum*, in Florida. Biocon Sci Technol 19:769–772

Braithwaite RW, Lonsdale W, Estbergs JA (1989) Alien vegetation and native biota in tropical Australia: the impact of *Mimosa pigra*. Biol Conserv 48:189–210

Briese DT (1997) Biological control of St. John's wort: past, present and future. Plant Prot Q 12:73–80

Briese DT, Cullen J (2012) *Hypericum perforatum* L. – St. John's wort. In: Julien M, Cruttwell McFadyen RE, Cullen J (eds) Biological control of weeds in Australia. CSIRO, Melbourne, pp 299–307

Brooks ML, Pyke DA (2001) Invasive plants and fire in the deserts of North America. In: Galley KEM, Wilson TP (eds) Proceedings of the invasive plant workshop: the role of fire in the control and spread of invasive species. Tall Timbers Research Station, Tallahassee, pp 1–14

Brooks ML, D'Antonio CM, Richardson DM et al (2004) Effects of invasive alien plants on fire regimes. BioScience 54:677–688

Brown CJ, Blossey B, Maerz JC et al (2006) Invasive plant and experimental venue affect tadpole performance. Biol Invasions 8:327–338

Buckingham GR (2002) Alligatorweed. In: Van Driesche R, Blossey B, Hoddle M et al (eds) Biological control of invasive plants in the eastern United States. FHTET-2002-04, USDA, Forest Service, Morgantown, pp 5–15

586   R. Van Driesche and T. Center

Byrne MJ, Witkowski ETF, Kalibbala FN (2011) A review of recent efforts at biological control of *Caesalpinia decapetala* (Roth) Alston (Fabaceae) in South Africa. Afr Entomol 19:247–257

Cagnotti C, McKay F, Gandolfo D (2007) Biology and host specificity of *Plectonycha correntina* Lacordaire (Chrysomelidae), a candidate for the biological control of *Anredera cordifolia* (Tenore) Steenis (Basellaceae). Afr Entomol 15:300–309

Carruthers RI, DeLoach CJ, Herr J et al (2008) Saltcedar areawide pest management in the western United States. In: Koul O, Cuperus G, Elliott N (eds) Areawide pest management: theory and practice. CABI, Cambridge, pp 271–299

Center TD, Grodowitz MJ, Cofrancesco AF et al (1997) Establishment of *Hydrellia pakistanae* (Diptera: Ephydridae) for the biological control of the submersed aquatic plant *Hydrilla verticillata* (Hydrocharitaceae) in the southeastern United States. Biol Control 8:65–73

Center TD, Hill MP, Cordo H et al (2002) Waterhyacinth. In: Van Driesche R, Blossey B, Hoddle M et al (eds) Biological control of invasive plants in the eastern United States. FHTET-2002-04, USDA, Forest Service, Morgantown, pp 41–64

Center TD, Pratt PD, Tipping PW et al (2007) Initial impacts and field validation of host range for *Boreioglycaspis melaleucae* Moore (Hemiptera: Psyllidae), a biological control agent of the invasive tree *Melaleuca quinquenervia* (Cav.) Blake (Myrtales: Myrtaceae: Leptospermoideae). Environ Entomol 36:569–576

Center TD, Purcell MF, Pratt PD et al (2012) Biological control of *Melaleuca quinquenervia*: an Everglades invader. Biol Control 57:151–165

Cline D, Juricek C, Lym RG et al (2008) Leafy spurge (*Euphorbia esula*) control with *Aphthona* spp. affects seedbank composition and native grass reestablishment. Invasions Plant Sci Manage 1:120–132

Coetzee JA, Hill MP (2012) The role of eutrophication in the biological control of water hyacinth, Eichhornia crassipes, in South Africa. Biol Control 57:247–261

Coetzee JA, Hill MP, Julien MH et al (2009) *Eichhornia crassipes* (Mart.) Solms-Laub. (Pontederiaceae). In: Muniappan R, Reddy GV, Raman A (eds) Biological control of tropical weeds using arthropods. Cambridge University Press, Cambridge, pp 183–210

Coetzee JA, Hill MP, Byrne MJ et al (2011a) A review of the biological control programmes on *Eichhornia crassipes* (C. Mart.) Solms (Pontederiaceae), *Salvinia molesta* D.S. Mitch. (Salviniaceae), *Pistia stratiotes* L. (Araceae), *Myriophyllum aquaticum* (Vell.) Verdc. (Haloragaceae) and *Azolla filiculoides* Lam. (Azollaceae) in South Africa. Afr Entomol 19:451–468

Coetzee JA, Bownes A, Martin GD (2011b) Prospects for the biological control of submerged macrophytes in South Africa. Afr Entomol 19:469–487

Cook GD, Setterfield SA, Maddison JP (1996) Shrub invasion of a tropical wetland: implications for weed management. Ecol Appl 6:531–537

Coombs EM, Radtke H, Isaacson DL et al (1996) Economic and regional benefits from the biological control of tansy ragwort, *Senecio jacobaea*, in Oregon. In: Moran VC, Hoffman JH (eds) Proceedings of the IX international symposium on biological control of weeds, 19–26 Jan 1996, Stellenbosch, South Africa. University of Cape Town, Cape Town, South Africa, pp 489–494

Coombs EM, McEvoy PB, Markin GP (2004) Tansy ragwort *Senecio jacobaea*. In: Coombs EM, Clark JK, Piper GL et al (eds) Biological control of invasive plants in the United States. Oregon State University Press, Corvallis, pp 335–336

Corn JG, Story JM, White LJ (2006) Impacts of the biological control agent *Cyphocleonus achates* on spotted knapweed, *Centaurea maculosa*, in experimental plots. Biol Control 37:75–81

Cornett MW, Bauman PJ, Breyfogle DD (2006) Can we control leafy spurge? Adaptive management and the recovery of native vegetation. Ecol Restor 24:145–150

Coulson JR (1977) Biological control of alligatorweed, 1959–1972. A review and evolution. ARS-USDA Technical bulletin no 1547

Crins WL (1989) The Tamaricaceae in the southeastern United States. J Arnold Arbor 70:403–425

Cruttwell Mcfadyen RE (1998) Biological control of weeds. Ann Rev Entomol 43:369–393
Cruttwell McFadyen RE (2012) *Ageratina adenophora* (Spreng.) King and Robinson. In: Julien M, Cruttwell McFadyen RE, Cullen J (eds) Biological control of weeds in Australia. CSIRO, Melbourne, pp 29–32
Cuda JP, Medal JC, Gillmore JL et al (2009) Fundamental host range of *Pseudophilothrips ichini* sensu lato (Thysanoptera: Phlaeothripidae), a candidate biological control agent of *Schinus terebinthifolius* (Sapindales: Anacardiaceae) in the USA. Environ Entomol 38:1642–1652
Davies JT, Ireson JE, Allen GR (2007) The impact of the gorse spider mite, *Tetranychus lintearius*, on the growth and development of gorse, *Ulex europaeus*. Biol Control 41:86–93
Day M (2012a) *Lantana camara* L. – lantana. In: Julien M, Cruttwell McFadyen R, Cullen J (eds) Biological control of weeds in Australia. CSIRO, Melbourne, pp 334–346
Day M (2012b) *Mikania micrantha* Kunth – mile-a-minute. In: Julien M, Cruttwell McFadyen R, Cullen J (eds) Biological control of weeds in Australia. CSIRO, Melbourne, pp 368–372
Day M (2012c) *Pistia stratiotes* L. – water lettuce. In: Julien M, Cruttwell McFadyen R, Cullen J (eds) Biological control of weeds in Australia. CSIRO, Melbourne, pp 472–476
Day MD, Bofeng I (2007) Biocontrol of *Chromolaena odorata* in Papua New Guinea. In: Lai P-Y, Reddy GVP, Muniappan R (eds) Proceedings of the 7th international workshop on the biological control and management of *Chromolaena odorata* and *Mikania micrantha*. National Pingtung University of Science and Technology, Taiwan, pp 53–67
Day M, Cruttwell McFadyen RE (2012) *Chromolaena odorata* (L.) King and Robinson – chromolaena. In: Julien M, Cruttwell McFadyen RE, Cullen J (eds) Biological control of weeds in Australia. CSIRO, Melbourne, pp 162–169
Day MD, Zalucki MP (2009) *Lantana camara* Linn. (Verbenaceae). In: Muniappan R, Reddy GVP, Raman A (eds) Biological control of tropical weeds using arthropods. Cambridge University Press, Cambridge, pp 211–246
DeBach P (1974) Biological control by natural enemies. Cambridge University Press, London
DEC (2006) NSW threat abatement plan: Invasion of native plant communities by *Chrysanthemoides monilifera* (bitou bush and boneseed). Department of Environment and Conservation, NSW, Hurstville, Australia
DeLoach CJ, Lewis PA, Herr JC et al (2003) Host specificity of the leaf beetle *Diorhabda elongata deserticola* (Coleoptera: Chrysomelidae) from Asia, a biological control agent for saltcedars (*Tamarix*: Tamaricaceae) in the western United States. Biol Control 27:117–147
DeLoach CJ, Moran PJ, Knutson AE et al (2008) Beginning success of biological control of saltcedars (*Tamarix* spp.) in the southwestern USA. In: Julien MH, Sforza R, Bon MC et al (eds) Proceedings of the XII international symposium of biological control of weeds, La Grande Motte, France, 22–27 Apr 2007. CAB International, Wallingford, pp 535–539
Dennill GB, Donnelly D (1991) Biological control of *Acacia longifolia* and related weed species (Fabaceae) in South Africa. Agric Ecosyst Environ 37:115–135
Dennill GB, Donnelly D, Stewart K et al (1999) Insect agents used for the biological control of Australian *Acacia* species and *Paraserianthes lophantha* (Willd.) Nielsen (Fabaceae) in South Africa. Afr Entomol Mem 1:45–54
Denoth M, Myers JH (2005) Variable success of biological control of *Lythrum salicaria* in British Columbia. Biol Control 32:269–279
Dhileepan K (2012) *Macfadyena unguis-cati* (L.) A.H. Gentry – cat's claw creeper. In: Julien M, Cruttwell McFadyen R, Cullen J (eds) Biological control of weeds in Australia. CSIRO, Melbourne, pp 351–359
Dhileepan K, Cruttwell McFadyen R (2012) *Parthenium hysterophorus* L. – parthenium. In: Julien M, McFadyen R, Cullen J (eds) Biological control of weeds in Australia. CSIRO, Melbourne, pp 448–462
Dhileepan K, Snow EL, Rafter MA et al (2007a) The leaf-tying moth *Hypocosmia pyrochroma* (Lep., Pyralidae), a host-specific biological control agent for cat's claw creeper *Macfadyena unguis-cati* (Bignoniaceae) in Australia. J Appl Entomol 131:564–568

Dhileepan K, Treviño M, Snow EL (2007b) Specificity of *Carvalhotingis visenda* (Hemiptera: Tingidae) as a biological control agent for cat's claw creeper *Macfadyena unguis-cati* (Bignoniaceae) in Australia. Biol Control 41:283–290

Diop O, Hill MP (2009) Quantitative post-release evaluation of biological control of floating fern, *Salvinia molesta* D.S. Mitchell (Salviniaceae), with *Cyrtobagous salviniae* Calder and Sands (Coleoptera: Curculionidae) on the Senegal River and Senegal River Delta. Afr Entomol 17:64–71

Dodd AP (1940) The biological campaign against prickly pear. Commonwealth Prickly Pear Board Bulletin, Brisbane

Downey PO (2006) The weed impact to native species (WINS) assessment tool – results from a trial for bridal creeper (*Asparagus asparagoides* (L.) Druce) and ground asparagus (*Asparagus aethiopicus* L.) in southern New South Wales. Plant Prot Q 21:109–116

Dray FA, Center TD (1992) Biological control of *Pistia stratiotes* L. (waterlettuce) using *Neohydronomus affinis* Hustache (Coleoptera: Curculionidae). Misc Paper A-92-1. U.S. Army Engineers Waterways Exper Sta, Vicksburg

Dudley TL, Bean DW (2012) Tamarisk biocontrol, endangered species risk and resolution of conflict through riparian restoration. Biol Control 57:331–347

Edwards PB, Adair RJ, Holtkamp RH et al (2009) Impact of the biological control agent *Mesoclanis polana* (Tephritidae) on bitou bush (*Chrysanthemoides monilifera* subsp. *rotundata*) in Eastern Australia. Bull Entomol Res 99:51–63

Esler KJ, Van Wilgen BW, Roller KS et al (2010) A landscape-scale assessment of the long-term integrated control of an invasive shrub in South Africa. Biol Invasion 12:211–218

Evans HC, Tomley AJ (1994) Studies on the rust, *Maravalia cryptostegiae*, a potential biological control agent of rubber-vine weed, *Cryptostegia grandiflora* (Asclepiadaceae: Periplocoideae), in Australia, III: Host range. Mycopathologia 126:93–108

Evans JM, Wilkie AC, Burkhardt J (2008) Adaptive management of nonnative species: moving beyond the "either-or" through experimental pluralism. J Agric Envrion Ethics 21:521–539

Fischer JR, Lewis-Weis LA, Tate CM et al (2006) Avian vacuolar myelinopathy outbreaks at a southeastern reservoir. J Wildlife Dis 42:501–510

Fourie A (2011) Preliminary attempts to identify pathogens as biological control agents against *Cestrum* species (Solanaceae) in South Africa. Afr Entomol 19:278–281

Gardner DE, Smith CW, Markin GP (1995) Biological control of alien plants in natural areas of Hawaii. In: Delfosse ES, Scott RR (eds) Proceedings of the 8th international symposium on biological control of weeds. CSIRO, Melbourne, Australia, pp 35–40

Gerber E, Krebs C, Murrell C et al (2008) Exotic invasive knotweeds (*Fallopia* spp.) negatively affect native plant and invertebrate assemblages in European riparian habitats. Biol Conserv 141:646–654

Goolsby JA, Moran PJ (2009) Host range of *Tetramesa romana* Walker (Hymenoptera: Eurytomidae), a potential biological control agent of giant reed, *Arundo donax* L. in North America. Biol Control 49:160–168

Goolsby JA, Kirk AA, Moran PJ et al (2011) Establishment of the armored scale, *Rhizaspidiotus donacis*, a biological control agent of *Arundo donax*. Southwest Entomol 36:373–374

Gordon AJ (1999) A review of established and new insect agents for the biological control of *Hakea sericea* Schrader (Proteaceae) in South Africa. In: Olckers T, Hill MP (eds) Biological control of weeds in South Africa (1990–1998). Afr Entomol Mem 1:35–43

Gordon AJ (2011) Biological control and endeavours against Australian myrtle, *Leptospermum laevigatum* (Gaertn.) F. Muell. (Myrtaceae), in South Africa. Afr Entomol 19:349–355

Gordon AJ, Fourie A (2011) Biological control of *Hakea sericea* Schrad. and J.C. Wendl. and *Hakea gibbosa* (Sm.) Cav. (Proteaceae) in South Africa. Afr Entomol 19:303–314

Grevstad FS (2006) Ten-year impacts of the biological control agents *Galerucella pusilla* and *G. calmariensis* (Coleoptera: Chrysomelidae) on purple loosestrife (*Lythrum salicaria*) in Central New York. Biol Control 39:1–8

Grodowitz MJ, Center TD, Cofrancesco AF et al (1997) Release and establishment of *Hydrellia balciunasi* (Diptera: Ephydridae) for the biological control of the submersed aquatic plant *Hydrilla verticillata* (Hydrocharitaceae) in the United States. Biol Control 9:15–23

Gruber E, Whytemare A (1997) The return of the native? *Sidalcea hirtipes* in coastal Oregon. In: Kaye TN, Liston A, Love RM et al (eds) Conservation and management of native plants and fungi. Proceedings of Oregon conference conservation and manage of native vascular plants, Bryophytes, and Fungi. Native Plant Society of Oregon, Corvallis, Oregon, USA, pp 121–124

Gutierrez AP, Pitcairn MJ, Ellis CK et al (2005) Evaluating biological control of yellow starthistle (*Centaurea solstitialis*) in California: a GIS based supply-demand demographic model. Biol Control 34:115–131

Hansen KL, Ruby EG, Thompson RL (1971) Trophic relationships in the waterhyacinth community. Quart J Florida Acad Sci 34:107–113

Harley KLS, Forno IW (1992) Biological control of weeds. A handbook for practioners and students. Inkata Press, Melbourne

Harley KLS, Kassulke RC, Sands DPA et al (1990) Biological control of water lettuce, *Pistia stratiotes* (Araceae) by *Neohydronomus affinis* (Coleoptera: Curculionidae). Entomophaga 35:363–374

Heard TA (2012) *Mimosa pigra* L. – mimosa. In: Julien M, Cruttwell McFadyen R, Cullen J (eds) Biological control of weeds in Australia. CSIRO, Melbourne, pp 378–397

Heard TA, Paynter Q (2009) *Mimosa pigra* (Leguminosae). In: Muniappan R, Reddy GV, Raman A (eds) Biological control of tropical weeds using arthropods. Cambridge University Press, Cambridge, pp 256–273

Heard TA, Dhileepan K, Bebawi F et al (2012) *Jatropha gossypiifolia* L. – bellyache bush. In: Julien M, Cruttwell McFadyen R, Cullen J (eds) Biological control of weeds in Australia. CSIRO, Melbourne, pp 324–333

Heystek F, Wood AR, Neser S et al (2011) Biological control of two *Ageratina* species (Asteraceae: Eupatoreiae) in South Africa. Afr Entomol 19:208–216

Higgins SI, Richardson DM, Cowling RM et al (1999) Predicting the landscape-scale distribution of alien plants and their threat to plant diversity. Conserv Biol 13:303–313

Hill MP, McConnachie AJ (2009) *Azolla filiculoides*. In: Muniappan R, Reddy GV, Raman A (eds) Biological control of tropical weeds using arthropods. Cambridge University Press, Cambridge, pp 74–87

Hill RL, Ireson J, Sheppard AW et al (2008) A global view of the future for biological control of gorse. In: Julien MH, Sforza R, Bon MC et al BG (eds) Proceedings of the XII international symposium of biological control of weeds, La Grande Motte, France, 22–27 Apr 2007. CAB International, Wallingford, pp 680–686

Hoffmann JH (1996) Biological control of weeds: the way forward, a South African perspective. In: Stirton CH (ed) Weeds in a changing world: Proceedings of the British Crop Protection Council international symposium, Brighton, England 20 Nov 1995. British Crop Protection Council, Farnham, pp 77–89

Hoffmann JH, Moran VC (1991) Biological control of *Sesbania punicea* (Fabaceae) in South Africa. Agric Ecosyst Environ 37:157–173

Hoffmann JH, Moran VC (1998) The population dynamics of an introduced tree, *Sesbania punicea*, in South Africa, in response to long-term damage caused by different combinations of three species of biological control agents. Oecologia 114:343–348

Holmes PM, Richardson DM, van Wilgen BW et al (2000) Recovery of South African fynbos vegetation following alien woody plant clearing and fire: implications for restoration. Aust Ecol 25:631–639

Holtkamp RH (2002) Impact of the bitou bush tip moth, *Comostolopsis germana*, on bitou bush in New South Wales. In: Jacob HS, Dodd J, Moore JH (eds) Proceedings of 13th Australian weed conference. Plant Protection Society of West Australia, Perth, Australia, pp 405–406

Holtkamp RH (2012) *Cylindropuntia imbricata* (Haw.) F.M. Knuth – rope pear and *Cylindropuntia rosea* (DC.) Beckeb. – Hudson pear. In: Julien M, Mcfadyen R, Cullen J (eds) Biological control of weeds in Australia. CSIRO, Melbourne, pp 198–202

Hosking JR (2012) *Opuntia* spp. In: Julien M, Mcfadyen R, Cullen J (eds) Biological control of weeds in Australia. CSIRO, Melbourne, pp 431–436

Hosking JR, Sheppard AW, Sagliocco J-L (2012) *Cytisus scoparius* (L.) Link – broom, Scotch broom or English broom. In: Julien M, McFadyen R, Cullen J (eds) Biological control of weeds in Australia. CSIRO, Melbourne, pp 201–210

Hough-Goldstein J, Lake E, Reardon R et al (2008) Biology and biological control of mile-a-minute weed. FHTET-2008-10, USDA Forest Service, Morgantown

Hough-Goldstein J, Mayer MA, Hudson W et al (2009) Monitored releases of *Rhinoncomimus latipes* (Coleoptera: Curculionidae), a biological control agent of mile-a-minute weed (*Persicaria perfoliata*), 2004–2008. Biol Control 51:450–457

Hough-Goldstein J, Lake E, Reardon R (2012) Status of an ongoing biological control program for the invasive vine, *Persicaria perfoliata* in eastern North America. Biol Control 57:181–189

Hudgeons JL, Knutson AE, Heinz KM et al (2007) Defoliation by introduced *Diorhabda elongata* leaf beetles (Coleoptera: Chrysomelidae) reduces carbohydrate reserves and regrowth of *Tamarix* (Tamaricaceae). Biol Control 43:213–221

Huffaker CB, Kennett CE (1959) A ten-year study of vegetational changes associated with biological control of Klamath weed. J Range Manage 12:69–82

Impson FAC, Moran VC, Hoffmann JH (2004) Biological control of an alien tree, *Acacia cyclops*, in South Africa: impact and dispersal of a seed-feeding weevil, *Melanterius servulus*. Biol Control 29:375–381

Impson FAC, Kleinjan CA, Hoffmann JH et al (2008) *Dasineura rubiformis* (Diptera: Cecidomyiidae), a new biological control agent for *Acacia mearnsii* in South Africa. S Afr J Sci 104:247–249

Impson F, Hoffmann J, Kleinjan C (2009) Australian *Acacia* species (Mimosaceae) in South Africa. In: Muniappan R, Reddy GVP, Raman A (eds) Biological control of tropical weeds using arthropods. Cambridge University Press, Cambridge, pp 38–62

Impson FAC, Kleinjan CA, Hoffmann JH et al (2011) Biological control of Australian *Acacia* species and *Paraserianthes lophantha* (Willd.) Nielsen (Mimosaceae) in South Africa. Afr Entomol 19:186–207

Ireson JE, Davies JT (2012) *Ulex europaeus* L. – gorse. In: Julien M, McFadyen R, Cullen J (eds) Biological control of weeds in Australia. CSIRO, Melbourne, pp 581–590

Julien MH (1981) Control of aquatic *Alternanthera philoxeroides* in Australia: another success for *Agasicles hygrophila*. In: Delfosse ES (ed) Proceedings of the Vth international symposium on biological control of weeds. 22–29 July 1980, Brisbane, Australia, pp 583–588

Julien M (2012a) *Eichhornia crassipes* (Martius) Solms-Laubach – water hyacinth. In: Julien M, McFadyen R, Cullen J (eds) Biological control of weeds in Australia. CSIRO, Melbourne, pp 227–237

Julien M (2012b) *Salvinia molesta* D.S. Mitchell – salvinia. In: Julien M, McFadyen R, Cullen J (eds) Biological control of weeds in Australia. CSIRO, Melbourne, pp 518–525

Julien MH, Griffiths MW (1998) Biological control of weeds: a world catalogue of agents and their target weeds. CABI Publishing, Wallingford

Julien MH, Hill MP, Tipping PW (2009) *Salvinia molesta* D.S. Mitchell (Salviniaceae). In: Muniappan R, Reddy GVP, Raman A et al (eds) Weed biological control with arthropods in the tropics. Cambridge University Press, Cambridge, pp 378–407

Julien M, Sosa A, Chan R et al (2012a) *Alternanthera philoxeroides* (Martius) Grisebach – alligator weed. In: Julien M, McFadyen R, Cullen J (eds) Biological control of weeds in Australia. CSIRO, Melbourne, pp 43–51

Julien M, Sosa A, Traversa G (2012b) *Phyla canescens* (Kunth) Greene – lippie. In: Julien M, McFadyen R, Cullen J (eds) Biological control of weeds in Australia. CSIRO, Melbourne, pp 463–471

Kaufmann S, McKey DB, Hossaert-Mckey M et al (1991) Adaptations for a two-phase seed dispersal system involving vertebrates and ants in a hemiepiphytic fig (*Ficus microcarpa*: Moraceae). Am J Bot 78:971–977

Killgore EM, Sugiyama LS, Barreto R et al (1999) Evaluation of *Colletrotrichum gloeosporioides* for biological control of *Miconia calvescens* in Hawaii. Plant Dis 83:964

King AM, Williams HE, Madire LG (2011) Biological control of cat's claw creeper, *Macfadyena unguis-cati* (L.) A.H. Gentry (Bignoniaceae), in South Africa. Afr Entomol 19:366–377

Knight TM, Dunn JL, Smith LA et al (2009) Deer facilitate invasive plant success in a Pennsylvania forest understory. Nat Areas J 29:110–116

Lake EC (2011) Biological control of mile-a-minute weed, *Persicaria perfoliata*, and integrating weed management techniques to restore invaded sites. PhD dissertation, University of Delaware, Newark, Delaware, USA

Lamb AJ, Klaussner E (1988) Response of the fynbos shrubs *Protea repens* and *Erica plukenetii* to low levels of nitrogen and phosphorus applications. S Afr J Bot 54:558–564

Landis DA, Sebolt DC, Haas MJ et al (2003) Establishment and impact of *Galerucella calmariensis* L. (Coleoptera: Chrysomelidae) on *Lythrum salicaria* L. and associated plant communities in Michigan. Biol Control 28:78–91

Madire LG, Wood AR, Williams HE et al (2011) Potential agents for the biological control of *Tecoma stans* (L.) Juss ex Kunth var. Stans (Bignonicaceae) in South Africa. Afr Entomol 19:434–442

Maerz JC, Brown CJ, Chapin CT et al (2005) Can secondary compounds of an invasive plant affect larval amphibians? Funct Ecol 19:970–975

Maerz JC, Nuzzo VA, Blossey B (2009) Declines in woodland salamander abundance associated with non-native earthworm and plant invasions. Consev Biol 23:975–981

Markin GP, Lai P-Y, Funusaki GP (1992) Status of biological control of weeds in Hawai'i and implications for managing native ecosystems. In: Stone CP, Smith CW, Tunison JT (eds) Alien plant invasions in native ecosystems of Hawai'i: management and research. University of Hawaii Press, Honolulu, pp 466–482

Mbati G, Neuenschwander P (2005) Biological control of three floating water weeds, *Eichhornia crassipes*, *Pistia stratiotes*, and *Salvinia molesta* in the Republic of Congo. Biol Control 50:635–645

McCaffrey JP, Campbell CL, Andres LA (1995) St. Johnswort. In: Nechols JR, Andres LA, Beardsley JW et al (eds) Biological control in the western United States: accomplishments and benefits of regional research project W-84, 1964–1989. University of California, Division of Agriculture and Natural Resources, Oakland, Pub No. 3361, pp 281–285

McConnachie AJ, Hill MP, Byrne MJ (2004) Field assessment of a frond-feeding weevil, a successful biological control agent of red waterfern, *Azolla filiculoides*, in southern Africa. Biol Control 29:326–331

McConnachie AJ, Retief E, Henderson L et al (2011) The initiation of a biological control programme against pompom weed, *Campuloclinium macrocephalum* (Less.) DC. (Asteraceae), in South Africa. Afr Entomol 19:258–268

McEvoy PB, Coz CS, Coombs EM (1991) Successful biological control of ragwort. Ecol Appl 1:430–432

Medal JC, Cuda JP (2010) Establishment and initial impact of the leaf-beetle *Gratiana boliviana* (Chrysomelidae), first biocontrol agent released against tropical soda apple in Florida. Florida Entomol 93:493–500

Medal J, Overholt W, Stansly P et al (2008) Establishment, spread, and initial impacts of *Gratiana boliviana* (Chrysomelidae) on *Solanum viarum* in Florida. In: Julien MH, Sforza R, Bon MC et al (eds) Proceedings of the XII international symposium of biological control of weeds, La Grande Motte, France, 22–27 Apr 2007. CAB International, Wallingford, UK, pp 591–596

Medeiros AC, Loope LL, Conant P et al (1997) Status, ecology, and management of the invasive plant *Miconia calvescens* DC (Melastomataceae) in the Hawaiian Islands. Bishop Mus Occas Pap 48:23–36

Meyer J-Y (1998) Observation on the reproductive biology of *Miconia calvescens* DC (Melastomataceae), an alien invasive tree on the island of Tahiti (South Pacific Ocean). Biotropica 30:609–624

Meyer J-Y, Florence J (1996) Tahiti's native flora endangered by the invasion of *Miconia calvescens* DC (Melastomataceae). J Biogeogr 23:775–781

Meyer J-Y, Taputuarai R, Killgore E (2008) Dissemination and impacts of the fungal pathogen, *Colletotrichum gloeosporioides* f. sp. *miconiae*, on the invasive alien tree, *Miconia calvescens*, in Tahiti (South Pacific). In: Julien MH, Sforza R, Bon MC et al (eds) Proceedings of the XII international symposium of biological control of weeds, La Grande Motte, France, 22–27 Apr 2007. CAB International, Wallingford, pp 594–600

Meyer J-Y, Fourdrigniez M, Taputuarai R (2009) Habitat restoration using a biocontrol agent: the positive effects of the fungal pathogen *Colletotrichum gloeosporioides* f. sp. *miconiae* on native plant recruitment in Tahiti (French Polynesia). Abstract of talk at Pacific Science International-Congress in Tahiti, February 2009

Michel A, Arias RS, Scheffler BE et al (2004) Somatic mutation-mediated evolution of herbicide resistance in the nonindigenous invasive plant hydrilla (*Hydrilla verticillata*). Mol Ecol 13:3229–3237

Mitchell DS (1978) Aquatic weeds in Australian waters. Australian Government Public Service, Canberra

Mo J, Treviño M, Palmer WA (2000) Establishment and distribution of the rubber vine moth, *Euclasta whalleyi* Popescu-Gorj and Constantinescu (Lepidoptera: Pyralidae) following its release in Australia. Aust J Entomol 39:344–350

Moran VC, Hoffmann JH, Zimmermann HG (2005) Biological control of invasive alien plants in South Africa: necessity, circumspection, and success. Front Ecol Environ 3:71–77

Morin L, Edwards PB (2006) Selection of biological control agents for bridal creeper – a retrospective review. Aust J Entomol 45:286–290

Morin L, Evans KJ (2012) *Rubus fruticosus* L. aggregate – European blackberry. In: Julien M, Cruttwell McFadyen R, Cullen J (eds) Biological control of weeds in Australia. CSIRO, Melbourne, pp 499–509

Morin L, Scott JK (2012) *Asparagus asparagoides* (L.) Druce – bridal creeper. In: Julien M, Cruttwell McFadyen R, Cullen J (eds) Biological control of weeds in Australia. CSIRO, Melbourne, pp 170–183

Morin L, Batchelor KL, Scott JK (2006a) The biology of Australian weeds: *Asparagus asparagoides* (L.) Druce. Plant Prot Q 21:46–62

Morin L, Neave M, Batchelor KL et al (2006b) Biological control: a promising tool for managing bridal creeper in Australia. Plant Prot Q 21:69–77

Nadel H, Frank JH, Knight RJ (1992) Escapees and accomplices: the naturalization of exotic *Ficus* and their associated faunas in Florida. Florida Entomol 75:29–38

Neuenschwander P, Julien MH, Center TD et al (2009) *Pistia stratiotes* L. (Araceae). In: Muniappan R, Reddy GV, Raman A (eds) Biological control of tropical weeds using arthropods. Cambridge University Press, Cambridge, pp 332–352

Newman RM (2004) Biological control of Eurasian watermilfoil by aquatic insects: basic insights from an applied problem. Arch Hydrobiol 159:145–184

Norambuena H (1995) Impact of *Apion ulicis* Forster (Coleoptera: Apionidae) on gorse *Ulex europaeus* L. (Fabaceae) in agricultural and silvicultural habitats in Southern Chile. Ph.D. Entomology Department, Washington State University, Pullman, Washington, USA

Norambuena H, Piper GL (2000) Impact of *Apion ulicis* on *Ulex europaeus* seed dispersal. Biol Control 17:267–271

Norambuena H, Martinez G, Carillo R et al (2007) Host specificity and establishment of *Tetranychus lintearius* (Acari: Tetranychidae) for biological control of gorse (*Ulex europaeus*). Biol Control 40:204–212

Oehrens E (1977) Biological control of the blackberry through the introduction of rust, *Phragmidium violaceum*. FAO Plant Prot Bull 25:26–28

Oehrens EB, Gonzalez SM (1977) Dispersion, ciclo biologico y daños causados por *Phragmidium violaceum* (Schultz) Winter en zarzamora (*Rubus constrictus* Lef. et M. y *R. ulmifolius* Schott.) en la zonas centro sur y sur de Chile. Agro Sur 5:73–85

Olckers T (2011a) Biological control of *Leucaena leucocephala* (Lam.) de Wit (Fabaceae) in South Africa: a tale of opportunism, seed feeders and unanswered questions. Afr Entomol 19:356–365

Olckers T (2011b) Biological control of *Solanum mauritianum* Scop. (Solanaceae) in South Africa: will perseverance pay off? Afr Entomol 19:416–426

Ortega YK, Pearson DE, McKelvey KS (2004) Effects of biological control agents and exotic plant invasion on deer mouse populations. Ecol Appl 14:241–253

Palmer B, Senaratne W (2012) *Anredera cordifolia* (Ten.) Steenis – Madeira vine. In: Julien M, Cruttwell McFadyen R, Cullen J (eds) Biological control of weeds in Australia. CSIRO, Melbourne, pp 60–64

Palmer B, Sims-Chilton N (2012) *Baccharis halimifolia* L. – groundel bush. In: Julien M, Cruttwell McFadyen R, Cullen J (eds) Biological control of weeds in Australia. CSIRO, Melbourne, pp 86–95

Palmer B, Vogler W (2012) *Cryptostegia grandifolia* (Roxb.) R. Br. – rubber vine. In: Julien M, Cruttwell McFadyen R, Cullen J (eds) Biological control of weeds in Australia. CSIRO, Melbourne, pp 190–197

Palmer WA, Yobo KS, Witt ABR (2008) Prospects for the biological control of the weedy sporobolus grasses in Australia. In: Anon. Proceedings of the 16th Australian weeds conference, Cairns Convention Centre, North Queensland, Australia, 18–22 May 2008. Queensland Weed Society, Queensland, Australia, pp 264–266

Palmer B, Lockett C, Dhileepan K (2012) *Acacia nilotica* subsp. *indica* (Benth.) Brenan – prickly acacia. In: Julien M, Cruttwell McFadyen R, Cullen J (eds) Biological control of weeds in Australia. CSIRO, Melbourne, pp 18–28

Paterson ID, Hoffmann JH, Klein H et al (2011) Biological control of Cactaceae in South Africa. Afr Entomol 19:230–246

Pearson DE, Callaway RM (2006) Biological control agents elevate hantavirus by subsidizing deer mouse populations. Ecol Lett 9:443–450

Pearson DE, McKelvey KS, Ruggiero LF (2000) Non-target effects of an introduced biological control agent on deer mouse ecology. Oecologia 122:121–128

Pemberton RW (2000) Predictable risk to native plants in weed biological control. Oecologia 125:489–494

Pemberton RW (2009) Proposed field release of *Lilioceris* sp. near *impressa* (Fabricius) (Coleoptera: Chrysomelidae), a leaf and bulbil feeder of air potato, *Dioscorea bulbifera* L. (Dioscoreaceae) in Florida. Petition to release a biological control agent, submitted 2 Feb 2009. On file at the USDA, ARS Biological Control Documentation Center, National Agricultural Library, Beltsville, Maryland, USA

Pemberton RW, Cordo HA (2001) Potential and risks of biological control of *Cactoblastis cactorum* (Lepidoptera: Pyralidae) in North America. Florida Entomol 84:513–526

Pemberton RW, Ferriter AP (1998) Old World climbing fern (*Lygodium microphyllum*), a dangerous invasive weed in Florida. Am Fern J 88:165–175

Pemberton RW, Liu H (2007) Control and persistence of native *Opuntia* on Nevis and St. Kitts 50 years after the introduction of *Cactoblastis cactorum*. Biol Control 41:272–282

Pemberton RW, Turner CE (1990) Biological control of *Senecio jacobaea* in northern California, an enduring success. Entomophaga 35:71–77

Pitcairn MJ, Woods DM, Popescu V (2005) Update on the long-term monitoring of the combined impact of biological control insects on yellow starthistle. In: Woods DM (ed) Biological control program annual summary, 2004. California Depart Food and Agric, Plant Health and Pest Prevention Serv, Sacramento, California, pp 27–30

Pratt PD, Center TD (2012) Biocontrol without borders: the unintended spread of introduced weed biological control agents. Biol Control 57:319–329

Pratt PD, Rayamajhi MB, Van TK et al (2005) Herbivory alters resource allocation and compensation in the invasive tree *Melaleuca quinquenervia*. Ecol Entomol 15:443–462

Racelis AE, Goolsby JA, Moran P (2009) Seasonality and movement of adventive populations of the arundo wasp (Hymenoptera: Eurytomidaea), a biological control agent of giant reed in the lower Rio Grande basin in south Texas. Southwest Entomol 34:347–357

Racelis AE, Goolsby JA, Penk R et al (2010) Development of an inundative aerial release technique for the arundo wasp, biological control agent of the invasive *Arundo donax* L. Southwest Entomol 35:495–502

Rayamajhi MB, Purcell MF, Van TK et al (2002) Australian paperbark tree (*Melaluca*). In: Van Driesche RG, Blossey B, Hoddle MS et al (eds) Biological control of invasive plants in the eastern United States. Forest Health Tech Enterprise Team, Morgantown, West Virginia, USA, pp 117–130

Rayamajhi MB, Van TK, Pratt PD (2007) *Melaleuca quinquenervia*-dominated forests in Florida: analyses of natural-enemy impacts on stand dynamics. Plant Ecol 192:119–132

Rayamajhi MB, Pratt PD, Van TK et al (2008) Aboveground biomass of an invasive tree melaleuca (*Melaleuca quinquenervia*), before and after herbivory by adventive and introduced natural enemies: a temporal case study in Florida. Weed Sci 56:451–456

Rayamajhi MB, Pratt PD, Van TK et al (2009) Decline in exotic tree density facilitates increased plant diversity: the experience from *Melaleuca quinquenervia* invaded wetlands. Wetlands Ecol Manage 17:455–467

Reichard SH, Hamilton CW (1997) Predicting invasions of woody plants introduced into North America. Conserv Biol 11:193–203

Richard DI, Small JW Jr, Osborne JA (1984) Phytoplankton responses to reduction and elimination of submerged vegetation by herbicides and grass carp in four Florida lakes. Aquatic Bot 20:307–319

Richardson DM, van Wilgen BW, Higgins SI et al (1996) Current and future threats to biodiversity on the Cape Peninsula. Biodivers Conserv 5:607–647

Richardson DM, Macdonald IAW, Hoffmann JH et al (1997) Alien plant invasion. In: Cowling RM, Richardson DM, Pierce SM (eds) Vegetation of southern Africa. Cambridge University Press, Cambridge, pp 535–570

Room PM, Harley KLS, Forno IW et al (1981) Successful biological control of the floating weed salvinia. Nature 294:78–80

Sainty G, McCorkelle G, Julien M (1998) Control and spread of alligator weed (*Alternanthera philoxeroides* (Mart.) Griseb.), in Australia: lesson for other regions. Wetlands Ecol Manage 5:195–201

Samuel L, Kirby DR, Norland JE et al (2008) Leafy spurge suppression by flea beetles in the Little Missouri drainage basin, USA. Rangeland Ecol Manage 61:437–443

Schooler SS, McEvoy PB, Hammond P et al (2009) Negative per capita effects of two invasive plants, *Lythrum salicaria* and *Phalaris arundinacea*, on the moth diversity of wetland communities. Bull Entomol Res 99:229–243

Schooler S, Palmer B, Morin L (2012a) *Ageratina riparia* (Regel) K. and R. – mistflower. In: Julien M, Cruttwell McFadyen R, Cullen J (eds) Biological control of weeds in Australia. CSIRO, Melbourne, pp 33–42

Schooler S, Cabrera-Walsh W, Julien M (2012b) *Cabomba caroliniana* Gray – cabomba. In: Julien M, Cruttwell McFadyen R, Cullen J (eds) Biological control of weeds in Australia. CSIRO, Melbourne, pp 108–117

Schroder D (1980) The biological control of thistles. Biocontr News Inform 1:9–26

Scott JK (2012) *Euphorbia paralias* L. – sea spurge. In: Julien M, Cruttwell McFadyen R, Cullen J (eds) Biological control of weeds in Australia. CSIRO, Melbourne, pp 259–262

Scott JK, Morin L (2012) *Moraea flaccida* Sweet – one-leaf Cape tulip and *Moraea miniata* Andrews – two-leaf Cape tulip. In: Julien M, Cruttwell McFadyen R, Cullen J (eds) Biological control of weeds in Australia. CSIRO, Melbourne, pp 398–403

Seixas CDS, Barreto RW, Freitas LG et al (2004) *Ditylenchus drepanocercus* (Nematoda), a potential biological control agent for *Miconia calvescens* (Melastomataceae): host-specificity and epidemiology. Biol Control 31:29–37

Serbesoff-King K (2003) Melaleuca in Florida: a literature review on the taxonomy, distribution, biology, ecology, economic importance and control measures. J Aquatic Plant Manage 41:98–112

Shafroth PB, Cleverly JR, Dudley TL et al (2005) Control of *Tamarix* in the western United States: implications for water salvage, wildlife use, and riparian restoration. Environ Manage 35:231–246

Shaw RH, Bryner S, Tanner R (2009) The life history and host range of the Japanese knotweed psyllid, *Aphalara itadori* Shinji: potentially the first classical biological weed control agent for the European Union. Biol Control 49:105–113

Sheppard AW, Henry K (2012) *Genista monspessulana*(L.) L. Johnson – Cape broom. In: Julien M, Cruttwell McFadyen R, Cullen J (eds) Biological control of weeds in Australia. CSIRO, Melbourne, pp 267–273

Simberloff D, von Holle B (1999) Positive interactions of nonindigenous species: invasional meltdown? Biol Invasion 1:21–32

Simelane DO, Fourie A, Mawela KV (2011) Prospective agents for the biological control of *Cardiospermum grandiflorum* Sw. (Sapindaceae) in South Africa. Afr Entomol 19:269–277

Simmonds FJ, Bennett FD (1966) Biological control of *Opuntia* spp. by *Cactoblastis cactorum* in the Leeward Islands (West Indies). Entomophaga 11:183–189

Smith L (2005) Host plant specificity and potential impact of *Aceria salsolae* (Acari: Eriophyidae), an agent proposed for biological control of Russian thistle (*Salsola tragus*). Biol Control 34:83–92

Smith L (2007) Physiological host range of *Ceratapion basicorne*, a prospective biological control agent of *Centaurea solstitialis* (Asteraceae). Biol Control 41:120–133

Smith L, Cristofaro M, de Lillo E et al (2009) Field assessment of host plant specificity and potential effectiveness of a prospective biological control agent, *Aceria salsolae*, of Russian thistle, *Salsola tragus*. Biol Control 48:237–243

Sorensen B, Jusaitis M (1995) The impact of bridal creeper on an endangered orchid. In: Cooke D, Choate J (eds) Weeds of conservation concern: seminar and workshop papers. Department of Environmental and Natural Resources and Animal and Plant Control Commission, Adelaide, Australia, pp 27–31

Stephens AEA, Krannitz PG, Myers JH (2009) Plant community changes after the reduction of an invasive rangeland weed, diffuse knapweed, *Centaurea diffusa*. Biol Control 51:140–146

Stock WD, Wienand KT, Baker AC (1995) Impacts of invading N2-fixing *Acacia* species on patterns of nutrient cycling in two Cape ecosystems: evidence from soil incubation studies and 15N natural abundance values. Oecologia 101:375–382

Story JM, Callan NW, Corn JG et al (2006) Decline of spotted knapweed density at two sites in western Montana with large populations of the introduced root weevil, *Cyphocleonus achates* (Fahraeus). Biol Control 38:227–232

Story JM, Smith L, Corn JG et al (2008) Influence of seed head-attacking biological control agents on spotted knapweed reproductive potential in western Montana over a 30-year period. Environ Entomol 37:510–519

Thomas J, Leys A (2002) Strategic management of bitou bush (*Chrysanthemoides monilifera* ssp. *rotundata* (L.) T. Norl.). In: Spafford Jacob H, Dodd J, Moore JH (eds) Proceedings of 13th Australian weeds. Plant Protection Society of West Australia, Perth, Australia, pp 586–590

Thomas PA, Room PM (1986) Successful control of the floating weed *Salvinia molesta* in Papua New Guinea: a useful biological invasion neutralizes a disastrous one. Environ Conserv 13:242–248

Tipping PW, Martin MR, Center TD et al (2008a) Suppression of *Salvinia molesta* Mitchell in Texas and Louisiana by *Cyrtobagous salviniae* Calder and Sands. Aquatic Bot 88:196–202

Tipping PW, Martin MR, Pratt PD et al (2008b) Suppression of growth and reproduction of an exotic invasive tree by two introduced insects. Biol Control 44:235–241

Tipping PW, Martin MR, Nimmo KR et al (2009) Invasion of a West Everglades wetland by *Melaleuca quinquenervia* countered by classical biological control. Biol Control 48:73–78

Toft JD, Simenstad CA, Cordell JR et al (2003) The effects of introduced waterhyacinth on habitat structure, invertebrate assemblages, and fish diets. Estuar Coasts 26:746–758

Tomley AJ (1995) The biology of Australian weeds No. 26. *Cryptostegia grandiflora*. R Br Plant Prot Q 10:122–130

Tracy JL, DeLoach CJ (1998) Suitability of classical biological control for giant reed (*Arundo donax*) in the United States. In: Bell CE (ed) Arundo and saltcedar management workshop proceedings, 17 June 1998, Ontario, California, University of California Cooperative Extension Service, Holtville, California, USA, pp 73–153

Turner CE, McEvoy PB (1995) Tansy ragwort. In: Nechols JR, Andres LA, Beardsley JW et al (eds) Biological control in the western United States: accomplishments and benefits of regional research project W-84, 1964–1989. University of California, Division of Agriculture and Natural Resources, Oakland, Pub No. 3361, pp 264–269

Turner PJ, Scott JK, Spafford H (2008a) The ecological barriers to the recovery of bridal creeper (*Asparagus asparagoides* [L.] Druce) infested sites: impacts on vegetation and the potential increase in other exotic species. Aust Ecol 33:713–722

Turner PJ, Scott JK, Spafford H (2008b) Implications of successful biological control of bridal creeper (*Asparagus asparagoides* (L.) Druce in south-west Australia. In: van Klinken RD, Osten VA, Panetta FD et al (eds) Proceedings of the 16th Australian weeds conference. Queensland Weed Society, Brisbane, Australia, pp 390–392

Urban AJ, Simelane DO, Retief E et al (2011) The invasive '*Lantana camara* L'. hybrid complex: a review of research into its identity and biological control in South Africa. Afr Entomol 19:315–348

Valentine LE, Roberts B, Schwarzkopf L (2007) Mechanisms driving avoidance of non-native plants by lizards. J Appl Ecol 44:228–237

van der Westhuizen L (2006) The evaluation of *Phenrica* sp. 2 (Coleoptera: Chrysomelidae: Alticinae), as a possible biological control agent for Madeira vine, *Andredera cordifolia* (Ten.) Steenis, in South Africa. M.S. Department Zoology and Entomology, Rhodes University, Grahamstown, South Africa

van der Westhuizen L (2011) Initiation of a biological control programme against Madeira vine *Anredera cordifolia* (Ten.) Steenis (Basellaceae), in South Africa. Afr Entomol 19:217–222

Van Driesche RG (2012) The role of biological control in wildlands. BioControl 57:131–137

Van Driesche RG, Hoddle M, Center T (2008) Control of pests and weeds by natural enemies: an introduction to biological control. Blackwell, Oxford

Van Driesche RG, Carruthers RI, Center T et al (2010) Classical biological control for the protection of natural ecosystems. Biol Cont Suppl 1:S2–S33

van Klinken RD (2012) *Prosopis* spp. – mesquite. In: Julien M, Cruttwell McFadyen R, Cullen J (eds) Biological control of weeds in Australia. CSIRO, Melbourne, pp 477–485

van Klinken RD, Campbell S (2009) Australian weeds series: *Prosopis* species. In: Panetta FD (ed) Australian weeds series, vol 3. RG and FJ Richardson, Melbourne, pp 238–273

van Klinken RD, Heard TA (2012) *Parkinsonia aculeate* L. – parkinsonia. In: Julien M, Cruttwell McFadyen R, Cullen J (eds) Biological control of weeds in Australia. CSIRO, Melbourne, pp 437–447

Vogler W, Lindsay A (2002) The impact of the rust fungus *Maravalia crytostegiae* on three rubber vine (*Cryptostegia grandiflora*) populations in tropical Queensland. In: Jacob HS, Dodd J, Moore JH (eds) 13th Australian weeds conference "Threats now and forever?" Perth, Western Australia, 8–13 Sept 2002. Victoria Park, Plant Prot Soc West Australia, pp 180–182

Walden D, van Dam R, Finlayson M et al (2004) A risk assessment of the tropical wetland weed *Mimosa pigra* in Northern Australia. In: Julien M, Flanagan G, Heard T et al (eds) Research and management of *Mimosa pigra*. CSIRO Entomology, Canberra, pp 11–21

Wang Y, Ding J, Wheeler G et al (2009) *Heterapoderopsis bicallosicollis* (Coleoptera: Attelabidae), a potential biological control agent for Chinese tallow (*Triadica sebifera*). Environ Entomol 38:1135–1144

Weed AS, Casagrande RA (2010) Biology and larval feeding impact of *Hypena opulent* (Christoph) (Lepidoptera: Noctuidae): a potential biological control agent for *Vincetoxicum nigrum* and *V. rossicum*. Biol Control 53:214–222

Weiss J, Sagliocco J-L (2012) *Marrubium vulgare* L. – horehound. In: Julien M, Cruttwell McFadyen R, Cullen J (eds) Biological control of weeds in Australia. CSIRO, Melbourne, pp 260–367

Wheeler GS, Center TD (2007) Hydrilla stems and tubers as hosts for three *Bagous* species: two introduced biological control agents (*Bagous hydrillae* and *B. affinis*) and one native species (*B. restrictus*). Environ Entomol 36:409–415

Wilde SB, Murphy TM, Hope CP et al (2005) Avian vacuolar myelinopathy AVM linked to exotic aquatic plants and a novel cyanobacterial species. Environ Toxicol 20:348–353

Willis AJ, McKay R, Vranjic JA et al (2003) Comparative seed ecology of the endangered shrub, *Pimelea spicata* and a threatening weed, bridal creeper: smoke, heat and other fire-related germination cues. Ecol Manage Restor 4:55–65

Wilson JRU, Ajuonu O, Center TD et al (2007) The decline of waterhyacinth on Lake Victoria was due to biological control by *Neochetina* spp. Aquatic Bot 87:90–93

Wood AR, Morris MJ (2007) Impact of the gall-forming rust fungus *Uromycladium tepperianum* on the invasive tree *Acacia saligna* in South Africa: 15 years of monitoring. Biol Control 41:68–77

Wu Y, Reardon RC, Ding J (2002) Mile-a-minute weed. In: Van Driesche R, Lyon S, Blossey B et al. (eds) Biological control of invasive plants in the eastern United States. USDA Forest Service Public FHTET-2002-04, pp 331–341

Yelenik SG, Stock WD, Richardson DM (2004) Ecosystem level impacts of invasive *Acacia saligna* in the South African fynbos. Restor Ecol 12:44–51

Zachariades C, Day MD, Muniappan R et al (2009) *Chromolaena odorata* (L.). King and Robinson (Asteraceae). In: Muniappan R, Reddy GV, Raman A (eds) Biological control of tropical weeds using arthropods. Cambridge University Press, Cambridge, pp 130–162

Zachariades C, Hoffmann JH, Roberts AP (2011a) Biological control of mesquite (*Prosopis* species) (Fabaceae) in South Africa. Afr Entomol 19:402–415

Zachariades C, Strathie LW, Retief E et al (2011b) Progress toward the biological control of *Chromolaena odorata* (L.) R.M. King and H. Rob (Asteraceae) in South Africa. Afr Entomol 19:282–302

Chapter 27
Restoration Within Protected Areas: When and How to Intervene to Manage Plant Invasions?

Loralee Larios and Katharine N. Suding

Abstract Despite the on-going efforts to set aside land for conservation, biodiversity is increasingly being threatened by factors such as invasive alien species that do not recognise these boundaries. Invasive species management programmes are widely incorporated into protected area management plans; however, the success of these programmes hinges on the ability to identify when a system will be able to recover after invader control and eradication efforts and when further intervention will be necessary to aide recovery. Invasive alien plants can alter ecosystem attributes to produce strong legacy effects that prevent the recovery of a system. Here we provide a framework for how to identify and incorporate recovery constraints into restoration efforts. Identifying recovery constraints can help improve how ecological theory – assembly rules, ecological succession, and threshold dynamics – can be used to guide restoration efforts.

Keywords Assembly rules • Ecological succession • Threshold dynamics • Recovery constraints • Invader impacts

27.1 Introduction

Protected areas (PAs) serve as the primary method to maintain and protect global biodiversity (UNEP-SCBD 2001). Therefore, an important goal in PAs is to minimise threats to biodiversity and maintain ecological communities in their natural states (Lockwood et al. 2006). Protected areas can manage certain threats such as deforestation or poaching, but even the most well managed reserves are still

L. Larios (✉) • K.N. Suding
Department of Environmental Science, Policy and Management, University of California
Berkeley, 137 Mulford Hall, Berkeley, CA 94720-3114, USA
e-mail: llarios@berkeley.edu; suding@berkeley.edu

L.C. Foxcroft et al. (eds.), *Plant Invasions in Protected Areas: Patterns, Problems and Challenges*, Invading Nature - Springer Series in Invasion Ecology 7, DOI 10.1007/978-94-007-7750-7_27, © Springer Science+Business Media Dordrecht 2013

susceptible to threats such as climate change, pollution, and invasive alien species that do not recognise these conservation boundaries and fence lines. We focus on managing one of these threats, invasive alien plant species (IAPs), in PAs. In response to this threat, many PAs have implemented large-scale invasive species management programmes that employ prevention, eradication, and control strategies aimed at slowing or stopping the process of invasion (Foxcroft and Richardson 2003; Doren et al. 2009b).

Increasingly, a challenge in this process is that simply removing the invasive species is not sufficient to restore native biodiversity. A recent review by Kettenring and Adams (2011) found that invasive removal successfully reduced the cover of invasive alien plants (IAP), but did not always result in native species recovery. Further intervention – with a focus on restoration – may be necessary to take into account the impacts an invader has on a system (D'Antonio and Meyerson 2002), as well as address recovery constraints of the native community. However, these additional intervention actions can be costly in terms of time and money and, in some cases, they have unintended consequences and actually slow recovery (Zavaleta et al. 2001; Hobbs and Richardson 2011). Integrating additional intervention efforts within an existing protected area management plan can be complicated by a variety of factors such as limited resources (e.g. staff and infrastructure), legal mandates under IUCN management categories, or differing agendas among the stakeholders in the governance group (Keenleyside et al. 2012). The isolated nature of PAs requires intervention efforts to be a concerted endeavour with agencies/land owners outside of the reserve, further complicating the success of management efforts.

In this chapter, we focus on this conundrum: when should we expect a system to recover without additional restoration efforts after invasive species control efforts? And when is further intervention necessary for recovery? Resources are often scarce for PAs, with eradication and control of invasive species often consuming a disproportionate amount of reserve budgets (D'Antonio and Meyerson 2002). Identifying necessary points of intervention prior to action is therefore critical for successful protected area management. We begin by providing an overview of invader impacts that may constrain and preclude the recovery of a system after IAP management. We then explore key ecological theories that can be used to guide restoration strategies. Finally, we discuss how land managers could adjust restoration efforts depending on the constraints present in the system.

In this chapter, we consider restoration to include both IAP control and eradication efforts as well as additional actions to aid native recovery. As emphasised elsewhere in this volume, invader management plans in PAs often include control and eradication efforts in tandem with native recovery efforts. Here, we focus on restoration after the invaders are removed or reduced. The key questions are thus: when will passive recovery following these efforts be sufficient to recover desired native communities, and when will active intervention (*sensu* Suding 2011) be needed?

27.2 Invader Impacts and Recovery Constraints

As PAs operate under the mandate to protect local biodiversity, the continuing and growing threat of IAP invaders on native biodiversity has made IAP management a priority for PAs (Macdonald et al. 1988; Vitousek et al. 1997; McNeely 2001). Understanding invader impacts on ecological communities is an important first step in understanding how native communities may recover following IAP control. We particularly focus on IAP legacy effects in PAs, where the impacts of invasion persist even after invader control or eradication. In these cases, removing the invader may not always lead to successful recovery of the degraded system (D'Antonio and Meyerson 2002); additional management and restoration actions may be necessary to put the native community on a path to recovery (Suding et al. 2004). Alternatively, if an IAP does not have strong legacy effects, additional efforts may not be necessary and native communities should be expected to passively recover following control efforts. Importantly, the impacts of invasion may occur either progressively with invader abundance or abruptly once the invader reaches a certain abundance threshold (D'Antonio and Chambers 2006; Didham et al. 2007). Consequently, whether active or passive restoration is necessary may depend on the pattern as well as the nature of legacy effects.

Native species recovery may often be limited by dispersal following IAP control (Galatowitsch and Richardson 2005; Traveset and Richardson 2006). Source populations of native species may be far from the restoration area (McKinney and Lockwood 1999) or seed dispersal networks may be altered in the invaded area (Traveset and Richardson 2006; McConkey et al. 2012). For example, in Australia, recovery of coastal dune communities invaded by *Chrysanthemoides monilifera* subsp. *rotundata* (the South African bitou bush) is limited by poor seed dispersal from existing native vegetation (French et al. 2011), and in New Zealand, native shrublands dominated by *Kunzea ericoides* (kanuka) have a different composition and a smaller abundance of the avian seed dispersers compared to *Ulex europaeus* (gorse) invaded stands (Williams and Karl 2002). Additionally, native seed bank at a restoration site could be diminished if natives have been absent or in low abundance, reducing the potential for recovery from in situ germination (D'Antonio and Meyerson 2002). In southern California, passive recovery of the native coastal sage scrub community is limited due to the depauperate native seedbank in long term invaded alien grassland sites (Cione et al. 2002; Cox and Allen 2008).

Plant invaders can alter disturbance regimes, which may create positive feedbacks that promote invader success (D'Antonio and Vitousek 1992; Mack and D'Antonio 1998). These feedbacks must be disrupted to allow the recovery of a system (Suding et al. 2004). A widespread example occurs when annual grass invaders increase the intensity and frequency of fire (D'Antonio and Vitousek 1992). In the Western United States, for example, alien annual grasses increase fuel loads, which promotes a fire frequency for which the resident community is not adapted (Whisenant 1990). Conversely, IAPs can also impact disturbance regimes by suppressing disturbances (Mack and D'Antonio 1998). *Schinus terebinthifolius* (pepper tree) invasion in Florida's Everglades National Park has suppressed fire intensity by decreasing fuel loads (i.e. understory vegetation), which enhances its own recruitment (Doren and Whiteaker 1990). In these cases, the disturbance regime may not recover following IAP control, and additional actions may be needed to re-establish the disturbance regime needed to support the native community (Davies et al. 2009).

Invasive alien plants can also impact the physical structure of soils by increasing erosion rates or sedimentation rates and directly by affecting substrate stability (D'Antonio et al. 1999), resulting in soil legacies (*sensu*, Corbin and D'Antonio 2004). For example, while increased sedimentation can promote succession and facilitate the establishment of native species in degraded forests in Algiers (Wojterski 1990), increased erosion rates can limit recovery by eliminating habitat for native species and promoting the establishment of introduced species in the South African fynbos (Macdonald and Richardson 1986).

Soil legacies can also influence belowground biological processes that promote IAP abundance and stall native species recovery (van der Putten et al. 2007; Inderjit and van der Putten 2010). An invader can be successful because it is able to escape soil pathogens (Klironomos 2002), and it may also alter pathogen incidence in the native community to reduce competitive effects and facilitate its spread (Eppinga et al. 2006; Mangla et al. 2008). Pathogen loads may slow the recovery rates of communities, as they continue to influence the performance of native species even after IAP removal (Malmstrom et al. 2005). Invasive alien plants can facilitate their invasion by allelopathy (i.e. the release of phytotoxins, which inhibit the growth of neighboring plants; Callaway and Ridenour 2004). For example, high impact invader *Centaurea maculosa* (spotted knapweed) releases a compound that inhibits root growth of its neighbouring plants (Bais et al. 2003). Additionally, *Alliaria petiolata* (garlic mustard), a widespread invader in North American forests, secretes compounds that inhibit the symbiotic mycorrhizal associations of native plants. These altered relationships can prevent the recovery of the community once the invader has been removed due to residual toxins (Perry et al. 2005). Other invaders can alter soil properties such as salt concentrations or soil pH, reducing the potential for subsequent colonization by native species (Vivrette and Muller 1977; Conser and Connor 2009).

Soil legacies also include invader impacts on biogeochemical cycles that alter resource availability (Mack et al. 2001; Ehrenfeld 2003). Nitrogen cycling rates are regularly increased by invaders by altering the microbial community (Hawkes et al. 2005), altering litter quality (Sperry et al. 2006), or directly by nitrogen-fixing species (Vitousek and Walker 1989; Le Maitre et al. 2011). Increased nitrogen availability can result in positive feedbacks that maintain the invaded state, thwarting recovery efforts (Clark et al. 2005). For example, in temperate grasslands in Australia, alien annual species that invade native perennial tussock grasslands can alter nitrogen cycling to favour their own growth. These nutrient changes are sufficient to push the system past a threshold, preventing the recovery of native grasses (Prober et al. 2009). Lastly, invaders can also alter the hydrology of a system via altered transpiration rates, rooting depths, phenology, and growth rates (Levine et al. 2003). *Tamarix* spp. (salt cedar) invasion in the south-western United States has resulted in higher transpiration rates and marginal water loss due to the salt cedar's deeper root system in this water limited system (Zavaleta 2000).

27.3   Ecosystem Models in Restoration

While it is clear that many IAPs have strong legacy effects that can influence the recovery of native communities, it is also important to put these effects in the context of ecological processes that guide the path to recovery (Young et al. 2005; Hobbs et al. 2007; Suding and Hobbs 2009). Conceptual models of ecosystem dynamics such as assembly theory, ecological succession, and threshold dynamics can guide restoration projects by providing insights into these ecological processes (Fig. 27.1). In the following paragraphs we explore these three concepts and how they can guide decisions about when and how to intervene in PAs following invasive plant control efforts. For each, we first present the basic framework, then a case study examining application to restoration in PAs.

27.3.1   Assembly Rules

Assembly theory focuses on how a suite of processes (e.g. dispersal, disturbance, environment, competition) influence which species are able to establish over time (Young et al. 2001; Temperton et al. 2004; White and Jentsch 2004; Hobbs et al. 2007). This framework integrates these processes into a series of filters (dispersal, environmental, and biotic) that act at varying spatial scales, which can explain which species from a regional species pool (large scale) are found in the local community (small scale, Weiher and Keddy 1995; Diaz et al. 1998, 1999). In the context of native species recovery following IAP control, recovery requires that filters at each scale allow native species to establish and persist (Fig. 27.1b). Additional intervention efforts would be focused on the filters that excluded the desired species from recovering (Fig. 27.1b, dashed arrow).

Three general types of filters are emphasised in assembly theory. The first filter that species must overcome is dispersal: species must have dispersal traits that allow them to arrive at a site (Levine and Murrell 2003). As discussed above, invasive plants can increase the dispersal limitation of native species in many ways, creating new barriers to the dispersal filter for some native species. If a species is able to colonise a site, the next filter acting upon it is the environmental filter. To successfully cross the environmental filter, a species must have the suite of traits that allow it to survive the given environmental conditions (Weiher and Keddy 1995; Diaz et al. 1998, 1999). Soil legacies of invasive plants, such as erosion and resource cycling impacts, can alter this filter. An extension to the environmental filter is the disturbance filter (White and Jentsch 2004), which invasive species may

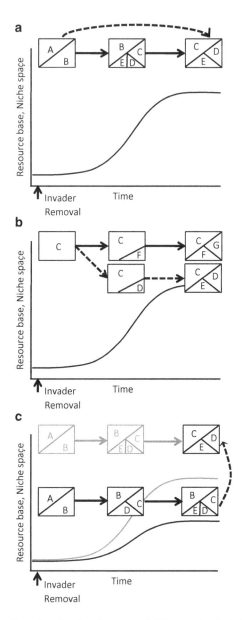

Fig. 27.1 Recovery trajectories after invader removal, (**a**) assuming little invader impact or, (**b**) and (**c**), a legacy of invader impacts. Species composition is symbolised by the *capital letters* and abundance by proportion of each square; the desired goal community is C/E/D. *Solid lines* indicate scenarios with passive restoration after IAP removal; *dotted arrows* indicate restoration interven-tion. We present scenarios consistent with each of the three ecosystem models of recovery. Successional theory (**a**) is most appropriate in systems where there is little expectation of strong invader impacts. In (**a**), successional theory assumes directional change in species composition over time. If the natural recovery takes too long, land managers can intervene to accelerate recovery (*dashed arrow* in **a**). In systems impacted by invader legacy affects (**b**, **c**), assembly theory and threshold theory may be most appropriate to guide restoration efforts. In (**b**), IAP legacies affect the order of species arrival. Active intervention can focus on adding species,

similarly alter. The final filter in assembly theory is the biotic filter, which restricts the community to those species that can coexist in the presence of interspecific interactions (MacArthur and Levins 1967; Tilman 1990; Chesson 2000). Under the biotic filter, competitive interactions would limit the co-occurrence of functionally equivalent species due to niche limitation resulting in limited similarity among species within community (MacArthur and Levins 1967). Under situations where invasive species have been controlled or eradicated in PAs, we would expect that this biotic filter would be less of a consideration compared to the other filters, but it would be important to manage were reinvasion possible.

The efficacy of active intervention efforts in restoration (e.g. species palette for planting, selection of planned disturbance to limit competitive interactions) can be assessed in this assembly filter framework by equating restoration actions with changes in assembly filters (Funk et al. 2008). For example, seed addition or planting of native species can be viewed as changing the dispersal filter at a site. Similarly, a trait-based approach could increase the success of restoration efforts areas where managers fear invasive species could re-invade following control efforts by identifying a suite of native species with traits similar to the IAP to enhance the invasion resistance of the community, thereby strengthening the biotic filter (Funk et al. 2008).

27.3.2 Case Study 1: California Grasslands

Protected areas such as county parks and reserves within California are often imbedded within a highly fragmented landscape (Greer 2005). In California PAs, alien annual grasses have the potential to gain access to the interior of natural areas by initially colonizing disturbed roadside areas (Gelbard and Belnap 2003). Roadsides can have large inputs of atmospheric nitrogen deposition (Pearson et al. 2000), which can interact with local grassland's N cycling to increase N availability (Sirulnik et al. 2007), and further promote these annual grasses (Padgett and Allen 1999). Furthermore, prolonged dominance of alien grasses within a site can reduce the seedbank of native species and prevent the recovery of a system once the grasses have been removed (Cione et al. 2002).

To evaluate if the biotic filter can be manipulated to slow or stop the re-invasion of aliens after control, Cleland et al. (2013) conducted a restoration experiment along a roadside edge of the Laguna Coast Wilderness Park in southern California

Fig. 27.1 (continued) affecting the order of species arrival, to guide the assembly process to arrive at the target community. In (c), recovery may result in a new undesired state due to invader legacy impacts, preventing the successional process that would occur naturally (*grey boxes*). A threshold model may be the most appropriate to apply in cases such as these, where multiple restoration activities would need to be done to overcome this feedback (*dashed arrow*, c) (Modified from White and Jentsch 2004)

where they manipulated nitrogen availability and added native seeds representing different functional groups (annual/perennial grasses, early/late forbs and N-fixing legumes). In the first year, they removed alien annual grasses and forbs. Then in the second year, they allowed alien species to colonise naturally. Native communities with low N availability and in which early forb seed was added best resisted re-invasion. Thus, they found that by altering resource availability and adding species that have similar phenology to the problematic invader they could manipulate the biotic filter to increase invasion resistance.

27.3.3 Ecological Succession

Successional dynamics, the changes in species composition within a community over time, have been a classic and focal question in ecology since the 1900s (Cowles 1899; Clements 1916; Gleason 1926). Succession traditionally describes the patterns of compositional change after a disturbance (Clements 1916; Pickett et al. 1989) but recent studies have gone beyond describing the patterns to identify the mechanisms, which influence these patterns (Connell and Slatyer 1977; Tilman 1988; Pickett et al. 2009). As successional theory has expanded to incorporate the possibility of multiple successional pathways versus a single climax community (Glenn-Lewin et al. 1992), comparing and analysing successional trajectories has been adopted to describe the temporal change in community composition (Hobbs and Mooney 1995). Once a disturbance occurs at a site, the availability of safe sites and propagules for colonization in conjunction with the impacts of established species determine subsequent successional dynamics (Pickett et al. 1987). In the context of whether to intervene following invasive species control, additional intervention activities can be viewed as either altering or initiating any of these recovery processes (del Moral et al. 2007; Fig. 27.1a).

Ecological restoration can take a variety of approaches to manage succession toward a desired target. The first and simplest approach is to allow succession to occur unaided (spontaneous succession, Prach et al. 2001) and should be a viable option if most abiotic and biotic functioning remain intact after invasive species control (Lockwood and Samuels 2004; Prach and Hobbs 2008). However, in the case of large-scale invasions, natural succession is unlikely to be a viable option as legacies from the invader may influence recovery (Zavaleta et al. 2001). When legacies are present another approach is to assist succession via manipulations to the physical environment and to biotic processes that may be important within the target system (technical reclamation, Prach et al. 2007). Technical reclamation may be necessary if invasion has resulted in the complete loss of any of the overarching processes governing succession (e.g. availability of safe sites, propagules, and species impacts; del Moral et al. 2007; Prach and Hobbs 2008). The third approach, assisted succession, is a combination of technical reclamation and spontaneous succession in which site conditions are initially modified to support native species but subsequent succession is allowed to occur naturally (Prach et al. 2007;

Fig. 27.1a, dashed line). This approach has been implemented within rangeland invasive plant management, by pairing removal efforts with post-removal restoration activities (Sheley et al. 2010). While this framework is similar to assembly theory in that it emphasises identification of processes that constrain recovery, it also emphasises trajectories of community development over time.

27.3.4 Case Study 2: South African Fynbos

The fynbos vegetation in the Cape Floristic Region of South Africa is highly impacted by alien trees and shrubs (*Acacia* spp; Macdonald 1984; Le Maitre et al. 2011). *Acacia spp.* are nitrogen (N)-fixing plants, which can increase soil fertility after an extended presence in an area (Yelenik et al. 2004). They also have a large impact on water resources, as they consume more water than the native vegetation (Le Maitre et al. 2000). Under the national 'Working for Water' program, *Acacia* spp. and other woody invasive plants have been targeted for removal (Turpie et al. 2008). Clearing of these invaders is often a combined effort of cutting down the tree/shrub and, for those species that resprout, applying herbicide to the stumps with the felled biomass left on site. It can also involve the removal of the felled material and/or burning (Macdonald 2004). Cleared sites are often allowed to recover spontaneously after treatment; however, the success of passive recovery is often dependent on the type of treatment (i.e. spontaneous succession was the most successful with clearing and removal and the least successful under burning; Blanchard and Holmes 2008). Blanchard and Holmes (2008) found that once the biomass was removed, native species had space to establish and assisted succession approaches were needed. For example, seeding after burn treatments to overcome dispersal constraints can increase the presence of native fynbos vegetation and enhance natural recovery; however, continuous eradication efforts are needed until the large *Acacia* seedbank is reduced as natural wildfires may continue to promote the establishment of *Acacia* after initial removal (Milton and Hall 1981).

27.3.5 Threshold Dynamics

Ecological thresholds are a breakpoint between two systems that, when crossed, result in an abrupt change in community states (Holling 1973). Thresholds occur due to positive feedback mechanisms, which make systems resistant to change (Folke et al. 2004; Suding et al. 2004). While successional models and recovery pathways apply to many situations of recovery following IAP control, threshold models can help explain why some systems are not able to recover once the invader has been removed (Prober et al. 2009). In the context of these 'stuck' systems, threshold models point to the importance of breaking these positive feedbacks in order to facilitate recovery (Fig. 27.1c).

A useful framework for incorporating ecological thresholds into management has been to divide thresholds into two stages. The first stage is the biotic threshold, which can be identified by changes in vegetative structure or composition (Friedel 1991; Whisenant 1999). The second stage is an abiotic threshold, which identifies changes in ecosystem functioning (Whisenant 1999). Because impacts on functioning are thought to lag behind biotic changes, a system is thought to first encounter the biotic threshold and subsequently the abiotic (Whisenant 1999; Hobbs and Harris 2001; Briske et al. 2005). Invasive alien plants that trigger biotic threshold changes may be easier to control than those that cause biotic and abiotic threshold changes. Invaders that cause the system to cross both thresholds (ecosystem engineers, *sensu* Jones et al. 1994) make the success of restoration efforts highly uncertain (Ehrenfeld et al. 2005; Kulmatiski 2006; Doren et al. 2009a). Once management has identified key variables that can indicate whether a threshold has been crossed, this knowledge can be incorporated into management to identify when and what management efforts are needed to increase the success of control and subsequent restoration efforts (Foxcroft and Richardson 2003; Doren et al. 2009b).

27.3.6 Case Study 3: Australian Subtropical Rainforests

One of the world's most notorious invaders, *Lantana camara* (lantana), has invaded and replaced much of the native vegetation in the subtropical forests in eastern Australia (Lowe et al. 2000; Bhagwat et al. 2012). *Lantana camara* was introduced as an ornamental shrub in the mid-nineteenth century (Swarbick 1986) but has rapidly spread to the detriment of native diversity, including PAs within Australia's national parks. Many of the national parks within eastern Australia are isolated within a highly disturbed system, a problem common to many PAs globally (Fox et al. 1997). Edges between the reserves and disturbed areas (e.g. old agricultural fields in Australia) make reserves vulnerable to weedy invaders such as *L. camara*, which readily spread across disturbed landscapes (Gentle and Duggin 1997; Stock 2004).

However, this landscape also provides an opportunity to investigate the dynamics that allow this invader to invade pristine habitats. Stock (2004) and Gooden et al. (2009) monitored *L. camara* and native plant abundance in national parks in eastern Australia and were able to identify two separate thresholds. After measuring *L. camara* cover and canopy cover in gaps in two national parks, Stock (2004) identified a first invasion threshold: forests whose canopy cover is 75 % native species can prevent the establishment of *L. camara*, because the woody invader is shade intolerant in those forests. If *L. camara* reaches 75 % cover, however, the system crosses a second biotic threshold identified by Gooden et al. (2009) in which native species richness falls dramatically, likely due to *L. camara* effects on soil fertility (Bhatt et al. 1994) and soil seed banks (Fensham et al. 1994). These thresholds, which identify when a community can resist invasion and when invader impacts begin to increase dramatically, are being used to guide an integrated management plan (Stock 2005).

27.4 Addressing Recovery Constraints in the Context of the Three Models

The ability of ecosystems to recover after IAP control greatly varies and is often contingent on the system's intrinsic rate of recovery, its level of degradation, and its surrounding matrix (Jones and Schmitz 2009; Holl and Aide 2011; Gaertner et al. 2012). This variability makes it difficult to assess when land managers should intervene and implement additional restoration practices or leave a system to recover naturally (passive restoration, *sensu* Suding 2011). For example, in the south-western United States, invader removal without any paired plantings of native flora can detrimentally affect local fauna (Zavaleta et al. 2001), and yet, active plantings in the tropics may prevent the establishment of native flora and slow natural recovery (Murcia 1997). Furthermore, within PAs, additional complicating factors such as land use and pollution are either well documented or less severe than in non-protected areas. Therefore, restoration efforts conducted within PAs after IAP removal can help improve our collective understanding of invader impacts and recovery constraints. In this section, we suggest a series of steps to decide when additional intervention following IAP control may be needed.

First, an understanding of the extent of IAP impacts and whether they will persist following invader removal is critical (Sheley et al. 2010). A holistic assessment should try to identify the causes of the invasion as well as impacts of the invasion (see invader impacts section above; James et al. 2010). If a holistic assessment was not initially available, small scale experiments can be used to identify restraints (Kettenring and Adams 2011). Simply observing the natural recovery of a system after control efforts would also aide the decision of whether or not to intervene when assessments are not available (Holl and Aide 2011). If monitoring indicates natural recovery, land managers can use successional theory to make inferences about the trajectory of the system (Sheley et al. 2006; Prach and Walker 2011). However, if monitoring identifies invader legacies, the success of management efforts is contingent on effectively prioritizing and addressing those recovery constraints (Fig. 27.2a; Suding et al. 2004).

Identification of constraints can be done through knowledge of natural history, experimentation (Gaertner et al. 2011; Kettenring and Adams 2011) or research from other sites (e.g. recent reviews of the effects of the invaders *Acacia* (Le Maitre et al. 2011) and *L. camara* (Bhagwat et al. 2012) on ecosystems). If one single factor seems to strongly constrain recovery, natural recovery should be fairly straightforward if land managers can address the single constraint

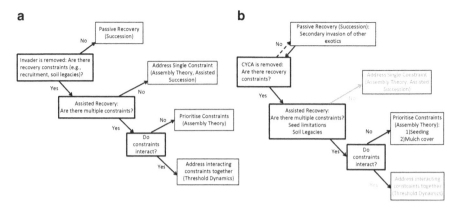

Fig. 27.2 (**a**) Decision Tree Model for assessing restoration activities following invader control efforts. Decision making nodes represent the assessment of identity, number of and interactions between recovery constraints (*bolded boxes*). At low and medium IAP abundances, control/removal efforts may be sufficient to return a community to its restored state (node 1: No intervention). However, at medium and high invader abundances management actions may not be sufficient to achieve the full recovery of the degraded system due to recovery constraints. A recovery constraint assessment can guide decisions for subsequent restoration actions (nodes 2 & 3). Ecological principles (*listed in parentheses*) can help inform which restoration tools to use. (**b**) Decision Tree model for the control and restoration efforts for *Cynara cardunculus*. Initially management efforts relied on passive recovery (*dashed line*); however after observing the ineffective recovery of native species, the land managers decided to implement efforts to overcome constraints. Evaluation of constraints (supporting citations listed in case study) indicated two potential constraints, and after determining that they likely do not interact to synergistically thwart recovery, constraints were prioritised. *Grey boxes* indicate paths that were not followed in this scenario

(Prach et al. 2001; Lockwood and Samuels 2004). Successional theory and assembly theory would be helpful in guiding restoration efforts with single constraints (Suding and Hobbs 2009). However, if multiple constraints are present, it is important to assess whether these constraints can be addressed independently or need to be addressed in tandem (Suding et al. 2004). If multiple constraints synergistically thwart the recovery of a system, it would be essential to address the constraints in tandem to disrupt any feedbacks that are preventing recovery (Fig. 27.2a; Suding et al. 2004).

Constraints can operate at multiple spatial and temporal scales (Suding and Hobbs 2009), and processes operating at one spatial or temporal scale can interact with processes operating at another scale to create strong internal feedbacks that prevent the recovery of the system (cross-scale interactions; Peters et al. 2007). In a hypothetical example, if an invader disrupted dispersal processes and produced soil legacies via allelopathy, successful restoration efforts would have to address both the soil condition as well as the dispersal constraint. Threshold models address strong internal feedbacks and nonlinear dynamics within ecosystems and would be helpful in guiding restoration efforts with interacting constraints

(Suding and Hobbs 2009). If the constraints do not interact, it would be important to prioritise constraints, and assembly theory could help elucidate which constraints and potential restoration approaches could be addressed based on how the degraded system has deviated from the historical environmental and biotic conditions (Lockwood and Samuels 2004). However, projects that incorporate several restoration actions are often more successful; therefore if resources are available, it would be wise to tackle multiple constraints (augmentative restoration; Bard et al. 2004; Buisson et al. 2008). While these approaches can help a land manager make better a priori decisions about what restoration activities to undertake, this approach is not fool proof and should always incorporate monitoring and re-assessment to ensure that the system is moving in the desired direction.

27.4.1 Case Study 4: Sustainable Control Efforts of Cynara cardunculus *(Artichoke Thistle) in Orange County, California*

Cynara cardunculus was introduced into southern California in the nineteenth century and has become a problematic invader across local grasslands (Thomsen et al. 1986). It is a perennial species with a deep taproot (about 1.5 m) and large inflorescences (up to 50 per rosette; Marushia and Holt 2006) filled with up to 800 wind dispersed seeds (Kelly 2000). It forms dense species-poor stands (Bowler 2008). Within the Nature Reserve of Orange County, it has invaded over 1,618 of the 14,973 ha of protected open space and its control has dominated the Reserve's budget for invasive species management (McAfee 2008). The primary control method since 1994 has been direct herbicide application with the assumption that native communities would passively recover. However, after 13 years native species did not recover in all treated areas; instead, the abundance of other alien plants increased (Seastedt et al. 2008). In an effort to implement more effective management activities for native grassland recovery, potential constraints were further identified using other published research studies. For example, seed limitation is often a constraint for native grass populations across California (Seabloom et al. 2003; Seabloom 2011). Additionally, Potts et al. (2008) identified that litter quantity and quality changes due to *C. cardunculus* invasion, which can negatively impact native recovery (Bartolome and Gemmill 1981; Coleman and Levine 2007). These findings can be integrated into a potential management plan for efficient and sustainable management of treated areas (Fig. 27.2b), where the passive recovery approach would be replaced with one where seed limitation and soil legacies constraints are both prioritised within the reserve.

27.5 Conclusions

Ecosystems globally are undergoing rapid changes due to global change drivers such as CO_2 enrichment, atmospheric nitrogen deposition, climate changes, land use, and biotic invasions (Sala et al. 2000). Among these drivers are invasive species, which are an increasing threat to natural and working landscapes as the globalization of trade and interactions with other global change drivers increase the opportunities for introductions (Levine and D'Antonio 2003; McNeely 2006). A small fraction of those invaders have the potential to trigger large changes in ecosystem functioning as they spread across a landscape (Williamson 1996) and can contribute to the degradation of native communities (McNeely 2001). Protected areas have the unprecedented burden of minimizing these negative invader impacts as they are tasked with the goal of protecting and maintaining the globe's biodiversity.

Here, we emphasised ways to determine whether additional intervention is needed for native recovery following IAP control in PAs. Multiple lines of evidence need to be weighed to best gauge when and where to invest in additional intervention approaches and when to stand back and allow the native system to recover naturally. This decision-making is not clear-cut but can be based on several ecological frameworks describing how communities are assembled and recover over time. Protected areas benefit from a holistic management approach, which addresses IAP detection, sources of invaders, potential external stressors, and management thresholds dictating when management efforts need to be initiated (Zavaleta et al. 2001; Foxcroft and Richardson 2003; Clewell and McDonald 2009). Worldwide efforts to evaluate the effectiveness of management within PAs (Hocking et al. 2000) provide a unique opportunity to assess the link between ecological theories that frame the process of recovery and restoration actions.

Acknowledgements We thank the UC Berkeley Range group, Suding lab members, and Marko Spasojevic for helpful comments on early drafts. We acknowledge support from the NSF Graduate Research Fellowship programme (to L.L.)

References

Bais HP, Vepachedu R, Gilroy S et al (2003) Allelopathy and exotic plant invasion: from molecules and genes to species interactions. Science 301:1377–1380
Bard EC, Sheley RL, Jacobsen JS et al (2004) Using ecological theory to guide the implementation of augmentative restoration. Weed Technol 18:1246–1249
Bartolome JW, Gemmill B (1981) The ecological status of *Stipa pulchra* (Poaceae) in California. Madrono 28:172–184
Bhagwat SA, Breman E, Thekaekara T et al (2012) A Battle lost? Report on two centuries of invasion and management of *Lantana camara* L. in Australia, India and South Africa. Plos One 7:e32407

Bhatt YD, Rawat YS, Singh SP (1994) Changes in ecosystem functioning after replacement of forest by *Lantana* shrubland in Kumaun Himalaya. J Veg Sci 5:67–70

Blanchard R, Holmes PM (2008) Riparian vegetation recovery after invasive alien tree clearance in the Fynbos Biome. S Afr J Bot 74:421–431

Bowler PA (2008) Artichoke thistle as an ecological resource and its utility as a precursor to restoration (California). Ecol Restor 26:7–8

Briske DD, Fuhlendor SD, Smeins EE (2005) State-and-transition models, thresholds, and range-land health: a synthesis of ecological concepts and perspectives. Rangel Ecol Manage 58:1–10

Buisson E, Anderson S, Holl KD et al (2008) Reintroduction of *Nassella pulchra* to California coastal grasslands: effects of topsoil removal, plant neighbour removal and grazing. Appl Veg Sci 11:195–204

Callaway RM, Ridenour WM (2004) Novel weapons: invasive success and the evolution of increased competitive ability. Front Ecol Environ 2:436–443

Chesson P (2000) Mechanisms of maintenance of species diversity. Ann Rev Ecol Syst 31:343–366

Cione NK, Padgett PE, Allen EB (2002) Restoration of a native shrubland impacted by exotic grasses, frequent fire, and nitrogen deposition in southern California. Restor Ecol 10:376–384

Clark BR, Hartley SE, Suding KN et al (2005) The effect of recycling on plant competitive hierarchies. Am Nat 165:609–622

Cleland EE, Larios L, Suding KN (2013) Strengthening invasion filters to reassemble native plant communities: soil resources and phenological overlap. Restor Ecol 21:390–398. doi:10.1111/j.1526-100X.2012.00896.x

Clements FE (1916) Plant succession, vol 242. Carnegie Institution, Washington, DC

Clewell A, McDonald T (2009) Relevance of natural recovery to ecological restoration. Ecol Restor 27:122–124

Coleman HM, Levine JM (2007) Mechanisms underlying the impacts of exotic annual grasses in a coastal California meadow. Biol Invasion 9:65–71

Connell JH, Slatyer RO (1977) Mechanisms of succession in natural communities and their role in community stability and organization. Am Nat 111:1119–1144

Conser C, Connor EF (2009) Assessing the residual effects of *Carpobrotus edulis* invasion, implications for restoration. Biol Invasion 11:349–358

Corbin JD, D'Antonio CM (2004) Effects of exotic species on soil nitrogen cycling: implications for restoration. Weed Technol 18:1464–1467

Cowles HC (1899) The ecological relations of the vegetation on the sand dunes of Lake Michigan. Bot Gazette 27:95–117

Cox RD, Allen EB (2008) Composition of soil seed banks in southern California coastal sage scrub and adjacent exotic grassland. Plant Ecol 198:37–46

D'Antonio CM, Chambers JC (2006) Using ecological theory to manage or restore ecosystems affected by invasive plant species. In: Falk DA, Palmer MA, Zedler JB (eds) Foundations of restoration ecology. Island Press, Washington, DC, pp 260–279

D'Antonio C, Meyerson LA (2002) Exotic plant species as problems and solutions in ecological restoration: a synthesis. Restor Ecol 10:703–713

D'Antonio CM, Vitousek PM (1992) Biological invasions by exotic grasses, the grass fire cycle, and global change. Ann Rev Ecol Syst 23:63–87

D'Antonio C, Dudley TL, Mack MC (1999) Disturbance and biological invasions: direct effects and feedbacks. In: Walker LR (ed) Ecosystems of disturbed ground. Elsevier, New York, pp 413–452

Davies KW, Svejcar TJ, Bates JD (2009) Interaction of historical and nonhistorical disturbances maintains native plant communities. Ecol Appl 19:1536–1545

del Moral R, Walker LR, Bakker JP (2007) Insights gained from succession for the restoration of landscpae structure and function. In: Walker LR, Walker J, Hobbs RJ (eds) Linking restoration and ecological succession. Springer, New York, pp 19–44

Diaz S, Cabido M, Casanoves F (1998) Plant functional traits and environmental filters at a regional scale. J Veg Sci 9:113–122

Diaz S, Cabido M, Casanoves F (1999) Functional implications of trait-environment linkages in plant communities. In: Weiher E, Keddy PA (eds) Ecological assembly rules. Cambridge University Press, Cambridge

Didham RK, Tylianakis JM, Gemmell NJ et al (2007) Interactive effects of habitat modification and species invasion on native species decline. Trends Ecol Evol 22:489–496

Doren RF, Whiteaker LD (1990) Effects of fire on different size individuals of *Schinus terbinthifolius*. Nat Areas J 10:107–113

Doren RF, Richards JH, Volin JC (2009a) A conceptual ecological model to facilitate understanding the role of invasive species in large-scale ecosystem restoration. Ecol Indic 9:S150–S160

Doren RF, Volin JC, Richards JH (2009b) Invasive exotic plant indicators for ecosystem restoration: an example from the Everglades restoration program. Ecol Indic 9:S29–S36

Ehrenfeld JG (2003) Effects of exotic plant invasions on soil nutrient cycling processes. Ecosystems 6:503–523

Ehrenfeld JG, Ravit B, Elgersma K (2005) Feedback in the plant-soil system. Ann Rev Environ Res 30:75–115

Eppinga MB, Rietkerk M, Dekker SC et al (2006) Accumulation of local pathogens: a new hypothesis to explain exotic plant invasions. Oikos 114:168–176

Fensham RJ, Fairfax RJ, Cannell RJ (1994) The invasions of *Lantana camara* L. in Forty-Mile Scrub National Park, north Queensland. Aust J Ecol 19:297–305

Folke C, Carpenter S, Walker B et al (2004) Regime shifts, resilience, and biodiversity in ecosystem management. Ann Rev Ecol Evol Syst 35:557–581

Fox BJ, Taylor JE, Fox MD et al (1997) Vegetation changes across edges of rainforest remnants. Biol Conserv 82:1–13

Foxcroft LC, Richardson DM (2003) Managing alien plant invasions in the Kruger National Park, South Africa. In: Child LE, Brock JH, Brundu G et al (eds) Plant invasions: ecological threats and management solutions. Backhuys Publishers, Leiden, pp 385–403

French K, Mason TJ, Sullivan N (2011) Recruitment limitation of native species in invaded coastal dune communities. Plant Ecol 212:601–609

Friedel MH (1991) Range condition assessment and the concept of thresholds – a viewpoint. J Range Manage 44:422–426

Funk JL, Cleland EE, Suding KN et al (2008) Restoration through reassembly: plant traits and invasion resistance. Trends Ecol Evol 23:695–703

Gaertner M, Richardson DM, Privett SDJ (2011) Effects of alien plants on ecosystem structure and functioning and implications for restoration: insights from three degraded sites in South African fynbos. Environ Manage 48:57–697

Gaertner M, Holmes PM, Richardson DM (2012) Biological invasions, resilience and restoration. In: van Andel J, Aronson J (eds) Restoration ecology: the new frontier. Wiley-Blackwell, Oxford, pp 265–280

Galatowitsch S, Richardson DM (2005) Riparian scrub recovery after clearing of invasive alien trees in headwater streams of the Western Cape, South Africa. Biol Conserv 122:509–521

Gelbard JL, Belnap J (2003) Roads as conduits for exotic plant invasions in a semiarid landscape. Conserv Biol 17:420–432

Gentle CB, Duggin JA (1997) *Lantana camara* L. invasions in dry rainforest open forest ecotones: the role of disturbances associated with fire and cattle grazing. Aust J Ecol 22:298–306

Gleason HA (1926) The individualistic concept of the plant association. Bull Torr Bot Club 53:7–26

Glenn-Lewin DC, Peet RK, Veblen TT (eds) (1992) Plant succession. Theory and prediction. Chapman & Hall, London

Gooden B, French K, Turner PJ et al (2009) Impact threshold for an alien plant invader, *Lantana camara* L., on native plant communities. Biol Conserv 142:2631–2641

Greer KA (2005) Habitat conservation planning in San Diego County, California: lessons learned after five years of implementation. Environ Pract 6:230–239

Hawkes CV, Wren IF, Herman DJ et al (2005) Plant invasion alters nitrogen cycling by modifying the soil nitrifying community. Ecol Lett 8:976–985

Hobbs RJ, Harris JA (2001) Restoration ecology: repairing the Earth's ecosystems in the new millennium. Restor Ecol 9:239–246

Hobbs RJ, Mooney HA (1995) Spatial and temporal variability in California annual grassland – results from a long-term study. J Veg Sci 6:43–56

Hobbs RJ, Richardson DM (2011) Invasion ecology and restoration ecology: parallel evolution in two fields of endeavour. In: Richardson DM (ed) Fifty years of invasion ecology: the legacy of Charles Elton. Blackwell Publishing, West Sussex, pp 61–69

Hobbs RJ, Jentsch A, Temperton VM (2007) Restoration as a process of assembly and succession mediated by disturbance. In: Walker LR, Walker J, Hobbs RJ (eds) Linking restoration and ecological succession. Springer, New York, pp 150–167

Hocking M, Stolton S, Dudley N (2000) Evaluating effectiveness: a framework for assessing the management of protected areas. Best practice protected area guidelines, vol 6. IUCN, Gland/Cambridge

Holl KD, Aide TM (2011) When and where to actively restore ecosystems? For Ecol Manage 261:1558–1563

Holling CS (1973) Resilience and stability of ecological systems. Ann Rev Ecol Syst 4:1–23

Inderjit, van der Putten WH (2010) Impacts of soil microbial communities on exotic plant invasions. Trends Ecol Evol 25:512–519

James JJ, Smith BS, Vasquez EA et al (2010) Principles for ecologically based invasive plant management. Invasion Plant Sci Manage 3:229–239

Jones HP, Schmitz OJ (2009) Rapid recovery of damaged ecosystems. Plos One 4:e5653

Jones CG, Lawton JH, Shachak M (1994) Organisms as ecosystem engineers. Oikos 69:373–386

Keenleyside KA, Dudley N, Cairns S et al (2012) Ecological restoration for protected areas: principles, guidelines and best practices. IUCN, Gland

Kelly M (2000) *Cynara cardunculus*. In: Bossard CC, Randall JM, Hoshovsky MC (eds) Invasive plants of California's wildlands. University of California Press, Berkeley, pp 139–145

Kettenring KM, Adams CR (2011) Lessons learned from invasive plant control experiments: a systematic review and meta-analysis. J Appl Ecol 48:970–979

Klironomos JN (2002) Feedback with soil biota contributes to plant rarity and invasiveness in communities. Nature 417:67–70

Kulmatiski A (2006) Exotic plants establish persistent communities. Plant Ecol 187:261–275

Le Maitre DC, Versfeld DB, Chapman RA (2000) The impact of invading alien plants on surface water resources in South Africa: a preliminary assessment. Water SA 26:397–408

Le Maitre DC, Gaertner M, Marchante E et al (2011) Impacts of invasive Australian Acacias: implications for management and restoration. Divers Distrib 17:1015–1029

Levine JM, D'Antonio CM (2003) Forecasting biological invasions with increasing international trade. Conserv Bio 17:322–326

Levine JM, Murrell DJ (2003) The community-level consequences of seed dispersal patterns. Ann Rev Ecol Evol Syst 34:549–574

Levine JM, Vilà M, D'Antonio CM et al (2003) Mechanisms underlying the impacts of exotic plant invasions. Proc R Soc Lond Ser B-Biol Sci 270:775–781

Lockwood JL, Samuels CL (2004) Assembly models and the practice of restoration. In: Temperton VM, Hobbs RJ, Nuttle T et al (eds) Assembly rules and restoration ecology: bridging the gap between theory and practice. Island Press, Washington, DC, pp 55–70

Lockwood M, Worboys GL, Kothari A (eds) (2006) Managing protected areas. Earthscan, London

Lowe SJ, Browne M, Boudjelas S et al (2000) 100 of the world's worst invasive alien species from the Global Invasive Species Database. Invasive Species Specialist Group (ISSG) a specialist group of the Species Survival Commission (SSC) of the World Conservation Union (IUCN), Auckland, New Zealand

MacArthur R, Levins R (1967) Limiting similarity convergence and divergence of coexisting species. Am Nat 101:377–385

Macdonald IAW (1984) Is the fynbos biome especially susceptible to invasion by alien plants – a re-analysis of available data. S Afr J Sci 80:369–377

Macdonald IAW (2004) Recent research on alien plant invasions and their management in South Africa: a review of the inaugural research symposium of the working for water programme. S Afr J Sci 100:21–26

Macdonald IAW, Richardson DM (1986) Alien species in terrestrial ecosystems of the fynbos biome. In: Macdonald IAW, Kruger FJ, Ferrar AA (eds) The ecology and management of biological invasions in Southern Africa. Oxford University Press, Cape Town, pp 77–91

Macdonald IAW, Graber DM, Debenedetti S et al (1988) Introduced species in nature reserves in Mediterranean-type climatic regions of the world. Biol Conserv 44:37–66

Mack MC, D'Antonio CM (1998) Impacts of biological invasions on disturbance regimes. Trends Ecol Evol 13:195–198

Mack MC, D'Antonio CM, Ley RE (2001) Alteration of ecosystem nitrogen dynamics by exotic plants: a case study of C-4 grasses in Hawaii. Ecol Appl 11:1323–1335

Malmstrom CM, Hughes CC, Newton LA et al (2005) Virus infection in remnant native bunch-grasses from invaded California grasslands. New Phytol 168:217–230

Mangla S, Inderjit, Callaway RM (2008) Exotic invasive plant accumulates native soil pathogens which inhibit native plants. J Ecol 96:58–67

Marushia RG, Holt JS (2006) The effects of habitat on dispersal patterns of an invasive thistle, *Cynara cardunculus*. Biol Invasion 8:577–593

McAfee L (2008) Nature reserve of orange county annual report. Nature Reserve of Orange County, Irvine

McConkey KR, Prasad S, Corlett RT et al (2012) Seed dispersal in changing landscapes. Biol Conserv 146:1–13

McKinney ML, Lockwood JL (1999) Biotic homogenization: a few winners replacing many losers in the next mass extinction. Trends Ecol Evol 14:450–453

McNeely JA (2001) Invasive species: a costly catastrophe for native biodiversity. Land Use Water Resour Res 1:1–10

McNeely JA (2006) As the world gets smaller, the chances of invasion grow. Euphytica 148:5–15

Milton SJ, Hall AV (1981) Reproductive biology of Australian acacias in the southwestern Cape-Province, South Africa. Trans Roy Soc S Afr 44:465–485

Murcia C (1997) Evaluation of Andean alder as a catalyst for the recovery of tropical cloud forests in Colombia. For Ecol Manage 99:163–170

Padgett PE, Allen EB (1999) Differential responses to nitrogen fertilization in native shrubs and exotic annuals common to Mediterranean coastal sage scrub of California. Plant Ecol 144:93–101

Pearson J, Wells DM, Seller KJ et al (2000) Traffic exposure increases natural (15)N and heavy metal concentrations in mosses. New Phytol 147:317–326

Perry LG, Johnson C, Alford ER et al (2005) Screening of grassland plants for restoration after spotted knapweed invasion. Restor Ecol 13:725–735

Peters DPC, Bestelmeyer BT, Turner MG (2007) Cross-scale interactions and changing pattern-process relationships: consequences for system dynamics. Ecosystems 10:790–796

Pickett STA, Collins SL, Armesto JJ (1987) A hierarchical consideration of causes and mechanisms of succession. Vegetation 69:109–114

Pickett STA, Kolasa J, Armesto JJ et al (1989) The ecological concept of disturbance and its expression at various hierarchical levels. Oikos 54:129–136

Pickett STA, Cadenasso ML, Meiners SJ (2009) Ever since Clements: from succession to vegetation dynamics and understanding to intervention. Appl Veg Sci 12:9–21

Potts DL, Harpole WS, Goulden ML et al (2008) The impact of invasion and subsequent removal of an exotic thistle, *Cynara cardunculus*, on CO2 and H2O vapor exchange in a coastal California grassland. Biol Invasion 10:1073–1084

Prach K, Hobbs RJ (2008) Spontaneous succession versus technical reclamation in the restoration of disturbed sites. Restor Ecol 16:363–366

Prach K, Walker LR (2011) Four opportunities for studies of ecological succession. Trends Ecol Evol 26:119–123

Prach K, Bartha S, Joyce CB et al (2001) The role of spontaneous vegetation succession in ecosystem restoration: a perspective. Appl Veg Sci 4:111–114

Prach K, Marrs R, Pyšek P et al (2007) Manipulation of succession. In: Walker LR, Walker J, Hobbs RJ (eds) Linking restoration and ecological succession. Springer, New York, pp 121–149

Prober SM, Lunt ID, Morgan JW (2009) Rapid internal plant-soil feedbacks lead to alternative stable states in temperate Australian grassy woodlands. In: Hobbs RJ, Suding KN (eds) New models for ecosystem dynamics and restoration. Island Press, Washington, DC, pp 156–168

Sala OE, Chapin FS, Armesto JJ et al (2000) Biodiversity – global biodiversity scenarios for the year 2100. Science 287:1770–1774

Seabloom EW (2011) Spatial and temporal variability in propagule limitation of California native grasses. Oikos 120:291–301

Seabloom EW, Borer ET, Boucher VL et al (2003) Competition, seed limitation, disturbance, and reestablishment of California native annual forbs. Ecol Appl 13:575–592

Seastedt TR, Hobbs RJ, Suding KN (2008) Management of novel ecosystems: are novel approaches required? Front Ecol Environ 6:547–553

Sheley RL, Mangold JM, Anderson JL (2006) Potential for successional theory to guide restoration of invasive-plant-dominated rangeland. Ecol Monogr 76:365–379

Sheley R, James J, Smith B et al (2010) Applying ecologically based invasive-plant management. Rangel Ecol Manage 63:605–613

Sirulnik AG, Allen EB, Meixner T et al (2007) Impacts of anthropogenic N additions on nitrogen mineralization from plant litter in exotic annual grasslands. Soil Biol Biochem 39:24–32

Sperry LJ, Belnap J, Evans RD (2006) *Bromus tectorum* invasion alters nitrogen dynamics in an undisturbed arid grassland ecosystem. Ecology 87:603–615

Stock D (2004) The dynamics of *Lantana camara* (L.) invasion of subtropical rainforest in Southeastern Australia. Griffith University, Gold Coast Campus, Brisbane

Stock D (2005) Management of *Lantana* in eastern Australia subtropical rainforests. In: Worboys GL, Lockwood M, De Lacy T (eds) Protected area management: principles and practices, 2nd edn. Oxford University Press, Melbourne

Suding KN (2011) Toward an era of restoration in ecology: successes, failures, and opportunities ahead. Ann Rev Ecol Evol Syst 42:465–487

Suding KN, Hobbs RJ (2009) Threshold models in restoration and conservation: a developing framework. Trends Ecol Evol 24:271–279

Suding KN, Gross KL, Houseman GR (2004) Alternative states and positive feedbacks in restoration ecology. Trends Ecol Evol 19:46–53

Swarbick JT (1986) History of the lantanas in Australia and origins of the weedy biotypes. Plant Protect Quart 1:115–121

Temperton VM, Hobbs RJ, Nuttle T et al (eds) (2004) Assembly rules and restoration ecology: bridging the gap between theory and practice. The science and practice of ecological restoration. Island Press, Washington, DC

Thomsen C, Barbe G, Williams W et al (1986) "Escaped" artichokes are troublesome pests. Calif Agric 40:7–9

Tilman D (1988) Monographs in population biology No 26. Plant strategies and the dynamics and structure of plant communities. Princeton University Press, Princeton

Tilman D (1990) Constraints and tradeoffs – toward a predictive theory of competition and succession. Oikos 58:3–15

Traveset A, Richardson DM (2006) Biological invasions as disruptors of plant reproductive mutualisms. Trends Ecol Evol 21:208–216

Turpie JK, Marais C, Blignaut JN (2008) The working for water programme: evolution of a payments for ecosystem services mechanism that addresses both poverty and ecosystem service delivery in South Africa. Ecol Econ 65:788–798

UNEP-SCBD (2001) Global biodiversity outlook. Paper presented at the UNEP Secretariat of the convention on biological diversity, Montreal, Canada

van der Putten WH, Klironomos JN, Wardle DA (2007) Microbial ecology of biological invasions. Isme J 1:28–37

Vitousek PM, Walker LR (1989) Biological invasion by *Myrica-Faya* in Hawaii – plant demography, nitrogen-fixation, ecosystem effects. Ecol Monogr 59:247–265

Vitousek PM, Dantonio CM, Loope LL et al (1997) Introduced species: a significant component of human-caused global change. N Z J Ecol 21:1–16

Vivrette NJ, Muller CH (1977) Mechanism of invasion and dominance of coastal grassland by *Mesembryanthemum crystallinum*. Ecol Monogr 47:301–318

Weiher E, Keddy PA (1995) Assembly rules, null models, and trait dispersion – new questions from old patterns. Oikos 74:159–164

Whisenant SG (1990) Changing fire frequencies on Idaho's Snake River Plains: ecological and management implications. In: Proceedings-symposium on cheatgrass invasion, shrub die-off, and other aspects of shrub biology and management, Las Vegas, pp 4–10

Whisenant S (1999) Repairing damaged wildlands: a process-oriented, landscape scale approach. Cambridge University Press, Port Chester

White PS, Jentsch A (2004) Disturbance, succession, and community assembly in terrestrial plant communities. In: Temperton VM, Hobbs RJ, Nuttle T et al (eds) Assembly rules and restoration ecology: bridging the gap between theory and practice. Island Press, Washington, DC, pp 342–366

Williams PA, Karl BJ (2002) Birds and small mammals in kanuka (*Kunzea ericoides*) and gorse (*Ulex europaeus*) scrub and the resulting seed rain and seedling dynamics. N Z J Ecol 26:31–41

Williamson M (1996) Biological invasions. Population and community biology. Chapman & Hall, Cornwall

Wojterski TW (1990) Degradation stages of the oak forests in the area of Algiers. Vegetation 87:135–143

Yelenik SG, Stock WD, Richardson DM (2004) Ecosystem level impacts of invasive *Acacia saligna* in the South African fynbos. Restor Ecol 12:44–51

Young TP, Chase JM, Huddleston RT (2001) Community succession and assembly: comparing, contrasting and combining paradigms in the context of ecological restoration. Ecol Restor 19:5–18

Young TP, Petersen DA, Clary JJ (2005) The ecology of restoration: historical links, emerging issues and unexplored realms. Ecol Lett 8:662–673

Zavaleta E (2000) The economic value of controlling an invasive shrub. Ambio 29:462–467

Zavaleta ES, Hobbs RJ, Mooney HA (2001) Viewing invasive species removal in a whole-ecosystem context. Trends Ecol Evol 16:454–459

Part IV
Conclusion

Chapter 28
Invasive Alien Plants in Protected Areas: Threats, Opportunities, and the Way Forward

Llewellyn C. Foxcroft, David M. Richardson, Petr Pyšek, and Piero Genovesi

Abstract The potential threats posed by biological invasions are widely appreciated, but the state of knowledge and level of management of invasive alien plants in protected areas differs considerably across the world. Research done on nature reserves as part of the international SCOPE programme on biological invasions in the 1980s showed the vulnerability of natural or undisturbed areas to invasions. Subsequent work, including the chapters in this book, shows the serious situation regarding plant invasions that prevails in many protected areas. Many invasive plants have, or have the potential to, greatly lessen the potential of protected areas to achieve the things they were proclaimed to do – provide refugia for species,

L.C. Foxcroft (✉)
Conservation Services, South African National Parks,
Private Bag X402, Skukuza 1350, South Africa

Centre for Invasion Biology, Department of Botany and Zoology,
Stellenbosch University, Private Bag X1, Stellenbosch 7602, South Africa
e-mail: Llewellyn.foxcroft@sanparks.org

D.M. Richardson
Centre for Invasion Biology, Department of Botany and Zoology,
Stellenbosch University, Private Bag X1, Stellenbosch 7602, South Africa
e-mail: rich@sun.ac.za

P. Pyšek
Department of Invasion Ecology, Institute of Botany, Academy of Sciences of the Czech
Republic, Průhonice CZ 252 43, Czech Republic

Department of Ecology, Faculty of Science, Charles University in Prague,
CZ 128 44 Viničná 7, Prague 2, Czech Republic
e-mail: pysek@ibot.cas.cz

P. Genovesi
ISPRA, Institute for Environmental Protection and Research,
Via V. Brancati 48, I-00144 Rome, Italy

Chair IUCN SSC Invasive Species Specialist Group, Rome, Italy
e-mail: piero.genovesi@isprambiente.it

L.C. Foxcroft et al. (eds.), *Plant Invasions in Protected Areas: Patterns, Problems and Challenges*, Invading Nature - Springer Series in Invasion Ecology 7, DOI 10.1007/978-94-007-7750-7_28, © Springer Science+Business Media Dordrecht 2013

habitats and the ecosystem services that they sustain. This brief synthesis discusses some emerging insights from protected areas of varying kinds and sizes, from the Azores, Australia, Chile, East and South Africa, Europe, Galapagos, India, Mediterranean Islands, New Zealand, Pacific Islands and Hawaii, Southern Ocean Islands, United States of America and the Western Indian Ocean Islands. Work in some protected areas has led to well-developed management and policy frameworks. In others, important insights have emerged on invasion mechanisms and the impacts of invasions. Although there is awareness of invasive alien plants in most of the 135 protected areas mentioned in this volume, better and more focused actions are urgently needed. This requires, among other things, improved capacity to prevent invasions and to react promptly to new incursions, and increasing efforts to manage well-established invasive species. Research to improve the understanding of invasion dynamics is essential. Full species lists are available only for a group of well-known protected areas. Updating species lists and distribution data is crucial for successful long-term management, as are collaborative networks, research groups, volunteers, and improved accessibility to resources such as online databases. Efforts to lessen the science-management divide are especially important in protected areas. One reason is that managers are usually required to implement invasive alien plant control programmes as part of general protected area management activities, and in many cases lack the knowledge and support for effective science-based management solutions. Overcoming this barrier is not trivial and will require partnerships between local, municipal, regional and national-level organizations and international non-profit NGOs and donor organisations.

Keywords Biological invasions • Impacts • Invasive alien plants • Non-native plants • Protected areas

28.1   Introduction

Many books (e.g. Cadotte et al. 2006; Nentwig 2007; Davis 2009; Richardson 2011) and more than 800 journal articles per year for the last 4 years (Fig. 1.2 in Chap. 1 Foxcroft et al. 2014b) attest to the huge interest in biological invasions. Indeed invasion biology has grown rapidly to become a strong and vibrant field within the ecological and conservation sciences. The main aim of this book was to provide a deeper understanding of the extent and dimensions of invasive alien plants (IAPs) in protected areas (PAs) – the pillars on which many conservation efforts are built and rely. Although conservation measures outside formally protected areas are being given more attention (e.g. Ervin et al. 2010), many conservation aims can only be achieved in areas that enjoy special levels of protection. Protected areas are therefore usually seen as core conservation areas and afforded the highest priority to, for example, maintain functional systems and populations from which species may disperse. Biological invasions are widely seen as one of the major threats to biodiversity in general, and as chapters in this book indicate, in many places also directly to PAs. In many cases, the extent of invasions and the levels of impacts due to invasive species have exploded in recent decades. Traditional measures of protection and intervention do not appear to limit the impact of invasions within PAs. Another problem is that PAs are increasingly becoming embedded in a matrix of human-modified landscapes which is creating an increasingly sharp interface between 'natural' and highly modified ecosystems. Among other things, this creates a continuous source of propagules for invasion. It is also not only the physical location of PAs that affects potential threats. Protected areas do not always enjoy high priority for the allocation of financial resources and are, in many cases, expected to generate their own resources, usually through tourism. Within PAs, limited resources are increasingly stretched across many aspects of management. Invasive alien plants often end up low on the list of priorities, notwithstanding their ability to cause impacts on basic ecosystem processes, for example biogeochemical cycling and fire (Chap. 2 Foxcroft et al. 2014a) and shifting PAs into alternative states that even active restoration may be unable to reverse (Chap. 27 Larios and Suding 2014).

The chapters in this book were grouped into three parts which sought to: (i) synthesise insights on plant invasions in PAs and integrate these with current models and theories of plant invasion ecology, (ii) determine the status of knowledge of IAPs in PAs, and (iii) determine key knowledge areas for informing the development of successful management strategies.

In the first part we asked authors to explore ways in which PAs could and have provided unique opportunities for gaining insights into broadly encompassing themes. Chapters in this part also show how work in PAs has led to advances in the field. These topics include the role of PAs for developing further understanding of plant invasions and succession in natural systems, impacts of IAPs, the invasion of mountain ecosystems and large scale monitoring programmes.

The 14 case studies in Part II aimed to capture experiences from PAs in different settings and regions. These focus on a range of island systems (e.g. Pacific, Western Indian, Southern Ocean; Mediterranean) and PAs in continental regions of Africa, Australia, Chile, Europe, the USA and other areas. The contributions synthesise diverse insights on IAPs from PAs of different kinds and sizes in many different environmental settings, and detail what has been learned from research and management experiences in these areas. The case studies also examined the specific contexts of the systems and their unique attributes, and explore the extent to which aspects such as modes and pathways of introduction and dispersal, impacts on biodiversity, the role of natural disturbance regimes and anthropogenic disturbance, define natural laboratories for examining questions that cannot easily be studied in other regions.

The management chapters round off the book by collating insights from general approaches, integrating them with experiences and context-specific examples. This section does not present a manual on how to clear IAPs, or lists of the techniques and herbicides to apply, as these are well documented elsewhere. Rather, the chapters provide in-depth syntheses of specific fields (e.g. biological control, Chap. 26 Van Driesche and Center 2014), examine the arguments for well-known but under-implemented approaches (e.g. eradication, Chap. 25 Simberloff 2014), while others offer novel methods for implementing actions within the PA context (rehabilitation, Chap. 27 Larios and Suding 2014; outcome-based planning, Chap. 23 Downey 2014; prevention, Chap. 21 Meyerson and Pyšek 2014). Two chapters discuss the challenges of managing IAPs in PAs more generally (Chap. 22 Genovesi and Monaco 2014; Chap. 24 Tu and Robison 2014). Also emphasised is the role of PAs beyond their borders, stressing the need to improve prevention and prompt response to new incursions. Collectively these chapters provide PA managers, conservation biologists, invasion biologists and others with collated knowledge on a wide variety of elements required for formulating comprehensive approaches to managing IAPs.

28.2 Key Outcomes

In this chapter we focus on three key themes: (i) impacts, (ii) management and (iii) the role of PAs as focal research sites. The outcomes of these chapters provide interesting examples on how the knowledge collected in PAs and the growing understanding of impacts caused by IAPs, can guide more effective management of invasions.

28.2.1 Impacts

Impacts of IAPs have been demonstrated in almost all regions and a large number of examples show the severity of change that is likely if invasions continue unabated. Although the quantification of impacts on species, communities, landscapes, habitats, ecosystem dynamics and services is difficult to elucidate, a surprising amount of work has been done in PAs (Chap. 2 Foxcroft et al. 2014a). In some cases, complete switches in ecosystem states have been observed that are unlikely to be reversed even if the invader is removed ('legacy effects' e.g. D'Antonio and Meyerson 2002). While exaggerated 'scare tactics' are certainly not appropriate, this kind of information has the potential to illustrate the severity of problems that could occur more widely if IAPs are not effectively managed. Finding ways of presenting such information accurately, but in ways that are also compelling to the public and decision makers, is a key challenge.

At the same time that invasion science has started moving away from a focus on single species to ecosystem level alterations (Chap. 25 Simberloff et al. 2014), conservation biologists have also started shifting their focus from single species or habitat conservation towards a more ecosystems based approach (Ostfeld et al. 1997). Many organisations or individual PAs were not necessarily designed within this context, and in some cases PAs were especially designated as species-specific reserves (e.g. Kaziranga National Park established for the one-horned rhinoceros, *Rhinoceros unicornis*, and now threatened by IAPs altering feeding areas, Chap. 12 Hiremath and Sundaram 2014). In such cases focusing research on individual IAPs most likely to impact the species of concern is appropriate. In some cases assessing the impacts of IAPs on particular taxa as indicator species (commonly beetles and spiders, see examples in Chap. 2 Foxcroft et al. 2014a) can serve as indicators of wider ecosystem-level changes. Research on less visible impacts, for example plant-soil nutrient dynamics, are required to understand whole-ecosystem energy budgets and function (Ehrenfeld 2010), the alteration of which can lead to profound changes to PAs. Integration of species- and ecosystem-based approaches (Likens and Lindenmayer 2012) in future may however provide interesting and novel approaches for understanding and managing biological invasions.

28.2.2 Management Approaches

Biological invasions are a good example of what has come to be known as 'wicked problems' (Conklin 2005). They are inherently complex and there is no single, easy or correct answer to management problems. There are numerous stakeholders, from local to global scales, with different perceptions and personal values, and economic incentives (e.g. increasing international trade), which greatly complicate the formulation of common goals (Chap. 4 McNeely 2014).

There has been a shift away from a focus on techniques for controlling IAPs to approaches for defining priorities, examining when and where active or passive restoration is required, defining the appropriate setting and deciding on how to achieve conservation outcomes. Further innovations are needed to guide the implementation of control measures with shrinking resources. Collaborations and knowledge sharing can lead to improved IAP management. Strategies need to focus on determining biodiversity areas most at risk and optimising approaches for conservation of these, rather than focussing on the control itself (Chap. 23 Downey 2014). Many hurdles need to be negotiated in formulating and implementing a successful IAP management programme; many sources of information are now available to assist in the development of effective plans. Once the foundations of a management approach have been laid, the process can evolve as more resources become available, and knowledge and management expertise is developed (Chap. 22 Genovesi and Monaco 2014; Chap. 24 Tu and Robison 2014).

Although IAP management programmes are often built into PA management plans, long-term success depends on whether the affected system can recover after control, or whether further intervention is needed to aid recovery (Chap. 27 Larios and Suding 2014). Preventing the introduction and spread of IAPs into PAs is an essential management strategy, even in the current era of global change. Management of invasion pathways and propagule pressure is an emerging preventative measure (Chap. 21 Meyerson and Pyšek 2014). While the literature abounds with information on eradication as the most effective strategy for managing alien plant invasions, many managers and scientists remain pessimistic about the feasibility of eradication. However, if well-funded, planned and implemented, once-off eradication campaigns can make projects that appear impossible a reality (Chap. 25 Simberloff 2014). In regions where sufficient resources are unlikely to be procured over the long term, biological control should be a key component of the overall management strategy. Once tested and released, biocontrol agents can be used over large areas and in places that are difficult to reach, at low cost (Chap. 26 Van Driesche and Center 2014). There is considerable scope for sharing of experiences with biological control across PAs.

Monitoring is one of the most important aspects of an overall IAP management programme, and probably also in PA management generally. Monitoring is needed to detect incursions by new species and populations (surveillance), track the status and extent of invasions, determine progress of control operations, and assess the outcome of eradication attempts and long-term maintenance programmes. Large PAs, however, present particular problems for effective monitoring, due to need for robust, cost effective, but rapid assessments. Protected areas provide opportunities for developing these concepts, and to test their efficacy in the field, for example using data from routine ranger patrols that are available in many PAs (Chap. 5 Hui et al. 2014).

28.2.3   Protected Areas as Science Hubs

There are many examples where PAs have formed the nucleus of science research programmes (e.g. Hawaii, Chap. 15 Loope et al. 2014; Galapagos, Chap. 16 Gardener et al. 2014; South African National Parks, Chap. 7 Foxcroft et al. 2014c), and where context-specific science can be of direct relevance to local PA management authorities. Motivations for forming research hubs with PAs range from, for example, (i) provision of good logistical support, (ii) employment of scientists within PAs actively carrying out studies, and forging collaborative programmes with other institutions, (iii) at the request of PA managers for support, or (iv) interest by scientists in a particular area and/or problem, and (v) in studying ecological processes in relatively undisturbed environments.

In Haleakala and Hawaii Volcanoes NPs (Chap. 15 Loope et al. 2014), species lists (more than 300 alien plants in Haleakala and 600 in Hawaii Volcanoes) and distribution data have been extensively assessed. Much research has been carried out in this PA, providing a very detailed understanding of a few species (*Morella faya*, faya tree; *Hedychium gardnerianum*, Kahili Ginger), a good practical understanding of impacts for the most threatening invaders, and helping to inform management efforts (Table 28.1). In another example, the Charles Darwin Research Station, Galapagos National Park and other institutions, have collaborated to develop a scientific basis to guide management of IAPs in Galapagos (Chap. 16 Gardener et al. 2014). The programme comprised a number of components including: baseline inventories, quarantine development, while the science programme focused on, for example, impacts of *Cinchona pubescens* (quinine tree) and *Rubus*

Table 28.1 Examples from selected protected areas as indicators of the status of plant invasions

Protected area	Species lists and distribution	Level of management	Level of scientific input/involvement	Understanding of impacts
Amani Nature Reserve, Tanzania (Chap. 8 Hulme et al. 2014)	Detailed species lists and distribution data.	No existing plans to manage invasive species in Amani Nature Reserve or the sources of invaders in Amani Botanical Garden.	Work on propagule pressure as major driver of invasion. Role of Amani Botanical Garden as a source of invasive species. Weed risk assessment tools (e.g. Australian Weed Risk Assessment).	Impact of *Maesopsis eminii* on native species recruitment and succession.
Biligiri Rangaswamy Temple Tiger Reserve, India (Chap. 12 Hiremath and Sundaram 2014)	Long-term monitoring of *Lantana camara* spread and changes in density.	Ad hoc small-scale *L. camara* removal by uprooting, and cutting-and-burning, but no systematic removal and restoration plan in place.	Work on drivers of invasive species spread. Experimental work on different weed removal techniques to assess relative efficacy and to determine barriers to native species restoration.	Effect of *L. camara* and *Chromolaena odorata* growth on species diversity, regeneration and stem density of tree and shrub layer.
Fiordland NP, New Zealand (Chap. 14 West and Thomson 2014)	Detailed species lists and distribution data exist for coastal locations and many specific locations within the NP.	Site-specific control until the mid-1990s when control to zero-density was initiated for a number of specific species, e.g. *Ammophila arenaria*, *Ulex europaeus*, *Cytisus scoparius*.	Field staff have access to advice from technical and scientific experts within the Department of Conservation as well as externally from research agencies and universities.	Impacts of some species are understood well, e.g. *U. europaeus*, *A. arenaria*, *Salix fragilis*, but understood less well for others, e.g. *Hypericum androsaemum*.
Galapagos NP Service, Galapagos (Chap. 16 Gardener et al. 2014)	Systematic once-off survey of inhabited areas adjacent to protected areas. Ad hoc collection in protected areas. To date, a total of about 871 alien species and 33,000 records.	On-going management since early 1980s. Between 2005 and 2010, 17 projects on five islands covering 11 species (~2,000 ha).	Since mid-1990s pragmatic research based approach used to: (i) understand the problem; (ii) developing management tools; and (iii) address the challenges of implementation.	Impacts of *Cinchona pubescens* and *Rubus niveus* on biodiversity and ecosystem function. Role of alien plants in pollination and seed-dispersal networks.

(continued)

Table 28.1 (continued)

Protected area	Species lists and distribution	Level of management	Level of scientific input/involvement	Understanding of impacts
Haleakala NP, Hawaii (Chap. 15 Loope et al. 2014)	Species and distribution data well developed (> 300 alien plants).	Modest within-Park control efforts, hindered by remoteness of high biodiversity rainforest. Substantial collaborative island-wide strategic proactive efforts to contain or eradicate encroaching invasions from outside the NP.	Research has helped inform and prioritise management efforts.	Good practical understanding of impacts for the most threatening invaders.
Hawaii Volcanoes NP, Hawaii (Chap. 15 Loope et al. 2014)	Species and distribution data well-developed and up-to-date (> 600 alien plants of which > 400 are naturalised).	Intensive strategic focus on alien plant control in 'Special Ecological Areas' as well as eradication or containment of selected species park wide.	Research has informed and helped to prioritise management efforts. Biological control research has been substantial; benefits are as yet small but opportunities may be large.	Impact of a few species exceptionally well understood through research (*Morella faya*, *Hedychium gardnerianum*, grasses), good practical understanding of others.
Kakadu NP, Australia (Chap. 9 Setterfield et al. 2014)	On-going surveys and collection of alien plant occurrence data.	Two dedicated alien plant control teams (*Mimosa pigra* and grassy weeds). Additional management undertaken by rangers.	Research undertaken by external agencies (universities, CSIRO) in collaboration with NPs. Major research project currently on predicting and managing spread of floodplain grassy weeds. Previous research on *Salvinia molesta* and *M. pigra*. Research Committee oversees research activity.	A major project has been completed in Kakadu NP on impact of *Urochloa mutica*.

Kruger NP, South Africa (Chap. 7 Foxcroft et al. 2014c)	On-going surveys, species list well developed (~350 alien plants). Distribution data covers entire park, with about 30,000 point records.	Before 1996, small but long-term control efforts. After 1996, substantial programme with at times several hundred employees, funded by Working for Water. Main focus on woody shrubs along main rivers, aquatic species and *Opuntia stricta*.	Since 2001, full time invasion biologist appointed. Work on biocontrol post release evaluation, risk analysis, ornamental plant invasions and boundaries as barriers to invasion to support management efforts.	Mostly anecdotal evidence. Some research done on impacts of *O. stricta* on biodiversity indicators.
Table Mountain NP, (Chap. 7 Foxcroft et al. 2014c)	On-going surveys, species lists well developed (~239 alien plants). Fine scale distribution data covers most of the park.	Long history of control efforts, although initially unsuccessful. In the last 15 years Working for Water has supported a substantial programme, investing about US$1,7 million annually.	Much research has been conducted in TMNP generally, with increasing focus on IAPs. Partnerships between academic researchers, managers, policy makers and funders facilitated an integrated strategic approach to implementation of control programmes.	Work has focused on IAPs changing fire behaviour, erosion, and potential extinction of rare and endangered species.
Western Indian Ocean Islands (Chap. 19 Baret et al. 2014)	Lists of all alien plant species present within terrestrial protected habitats.	Varies according to countries.	From early detection and rapid response, to on-going eradication of large propagation plants.	Variable between country and type of protected areas.

niveus (Ceylon raspberry) on biodiversity and ecosystem function and the role of IAPs in pollination and seed-dispersal networks (Table 28.1).

In general, international collaborative networks provide opportunities for exploring the impacts and role of IAPs in different settings, and achieving a broader synthetic understanding of invasions. Some of these recent efforts have been conducted with a strong conservation focus. For example, many mountain systems worldwide have been designated as national parks and reserves (Spehn et al. 2006), and one third of the world's PAs are in mountainous regions (Chap. 6 Kueffer et al. 2014). Mountains are important for people, biodiversity and for providing ecosystem services, but they are often considered as having a low risk of invasion. However, a global network (MIRN) examining IAPs in 11 mountain ranges showed that mountains are not inherently more resistant to invasion. Moreover, future risks will increase with global warming and continuing human land use and expansion (Chap. 6 Kueffer et al. 2014). It is important to link these focused projects to the global efforts to increase the inter-operability of information systems on IAS, such as the Global Invasive Alien Species Information Partnership, launched by the Convention on Biological Diversity (Chap. 22 Genovesi and Monaco 2014).

28.3   Glaring Gaps and Serious Shortcomings

Useful insights and experiences in at least some facets of policy, management, or scientific investigation of IAPs emerged from PAs in all regions. However, some aspects emerged as fairly general problems or gaps, which are impeding the implementation of effective IAP management programmes. Such issues relate to incomplete data on IAPs, poor implementation of research outputs in management programmes, and the shortage of financial resources.

28.3.1   Species Lists and Distribution Maps: The Urgency of Systematic Surveys

Inventories of IAPs and of the key correlates of invasions, including geographic data, are crucial requirements for robust management plans. While the situation is improving, some areas still lack even basic information on invasive plants in PAs, for example in many African PAs (Chap. 7 Foxcroft et al. 2014c). This is in accordance with global distribution of knowledge of invasions in general. A recent assessment of regional contributions to invasion ecology found a low representation of developing countries, and Asia and Africa (with the exception of South Africa) were found to be markedly understudied (Pyšek et al. 2008). In a literature review of ecology and biodiversity conservation, only 15.8 % of all published papers related to alien species had authors from developing countries, and only 6.5 %

had authors solely from developing countries (Nuñez and Pauchard 2010). In general, the number of documented invasive species also gives a significant under-estimate of the magnitude of the problem, because its value is negatively affected by country development status (McGeoch et al. 2010). However, surprisingly also in Europe, one of the most intensively studied continents (Pyšek et al. 2008), systematically collected information on IAPs in PAs does not exist (Chap. 11 Pyšek et al. 2014).

The paucity of essential baseline information, including species lists and distri-bution data impedes development of both local and regional scale prevention, management and monitoring approaches (McGeoch et al. 2012). Potential errors common in compiling lists of alien species are often unrecognised (McGeoch et al. 2012) and need to be taken into account as these data start being collected and compiled. This uncertainty can be compounded as the lists are used at different scales and for different purposes, for example, local scale management vs. reporting globally under the CBD Biodiversity targets across regions (Pyšek et al. 2008). While many areas are currently developing or expanding such datasets, cognisance of the potential errors and future use of the data from the outset will greatly increase the value of the data. A number of potential errors that can lead to uncertainty in IAP data include obtaining globally comparable data for individual PAs, non-systematic research, human error and taxonomic issues resulting in species misidentification, and variations in criteria applied to designate species as invasive (McGeoch et al. 2012). While some level of uncertainty is likely to be unavoidable, when determining the primary use of the data, strategies for reducing the level of errors need to be included.

While many regions have species lists, albeit usually incomplete and often only considering the most problematic and visible invasive species, there is generally an absence of distribution data, or where present, it is poor. Such information is essential for developing management and monitoring programmes. A number of tools that facilitate collection of spatial data are becoming more readily available, including free software that can be customised to the needs of a particular PA (e.g. CyberTracker, Kruger and MacFadyen 2011, and Geoweed, http://geoweed.org).

Not only is collection of data important, but data warehousing – a complex process – needs to be carefully managed to ensure that data are updated and readily available. A number of approaches are already in place in various regions and could serve as model systems for PAs, or by including specific PA-related modules into existing structures. Some examples are the DASIE project (Delivering Alien Invasive Species Inventories for Europe; DAISIE 2009; http://www.europe-aliens.org), CAB Interna-tional's (CABI) Invasive Species Compendium (www.cabi.org/ISC), the IUCN Inva-sive Species Specialist Group Global Database (http::/http://www.issg.org/database), the South African Plant Invader Atlas (SAPIA, Henderson 1998; http://www.agis.agric.za/wip), and a number of freely available, customisable databases (e.g. Pl@ntnote; http://www.plantnet-project.org/page:tools?langue=en). However while these tools can be effective in developed countries where the required infra-structure is available, in developing countries access to internet based data is often problematic and approaches for overcoming these challenges need to be sought. These may be alleviated, at least to some extent, through collaborative networks.

28.3.2 Bridging the Science–Management Divide

While the scientific aspects of biological invasions are comparatively well understood, the extent to which this science is problem focused, and relevant for policy makers and managers, has been questioned (Esler et al. 2010). Although it is generally acknowledged that scientists and managers must work more closely together, this is perhaps especially important for PAs, because the managers are often based in remote areas, without the necessary skills and awareness, have many other responsibilities, or are trying to implement management programmes without adequate professional support.

Collaborative networks can help to consolidate information in suitable repositories or to facilitate links and access between data and information repositories, and also between members to share ideas and help each other. A number of agencies and organisations are already involved in various management efforts in different regions, for example the Mountain Invasion Research Network (Chap. 6 Kueffer et al. 2014) and Global Island Plant Conservation Network (Chap. 19 Baret et al. 2014). The development of a regional inter-agency collaborative network focusing on invasive species in PAs is likely to be better positioned to attract funding and enlist support from experts. Perhaps formal networks (e.g. the IUCN SSC Invasive Species Specialist Group and the IUCN World Commission on Protected Areas) can help stimulate or form collaborative networks where they would be useful, even if only through bringing role players together. Similarly, the coordinated research programmes launched by the European Union may present opportunities to overcome the lack of data from PAs across the continent, by including such a theme into European-wide proposal calls, and forming an international collaborative network.

28.3.3 Resources: The Ubiquitous Problem

The resources available for IAP management in PAs are, in most cases, inadequate to implement actions that are deemed necessary to reduce the abundance and density of invasive plants and manage their impacts. This is partly due to the many demands being placed on the limited resources available for PA management in general. Although not unique to PAs, but perhaps felt more acutely in the conservation sector, is the availability of resources across countries and regions of the world. For example, Africa (with the exception of South Africa) is probably at this stage unable, and is in the future unlikely to be able, to provide the resources for managing IAPs at anywhere near the levels that will be necessary to prevent many invasive species from becoming widespread and having major impacts.

Overcoming these constraints in a time of global financial austerity is a huge challenge, but the case studies reported in this book show how improved, science-based planning can help to make the best use of available resources. In many cases, solutions lie in sagacious project management, for example by applying objective decision-making tools for prioritizing action, and clever integration of prevention and other forms of management into overall management plans for PAs. Also, partnerships between local, municipal, regional and national governments and NGOs, and international non-profit or donor organisations, will need to be included in any management programme.

The availability and use of technology, whether for management or science, varies considerably, although not always necessarily in regions or countries as a whole. Fortunately, additional support has been provided for management in some PAs through interventions of non-profit organisations. Science-based associations have often developed through academic – government partnerships, or international – local scientific collaboration, where the country does not have the expertise and is otherwise unable to implement various technologies. This also fulfils the role of knowledge transfer.

It is important that case studies on management programmes (prevention, eradication, control, mitigation of impacts, etc.) are compiled and assessed, and that key lessons are circulated. Such insights are unlikely to appear in scientific journals because of the lack of rigorous experimental designs, non-standard methods, and the lack of incentives to spend time to publish such work. For this reason these data and information should be organised and made available online, for example integrating the different IUCN knowledge products such as the Red List, the Global Invasive Species Database, and the World Database on Protected Areas (www.wdpa.org). Integrating data on native and invasive species, as well as on the features of PAs, would provide extremely helpful information to managers.

28.4   Protected Areas as Natural Laboratories: Unique Opportunities

Protected areas have many features that make them ideal outdoor laboratories. Some aspects are unique to PAs due to the kinds of management interventions they experience, for example, from actively patrolled to unprotected/extensively utilised, and intense to laissez-faire management approaches. We highlight some features associated with PAs that may provide for interesting research opportunities.

- The range in sizes of PAs, from small nature reserves covering a few hectares to those covering millions of hectares. Also, the global distribution and variety of geographic and environmental features of PAs, and from islands to mountains, all provide many opportunities for comparative work across multiple scales. These attributes can provide opportunities for identifying general determinants of invasions in relatively undisturbed environments.

- The management approaches of PAs provide many unique settings. For example, many PAs in East Africa are unfenced, have no (or little) artificial water provision, and some have been studied for decades (e.g. Serengeti NP, Sinclair et al. 2008). In contrast, Kruger NP is fenced, has numerous artificial water points, and has also been well studied for several decades (Du Toit et al. 2003). Protected areas have been proclaimed at different times, have different histories and fall in different categories (e.g. one of the six IUCN PA classification categories). Such features provide outdoor laboratories for research of processes that may not be available elsewhere, where causes of local patterns and specific mechanisms of particular invasions can be addressed.
- It is reasonable to assume that PAs should have had significantly lower rates of intentional introductions of ornamental alien plants. In one of the world's oldest national parks (Kruger NP, South Africa), ornamental plants accounted for a large number of the alien plant species in the park. Ornamental plant use in and adjacent to PAs could thus be an important pathway of invasion and an important area to focus management attention (Foxcroft et al. 2008).
- The interaction of fire with many IAPs has induced radical ecosystem-level changes in many regions, including PAs (Brooks et al. 2004). Fire is an integral part of many, although not all, grassland and savanna ecosystems; interactions between fire and other factors have evolved over millennia, continuously adapting the structure, biodiversity and function of these systems (Bond et al. 2005; van Wilgen et al. 2003). The interaction of fire with certain alien plant species can create self-reinforcing feedback loops which drives rapid invasion, as has happened in northern Australia with *A. gayanus* (Rossiter et al. 2003; Setterfield et al. 2010; Chap. 2 Foxcroft et al. 2014b). In another example, a large-scale experiment in Hluhluwe-iMfolozi game reserve, fire interacted with conventional control actions, inducing a shift from woodland dominated, to grass-dominated systems (te Beest et al. 2012). Changing fire regimes (e.g. due to increased ignition events caused by human activities, changes in climate and vegetation) could alter the trajectories of invasions in many ways. Changing perceptions of the role of fire as a management tool will also change options for managing IAPs in many PAs.
- Tourism has been shown to be correlated with increased plant invasions (e.g. Usher 1988; Macdonald et al. 1989; Lonsdale 1999; Chap. 18 Brundu 2014). Many regions rely on tourism and associated economic benefits for financial support beyond PA boundaries (Christie and Crompton 2001; Eagles et al. 2002; Naughton-Treves et al. 2005). Much work remains to be done to improve our understanding, and where necessary, reduce the facilitative role of certain categories of tourism in the spread of IAPs. Tourism can also play a positive role, by making visitors aware of the threats of IAPs and the value of non-invaded, well-functioning ecosystems. Conservation agencies are increasingly required to become financially independent as government grants are redirected to other

areas of the economy. The different models and numbers of tourists in different PAs provide an exciting study system for exploring these questions.

- The transformation of habitat by IAPs from, for example, open woodlands and grassland to dense patches of alien woody shrubs (or vice versa), may potentially impact herbivore migration patterns. Could such changes, together with the combination of human encroachment on open boundary PAs and landscape fragmentation, impact on, for example, the iconic migrations of wildebeest (*Connochaetes taurinus*) and zebra (*Equus zebra*) in Serengeti NP and Tarangire NP and white-eared kob (*Kobus kob leucotis*) in southern Sudan? Similarly, the transformation of grasslands to impenetrable thickets of IAPs is impacting on feeding areas of the one-horned rhinoceros (Lahkar et al. 2011) and replacement of woody thickets by *Chromolaena odorata* (siam weed) displacing black rhino (*Diceros bicornis*, Howison 2009).

- A characteristic feature of many PAs is the presence of large numbers of mammalian herbivores. Large herbivores, in high numbers, have either been present up until recent decades or are still present. High populations of large mammals create disturbances of varying kinds and intensities, which could conceivably allow for invasion into natural areas, while alternatively certain species may be suppressed. While there have been some studies on the role that these animals play in preventing or promoting alien plant invasions (e.g. Constible et al. 2005; Rose and Hermanutz 2004) an improved understanding of interactions between such key elements of the native biota and IAPs is needed as a tool to facilitate the development of integrated PA management plans.

28.5   Conclusion

The chapters in this book show that the extent of invasions of alien plants is growing rapidly in PAs globally. The effects on biodiversity of PAs are already dramatic in many cases, and the overall impacts attributed to IAPs are increasing rapidly. Future trajectories are difficult to predict, as invasive species interact in complex ways with other factors of global change such as climate change, habitat loss and many forms of human pressure. The impact of IAS on PAs has long been underestimated, and the concerns raised by scientists more than 20 years ago that this threat was going to increase (Macdonald et al. 1989; Usher 1988) were largely ignored by national and international institutions.

Management strategies for IAPs (and invasive alien species in general) in PAs require urgent attention to avoid a rapid and irreversible escalation of impacts of the sort that are described on many pages of this book. Assuming that IAPs in PAs will look after themselves due to some natural processes (following the laissez-faire approach to PA management by some agencies) is not an option (Chap. 3 Meiners and Pickett 2014; Chap. 21 Meyerson and Pyšek 2014) and active management of this threat is crucial. Evidence-based policy and management, developed through rigorous science, will allow PAs to respond appropriately to this growing threat at all scales.

Protected areas can and should play a major role in combating invasions, not only by improving the efficacy of IAS management within their borders, but also monitoring patterns of invasions, raising awareness at all levels, improving the capacity of practitioners to deal with invaders, implementing site-based prevention efforts, enforcing early detection and rapid response frameworks, and catalysing action beyond the park boundaries. Protected areas must be more active in preventing and mitigating the global effects of invasions by being: reservoirs of the heritage of native species and ecosystems; sentinels of incursions to speed up response at all levels; champions of information and awareness with the different sectors of the society; and catalysts of action at all scales.

Acknowledgements LCF thanks South African National Parks for supporting work on this book and for general support. LCF and DMR thank the DST-NRF Centre of Excellence for Invasion Biology, Stellenbosch University, and the National Research Foundation for support. PP was supported by long-term research development project no. RVO 67985939 (Academy of Sciences of the Czech Republic), institutional resources of Ministry of Education, Youth and Sports of the Czech Republic, and acknowledges the support by Praemium Academiae award from the Academy of Sciences of the Czech Republic. We thank Zuzana Sixtová for technical assistance with editing, and Phil Hulme, Ankila Hiremath, Carol West, Mark Gardener, Lloyd Loope, Samantha Setterfield and Stéphane Baret for contributing information for Table 28.1.

References

Baret S et al (2014) Chapter 19: Threats to paradise? Plant invasions in protected areas of the western Indian Ocean islands. In: Foxcroft LC, Pyšek P, Richardson DM, Genovesi P (eds) Plant invasions in protected areas: patterns, problems and challenges. Springer, Dordrecht, pp 423–447

Bond WJ, Woodward FI, Midgley GF (2005) The global distribution of ecosystems in a world without fire. New Phytol 165:525–538

Brooks ML, D'Antonio CM, Richardson DM et al (2004) Effects of invasive alien plants on fire regimes. BioScience 54:677–688

Brundu G (2014) Chapter 18: Invasive alien plants in protected areas in Mediterranean islands: knowledge gaps and main threats. In: Foxcroft LC, Pyšek P, Richardson DM, Genovesi P (eds) Plant invasions in protected areas: patterns, problems and challenges. Springer, Dordrecht, pp 395–422

Cadotte MW, McMahon SM, Fukami T (2006) Conceptual ecology and invasion biology: reciprocal approaches to nature. Springer, Berlin

Christie IT, Crompton DE (2001) Tourism in Africa. Africa Region working paper series no 12. The World Bank Group

Conklin J (2005) Dialogue mapping: building shared understanding of wicked problems. John Wiley & Sons, New York

Constible JM, Sweitzer RA, Van Vuren DH et al (2005) Dispersal of non-native plants by introduced bison in an island ecosystem. Biol Invasions 7:699–709

D'Antonio C, Meyerson LA (2002) Exotic plant species as problems and solutions in ecological restoration: a synthesis. Restor Ecol 10:703–713

DAISIE (2009) Handbook of alien species in Europe. Springer, Berlin

Davis MA (2009) Invasion biology. Oxford University Press, Oxford

Downey PO (2014) Chapter 23: Protecting biodiversity through strategic alien plant management: an approach for increasing conservation outcomes in protected areas. In: Foxcroft LC, Pyšek P, Richardson DM, Genovesi P (eds) Plant invasions in protected areas: patterns, problems and challenges. Springer, Dordrecht, pp 507–528

Du Toit JT, Rogers KH, Biggs HC (eds) (2003) The Kruger experience. Ecology and management of savanna heterogeneity. Island Press, Washington, DC

Eagles PFJ, McCool SF, Haynes CDA (2002) Sustainable tourism in protected areas: guidelines for planning and management. IUCN, Gland/Cambridge, xv + 183 pp

Ehrenfeld JG (2010) Ecosystem consequences of biological invasions. Annu Rev Ecol Evol Syst 41:59–80

Ervin J, Mulongoy KJ, Lawrence K et al (2010) Making protected areas relevant: a guide to integrating protected areas into wider landscapes, seascapes and sectoral plans and strategies, CBD technical series no 44. Convention on Biological Diversity, Montreal, 94 pp

Esler KJ, Prozesky H, Sharma GP et al (2010) How wide is the "knowing-doing" gap in invasion biology? Biol Invasions 12:4065–4075

Foxcroft LC, Richardson DM, Wilson JRU (2008) Ornamental plants as invasive aliens: problems and solutions in Kruger National Park, South Africa. Environ Manage 41:32–51

Foxcroft LC, Pyšek P, Richardson DM et al (2014a) Chapter 2: Impacts of alien plant invasions in protected areas. In: Foxcroft LC, Pyšek P, Richardson DM, Genovesi P (eds) Plant invasions in protected areas: patterns, problems and challenges. Springer, Dordrecht, pp 19–41

Foxcroft LC, Richardson DM, Pyšek P et al (2014b) Chapter 1: Plant invasions in protected areas: outlining the issues and creating the links. In: Foxcroft LC, Pyšek P, Richardson DM, Genovesi P (eds) Plant invasions in protected areas: patterns, problems and challenges. Springer, Dordrecht, pp 3–18

Foxcroft LC, Witt A, Lotter WD (2014c) Chapter 7: Icons in peril: invasive alien plants in African protected areas. In: Foxcroft LC, Pyšek P, Richardson DM, Genovesi P (eds) Plant invasions in protected areas: patterns, problems and challenges. Springer, Dordrecht, pp 117–143

Gardener MR, Trueman M, Buddenhagen C et al (2014) Chapter 16: A pragmatic approach to management of plant invasions in Galapagos. In: Foxcroft LC, Pyšek P, Richardson DM, Genovesi P (eds) Plant invasions in protected areas: patterns, problems and challenges. Springer, Dordrecht, pp 349–374

Genovesi P, Monaco A (2014) Chapter 22: Guidelines for addressing invasive species in protected areas. In: Foxcroft LC, Pyšek P, Richardson DM, Genovesi P (eds) Plant invasions in protected areas: patterns, problems and challenges. Springer, Dordrecht, pp 487–506

Henderson L (1998) Southern African plant invaders atlas (SAPIA). Appl Plant Sci 12:31–32

Hiremath AJ, Sundaram B (2014) Chapter 12: Invasive plant species in Indian protected areas: conserving biodiversity in cultural landscapes. In: Foxcroft LC, Pyšek P, Richardson DM, Genovesi P (eds) Plant invasions in protected areas: patterns, problems and challenges. Springer, Dordrecht, pp 241–266

Howison RA (2009) Food preferences and feeding interactions among browsers, and the effect of an exotic invasive weed *Chromolaena odorata* on the endangered Black Rhino in an African savanna. University of Kwazulu-Natal, Westville

Hui C et al (2014) Chapter 5: A cross-scale approach for abundance estimation of invasive alien plants in a large protected area. In: Foxcroft LC, Pyšek P, Richardson DM, Genovesi P (eds) Plant invasions in protected areas: patterns, problems and challenges. Springer, Dordrecht, pp 73–88

Hulme PE, Burslem DFRP, Dawson W et al (2014) Chapter 8: Aliens in the Arc: are invasive trees a threat to the montane forests of East Africa? In: Foxcroft LC, Pyšek P, Richardson DM, Genovesi P (eds) Plant invasions in protected areas: patterns, problems and challenges. Springer, Dordrecht, pp 145–165

Kruger JM, MacFadyen S (2011) Science support within the South African National Parks adaptive management framework. Koedoe 53(2): Art. #1010. doi:10.4102/koedoe.v53i2.1010

Kueffer C, McDougall K, Alexander J et al (2014) Chapter 6: Plant invasions into mountain protected areas: assessment, prevention and control at multiple spatial scales. In: Foxcroft LC, Pyšek P, Richardson DM, Genovesi P (eds) Plant invasions in protected areas: patterns, problems and challenges. Springer, Dordrecht, pp 89–113

Lahkar BP, Talukdar BK, Sarma P (2011) Invasive species in grassland habitat: an ecological threat to the greater one-horned rhino (*Rhinoceros unicornis*). Pachyderm 49:33–39

Larios L, Suding KN (2014) Chapter 28: Restoration within protected areas: when and how to intervene to manage plant invasions? In: Foxcroft LC, Pyšek P, Richardson DM, Genovesi P (eds) Plant invasions in protected areas: patterns, problems and challenges. Springer, Dordrecht, pp 621–639

Likens GE, Lindenmayer DB (2012) Integrating approaches leads to more effective conservation of biodiversity. Biodivers Conserv 21:3323–3341

Lonsdale WM (1999) Global patterns of plant invasions and the concept of invasibility. Ecology 80:1522–1536

Loope LL, Hughes RF, Meyer J-Y et al (2014) Chapter 15: Plant invasions in protected areas of tropical Pacific Islands, with special reference to Hawaii. In: Foxcroft LC, Pyšek P, Richardson DM, Genovesi P (eds) Plant invasions in protected areas: patterns, problems and challenges. Springer, Dordrecht, pp 313–348

Macdonald IAW, Loope LL, Usher M et al (1989) Wildlife conservation and the invasion of nature reserves by introduced species: a global perspective. In: Drake JA, Mooney H, di Castri F et al (eds) Biological invasions. A global perspective, Scope 37. Wiley, Chichester, pp 215–256

McGeoch MA, Butchart SHM, Spear D et al (2010) Global indicators of biological invasion: species numbers, biodiversity impact and policy responses. Divers Distrib 16:95–108

McGeoch MA, Spear D, Kleynhans E et al (2012) Uncertainty in invasive alien species listing. Ecol Appl 22:959–971

McNeely J (2014) Chapter 4: Global efforts to address the wicked problem of invasive alien species. In: Foxcroft LC, Pyšek P, Richardson DM, Genovesi P (eds) Plant invasions in protected areas: patterns, problems and challenges. Springer, Dordrecht, pp 61–71

Meiners SJ, Pickett STA (2014) Chapter 3: Plant invasion in protected landscapes: exception or expectation? In: Foxcroft LC, Pyšek P, Richardson DM, Genovesi P (eds) Plant invasions in protected areas: patterns, problems and challenges. Springer, Dordrecht, pp 43–60

Meyerson LA, Pyšek P (2014) Chapter 21: Manipulating alien species propagule pressure as a prevention strategy in protected areas. In: Foxcroft LC, Pyšek P, Richardson DM, Genovesi P (eds) Plant invasions in protected areas: patterns, problems and challenges. Springer, Dordrecht, pp 473–486

Naughton-Treves L, Holland MB, Brandon K (2005) The role of protected areas in conserving biodiversity and sustaining local livelihoods. Annu Rev Environ Res 30:219–252

Nentwig W (ed) (2007) Biological invasions. Springer, Berlin

Nuñez MA, Pauchard A (2010) Biological invasions in developing and developed countries: does one model fit all? Biol Invasions 12:707–714

Ostfeld RS, Pickett STA, Shachak M et al (1997) Defining the scientific issues. In: Pickett STA, Ostfeld RS, Shachak M et al (eds) The ecological basis of conservation. Heterogeneity, ecosystems, and biodiversity. Chapman and Hall, New York, pp 3–10

Pyšek P, Richardson DM, Pergl J et al (2008) Geographical and taxonomic biases in invasion ecology. Trends Ecol Evol 23:237–244

Pyšek P, Genovesi P, Pergl J et al (2014) Chapter 11: Invasion of protected areas in Europe: an old continent facing new problems. In: Foxcroft LC, Pyšek P, Richardson DM, Genovesi P (eds) Plant invasions in protected areas: patterns, problems and challenges. Springer, Dordrecht, pp 209–240

Richardson DM (ed) (2011) Fifty years of invasion ecology: the legacy of Charles Elton. Wiley-Blackwell, Oxford

Rose M, Hermanutz L (2004) Are boreal ecosystems susceptible to alien plant invasion? Evidence from protected areas. Oecologia 139:467–477

Rossiter NA, Setterfield SA, Douglas MM et al (2003) Testing the grass-fire cycle: alien grass invasion in the tropical savannas of northern Australia. Divers Distrib 9:169–176

Setterfield SA, Rossiter-Rachor NA, Hutley LB et al (2010) Turning up the heat: the impacts of *Andropogon gayanus* (gamba grass) invasion on fire behaviour in northern Australian savannas. Divers Distrib 16:854–861

Setterfield SA, Douglas MM, Petty AM et al (2014) Chapter 9: Invasive plants in the floodplains of Australia's Kakadu National Park. In: Foxcroft LC, Pyšek P, Richardson DM, Genovesi P (eds) Plant invasions in protected areas: patterns, problems and challenges. Springer, Dordrecht, pp 167–189

Simberloff D (2014) Chapter 25: Eradication – pipe dream or real option? In: Foxcroft LC, Pyšek P, Richardson DM, Genovesi P (eds) Plant invasions in protected areas: patterns, problems and challenges. Springer, Dordrecht, pp 549–559

Simberloff D, Martin J-L, Genovesi P et al (2014) Impacts of biological invasions: what's what and the way forward. TREE 28:58–66

Sinclair ARE, Packer C, Mduma SAR et al (eds) (2008) Serengeti III: human impacts on ecosystem dynamics. University of Chicago Press, Chicago

Spehn EM, Libermann M, Körner C (eds) (2006) Land use change and mountain biodiversity. CRC Press, Andover

te Beest M, Cromsigt JPGM, Ngobese J et al (2012) Managing invasions at the cost of native habitat? An experimental test of the impact of fire on the invasion of *Chromolaena odorata* in a South African savanna. Biol Invasions 14:607–618

Tu M, Robison RA (2014) Chapter 24: Overcoming barriers to the prevention and management of alien plant invasions in protected areas: a practical approach. In: Foxcroft LC, Pyšek P, Richardson DM, Genovesi P (eds) Plant invasions in protected areas: patterns, problems and challenges. Springer, Dordrecht, pp 529–547

Usher MB (1988) Biological invasions of nature reserves: a search for generalizations. Biol Conserv 44:119–135

Van Driesche R, Center T (2014) Chapter 26: Biological control of invasive plants in protected areas. In: Foxcroft LC, Pyšek P, Richardson DM, Genovesi P (eds) Plant invasions in protected areas: patterns, problems and challenges. Springer, Dordrecht, pp 561–597

van Wilgen B, Trollope WSW, Biggs HC et al (2003) Fire as a drive of ecosystem variability. In: du Toit JT, Rogers KH, Biggs HC (eds) The Kruger experience. Ecology and management of savanna heterogeneity. Island Press, Washington, DC, pp 149–170

West CJ, Thomson AM (2014) Chapter 14: Small, dynamic and recently settled: responding to the impacts of plant invasions in the New Zealand (Aotearoa) archipelago. In: Foxcroft LC, Pyšek P, Richardson DM, Genovesi P (eds) Plant invasions in protected areas: patterns, problems and challenges. Springer, Dordrecht, pp 285–311

Plant Invasions Index

A

Abundance
  abundance-effort relationship, 78
  estimation, 73–86
  occupancy-abundance relationship, 85
Acid deposition, 48–49
Africa
  East Africa, 120, 122–128, 134,
    136, 145–162, 580, 634
  Eastern Arc Mountains, 134,
    146–150, 158–162
  East Usambara Mountains, 147,
    150–152, 154, 155, 158, 159
  Kenya, 25, 121, 123–127, 134, 146,
    158, 573
  South Africa, 9, 15, 21, 25, 29, 35, 75, 91,
    120–122, 125, 128–133, 135–137, 260,
    403, 411, 453, 458, 478, 490, 492, 509,
    510, 514, 520, 551, 562, 563, 572–578,
    582, 629, 630, 632
  Tanzania, 62, 121–124, 126–128, 135, 146,
    150–154, 158–160, 627
  Uganda, 122, 123, 158
  West Africa, 122, 158, 574
  Zambia, 134
  Zimbabwe, 128, 573
Agroforestry, 158, 160, 161
Alpine, 92, 95–97, 99, 101, 103, 107, 248, 288
Altitude, 97, 101, 108, 147, 148, 154, 195, 218,
    220, 288, 377, 379, 438, 452
America
  North, 48, 53, 64, 91, 99, 213, 230, 493,
    562–564, 573, 575, 577, 581
  South, 47, 92, 249, 274, 321, 575, 576
Arctic, 95–97, 108
Argentina, 24, 250, 254
Assembly rules, 603–605

Australia, 4, 24, 64, 91, 136, 168, 221, 249,
    278, 285, 332, 364, 403, 453, 478, 492,
    514, 551, 562, 601, 623
Austria, 213, 215, 226, 230. *See also* Europe
Awareness, 11, 14, 28, 97–102, 105, 108, 126,
    135, 161, 182, 183, 231, 232, 235, 259,
    268, 278, 335, 352, 358, 367, 369, 385,
    387, 389, 404, 412, 424, 427, 429, 431,
    438–441, 462, 488–493, 495–497, 500,
    503, 530, 531, 543, 622, 632, 636
Azores, 331, 360, 376–392, 398, 400, 403,
    491, 622

B

Biodiversity, 4, 20, 62, 75, 90, 118, 168,
    192, 210, 244, 270, 287, 314, 352,
    376, 405, 424, 453, 476, 488, 508,
    536, 562, 599, 622
  hotspot, 314, 424
Biogeochemistry, 23, 30–31
Biological control, 66, 102, 122, 161, 175,
    275, 316, 364, 434, 511, 544, 555,
    562, 623
Biomes, 5, 7, 28, 121, 122, 131, 133, 146, 490,
    518, 572, 574, 576, 578
Biosphere reserve, 243, 247, 248, 259, 386, 398
Botanic gardens, 155, 157

C

Canada, 21, 34, 219, 221, 576
Capacity building, 160, 161, 352, 530
CBD. *See* Convention on Biological Diversity
    (CBD)
Charles Darwin Research Station, 352,
    358, 626

L.C. Foxcroft et al. (eds.), *Plant Invasions in Protected Areas: Patterns, Problems
and Challenges*, Invading Nature - Springer Series in Invasion Ecology 7,
DOI 10.1007/978-94-007-7750-7, © Springer Science+Business Media Dordrecht 2013

Mauritius, 67, 294, 425–428, 433–437, 439, 440, 573, 574. *See also* Islands
Mediterranean endemics, 396, 403–405, 411, 412
Minimum sampling effort, 78, 83, 84
MIREN. *See* Mountain Invasion Research Network (MIREN)
Monitoring, 13, 31, 51, 52, 56, 75, 82–86, 100–102, 104, 106–107, 121, 123, 135, 161, 181, 184, 194, 204–206, 223, 224, 230, 231, 234, 251, 259, 260, 308, 324, 330, 336, 368, 385, 387, 391, 437, 454, 481, 488, 491, 493, 495, 498, 500–501, 503, 510, 512, 515, 518, 521–523, 530–532, 534–538, 540, 543, 545, 552–554, 556, 569, 609, 623, 626, 627, 631, 636
Mountain Invasion Research Network (MIREN), 95, 104–106, 501, 632
Mountains, 25, 62, 89–108, 119, 146, 193, 220, 273, 378, 396, 430, 501, 508, 539, 623
Mutualisms, 20, 27, 352, 360–361

N
Natura 2000 Network, 210, 213, 214, 218, 380, 398, 412
Neophytes, 215, 217–221
Nepal, 63, 333
New Zealand, 24, 67, 91–93, 96, 99, 221, 223, 269, 285–308, 314, 332, 359, 453, 458, 463, 474, 492, 494, 496, 498, 552, 556, 572–574, 576, 577, 579, 601, 622, 627
Nutrient
cycling, 20, 24, 30, 31, 35, 55, 260, 475
deposition, 48–49, 56, 288

O
Ornamental plant trade, 92. *See also* Invasive plants
horticulture, 108

P
Partnerships, 11, 133, 161, 162, 298, 335, 336, 388, 389, 546, 566–467, 622, 629, 633
Pollination, 27, 35, 228, 360, 361, 368, 627, 630
Polynesia, 274, 286, 314, 317–319, 321, 322, 337–342, 497
Portugal, 214, 215, 226, 377, 388, 398, 400. *See also* Europe

Precautionary principle, 97, 367
Presence-absence records, 74
Preservation, 5, 14, 44, 48, 305, 315, 316, 384, 430, 481, 491, 533
Prevention, 11, 14, 48, 66, 67, 89–108, 122, 214, 223, 224, 234, 277–278, 337, 363, 366, 368, 402, 432, 473–483, 490, 492–497, 499, 503, 511, 529–545, 600, 624, 631, 633, 636
Prioritise, 33, 135, 158, 161, 287, 350, 352, 361–362, 368, 511, 515, 516, 518, 520, 521, 528, 541
Propagule pressure, 45, 55, 56, 91, 94, 95, 103, 104, 108, 146, 153, 155–157, 159, 160, 193, 205, 218, 231, 253, 450, 458, 459, 462, 473–483, 625, 627
Protection, 5, 14, 15, 44, 51, 56, 75, 101, 148, 160, 172, 176, 192, 210, 213, 214, 229, 231, 235, 243, 244, 254, 260, 287, 298, 315–317, 335, 338, 364, 369, 388, 399, 409, 410, 424–428, 436, 439, 453, 464, 474, 476–478, 482, 489, 503, 509–513, 517–519, 521, 523, 562, 622, 623

Q
Quarantine, 330, 352, 361–363, 453, 462, 496, 573, 626

R
Ramsar Convention, 11, 68, 398
Recovery constraints, 600–603, 609–611
Residence time, 156, 456, 457
Restoration
active, 30, 623
passive, 30, 601, 604, 609
Réunion (La Réunion, France), 425–428, 434, 436–437, 440. *See also* Islands
Rodrigues (Îles Rodrigues, Mauritius), 425, 428, 437. *See also* Islands

S
Samoa, 15, 328, 329, 335
Sampling effort, 77–79, 82–85, 356
Scale, 5, 30, 52, 63, 74, 92, 129, 151, 176, 193, 214, 253, 289, 316, 364, 379, 398, 429, 460, 489, 513, 535, 562, 603, 623
Scaling pattern, 73, 80, 85, 86
Scientific Committee on Problems of the Environment (SCOPE), 6–8, 12, 21, 22, 35, 121, 512
SEAs. *See* Special Ecological Areas (SEAs)

Protected Areas Index

L.C. Foxcroft et al. (eds.), *Plant Invasions in Protected Areas: Patterns, Problems and Challenges*, Invading Nature - Springer Series in Invasion Ecology 7, DOI 10.1007/978-94-007-7750-7, © Springer Science+Business Media Dordrecht 2013

Species Index

L.C. Foxcroft et al. (eds.), *Plant Invasions in Protected Areas: Patterns, Problems and Challenges*, Invading Nature - Springer Series in Invasion Ecology 7, DOI 10.1007/978-94-007-7750-7, © Springer Science+Business Media Dordrecht 2013

CPSIA information can be obtained at www.ICGtesting.com
Printed in the USA
LVOW02*1555021114

411666LV00001B/10/P